Lectures on Euclidean Geometry - Volume 1

Paris Pamfilos

Lectures on Euclidean Geometry - Volume 1

Euclidean Geometry of the Plane

 Springer

Paris Pamfilos
Department of Mathematics
& Applied Mathematics
University of Crete
Heraklion, Greece

ISBN 978-3-031-48908-2   ISBN 978-3-031-48906-8   (eBook)
https://doi.org/10.1007/978-3-031-48906-8

Mathematics Subject Classification (2020): 51-01, 51M04, 51M15, 51M25

© The Editor(s) (if applicable) and The Author(s), under exclusive license to Springer Nature Switzerland AG 2024

This work is subject to copyright. All rights are solely and exclusively licensed by the Publisher, whether the whole or part of the material is concerned, specifically the rights of reprinting, reuse of illustrations, recitation, broadcasting, reproduction on microfilms or in any other physical way, and transmission or information storage and retrieval, electronic adaptation, computer software, or by similar or dissimilar methodology now known or hereafter developed.

The use of general descriptive names, registered names, trademarks, service marks, etc. in this publication does not imply, even in the absence of a specific statement, that such names are exempt from the relevant protective laws and regulations and therefore free for general use.

The publisher, the authors, and the editors are safe to assume that the advice and information in this book are believed to be true and accurate at the date of publication. Neither the publisher nor the authors or the editors give a warranty, expressed or implied, with respect to the material contained herein or for any errors or omissions that may have been made. The publisher remains neutral with regard to jurisdictional claims in published maps and institutional affiliations.

Cover illustration: A visual proof of Pappus' theorem generalizing the Pythagorean theorem.

This Springer imprint is published by the registered company Springer Nature Switzerland AG
The registered company address is: Gewerbestrasse 11, 6330 Cham, Switzerland

Paper in this product is recyclable.

To Kostas, to Michalis, to Iasonas, to Odysseas, to Danae

Preface

> If I had to live my life again, I would have made a rule to read some poetry and listen to some music at least once every week; for perhaps the parts of my brain now atrophied would thus have been kept active through use. The loss of these tastes is a loss of happiness, and may possibly be injurious to the intellect, and more probably to the moral character, by enfeebling the emotional part of our nature.
>
> C. Darwin

> All things are mutually intertwined, and the tie is sacred, and scarcely anything is alien the one to the other. For all things have been ranged side by side, and together help to order one ordered Universe.
>
> Marcus Aurelius, To himself VII, 9

This book is the result of processed notes from courses in *Geometry, Euclidean Geometry, Geometry at School* and *Geometry and Computers*, which I repeatedly taught during the last twenty five years at the University of Crete in Herakleion Greece. Although the book is intended for school teachers and courses, the material it contains is much more extended than what can be naturally taught in school classes. The material however is developed gradually from simple and easy stuff to more complex and difficult subjects. This way, in the beginning chapters, I even avoid using negative numbers and the notion of transformation, so that the book can be used in all school courses.

In its content and organization the whole work complies with the philosophy of having one book of reference for every school course on a specific subject: the book of Geometry, the book of Physics, the book of Chemistry, etc. If not for the student, at least for the teacher. The book's intention is to offer a solid and complete foundation to both student and teacher, so that both can consult it for studying, understanding and examining various problems and extensions of elementary geometry.

The book's intention is not to develop a structural exposition involving mathematical structures such as vectors, groups, etc., but to proceed in the traditional, synthetic method and familiarize the reader with the basic notions and the problems related to them. Mathematics has definitely passed from the art of calculations to the discovery and investigation of structures. Although this is a long term accepted development, I do not necessarily consider this path appropriate for a beginning and an introduction to Geometry, its related notions, the elementary simple shapes and the problems they lead to. I reckon that the pupil must first have some elementary experience and incoming impressions of the simplest possible kind, without the interjection of notions of abstract structures, which in my opinion at a beginning stage would make the student's approach more difficult.

Thus, I spend a minimal time with axioms so that the reader has a reference point, and proceed quickly towards their logical consequences, so as to speed up the reader's contact with more complex and more interesting shapes. The proposed axioms represent very basic properties, some of which could conceivably be proved from others, even simpler. Such a practice however would cause a slightly more involved discussion on trivial consequences and conclusions, which may appear boring and repel the pupil from the course. In my opinion, a discussion of the foundations of geometry is the work of a late wisdom and must be done after one starts to love the material. In first place one has to examine what exactly is this stuff, practice and, slowly, depending on one's interest and capabilities, proceed to theory. This book therefore has an elementary character, avoiding the use of complex mathematical structures, yet exposing the reader to a wide range of problems.

I think that a teacher can use this book as a reference and road-map of Geometry's material in several geometry courses. Individual courses for specific classes, must and should be supported with companion aids of instructional character (practical exercises, additional exercises, drawing exercises for consolidation of ideas, spreadsheets, drawing software, etc.). The book contains many (over 1400) exercises, most of them with solutions or hints.

In writing this book, my deeper desire is to see Geometry return to its ancient respectable place in the school. This, because, Geometry, with all its beautiful shapes, offers a strong motivation and help for inductive thinking and many tangible pedagogical benefits. I will mention four main ones.

The first is the realization that there are things in front of you, which you do not see. Simple things, simple relations exposed to public view, initially invisible, which start to reveal themselves after lots of work and effort. Attention therefore increases, as does capability for correct observation. To anyone who asks "did I miss something" the answer is everywhere and always, "many (things)". Not asking this question, avoiding to answer it or answering carelessly, is completely incompatible with Geometry's pedagogy.

The second is the all encompassing power of detail, in other words, the accuracy of thought. Real, creative work, means involving yourself with details. With Geometry, this is accomplished with the exercises. Good intentions, visions and abstractions are void of content, when they are not emerging from the sea of details. Speaking on generalities is characteristic of rhetorical speech, the art of words. It is not a coincidence that Geometry has been pushed aside while flourishing the rhetorical and political speech.

The third and most important is the very essence of thought, as well as the human's character, if there is one, consistency. Mathematics and Geometry especially, with the assistance of its figures, is the great teacher of consistency. You begin from certain notions and basic properties (axioms) and start building, essentially, ad infinitum, without ever diverging from the initial principles and the simple rules of logic. This way the work done is always additive towards a building of absolute validity, which has nothing to do with lame improvisations and products, which are the result of continuously changing rules. Man, often, to please his desire, changes arbitrarily the rules of the game. He creates thus a certain culture in which, he is either an abuser, when he himself changes the rules, or a victim, when he suffers from external arbitrary rule changes. This way the work done is sometimes additive and other times negative, canceling the previous work done. This mode is the widespread culture of non-thought, since clear thought is virtually synonymous to mathematical thought and mathematical prototyping, the discovery and the respect of accepted rules.

The fourth and very crucial benefit is the acquaintance with the general problem of knowledge, and the balance between the quantity and quality. In the process of learning Geometry we realize with a particular intention the infinity of the directions towards it extends, but also the unity and the intimate relations of its parts. There results a question of approach, a question of psychology, a question of philosophy. How you can approach this whole, an immense body of knowledge? The question is crucial and posed at an early age. A correct or wrong attitude sets the foundation for a corresponding evolution of the whole life of the student. A big part of the failure of the learning is due to the misconceptions about its nature. If we look at the established practice, we realize that the mainstream behavior is that of the hunter. You target the language, history, chemistry, etc. you find the crucial paths and passages and you shoot. Unfortunately though, knowledge consists not of woodcocks. Any subject we may consider consists not of isolated entities. It is not a large herd of birds containing some rare and some exotic individuals. It is rather a connected continuous and consistent body, accepting only one approach, through the feeling. You cannot learn something if you do not approach it with positive feeling. Besides, every subject of knowledge is similar to a musical instrument. And as you cannot learn 10 instruments at the same time, so you cannot learn 10 subjects at the same time. The 10 instruments you can touch, put out of their cases, taste their sounds. You have to choose though and indulge into one. This is the characteristic property of the thinking man. He can indulge into his subject withdrawing from everything else. Like the virtuoso of the musical instrument absorbed to its music brings to unity the instrument, the music and his existence.

If we were asked to set the targets of education, one of the principals would be the ability to indulge into a subject. This, though it is developed by a mental process, it has an emotional basis. The teacher at the school, the college, the university, meets invariably the same stereotyped problem. The failure of the student is not due to missing mental forces. It is due to underdeveloped or totally missing emotional basis for its subject. The failure of the teacher is not due to what he teaches or lefts aside. It results from not recognizing the role of the emotional basis, not highlighting it, not cultivating it.

Motivated by these ideas I proceed with my proposition, proposing this foundation for the organization of lessons as simple as possible or, at any rate, as simple as the subject allows. I wish that the teacher and the studious pupil will read the book with the same and even greater pleasure than I had while studying for a long time the exquisite material. A material perfect and timeless, the only responsible for imperfections and mistakes being myself.

At this point I wish to thank the colleagues Georgia Athanasaki, Ioanna Gazani for the correction of many mistakes and also many suggestions for improvements. I thank also Dimitris Kontokostas for his numerous interventions on the first chapters of the book and the supervisors of the two greek editions, John Kotsopoulos and John Papadogonas. I am also very indebted to my collegue Manolis Katsoprinakis for many corrections of formulas as well as of figures. I thank also my colleagues Stylianos Negrepontis and George Stamou for their encouragement to continue my work and the colleague Antonis Tsolomitis for his help on "latex". I also express my gratitude to my colleague Giannis Galidakis for his assistance in the translation from the Greek. The suggestion to publish the English text with Springer came from my colleague Athanase Papadopoulos from Strasbourg and the heavy work of organization and management of the editorial work has been done by Springer's senior editor Elena Griniari. To them both I would like to express my deepest gratitude.

Regarding my sources, I have included all bibliographical references, which existed in my notes and were used to backtrack and complete the material. I believe they will be useful to those who wish to dig deeper and compare with other sources. It is especially interesting to search and enter in discussion with great minds, who came before and created material in the subject.

Looking back at the times of school, when I was initiated to Geometry by my excellent teachers, like the late Papadimitriou, Kanellos and Mageiras, I wish to note that the book binds, hopefully worthily, to a tradition we had on the area, which was cultivated, at that time, by the strong presence of Geometry in high schools and gymnasiums. Many of the problems herein are the ones I encountered in books and notes of the teachers I mentioned, as well as these found in Papanikolaou, Ioannidis, Panakis, Tsaousis. Several other problems were collected from classical Geometry texts, such as Catalan [9], Lalesco [28], Legendre [30], Lachlan [27], Coxeter [12], [13], Hadamard [20], F.G.M. [17] and many others, which are too numerous to list.

Concluding the preface, I mention, for those interested, certain books which contain historical themes on the subject [11], [16], [21], [22], [14], [26], [7], [34], [10].

In the second edition of the book there are substantial changes having to do with the material's articulation and expansion, the addition of exercises, figures, and the return of epigrams, which were omitted in the first edition, because of some mistaken reservations concerning the total volume of the book. Going through the first edition, I felt somewhat guilty and had a strong feeling that these small connections with the other non-mathematical directions of thinking and their creators are necessary and very valuable to be omitted for the benefit of a minimal space reduction. In my older notes I used to put them also at the end or even the middle of lengthy proofs. I had thus the feeling to be a member of the international and timeless university and the ability to ask and discuss with these great teachers of all possible subjects. Perhaps some of the readers will eventually keep in mind some of the many figures of the book and some of the epigrams, which are a sort of figures in this "invisible geomery", as the great poet says.

Inatos, July 2003 *Paris Pamfilos*

Symbol index

$(AB\Gamma)$	Circle going through the points A, B, Γ
$\kappa = (AB\Gamma\Delta)$	Circle going through the points A, B, Γ, Δ
$X = (AB, \Gamma\Delta)$	Intersection point X of the lines $AB, \Gamma\Delta$
$(AB; \Gamma\Delta)$	Cross ratio $\frac{\Gamma A}{\Gamma B} : \frac{\Delta A}{\Delta B}$
$(AB; \Gamma\Delta) = -1$	Harmonic quadruple of four collinear points
$(A, B) \sim (\Gamma, \Delta) \Leftrightarrow (AB; \Gamma\Delta) = -1$	Harmonic pairs
$\Delta = \Gamma(A, B) \leftrightarrow (AB; \Gamma\Delta) = -1$	Δ harmonic conjugate to Γ w.r.t. (A, B)
$\|AB\|$	Length of AB
$\kappa(O, \rho)$	Circle κ, center O, radius ρ
$\kappa(O)$	Circle κ with center O
$\kappa(\rho)$	Circle κ of radius ρ
$O(\rho)$	Circle: center O, radius ρ
$p(X)$ or $p(X, \kappa)$ or $p_\kappa(X)$	Power of point X relative to circle κ
$\|\widehat{AB\Gamma}\|$	Angle measure $\widehat{AB\Gamma}$
$O(A, B, \Gamma, \Delta)$	Pencil of four lines $OA, OB, O\Gamma, O\Delta$ through O
$\varepsilon(AB\Gamma)$	Area of triangle $AB\Gamma$
$\varepsilon(AB\Gamma...)$	Area of polygon $AB\Gamma...$
$o(AB\Gamma...)$	Volume of polyhedron $AB\Gamma...$
τ	Half perimeter of triangle
α, β, γ	Measures of angles of triangle $AB\Gamma$
R	Radius(circumradius) of $AB\Gamma$
v_A, v_B, v_Γ	Altitudes of triangle $AB\Gamma$
μ_A, μ_B, μ_Γ	Medians of triangle $AB\Gamma$
$\delta_A, \delta_B, \delta_\Gamma$	Inner bisectors of triangle $AB\Gamma$
a, b, c	Lengths of sides of triangle $AB\Gamma$
r, r_A, r_B, r_Γ	Radius of inscribed/escribed circles of $AB\Gamma$
$\frac{AB}{\Gamma\Delta}$	Signed ratio of segments of the same line
$\phi = \frac{\sqrt{5}+1}{2} =\sim 1.61803398874989484820...$	Golden section ratio
$\mathbb{N}$	Set $\{1, 2, 3, ...\}$ of natural numbers
$\mathbb{Z}$	Set of integers (positive, negative, 0)
$\mathbb{Q}$	Set (field) of rational numbers
$\mathbb{R}$	Set (field) of real numbers

Contents

Part I Euclidean Geometry of the plane

1   The basic notions 3
    1.1   Undefined terms, axioms 3
    1.2   Line and line segment 6
    1.3   Length, distance 9
    1.4   Angles 12
    1.5   Angle kinds 16
    1.6   Triangles 19
    1.7   Congruence, the equality of shapes 23
    1.8   Isosceles and right triangle 25
    1.9   Triangle congruence criteria 27
    1.10 Triangle's sides and angles relations 33
    1.11 The triangle inequality 40
    1.12 The orthogonal to a line 42
    1.13 The parallel from a point 45
    1.14 The sum of triangle's angles 48
    1.15 The axiom of parallels 52
    1.16 Symmetries 61
    1.17 Ratios, harmonic quadruples 66
    1.18 Comments and exercises for the chapter 72
    References 77

2   Circle and polygons 79
    2.1   The circle, the diameter, the chord 79
    2.2   Circle and line 84
    2.3   Two circles 88
    2.4   Constructions using ruler and compass 91
    2.5   Parallelograms 101
    2.6   Quadrilaterals 105
    2.7   The middles of sides 110

	2.8 The triangle's medians	114
	2.9 The rectangle and the square	118
	2.10 Other kinds of quadrilaterals	123
	2.11 Polygons, regular polygons	127
	2.12 Arcs, central angles	135
	2.13 Inscribed angles	139
	2.14 Inscriptible or cyclic quadrilaterals	151
	2.15 Circumscribed quadrilaterals	158
	2.16 Geometric loci	160
	2.17 Comments and exercises for the chapter	164
	References	180
3	**Areas, Thales, Pythagoras, Pappus**	181
	3.1 Area of polygons	181
	3.2 The area of the rectangle	183
	3.3 Area of parallelogram and triangle	187
	3.4 Pythagoras and Pappus	196
	3.5 Similar right triangles	205
	3.6 The trigonometric functions	213
	3.7 The theorem of Thales	221
	3.8 Pencils of lines	227
	3.9 Similar triangles	237
	3.10 Similar polygons	246
	3.11 Triangle's sine and cosine rules	258
	3.12 Stewart, medians, bisectors, altitudes	268
	3.13 Antiparallels, symmedians	277
	3.14 Comments and exercises for the chapter	284
	References	309
4	**The power of the circle**	311
	4.1 Power with respect to a circle	311
	4.2 Golden section and regular pentagon	316
	4.3 Radical axis, radical center	321
	4.4 Apollonian circles	327
	4.5 Circle pencils	333
	4.6 Orthogonal circles and pencils	344
	4.7 Similarity centers of two circles	352
	4.8 Inversion	360
	4.9 Polar and pole	372
	4.10 Comments and exercises for the chapter	379
	References	412

5 From the classical theorems ... 413
- 5.1 Escribed circles and excenters ... 413
- 5.2 Heron's formula ... 419
- 5.3 Euler's circle ... 422
- 5.4 Feuerbach's Theorem ... 427
- 5.5 Euler's theorem ... 429
- 5.6 Tangent circles of Apollonius ... 436
- 5.7 Theorems of Ptolemy and Brahmagupta ... 443
- 5.8 Simson's and Steiner's lines ... 452
- 5.9 Miquel point, pedal triangle ... 457
- 5.10 Arbelos ... 466
- 5.11 Sangaku ... 472
- 5.12 Fermat's and Fagnano's theorems ... 477
- 5.13 Morley's theorem ... 482
- 5.14 Signed ratio and distance ... 485
- 5.15 Cross ratio, harmonic pencils ... 493
- 5.16 Theorems of Menelaus and Ceva ... 501
- 5.17 The complete quadrilateral ... 513
- 5.18 Desargues' theorem ... 521
- 5.19 Pappus' theorem ... 526
- 5.20 Pascal's and Brianchon's theorems ... 530
- 5.21 Castillon's problem, homographic relations ... 540
- 5.22 Malfatti's problem ... 549
- 5.23 Calabi's triangle ... 553
- 5.24 Comments and exercises for the chapter ... 557
- References ... 587

Index ... 589

Part I
Euclidean Geometry of the plane

Chapter 1
The basic notions

1.1 Undefined terms, axioms

> This is valid even for our own ego: we understand it only through its properties, not as something which can exist in and of itself.
>
> Thomas Mann, Schopenhauer

The notions, at least the mathematical, are like the forms of matter, which are split in molecules, these in atoms, these in elementary particles etc. In geometry the reduction to simpler and more elementary notions comes eventually to the so called **undefined terms**. These are notions, simple and familiar from our experience, for which it is difficult to find even simpler ones, with which to describe them ([24]). Such notions in geometry are the **point**, the **plane**, the **space**, the **line**, the notion of the point **between** two other points and the notion of **equality** of two *shapes*.

We learn to handle these notions using their **properties** or **axioms**, which describe some of their characteristics, which *we accept without proof*. We begin therefore with the undefined terms. We describe their basic properties with axioms, and from there on, by combining these with logic, we deduce other properties, theorems or propositions and corollaries (direct logical consequences of theorems). Axioms and theorems, up to a point, are used to deduce new properties, in other words, new theorems.

Proceeding in this way, we build gradually a well-organized and structured spiritual edifice, which represents our knowledge in geometry and is, potentially, infinitely extendable. If at some point we accept a hypothesis, for example $A = B$, and, relying on logic, we conclude that this leads to a contradictory result against some axiom or some already proved theorem, then we say that our hypothesis has lead us to a contradiction, and we are then obliged to accept that the negation of the hypothesis must necessarily hold (in this case $A \neq B$). This method of reasoning is commonly called **re-**

duction to contradiction (Latin: *Reductio ad absurdum*) and is used very often in geometry.

Euclidean Geometry examines the properties of shapes in space and on the plane and, mainly, properties associated with measurement. As **shape** we consider any collection of points on the plane (plane shape) or space (space shape). We measure lengths, angles and areas. In space we also measure volumes. Usually the lesson is divided into two parts. In the first part, called **plane geometry**, we examine properties of plane figures, such as the triangle, the square, the circle etc. In the second part, called **space geometry**, we examine properties of space figures, such as the cube, the pyramid, the cylinder, the sphere etc.

Remark 1.1. The axioms, we'll choose as basic properties and starting point of our study, are not really independent from each other. Some of them are consequences of others, therefore we could conceivably start with fewer independent axioms, which would be sufficient in proving all the rest as theorems. Such a practice however would have the disadvantage of spending considerable time on very simple properties, which could otherwise be proved using our very few axioms.

I have therefore decided to incorporate some of these properties into the axioms, conforming to the philosophy, that the revelation of the more hidden properties of shapes creates stronger motivation than the discussion on the obvious ones. For an alternative course, where the subject of axioms is examined in detail, one may consult the well known book by Hilbert (1862-1943) on the *Foundations of Geometry* [25], which is devoted solely to the discussion of axioms (initially 22 of them), their independence and their non-contradictory nature, their so-called *consistency*. Most line axioms I formulate further ahead are taken from this book. I have replaced however some of these axioms with some from Birkhoff's (1884-1944) system ([5]), which guarantee that lines are, essentially, copies of the set $\mathbb{R}$ of real numbers.

In any case, let it be noticed that Euclidean Geometry's foundations can be done using very few axioms. Hilbert in the aforementioned book, as well as Cairns (1904-1982) ([8]) present systems with only four axioms. Bachmann (1909-1982) ([3]) presents a system with five axioms, which however involve some complex mathematical structures (topological spaces, transformations, groups etc.).

Remark 1.2. Euclid's Elements (approximately 325-265 B.C.) ([23], [21]) begin with the presentation of 23 definitions, 4 of which and the last are the following:

(1) A point is that of which there is no part.
(2) And a line is a length without breadth.
(3) And the extremities of a line are points.
(4) A straight-line is (any) one which lies evenly with points on itself.
...

1.1. UNDEFINED TERMS, AXIOMS

(23) Parallel lines are straight-lines which, being in the same plane, and being produced to infinity in each direction, meet with one another in neither (of these directions).

Immediately after the 23 definitions follow 5 **Postulates** which we name **axioms**:

1. Let it have been postulated to draw a straight-line from any point to any point.
2. And to produce a finite straight-line continuously in a straight-line.
3. And to draw a circle with any center and radius.
4. And that all right-angles are equal to one another.
5. And that if a straight-line falling across two (other) straight-lines makes internal angles on the same side (of itself whose sum is) less than two right-angles, then the two (other) straight-lines, being produced to infinity, meet on that side (of the original straight-line) that the (sum of the internal angles) is less than two right-angles (and do not meet on the other side).

These definitions are examples of notions we have called *undefined* (1, 2, 4), as well as, normal definitions in the modern sense (3, 23). Euclid's five axioms unfortunately are not sufficient in proving all the propositions which follow in his book. Euclid often tacitly supposes some additional properties, which are not deduced from these axioms, but are correct nevertheless. What is needed, is the addition of more axioms, so that the final set becomes, what we today call a *complete axiomatic system,* a system which is capable of supporting the proofs of all the properties of geometric shapes we discover using logical deduction ([39, p.36]).

In this book I conform to Euclid's choice to devote relatively little time on definitions and axioms, because I have often observed, that when a student is exposed to lots of explanations and detailed analyses of naturally intuitive notions, he usually starts doubting familiar to him notions and ends up being confused instead of illuminated. Therefore caution is needed, so that the student's natural intuition about previous empirical knowledge is fortified rather than challenged. Following then Euclid, I will not spend too much time on abstract notions and axioms. I will give a complete system capable of supporting all subsequent propositions and theorems. Trusting the reader's intuition however, I will not delve in the discussion of interdependence of these axioms or the undefined terms the axioms refer to ([39, p.165], [31]).

1.2 Line and line segment

> I think simplicity, speaking accurately, resembles very much the non-being. The great inability of my mind consists in that it does not have an instrument fine enough with which to grasp this. Someone may claim that the mathematical point is simple. In reality however, the mathematical point does not exist.
>
> Voltaire, *The Ignorant Philosopher*

The **plane** consists of **points**, which we denote by greek capital letters A, B, Γ, ..., A', B', Γ', ... A_1, A_2, ... etc. One of the simplest shapes of the plane is the **line** (See Figure 1.1) which we denote with greek letters ε, ζ, ..., ε', ζ', ... ε_1, ε_2, ... etc. For lines we accept the following initial properties (axioms).

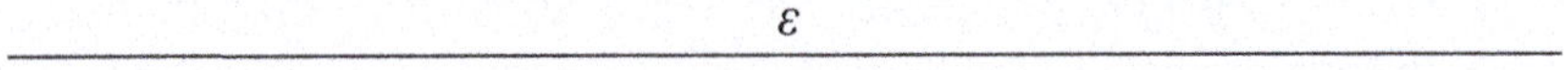

Fig. 1.1: Line ε

Axiom 1.1 *Two different points A, B define exactly one line denoted by AB (See Figure 1.2).*

Fig. 1.2: Line AB

Axiom 1.2 *Every line contains infinitely many points. For every line there are infinitely many points which do not belong to the line. For every point there are infinitely many lines which do not pass through that point.*

Fig. 1.3: Half-planes or sides defined by a line

Axiom 1.3 *Every line divides the plane into two parts called* **half-planes**, *which have no common points with that line. A line ε', which has two points A and B lying in different half-planes of the line ε, intersects the line ε (the first theorem below establishes that there exists exactly one intersection point). We often use the word* **side** *of line, meaning one of the two half-planes defined by it (See Figure 1.3).*

Axiom 1.4 *Two points A, B of a line ε define a* **line segment** *denoted also by AB. AB consists of A, B as well as all the points* **between** *A and B. A and B are called*

1.2. LINE AND LINE SEGMENT

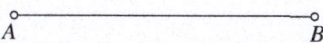

Fig. 1.4: Line segment AB

endpoints of the line segment (See Figure 1.4). The points of the line segment, excluding the endpoints, are called **interior** points of the segment. The set of all interior points is called the **interior** of the line segment. The points of the line AB not belonging to the segment AB are called **exterior** points of the segment.

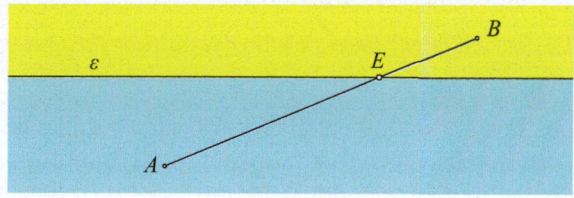

Fig. 1.5: A and B on different half-plane of the line ε

Axiom 1.5 *If the two points A and B belong to the same half-plane defined by the line ε, then all points of the line segment AB are contained in the same half-plane. If the points A and B belong to different half-planes defined by the line ε, then the intersection point E, of the line ε and the line AB, lies **between** the points A and B (See Figure 1.5).*

Remark 1.3. In Axiom 1.4 the word **between** is temporarily ambiguous. It will become clear in the next section with the help of the notion of length of a line segment.

Remark 1.4. The simultaneous usage of the symbol AB for both, the line segment it represents, and the line defined by points A and B, should not confuse us. Each time the meaning of the symbol will be clear from context. Often we'll denote the line $\varepsilon = AB$, considering that this symbol represents the phrase **the line ε defined by the points A and B**. We may also consider the line segment AB as defining a **direction** along the line AB with A being the **start** and B being the **end** of the segment AB.

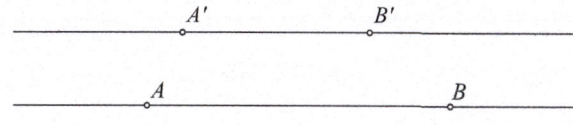

Fig. 1.6: Parallels AB and $A'B'$

Parallel we call two lines which do not intersect (See Figure 1.6). A line containing a segment AB is called **carrier** of the line segment. We call two line segments **parallel** when their corresponding carriers are parallel lines. **Secant** of a line ε, is called a line ε', different from ε, which intersects ε (See Figure 1.7).

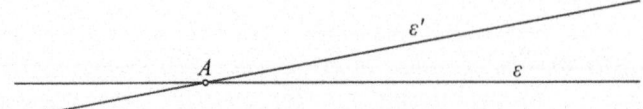

Fig. 1.7: Intersecting lines ε and ε'

Proposition 1.1. *Two different lines are either parallel or they intersect at exactly one point.*

Proof. In Proposition 1.9 we'll see that there do exist parallel lines. If the two lines ε and ε' do not intersect, then they are by definition parallel. If they do intersect they will have only one common point A. If they had a second intersection point B, different from A, we would have two different lines ε and ε' passing through the two points A and B, which is impossible, as this contradicts Axiom 1.1

Exercise 1.1. Given a line ε, show that if the line segment AB does not intersect the line ε, then the points A and B are contained in the same half-plane defined by ε (See Figure 1.8).

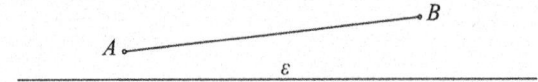

Fig. 1.8: A, B on the same side of ε

Hint: Use reduction to contradiction. Suppose that AB does not intersect the line ε and points A, B are contained in different half-planes of ε. Then, according to Axiom 1.5 the line segment AB will intersect line ε at a point E, contradicting the assumption.

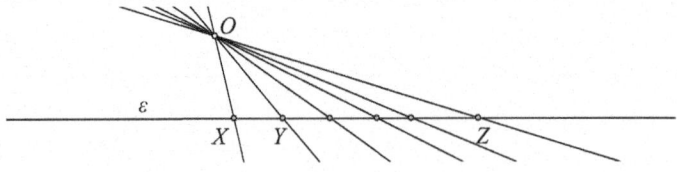

Fig. 1.9: Infinitely many lines through O

Exercise 1.2. Show that for each point O of the plane, there are infinitely many lines passing though it.

1.3. LENGTH, DISTANCE

Hint: Consider a line ε not passing through O. According to Axiom 1.2 there is at least one such line. Next, define the lines OX, OY, ... etc., which pass through O and through the points respectively X, Y,..., Z of ε (See Figure 1.9). Each of these points defines a different line through O.

1.3 Length, distance

> No one can read two thousand books. In the four hundred years I have lived, I've not read more than half a dozen. And in any case, it is not the reading that matters, but the rereading. Printing, which is now forbidden, was one of the worst evils of mankind, for it tended to multiply unnecessary texts to a dizzying degree.
>
> *J. L. Borges, A Weary Man's Utopia*

The axioms of this section relate lines with **real numbers** through the notion of *distance* of two points. They clarify the notion of *point between two other points*, as well as the notion of the line segment AB, which consists of all the points between A, B.

Axiom 1.6 *Every pair of points A and B defines a real number $|AB| \geq 0$, which we call **distance** of the points and which satisfies the properties $|AB| = |BA|$ and $|AB| = 0$, if and only if the two points coincide.*

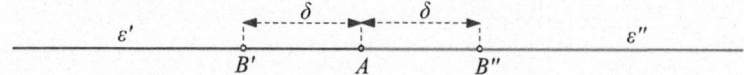

Fig. 1.10: $|AB| = |AE| + |EB|$

Axiom 1.7 *For every three different points A, B and E on the same line, one of the three necessarily lies between the other two. If E lies between A and B then $|AB| = |AE| + |EB|$ (See Figure 1.10). Conversely, if the preceding relation holds then E lies between A and B.*

Fig. 1.11: Points at distance δ from the startpoint of opposite half-lines

Axiom 1.8 *A point A on line ε separates the line in two parts ε' and ε'', which have only the point A in common and are called **half-lines** with A as **start-point**. For every positive number δ, there is exactly one point B' on ε' with $|B'A| = \delta$ and exactly one point B'' on ε'' with $|B''A| = \delta$. A is the **middle** of the segment $B'B''$.*

If the points A, B' and B'' are contained in the same line ε and ε', ε'' denote the half-lines of ε with startpoint A, we say that B, B' are in **different sides** of A, when one is contained in ε' and the other in ε'' (See Figure 1.11). We say that B and B'' are in the **same side** of A when they are both contained in one of ε' or ε''.

We call **length** of the segment AB the distance $|AB|$ of its endpoints. We call two line segments AB and $\Gamma\Delta$ of the same or different lines **equal**, when they have the same length.

The two half-lines which are defined by a point A on the line ε are called **opposite**. We call two half-lines **parallel** when they are contained in parallel lines.

Remark 1.5. Line Axiom 1.8 means that we can construct a line segment of any length we desire. We'll consider often the problem of practically constructing a line segment of given length, limiting ourselves to the usage of only two tools: the straight edge and the compass. For example, the construction of the middle point M of a given line segment AB, using a straight edge and compass, requires knowledge of the properties of the circle, which we do not have as of yet. Note however that the proof of the existence of M relying on the aforementioned properties is simple.

Exercise 1.3. Let B and E be two points on the same half-line AX with startpoint A (See Figure 1.12). Show that if $|AE| > |AB|$ then B lies between A and E. Conversely, if B lies between A and E then the preceding relation holds.

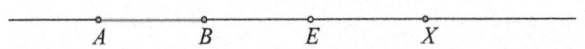

Fig. 1.12: B between A and E

Hint: Suppose B is not between A and E. Then either B is identified with E and therefore $|AB| = |AE|$ which is a contradiction, or E lies between A and B, which immediately implies $|AB| > |AE|$, contrary to the hypothesis.

Exercise 1.4. (Doubling the length of line segment) Given a line segment AB, show that on the line AB there exist points E and Z such that B is the middle of AE and A is the middle of ZB (See Figure 1.13).

Fig. 1.13: Doubling the length of AB

Hint: Take E on the half-line with start-point B which does not contain A at a distance $|AB|$ from B. Similarly work for Z.

Exercise 1.5. Show that for every line segment AB there exists exactly one point M (the middle of AB), such that $|AM| = |MB|$.

1.3. LENGTH, DISTANCE

Hint: If $|AB| = \lambda$ then the point M at distance $\lambda/2$ from A on the side of B exists using Axiom 1.8 and this is what we want.

Exercise 1.6. Show that if two points A and B are on the same side of line ε, then the line segment AB does not intersect the line ε. Is this the same with exercise 1.1?

Hint: If AB intersected ε, then the intersection point Γ would be different from A and B, therefore it would be between them and we have a contradiction in Axiom 1.5.

Exercise 1.7. Show that a line ε' is parallel to ε, if and only if one of the two half-planes of ε contains every pair of different points of ε'.

Hint: If we suppose there are two points A and B of ε' contained on different half-planes of ε, then according to Axiom 1.5, ε' would intersect ε. Conversely, if one ha-lf plane of ε contains all possible pairs of points of ε' then the latter cannot intersect ε, for if it did at A, then A would define two opposite half-lines on ε'. Choosing one point on each half-line we would find two points of ε' on different half-planes of ε.

Exercise 1.8. Show that points B and Γ of line ε are on the same side of point A of ε, if and only if $|B\Gamma| = ||AB| - |A\Gamma||$.

Hint: If points B, Γ are on the same half-line of A, then either B will be between A and Γ so that $|A\Gamma| = |AB| + |B\Gamma|$, or Γ will be between A and B so that $|AB| = |A\Gamma| + |\Gamma B|$. Therefore in both cases $|B\Gamma| = ||AB| - |A\Gamma||$, which is what we want. A similar argument proves the converse.

Exercise 1.9. Let M be the middle of the line segment AB. Show that if point Γ is on the interior of AB, then $|\Gamma M| = \frac{1}{2}||\Gamma A| - |\Gamma B||$. If Γ is on line AB but outside of segment AB, then $|\Gamma M| = \frac{1}{2}(|\Gamma A| + |\Gamma B|)$.

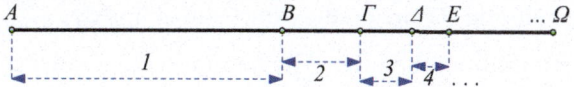

Fig. 1.14: Zeno's paradox

Exercise 1.10. In a race from A to Ω, the quickest runner (α :Achilles) can never overtake the slowest (τ : tortoise), which starts at the midpoint B. In fact, when α reaches B, τ has advanced to Γ. When α reaches Γ, τ has advanced to Δ, and so on (See Figure 1.14). Thus, Achilles will never catch the tortoise (Zeno's paradox 490-430 BC). How you resolve this paradox?

1.4 Angles

> The Master said, "I will not open the door for a mind that is not already striving to understand, nor will I provide words to a tongue that is not already struggling to speak. If I hold up one corner of a problem, and the student cannot come back to me with the other three, I will not attempt to instruct him again."
>
> *Confucius, Analects 7.8*

Two half-lines OX, OY, with common start-point O, and not contained in the same line, separate the plane in two parts and define an **acute angle** or simply **angle** and a **non-convex angle**.

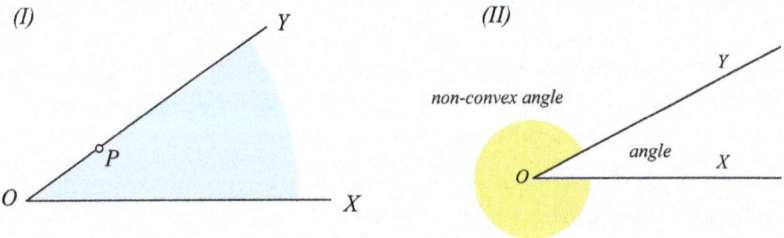

Fig. 1.15: Angle $\widehat{XOY}$ Non-convex angle $\widehat{XOY}$

Convex angle or simply **angle** is called the figure denoted by $\widehat{XOY}$ and consists of the two half-lines OX and OY together with one part of the plane we call **interior** of the angle. The angle's **interior** (See Figure 1.15-I) is the part of the plane whose points P satisfy the two properties:

1. P and the half-line OY are on the same side of the line OX,
2. P and the half-line OX are on the same side of line OY.

Point O is called the angle's **vertex**. The half-lines OX, OY are called **sides of the angle**. **Non-convex angle** is called the figure again defined by the half-lines OX and OY but this time consisting of the rest of the plane minus the interior of the acute angle $\widehat{XOY}$ and the half-lines which define it (See Figure 1.15-II). This part of the plane we call **interior** of the non-convex angle $\widehat{XOY}$, or **exterior** of the convex angle $\widehat{XOY}$.

Often we'll talk about angles without making precise whether these are convex or not. The exact meaning of the case will then result from the context. In the case where the two half-lines, which define an angle, are contained in the same line we define the following special angles.

Flat angle or **straight angle** we call the shape consisting of two opposite half-lines (See Figure 1.16-I). Any of the the two half-planes, defined by the line OX, can be considered as the *interior* or *exterior* of the flat angle.

1.4. ANGLES

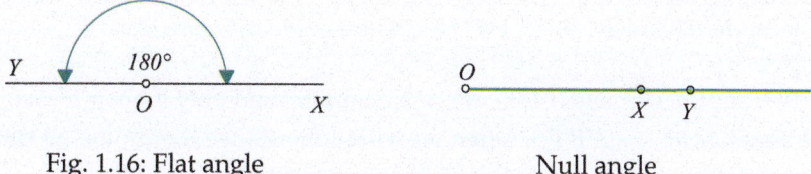

Fig. 1.16: Flat angle   Null angle

We call **Null angle** the shape consisting of two identical half-lines OX and OY (See Figure 1.16-II). In this case the angle has no interior, and its exterior coincides with the whole plane without the half-line OX.

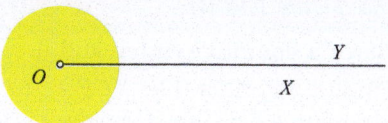

Fig. 1.17: Full turn

We call **Full turn** or **complete turn** or **full revolution** the shape consisting of two identical half-lines OX and OY (See Figure 1.17). In this case the angle has no exterior, and its interior coincides with the whole plane without the half-line OX. The basic properties (axioms) of angles are the following:

Axiom 1.9 *To every angle $\widehat{XOY}$ (convex or not) we associate a number $|\widehat{XOY}| = |\widehat{YOX}| \geq 0$ called* **angle measure in degrees** *. It holds $|\widehat{XOY}| = 0$ if and only if the angle is the null angle.*

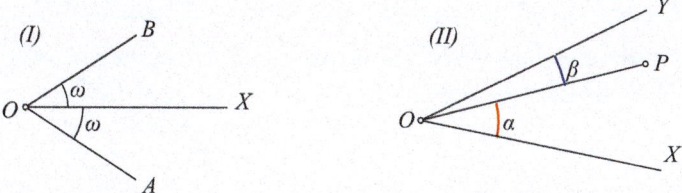

Fig. 1.18: Equal angles on sides of OX $|\widehat{XOY}| = |\widehat{XOP}| + |\widehat{POY}|$

Axiom 1.10 *For every number ω with $0 < \omega < 180°$, there exist exactly two half-lines OA, OB on the two sides of the line OX such that the angles $\widehat{XOA}$ and $\widehat{XOB}$ satisfy $|\widehat{XOA}| = |\widehat{XOB}| = \omega$ (See Figure 1.18-I). A flat angle has measure 180 degrees.*

Traditionally the measure ω in degrees is denoted as $\omega°$. Thus, $30°$ means angle of 30 degrees. A 1/60-th of a degree is called **minute** of a degree and is denoted by a prime on the number of minutes. A 1/60-th of a minute is called **second** of a degree and is denoted by two primes on the number of seconds. This way, $30°23'11''$ denotes an angle measure of $30 + \frac{23}{60} + \frac{11}{3600}$ degrees.

Two angles $\widehat{AB\Gamma}$ and $\widehat{A'B'\Gamma'}$ are called **equal** if and only if, their measures are equal: $|\widehat{AB\Gamma}| = |\widehat{A'B'\Gamma'}|$. Often, in what follows, we'll omit the absolute values and for two equal angles we'll simply write $\widehat{AB\Gamma} = \widehat{A'B'\Gamma'}$.

Axiom 1.11 *For every point P in the interior of angle $\widehat{XOY}$ (convex or not) the measures of angles $\widehat{XOY}$, $\widehat{XOP}$ and $\widehat{POY}$ satisfy the equation $|\widehat{XOY}| = |\widehat{XOP}| + |\widehat{POY}|$. In all such cases we say that angle $\widehat{XOY}$ is the **sum** of angles $\widehat{XOP}$ and $\widehat{POY}$. Often to declare such a relation we'll omit the absolute values and we'll write $\widehat{XOY} = \widehat{XOP} + \widehat{POY}$ (See Figure 1.18-II).*

Two angles that have a common vertex and a common side and non-intersecting corresponding interiors (like $\widehat{XOP}$ and $\widehat{POY}$ of figure 1.18) are called **adjacent**. By drawing a half-line, from the vertex to an interior point of the angle, and using the preceding axiom, we see that non-convex angles have a measure $\omega > 180°$. The axiom guaranties also the existence of the **bisector**, which is a half-line through the vertex of the angle, dividing it in two equal (adjacent) angles.

Exercise 1.11. (Existence of bisector) Show that, for every angle $\widehat{XOY}$, there exists exactly one half-line OZ in its interior, which divides $\widehat{XOY}$ into two equal angles $\widehat{XOZ}$, $\widehat{ZOY}$ with $|\widehat{XOZ}| = |\widehat{ZOY}| = |\widehat{XOY}|/2$.

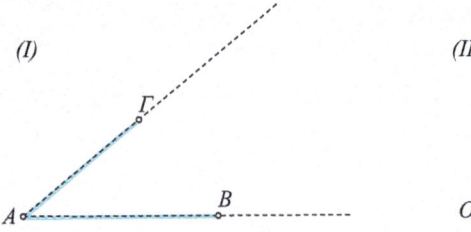
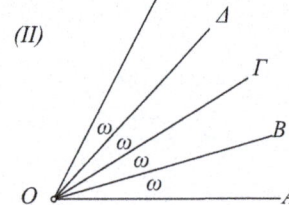

Fig. 1.19:   Angle of AB and $A\Gamma$ Sum of equal angles

Angle of two line segments AB and $A\Gamma$ having a common endpoint A, we call the angle formed by the corresponding half-lines AB and $A\Gamma$ (See Figure 1.19-I).

Exercise 1.12. Find the bisector of a flat angle $\widehat{XOY}$. Show that the measure of a full turn is 360 degrees.

Exercise 1.13. Starting with an angle $\widehat{AOB}$ of measure ω, we construct equal and successive adjacent angles towards the same side $\widehat{BO\Gamma}$, $\widehat{\Gamma O\Delta}$, etc. For

1.4. ANGLES

which measures ω does this procedure applied ν times defines an angle $\widehat{AO\Omega}$, whose side $O\Omega$ coincides with the initial segment OA (See Figure 1.19-II)?

Exercise 1.14. Consider an angle $\widehat{XOY}$ with measure $|\widehat{XOY}| = \alpha$ and a point P in its interior. Show that $|\widehat{XOP}| < \alpha$. Conversely show that for every positive $\beta < \alpha$ there exists a point P in the angle's interior such that $|\widehat{XOP}| = \beta$. Show also that all these P are contained in a half-line with endpoint O.

Exercise 1.15. Let the points A and B be contained in the interior of the convex angle $\widehat{XOY}$. Show that every point of the line segment AB is contained in the interior of the angle $\widehat{XOY}$. Show that the corresponding property does not hold for non-convex angles.

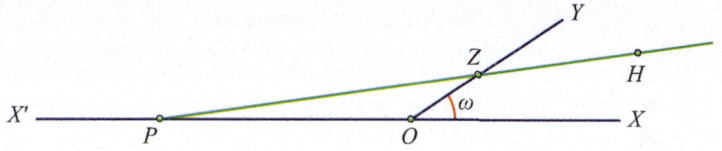

Fig. 1.20: Points in the interior of angle $\widehat{XOY}$

Exercise 1.16. Given is a convex angle $\widehat{XOY}$ and a point P on the opposite half-line OX' of OX, as well as also point Z of the half-line OY. Show that every point H of the half-line PZ, lying outside the line segment PZ, is contained in the interior of the angle $\widehat{XOY}$ (See Figure 1.20).

Hint: By construction, point H is contained in the side of the line OY, in which OX is also contained. Also points Z, H are contained in the same side of the line OX, because the intersection point P of ZH with OX is on the exterior of the line segment ZH.

Remark 1.6. Angle Axiom 1.10 means that we can construct any angle we desire on either sides of a half-line. Like with line segments however, so with angles, the practical construction of an angle with specific measure, using only the ruler and the compass, is a different problem. Thus, for example, for the construction (by ruler and compass) of the angle of 60 degrees, we'll need again properties of the circle, to be discussed later.

Remark 1.7. It is worthwhile to observe some common properties between angles and line segments, especially these that concern the notions of *between*, *adjacent* and *measure*. Figure 1.21 illustrates this. Point O is fixed and does not belong to the fixed line ε. For every line segment AB belonging

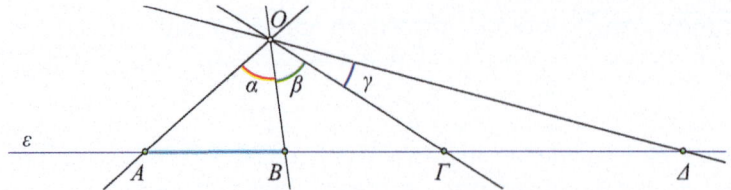

Fig. 1.21: Correspondence of line segments - angles

to the line one may construct the angle $\widehat{AOB}$. Through this correspondence between line segments and angles the notions I mentioned are transferred from the line to the angles with vertex O. This way $A\Gamma$ is the sum of AB and $B\Gamma$ and the corresponding angle $\widehat{AO\Gamma}$ is the sum of $\widehat{AOB}$ and $\widehat{BO\Gamma}$. B is between A and Γ and similarly OB is between OA and $O\Gamma$, adjacent line segments correspond to adjacent angles and so on.

On the occasion of figure 1.21, we can formulate immediately two problems, but in order to solve them, we must first learn to use some tools (for their solution see Problem 3.93 and 3.83).

Problem 1.1. Suppose that in figure 1.21 the angle $\widehat{AOB}$ has constant measure $|\widehat{AOB}| = \alpha$ and rotates about O. Which position of the angle $\widehat{AOB}$ makes $|AB|$ minimal?

Problem 1.2. Suppose that in figure 1.21 the line segment AB glides on line ε keeping its length constant. Which position of AB maximizes the corresponding angle $\widehat{AOB}$?

1.5 Angle kinds

> Many ignore the figure's accuracy, they presuppose it, and pay attention to the proof. Contrary to this, we'll never talk about proofs: the most important job would be to draw completely straight lines, correct, even. To construct a perfect square, to carve a fully round circle.
>
> *Jean-Jacques Rousseau, Émile, or On Education*

Two lines OX and OY intersecting at O define four angles and two pairs of **vertical** angles, i.e. angles, such that the sides of one are extensions of the sides of the other (See Figure 1.22). For the flat angles $\widehat{XOX'}$ and $\widehat{YOY'}$ we have $180° = |\widehat{XOX'}| = |\widehat{XOY}| + |\widehat{YOX'}|$. Also $180° = |\widehat{YOY'}| = |\widehat{YOX}| + |\widehat{XOY'}|$. Because $|\widehat{XOY}| = |\widehat{YOX}|$, we conclude that the opposite vertical angles $\widehat{YOX'}$

1.5. ANGLE KINDS

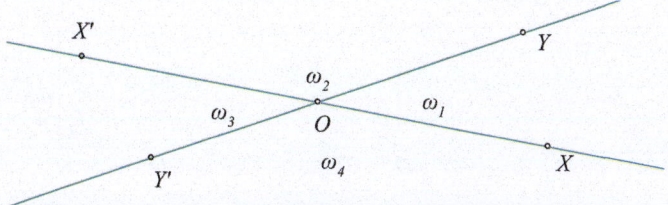

Fig. 1.22: Angles of two lines

and $\widehat{XOY'}$ are equal. Similarly we can also show that $\widehat{XOY}$ and $\widehat{X'OY'}$ are equal. We have therefore proved:

Proposition 1.2. *Vertical angles are equal.*

Two angles adding up to 180° are called **supplementary**. In the preceding figure every pair of adjacent angles consists of supplementary angles. An angle measuring 90° is called a **right** angle. Obviously a right angle is equal to its supplementary. By extending the sides of a right angle at O, that is, by considering the opposite half lines of the sides of the right angle, we define four right angles around this point, every two of which are either supplementary or vertical. Thus, two lines which intersect at O and form

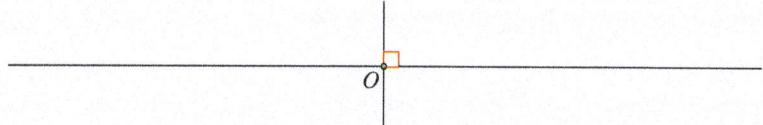

Fig. 1.23: Orthogonal lines

at least one right angle (from the resulting four), will necessarily have the other three angles also right. Two such lines are called **orthogonal** or **vertical** or **perpependicular** (See Figure 1.23). We call an angle $\widehat{XOY}$ **acute** when

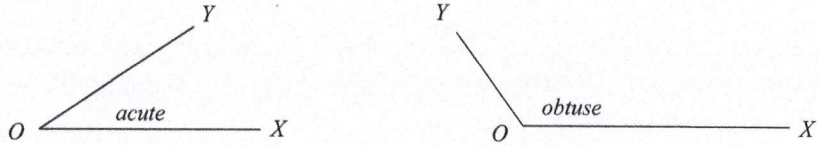

Fig. 1.24: Acute and obtuse angle

its measure is $|\widehat{XOY}| < 90°$ (See Figure 1.24). We call an angle **obtuse** when its measure is $|\widehat{XOY}| > 90°$. Obviously if an angle is acute its supplementary angle will be obtuse and vice-versa. We say that angle α is less/greater than angle β when their measures satisfy: $|\alpha| < |\beta|$ (resp. $|\alpha| > |\beta|$). Obviously every obtuse is greater than a right angle which in turn is greater than every

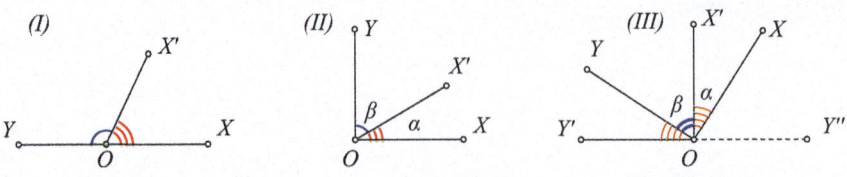

Fig. 1.25: Supplementary,   Complementary,   Orthogonal sides

acute. Two angles whose measures α, β satisfy $\alpha + \beta = 90°$ are called **complementary**. Obviously two complementary angles are both acute. A point X' in the interior of a right angle $\widehat{XOY}$ defines two complementary angles $\alpha = \widehat{XOX'}$, $\beta = \widehat{X'OY}$ (See Figure 1.25-II).

Proposition 1.3. *Two angles $\widehat{XOX'}$ and $\widehat{YOY'}$ which have sides respectively orthogonal are either equal or supplementary (See Figure 1.25-III).*

Proof. If OY and OY' are on the same side of OX', then angles $\alpha = \widehat{XOX'}$ and $\alpha' = \widehat{YOY'}$ are equal, having a common complementary angle β. If OY and OY'' are on opposite sides of OX', then the opposite half line OY' of OY'' forms a supplementary of $\alpha' = \widehat{YOY''}$ and it is on the same of OX' with OY, therefore according to the preceding case $180° - \alpha' = \alpha$.

Corollary 1.1. *From a point A on the line ε passes exactly one line ζ orthogonal to ε (See Figure 1.26).*

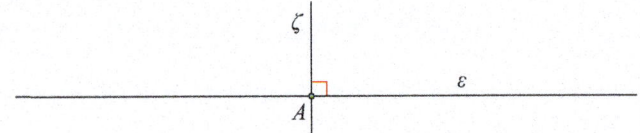

Fig. 1.26: Line ζ orthogonal to ε

Proof. Immediate consequence of Axiom 1.10, according to which there exists exactly one angle of 90 degrees with vertex at A, with one side on ε and contained in one of the half planes of ε.

1.6 Triangles

> The great trick of regarding small departures from the truth as the truth itself - on which is founded the entire integral calculus - is also the basis of our witty speculations, where the whole thing would often collapse if we considered the departures with philosophical rigour.
>
> G.J. Lichtenberg, Aphorisms

Triangles, after lines and angles, are the simplest figures of the plane. In spite of their simplicity they have infinite many properties and are an object of study by well known Mathematicians throughout the ages. The up to date properties and conclusions are so many, that they form a special branch of geometry, the so called *Geometry of the triangle* ([28], [19], [40]).

A **triangle** is a figure defined by three points A, B and Γ, not contained in a line, as well as the line segments which join these points together. The

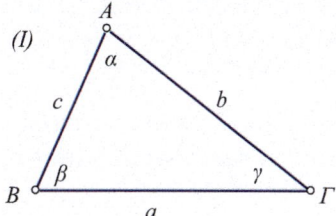

Fig. 1.27: Triangle   Interior and exterior of triangle

three points are called **vertices** of the triangle. The line segments defined by pairs of vertices are called **triangle sides** (See Figure 1.27). The **Angles** of the triangle are the (convex) angles formed by the triangle's sides. The lengths of the sides of a triangle are usually denoted by the Latin letters

$$a = |B\Gamma|, \ b = |\Gamma A|, \ c = |AB|$$

and the measures of the angles of a triangle by small greek letters

$$\alpha = |\widehat{BA\Gamma}|, \ \beta = |\widehat{\Gamma BA}|, \ \gamma = |\widehat{A\Gamma B}| \ \text{ or simply } \ \alpha = \widehat{A}, \ \beta = \widehat{B}, \ \gamma = \widehat{\Gamma}.$$

I'll use this notation often in subsequent chapters.

We say that the angles $\widehat{AB\Gamma}$, $\widehat{B\Gamma A}$, $\widehat{\Gamma AB}$ are respectively **opposite** to the sides $A\Gamma$, BA and ΓB. The sum of the lengths

$$\sigma = a + b + c,$$

is called the triangle's **perimeter**. We often also use the **half-perimeter** or **semi-perimeter**, denoted by $\tau = \sigma/2$.

A triangle is called (1) **acute**, (2) **obtuse**, (3) **scalene**, when it has respectively, (1) all its angles acute, (2) one obtuse angle, (3) sides with different lengths.

Two triangles $AB\Gamma$ and $A'B'\Gamma'$ are called **congruent** or **isometric** or **equal**, when their corresponding sides are equal ($a = a'$, $b = b'$, $c = c'$) and they have respectively equal angles ($\alpha = \alpha'$, $\beta = \beta'$, $\gamma = \gamma'$).

The basic properties (axioms) of the triangle are:

Axiom 1.12 *Every triangle divides the plane into two parts, the **interior** and **exterior**. Two points X and Y contained in the interior of the triangle define a line segment XY, which is contained entirely in the triangle's interior (See Figure 1.27-II). Two points X and Z, one of which is contained in the interior and the other in the exterior of a triangle define a line segment XZ which either contains a triangle vertex or intersects exactly one side of the triangle in an interior point Ω of this side.*

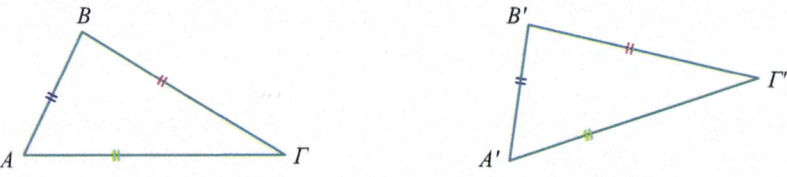

Fig. 1.28: Triangles with corresponding sides equal

Axiom 1.13 (of triangle congruence) *Two triangles $AB\Gamma$ and $A'B'\Gamma'$ having their corresponding sides equal ($|AB| = |A'B'|$, $|B\Gamma| = |B'\Gamma'|$, $|\Gamma A| = |\Gamma'A'|$) are congruent. In other words their corresponding angles are also equal in measure. Moreover opposite of respectively equal sides, the corresponding angles are equal in measure (See Figure 1.28).*

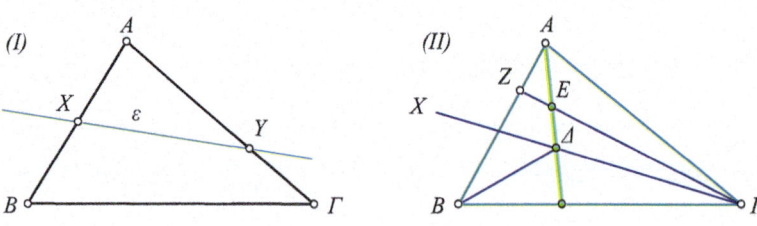

Fig. 1.29: Pasch's axiom   Intersection on opposite side

Axiom 1.14 (of Pasch (1843-1940)) *If a line ε intersects the side AB of a triangle and does not contain any of the triangle's vertices, then it also intersects one of the other sides (See Figure 1.29-I).*

Remark 1.8. The axiom for the interior and exterior of a triangle is one of the cases I mentioned in the beginning of the chapter. It can be inferred from

1.6. TRIANGLES

the rest of the axioms, consequently it could be proved as a theorem. The proof however contains details I don't consider interesting for the student to involve himself with at this stage. Hence its presentation here as an axiom.

The last axiom, Axiom 1.14 seems to be self-evident. However the property it expresses cannot be inferred from the preceding axioms. Its usefulness can be seen from the next proposition as well as from the exercise which follows. These two propositions are presented here in order to give a taste of the details one should pay attention to, if he insists to prove *all* obvious properties, relying on the axioms. A plethora of similar "self-evident" propositions can be seen in [15, pp.42-84], [4].

Proposition 1.4. *If Δ is an interior point of a triangle, then $A\Delta$ intersects the opposite side $B\Gamma$ of the triangle (See Figure 1.29-II).*

Proof. Consider a point E on the segment $A\Delta$. Next consider the triangle $AB\Delta$ and the secant line ΓE. According to Axiom 1.14, ΓE will also meet a second side of the triangle $AB\Delta$. Side $B\Delta$ of this triangle is outside of angle $\widehat{X\Gamma A}$, so ΓE will meet AB at a point Z. Consider then the triangle $B\Gamma Z$ and the line AE, which intersects its side of ΓZ. According to the Axiom 1.14, AE will also meet another side of the triangle, which cannot be BZ, because then AE would coincide with BZ. Therefore AE, which is the same as line $A\Delta$, will meet side $B\Gamma$ of triangle $B\Gamma Z$, which is also a side of the triangle $AB\Gamma$.

Exercise 1.17. Given a line ε, show that the relation between two points A and B: "A and B are contained in the same half plane of ε" is transitive. In other words if A and B are contained in the same half plane and B and Γ are also contained in the same half plane, then A and Γ are also contained in the same half plane.

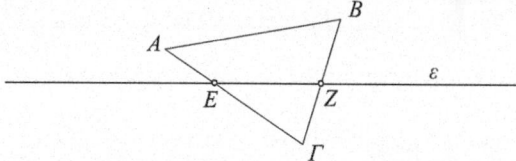

Fig. 1.30: Meaning of axiom 1.14

Hint: Let points A, B be in the same half plane and that B, Γ are also in the same half plane, but A, Γ are not (See Figure 1.30). Then, by Axiom 1.5 there exists a point E of line ε belonging to $A\Gamma$ and lying between A and Γ. In other words ε intersects $A\Gamma$. Because ε does not contain A, B, Γ and intersects one side of the triangle ($A\Gamma$) it will also intersect, by Axiom 1.14, another side. If it intersects $B\Gamma$ at point Z, we have a contradiction, because then B, Γ will be contained in different half planes of ε. If it intersects AB we get a similar contradiction. Therefore $A\Gamma$ cannot intersect ε.

Median of the triangle is called the line segment which joins any vertex with the midpoint of the opposite side. **Bisector** of the triangle is called the line segment which joins any vertex with the opposite side and divides the

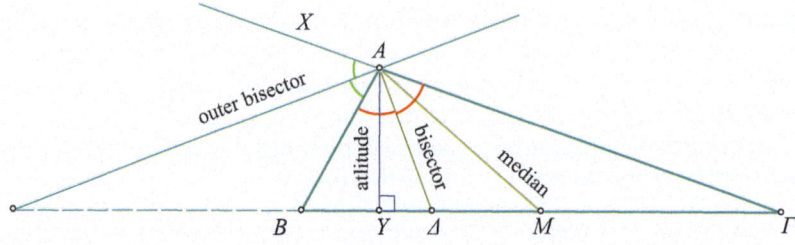

Fig. 1.31: Median, bisector, altitude

angle at this vertex into two equal parts. We often call also "bisector" the half line or the entire line, which divides the corresponding angle. **Exterior angle** of a triangle is called the angle which is supplementary to a given triangle angle, for example to angle $\widehat{A}$, and which results by extending one of the sides of the triangle, like $A\Gamma$ (by considering the line $A\Gamma$), resulting in angle $\widehat{BAX}$ (See Figure 1.31). **External bisector** is called the bisector of an external triangle angle. **Altitude** of a triangle is called the line segment, which joins a vertex with a point belonging to the opposite side, to which it is also perpendicular (we'll guarantee the existence of the altitude later in § 1.12). As we'll see later, the three triangle medians meet in point (Theorem 2.15), the three bisectors meet in another common point (Theorem 2.3) and finally the three altitudes meet in a third point (Theorem 2.16).

A triangle's medians, altitudes and bisectors (internal and external) are commonly referred to as **secondary elements** of the triangle.

Remark 1.9. Often knowledge of the length of three of these elements is sufficient for the accurate construction of a triangle. For example, in Exercise 2.151, we'll see that a triangle can be constructed easily when we know the three lengths $|AY|$, $|A\Delta|$ and $|AM|$ corresponding to altitude, bisector and median from a given vertex. Usually we require the exclusive use of ruler and compass for the construction of triangles. A relatively complex problem is to prove, that a certain geometric construction is impossible (using only the ruler and the compass). For example, to construct the triangle from the altitude $|AY|$, the median $|AM|$ from the same vertex but the bisector from a vertex different than A. The construction of the corresponding triangle from these elements has been proved impossible ([18, p.38]). Notice however that non-constructibility, using exclusively the ruler and compass does not mean that the triangle is not constructible by using other means. Thus, for example, given three positive numbers there exists exactly one triangle having these numbers as its bisector lengths. However, this triangle cannot be constructed using only ruler and compass ([32], [35]).

Exercise 1.18. Show that the internal and external bisectors of a given triangle vertex are orthogonal.

1.7 Congruence, the equality of shapes

> It is a mistake that equality is the law of nature. Nature does not produce equality. The supreme law is ordering and dependence.
>
> *Vauvenargues, Reflections and Maxims*

A point, a line, a half-line, a line segment, a triangle, are **shapes**. More generally, we call (plane) shape any specific set of points on the plane. The ones we examined, up to now, are the simplest shapes. In the next lessons we'll encounter other more complex shapes and we'll study properties which hold for each of them and are the same for the so called **congruent shapes** or **isometric shapes**.

Every shape has a rule, which determines when it is equal to another. Line segments have their length. They are equal (or congruent or isometric) exactly when they have the same length. The same for angles. They have their measure. They are equal when they have the same measure. With triangles the notion of equality contains more ingredients. The definition of congruence in this case requires two triangles to have equal respective sides and respective angles. Axiom 1.13 gives the basic criterion for triangle congruence. It says that, when two triangles have respective sides equal, then they are congruent. In other words, their respective angles (the ones opposite to corresponding equal sides) will also be equal. Later (§ 1.9) we'll see more criteria for triangle congruence. The more complex the shape, the more elements of it we must compare to conclude that it is congruent to another shape.

Euclid in his *Elements* does not waste any time analyzing the notion of congruence. He adopts a simple notion of equality, under which two shapes are equal, if and only if we can displace one of the shapes and place it upon the other, so that the two shapes become exactly coincident. But what does it mean to *displace*? This notion of displacement is complex. It can be founded upon the more general notion of *transformation* and more specifically on that of *isometry* or *congruence*, which we'll discuss later (II-§ 2.1).

We initially define *congruence* by giving each shape its rule, which dictates when it is congruent to another. It doesn't hurt however to think in Euclid's terms. Two shapes of the plane, which are congruent according to their rule of congruence, are indeed equal according to Euclid's definition, using displacement and coincidence. The converse is also true: If two shapes can be moved around or displaced until they actually become coincident, then they are congruent according to the rules we give for each case. The problem is, that in order to actually prove this sort of equivalence, we must study various additional topics, whose presentation here, at this point, would create some inconvenience. We therefore limit ourselves to Euclid's rule. For practical purposes this means, that we conceive the plane as a transparent plastic medium and that the shapes can be cut from their initial place, displaced to the position of the second shape and placed on top of it until coincidence is achieved. The actual precise foundation of the notion of congruence is done much later in (II-§ 2.5), which I recommend for a second reading. I

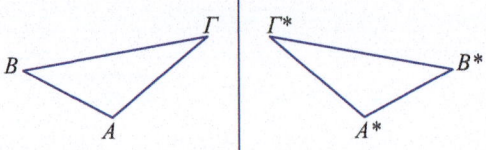

Fig. 1.32: Congruent but with opposite orientation

note a peculiarity of the notion of congruence which reveals itself in figure 1.32 and has to do with the so called **orientation** of shapes. The two triangles are congruent with our notion of equality. They are peculiar however, in that the direction of succession $A \to B \to \Gamma$ is clockwise, while that of succession $A^* \to B^* \to \Gamma^*$ is counterclockwise. Triangle $AB\Gamma$ is said to be **negatively oriented**, while $A^*B^*\Gamma^*$ is said to be **positively oriented**. In order to achieve total coincidence between triangles $AB\Gamma$ and $A^*B^*\Gamma^*$ under the notion of displacement, we must cut one triangle and turn it upside down, in the same way we turn a page and read the opposite side. Things become slightly more complex in space, where a similar phenomenon occurs yet in this case there's no outside space to perform this turning of the page. There the notion of congruence the way we defined it, for each figure separately, is not equivalent to the notion of coincidence (see for example the remark II-4.1 and on a second reading the full description of the notion of congruence in II-§ 2.5 and II-§ 7.5). Therefore, the way we handle congruence is safer than that of *displacement*, as long as we refrain from going into the details of the exact definition of this notion.

Remark 1.10. For some shapes, congruence, under the notion of displacement, is obvious. Thus, for example, any two lines α and β are equal, in the sense that α can clearly be displaced and be put onto β, so that the two lines become coincident. Likewise, two intersecting lines α and β forming an angle of measure ω make a shape which is congruent with the shape of two other lines α' and β' making an angle of equal measure ω.

Remark 1.11. A remark on notation: Often, with the congruence of two shapes, which have angles, vertices or other similar characteristics, we create correspondences between their vertices, by denoting corresponding vertices with the same letter and adding an index, prime, star or some other such symbol to the second vertex to show the implied correspondence. This way, when we say that triangles $AB\Gamma$ and $A'B'\Gamma'$ are congruent because they have their corresponding sides equal, we mean that side AB is respectively equal to $A'B'$, $B\Gamma$ is equal to $B'\Gamma'$ etc. We follow this rule even with space figures. Creating this notational correspondence in shapes, which are candidates for congruence, becomes the first important step in trying to prove their congruence, which usually reduces to equality of their respective simpler elements.

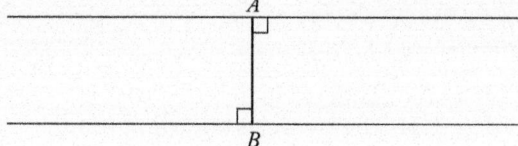

Fig. 1.33: A simple shape

Exercise 1.19. The shape of figure 1.33 consists of a line segment AB of length δ and the lines which are orthogonal at each of its endpoints. Show that every line segment $A'B'$ of same length δ, defines by analogy a shape congruent to the first shape under the notion of displacement.

Hint: Because of length equality, line segment $A'B'$ can be displaced until it becomes coincident with AB. Then, according to Axiom 1.10, the orthogonal lines at the endpoints of $A'B'$ will necessarily coincide with those of AB at its endpoints.

1.8 Isosceles and right triangle

> 'Tis the perception of the beautiful,
> A fine extension of the faculties,
> Platonic, universal, wonderful,
> Drawn from the stars, and filter'd through the skies,
> Without which life would be extremely dull;
> In short, it is the use of our own eyes,
> With one or two small senses added, just
> To hint that flesh is form'd of fiery dust.
>
> Byron, Don Juan Canto II, 212

In Geometry, as in all of Mathematics, after a general categorical definition, it is useful to examine some special cases. It is not rare, for some general property we want to prove, to establish the result in a special case and find through it the way to the general case. Other times, again, the properties of the special case help in expressing the proof and properties of the general case or the rejection of some general conjecture. The isosceles and right triangles are special cases of triangles, which we meet in the formulations and proofs of a plethora of more general properties in all geometry chapters.

Fig. 1.34: The isosceles and right triangle

Isosceles triangle is called a triangle which has two equal sides (See Figure 1.34). The two sides, which are equal in length, are called **legs** of the isosceles. The third side is called **base** of the isosceles. The vertex, formed by the two equal sides, is usually called **apex** of the isosceles.

Right triangle (or right-angled triangle) is called a triangle which has one angle equal in measure to 90 degrees. The sides which define this angle are called **orthogonal sides**. The side opposite of the right angle is called **hypotenuse** ($A\Gamma$ in right figure 1.34) of the right triangle.

Theorem 1.1. *In every isosceles triangle ($|AB| = |A\Gamma|$) the two base angles ($\widehat{B}$ and $\widehat{\Gamma}$) are equal.*

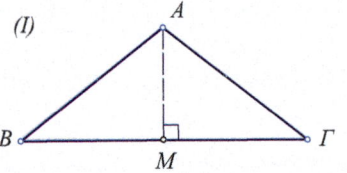
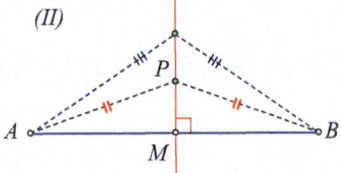

Fig. 1.35: Median AM of isosceles   Medial line of AB

Proof. Consider the two triangles ABM and $AM\Gamma$, which are formed by drawing AM, where M is the middle of the base $B\Gamma$ (See Figure 1.35-I). The two triangles have their sides respectively equal: $|AB| = |A\Gamma|$ by hypothesis, $|BM| = |M\Gamma|$ because M is the middle of $B\Gamma$ and finally AM is a common side of both triangles. According to triangle Axiom 1.13 these two triangles will be congruent, therefore their angles at B and Γ will be respectively equal.

Corollary 1.2. *In every isosceles triangle ($|AB| = |A\Gamma|$) the line which joins its apex A with the middle M of the opposite side bisects the apex angle at A (See Exercise 1.29 for the converse).*

Corollary 1.3. *In every isosceles triangle ($|AB| = |A\Gamma|$) the line which joins its vertex A with the middle M of the opposite side is orthogonal to the base and divides the triangle into two congruent right triangles (AMB and AMΓ) (See Exercise 1.31 for the converse).*

Medial line or **Line-bisector** of the segment AB we call the line, which is orthogonal to the segment at its middle (See Figure 1.35-II). The preceding corollary can also be expressed in the following form.

Corollary 1.4. *In every isosceles triangle ABΓ with base AB, its apex Γ lies on the medial line of the segment AB.*

The following formulation is also equivalent:

Corollary 1.5. *Every point P, which is equidistant from points A and B, lies on the medial line of the segment AB.*

Exercise 1.20. For every isosceles triangle $AB\Gamma$ with base AB, its apex Γ lies on the medial line of AB.

Corollary 1.6. *Every point P which is equidistant from the points $\{A, B\}$ lies on the medial line of AB.*

Exercise 1.21. Assuming that two lines which are orthogonal to the same line do not intersect (we'll show this later in corollary 1.15), show that: Given three different points $\{A, B, \Gamma\}$ on the same line ε, there is no point X, which is equidistant from these points.

1.9 Triangle congruence criteria

> Laws, taken in the broadest meaning, are the necessary relations deriving from the nature of things; and in this sense, all beings have their laws.
>
> Montesquieu, *The Spirit of the Laws*

Besides the basic Axiom 1.13 of triangle congruence, which is also referred to as *SSS-criterion* (side-side-side criterion) of congruence, there are two more congruence criteria, which result as theorems relying on the SSS-criterion. These are referred to as SAS-criterion (side-angle-side criterion) and ASA-criterion (angle-side-angle criterion).

Proposition 1.5 (SAS-criterion). *Two triangles $AB\Gamma$, $A'B'\Gamma'$, which have two equal respective sides ($|AB| = |A'B'|$, $|A\Gamma| = |A'\Gamma'|$) and the contained in them angles also equal ($|\widehat{BA\Gamma}| = |\widehat{B'A'\Gamma'}|$), are congruent.*

Proof. Place angle $\widehat{A'}$ onto $\widehat{A}$ so that the half lines AB and $A'B'$ and $A\Gamma$ and $A'\Gamma'$ become coincident. This is possible because of the supposed equality of the angles respectively at A and A' (See Figure 1.36). Because of the also supposed equality of lengths $|AB| = |A'B'|$, we'll also have coincidence of B and B' (according to Axiom 1.8), and, for the same reason, we'll have

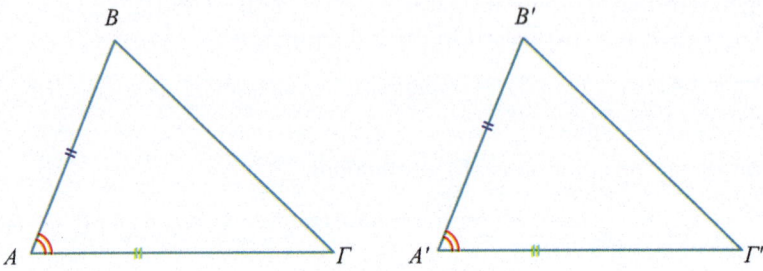

Fig. 1.36: SAS-criterion

coincidence of Γ and Γ'. As a result, we have coincidence of the sides $B\Gamma$ and $B'\Gamma'$ and therefore their lengths will be equal $|B\Gamma| = |B'\Gamma'|$. The truth of the proposition results by applying the SSS-criterion.

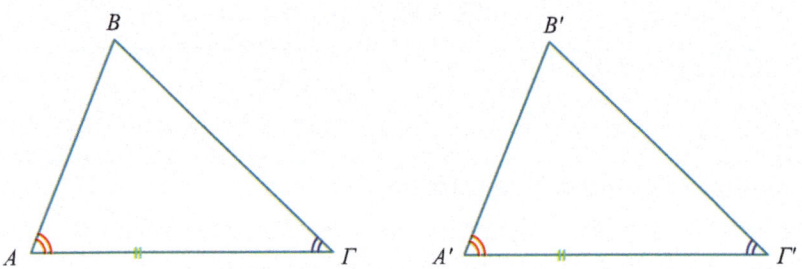

Fig. 1.37: ASA-criterion

Proposition 1.6 (ASA-criterion). *Two triangles $AB\Gamma$, $A'B'\Gamma'$, which have two respective angles equal ($|\widehat{AB\Gamma}| = |\widehat{A'B'\Gamma'}|$ and $|\widehat{B\Gamma A}| = |\widehat{B'\Gamma'A'}|$) and the contained in them sides also equal ($|B\Gamma| = |B'\Gamma'|$), are congruent.*

Proof. The proof is similar to the preceding one. Place the triangles so that $B\Gamma$ and $B'\Gamma'$, as well as the angles at B, B' and Γ, Γ' become coincident (See Figure 1.37). This is possible because of Axiom 1.10. Then we get coincidence of lines BA, $B'A'$ as well as ΓA, $\Gamma'A'$, consequently we get coincidence of their respective intersections which define the points A and A'. From this last coincidence we get $|BA| = |B'A'|$ and $|\Gamma A| = |\Gamma'A'|$. The truth of the proposition results by applying the SSS-criterion.

1.9. TRIANGLE CONGRUENCE CRITERIA

Proposition 1.7. *If a triangle has two of its angles equal then it is an isosceles.*

Proof. Consider triangle $A'B'\Gamma'$ congruent to $AB\Gamma$ and apply the ASA criterion (See Figure 1.38). The two triangles have the sides $B\Gamma$ and $B'\Gamma'$ equal respectively and angles $\widehat{AB\Gamma}$ and $\widehat{A'\Gamma'B'}$ equal, as well as $\widehat{A\Gamma B}$ and $\widehat{A'B'\Gamma'}$

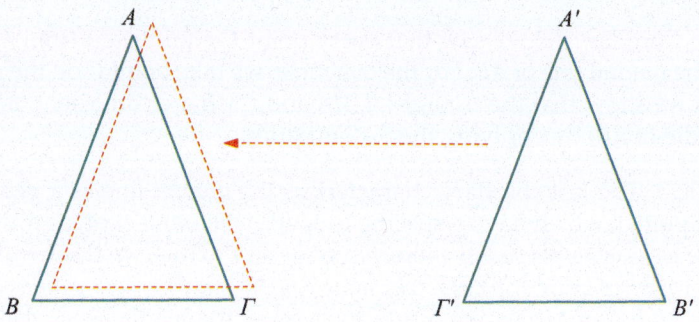

Fig. 1.38: Two equal angles produce an isosceles

equal, therefore they are congruent. Side $A\Gamma$ opposite to angle $\widehat{AB\Gamma}$ will be equal to the side $A'B'$ opposite to angle $\widehat{A'\Gamma'B'}$, which is equal to the preceding angle. But, by construction $A'B'$ is equal to AB, therefore finally AB and $A\Gamma$ are equal.

Remark 1.12. This proof (due to Pappus) contains a subtle and paradoxical point, where two congruent triangles are proved again equal. The resulting wordplay has to do with the triangle's **orientation**. It is true that $A'B'\Gamma'$ is equal to $AB\Gamma$, yet it has been placed against $AB\Gamma$ after reversing its orien-

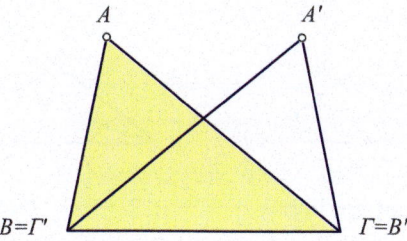

Fig. 1.39: Placement of congruent triangles with opposite orientation

tation. Figure 1.39 illustrates the same procedure of repositioning the same triangle on its base $B\Gamma$ for a non-isosceles $AB\Gamma$. Note that, although the two triangles are congruent, this particular placement does not make them coincident. The deeper meaning of the proposition is, that the two congruent triangles, placed this way, become coincident, when and only when they are equal isosceli.

Corollary 1.7. *Point Γ belongs to the medial line ε of line segment AB, if and only if it is equidistant from A and B.*

Proof. In Corollary 1.5 we saw that every point Γ equidistant from A and B lies on the medial line of AB. For the converse, we take point Γ on the medial line and we show that the triangles ΓMA and ΓMB are congruent (M being the middle of AB) by applying the SAS-criterion.

Remark 1.13. The last corollary characterizes the medial line as a **geometric locus** of points, which have a specific property. We often express this as: *the geometric locus of points having property x is the set Y*. Thus, we'll say from now on: *the geometric locus of points, which are equidistant from two points A and B is the medial line of AB*. As we did in the case of the medial line, also in the general case of a geometric locus, we must show two things: a) every point of the geometric locus Y has property x, b) if a point has property x then it necessarily belongs to the geometric locus Y (more in § 2.16).

Exercise 1.22. Two right triangles which have respective perpendicular sides of the same length are congruent.

Hint: Apply the SAS-criterion with respective angles the right angles of the two triangles.

Exercise 1.23. Two right triangles which have an orthogonal side and the adjacent acute angle respectively equal are congruent.

Hint: Apply the ASA-criterion.

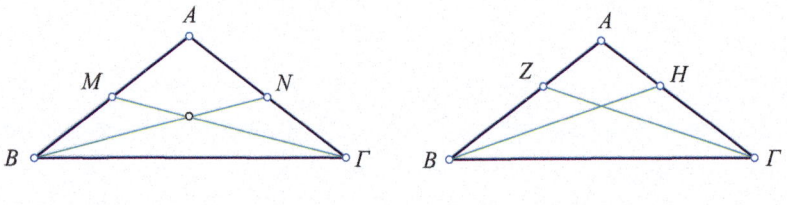

Fig. 1.40: Equal medians   Equal bisectors

Exercise 1.24. Let ABΓ be an isosceles triangle with equal angles at B and Γ. Show that the medians through these vertices are equal. Also show that the bisectors through these vertices are equal.

Hint: Let M and N be the middles of BA and ΓA respectively (See Figure 1.40). Triangles BMΓ and BNΓ are congruent as having a) BΓ common, b)

1.9. TRIANGLE CONGRUENCE CRITERIA

BM and ΓN equal as halves of equal sides, c) angles at B and Γ equal. The SAS-criterion therefore is applicable. The proof for bisectors is similar, only this time the ASA-criterion is applicable. Indeed, suppose BH and ΓZ are the bisectors of the angles at B and Γ, respectively. Then triangles $B\Gamma H$ and $B\Gamma Z$ are congruent as having a) $B\Gamma$ common, b) angles at B and Γ equal, c) angles $|\widehat{HB\Gamma}| = |\widehat{Z\Gamma B}|$ equal as halves of equal angles.

Remark 1.14. The converse of the proposition in the last exercise, also holds, but in the case of medians we need a property to be discussed later (see Exercise 2.83). In the case of bisectors the proof of the converse, which we give in the next section (Theorem 2.8), is referred to as *theorem of Steiner-Lehmus* and is unexpectedly difficult. For a computational proof see Exercise 3.139.

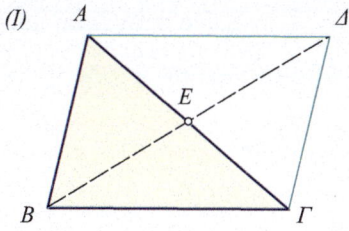
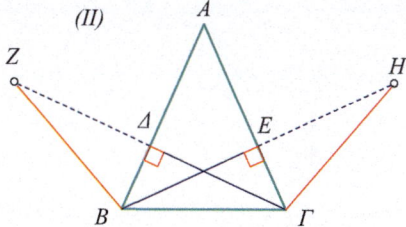

Fig. 1.41: Extension of the median   Equal altitudes of isosceles

Exercise 1.25. Let E be the middle of side $A\Gamma$ of the triangle $AB\Gamma$. Extend BE (median) so as to double its length until Δ. Show that triangles $A\Gamma\Delta$ and $A\Gamma B$ are congruent.

Hint: Using the SAS-criterion first show that triangles AEB and $\Gamma E\Delta$ are congruent (See Figure 1.41-I). Similarly, also show that $BE\Gamma$ and $AE\Delta$ are congruent. Conclude next, using the SSS-criterion, that $AB\Gamma$ and $A\Gamma\Delta$ are congruent.

Exercise 1.26. Let $AB\Gamma$ be an isosceles triangle with equal angles at B and Γ. Show that the altitudes through these vertices are equal.

Hint: Let BE and $\Gamma\Delta$ be the altitudes respectively from B and Γ (See Figure 1.41-II). Extend BE by doubling its length to point H and $\Gamma\Delta$ by doubling its length to point Z. Triangles $BE\Gamma$ and $HE\Gamma$ are congruent having (i) $E\Gamma$ common, (ii) angles at E right and (iii) sides BE and EH equal by construction (SAS-criterion). Consequently triangle $B\Gamma H$ is isosceles. Similarly we show that triangle $B\Gamma Z$ is isosceles. These two triangles are congruent, having (i) $B\Gamma$ common, (ii) BZ equal to ΓH and (iii) their angles at B and Γ equal to 2β and 2γ respectively. Thus ΓZ and BH, which are twice the altitudes, are equal. For the converse of this property see Exercise 1.33.

Remark 1.15. We'll see that there is one more triangle congruence criterion, which could be called AAS-criterion. According to this if two triangles $AB\Gamma$ and $A'B'\Gamma'$ have their angles $\alpha = \alpha'$, $\beta = \beta'$ and sides $a = |B\Gamma| = |B'\Gamma'| = a'$, then they are congruent. In this case it is supposed that the triangles have two angles equal and a side respectively equal to another, but this side is *opposite* to α and not adjacent to α (like in the ASA-criterion). This criterion however reduces to the ASA-criterion, because from the equality of the two angles and the relation $\alpha + \beta + \gamma = 180°$, which we'll show later, results the equality of *all* angles of the two triangles.

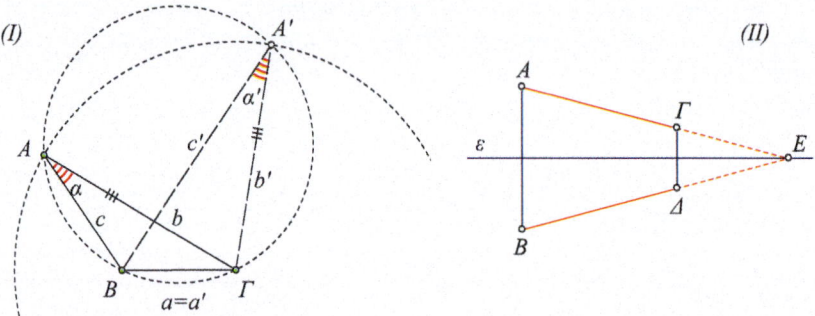

Fig. 1.42: Ambiguous case   Common medial line

Figure 1.42-I shows, that a conceivable SSA-criterion is not valid. In general (for a not right triangle), there are two triangles $AB\Gamma$ and $A'B'\Gamma'$ for which holds $a = a'$, $b = b'$ and $\widehat{A} = \widehat{A'}$. We'll analyze this figure later, when we'll have sufficient knowledge about circles and their properties.

Exercise 1.27. Show that, if the triangles $AB\Gamma$ and $A'B'\Gamma'$ are congruent, then the medians/bisectors of $AB\Gamma$ are respectively equal to those of $A'B'\Gamma'$.

Exercise 1.28. Suppose that the line segments AB and $\Gamma\Delta$ have a common medial line ε and that $A\Gamma$ meets ε in E. Show that $B\Delta$ also meets ε in E (See Figure 1.42-II).

Exercise 1.29. Show that, if the median AM of triangle $AB\Gamma$ bisects the angle $\widehat{BA\Gamma}$, then the triangle is isosceles (converse of Corollary 1.2).

Exercise 1.30. Show that two congruent triangles $AB\Gamma$ and $A'B'\Gamma'$ have equal respective altitudes.

Hint: Place the triangles so that their bases $B\Gamma$ and $B'\Gamma'$ coincide and vertices A, A' are on either side of $B\Gamma$. Then ABA' is an isosceles triangle, etc.

Exercise 1.31. Show that, if the median AM of triangle $AB\Gamma$ is perpendicular to the base $B\Gamma$, then the triangle is isosceles (converse of Corollary 1.3).

1.10 Triangle's sides and angles relations

> Vain would be the attempt of telling all the figures of them circling as in a dance, and their juxtapositions, and the return of them in their revolutions upon themselves, and their approximations...
>
> *Plato, Timaeus 40*

Next properties of the triangle rely on the axioms we have accepted up to now and prepare the ground for the all important *triangle inequality* of the next section.

Theorem 1.2. *The supplementary of a triangle's angle is greater than each of the other two angles.*

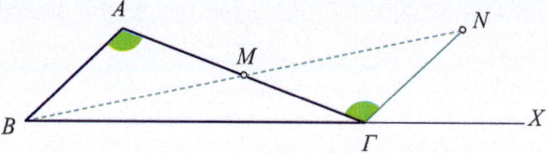

Fig. 1.43: Triangle angles comparison

Proof. In the triangle $AB\Gamma$ (See Figure 1.43) we show that $\widehat{X\Gamma A}$, which is the supplementary angle of $\widehat{\Gamma}$, is greater than $\widehat{A}$. In fact, let M be the middle of $A\Gamma$ and suppose that N lies on the half line BM, so that $|BM| = |MN|$. Triangles ABM and ΓNM have: (i) the angles $\widehat{AMB}$ and $\widehat{\Gamma MN}$ equal, since they are vertical, (ii) the sides AM and $M\Gamma$ equal, because M is the middle of $A\Gamma$, (iii) the sides MB and MN equal by construction. Consequently, by the SAS-criterion Proposition 1.5, the triangles are congruent. From this results that angle $\widehat{BA\Gamma}$ is equal to $\widehat{A\Gamma N}$. The last one however is smaller than $\widehat{A\Gamma X}$ because N belongs in the interior of $\widehat{A\Gamma X}$ (Exercise 1.16). Similarly we show that angle $\widehat{AB\Gamma}$ is smaller than $\widehat{A\Gamma X}$.

Remark 1.16. Angle $\widehat{A\Gamma X}$ is referred to as an **external angle** (see § 1.6) of the triangle and the theorem takes the form:

An external angle of a triangle is greater than each of the opposite internal angles.

Proposition 1.8. *Let the point Δ be in the interior of triangle $AB\Gamma$. Then angle $\widehat{B\Delta\Gamma}$ is greater than $\widehat{BA\Gamma}$.*

Proof. Extend one of the sides of the internal angle, for example $B\Delta$ and define the intersection point E with $A\Gamma$ (See Figure 1.44-I). Angle $\varepsilon = |\widehat{BE\Gamma}| > \alpha = |\widehat{BA\Gamma}|$ as an external and opposite of α in triangle ABE. Similarly also $\delta = |\widehat{B\Delta\Gamma}| > \varepsilon$ and by combining the inequalities, $\delta > \alpha$.

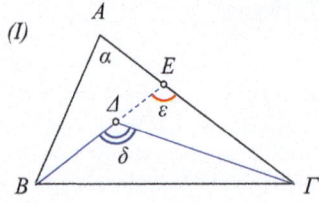 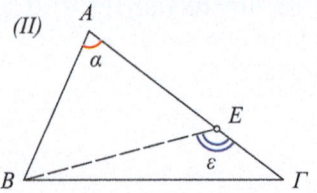

Fig. 1.44: $\alpha < \varepsilon < \delta$ Greater angle opposite greater side

Theorem 1.3. *In every triangle* $AB\Gamma$ *opposite to a greater side lies a greater angle.*

Proof. Let the side $A\Gamma$ be greater than AB (See Figure 1.44-II). Then, there is a point E between A and Γ, such that $|AE| = |AB|$ (Axiom 1.8, Exercise 1.3). The triangle ABE is isosceles and, according to the preceding theorem, $|\widehat{ABE}| = |\widehat{AEB}| > |\widehat{A\Gamma B}|$. Also $|\widehat{AB\Gamma}| > |\widehat{ABE}|$ because E belongs in the interior of angle $\widehat{AB\Gamma}$. Combining the inequalities we get $|\widehat{AB\Gamma}| > |\widehat{A\Gamma B}|$.

Theorem 1.4. *In every triangle* $AB\Gamma$ *opposite to a greater angle lies a greater side.*

Proof. Using reduction to contradiction. Suppose that in the triangle $AB\Gamma$ the angle $\alpha > \beta$ but the side $a \leq b$. The equality $a = b$ is not possible, since the triangle would be isosceles and the angles $\alpha = \beta$ (Theorem 1.1), contrary to our assumption. The inequality $a < b$ is also not possible, since, if this was the case, then according to the preceding proposition, we would obtain $\alpha < \beta$, which is again contrary to our assumption. Therefore $a \leq b$ cannot be true and we must have $a > b$

Corollary 1.8. *In every triangle the sum of any two of its angles is less than 180 degrees.*

Proof. Denote the angles by α, β and γ. According to Theorem 1.2, each of α, β is less than $180° - \gamma$. Therefore $\alpha + \gamma < 180°, \beta + \gamma < 180°$. Similarly one proves $\alpha + \beta < 180°$.

Corollary 1.9. *A triangle has at most one obtuse angle.*

Proof. If it had two obtuse angles, for example α and β, then it would be $\alpha + \beta > 180°$, which contradicts the preceding corollary.

Corollary 1.10. *In every isosceles triangle its two equal angles are acute.*

Corollary 1.11. *Every right triangle has its other two angles acute.*

Corollary 1.12. *In a right triangle each one of the orthogonal sides is smaller than the hypotenuse .*

1.10. TRIANGLE'S SIDES AND ANGLES RELATIONS

Proof. The hypotenuse is opposite to the right angle which is greater than each of the other angles which are acute (Corollary 1.11), therefore the two orthogonal sides, which are opposite to the acute angles, will be less than the hypotenuse (Theorem 1.4).

Corollary 1.13. *Let $A\Delta$ be orthogonal to XY, where Δ is a point of XY. Then, points B and Γ satisfy $|B\Delta| < |\Gamma\Delta|$, if and only if $|BA| < |\Gamma A|$.*

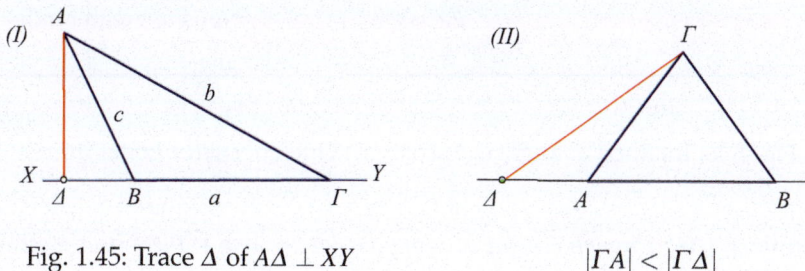

Fig. 1.45: Trace Δ of $A\Delta \perp XY$ $\quad\quad\quad |\Gamma A| < |\Gamma \Delta|$

Proof. Angle $\widehat{ABA}$ is acute (See Figure 1.45-I) because it is an angle different from the right one of triangle $AB\Delta$ (Corollary 1.11). Similarly angle $\widehat{A\Gamma A}$ is acute. Suppose now that $|B\Delta| < |\Gamma\Delta|$. Then in triangle $AB\Gamma$, Γ lies in the extension of ΔB towards B and the angle at B is obtuse and at Γ acute. Therefore the opposite to the obtuse side $A\Gamma$ will be greater than the opposite to the acute side AB. For the converse, suppose that $|BA| < |\Gamma A|$ but $|B\Delta| > |\Gamma\Delta|$. There is a contradiction here, because, according to the preceding part of the proof, $|\Gamma\Delta| < |B\Delta|$ implies $|\Gamma A| < |BA|$, contrary to the hypothesis. Likewise there is a contradiction if we suppose $|BA| < |\Gamma A|$ and $|B\Delta| = |\Gamma\Delta|$, because the latter implies that triangles $A\Delta B$ and $A\Delta\Gamma$ are congruent. Therefore, when $|BA| < |\Gamma A|$, it must also be $|B\Delta| < |\Gamma\Delta|$.

Remark 1.17. The last corollary is equivalent to the fact that the length of the hypotenuse $|AB|$ of the right triangle $A\Delta B$ is an increasing function of the length of the orthogonal side $|\Delta B|$, when the other orthogonal side $A\Delta$ remains fixed.

Corollary 1.14. *Let the isosceles triangle $AB\Gamma$ have equal angles at A and B. Consider a point Δ in the extension of line segment AB. Then $\Gamma\Delta$ is greater than ΓA (See Figure 1.45-II).*

Corollary 1.15. *If lines α, β intersect line ε and they form equal angles on the same side of the line and to the same direction of it, then they are parallel. Specifically two lines perpendicular on this line are parallel.*

Proof. If they were not parallel, they would intersect at say, Γ (See Figure 1.46-I), and then they would form a triangle with sum of angles at the vertices different from Γ: $\omega + (180° - \omega) = 180°$, which is a contradiction (Corollary 1.8).

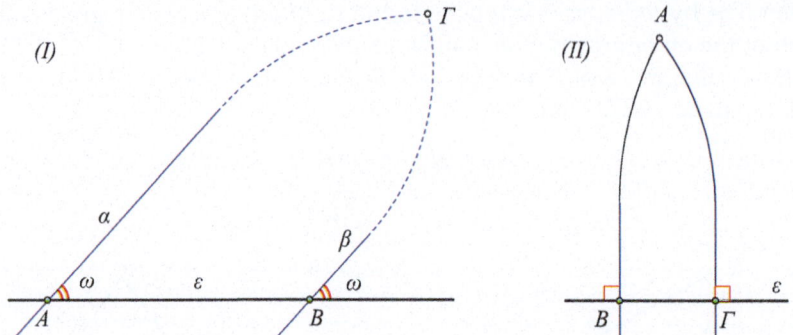

Fig. 1.46: Inclined at angle ω to ε   One only orthogonal from A

Corollary 1.16. *Through a point A, lying outside the line ε, there is at most one line orthogonal to ε.*

Proof. If there were two different orthogonals from A (See Figure 1.46-II), then they would form a triangle and its two angles different from A would sum up to $90° + 90° = 180°$, which is a contradiction (Corollary 1.8).

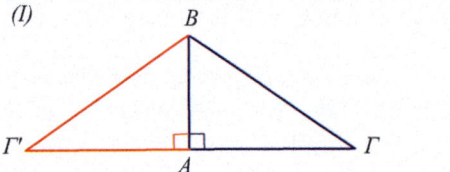 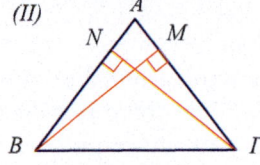

Fig. 1.47: Congruent right triangles   Equal altitudes

Exercise 1.32. Two right triangles with equal hypotenuses and two respective orthogonal sides equal, are congruent.

Hint: Place the two triangles so, that the two orthogonal sides coincide with the line segment AB and the triangles ABΓ and ABΓ' lie on different sides of AB with the right angle at A (See Figure 1.47-I). Then, the vertices Γ, Γ' and A lie all on a line and B, by hypothesis, is equidistant from Γ and Γ'. Thus, B belongs to the medial line of ΓΓ' and A is the middle of ΓΓ'.

Exercise 1.33. If two altitudes in a triangle are equal, then the triangle is isosceles.

Hint: Let BM and ΓN be the two altitudes (See Figure 1.47-II). Then triangles BMΓ and BNΓ are right triangles with common hypotenuse BΓ and the orthogonal sides BM and ΓN by hypothesis equal. According to the preceding exercise, the right triangles are congruent and therefore their angles at B and Γ are equal. Hence the triangle ABΓ will be isosceles (Corollary 1.7). Note that this property is the converse of Exercise 1.26.

Theorem 1.5. *If two triangles $AB\Gamma$ and $A'B'\Gamma'$ have equal sides respectively adjacent to A and A' ($|AB| = |A'B'|$, $|A\Gamma| = |A'\Gamma'|$) and the angles at A and A' are unequal ($\widehat{A} > \widehat{A'}$), then their sides are respectively unequal ($|B\Gamma| > |B'\Gamma'|$).*

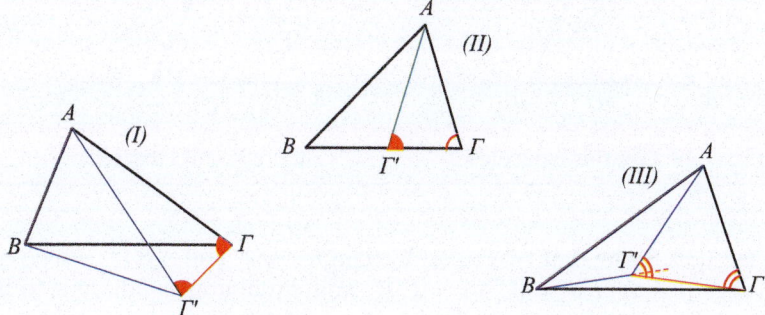

Fig. 1.48: Greater side opposite greater angle

Proof. Place the triangles so that sides AB and $A'B'$ become coincident and $A\Gamma'$ lies in the interior of $\widehat{A}$. The possible cases are three, depending on the position of Γ' relative to the line $B\Gamma$ (I-III in figure 1.48). Let us see the first case (See Figure 1.48-I) and leave the other two as exercises. In this case, we suppose that Γ' and A are contained in different sides (half planes) of $B\Gamma$. Then comparing the angles of the triangle $B\Gamma'\Gamma$ we have $|\widehat{B\Gamma'\Gamma}| > |\widehat{A\Gamma'\Gamma}| = |\widehat{A\Gamma\Gamma'}| > |\widehat{B\Gamma\Gamma'}|$, where the last holds because $|A\Gamma'| = |A'\Gamma'| = |A\Gamma|$. But according to theorem 1.4, the inequality $|\widehat{B\Gamma'\Gamma}| > |\widehat{B\Gamma\Gamma'}|$ implies $|B\Gamma| > |B\Gamma'|$

Corollary 1.17. *If two triangles $AB\Gamma$ and $A'B'\Gamma'$ have equal sides respectively adjacent to A and A' ($|AB| = |A'B'|$, $|A\Gamma| = |A'\Gamma'|$) and the third sides are unequal ($|B\Gamma| > |B'\Gamma'|$), then their angles at A and A' are respectively unequal ($\widehat{A} > \widehat{A'}$).*

Proof. If, in addition to the corollary's hypotheses we had $|\widehat{BA\Gamma}| \leq |\widehat{B'A'\Gamma'}|$ then we would get an contradiction. Indeed, if $|\widehat{BA\Gamma}| = |\widehat{B'A'\Gamma'}|$, then according to the SAS-criterion the triangles would be congruent and we would have $|B\Gamma| = |B'\Gamma'|$, a contradiction. Similarly, if in addition to the hypothesis we had $|\widehat{BA\Gamma}| < |\widehat{B'A'\Gamma'}|$, then according to theorem 1.5, we would obtain $|B\Gamma| < |B'\Gamma'|$, also a contradiction. Hence, under the hypothesis only $\widehat{A} > \widehat{A'}$ is possible

Exercise 1.34. Show that between two isosceli triangles $AB\Gamma$ and $\Delta B\Gamma$ with the same base, the one which has the greater angle at its apex is the one with the smaller legs and conversely.

Hint: Show first that their corresponding third vertices, which are contained in the medial line of the base $B\Gamma$ (See Figure 1.49-I), are closer to its middle M, when the corresponding triangle leg becomes smaller (Corollary 1.13).

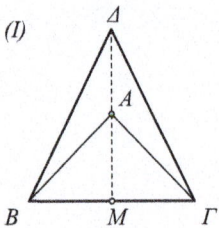 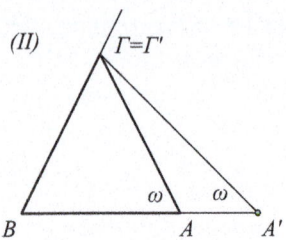

Fig. 1.49: Greater isosceles　　　　　Congruent isosceli

Exercise 1.35. Show that two isosceli triangles $AB\Gamma$ and $A'B'\Gamma'$, which have equal legs $|\Gamma A| = |\Gamma B| = |\Gamma'A'| = |\Gamma'B'|$ and equal angles adjacent to their bases, are congruent.

Hint: Place the triangles so that their angles at B and B' (See Figure 1.49-II), as well as $B\Gamma$ and $B'\Gamma'$, because of equality of lengths, become coincident. If A, A' were different and A' was farther than B than A, we'd have an external angle ω of triangle $AA'\Gamma$ at A greater than the internal ω at A' ($\omega > \omega$), which is a contradiction. Therefore the vertices A, A' coincide.

Exercise 1.36. Given a point A not in line ε and an angle ω ($0 < \omega < 90$), show that there is, at most, one isosceles triangle with vertex at A, base at ε and base angles equal to ω.

Exercise 1.37. Show that in every triangle $AB\Gamma$ the sum of its altitudes is less than its perimeter.

Exercise 1.38. Show that two right triangles $AB\Gamma$ and $A'B'\Gamma'$, which have equal hypotenuses and one angle $\omega \neq 90°$ equal, are congruent.

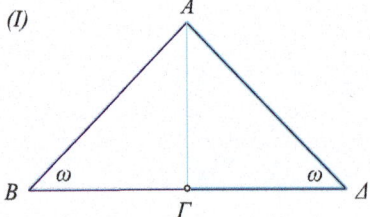 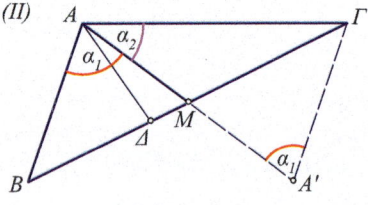

Fig. 1.50: Congruent right triangles　　　Median slope

Hint: Extend the adjacent to ω orghogonal $B\Gamma$ by doubling its length until Δ (See Figure 1.50-I). The resulting triangle $AB\Delta$ is isosceles. Similarly results the isosceles $A'B'\Delta'$ from right triangle $A'B'\Gamma'$. The two isosceli triangles $AB\Delta$ and $A'B'\Delta'$ have equal legs respectively and equal angles adjacent to their base, therefore they are congruent (Exercise 1.35). Then their halves, i.e. the right triangles $AB\Gamma$ and $A'B'\Gamma'$ are congruent.

1.10. TRIANGLE'S SIDES AND ANGLES RELATIONS

Exercise 1.39. Show that the median AM of triangle $AB\Gamma$ with unequal sides $AB, A\Gamma$, is inclined towards the smaller side. Also, between angles $\widehat{BAM}$ and $\widehat{MA\Gamma}$ the one adjacent to the smaller side is greater. Conclude that between vertices B and Γ, the trace Δ of the bisector $A\Delta$ is closer to the vertex which belongs to the smaller side (See Figure 1.50-II).

Exercise 1.40. Show the converse of the preceding exercise, i.e. if the median AM is inclined towards B, then $|AB| < |A\Gamma|$ and $\widehat{BAM} > \widehat{MA\Gamma}$.

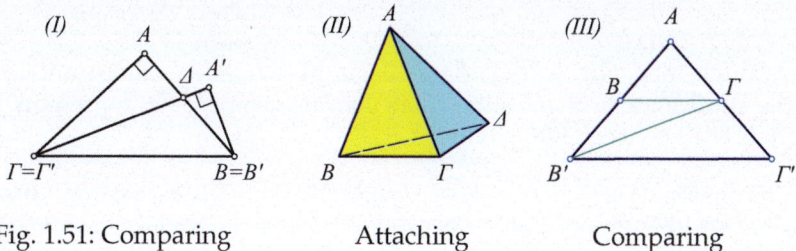

Fig. 1.51: Comparing   Attaching   Comparing

Exercise 1.41. The right angled triangles $\{AB\Gamma, A'B'\Gamma'\}$ have equal hypotenuses $|B\Gamma| = |B'\Gamma'|$ and angles $\phi = \widehat{AB\Gamma} < \omega = \widehat{A'B'\Gamma'}$. Show that $|A\Gamma| < |A'\Gamma'|$ (See Figure 1.51-I).

Hint: Referring to the figure, show first by contradiction using proposition 1.8 that Δ is between A, B.

Exercise 1.42. Let $BA\Gamma$ and $\Gamma A\Delta$ be isosceli triangles with apex at A and a common side $A\Gamma$. Show that sides AB and $A\Delta$ are either contained in the same line or they form an isosceles triangle $BA\Delta$ (See Figure 1.51-II).

Exercise 1.43. Extend the sides $\{AB, A\Gamma\}$ of the triangle $AB\Gamma$, whose angle $\widehat{B}$ is acute, to their double $\{AB', A\Gamma'\}$. Show that $|B\Gamma| < |\Gamma B'| < |B'\Gamma'|$ (See Figure 1.51-III).

Exercise 1.44. Continuing the last exercise, show that the angles of the triangle $AB'\Gamma'$ at $\{B', \Gamma'\}$ are not greater than the respective angles at $\{B, \Gamma\}$ of the triangle $AB\Gamma$. Later, after introducing the axiom of parallels (§ 1.15), we'll see that the angles are equal $\{\widehat{B} = \widehat{B'}, \widehat{\Gamma} = \widehat{\Gamma'}\}$.

1.11 The triangle inequality

> The third, to conduct my thoughts in such order that, by commencing with objects the simplest and easiest to know, I might ascend by little and little, and, as it were, step by step, to the knowledge of the more complex; assigning in thought a certain order even to those objects which in their own nature do not stand in a relation of antecedence and sequence.
>
> *Descartes, Discourse on Method II*

The triangle inequality is of fundamental importance in geometry and leads to the confirmation of intuition, that the line segment AB represents the shortest path between two points $\{A, B\}$.

Theorem 1.6. *The sum of the lengths of two sides of a triangle is greater than the length of the third side.*

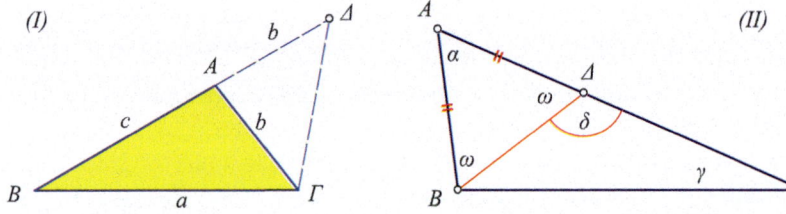

Fig. 1.52: Triangle inequality   Inequality for the difference

Proof. Let $a = |B\Gamma|$ be the greatest of all sides of the triangle $AB\Gamma$ (See Figure 1.52-I). It suffices to show that $a < b + c$. For this, extend AB by the segment $A\Delta$ equal to $A\Gamma$. In the resulting triangle $\Delta B\Gamma$, angle $\widehat{B\Gamma\Delta} = \widehat{B\Gamma A} + \widehat{A\Gamma\Delta}$ is greater than $\widehat{\Gamma\Delta B}$ therefore (Theorem 1.4) also side $B\Delta$, for which $|B\Delta| = b + c$, will be greater than $B\Gamma$ with $|B\Gamma| = a$.

Theorem 1.7. *The difference of lengths of two sides of a triangle is less than the length of the third side.*

Proof. Let $A\Gamma$ in the triangle $AB\Gamma$ be greater than AB ($b > c$) (See Figure 1.52-II). It suffices to show that $a > b - c$. For this, consider on $A\Gamma$ the segment $A\Delta$ equal to AB. In the resulting triangle $\Delta B\Gamma$, angle $\widehat{BA\Gamma}$ is greater than $\widehat{AB\Gamma}$. This, because as external of the basis angle ω of the isosceles triangle $BA\Delta$, it is obtuse (Corollary 1.10). And if a triangle has one obtuse angle, this angle is greater than the other two, which must be acute (Corollary 1.9). It follows that side $B\Gamma$, which is opposite to the obtuse $\delta = \widehat{BA\Gamma}$ is greater than $\Delta\Gamma$, which has length $|\Delta\Gamma| = |A\Gamma| - |AB|$.

1.11. THE TRIANGLE INEQUALITY

Remark 1.18. The two theorems together mean that in every triangle the following inequalities hold:

$$|a-b| < c < a+b, \quad |b-c| < a < b+c, \quad |c-a| < b < c+a.$$

It is easy to see that all these inequalities are equivalent to the inequality $a < b+c$, where a is the triangle's greatest side.

A **broken line** is a shape which consists of a sequence of line segments $AB, B\Gamma,..., XY, Y\Omega$ in which every pair of successive segments has exactly

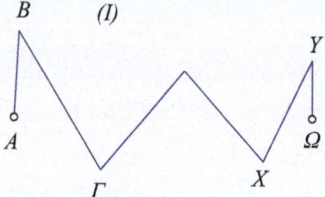

Fig. 1.53: Broken line   Shortening of broken line

one common endpoint (See Figure 1.53-I). This broken line is denoted by $AB\Gamma...Y\Omega$ and we say that it **joins** points A and Ω. The broken line is called **closed**, when points A and Ω coincide. The line segments $AB, B\Gamma, ...$ are called **sides** and the sum of their lengths $|AB| + |B\Gamma| + ... + |Y\Omega|$ is called **length** of the broken line.

Corollary 1.18. *The line segment $A\Omega$ has length $|A\Omega|$, less than the length of any broken line which joins A with Ω.*

Proof. Given the broken line $AB...Y\Omega$ and replacing $\{AB, B\Gamma\}$ with $A\Gamma$ we create a new shorter broken line $A\Gamma\Delta E...XY\Omega$ with one side less (See Figure 1.53-II). This follows by applying theorem 1.6 to the triangle $AB\Gamma$. Continuing this way and replacing $\{A\Gamma, \Gamma\Delta\}$ with $A\Delta$, ... we reduce successively the length μ of the given broken line and its number of sides. The process continues until we reach the line segment $A\Omega$ and the corresponding reduction of length gives the inequality $|A\Omega| < \mu$.

Exercise 1.45. Let the point Δ be in the interior of triangle $AB\Gamma$. Show that $|\Delta B| + |\Delta \Gamma| < |AB| + |A\Gamma|$ (See Figure 1.54-I).

Exercise 1.46. Let point Δ be in the interior of triangle $AB\Gamma$. Show that the sum of the distances of Δ from the triangle's vertices is less than the triangle's perimeter and greater than the half perimeter.

Exercise 1.47. Let the point Δ be contained in the exterior of triangle $AB\Gamma$. Show that the sum of the distances of Δ from the triangle's vertices is greater than the triangle's half perimeter.

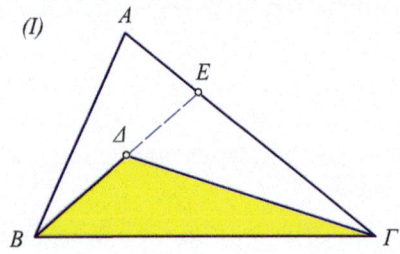
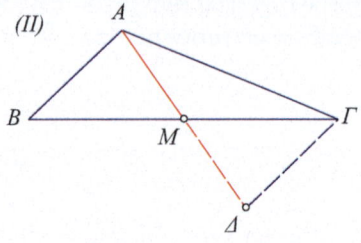

Fig. 1.54: $|\Delta B|+|\Delta \Gamma|<|AB|+|A\Gamma|$ $\frac{||AB|-|A\Gamma||}{2}<|AM|<\frac{|AB|+|A\Gamma|}{2}$

Exercise 1.48. Show that the median AM of triangle $AB\Gamma$ has length satisfying the inequalities $|A\Gamma|-|AB|<2|AM|<|AB|+|A\Gamma|$, assuming $|AB| \leq |A\Gamma|$.

Hint: Extend AM by doubling it to Δ (See Figure 1.54-II). Triangles AMB and $\Delta M\Gamma$ are congruent, having $|AM|=|M\Delta|$, $|BM|=|M\Gamma|$ and contained angles $\widehat{AMB}$ and $\widehat{\Gamma M\Delta}$ equal. Consequently, from the triangle inequality, we have $2|AM|=|A\Delta|<|A\Gamma|+|\Gamma\Delta|=|A\Gamma|+|AB|$.

Notice also the relation $|AM|>\frac{1}{2}(|AB|+|A\Gamma|-|B\Gamma|)$ (Corollary 3.26).

Exercise 1.49. Show that in every triangle the sum of the lengths of the medians is less than the perimeter of the triangle.

Exercise 1.50. Show that in an isosceles triangle $AB\Gamma$ with altitude from apex, $\upsilon = |A\Delta|$, greater from the basis length, $b = |B\Gamma|$, the sum of the lateral sides, $2\lambda = |BA|+|A\Gamma|$, is greater than $b+\upsilon$. Later (using Pythagora's theorem 3.4) we'll see that $2\lambda \geq b+\upsilon$, precisely when $\upsilon \geq \frac{2}{3}b$.

1.12 The orthogonal to a line

> Let us honour if we can
> The vertical man
> Though we value none
> But the horizontal one.
>
> W. H. Auden, To C. Isherwoods

From Corollary 1.16 we know that, if from point A not lying on line ε, there is a line ε' orthogonal to ε, then it will be unique. Next theorem proves the *existence* of such a line ε'. For this we use the fundamental capability of constructing a specific angle (Axiom 1.10) having its vertex and one of its sides on a given line. From the same fundamental capability results also the existence of the orthogonal ε' to line ε from a point A **on** the line ε (Corollary

1.12. THE ORTHOGONAL TO A LINE

1.1). The practical construction of the orthogonal ε', using ruler and compass depending on properties of the circle will be discussed in § 2.4.

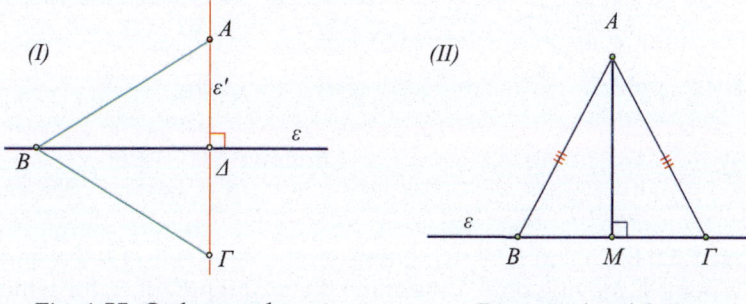

Fig. 1.55: Orthogonal to ε Distance $|AM|$ from ε

Theorem 1.8. *From a point A lying outside the line ε precisely one orthogonal to the line can be drawn.*

Proof. We construct the orthogonal ε' to line ε using an isosceles triangle as follows (See Figure 1.55-I). We take an arbitrary point B in ε and draw the line BA. In the half plane of ε, which does not contain point A, we define the line segment $B\Gamma$ which forms with ε the same angle as BA (Axiom 1.10). On this line we consider the point Γ such that $|B\Gamma| = |BA|$. Line $A\Gamma$ is the wanted one. Indeed, triangle $AB\Gamma$ is isosceles with base $A\Gamma$ by construction and ε coincides with the bisector of its apex. Consequently ε will be orthogonal to the base $A\Gamma$ of the triangle (Corollary 1.3).

Corollary 1.19. *(Distance of point from line) Let A be a point not lying on the line ε and AM be orthogonal to it at the point M belonging to ε. For every other point B of the line, AB is greater than AM.*

Proof. Any other point B (See Figure 1.55-II) makes a right triangle AMB with hypotenuse AB, which is always greater than the orthogonal AM (Corollary 1.12).

Point M is called (orthogonal or vertical) **projection** of A onto the line ε. We call the length $|AM|$ of AM **distance** of the point A from the line ε.

Exercise 1.51. Extend the altitude v_α of the triangle $AB\Gamma$ from A, towards $B\Gamma$, to its double until point E. The created triangle $BE\Gamma$ is congruent to $AB\Gamma$ (See Figure 1.56-I).

Hint: By applying the SAS-criterion, show first that triangles $B\Delta A$ and $B\Delta E$ are congruent and triangles $A\Delta\Gamma$ and $E\Delta\Gamma$ are also congruent. Then, by applying the SSS-criterion, show that $AB\Gamma$ and $EB\Gamma$ are congruent.

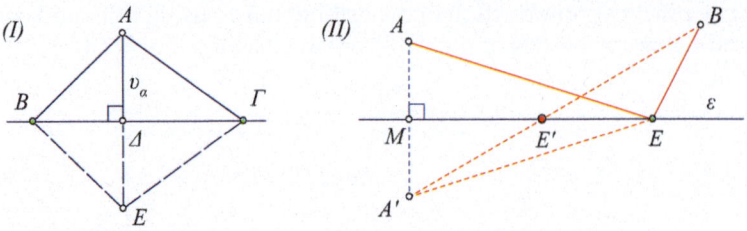

Fig. 1.56: Extending $|A\Delta|$ Minimize $|EA| + |EB|$

Exercise 1.52. Let A and B be two points on the same side of line ε and consider a point E moving on ε. Determine the position E' of E, for which the sum of the distances $|EA| + |EB|$ becomes minimal (See Figure 1.56-II).

Hint: Project point A onto ε to point M and extend AM towards M to its double until point A'. By construction then, ε is the medial line of AA' and for any point E of ε line segments EA and EA' are equal. Consequently $|EA| + |EB| = |EA'| + |EB|$. Suppose E' is the intersection of $A'B$ and ε. In triangle EBA', for points E of ε different from E', we have always $|EA'| + |EB| \geq |A'B|$ with equality valid only when E coincides with E'. This is the point at which the expression $|EA| + |EB|$ becomes minimal.

Exercise 1.53. Let A and B be points on the same side of line XY and consider a point E moving on line XY. Determine the position Z of E for which angles $\widehat{AZX}$ and $\widehat{BZY}$ are equal.

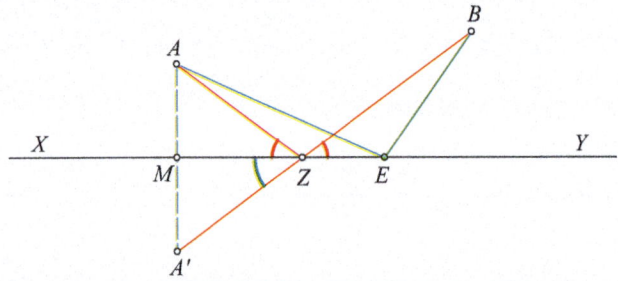

Fig. 1.57: Point of reflection Z from A to B.

Hint: Suppose that the problem has been solved and the position of Z has been determined (See Figure 1.57). Take on the extension of ZB the segment ZA' equal to ZA. The triangle ZAA' is isosceles and XY, by construction, is a bisector of the angle at the vertex Z. Consequently AA' is orthogonal to XY and intersects it at the middle M of AA' (Corollary 1.3). Consequently A' is determined from the given data by drawing an orthogonal AM from A to XY and extending it to its double towards M. The intersection of $A'B$ with XY determines the desired point.

1.13. THE PARALLEL FROM A POINT

Remark 1.19. It is noteworthy that the solutions of the last two exercises define the same point. In physics this corresponds to the law of reflection, according to which a ray emitted from A, gets reflected onto XY (flat mirror) and the reflected ray which passes through B has two simultaneous properties: a) the length $|AZ| + |ZB|$ is the least possible, b) the angle $\widehat{AZX}$ (its complementary is called **incidence angle**) (See Figure 1.57) is equal to angle $\widehat{YZB}$ (its complementary is called **reflection angle**).

Exercise 1.54. Let A and B be two points on the same side of line ε. Determine a point Γ on ε, for which the difference $||\Gamma A| - |\Gamma B||$ becomes maximal. Examine the same problem in the case when A and B are on different sides of ε.

Exercise 1.55. Show that, if $|AB| < |A\Gamma|$, then the projection Δ of vertex A onto side $B\Gamma$ of the triangle $AB\Gamma$ is closer to B than Γ ($|B\Delta| < |\Delta\Gamma|$).

1.13 The parallel from a point

> As lines, so loves oblique, may well
> Themselves in every angle greet;
> But ours, so truly parallel,
> Though infinite, can never meet.
>
> A.A. Cooper, Love Definition st. 7

The following proposition guaranties the *existence* of a parallel line. It says nothing, however, about the *uniqueness* of this parallel line. We'll see later that here a basic hypothesis (axiom) is needed, one according to which we accept that there is no other parallel different from the constructed one. The propositions we have proved so far did not need anywhere the hypothesis of uniqueness of a parallel. They depended exclusively on properties (axioms) for lines, angles and triangles, accepted without proofs (in sections 1.2-1.6).

Theorems, not depending on the axiom of parallels, make up the subject of the so called **Absolute Geometry**. The propositions of Absolute Geometry, to which belong all the propositions we have proved so far, are valid not only in Euclidean Geometry, where we accept the uniqueness of the parallel, but also in the so called **Hyperbolic Geometry** (a short account of it is given in II-§ 2.9), where we accept that there exist more than one parallels to line ε from a point A not lying on ε. In the next section we prove two last propositions of Absolute Geometry and pass over to Euclidean Geometry, accepting the uniqueness of the parallel.

Proposition 1.9. *From a point A, lying outside the line ε, a parallel δ to ε can be drawn (See Figure 1.58).*

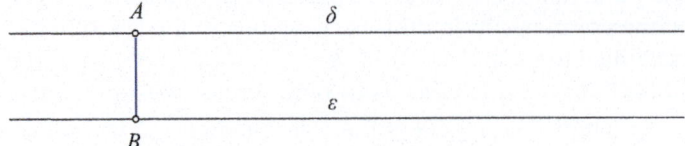

Fig. 1.58: Parallel from a point

Proof. According to Theorem 1.8, there exists an orthogonal segment *AB* to ε. The orthogonal line δ to *AB* at *A* (Corollary 1.1) is also constructible. Lines ε and δ, according to Corollary 1.15, are parallel.

Exercise 1.56. At the points *A* and *B* of a line and on the same side draw two equal orthogonals *AΔ* and *BΓ*. Show that *ΓΔ* is parallel to line *AB*.

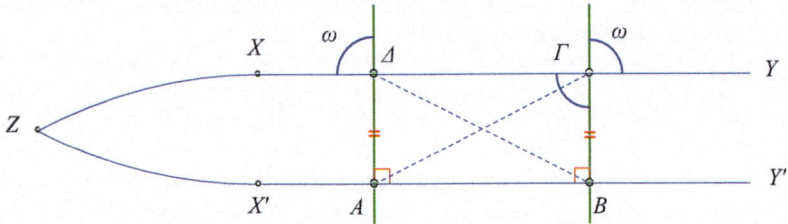

Fig. 1.59: Saccheri trapezium

Hint: The triangles *ABΔ* and *ABΓ* are congruent (See Figure 1.59), having *AB* common, *AΔ* and *BΓ* equal by construction and the contained in them angles equal as right (SAS-criterion). It follows that *AΓ* and *BΔ* are equal and from this that the triangles *AΔΓ* and *BΔΓ* are congruent (SSS-criterion). From the congruence of the last triangles follows also the equality of angles $\widehat{X\Delta A}$ and $\widehat{Y\Gamma B}$. Using reduction to contradiction, we proceed now to the proof that *AB* and *ΓΔ* do not intersect. In fact, suppose that the two lines intersect at *Z*. Then, in the triangle *ΔAZ* the external angle at *Δ* will be $\omega > 90°$, whereas in *BΓZ* the external angle at *B* will be $90° > \omega$. From the last two expressions it follows that $\omega > \omega$, which is a contradiction. Therefore *AB* and *ΓΔ* are parallel.

Remark 1.20. The shape *ABΓΔ* in the preceding exercise is called *Trapezium of Saccheri (1667-1733)* and played an important role in the history of Hyperbolic Geometry ([36, p.5]). After accepting the axiom of parallels, we'll be able to see that angle ω is right. Currently, however, where we have not supposed anything about the uniqueness of parallels, we cannot prove something like that.

1.13. THE PARALLEL FROM A POINT

Exercise 1.57. From points A and A' of line ε and towards the same side draw two equal line segments AB, $A'B'$ having the same inclination ω to ε. Show that BB' is parallel to ε.

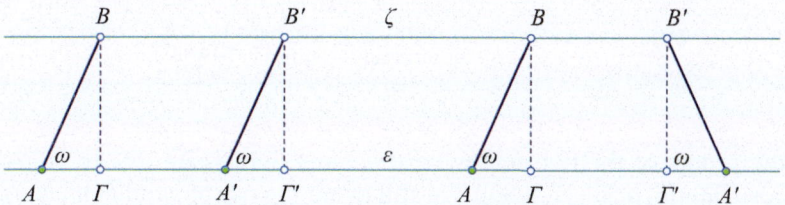

Fig. 1.60: Equal line segments defining a parallel

Hint: Draw the orthogonals $B\Gamma$ and $B'\Gamma'$ towards ε (See Figure 1.60). The right triangles $AB\Gamma$ and $A'B'\Gamma'$ have equal hypotenuse and one acute angle (ω), equal, consequently they are congruent (Exercise 1.38). Consequently the orthogonals $B\Gamma$ and $B'\Gamma'$ are equal and the result follows from Exercise 1.56.

Exercise 1.58. Suppose that the two angle bisectors of angle $\widehat{A}$ of triangle $AB\Gamma$ intersect the opposite side at points: Δ (internal) and E (external). Project vertex B to Z on AE and double the segment BZ in the direction of Z until H (See Figure 1.61). Show that H belongs to line $A\Gamma$. Also show that HB is parallel to $A\Delta$.

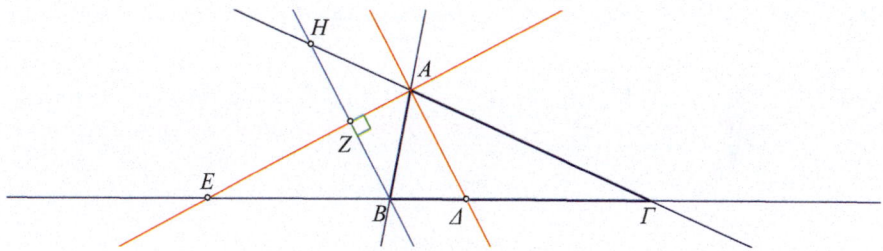

Fig. 1.61: Parallel of a bisector

Hint: The right triangles ABZ and AHZ are congruent by construction (SAS-criterion), therefore their angles at A are equal and because AZ is a bisector of the supplementary of angle A, AH will coincide with line $A\Gamma$. The lines $A\Delta$ and ZB are perpendicular to AE, hence they do not intersect (Theorem 1.8).

1.14 The sum of triangle's angles

> It is not always by plugging away at a difficulty and sticking at it that one overcomes it; but, rather, often by working on the one next to it. Certain people and certain things require to be approached on an angle.
>
> *Andre Gide, Journals, Oct. 26, 1924*

In the proofs of the two following theorems (which are due to Legendre (1752-1833) ([6, p.55], [1, p.80]) we still avoid the use of parallel lines. These theorems essentially lie close to the boundary between Absolute Geometry and other Geometries. The second theorem reveals, that when the axioms of lines, angles and triangles which we accepted in sections 1.2-1.6 hold, and we find on the plane some triangle $AB\Gamma$ with angle sum $\alpha + \beta + \gamma = 180°$, then every other triangle on the same plane will have angle sum $180°$ and the uniqueness of the parallel from a point not lying on the line holds. In other words, on this plane Euclidean Geometry is valid. Similarly, if we find a triangle $AB\Gamma$ on the plane, with angle sum $\alpha + \beta + \gamma < 180°$, then in every other triangle on the same plane its angle sum will again be less than $180°$. In the latter case, another kind of geometry is valid, different from Euclidean. This is the Hyperbolic Geometry or the Geometry of Bolyai 1802-1860 and Lobatsevsky 1792-1856 ([36], [1, p.98]), which was mentioned in the preceding section and of which a model is discussed in II-§ 2.9.

Theorem 1.9 (Saccheri-Legendre). *In every triangle $AB\Gamma$ the sum of its angles is less than or equal to 180 degrees ($\alpha + \beta + \gamma \leq 180°$).*

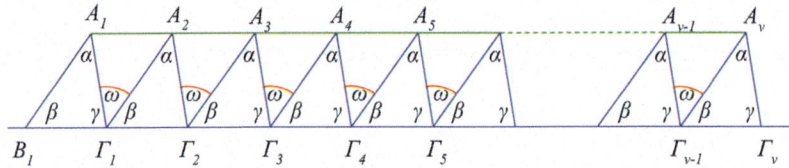

Fig. 1.62: A sequence of congruent triangles

Proof. Using reduction to contradiction. We start using the hypothesis that there exists a triangle $AB\Gamma$ with $\alpha + \beta + \gamma > 180°$ and we show that this leads to something contradictory. We place ν copies of this triangle next to each other so that their bases $B_1\Gamma_1$, $\Gamma_1\Gamma_2$,... become successive equal line segments of the same line (Axiom 1.8), as in the figure 1.62. Additionally we suppose that $|A\Gamma| \leq |B\Gamma|$. We construct, thus, triangles $A_1\Gamma_1A_2$, $A_2\Gamma_2A_3$, ... which according to the SAS-criterion are all congruent. If $\omega = |\widehat{A_1\Gamma_1A_2}|$, then the expression $\beta + \omega + \gamma = 180°$ along with the hypothesis $\alpha + \beta + \gamma > 180°$ implies $\omega < \alpha$. Consequently, comparing the triangles $A_1B_1\Gamma_1$ and $A_1\Gamma_1A_2$, which

1.14. THE SUM OF TRIANGLE'S ANGLES

have their sides on vertices A_1 and Γ_1 respectively equal and the contained angles unequal ($\omega < \alpha$), according to Theorem 1.5, we conclude that their third sides will be respectively unequal $|A_1 A_2| < |B_1 \Gamma_1|$. We compare now the lengths of the two lines which join B_1 and Γ_ν. According to Corollary 1.18, the length of the line segment $|B_1 \Gamma_\nu| = \nu |B_1 \Gamma_1|$ will be less than the length of the broken line $B_1 A_1 A_2 A_3 ... A_\nu \Gamma_\nu$, which is $|B_1 A_1| + (\nu - 1)|A_1 A_2| + |A_1 \Gamma_1|$. From the inequality

$$\nu |B_1 \Gamma_1| < |B_1 A_1| + (\nu - 1)|A_1 A_2| + |A_1 \Gamma_1|$$

because of $|A_1 \Gamma_1| \leq |B_1 \Gamma_1|$, which we supposed previously, follows that

$$\nu |B_1 \Gamma_1| < |B_1 A_1| + (\nu - 1)|A_1 A_2| + |B_1 \Gamma_1|.$$

From this, using again simple calculations, follows

$$(\nu - 1)(|B_1 \Gamma_1| - |A_1 A_2|) < |B_1 A_1|.$$

This last inequality is contradictory, because, as we noted above, its left side is positive and increases beyond any bound as ν increases, whereas its right side is constant. The contradiction, which resulted by assuming $\alpha + \beta + \gamma > 180°$, implies that we must have $\alpha + \beta + \gamma \leq 180°$.

Theorem 1.10 (Legendre). *If there exists a triangle $AB\Gamma$ of the plane ε, for which the sum of its angles is $180°$, then for every other triangle $A'B'\Gamma'$ of ε the sum of its angles will also be $180°$.*

We divide the proof, which is somewhat extended, into three lemmata.

Lemma 1.1. *If the triangle $AB\Gamma$ has sum of angles $180°$, then there exists a triangle $A'B'\Gamma'$ with the same angles and sides $|A'B'| = 2^\nu \cdot |AB|$, $|A'\Gamma'| = 2^\nu \cdot |A\Gamma|$, for any natural number ν.*

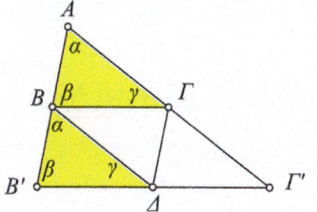 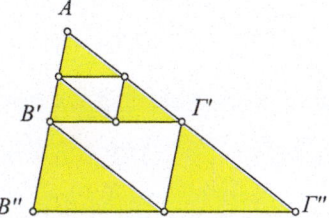

Fig. 1.63: Same angles but large sides

Proof. Suppose that triangle $AB\Gamma$ has angle sum $\alpha + \beta + \gamma = 180°$. We extend AB and we construct triangle $BB'\Delta$ congruent to $AB\Gamma$ (See Figure 1.63). Then, triangle $B\Delta\Gamma$ will be congruent to $BA\Gamma$, because the two triangles have the side $B\Gamma$ in common, $|B\Delta| = |A\Gamma|$ and the contained angle equal to γ. Then we extend $A\Gamma$ by doubling it and we define Γ' as $|A\Gamma'| = 2|A\Gamma|$. Then triangle $\Gamma\Delta\Gamma'$ is also congruent to $AB\Gamma$. Also, because at Δ the angles add up to 180°, points B', Δ, Γ' belong to the same line. Concluding therefore, triangle $AB'\Gamma'$ has the same angles as $AB\Gamma$ but sides of corresponding double length. The proof of the lemma results by repeating ν times the preceding construction.

Lemma 1.2. *If the triangle $AB\Gamma$ of the plane ε has sum of angles 180° and the triangle $A'B'\Gamma'$ has one of its angles equal to an angle of $AB\Gamma$, then triangle $A'B'\Gamma'$ will have sum of angles 180° too.*

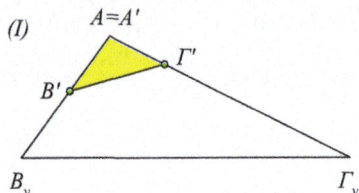
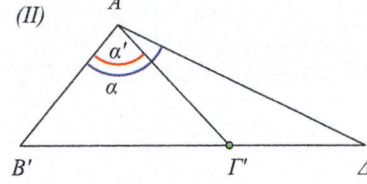

Fig. 1.64: Common angle ... and case $\alpha' < \alpha$

Proof. Suppose that the two triangles have equal angles at A and A', respectively. Then we place them in such a way, so that their equal angles coincide (See Figure 1.64-I), points A, B', B become collinear and A, Γ', Γ become also collinear. If needs be, by increasing $AB\Gamma$, the same way we did in the preceding lemma, we can consider that $AB_\nu\Gamma_\nu$ has the same angles as $AB\Gamma$ and is so large, that B' lies between A, B_ν and Γ' lies between A and Γ_ν. Drawing then the line $B'\Gamma_\nu$, we divide $B'\Gamma'\Gamma_\nu B_\nu$ into two triangles. By applying to each of these triangles the inequality of the preceding theorem and adding pairwise the two resulting inequalities, we get

$$\beta + \gamma + (180° - \beta') + (180° - \gamma') \leq 2 \cdot 180°,$$

which is equivalent to

$$\beta + \gamma \leq \beta' + \gamma'.$$

Adding on both sides of the inequality the common angle α we have

$$180° = \alpha + \beta + \gamma \leq \alpha + \beta' + \gamma' \leq 180°,$$

which implies that $\alpha + \beta' + \gamma' = 180°$.

Lemma 1.3. *If triangle $A'B'\Gamma'$ has one angle less than an angle of $AB\Gamma$, which has sum of angles 180°, then triangle $A'B'\Gamma'$ also has angle sum 180°.*

Proof. Extend $B'\Gamma'$ towards Γ' and consider point Δ, so that angle $\widehat{B'A\Delta}$ has measure α (See Figure 1.64-II). If the triangles $AB'\Gamma'$, $\Gamma'A\Delta$ have the angle sums Σ_1, Σ_2 respectively, then, according to the preceding lemma applied on $AB'\Delta$, and taking into account that $\widehat{B'\Gamma A} + \widehat{A\Gamma\Delta} = 180°$, we have

1.14. THE SUM OF TRIANGLE'S ANGLES

$$\Sigma_1 + \Sigma_2 = 2 \cdot 180°.$$

But, according to Theorem 1.9, it must be $\Sigma_1 \leq 180°$ and $\Sigma_2 \leq 180°$. Therefore, if the preceding equality is to hold, we must have $\Sigma_1 = 180°$ and $\Sigma_2 = 180°$.

Proof. (of theorem 1.10) Suppose that the triangle $AB\Gamma$ has angle sum $\alpha + \beta + \gamma = 180°$ and $A'B'\Gamma'$ is another triangle with angles α', β', γ'. It is impossible for the three following inequalities to hold simultaneously $\alpha < \alpha'$, $\beta < \beta'$, $\gamma < \gamma'$, because then, by adding them pairwise we get

$$180° = \alpha + \beta + \gamma < \alpha' + \beta' + \gamma',$$

which contradicts Theorem 1.9. Therefore, at least one of them must be invalid, say the first, which means that

$$\alpha' \leq \alpha.$$

Then, by applying one of the two last lemmata, we'll have that the angle sum of the triangle $A'B'\Gamma'$ is $\alpha' + \beta' + \gamma' = 180°$.

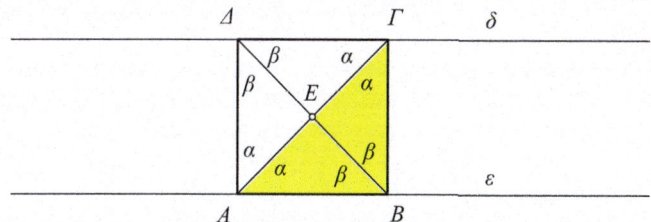

Fig. 1.65: Construction from a right isosceles

Exercise 1.59. Let $AB\Gamma$ be a right at B isosceles triangle and BE be its median. Extend BE towards E by doubling it until Δ. Show that the resulting triangles: ABE, $BE\Gamma$, $\Gamma E\Delta$ and ΔEA are congruent. Also show that the lines AB and $\Delta\Gamma$ are parallel (See Figure 1.65).

Remark 1.21. Note, that using the means available to us until now, we cannot prove, that besides the angles at B, Δ, the angles at Γ and A of $AB\Gamma\Delta$ are also right. Our knowledge however is sufficient to prove that AB and $\Gamma\Delta$, as well as $B\Gamma$ and $A\Delta$, are parallel.

1.15 The axiom of parallels

> All power is of one kind, a sharing of the nature of the world. The mind that is parallel with the laws of nature will be in the current of events, and strong with their strength.
>
> *Emerson, Power, The Conduct of Life*

As we noticed in the two preceding sections, the propositions we proved so far are inside the borders of *absolute geometry*, where we do not use properties (axioms) of *uniqueness* of the parallel to a line from a point. In this section we cross the boundary of absolute geometry and enter the domain of Euclidean geometry and the study of shapes, exploring also properties, which depend on the behavior of parallel lines.

Besides the properties of lines, angles and triangles which we accepted in sections 1.2-1.6, Euclidean Geometry also accepts the validity of the **uniqueness** of the parallel, in other words:

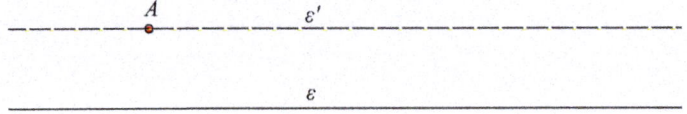

Fig. 1.66: ε' unique parallel of ε from A

Axiom 1.15 *Through a point A not lying on the line ε, one and only one parallel ε' to ε can be drawn (See Figure 1.66).*

Corollary 1.20. *If the line α intersects line β, then a parallel α' to α also intersects line β.*

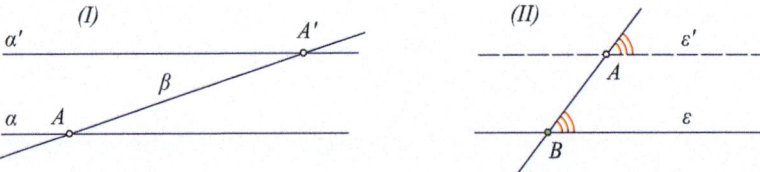

Fig. 1.67: If α then also α' intersects   Parallel from A

Proof. If α' did not intersect β (See Figure 1.67-I), then, from the intersection point A of α and β we would have two different parallels to α': α and β, which is a contradiction. Therefore α' intersects β

In section 1.13 we saw one method of construction of a parallel. More generally, we may consider a secant AB of ε, from A, not necessarily orthogonal to ε (Corollary 1.15) and we can construct the parallel forming at A the

1.15. THE AXIOM OF PARALLELS

same angle ω that is formed by AB and ε (at B) (See Figure 1.67-II). Because of the basic assumption of uniqueness, this will be the unique parallel to ε that can be drawn through point A.

Theorem 1.11. *From the assumption of uniqueness of the parallel, follows that the sum of the measures of the angles of a triangle is 180 degrees:* $\alpha + \beta + \gamma = 180°$.

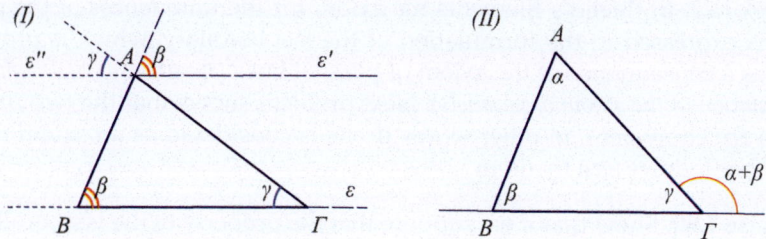

Fig. 1.68: Triangle angle sum   External angle of triangle

Proof. We make twice the construction of the parallel to the base $\varepsilon = B\Gamma$ of triangle from its vertex A (See Figure 1.68-I). The first time we consider line AB as a secant of $B\Gamma$ and we draw ε', which makes at A an angle equal to β. The second time we repeat the construction using line $A\Gamma$ as the secant of $B\Gamma$ and we draw ε'' which makes at A an angle equal to γ. From the uniqueness of the parallel axiom, it follows that the parallels ε', ε'' of ε coincide and therefore, at A are formed all three angles of the triangle and these sum up to $180°$.

Corollary 1.21. *Every external angle of the triangle has measure equal to the sum of the measures of the two opposite internal angles (See Figure 1.68-II).*

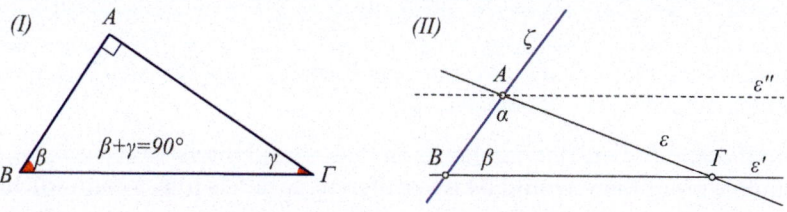

Fig. 1.69: Angles of right triangle   Euclid's parallel postulate

Corollary 1.22. *in every right triangle the sum of its acute angles equals a right angle (See Figure 1.69-I).*

Corollary 1.23. *If the line ζ intersects two other lines $\{\varepsilon, \varepsilon'\}$, respectively, at points $\{A, B\}$, making there internal and adjacent angles α and β with $\alpha + \beta < 180°$, then the lines ε and ε' intersect on the side of ζ containing the angles α and β (See Figure 1.69-II).*

Proof. First of all, it is impossible for ε, ε' to not intersect, since then, being parallel, we would have $\alpha + \beta = 180°$, which is contrary to our hypothesis. It is also impossible that they intersect on the other side because then, along with their intersection point Γ, they would form a triangle with angle sum greater than $180°$, which is a contradiction.

Remark 1.22. In Euclid's Elements the axiom for the uniqueness of the parallel is expressed in the formulation of the last corollary, which is proved here as a consequence of the axiom of parallel lines (referred to as *Playfair's formulation of the axiom of parallels*). Next problem shows that the two properties are equivalent, in other words if one is considered as an axiom then the other follows as a theorem.

Exercise 1.60. Show that if we suppose that the property of the last corollary holds, then we can prove, that from point A not lying on the line ε, one and only one parallel to ε can be drawn.

Hint: Referring to figure 1.69-II, consider the parallel ε'' of ε' from A, which is constructed by drawing the secant ε'' of ζ from A making with it an angle α'', such that $\alpha'' + \beta = 180°$. A different from ε'', line ε, will form at A an angle different from α'' and, consequently, on one side of the line ζ we'll have two internal and adjacent angles with $\alpha + \beta < 180°$, therefore the lines will intersect on this side. Therefore there exists one and only one parallel of ε' from A.

Exercise 1.61. Show the transitive property of parallels, that is: if line β is parallel to α and line γ is parallel to β, then γ is also parallel to α, or identical to it.

Hint: If γ were not parallel to α, then they would intersect. But then, (Corollary 1.20), β, which is parallel to α, would also intersect γ, a contradiction.

Exercise 1.62. Show that if two different lines α and β are, each one, parallel to line ε, then they are themselves parallel.

Hint: If α and β were not parallel, then they would intersect at a point A, and from this point there would be two different parallels to ε, a contradiction.

Corollary 1.24. *Line AB incident to parallels ε and ε' has:*
(i) The internal - external and on the same side angles equal (figure 1.70-α),
(ii) The internal and alternate angles equal (figure 1.70-β),
(iii) The internal and on the same side angles supplementary (figure 1.70-γ).

Proposition 1.10. *For every line segment AB and every pair of angles (β, γ) with measures $\beta + \gamma < 180°$, there exists a triangle $AB\Gamma$ with angles at B and Γ respectively of measure β and γ.*

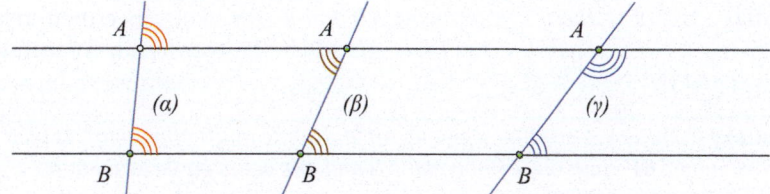

Fig. 1.70: Line incident to two parallels

Proof. According to Axiom 1.10 we can construct the angles β and γ on the same side of $B\Gamma$ (See Figure 1.71-I), with vertex at B and Γ respectively, which have one of their side the half line $B\Gamma$ and ΓB respectively. What we ensure now, is that the second sides of these angles will intersect and will define the third triangle vertex Γ. This, because if they did not intersect, that is, if they were parallel, then we would have two parallels which would form with the intersecting them line AB angles internal and on the same side with a sum less than 180°, which is a contradiction.

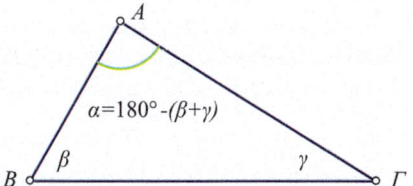 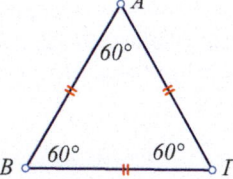

Fig. 1.71: ASA construction   Equilateral triangle

Corollary 1.25. *(Existence of equilateral triangle) For every positive number δ there exists a triangle $AB\Gamma$ which has all its sides equal to δ and all of its angles equal to 60°. Such a triangle is called **equilateral**. Conversely, every equilateral has angles of the same measure which is 60 degrees.*

Proof. Construct the triangle with base $B\Gamma$ of length δ and adjacent angles $\alpha = 60°$ and $\beta = 60°$ (Proposition 1.10). Because $\alpha + \beta + \gamma = 180°$, the third angle will have a measure of 60 degrees, hence the triangle will be isosceles relative to any of its sides as base, hence equilateral (See Figure 1.71-II). Conversely, every equilateral is isosceles relative to any of its sides as base, therefore all its angles are equal to $\alpha = 180°/3$.

Corollary 1.26. *For every line segment AB and an angle of measure $\alpha < 180°$ there is an isosceles triangle $AB\Gamma$ with base $B\Gamma$ and angle of measure α at its apex.*

Proof. The angles at the isosceles base must be $\omega = \frac{180° - \alpha}{2}$ and such a triangle is constructible according to Proposition-1.10.

Corollary 1.27. *For every line segment AB and a pair of angles of measures α, β with α + β = 90° there is a right triangle ABΓ with hypotenuse AB and acute angles α and β.*

Corollary 1.28. *For every line segment AB and acute angle of measure ω there is a right triangle ABΓ with one orthogonal AB and one acute angle equal to ω.*

Proof. The way the corollary is formulated leaves an ambiguity on AB, being opposite to ω or adjacent to it. There are therefore two triangles with these specifications. One has adjacent to AB angles ω and 90° and the other has adjacent to AB the angles 90° − ω and 90°. The existence of both follows from Proposition 1.10.

Exercise 1.63. A triangle ABΓ has at least one angle greater or equal to 60 degrees, as well as two angles with sum greater than 90 degrees.

Hint: If all angles were strictly less than 60°, then their sum would be $α + β + γ < 3 \cdot 60° = 180°$, which is a contradiction. The second assertion is proved similarly.

Exercise 1.64. *Every triangle ABΓ has at least one angle less than or equal to 60 degrees.*

Two half lines AX and BY are called **equal oriented**, when either (i) they are coincident, or (ii) one contains the other, or (iii) they are parallel and

Fig. 1.72: Equally oriented and opposite oriented half lines

the line AB which joins their start points leaves them on the same side (See Figure 1.72). The half lines are called **opposite oriented**, when either (i) they are contained in the same line but they are not equal oriented, or (ii) they are parallel and they are on different sides of the line AB. Similarly, two line segments AB and ΓΔ are called *equal/opposite oriented* when the half lines that contain them, with start points A and Γ, are respectively equal/opposite oriented. From Corollary 1.24 we get immediately the following.

Corollary 1.29. *Two parallel and equal oriented half lines AX and BY make with AB equal internal-external and on the same side angles. Two parallels and opposite oriented half lines make internal and alternate angles equal (See Figure 1.72).*

1.15. THE AXIOM OF PARALLELS

Corollary 1.30. *Two angles, which have their sides parallel, are equal or supplementary.*

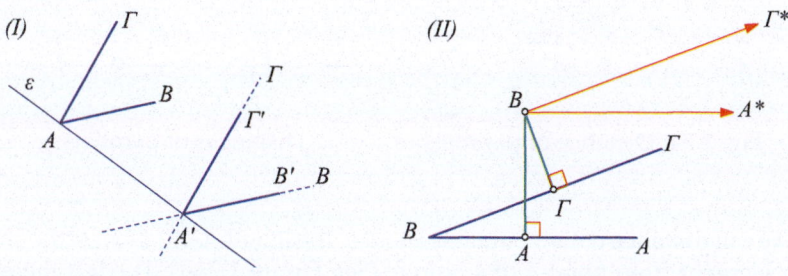

Fig. 1.73: Parallel sides   Orthogonal sides

Proof. Suppose that the angles $\widehat{BA\Gamma}$ and $\widehat{B'A'\Gamma'}$ have their sides parallel (See Figure 1.73-I). If their vertices at A and A' coincide, then the lines which contain their sides also coincide and the conclusion is obvious. If the vertices do not coincide, then from A' we draw equal oriented half lines in the direction of the sides of $\widehat{BA\Gamma}$ which, on one hand, form an angle equal to $\widehat{BA\Gamma}$ (Corollary 1.29), and on the other, their sides belong to the same lines as those of angle $\widehat{B'A'\Gamma'}$, consequently they make an angle either equal or supplementary to it.

Corollary 1.31. *Two angles, which have their sides respectively orthogonal, are equal or supplementary.*

Proof. Let the angles $\widehat{AB\Gamma}$ and $\widehat{A'B'\Gamma'}$ have their sides orthogonal (See Figure 1.73-II), i.e. line AB is orthogonal to $A'B'$ and $B\Gamma$ is orthogonal to $B'\Gamma'$. From B' we draw the parallel and equal oriented half lines to the sides of $\widehat{AB\Gamma}$. Then an angle $\widehat{A''B''\Gamma''}$, equal to $\widehat{AB\Gamma}$ is formed (Corollary 1.30) with sides orthogonal to those of $\widehat{AB\Gamma}$. The result follows from Proposition 1.3

Theorem 1.12. *Let ε and ε' be two parallel lines. For every point A of ε the segment AA' orthogonal to line ε', is also orthogonal to ε and its length $|AA'|$ is independent of the position of the point A on line ε.*

Proof. Consider a point B, different from A, on line ε and BB' also orthogonal to ε' (See Figure 1.74-I). By Corollary 1.24, lines AA', BB' make with ε, ε' the same angles, therefore they will also be orthogonal to ε. By the same corollary also the two right triangles $AA'B'$ and $B'BA$ will have respectively equal angles $\widehat{A'AB'} = \widehat{AB'B}$, $\widehat{A'B'A} = \widehat{B'AB}$ and common hypotenuse AB'. Consequently, by the ASA-criterion, the triangles will be congruent, with $|AA| = |BB'|$.

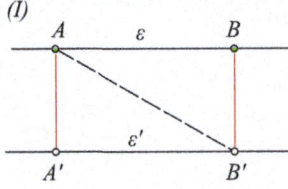 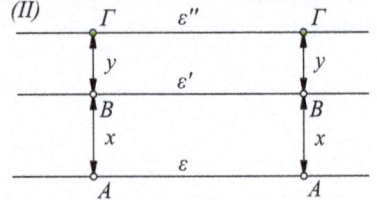

Fig. 1.74: Distance of parallels   Distances of parallels

We call **distance** of two parallels ε and ε' the length $|AA'|$ of an orthogonal line segment from point A of ε onto ε' (See Figure 1.74-I). The last theorem shows that the position of A on ε is immaterial.

Corollary 1.32. *Given three parallel lines ε, ε', ε'', every line orthogonal to one of them is orthogonal to all and intersects them respectively at points A, B, Γ, such that the distances $x = |AB|$, $y = |B\Gamma|$ are equal to the respective distances of the parallels and the distances are independent of the position of the orthogonal (See Figure 1.74-II).*

Corollary 1.33. *Given a line ε and distance δ, there is exactly one parallel line at distance δ on each side of ε.*

Proof. From an arbitrary point A of ε raise a orthogonal AB of length δ towards a specific side of ε. Next draw the orthogonal ε' of AB at B. Lines ε, ε' are parallel and at distance δ. Had there been another ε'' on the same side of ε as ε' also at distance δ, then ε'' would intersect AB at B', so that $|AB'| = \delta$, therefore B' would coincide with B and consequently lines ε', ε'' would also coincide.

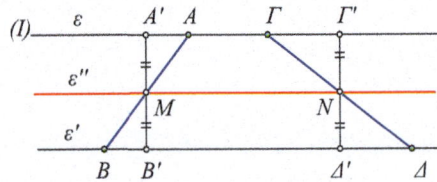 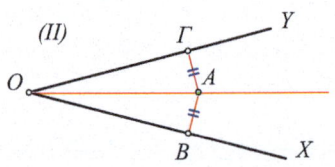

Fig. 1.75: The middle-parallel   Property of bisector

Corollary 1.34. *If points A, B belong respectively to parallels ε, ε' and we draw from the middle M of AB parallel ε'' to ε, then ε'' contains the middle N of every line segment $\Gamma\Delta$ with endpoints belonging to ε and ε' respectively.*

Proof. From the middle M of AB draw an orthogonal to the parallels, which intersects them at A', B' (See Figure 1.75-I). The right triangles $MA'A$, $MB'B$

1.15. THE AXIOM OF PARALLELS

are congruent, because they have equal angles and equal hypotenuse. Therefore M belongs to the parallel ε'' of ε, ε', which is characterized by the property of being equidistant from ε and ε'. Conversely, if line segment AB intersects this parallel at M and $A'M$, $B'M$ are orthogonal to ε, ε', then the right triangles $MA'A$, $MB'B$ are congruent, because the angles at M are orthogonal and $|MA'| = |MB'|$. Therefore their hypotenuses will also be equal, $|MA| = |MB|$.

We call the line ε'', which is defined in the preceding corollary, **middle-parallel** of the parallel lines ε and ε'.

Corollary 1.35. *For every point A on the bisector of the angle $\widehat{XOY}$, the distances AB and $A\Gamma$ from the sides of the angle are equal. Conversely: if a point A is equidistant from the sides of the angle $\widehat{XOY}$ then it belongs to the bisector of this angle.*

Proof. The right triangles $OA\Gamma$ and OAB have OA common and the adjacent angles to OA equal (See Figure 1.75-II), therefore (SAS-criterion) they are congruent and consequently $|A\Gamma| = |AB|$. Conversely: if the preceding equality holds, then triangles ABO and $A\Gamma O$ are congruent. This can be seen by placing them so that their right angles at B and Γ and their sides ΓA and BA become coincident. Then their respective hypotenuses, which are equal to OA must also coincide (Corollary 1.14) and the triangles will have all three sides respectively equal.

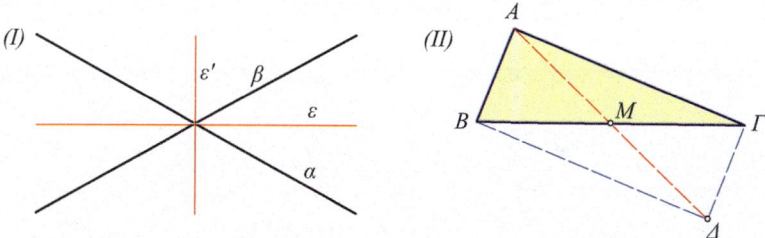

Fig. 1.76: Angle bisectors   Median of right triangle

Corollary 1.36. *The geometric locus of the points which are equidistant from two different lines α and β intersecting at O, consists of two orthogonal lines passing through the intersection point of α and β. These lines coincide with the bisectors of the angles which are formed by α and β at O (See Figure 1.76-I).*

Corollary 1.37. *in every right triangle the median to the hypotenuse is equal to half the hypotenuse and divides the right triangle into two isosceli triangles.*

Proof. Extend the median AM by doubling it until Δ: $|AM| = |M\Delta|$ (See Figure 1.76-II). Applying the SAS-criterion, we easily see that triangles AMB and $\Gamma M\Delta$ are congruent, as well as triangles $AM\Gamma$ and $BM\Delta$. Then, applying the SSS-criterion, we see that triangles $AB\Gamma$ and $\Delta\Gamma B$ are also congruent and the angle $\widehat{A\Gamma\Delta}$ is right ($\widehat{A\Gamma\Delta} = \widehat{A\Gamma B} + \widehat{B\Gamma\Delta} = \widehat{A\Gamma B} + \widehat{\Gamma BA} = 90°$). It follows that triangles $AB\Gamma$ and $A\Gamma\Delta$ are congruent, as right triangles with equal orthogonal sides (SAS-criterion). From this follows that their medians towards M will be equal ($|AM| = |\Gamma M|$), which is what we wanted. The two isosceles triangles mentioned in the corollary are AMB and $AM\Gamma$.

Exercise 1.65. Show the converse of the preceding corollary. If the median AM of triangle $AB\Gamma$ is half the side $B\Gamma$, then the triangle is right angled at A.

Exercise 1.66. Show that the median AM from acute/obtuse angle A of the triangle $AB\Gamma$ towards the opposite side $B\Gamma$ has length greater/less than $\frac{|B\Gamma|}{2}$.

Hint: Use figure 1.76-II and Theorem 1.5.

Exercise 1.67. Show the inverse to the preceding exercise, i.e. if the median AM of the triangle $AB\Gamma$ is greater/less of $|B\Gamma|/2$, then the angle α is acute/obtuse.

Exercise 1.68. Suppose that the interior bisectors of angles B and Γ of triangle $AB\Gamma$ intersect at Δ. Show that the measure of the angle $\widehat{B\Delta\Gamma}$ is equal to $\frac{\alpha}{2} + 90$, where $\alpha = |\widehat{BA\Gamma}|$ (also see Exercise 2.16).

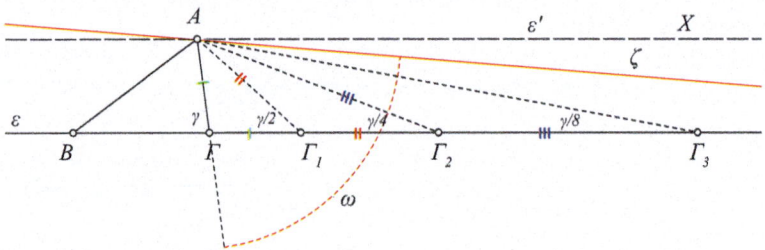

Fig. 1.77: Axiom of parallels equivalent to $\alpha + \beta + \gamma = 180°$

Theorem 1.13. *Assuming that all other axioms are valid, the axiom of the parallel is equivalent with the fact that the sum of the angles of a triangle is 180°.*

Proof. That the axiom of the parallel implies $\alpha + \beta + \gamma = 180°$ for the angles of triangle $AB\Gamma$, was proved in Theorem 1.11. For the converse, we show that, if the sum of the angles of a triangle is 180°, then there exists one and only one parallel to the line ε from a point A not lying on it. Indeed, let us consider the triangle $AB\Gamma$ with B, Γ belonging to ε and line $\varepsilon' = AX$, which makes an angle $|\widehat{XA\Gamma}| = \gamma$. We know that this is a parallel of ε. We'll see that there is no other, by proving that every other line $\zeta \neq \varepsilon'$, making an angle $\omega < \gamma$ at A, will necessarily intersect ε (See Figure 1.77). In fact, suppose that there is such a line ζ making an angle $\omega < \gamma$ with $A\Gamma$ and not intersecting line ε. This leads to a contradiction as follows. We construct successive points Γ_1,

$\Gamma_2, \Gamma_3, \ldots$ belonging to ε, so that $|\Gamma\Gamma_1| = |A\Gamma|$, next $|\Gamma_2\Gamma_1| = |A\Gamma_1|$, then $|\Gamma_3\Gamma_2| = |A\Gamma_2|$, etc. We see immediately that the triangles $A\Gamma\Gamma_1$, $A\Gamma_1\Gamma_2$, $A\Gamma_2\Gamma_3$, ... are isosceli and the angles at their base have measures, respectively $\frac{\gamma}{2}, \frac{\gamma}{4}, \frac{\gamma}{8}$, etc. Consequently after ν steps the angle at the base of the isosceles $A\Gamma_{\nu-1}\Gamma_\nu$ will be $\frac{1}{2^\nu}\gamma$ and therefore, for a big ν, the angle will become

$$\widehat{\Gamma A \Gamma_\nu} = \gamma - \frac{1}{2^\nu}\gamma > \omega.$$

For this, it suffices to take sufficiently big ν, so that

$$\gamma - \omega > \frac{1}{2^\nu}\gamma \quad \Leftrightarrow \quad \frac{\gamma - \omega}{\gamma} > \frac{1}{2^\nu}.$$

But then, line ζ will be found in the interior of angle A of triangle $\Gamma A \Gamma_\nu$ and consequently will intersect the opposite side, contrary to the hypothesis.

Exercise 1.69. Show that every triangle, for which one angle is the sum of the other two, can be partitioned into two isosceli triangles. Show the same thing for a triangle in which one of its angles is three times another one.

1.16 Symmetries

> It is true what they propose and name order and symmetry, analogies and correct reasoning. When you have seen them, you cannot deny them, anymore.
>
> J.L. Siesling, *The painter of Tour di Pen*

There are many kinds of symmetries that give pleasure to Mathematicians, for aesthetic reasons as well as for their contribution to the simplification of problems. Two of the simplest kinds are the *axis symmetry* and the *point symmetry*. The first is completely determined by a line and the second by a point. Given therefore a line ε, we call **symmetry relative to line ε** or

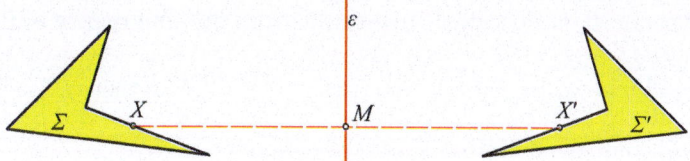

Fig. 1.78: Axial Symmetry

symmetry relative to axis ε the correspondence of X' to the point X not lying on ε, so that the line segment XX' is orthogonal to ε and the middle M of XX' belongs to ε. In other words, ε is the medial line of XX'. Point X' is called

symmetric or **reflected** of X **relative to** ε. When X belongs to ε, then we consider that X' coincides with X and conversely when X and X' coincide, then X belongs to line ε.

The symmetry relative to ε is the mathematical description of the action of folding the plane along ε, as if it was made of paper. Points which coincide after the folding are exactly symmetrical relative to ε. By the folding, the points of line ε stay fixed and make the edge of the folding. Often ε is called **axis of symmetry** and this symmetry is characterized as **axial symmetry**. From the definition follows that the symmetric of the symmetric of X is the original point X. The points of the axis are the **fixed points** of the symmetry.

We call two shapes Σ and Σ' **symmetric relative to an axis**, when there exists an axial symmetry so that every point X of Σ has a symmetric in Σ' (See Figure 1.78). A shape Σ is called **symmetric relative to an axis** when Σ is symmetrical to itself relative to that axis. This means that for every X of Σ its symmetric X' is contained in Σ too.

Proposition 1.11. *Every line ε, orthogonal to the line α, is an axis of symmetry of the line α.*

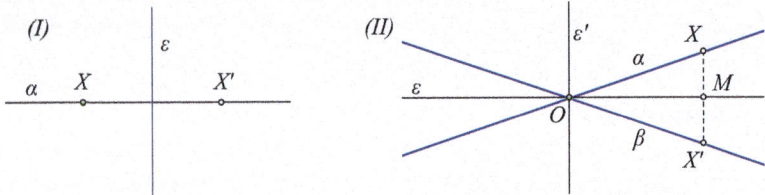

Fig. 1.79: α symmetric w.r.t. ε \quad\quad $\{\alpha, \beta\}$ symetric w.r.t. ε

Proof. By definition XX' will be orthogonal to ε (See Figure 1.79-I), therefore line XX' will coincide with α.

Proposition 1.12. *The shape which consists of two intersecting lines α and β is symmetric with respect to each one of the bisectors ε and ε' of the angles they form.*

Proof. If X is any point of α and X' the symmetric of X relative to the bisector ε (See Figure 1.79-II), then the triangle $XX'O$, where O is the intersection point of α and β, is isosceles. This results from the congruence of the right triangles OMX and OMX', which have OM common and $|MX| = |MX'|$, from the definition of symmetry. Consequently OM will bisect the angle at O. Because, by the hypothesis ε is also a bisector of the angle of α and β, lines OX' and β will coincide, therefore X' will be contained in β.

Remark 1.23. Axial symmetry is the key to the art of folding a sheet of paper, in such a way, that after a number of folds the resulting figure is interesting. This art, widely spread in Japan, is called Origami ([37], [2], [29]). Next exercise ([38, p.11]) shows, how we can construct angles of 30 and 60 degrees, using an appropriate folding of a square piece of paper.

1.16. SYMMETRIES

Exercise 1.70. Fold a square piece of paper $AB\Gamma\Delta$ along its middle. Next unfold it so that the trace of the line α along which it was folded (the crease) becomes visible. Next fold it again along angle B, this time until B coincides with a point Z of α, while keeping A fixed. Next unfold it so that the trace of line β, along which it was folded the second time becomes visible. Show that the angle between α and β is 60 degrees (See Figure 1.80).

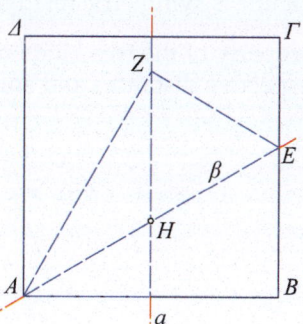

Fig. 1.80: Origami construction of angle of 30 degrees

Hint: During the second folding, AB overlaps AZ, therefore $|AZ| = |AB|$. However α is the medial line of AB, therefore $|ZA| = |ZB|$ (ZB not drawn in figure 1.80). Consequently triangle ABZ is equilateral, therefore the angle between α and β is 60 degrees.

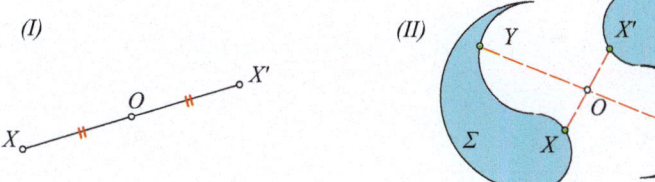

Fig. 1.81: Symmetry relative to O Symmetric relative to O

Given a point O, we call **symmetry relative to point** O, the correspondence between any point X different from O, and a point X', so that the line segment XX' has its middle at the point O (See Figure 1.81-I). X' is called **symmetric of X relative to O**. We consider the symmetric of O to be itself. Conversely, if X coincides with its symmetric point relative to O then X coincides with O. Point O is called **center of symmetry** and this symmetry is characterized as a **point symmetry**. From the definition follows that the symmetric of the symmetric of a point X is the original point X. Point O is the unique **fixed point** of the point symmetry.

We say that two shapes Σ and Σ' are **symmetric relative to a point**, when there exists a point symmetry so that every point X of Σ has its symmetric in Σ' and each point X' of Σ' has its symmetric in Σ (See Figure 1.81-II). A shape Σ is called **symmetric relative to a point** when Σ is symmetric to itself relative to a point. This means that for every X of Σ its symmetric X' is again contained in Σ.

Exercise 1.71. Show that a line is symmetric relative to any of its points, O.

The most important property of the two kinds of symmetry we have defined is the equality of respective distances and angles.

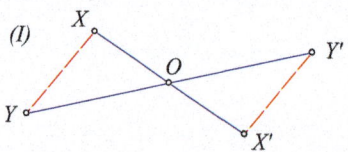
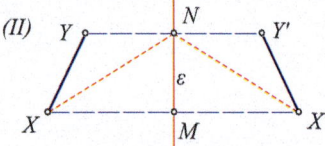

Fig. 1.82: Symmetry rel. to O Symmetry rel. to ε

Theorem 1.14. *If the shapes Σ and Σ' are symmetric relative to an axis or a point, then the distances of any two points $|XY|$ and the distances of their respective symmetric points $|X'Y'|$ are equal.*

Proof. The case of symmetry relative to point is simple. If (X,X') and (Y,Y') are symmetric relative to O (See Figure 1.82-I), then by definition the following equalities hold:

$$|XO| = |OX'|, \text{ and } |YO| = |OY'|.$$

Then, triangles XOY and $X'OY'$ are congruent (SAS-criterion), therefore also $|XY| = |X'Y'|$. Now, if (X,X') and (Y,Y') are symmetric relative to axis, then the middle points M and N respectively of XX' and YY' belong to the axis ε and N belongs to the medial line of XX' (See Figure 1.82-II). Therefore the triangle $XX'N$ is isosceles, has $|NX| = |NX'|$ and $|\widehat{YNX}| = |\widehat{Y'NX'}|$. Then, the triangles XNY and $X'NY'$ are congruent, having their sides $|XN| = |NX'|$, $|YN| = |NY'|$ and the angles $|\widehat{XNY}| = |\widehat{X'NY'}|$.

Corollary 1.38. *If the shapes Σ and Σ' are symmetric relative to an axis or a point, then the triangles which are formed by any three non-collinear points X, Y and Z of Σ and their respective symmetric points X', Y' and Z' of Σ', are congruent.*

Proof. It follows directly from the preceding proposition, since then $|XY| = |X'Y'|$, $|YZ| = |Y'Z'|$, $|ZX| = |Z'X'|$ and the result follows by applying the SSS-criterion for congruent triangles.

1.16. SYMMETRIES

Corollary 1.39. *If the shapes Σ and Σ' are symmetric relative to an axis or a point, then the angles $\widehat{XYZ}$ and $\widehat{X'Y'Z'}$ formed by any three points X, Y and Z of Σ and their corresponding points X', Y' and Z' of Σ', are equal.*

Proof. It follows directly from the preceding proposition, because of the equality of the corresponding angles of congruent triangles XYZ and $X'Y'Z'$.

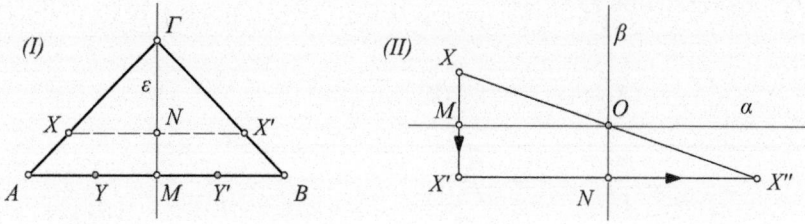

Fig. 1.83: Axis of isosceles   Point symmetry from axials

The classic example of a symmetric shape relative to an axis is the isosceles triangle (See Figure 1.83-I). Its axis of symmetry is its median to its base which is also the bisector of its apical angle and simultaneously its altitude to the base (Corollary 1.3).

Proposition 1.13. *The line ε, which joins the apex of the isosceles with the middle of its base, is an axis of symmetry for the triangle.*

Proof. The symmetry of base points Y, Y' (See Figure 1.83-I) is obvious. On the other hand every parallel to the base intersects the triangle's legs at points X, X' symmetric relative to the median ΓM, since also $X\Gamma X'$ is isosceles ($|\widehat{\Gamma XX'}| = |\widehat{\Gamma X'X}|$, because these angles have sides which are parallel to $\widehat{\Gamma AM}$ and $\widehat{\Gamma BM}$ respectively) and its median coincides with ΓM and is orthogonal to XX' (Corollary 1.3).

Theorem 1.15. *If a shape is symmetric relative to two axes α and β, which are orthogonal, then it is symmetric relative to their intersection point O.*

Proof. Indeed, let X' be the symmetric of point X relative to the axis α and X'' be the symmetric of X' relative to axis β (See Figure 1.83-II). Because of the axial symmetries, $XX'O$ and $X'OX''$ are isosceli triangles, therefore $|OX| = |OX''|$. Also thee angles $\widehat{XOM}$ and $\widehat{MOX'}$ are equal and the angles $\widehat{X'ON}$ and $\widehat{NOX''}$ are also equal. But the angle $\widehat{MON}$ is by definition right, therefore its double will be flat, in other words points X, O and X'' will be collinear.

Exercise 1.72. Show that if a triangle has an axis of symmetry then it is isosceles.

Hint: If ε is the axis of symmetry, then at least one triangle vertex, B say, is not contained in the axis. The symmetric Γ relative to ε will be again a triangle vertex. Then the angle at B will correspond by symmetry to an equal angle at Γ and the triangle will be isosceles.

Exercise 1.73. Show that a triangle is isosceles, if and only if it has an axis of symmetry.

Hint: Combine the preceding exercise with Proposition 1.13.

Exercise 1.74. Show that if the points Y, Y' are symmetric of X, X' (relative to axis or point) then the symmetric of every point Z of the line XX' is a point Ω belonging to line YY'.

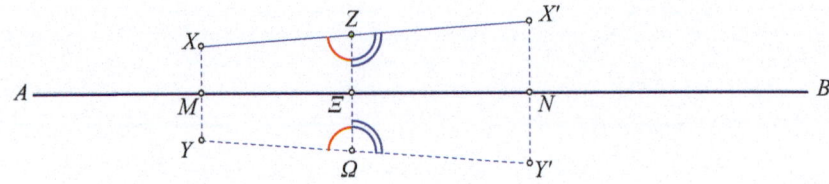

Fig. 1.84: Symmetric of line relative to axis

Hint: For the axial symmetry. Let Y, Y' be the symmetrics of X, X' relative to axis AB (See Figure 1.84). Let also Ω be the symmetric of a point Z of XX' relative to AB. The angles $\widehat{XZ\Xi}$ and $\widehat{\Xi\Omega Y}$ are equal (Corollary 1.39). Similarly the angles $\widehat{\Xi ZX'}$ and $\widehat{\Xi\Omega Y'}$ are equal. However $\widehat{XZ\Xi}$ and $\widehat{\Xi ZX'}$ are supplementary, therefore $\widehat{Y\Omega\Xi}$ and $\widehat{\Xi\Omega Y'}$ will also be supplementary and the three points Y, Ω and Y' will be collinear. The property for the corresponding point symmetry is proved similarly.

Exercise 1.75. Show that the figure consisting of two lines α and β always has a center of symmetry as well as axes of symmetry. When is there exactly one center of symmetry? When do the two lines have more than two axes of symmetry or more than two points of symmetry?

Remark 1.24. The point/axial symmetry is a special case of *isometry* (II-§ 2.1, II-§ 7.1), which in turn is a special kind of *transformation*.

1.17 Ratios, harmonic quadruples

> The harmony of the world is made manifest in Form and Number, and the heart and soul and all the poetry of Natural Philosophy are embodied in the concept of mathematical beauty.
>
> W. Thompson, *On Growth and Form*

In this section we examine, how the ratio of distances of a point X on the line ε, from two other fixed points A and B of the line, determines the position

1.17. RATIOS, HARMONIC QUADRUPLES

of X. It can be proved, that for two fixed points A and B on ε and a given ratio $t \neq 1$ there are precisely two points X, X' on the same line, which have ratio of distances t relative to A, B. This way we arrive at the definition of the *harmonic quadruple* (A, B, X, X') of four points on a line. A notion with many uses in Geometry. Setting $x = |AX|$, we easily see that the ratio $t = \frac{|XA|}{|XB|}$ is less

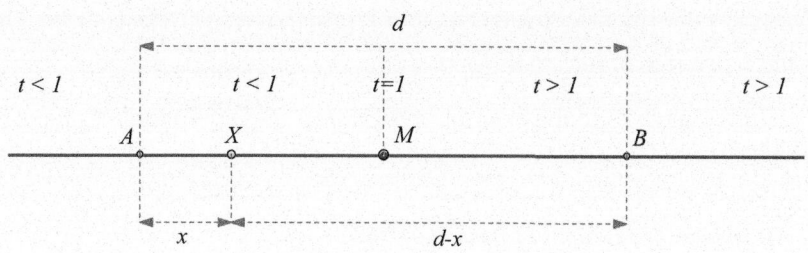

Fig. 1.85: Ratio $t = \frac{|XA|}{|XB|}$

than when X belongs to the half line with start point the middle M of AB, which contains A and is greater than 1, on the opposite half line, starting at the middle M, which contains B (See Figure 1.85). If $d = |AB|$ is the length of AB, then we have for points X in AB:

$$t = \frac{x}{d-x} \Leftrightarrow x = \frac{dt}{1+t} \text{ for } t > 0. \tag{1.1}$$

For the points outside of the line segment AB and on A's side:

$$t = \frac{x}{d+x} \Leftrightarrow x = \frac{dt}{1-t} \text{ for } t < 1. \tag{1.2}$$

Similarly for the points outside the line segment AB and on B's side:

$$t = \frac{x}{x-d} \Leftrightarrow x = \frac{dt}{t-1} \text{ for } t > 1. \tag{1.3}$$

We see that, besides the ratio value $t = 1$, which corresponds to exactly one point, the middle M of AB, for all other values of $t > 0$, there exist two points X, X', which have ratio $\frac{|XA|}{|XB|} = t$. The first point (X) is inside the segment and we find it from equation (1) and the second (X') is outside AB and we find it from equation (2) or (3), depending on whether $t < 1$ or $t > 1$.

For example, if $t = 2$, i.e. $\frac{|XA|}{|XB|} = 2$, then X belongs to the half line of M on the side of B. We find the point X of AB from equation (1): $x = \frac{2d}{3}$ and the point X' outside of AB from equation (3): $x = \frac{2d}{1} = 2d$.

We have therefore proved the next theorem.

Theorem 1.16. *For every positive number $t \neq 1$, there are precisely two points X, X', on the line AB, such that $\frac{|XA|}{|XB|} = t$. One of them lies in the interior of the segment AB, and the other lies outside this segment. For $t = 1$ there is just one point, which is the middle of AB.*

If we limit ourselves to the interior points of AB, then we obviously have the corollary:

Corollary 1.40. *For every positive number t, there is precisely one point X on line AB and between A, B, such that $\frac{|XA|}{|XB|} = t$.*

If we limit ourselves to the exterior of AB instead, then we similarly have the corollary:

Corollary 1.41. *For every positive number $t \neq 1$, there is precisely one point X on line AB and on the exterior of AB, so that $\frac{|XA|}{|XB|} = t$.*

An immediate consequence of the two corollaries, with many applications, is also the following corollary, which shows that the ratio $\frac{|XA|}{|XB|}$, coupled with the information on whether X is on the interior or the exterior of AB, determines X's position uniquely.

Corollary 1.42. *If two points X and Y of the line AB form the same ratio $\frac{|XA|}{|XB|} = \frac{|YA|}{|YB|}$ and they both belong to the interior of AB or both belong to the exterior of AB, then they coincide.*

Given two points A, B, we call two different points X and X' **harmonic conjugate relative to** A, B and write $X' = X(A,B)$ or equivalently $X = X'(A,B)$, when they form the same ratio $t = \frac{|XA|}{|XB|} \neq 1$. As we saw, one of them is interior to the segment AB and the other is outside of AB. We say that four points of the same line: A, B, X, X', of which the last two are harmonic conjugate relative to the first two, define a **harmonic quadruple** or **harmonic division** of points. We denote this property by writing $(AB;XX') = -1$.

Proposition 1.14. *The points X, X' are harmonic conjugate relative to A, B, which are at distance $|AB| = d$ (See Figure 1.86), if and only if their distances $x = |XA|$ and $x' = |X'A|$ from A satisfy one of the following relations*

$$2x \cdot x' = d \cdot (x' - x), \quad 2x \cdot x' = d \cdot (x' + x).$$

The first corresponds to a ratio $t = \frac{|XA|}{|XB|} < 1$ and the other to a ratio $t > 1$.

Proof. In the first case the equality of ratios $\frac{|XA|}{|XB|} = \frac{|X'A|}{|X'B|}$, gives according to equations (1) and (2): $\frac{x}{d-x} = \frac{x'}{d+x'}$. In the second case again equations (1) and (3) of this section give $\frac{x}{d-x} = \frac{x'}{x'-d}$. These two equations are equivalent respectively to the aforementioned.

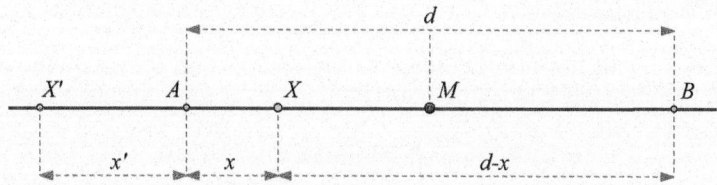

Fig. 1.86: $\{X, X'\}$ harmonic conjugate of $\{A, B\}$

Corollary 1.43. *If t is positive $t \neq 1$, then the two points X, X' of line AB which have ratio $\frac{|XA|}{|XB|} = t$, are in a distance*

$$|XX'| = \begin{cases} \frac{2dt}{1-t^2} & \text{for } t < 1, \\ \frac{2dt}{t^2-1} & \text{for } t > 1, \end{cases}$$

where $d = |AB|$.

Proof. For $t < 1$, both points belong to the half line of M (middle of AB), which contains A and we have $|XX'| = x + x'$, where $x = |AX| = \frac{dt}{1+t}$ and $x' = |AX'| = \frac{dt}{1-t}$ and the first expression follows from some obvious calculation. For $t > 1$, both points belong to the half line of M which contains B and we have $|XX'| = x' - x$, where $x = |AX| = \frac{dt}{1+t}$ and $x' = |AX'| = \frac{dt}{t-1}$ and the second expression follows from also from some obvious calculation.

Corollary 1.44. *Points X of the interior of AB and X' of the exterior of AB are harmonic conjugate to A and B, if and only if,*

$$|X'A||X'B| - |XA||XB| = |XX'|^2.$$

Proof. Simple calculations relying on the preceding expressions.

Exercise 1.76. Show that the harmonic conjugates X and X' of A and B always belong to the same half line from the two defined by the middle M of line segment AB.

Hint: As it is seen in figure 1.86, the harmonic conjugates belong to one or the other half line of M, depending on the value of the ratio $t = \frac{|XA|}{|XB|}$. If $t < 1$, then both belong to the half line containing A. If $t > 1$, then both belong to the half line containing B.

Proposition 1.15. *Points X and X' are harmonic conjugate relative to A and B, if and only if*

$$|MX||MX'| = \left(\frac{|AB|}{2}\right)^2,$$

where M is the middle of AB.

Proof. Let X, X' be harmonic conjugate relative to A and B. We use the expressions of Proposition 1.14, which we solve for x', assuming that X belongs to the interior of AB:

$$x' = \begin{cases} \frac{dx}{d-2x} & \text{for } t < 1, \\ \frac{dx}{2x-d} & \text{for } t > 1. \end{cases}$$

Also

$$|MX||MX'| = \begin{cases} \left(\frac{d}{2}-x\right)\left(\frac{d}{2}+x'\right) & \text{for } t < 1, \\ \left(x-\frac{d}{2}\right)\left(x'-\frac{d}{2}\right) & \text{for } t > 1. \end{cases}$$

In both cases, replacing x' from the preceding equations and simplifying, we arrive at $|MX||MX'| = \left(\frac{d}{2}\right)^2$. Conversely, if we suppose the relation, this translates to $\left(\frac{d}{2}\right)^2 = |MX||MX'|$, where again $|MX||MX'|$ is given by the preceding expressions, and leads to

$$\left(\frac{d}{2}\right)^2 = \begin{cases} \left(\frac{d}{2}-x\right)\left(\frac{d}{2}+x'\right) & \text{for } t < 1, \\ \left(x-\frac{d}{2}\right)\left(x'-\frac{d}{2}\right) & \text{for } t > 1. \end{cases}$$

Simplifying these, we arrive at the equations of Proposition 1.14, which characterize the harmonic conjugate points.

Exercise 1.77. Show that if points X, X' are harmonic conjugate relative to A, B, then A, B are also harmonic conjugate relative to X and X'.

Hint: When the X, X' are on the side of A (relative to middle M), then the ratios $\frac{|AX|}{|AX'|} = \frac{x}{x'}$ and $\frac{|BX|}{|BX'|} = \frac{d-x}{d+x'}$ are equal according to Proposition 1.14 (See Figure 1.86). When the X, X' are on the side of B, then $\frac{|AX|}{|AX'|} = \frac{x}{x'}$ and $\frac{|BX|}{|BX'|} = \frac{d-x}{x'-d}$ and the equality of the ratios results from the second case of Exercise 1.14.

Exercise 1.78. Using the preceding notation, show that the middle N of the interval XX' will be found at distance $|AN| = d\frac{t^2}{1-t^2}$ for $t < 1$ and $|BN| = d\frac{t^2}{t^2-1}$ for $t > 1$.

Hint: See first that, for example, for $t < 1$ holds $x' > x$, since $\frac{x}{x'} = \frac{1-t}{1+t} < 1$. Consequently, if $c = |AN|$, then the fact that N is the middle of XX', implies $x' - c = c + x \Rightarrow c = \frac{x'-x}{2}$. Now substitute in this expression x, x' from expressions (1) and (2) respectively. The other case, for $t > 1$. is handled similarly.

Exercise 1.79. For the points $\{A, B, X, X', M\}$ of figure 1.86, where $(AB; XX')$ is a harmonic quadruple and M is the middle of the segment AB, show the relations:

1.17. RATIOS, HARMONIC QUADRUPLES

$$\frac{2}{|AB|} = \frac{1}{|XB|} + \frac{1}{|X'B|} \quad \text{and} \quad |X'A| \cdot |X'B| = |AB| \cdot |XM|.$$

To investigate how these relations change when the ordering of the points is different from that of figure 1.86.

Exercise 1.80. Let $\{A, B, X, X'\}$ be points of the line ε for which we know the ratios $\kappa = \frac{|XA|}{|XB|}$, and $\lambda = \frac{|X'A|}{|X'B|}$. To find the distance $|XX'|$ and $|YY'|$ of the harmonic conjugates $\{Y, Y'\}$ of $\{X, X'\}$ with respect to $\{A, B\}$. To distinguish between the relative positions of $\{X, X'\}$ with respect to $\{A, B\}$.

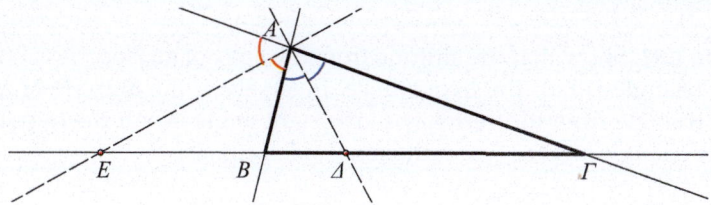

Fig. 1.87: Δ and E harmonic conjugate relative to B and Γ

Remark 1.25. The most famous harmonic conjugate points are the traces Δ and E, in $B\Gamma$, of the two bisectors (that is, the intersections of the two bisectors with $B\Gamma$), internal and external, from vertex A of triangle $AB\Gamma$, which *does not* have its adjacent sides (AB and $A\Gamma$) equal (See Figure 1.87). We'll talk about these in a subsequent chapter (Theorem 3.3 and Exercise 3.12). The exercises of this section (in particular the last exercise) must be repeated after § 5.14 on *signed* ratios.

Exercise 1.81. Point Γ is found in line AB on the side of B and holds $|\Gamma A| = v|AB|$. Find the ratio $t = \frac{|\Gamma A|}{|\Gamma B|}$. Same question when Γ is found on the side of point A.

Hint: In the first case $|\Gamma B| = |\Gamma A| - |AB|$, therefore $t = \frac{|\Gamma A|}{|\Gamma B|} = \frac{|\Gamma A|}{|\Gamma A| - |AB|} = \frac{v|AB|}{v|AB| - |AB|} = \frac{v}{v-1}$. In the second case $|\Gamma B| = |\Gamma A| + |AB|$, etc.

Exercise 1.82. The line segment AB is divided by point A_1 in parts with ratio p/q, where $\{p < q\}$ are positive integers. It is also divided by point B_1 in parts with ratio q/p. Find the length of the segment A_1B_1. Repeat the preceding construction for the segment A_1B_1 and find analogously the segment A_2B_2. Repeat the procedure for the last segment and find the new segment A_3B_3. Continuing that way, we find after n steps the segment A_nB_n. What is the length of this segment?

1.18 Comments and exercises for the chapter

> Digressions, incontestably, are the sunshine - they are the life, the soul of reading; take them out of this book for instance - you might as well take the book along with them.
>
> *Laurence Sterne, Tristram Shandy, I, 22*

Exercise 1.83. Determine the angles of an isosceles and right-angled triangle. Is it possible for such a triangle to have the equal legs enclosing an other than a right angle?

Exercise 1.84. Show that the altitude from the apex of an isosceles and right-angled triangle is half the hypotenuse. Inversely, if the altitude of a right-angled triangle from the vertex with the right angle is half the hypotenuse, then the triangle has the two orthogonal sides equal.

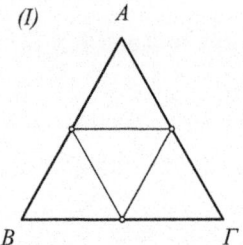 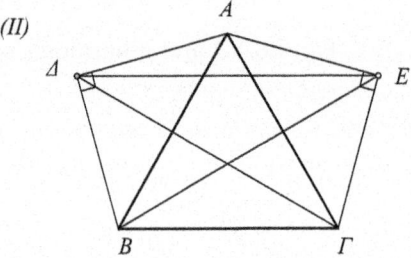

Fig. 1.88: Equilateral   Equality of sides

Exercise 1.85. Show that the middles of the sides of an equilateral triangle $AB\Gamma$ define four equal equilateral triangles (See Figure 1.88-I).

Exercise 1.86. On the sides of an equilateral triangle and outside we construct equal right-angled triangles (See Figure 1.88-II). Show that in such a construction it is impossible for the five resulting outer sides to be equal.

Exercise 1.87. If in the figure 1.88-II we suppose that the lengths $|A\Delta|, |\Delta B|$, $|\Gamma E|, |EA|$ are equal, then determine the angles at $\{A, B, \Gamma, \Delta, E\}$.

Exercise 1.88. If in a right triangle one of the acute angles is double the other, then the triangle has angles measuring $90°$, $60°$ and $30°$ and one of the orthogonal sides is half the hypotenuse.

Exercise 1.89. Given is a right-angled triangle $AB\Gamma$ with orthogonal sides $|A\Gamma| > |AB|$, median AM and altitude AY from A. Show that the $\widehat{YAM} = \beta - \gamma$ (see exercise 2.178 for the converse).

1.18. COMMENTS AND EXERCISES FOR THE CHAPTER

Exercise 1.90. In an arbitrary triangle $AB\Gamma$ consider the points Δ, E on the base $B\Gamma$, such that $\widehat{BA\Delta} = \widehat{\Gamma}$ and $\widehat{EA\Gamma} = \widehat{B}$. Show that the triangle ΔAE is isosceles.

Exercise 1.91. Given two intersecting lines α, β and a point P not lying on them, draw from P a line ε, so that the three lines α, β, ε form an isosceles triangle. Find cases in which there is only one solution.

Exercise 1.92. Given two points A, B and a line ε, draw two lines α, β, respectively, through A and B, so that the three lines α, β, ε form an equilateral triangle.

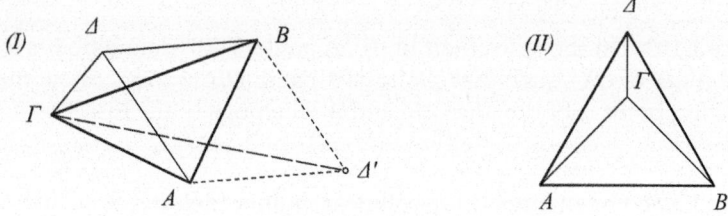

Fig. 1.89: Computation of angles

Exercise 1.93. On the orthogonal side AB of an isosceles right-angled triangle $AB\Gamma$ and towards Γ we construct an equilateral triangle $AB\Delta$ (See Figure 1.89-I). Find the measure of the angle $\widehat{BA\Gamma}$. Similarly determine the measure of the angle $\widehat{BA'\Gamma}$, where Δ' is the symmetric of Δ with respect to AB. Finally determine the angle of the lines $\{A\Gamma, B\Delta\}$.

Exercise 1.94. On the hypotenuse AB of an isosceles and right-angled triangle $AB\Gamma$ and towards the right angle we construct an equilateral triangle $AB\Delta$ (See Figure 1.89-II). Find the measure of angle $\widehat{BA\Gamma}$.

Exercise 1.95. Show that if the three points A, B, Γ satisfy the relation $|B\Gamma| = |BA| + |A\Gamma|$ then they are contained in a line (they are collinear).

Exercise 1.96. Show that for every line ε, point A, not lying on it and a positive number δ, there exist at most two line segments AB with B on ε, of length $|BA| = \delta$.

Exercise 1.97. Can a scalene triangle be divided, using a line, into two congruent triangles?

Exercise 1.98. In how many ways can an equilateral triangle be divided, by a line, into two equal triangles?

Exercise 1.99. In a triangle with base $a = |B\Gamma|$, show that $b + c - a < 2|A\Delta|$, where Δ is any point of the base.

Exercise 1.100. Show that any line segment EZ, contained completely in the interior of a triangle $AB\Gamma$, is less than the greater side of the triangle.

Exercise 1.101. $A\Delta$ is the bisector of the triangle $AB\Gamma$, which has sides $|AB| > |A\Gamma|$. Show that $|A\Delta| < |AB|$. Subsequently consider the point E of AB, such that $|AE| = |A\Delta|$. Show that the angle $\widehat{\Delta EB}$ is always obtuse and determine its measure as a function of the angles of the triangle. Finally show that $|B\Delta| > |\Gamma\Delta|$.

Exercise 1.102. In the right, at A, triangle $AB\Gamma$, with unequal orthogonals, AY, $A\Delta$ and AM are respectively the altitude, the bisector and the median. Show that $A\Delta$ is also the bisector of the angle $\widehat{YAM}$.

Exercise 1.103. On the sides (half lines) of angle $\widehat{XOX'}$ define points A, B of OX and A', B' of OX', such that $|OA| = |OA'|$ and $|OB| = |OB'|$. Show that the intersection point Δ of the lines AB' and $A'B$ belongs to the bisector of angle $\widehat{XOX'}$.

Exercise 1.104. Given three non collinear points, determine a line at the same distance from the three points (three solutions).

Exercise 1.105. Given two equal line segments AB, $A\Gamma$, find a point Δ on a given line ε, such that angles $\widehat{A\Delta B}$ and $\widehat{A\Delta\Gamma}$ are equal.

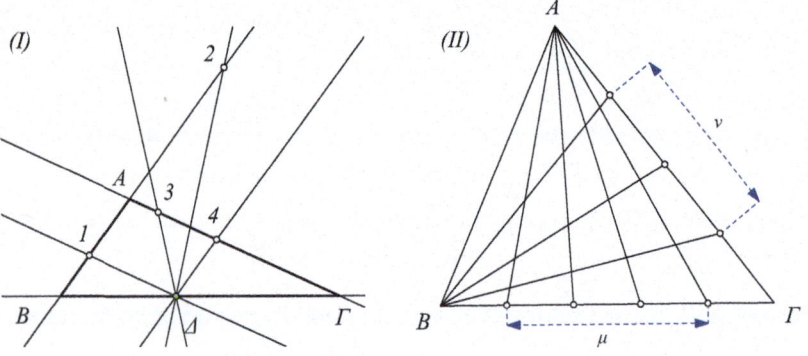

Fig. 1.90: $B\Delta 1$, $B\Delta 2$, $\Delta\Gamma 3$, $\Delta\Gamma 4$ How many triangles?

Exercise 1.106. Show that through every point Δ on the side $B\Gamma$ of the triangle $AB\Gamma$, different from B and Γ, passes a line ΔX which intersects the other sides of the triangle and forms with them a triangle with angles equal to those of $AB\Gamma$. Show that for every such Δ there exist four different lines ΔX which have this property (See Figure 1.90-I).

Exercise 1.107. We join vertex A of the triangle $AB\Gamma$ with μ different points of the opposite side. We also join vertex B with ν different points of the opposite side (See Figure 1.90-II). How many triangles are contained in the resulting shape?

Hint: $\frac{(\mu+1)(\nu+1)(\mu+\nu+2)}{2}$.

Exercise 1.108. The triangle $AB\Gamma$ has the sides $|A\Gamma| > |AB|$ and on the side $A\Gamma$ we take the point B', such that $|AB'| = |AB|$. Show that angle $\widehat{B'B\Gamma} = \frac{\beta-\gamma}{2}$. Show also that this angle is equal to the angle between the bisector and the altitude from A.

Exercise 1.109. Let X be a point not contained on either of the intersecting at O lines $\{\alpha, \beta\}$. Show that every line γ passing through X intersects at least one of $\{\alpha, \beta\}$.

Exercise 1.110. If Y, Δ, M are respectively labels for the traces on $B\Gamma$ of the altitude, bisector and median in triangle $AB\Gamma$, show that Δ lies always between Y and M. Also show that if two of these traces coincide, then all three traces coincide and the triangle is isosceles.

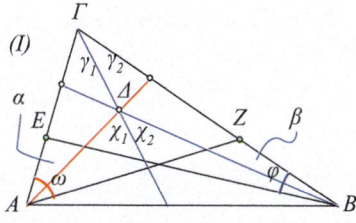
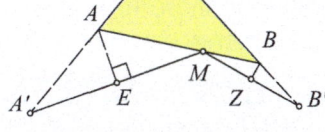

Fig. 1.91: $\widehat{A\Delta B} = \alpha + \beta$ Construction of point M

Exercise 1.111. On the sides ΓA, ΓB of the triangle $AB\Gamma$ we select respectively points E and Z and we draw the bisectors of the angles $\omega = \widehat{\Gamma AZ}$ and $\phi = \widehat{EB\Gamma}$, which intersect at Δ. Show that for angles $\alpha = \widehat{AEB}$ and $\beta = \widehat{AZB}$, holds $\alpha + \beta = 2\widehat{A\Delta B}$.

Hint: Suppose that $\Gamma \Delta$ divides angles $\gamma = \widehat{A\Gamma B}$ and $\widehat{A\Delta B}$, respectively, into two parts γ_1, γ_2 and χ_1, χ_2 (See Figure 1.91-I). Observe that $\alpha + \beta = 2\gamma + \omega + \phi$, $\chi_1 = \gamma_1 + \frac{\omega}{2}$, $\chi_2 = \gamma_2 + \frac{\phi}{2}$.

Exercise 1.112. Find a point M on the side AB of the triangle $AB\Gamma$, such that $|MA| + |A\Gamma| = |MB| + |B\Gamma|$.

Hint: Consider AA' equal to AM on the extension of ΓA and BB' equal to BM on the extension of ΓB (See Figure 1.91-II). The length $|\Gamma A'| = |\Gamma B'|$ is known and equal to half the perimeter of the triangle. Triangles $AA'M$ and $BB'M$ are isosceli and can be constructed.

Exercise 1.113. Given a point D inside the triangle ABC locate the closed billiard path returning to D after successive reflections on the sides $\{CA, AB, BC\}$ (See Figure 1.92). How many solutions can you find?

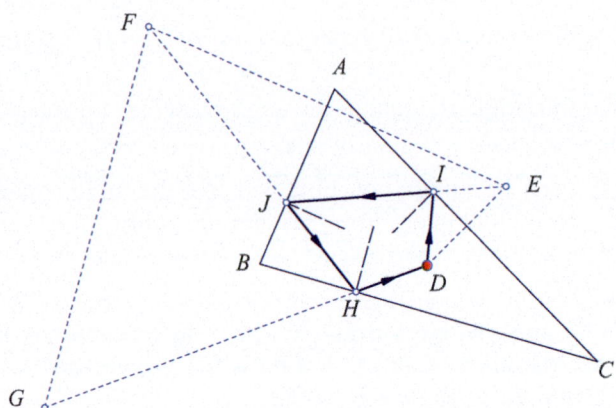

Fig. 1.92: Closed billiard ball path in triangle

Hint: Reflect D in AC to E, then reflect this in AB to F and this in BC to G. The closed path is determined from the intersection $H = (GD, BC)$.

Exercise 1.114. Show that if the triangle $AB\Gamma$ is contained entirely inside the triangle $A'B'\Gamma'$, then each side of $AB\Gamma$ is less than the greater side of $A'B'\Gamma'$ and each side of $A'B'\Gamma'$ is greater than the smaller side of $AB\Gamma$.

Exercise 1.115. Is there any line intersecting all side-lines of the triangle $AB\Gamma$ under the same angle ω?

Equal equilateral triangles are used as building elements (tiles) in some board games, in which the end of the game is to build shapes using these elements. Figure 1.93 shows the 24 building blocks of one such game, published by Robert Laffont and called "Trioker" ([33]). The unique rule of the game is the

match of number of points at concurring vertices.

The game is reminiscent of the classical "Stomachion of Archimedes" and "Tangram" discussed in § 3.14. The game's end results are shapes composed by (some or all of) these 24 tiles, characterized by the integers at the common vertices of the tiles building the shape (See Figure 1.94).

Exercise 1.116. Try to build the shaded hexagon in figure 1.94 using some of the trioker tiles.

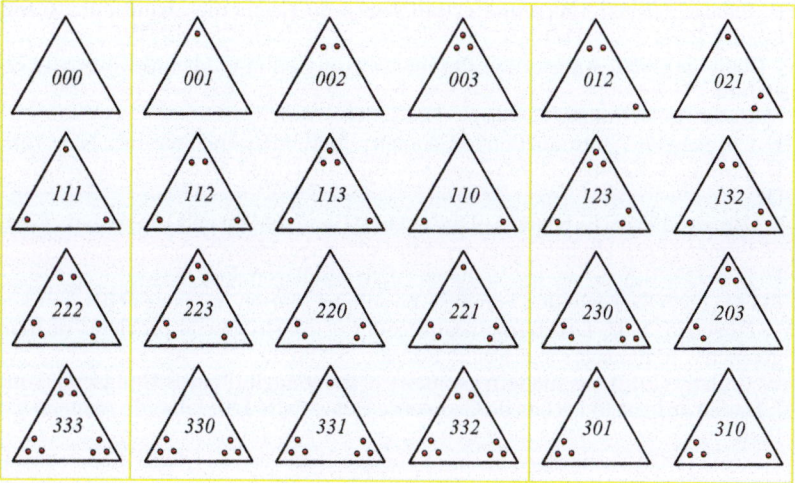

Fig. 1.93: Trioker tiles

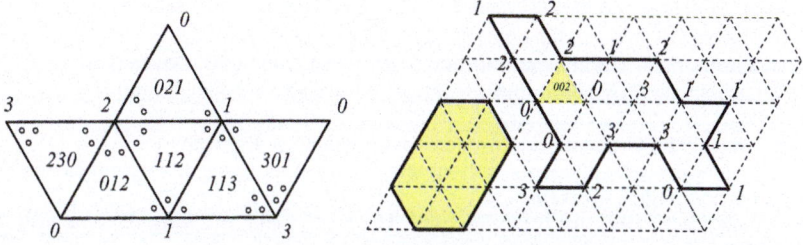

Fig. 1.94: Shapes build with trioker tiles

References

1. E. Agazzi, D. Palladino (1988) Le Geometrie non Euclidee e i Fondamenti della Geometria. La scuola, Brescia
2. R. Alperin, 2000. A Mathematical Theory of Origami Constructions and Numbers, *New York Journal of Mathematics*, 6:119-133
3. F. Bachmann (1973) Aufbau der Geometrie aus dem Spiegelungsbegriff. Springer, Heidelberg
4. O. Belyaev (2007) Fundamentals of Geometry.
   http://polly.phys.msu.ru/ belyaev/geometry.pdf
5. G. Birkhoff, 1932. A set of postulates for plane Geometry, Based on Scale and Protractor, *Annals of Mathematics*, 33:329-345
6. R. Bonola (1912) Non-Euclidean Geometry. The Open Court Publishing Company, Chicago
7. C. Boyer (1991) A History of Mathematics, 2nd Edition. John Wiley, New York
8. S. Cairns, 1933. An axiomatic basis for plane Geometry, *Transactions of the American Mathematical Society*, 35:234-244

9. E. Catalan (18582) Theoremes et problemes de Geometrie Elementaire. Carilian-Coeury, Paris
10. J. Coolidge (1980) A history of the Geometrical Methods. Oxford University Press, Oxford
11. N. Court (1980) College Geometry. Dover, New York
12. H. Coxeter (1961) Introduction to Geometry. John Wiley and Sons Inc., New York
13. H. Coxeter, L. Greitzer (1967) Geometry Revisited. Math. Assoc. Amer. Washington DC
14. T. Dantzig (1955) The bequest of the Greeks. George Allen and Unwin Ltd., London
15. N. Efimov (1980) Higher Geometry. Mir Publishers, Moscow
16. H. Eves (1963) A survey of Geometry. Allyn and Bacon, Inc., Boston
17. F.G.M (1920) Exercises de Geometrie, 6e edition. Maison A. Mame et fils, Tours
18. V. Fursenko. 1937. Lexicographical account of constructional problems of triangle geometry, *Mathematics in school (in Russian)*, 5:4-30
19. W. Gallatly (1913) The modern geometry of the triangle. Francis Hodgsonn, London
20. J. Hadamard (1905) Lecons de Geometrie elementaire I, II. Librairie Armand Colin, Paris
21. T. Heath (1908) The thirteen books of Euclid's elements vol. I, II, III. Cambridge University Press, Cambridge
22. T. Heath (1931) A manual of Greek Mathematics. Oxford University Press, Oxford
23. J. Heiberg (1885) Euclidis Elementa. Teubner, Leipzig
24. H. Helmholtz. 1876. The Origin and Meaning of Geometrical Axioms, *Mind*, 1:301-321
25. D. Hilbert (1903) Grundlagen der Geometrie. Teubner, Leipzig
26. W. Knorr (1993) The ancient tradition of geometric problems. Dover, New York
27. R. Lachlan (1893) Modern Pure Geometry. Macmillan and Co., London
28. T. Lalesco (1952) La Geometrie du triangle. Librairie Vuibert, Paris
29. R. Lang (1996) Origami and Geometric Constructions. Preprint
30. A. Legendre (1837) Elements de Geometrie suivis d' un traite de Trigonometrie. Langlet et compagnie, Bruxelles
31. D. Logothetti, 1980. An Interview with H.S.M. Coxeter, the King of Geometry, *The Two-Year College Mathematics Journal*, 11:2-19
32. P. Mironescu, L. Panaitopol, 1994. The existence of a triangle with prescribed angle bisector lengths, *Amer. Math. Monthly*, 101:58-60
33. Y. Odier, Y. Roussel (1979) Trioker mathematisch gespielt. Friedr. Vieweg, Braunschweig
34. A. Ostermann, G. Wanner (2012) Geometry by its history. Springer, Berlin
35. V. Oxman, 2008. A Purely Geometric Proof of the Uniqueness of a Triangle With Prescribed Angle Bisectors, *Forum Geometricorum*, 8:197-200
36. A. Papadopoulos (2010) Nikolai I. Lobachevsky, Pangeometry. European Mathematical Society, Zürich
37. J. Rourke (2011) How to fold it. Cambridge University Press, Cambridge
38. S. Row (1917) Geometric Exercises in Paper Folding. The open court publishing company, Chicago
39. J. Young (1917) Lectures on Fundamental Concepts of Algebra and Geometry. Macmillan, New York
40. P. Yiu (2013) Introduction to the Geometry of the Triangle. http://math.fau.edu/Yiu/Geometry.html

Chapter 2
Circle and polygons

2.1 The circle, the diameter, the chord

> Most people live, whether physically, intellectually, or morally, in a very restricted circle of their potential being. They make use of a very small portion of their possible consciousness, and of their soul's resources in general, much like a man who, out of his whole bodily organism, should get into a habit of using and moving only his little finger.
>
> William James, Letter to Lutoslawski

After the line and the related shapes produced by it, i.e. line segments, broken lines and polygons, the circle is the simplest curve which, like the line, has a certain homogeneity and shows the same behavior in all of its points and parts but, contrary to the line, it does not extend to infinity. Also, the line and the circle are the shapes for which we have the simplest drawing tools, the *ruler* and the *compass*. If we had a third, equally simple, tool for drawing a third category of curves, we would certainly include these curves also in the domain of elementary euclidean geometry.

Circle of radius ρ is called the shape of the plane consisting of all points X, whose distance from a fixed point O is ρ. Point O is called **center** of the circle (See Figure 2.1-I). **Radius** OX of the circle is also called the line segment with endpoints the center O and the circle point X. Often we also denote this circle with $O(\rho)$ or with $\kappa(O, \rho)$. **Interior** of the circle we call the set of points Y, whose distance from the center O is less than the radius: $|OY| < \rho$. **Exterior** of the circle we call the set of points Z, whose distance from the center O is greater than the radius: $|OZ| > \rho$. From Axiom 1.8 of lines, it follows that every line ε, which passes through a circle's center O, will meet the circle exactly at two points A and B. Such points of the circle are called **diametrically opposite** or **diametrical** or **diametral** and the line segment AB, which they define is called **diameter** of the circle (See Figure 2.1-II). Obviously the middle of every diameter is the center O and the diameter's length is twice

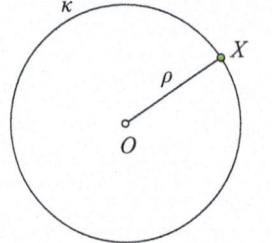 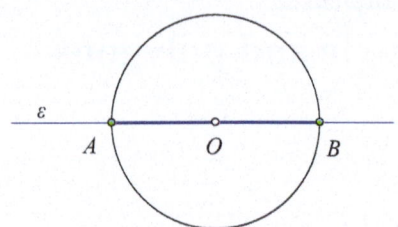

Fig. 2.1: Circle $\kappa(O,\rho)$,   Diameter AB

that of the radius $|AB| = 2\rho$. Two circles are called **congruent** or **equal**, when their radii are equal. **Chord** of a circle is called a line segment AB whose endpoints lie on the circle. The diameter is just a special case of chord. Every chord AB, which is not a diameter, defines, along with the center of the circle O, an isosceles triangle AOB (See Figure 2.2-I), whose legs have length ρ.

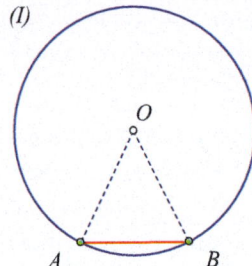 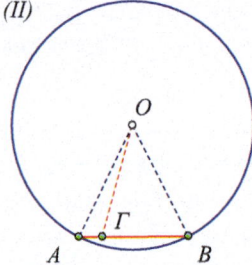

Fig. 2.2: Circle chord,   Chord-circle common points

Proposition 2.1. *A chord AB does not contain any points of the circle, other than A and B.*

Proof. If the chord contained one more circle point Γ (See Figure 2.2-II), then triangle $BO\Gamma$ would be isosceles and A would belong to the extension of the triangle's base, therefore, according to Corollary 1.14 it would be $\rho = |OA| < |O\Gamma| = \rho$, a contradiction.

Corollary 2.1. *The center of a circle is always contained in the medial line of its chords.*

Proof. Obviously, since the center is equidistant from the chord endpoints (See Figure 2.3-I), it will be contained in the medial line of the chord (Corollary 1.7).

Corollary 2.2. *The orthogonal line, from the center to a chord of a circle, passes through the chord's middle.*

2.1. THE CIRCLE, THE DIAMETER, THE CHORD

Corollary 2.3. *The middle points of parallel chords of a circle are contained in the diameter, which is orthogonal to these chords.*

Proof. From the center of the circle we draw an orthogonal line to one such chord (See Figure 2.3-II). This line will pass through the chord's middle and will be orthogonal also to any other parallel chord, therefore it will pass through the middle of any such chord (Corollary 2.2).

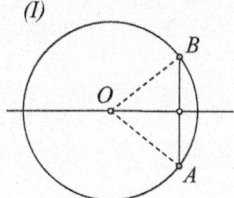 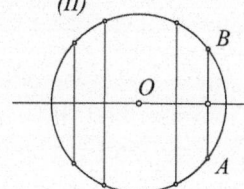

Fig. 2.3:   Chord middle   Middles of parallel chords

Corollary 2.4. *Every diameter of the circle is also an axis of symmetry of the circle.*

Corollary 2.5. *The center of a circle is also a center of symmetry of the circle.*

Exercise 2.1. Show that an axis of symmetry of the circle, must necessarily pass through its center. Show also, that a center of symmetry of the circle must coincide with its center.

Proposition 2.2. *The chord length, in a circle of radius ρ, is less than or equal to the length 2ρ of the circle's diameter. If the chord has length 2ρ then it coincides with a diameter of the circle.*

Proof. If O is the circle's center, $\delta = |AB|$ is the length of the chord and the points A, B, O are not collinear, then they define a triangle ABO. This triangle is isosceles and its legs have length ρ. Consequently, using the triangle inequality, $\delta = |AB| < |OA| + |OB| = 2\rho$. The last inequality also implies that the chord has length 2ρ precisely when points A, B and O are collinear.

Theorem 2.1. *For every triple of non-collinear points A, B and Γ, there exists a unique circle passing through them.*

Proof. The medial lines of line segments AB and $B\Gamma$ respectively, intersect at point O (See Figure 2.4-I). Had they not intersected, then they would be parallel and from B we would have verticals BM, BN to two parallels, therefore M, N and B would be collinear, a contradiction. From the first medial line we have (Corollary 1.7) $|OA| = |OB|$ and from the second $|OB| = |O\Gamma|$.

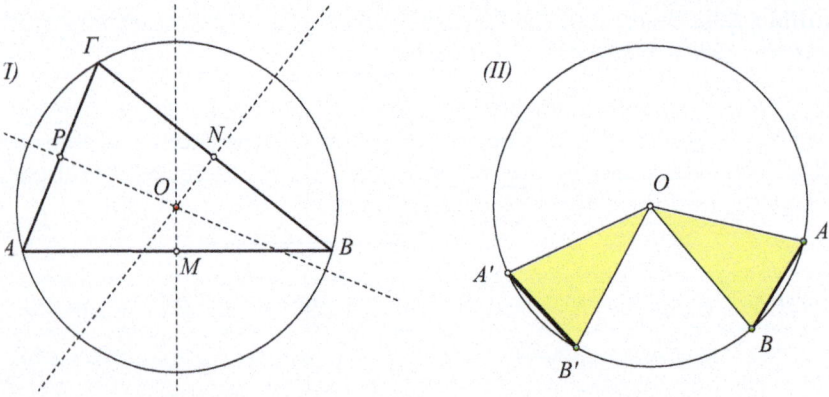

Fig. 2.4: Circumscribed circle (ABΓ)   Congruent isosceli

It follows then that $|OA| = |O\Gamma|$, therefore O also belongs to the medial line of segment $A\Gamma$. In other words, we proved that the three medial lines of the sides of triangle $AB\Gamma$ pass through the same point O, which is equidistant from the three points $|OA| = |OB| = |O\Gamma| = \rho$. Consequently, the circle with center O and radius ρ passes through these points. The fact, that this circle is unique, follows using the same argument. Every other circle, that would pass through the vertices of the triangle $AB\Gamma$, would have its center on the medial lines of triangle's $AB\Gamma$ sides, since these sides would be circle chords (Corollary 2.1). Therefore the circle's center and radius would coincide with those of the circle constructed previously.

Corollary 2.6. *For every triangle $AB\Gamma$ there exists a unique circle passing through the triangle's vertices.*

We call the circle of the preceding corollary **circumscribed** or **circumcircle** of the triangle. The center of this circle is called **circumcenter** and the radius **circumradius** of the triangle. Often the circle which passes through three non collinear points A, B and Γ is denoted by $(AB\Gamma)$.

Exercise 2.2. Show that equal chords AB and $A'B'$, on the same circle with center O, define congruent (isosceli) triangles AOB and $A'OB'$ (See Figure 2.4-II).

Corollary 2.7. *If a point A sees the line segment $B\Gamma$ under a right angle, that is, if $|\widehat{BA\Gamma}| = 90°$, then it is contained in the circle with diameter $B\Gamma$ (See Figure 2.5-I).*

Proof. If the point A is viewing the segment $B\Gamma$ under a right, then $AB\Gamma$ is a right-angled triangle and, according to Corollary 1.37, the median from A to $B\Gamma$ has length half of $B\Gamma$, consequently the points A, B and Γ will be on the circle with diameter $B\Gamma$.

Corollary 2.8. *For every point A of the circle, with diameter BΓ, the angle $\widehat{BA\Gamma}$ is a right one. Equivalently: a diameter is seen from a point on the circle (except its endpoints) under a right angle. Conversely, if a point of the circle is viewing a chord under a right angle, then this chord is a diameter.*

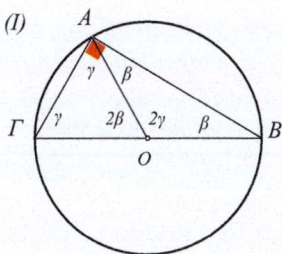
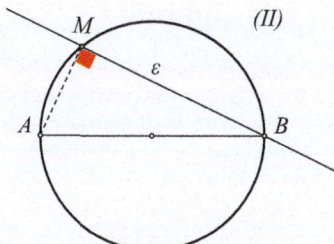

Fig. 2.5: *A "viewing" a diameter,*   Projection of A on ε

Proof. Indeed, if BΓ is a diameter and A is a point of the circle (different from B and Γ), then the triangles OAB and OAΓ are isosceli with base angles β and γ respectively, of triangle ABΓ (See Figure 2.5-I). The sum of the (external) angles of the two isosceli at O is $2\beta + 2\gamma = 180°$ (Corollary 1.21), therefore $\beta + \gamma = 90°$, which shows that ABΓ is right angled at A. Conversely, if the chord BΓ is seen under a right angle from A, then, according to the preceding corollary, A will belong to the circle with diameter BΓ, which, having three common points (A, B and Γ) with the given circle, must coincide with it.

Exercise 2.3. Show that between two circle chords α and β, the longer one lies at a lesser distance from the center.

Exercise 2.4. Show that if three points X, Y and Z lie on the circle κ, then their symmetric X', Y', Z' (relative to an axis or a point) will lie on circle κ' which is congruent to κ.

Exercise 2.5. What is the geometric locus of centers of circles κ which pass through two fixed points A and B?

Exercise 2.6. Let A and B be two fixed points and ε be a line passing through B. Let also M be the projection of A onto ε. Find the geometric locus of M as line ε rotates about B.

Hint: When line ε does not coincide with line AB or its orthogonal at B, then triangle ABM (See Figure 2.5-II) is a right triangle with fixed hypotenuse AB, therefore according to the preceding corollary point M will belong to the circle with diameter AB.

Exercise 2.7. Let α and β be two concentric circles. Then a line ε, which intersects both circles, defines line segments AB, ΓΔ contained between the circles, which are equal.

Exercise 2.8. Let ε be a line A, B be two points not contained in ε. Find a point $Γ$ of the line, which is equidistant from A and B. Has this problem always a solution?

2.2 Circle and line

> The heavens call to you, and circle around you, displaying to you their eternal splendors, and your eye gazes only to earth.
>
> *Dante, Purgatorio, 14,*

In this section we examine the relative positions of a line and a circle. The key is the distance between the center of the circle and the line. This distance, related to the radius of the circle, determines whether the two shapes have one, two or no points in common.

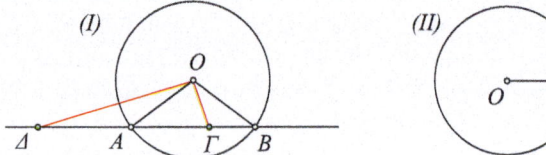

Fig. 2.6: Line-circle intersection   Circle-Half-line intersection

Proposition 2.3. *A line AB contains at most two points of a given circle.*

Proof. We show that a line cannot contain three or more points of a given circle κ. Indeed, if the line had three points A, B and $Γ$ in common with κ, then one of the three would lie between the other two which define a circle chord anyway (See Figure 2.6-I). Thus, we would have a chord containing a third circle point, a contradiction to Proposition 2.1.

Corollary 2.9. *A half line with start point at the center of a circle, intersects the circle at exactly one point (See Figure 2.6-II).*

Tangent of a circle is called a line, which has exactly one common point with the circle, called **contact point** of the line to the circle.

Theorem 2.2. *A line which has only one common point A with a circle, is orthogonal at A to the radius OA of the circle.*

Proof. We use a reduction to contradiction. If the line is not orthogonal to OA, then suppose OB is the orthogonal to the line and define point $Γ$ on the line, so that B is the middle of $AΓ$ (See Figure 2.7-I). Then triangles OBA and $OBΓ$ are right and congruent, having OB common, BA and $BΓ$ equal and the contained angles $\widehat{OBA}$ and $\widehat{OB\Gamma}$ equal as right. Then triangle $OA\Gamma$ is isosceles and we have $|O\Gamma| = |OA| = \rho$, in other words Γ belongs to the circle and the line has one more common point with the circle other than A, a contradiction to the hypothesis.

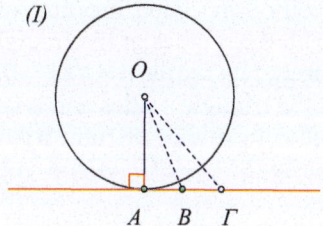 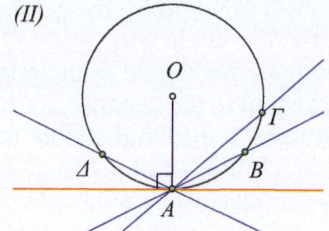

Fig. 2.7: Circle tangent   Lines through A

Corollary 2.10. *Every line which passes through a point A of a circle, except the orthogonal to the radius OA, intersects the circle at a second point B. The excepted orthogonal to OA at A is the unique tangent to the circle at A (See Figure 2.7-II).*

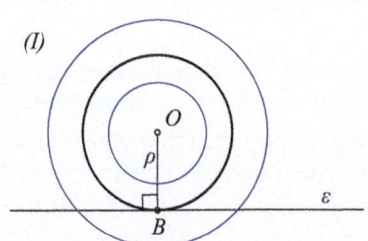 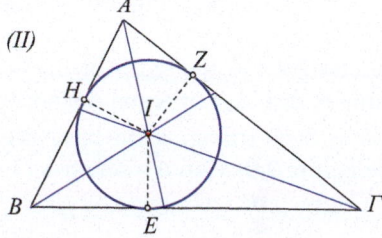

Fig. 2.8: Circle-line intersection   Inscribed circle

Corollary 2.11. *Given a line ε and a point O, not lying on it, the circle $O(\rho)$ intersects the line, if and only if $\rho > |OB|$ where B is the projection of O onto ε ($|OB|$ is the distance of point O from the line ε). When $\rho = |OB|$, then the circle is tangent to the line at B. Finally when $\rho < |OB|$ the circle does not intersect the line (See Figure 2.8-I).*

Theorem 2.3. *The three angle bisectors of triangle $AB\Gamma$ pass through a common point I, which is the center of a circle $I(r)$ simultaneously tangent to the three sides of the triangle.*

Proof. Let I be the intersection point of two angle bisectors, of angles $\widehat{B}$ and $\widehat{\Gamma}$ say (See Figure 2.8-II). We show that the bisector of the third angle $\widehat{A}$ passes through I too. Indeed, by Proposition 1.35, I's distances from the sides of angle $\widehat{B}$: IE and IH are equal. Similarly I's distances from the sides of angle $\widehat{\Gamma}$: IE and IZ are equal. Consequently the three distances IE, IH and IZ are equal, hence they are radii of the circle with center I and radius $r = |IE|$. The orthogonality of the sides to these radii at their endpoints shows (Corollary 2.10) that the circle is tangent to all three sides of the triangle.

Point I, ensured by the preceding proposition, is called **incenter** of the triangle. The circle with center I, tangent to all three sides of the triangle, is called **inscribed circle** or **incircle** of the triangle. Its radius $r = |IE| = |IZ| = |IH|$ is equal to the distance of I from any of the triangle's sides and is called **inradius** of the triangle. Note that from the congruence of right triangles $EI\Gamma$, $ZI\Gamma$ follows that $|\Gamma E| = |\Gamma Z|$ and similarly $|BE| = |BH|$, $|AH| = |AZ|$.

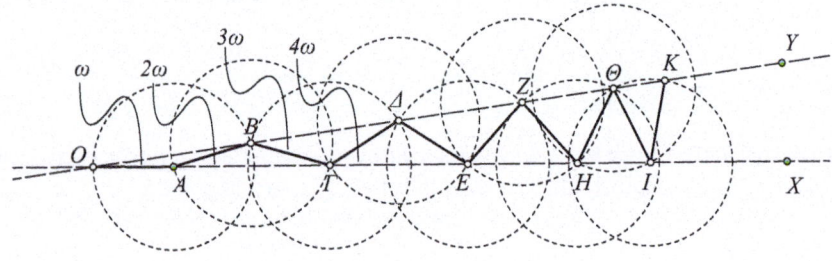

Fig. 2.9: Multiples of the angle ω

Exercise 2.9. On the sides of the acute angle $\widehat{XOY}$ we define alternatively points A, B, Γ, ... at equal distances $|OA| = |AB| = |B\Gamma| = ...$. Show that, in these isosceli triangles, the base angles are ω, 2ω, 3ω, ..., where $\omega = |\widehat{XOY}|$, (See Figure 2.9). Can we construct infinitely many such triangles?

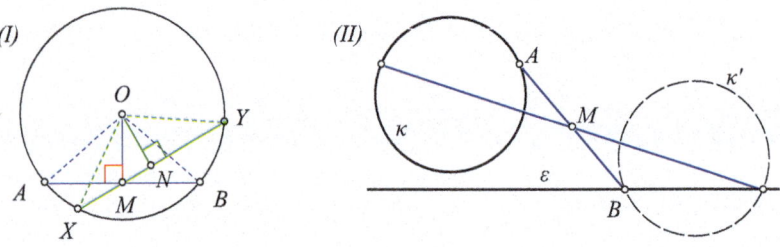

Fig. 2.10: Minimal chord AB Application of symmetry

Exercise 2.10. Show that the chord XY of least length of the circle $\kappa(O, \rho)$, which passes through a fixed inner point $M \neq O$, is the orthogonal chord AB to OM (See Figure 2.10-I).

Hint: Triangle *OMN* has hypotenuse $|OM| > |ON|$, therefore the isosceles *OAB* has a smaller base than that of the isosceles *OXY* (Exercise 2.3).

Often the distance $|OM|$ of the middle of the chord *AB* from the center is called **apothem** and $\rho - |OM|$ **sagitta** of the chord.

Exercise 2.11. Show that the symmetric X' of the points X of the circle κ, relative to a fixed point M, are contained in a circle κ' congruent to κ.

Exercise 2.12. Given is a circle κ, a point M and a line ε. Construct a line segment *AB*, with endpoints respectively on κ and ε, and having its middle at M (See Figure 2.10-II). When does the problem have (or doesn't have) a solution?

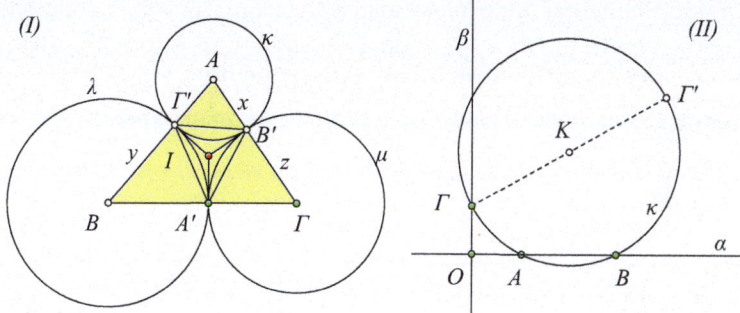

Fig. 2.11: Three tangent circles   Minimization of diameter

Exercise 2.13. Construct circles κ, λ, μ, which are pairwise tangent and have their centers respectively at the triangle vertices A, B and Γ (See Figure 2.11-I). Show that the half-perimeter τ of the triangle, and the respective radii x, y, z of these circles, satisfy the equalities (with $a = |B\Gamma|, b = |\Gamma A|, c = |AB|$):

$$\tau = \tfrac{1}{2}(a+b+c), \qquad y = \tfrac{1}{2}(c+a-b) = \tau - b,$$
$$x = \tfrac{1}{2}(b+c-a) = \tau - a, \qquad z = \tfrac{1}{2}(a+b-c) = \tau - c.$$

Exercise 2.14. In the preceding exercise, show that the circle contact points A', B', Γ' lie on the sides of triangle $AB\Gamma$, whose incenter I coincides with the circumcenter of triangle $A'B'\Gamma'$.

Exercise 2.15. The lines $\{\alpha, \beta\}$ are orthogonal at O, points $\{A, B\}$ are fixed on α and point Γ varies on β (See Figure 2.11-II). In the circle $\kappa = (AB\Gamma)$ consider the diametrically opposite Γ' of Γ. Find the geometric locus of Γ' and the position of the circle which minimizes $|\Gamma\Gamma'|$.

Exercise 2.16. Show that point I, on the bisector $A\Delta$ of triangle $AB\Gamma$, is the intersection point of the angle bisectors, if and only if $|\widehat{BI\Gamma}| = 90° + \frac{|\widehat{BA\Gamma}|}{2}$.

2.3 Two circles

> Little-minded people's thoughts move in such small circles that five minutes' conversation gives you an arc long enough to determine their whole curve. An arc in the movement of a large intellect does not sensibly differ from a straight line.
>
> — O.W. Holmes, *The Autocrat of the Breakfast Table, 1*

We examine here the relative positions of two circles. The key to the subject is the distance between the centers of the circles in relation to the magnitude of their radii. Next theorem shows that there are three possibilities: (i) two different intersection points, (ii) one intersection point, (iii) no intersection point. Of fundamental importance also is theorem 2.6, which completes the triangle inequality (§ 1.11), showing the *existence* of a triangle with given lengths of sides $\{a, b, c\}$, which satisfy this inequality.

Theorem 2.4. *Two different circles have at most two common points.*

Proof. Because if they had three or more, they would be coincident (Theorem 2.1).

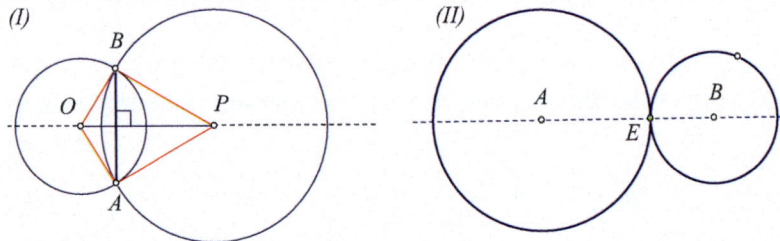

Fig. 2.12: Common chord   Contact point on centerline

The line segment AB of the common points A and B of two circles, which intersect at two different points (See Figure 2.12-I), is called **common chord** of the two circles. Two circles which have the same center are called **concentric**. The line which joins the centers O and P of two non-concentric circles is called **center-line** of the circles. We often also call *center-line* the line segment OP, which joins the centers. What exactly we mean in each case, will be made clear from context.

Corollary 2.12. *The center-line OP of two circles, which intersect at two different points A and B, coincides with the medial line of their common chord AB.*

Proof. According to Corollary 2.1 the centers of the two circles will belong to the medial line of their chord AB.

Two circles are called **tangent** when they have exactly one common point A (See Figure 2.12-II). This point is called **contact point** of the two circles.

2.3. TWO CIRCLES

Proposition 2.4. *The contact point of two tangent circles belongs to their center-line. Conversely, if two different circles have one intersection point, belonging to their center-line, then this point is unique and the circles are tangent.*

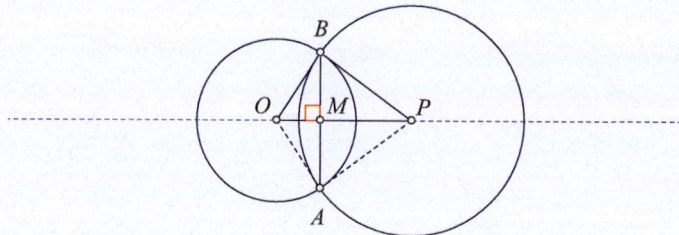

Fig. 2.13: Symmetry relative to center-line

Proof. We use reduction to contradiction. Suppose that the contact point A of the two circles does not belong to their center-line OP (See Figure 2.13). Draw then the orthogonal AM from A to the center-line OP and extend it to its double to point B. The triangles OMA and OMB are congruent (SAS-criterion). Similarly triangles PMA and PMB are congruent. It follows, that $|OB| = |OA| = \rho$ and $|PB| = |PA| = \rho'$, where ρ and ρ' are the radii of the two circles. Consequently, besides A, we find another point B common to the two circles, a contradiction. Conversely, if the circles have two different intersection points A and B, then their center-line coincides with the medial line of AB (Corollary 2.12) and none of the two points may belong to the center-line. Hence, if there is an intersection point on the center-line, then it is unique.

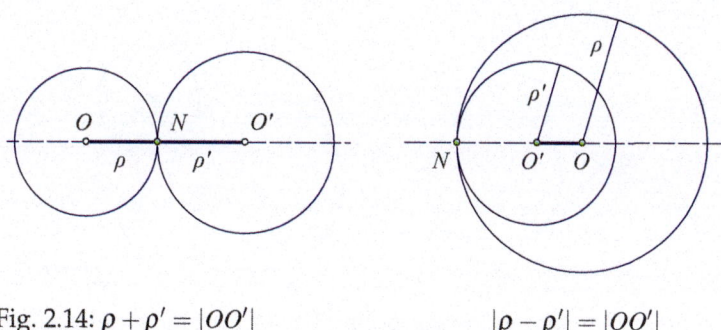

Fig. 2.14: $\rho + \rho' = |OO'|$ $\qquad\qquad |\rho - \rho'| = |OO'|$

Corollary 2.13. *Two circles $O(\rho)$ and $O'(\rho')$ have exactly one common point (they are tangent), if and only if one of the following equations hold:*

$$\rho+\rho'=|OO'|, \quad |\rho-\rho'|=|OO'|.$$

In the first case we say that the circles are tangent **externally** and in the second we say they are tangent **internally** (See Figure 2.14).

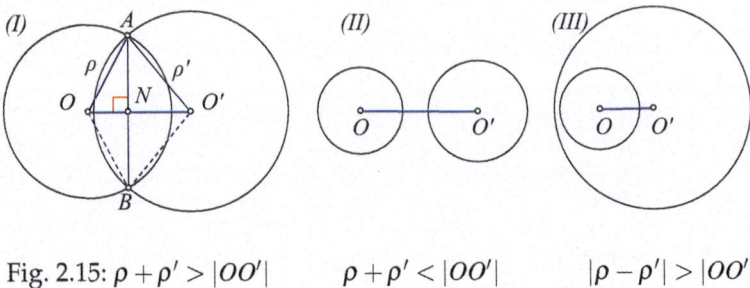

Fig. 2.15: $\rho+\rho'>|OO'|$ $\qquad$ $\rho+\rho'<|OO'|$ $\qquad$ $|\rho-\rho'|>|OO'|$

Theorem 2.5. *Two circles $O(\rho)$ and $O'(\rho')$ intersect at two different points, if and only if their center-line $|OO'|$ and their radii satisfy the triangle inequalities*

$$|\rho-\rho'|<|OO'|<\rho+\rho'.$$

Proof. If A is one of the two intersection points (See Figure 2.15-I), then triangle OAO', has sides of length ρ, ρ' and $|OO'|$ (Proposition 2.4). By Theorem 1.6 and Theorem 1.7, the triangle inequalities are satisfied. Conversely, if these inequalities are satisfied, then the two circles cannot be tangent, because then we would have $\rho+\rho'=|OO'|$ or $|\rho-\rho'|=|OO'|$. Also it is impossible for the circles to not intersect, because then, measuring distances on their center-line, we would have either $\rho+\rho'<|OO'|$ (See Figure 2.15-II) or $|\rho-\rho'|>|OO'|$ (See Figure 2.15-III). Therefore, if the inequalities are satisfied, the circles will intersect at two points.

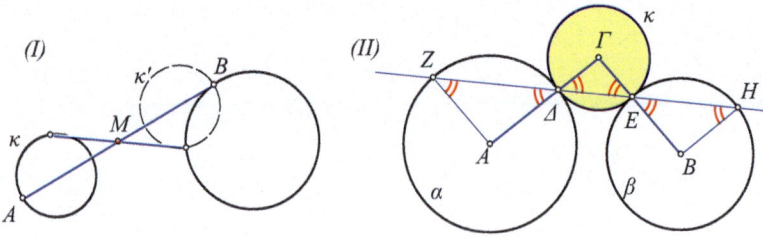

Fig. 2.16: Symmetry application $\qquad$ Circle tangent to two

Exercise 2.17. Given two circles κ and κ' and a point M, to find a line segment AB with its endpoints on the two circles respectively and having middle M (See Figure 2.16-I).

2.4. CONSTRUCTIONS USING RULER AND COMPASS

Exercise 2.18. Show that if a circle κ is tangent to two other circles α and β, then the line ΔE, which joins κ's contact points intersects a second time the other two circles at points Z and H, so that the radii AZ and BE are parallel and $A\Delta$ and BH are also parallel (See Figure 2.16-II).

Hint: Triangles formed by radii are isosceli. The angles of such isosceli at Z and E are equal as internal - external and on the same side for line ΔE intersecting lines AZ and $B\Gamma$.

Exercise 2.19. Show that the shape, consisting of two circles, is symmetric relative to the axis, which is coincident with their center-line. If the two circles are congruent, show that there is a second axis of symmetry orthogonal to the preceding one, as well as, a center of symmetry of the shape.

Exercise 2.20. Divide a chord AB of circle $\kappa(O,\rho)$ into three equal parts $A\Gamma$, $\Gamma\Delta$ and ΔB. Show that angles $\widehat{AO\Gamma}$, $\widehat{\Gamma O\Delta}$ and $\widehat{\Delta OB}$ are not equal. Specifically, $\widehat{AO\Gamma}$ and $\widehat{\Delta OB}$ are equal but smaller than $\widehat{\Gamma O\Delta}$.

Exercise 2.21. Let ε and ζ be two parallel lines and and A a point, not lying on them. Construct a circle, passing through A, and tangent to the two lines. When is this possible?

Exercise 2.22. Let κ be a circle and two points A and B, not lying on it. Construct a right angle $\widehat{XOY}$, whose vertex O is on the circle κ and its sides OX, OY respectively pass through A and B. When is this possible?

Exercise 2.23. Construct a circle κ, passing through two given points A and B and intersecting a given circle λ along a chord $\Gamma\Delta$, parallel to a given line ε.

Exercise 2.24. Which is the maximal number of circles of radius r, which can be placed externally and tangent to a circle κ of the same radius, so that, by two, they are tangent or non-intersecting?

2.4 Constructions using ruler and compass

> No doubt about it: error is the rule,
> truth is the accident of error.
>
> G. Duhamel, *Le Notaire du Havre*

With this term we mean a sequence of constructions of circles or/and lines and their intersection points. The ruler is supposed to give us the ability to define the line (and the line segment) defined by two points A and B. The compass gives us the ability to construct the circle with any center and any radius. We also have the capability to find the intersection points between (i) line and line, (ii) line and circle, (iii) circle and circle ([52]).

Constructions achieved with the exclusive use of ruler and compass play an important role in Euclidean geometry right from its beginnings. Since then are known the three classical problems: (i) The doubling of the volume of the cube (Delian problem), (ii) the trisection of an angle, (iii) the circle quadrature. The proof of the impossibility of the solution of these problems with the exclusive use of rule and compass is possible through the use of algebraic structures (fields) and is a relatively recent achievement ([54], [41], [16, p.183], [51]).

Theorem 2.6. *There exists a triangle with side lengths three given positive numbers a, b, c, if and only if these numbers satisfy the triangle inequalities, equivalently the greater of them, a say, satisfies:*

$$a < b+c.$$

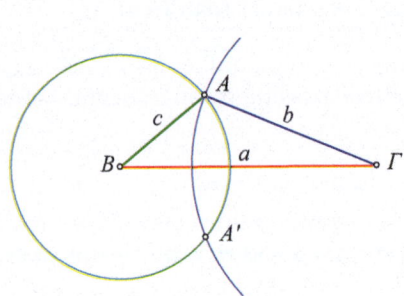

Fig. 2.17: Triangle construction from its sides a, b and c

Proof. If a, b, c are the lengths of triangle $AB\Gamma$, then these numbers satisfy the triangle inequalities (§ 1.11). Conversely, if these three numbers satisfy these inequalities, then a triangle having these numbers as side lengths can be constructed. Indeed, suppose $B\Gamma$ is a line segment of length a. We draw circles with centers B and Γ and radii c and b respectively (See Figure 2.17). These circles will intersect at two points A and A' not belonging to $B\Gamma$, because the inequalities satisfied by their radii are exactly the inequalities of Corollary 2.5. Triangles $AB\Gamma$ and $A'B\Gamma$ are congruent (SSS-criterion) and their sides have the lengths a, b and c.

Exercise 2.25. Show that the three inequalities $|b-c| < a < b+c$, $|c-a| < b < c+a$, $|a-b| < c < a+b$ are equivalent to the one $a < b+c$, where a is the biggest of $\{a,b,c\}$.

Construction 2.1 *Construction of the middle M of a line segment AB and its medial line.*

2.4. CONSTRUCTIONS USING RULER AND COMPASS

Construction: With center A and radius $\rho = |AB|$ we construct circle $A(\rho)$ (See Figure 2.18-I). Similarly with center B and the same radius we construct the circle $B(\rho)$. The radii of the two circles satisfy the triangle inequalities ($0 < \rho < \rho + \rho$), therefore according to Theorem 2.5, the two circles will intersect at two points Γ and Δ. The middle is the intersection point M of lines AB and $\Gamma\Delta$. Indeed, Corollary 2.12 implies that AB is orthogonal to $\Gamma\Delta$, and since by construction $AB\Gamma$ is isosceles, line $\Gamma\Delta$ passes through the middle of AB and coincides with the medial line of AB (Corollary 1.4).

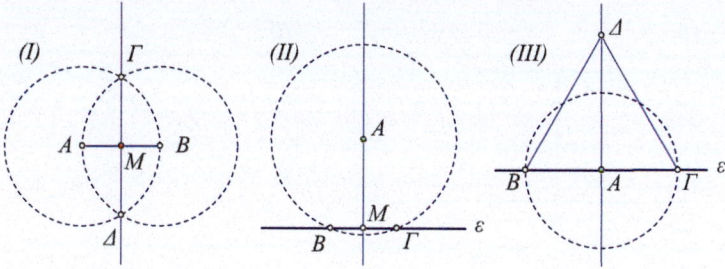

Fig. 2.18: Middle constr.　　A not lying on ε　　A on ε

Construction 2.2 *Construction of an equilateral triangle with side equal to a given line segment AB.*

Construction: Same construction with the preceding one (and same figure). Triangle $AB\Gamma$, of the preceding construction is the requested equilateral.

Construction 2.3 *Construction of the angles of 30 and 60 degrees.*

Construction: The right triangle $AM\Gamma$, which results in construction 2.2 has acute angles $\alpha = 60°$ and $\gamma = 30°$.

Construction 2.4 *Construction of the orthogonal line to the given line ε, from a given point A, not lying on ε.*

Construction: We choose an arbitrary point B on ε and with radius $\rho = |AB|$ we draw the circle $A(\rho)$ (See Figure 2.18-II). If this circle does not intersect ε at another point, then ε is tangent to it and therefore orthogonal to the radius AB (Theorem 2.2). If the circle intersects ε at a second point Γ, then we join the middle M of $B\Gamma$ (Construction 2.1) with A and define the requested orthogonal line (Corollary 1.4).

Construction 2.5 *Construction of the orthogonal line to the given line ε from a given point A on ε.*

Construction: We choose an arbitrary point B on ε (See Figure 2.18-III), different from A and we draw a circle $A(\rho)$, with center A and radius $\rho = |AB|$, which intersects ε at a second point Γ. We construct the equilateral triangle $B\Gamma\Delta$ with one side being $B\Gamma$ (Construction 2.2) and we draw the line $A\Delta$ which is the wanted one.

Construction 2.6 *Construction of the parallel δ to a given line ε, from a given point A, not lying on ε (See Figure 2.19-I).*

Construction: We construct the orthogonal AB to ε (Construction 2.4) and subsequently the orthogonal to AB at A (Construction 2.5), which is the requested line (Corollary 1.15).

Construction 2.7 *Construction of the bisector of an angle $\widehat{XOY}$ (See Figure 2.19-II).*

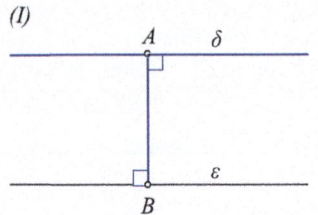
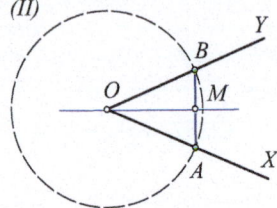

Fig. 2.19: Parallel to ε from A ⟶ Angle bisector of $\widehat{XOY}$

Construction: We consider an arbitrary point A on OX and construct the circle with center O and radius $\rho = |OA|$, intersecting the angle's side OY at B. If M is the middle of AB Construction 2.1, then OM is the requested line (Corollary 1.2).

Construction 2.8 *Construction of the tangents to a given circle, from a given point, lying outside it.*

Construction: Suppose the point A is outside the circle $O(\rho)$, that is $|OA| > \rho$. We construct the circle $O'(\rho')$, with center at the middle O' of OA and

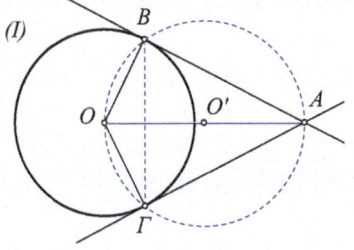
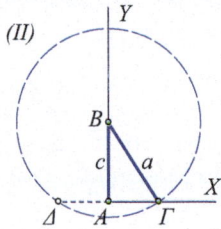

Fig. 2.20: Tangents from A ⟶ Hypotenuse and orthogonal

radius $\rho' = |OA|/2$, i.e. the circle with diameter OA (See Figure 2.20-I). It holds $\rho' - \rho < |OO'| = \rho' < \rho + \rho'$, therefore, according to Theorem 2.5, the two circles intersect at two points B and Γ. In the resulting triangles OAB and $OA\Gamma$, vertices B and Γ see the diameter OA under a right angle (Corollary 2.7), therefore AB and $A\Gamma$ are respectively orthogonal to the radii OB and $O\Gamma$ of the circle $O(\rho)$, hence they are tangent to it (Theorem 2.2).

2.4. CONSTRUCTIONS USING RULER AND COMPASS

We also call the line segments AB and $A\Gamma$ **tangents** to the circle $O(\rho)$ from A. The distinction between tangent *lines* and tangent *line segments* will always be clear from context.

Corollary 2.14. *From a point A, lying outside the circle $O(\rho)$, there are exactly two tangents to it, which are equal.*

Proof. The two tangents in the preceding construction are equal, as respective sides of congruent triangles OAB and $OA\Gamma$. These triangles are congruent, because they have OA common, OB and $O\Gamma$ equal and their angles at O equal, since $OB\Gamma$ is isosceles and OA is the bisector of $\widehat{OB\Gamma}$ (Corollary 2.12). Besides the two tangents from A of the preceding construction, there are no others. Indeed if there was one more $A\Delta$, where Δ is the contact point on the circle $O(\rho)$, then the triangle $OA\Delta$ would also be right at Δ and Δ would see OA under a right angle, therefore Δ would be on the circle with diameter OA (Corollary 2.7) and would coincide with one of B and Γ.

Construction 2.9 *Construct a right triangle $AB\Gamma$, for which are given one orthogonal side and the hypotenuse (See Figure 2.20-II).*

Construction: Analysis: Suppose that the requested triangle $AB\Gamma$, right at A, has been constructed and has a given orthogonal side $|AB| = c$ and hypotenuse $|B\Gamma| = a$. Point Γ is an intersection point of side OX of the right angle $\widehat{YOX}$ and the circle with center B on side OY of the angle and of radius equal to the given length $|B\Gamma|$ of the hypotenuse.

Synthesis: Consider two vertical lines AX, AY and define on AY the line segment AB equal to the given right side. Next, with center B and radius $|B\Gamma|$, equal to the given length of the hypotenuse, draw a circle which intersects OX at Γ and Δ. Triangles $AB\Gamma$ and $AB\Delta$ are congruent and satisfy the problem's requirements.

Discussion: The triangle can be constructed, if and only if the given length $a = |B\Gamma|$ of the hypotenuse is greater than the length of the vertical $c = |AB|$. There exists then exactly one solution ($AB\Gamma$ and $AB\Delta$ are congruent triangles).

Besides elementary constructions, in which, with the help of the ruler and compass, we find angles, orthogonals or perpendiculars and tangents according to the pattern of the preceding examples, there are more complex constructions, in which some specific requirements on a shape are formulated ([41], [42]) and wanted is a shape satisfying these requirements. In particular, the construction of the simplest plane figure, the triangle, constitutes an especially important field of problems in Geometry. In general, when the construction of a complex shape, from certain data is requested, we resort to the triple: **analysis, synthesis, discussion**.

In **analysis** we begin with the phrase: *Suppose that the requested shape has been constructed*, we suppose therefore that the construction has been completed and combining the constructed elements, we proceed to find out interconnections with the given data.

In **synthesis** we go backwards. We use the steps, we performed in analysis, in reverse order and we pass from the given data to the requested ones.

Finally, in the **discussion**, we examine, under what conditions, the given elements, lead to a solution and how many solutions there are. With the problems, I discuss in the book, I often restrict myself to the *synthesis*, leaving the other two stages as exercises. In particular, when the construction is very simple, the analysis is redundant (see also relevant remarks 2.2, 2.3).

Construction 2.10 *Construction of the common tangents between two external to each other circles.*

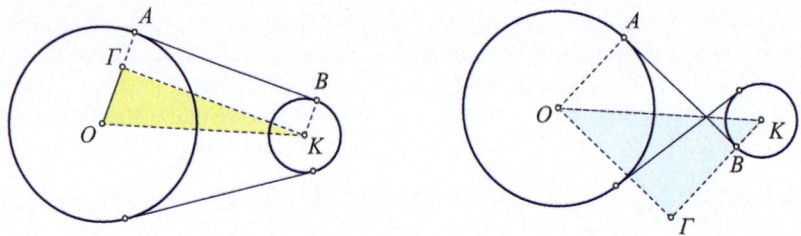

Fig. 2.21: Construction of common tangents of two circles

Construction: Analysis: Let the circles $O(\rho)$ and $K(\rho')$ have radii $\rho \geq \rho'$ and suppose that the common tangent AB has been constructed (See Figure 2.21). The tangent may have the circles on the same or on different sides. In both cases the lines OA and KB are orthogonal to AB and are contained in the same or different sides of AB, respectively. In the first case we project K onto OA. In the second we project O onto KB. This defines a triangle $OK\Gamma$, right at Γ and constructible from the given data. This, because $|O\Gamma| = \rho - \rho'$ in the first case and $|K\Gamma| = \rho + \rho'$ in the second. Therefore the right triangle $OK\Gamma$ can be constructed since we know its hypotenuse $|OK|$ and one orthogonal side (Construction 2.9).

Synthesis: Construct the right triangle $O\Gamma K$ having hypotenuse the center-line OK and one orthogonal $O\Gamma$ of length equal to the difference $\rho - \rho'$. Extend $O\Gamma$ until A onto $O(\rho)$ and draw the parallel ray KB of $K(\rho')$. AB then is the requested tangent. Similarly, construct the right triangle $O\Gamma K$, having hypotenuse OK and one orthogonal $O\Gamma$ of length $\rho + \rho'$. Determine the intersection B of $K\Gamma$ with $K(\rho')$ and draw the parallel radius OA to KB of circle $O(\rho)$, so that the angle $\widehat{AOK}$ is equal to $\widehat{OKB}$. The first construction gives the tangent AB, which leaves the two circles on the same side of it. The second leaves the two circles on opposite sides.

Discussion: The figure is symmetric relative to the center-line OK and in both cases the preceding method, constructs two equal tangents.

Exercise 2.26. Construct a triangle $AB\Gamma$ for which are given the lengths of two sides $b = |A\Gamma|$, $c = |AB|$ and the length of the contained in them median $\mu_A = |A\Delta|$ (See Figure 2.22-I).

2.4. CONSTRUCTIONS USING RULER AND COMPASS

Hint: Analysis: Suppose that the triangle $AB\Gamma$ has been constructed. Extend the median $A\Delta$ to its double until E. Triangles $A\Delta B$ and $E\Delta\Gamma$ are congruent (SAS-criterion). Triangle $A\Gamma E$ is constructible, because we know the lengths of of its three sides $|A\Gamma| = b$, $|E\Gamma| = c$ and $|AE| = 2\mu_A$.

Synthesis: We construct the triangle $A\Gamma E$, with sides $|A\Gamma| = b$, $|E\Gamma| = c$ and $|AE| = 2\mu_A$. We draw the median $\Gamma\Delta$ of this triangle and extend it to its double ΓB. Triangle $AB\Gamma$ is the requested one. This, because triangles $A\Delta B$ and $E\Delta\Gamma$ are congruent, therefore $|AB| = |E\Gamma| = c$, $|A\Gamma| = b$ by construction and, for the median, we have $|A\Delta| = \frac{|AE|}{2} = 2\mu_A$. Triangle $AB\Gamma$ therefore has the requested properties.

Discussion: The given lengths b, c and $2\mu_A$ must satisfy the triangle inequality. There exists then only one solution.

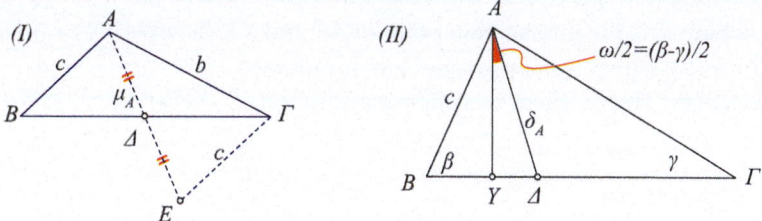

Fig. 2.22: Triangle from $\{b, c, \mu_A\}$ Triangle from $\{\delta_A, c, \omega = |\beta - \gamma|\}$

Exercise 2.27. Construct a triangle $AB\Gamma$, for which are given the length of the bisector $\delta_A = |A\Delta|$ from A, the side $c = |AB|$, and the difference $\omega = |\beta - \gamma|$.

Hint: Analysis: Suppose the requested triangle has been constructed, with $b > c$ and so also $\beta > \gamma$, hence $\omega = \beta - \gamma$ (See Figure 2.22-II). The key to the construction is the angle between altitude AY and bisector $A\Delta$, which has measure $|\widehat{YA\Delta}| = (\beta - \gamma)/2 = \omega/2$. This results easily, by considering the angles of the right triangle $AY\Delta$. $|\widehat{A\Delta Y}| = \gamma + \frac{\alpha}{2}$, as external of triangle $A\Delta\Gamma$. Then $|\widehat{YA\Delta}| = 90° - |\widehat{Y\Delta A}| = \frac{\alpha+\beta+\gamma}{2} - (\gamma + \frac{\alpha}{2}) = \frac{\beta-\gamma}{2}$. Consequently the right triangle $AY\Delta$ can be constructed from the given data, since we know the hypotenuse δ_A and its angles.

Synthesis: We construct the right triangle $AY\Delta$, for which we know its hypotenuse δ_A and its angles. Next we construct the right triangles AYB and $AY\Gamma$. The first, because we know the hypotenuse AB and the orthogonal AY from the preceding step. From the construction of AYB, we find the angle β and by subtracting $\beta - \omega = \gamma$ we find γ. The right triangle $AY\Gamma$ therefore is constructible, because we know the orthogonal AY and its angles.

Discussion: The construction is possible, when the given length c satisfies $c = |AB| > |AY|$. One solution then results. Later (§ 3.6) we'll see that the length $|AY|$, which results during the construction process, can be expressed with the help of the initial data ($|AY| = \delta_A \cdot \cos(\omega/2)$). Therefore, then, we'll be able to express the condition of existence of a solution using an inequality which must be satisfied by the given data: $c > \delta_A \cdot \cos(\omega/2)$.

Exercise 2.28. Let A, B be points on either side of line XY. Find a point Γ on the line, such that $|\widehat{X\Gamma A}| = 2|\widehat{B\Gamma Y}|$ (See Figure 2.23-I).

Hint: Analysis: Suppose Γ on line XY has been constructed. We extend $A\Gamma$ towards Γ and let B be the projection of B on $A\Gamma$. Then ΓB will be a bisector of angle $\widehat{E\Gamma Y}$ and B will be equidistant from lines XY and $A\Gamma$ coinciding with the center of the circle $\kappa(B, |B\Delta|)$. Point Δ can be determined from the given data and therefore the circle κ is constructible.

Synthesis: We project B to Δ on XY and construct the circle $\kappa(B, |B\Delta|)$. We draw the tangents from A to κ. From the two tangents we choose AE, which, together with the given XY, contains the circle in the interior of the angle $\widehat{Y\Gamma E}$, where Γ is the intersection point of AE and XY. Then $|\widehat{A\Gamma X}| = |\widehat{Y\Gamma E}| = 2|\widehat{Y\Gamma B}|$ and Γ satisfies the problem's requirements.

Discussion: The problem has always a solution.

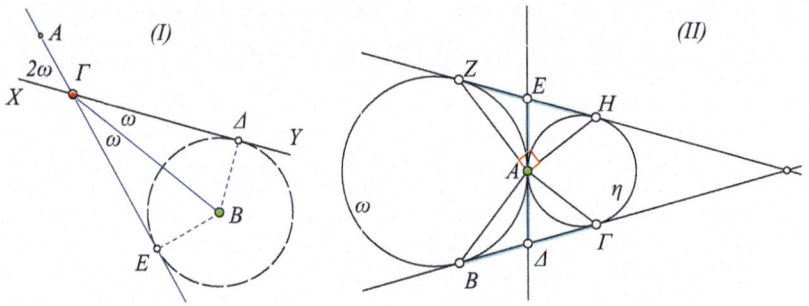

Fig. 2.23: $\Gamma : |\widehat{X\Gamma A}| = 2|\widehat{B\Gamma Y}|$ Three common tangents

Exercise 2.29. Show that two circles ω and η have in general 0, 1, 2, 3 or 4 common tangents. In the case where the circles have three common tangents, these tangents are equal. In figure 2.23-(II) $|B\Gamma| = |\Delta E| = |ZH|$. Also show, that in figure 2.23-(II), triangles $BA\Gamma$ and ZAH, which are formed by the contact points, are congruent right triangles.

Exercise 2.30. Construct a triangle $AB\Gamma$ for which are given the side $|B\Gamma|$, its median $\mu_A = |A\Delta|$ and the altitude $v_A = |AY|$.

Exercise 2.31. Construct an isosceles triangle, for which are given the traces of the two equal altitudes on its legs and a point of its base.

Exercise 2.32. Construct triangle $AB\Gamma$, for which are given the median $\mu_A = |A\Delta|$ as well as the angles, which the median forms with sides AB and $A\Gamma$.

2.4. CONSTRUCTIONS USING RULER AND COMPASS

Exercise 2.33. Construct an isosceles triangle, for which are given the middles of its two equal legs and a point of its base.

Exercise 2.34. Construct an isosceles triangle, for which are given the traces of the two angle bisectors on its legs and a point of its base.

Exercise 2.35. Find a point in the extension of a circle's diameter, from which a tangent can be drawn whose length is equal to the radius of the circle.

Exercise 2.36. Construct a circle κ with given radius ρ, passing through a given point A and having a tangent from a given point with given length.

Exercise 2.37. Given are three congruent circles, whose centers are not all collinear. Find a point, from which the tangents to these circles are equal.

Exercise 2.38. Construct a triangle $AB\Gamma$, for which are given two angles and the perimeter.

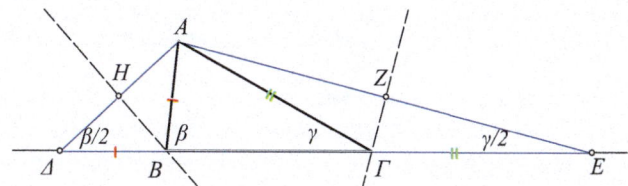

Fig. 2.24: Triangle construction from angles and perimeter

Hint: Analysis: To begin with, it is immaterial which of the three angles have been given. Let us suppose for example that the perimeter p and the angles β and γ are given, so that $\alpha = 180° - (\beta + \gamma)$. Suppose, then, that the triangle $AB\Gamma$ has been constructed. Extend side $B\Gamma$ and take segments $|B\Delta| = |BA|$ and $|\Gamma E| = |\Gamma A|$ (See Figure 2.24). This forms triangle $A\Delta E$, which has angles at Δ and E, $\beta/2$ and $\gamma/2$, respectively and $|\Delta E| = |AB| + |B\Gamma| + |\Gamma A| = p$, therefore it can be constructed from the given data.

Synthesis: Construct the triangle $A\Delta E$, which has $|\Delta E| = p$, equal to the given perimeter, and angles at Δ and E, respectively equal to $\beta/2$ and $\gamma/2$. After the construction of $A\Delta E$, draw the medial lines HB, $Z\Gamma$, which define the points B and Γ as intersections with the side ΔE. Triangle $AB\Gamma$ has the requested properties. This, because from the medial line HB of $A\Delta$, we have $|AB| = |\Delta B|$. Similarly from the medial line $Z\Gamma$ of AE, we have $|A\Gamma| = |\Gamma E|$. Therefore the perimeter, $|AB| + |B\Gamma| + |\Gamma A| = |\Delta E|$. We also have $|\widehat{AB\Gamma}| = 2\frac{\beta}{2} = \beta$, the angle being external of the isosceles $AB\Delta$ at B. Similarly, also $|\widehat{A\Gamma B}| = \gamma$.

Discussion: The problem has always exactly one solution, provided the angles satisfy $\beta + \gamma < 180°$.

Exercise 2.39. Construct a circle of radius r, which is tangent to two other given circles κ and λ.

Exercise 2.40. Let $\{\kappa_1, \kappa_2\}$ be two circles tangent to the sides of the angle $\widehat{XOZ}$ correspondingly at the points $\{A_1, B_1\}$ (See Figure 2.25-I). Show that a line, tangent to the circles correspondingly at $\{A_0, B_0\}$, passes through O, if and only if
$$||OA_1| - |OB_1|| = |A_0B_0|.$$

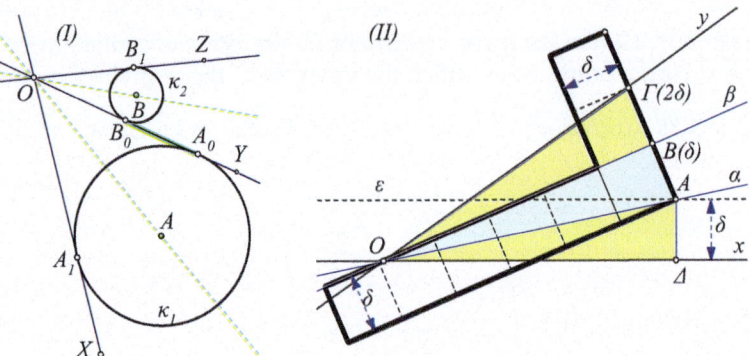

Fig. 2.25: Tangent circles   Carpenter's square trisecting

As we noticed at the beginning of the section, the trisection of an arbitrary angle with exclusive use of ruler and compass is in general impossible. The following exercise ([57]) shows that the trisection is possible using some other simple instruments, like the carpenter's square.

Exercise 2.41. Trisection of the angle $\widehat{xOy}$ using carpenter's square of width δ. To trisect the angle place the carpenter's square so that the inner edge of its larger arm is on the point O, its outer corner A lies on the parallel ε to Ox at distance δ from it, and the mark $\Gamma(2\delta)$ is on Oy (See Figure 2.25-II). Then the lines through the marks $\{\alpha = OA, \beta = OB(\delta)\}$ trisect the angle $\widehat{xOy}$.

Hint: If Δ is the projection of A on Ox, then triangles $\{AO\Delta, AOB, BO\Gamma\}$ are equal.

Exercise 2.42. Show that for a convex polygon p, which has an axis of symmetry, the folding along this axis, creates a new polygon p', with the same number and measures of angles with p, if and only if, p is a right isosceles triangle or a rectangle. For the rectangle's definition see § 2.9.

Exercise 2.43. Show that the only triangle which can be divided into two equal triangles, which have also the same angles with the original triangle, is the right-angled isosceles.

2.5 Parallelograms

> This is something the connoisseurs of exotic cooking first learn as soon as they attempt to eat with chopsticks: the simple tool requires greater dexterity than the complex one.
>
> *Claude Levi-Strauss, Listening to Rameau*

Before to proceed to general quadrilaterals, we examine the special case of the parallelogram. This simple figure, which can be divided into two congruent triangles, has many properties, which are not only interesting in and of themselves, but are useful also in the detection of properties of other, more complex quadrilaterals and general figures. Next theorem describes one of their simplest constructions.

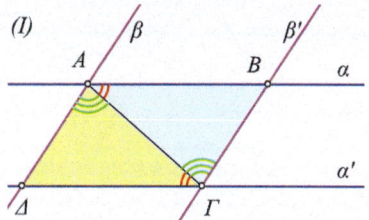
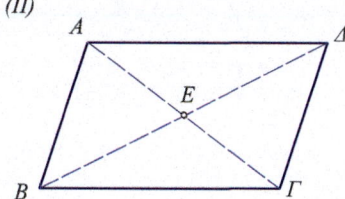

Fig. 2.26: Pairs of parallels   Common middle E

Theorem 2.7. *Two parallel lines α and α' cut on two other parallels β and β', respectively, line segments $A\Delta$ and $B\Gamma$, which are equal (See Figure 2.26-I).*

Proof. Drawing the line $A\Gamma$, creates triangles $AB\Gamma$ and $A\Gamma\Delta$. These triangles have a common side $A\Gamma$. They also have angles $\widehat{\Delta A\Gamma}$ and $\widehat{A\Gamma B}$ equal, as internal and alternate between parallels (β and β') intersected by $A\Gamma$. Angles $\widehat{A\Gamma\Delta}$ and $\widehat{\Gamma AB}$ are also equal as alternate internal between parallels (α and α'). Applying the ASA-criterion for triangles, we conclude that the aforementioned triangles are congruent. From the congruence of the triangles, follows the equality of $A\Delta$ and $B\Gamma$.

Parallelogram is called the figure $AB\Gamma\Delta$, which is defined from two pairs of parallel line segments $(AB, \Gamma\Delta)$ and $(B\Gamma, \Delta A)$, like in the preceding proposition (See Figure 2.26-I). Points A, B, Γ and Δ are called **vertices** of the parallelogram. The line segments AB, $B\Gamma$, $\Gamma\Delta$ and ΔA are called **sides** of the parallelogram and $A\Gamma$ and $B\Delta$ are called **diagonals** and their intersection **center** of the parallelogram.

Corollary 2.15. *Every parallelogram has the following properties:*

1. Each diagonal divides the parallelogram into two congruent triangles.
2. The opposite sides in a parallelogram are equal.
3. The opposite angles in a parallelogram are equal.
4. Two adjacent angles in a parallelogram are supplementary.

Proposition 2.5. *The diagonals of a parallelogram are bisected at their intersection point (See Figure 2.26-II).*

Proof. The proof results from the congruence of triangles $AE\Delta$ and $BE\Gamma$. The latter results from the fact that sides $A\Delta$ and $B\Gamma$ are equal and that angles $\widehat{EA\Delta}$ and $\widehat{E\Gamma B}$ are equal, as well as, that $\widehat{A\Delta E}$ and $\widehat{EB\Gamma}$ are also equal (as internal and alternate etc.).

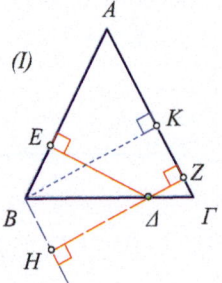 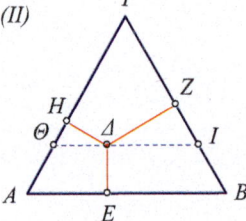

Fig. 2.27: Isosceles basis property   Equilateral property

Exercise 2.44. Show that every point Δ of the base $B\Gamma$ of the isosceles triangle $AB\Gamma$ has sum of distances $|\Delta E| + |\Delta Z|$ from the legs constant and equal to the altitude $|BK|$ from B.

Hint: Draw from B the parallel to $A\Gamma$ and extend ΔZ until it intersects it at H (See Figure 2.27-I). The angle at H is right and the right triangles $BE\Delta$ and $BH\Delta$ have $B\Delta$ common and their adjacent angles equal, therefore they are congruent. Consequently $|\Delta E| + |\Delta Z| = |\Delta H| + |\Delta Z| = |BK|$. The last equality holds because $ZHBK$ is a parallelogram (also see Exercise 2.100).

Exercise 2.45. Show that every point Δ, in the interior of an equilateral triangle $AB\Gamma$, has sum of distances from its sides $|\Delta E| + |\Delta Z| + |\Delta H|$ constant and equal to the altitude of the equilateral.

Hint: Draw from Δ the parallel ΘI to AB and apply the preceding exercise on the (equilateral) triangle $\Theta\Gamma I$ (See Figure 2.27-II). To begin with, $|\Delta Z| + |\Delta H|$ is equal to the altitude of $\Theta\Gamma I$ from Θ. Because the triangle $\Theta\Gamma I$ is also equilateral, this altitude is equal to the altitude from Γ and adding to this $|\Delta E|$ we get the altitude of the original triangle from Γ.

Exercise 2.46. Construct a parallelogram $AB\Gamma\Delta$ with given side lengths $|AB|$, $|B\Gamma|$ and the angle $\widehat{AB\Gamma}$ between them. Similarly, construct the parallelogram from the same side lengths and the diagonal $|A\Gamma|$.

2.5. PARALLELOGRAMS

Exercise 2.47. Construct a parallelogram for which are given the lengths of the sides and the distance between two of its parallel sides.

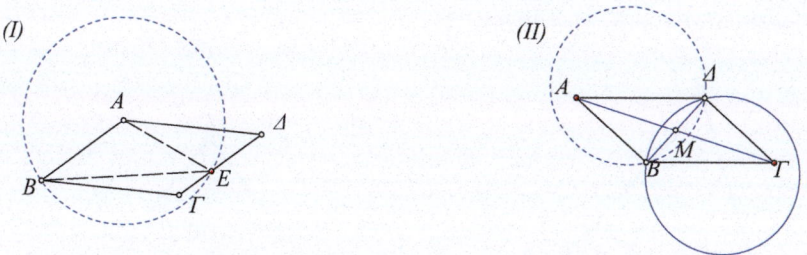

Fig. 2.28: Constructions of parallelograms

Exercise 2.48. Find a point E, belonging to the side $\Gamma\Delta$ of a parallelogram $AB\Gamma\Delta$, such that angle $\widehat{AEB}$ is equal to angle $\widehat{BE\Gamma}$ (See Figure 2.28-I).

Exercise 2.49. Construct a parallelogram $AB\Gamma\Delta$, which has its vertices A and Γ at given points and vertices B, Δ on a given circle (See Figure 2.28-II).

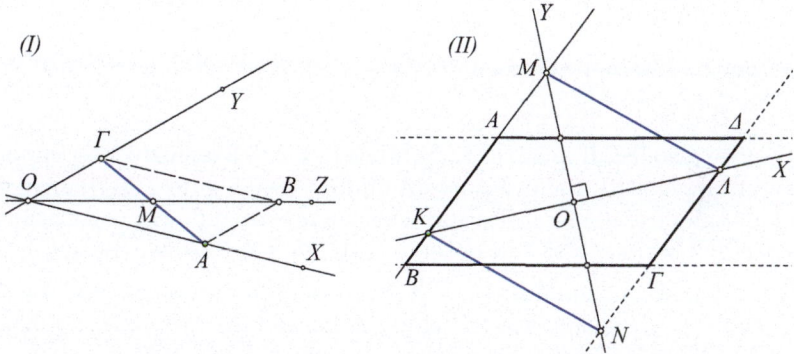

Fig. 2.29: Middle M on OZ Parallelogram and right angle

Exercise 2.50. From a point A of the side OX of the angle $\widehat{XOY}$, to draw a line segment $A\Gamma$ with Γ on OY, such that its middle M is contained in a given line OZ in the interior of angle $\widehat{XOY}$ (See Figure 2.29-I).

Exercise 2.51. From the intersection point O of the diagonals of a parallelogram $AB\Gamma\Delta$, draw two orthogonal lines XOY. If OX intersects AB and $\Gamma\Delta$ at points K and Λ and OY intersects AB and $\Gamma\Delta$ at M and N respectively, show that $KM\Lambda N$ is a parallelogram which has equal sides (See Figure 2.29-II).

Exercise 2.52. Construct a triangle $AB\Gamma$, whose side $|A\Gamma|$ and median $|A\Delta|$ are equal to d and the angle $\widehat{\Delta A\Gamma}$ is equal to $120°$.

Exercise 2.53. Construct a segment ΔE parallel and equal oriented to the base $B\Gamma$ of triangle $AB\Gamma$, such that $|\Delta E| = |B\Delta| + |E\Gamma|$ and the points $\{\Delta, E\}$ respectively on the sides $\{AB, A\Gamma\}$.

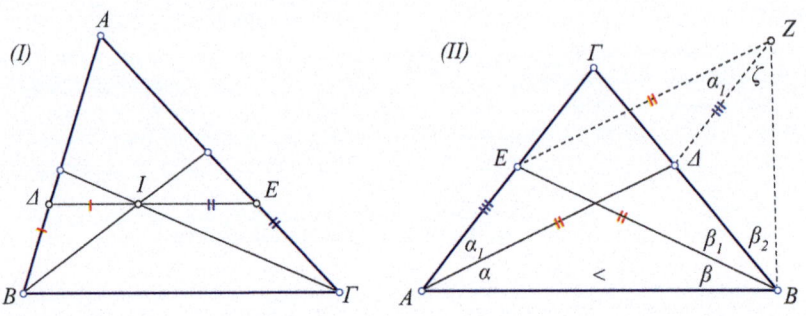

Fig. 2.30: Bisectors property   Steiner-Lehmus theorem

Hint: If I is the intersection point of the bisectors at B and Γ (See Figure 2.30-I), then the parallel ΔE of $B\Gamma$ from I has this property and conversely, one parallel having the properties of the exercise passes through the intersection point of the bisectors I.

Theorem 2.8 (Steiner-Lehmus). *A triangle which has two internal equal bisectors is isosceles.*

Proof. (See also [58, II, p.321], [13, p.14], [11, p.72], Exercise 3.139) Suppose that in triangle $AB\Gamma$ the bisectors $A\Delta$ and BE (See Figure 2.30-II) are equal but the triangle is not isosceles and we have $\alpha = \alpha_1 < \beta = \beta_1$ (∗). Construct the parallelogram $A\Delta ZE$ and observe that triangle EZB is isosceles, hence the basis angles are equal $\widehat{\alpha_1 + \zeta} = \widehat{\beta_1 + \beta_2}$ (∗∗). But in triangle ΔBZ the side $|\Delta B| < |\Delta Z| = |AE|$, since the triangles $AEB, \Delta AB$ have two sides equal and respective unequal angles $\widehat{\alpha} < \widehat{\beta}$ (Theorem 1.5). Consequently (Theorem 1.3) $\widehat{\zeta} < \widehat{\beta_2}$ and, because of (∗∗), the angle $\widehat{\alpha_1} > \widehat{\beta_1}$, contradicting (∗).

Exercise 2.54. Show that the altitude towards a greater side of a triangle is less than the altitude towards a smaller side.

Exercise 2.55. Construct a parallelogram whose are given the two altitudes (distances of opposite sides) and one diagonal.

Exercise 2.56. Inscribe in an equilateral triangle another equilateral whose sides are orthogonal to the first one.

2.6 Quadrilaterals

> If even small upon the small you place and
> do this oft, the whole will soon be great.
>
> Hesiod, Works and Days 361

Four successive line segments AB, $B\Gamma$, $\Gamma\Delta$, ΔA, which form a closed broken line define a **quadrilateral** or **Quadrangle**. Points A, B, Γ and Δ are called

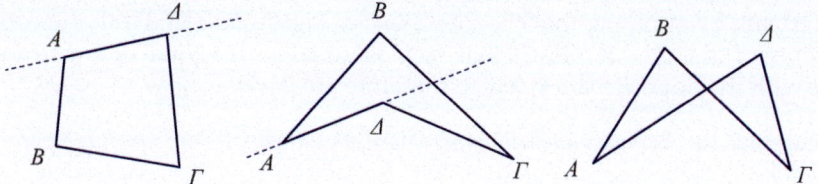

Fig. 2.31: Convex, non-convex, and self-intersecting quadrilateral

vertices of the quadrilateral. The first quadrilateral of figure 2.31 is **convex**, that is, extending any side of it leaves the quadrilateral on the same side of the extended line. The second quadrilateral of the figure is **non-convex**. Extending, for example, $A\Delta$ we have parts of the quadrilateral on both sides of the line $A\Delta$. The third quadrilateral is **self-intersecting**, in other words it has two sides which intersect at a point different from its vertices. In these lessons we'll deal exclusively with convex quadrilaterals, an example of which are the parallelograms. Two sides/angles of the quadrilateral, which have no common vertex/side are called **opposite**. The two line segments which are defined by pairs of vertices which are not contained in the same side are called **diagonals**. We say that two quadrilaterals $AB\Gamma\Delta$ and $A'B'\Gamma'\Delta'$ are **congruent** when they have corresponding sides equal and corresponding angles also equal.

Proposition 2.6. *The sum of the measures of the angles in a convex quadrilateral is 360 degrees.*

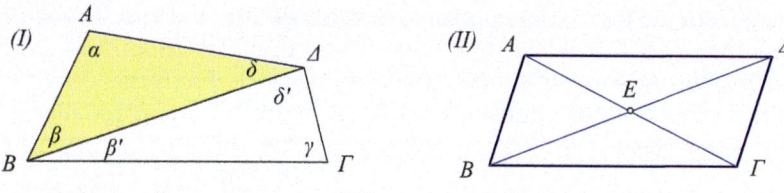

Fig. 2.32: Division into triangles   Diagonals bisected

Proof. The proof results immediately by dividing the quadrilateral into two triangles, using one of its diagonals (See Figure 2.32-I). The sum of the angles of the quadrilateral is equal to the sum of angles of the two triangles $\alpha + \beta + \delta = 180°$ and $\beta' + \gamma + \delta' = 180°$ (Theorem 1.11).

Next properties are the converses of properties we proved in § 2.5 and show that, among all quadrilaterals, they are characteristic of parallelograms.

Theorem 2.9. *A convex quadrilateral, whose opposite sides are equal, is a parallelogram.*

Proof. We draw a diagonal and we prove, that the two created triangles are congruent, by applying the SSS-criterion (See Figure 2.32-II). From the congruence of triangles, it follows that each diagonal makes equal alternate angles with the opposite sides, which therefore are parallel.

Theorem 2.10. *A convex quadrilateral, whose opposite angles are equal, is a parallelogram.*

Proof. First we show that successive angles are supplementary. Indeed since the angles are equal in pairs then their measures will successively be α, β, α, β and the sum must be $2\alpha + 2\beta = 360°$. Consequently $\alpha + \beta = 180°$. It therefore follows that the quadrilateral has opposite sides parallel, therefore it is a parallelogram.

Theorem 2.11. *A quadrilateral, whose diagonals are bisected at their intersection point, is a parallelogram.*

Proof. Here also it suffices to consider the triangles $AE\Delta$ and $BE\Gamma$, which are created from two opposite sides $A\Delta$, $B\Gamma$ and the intersection point of the diagonals E (See Figure 2.32-II). These triangles are proved to be congruent using the SAS-criterion of triangle congruence. From this congruence results the fact that the opposite sides of the quadrilateral are parallel.

Theorem 2.12. *A convex quadrilateral, whose two opposite sides are equal and parallel, is a parallelogram.*

Proof. Here also it suffices to consider the triangles $AE\Delta$ and $BE\Gamma$, which are created from the two equal opposite sides $A\Delta$, $B\Gamma$ and the intersection point of the diagonals E (See Figure 2.32-II). These triangles are proved congruent using the ASA-criterion of triangle congruence. From this congruence results the fact that the opposite sides of the quadrilateral are parallel.

Exercise 2.57. Show that, if in a quadrilateral each pair of successive angles is a pair of supplementary angles, then the quadrilateral is a parallelogram.

Hint: Single out an angle α of the quadrilateral. Since its preceding and next are supplementary to it (equal to $180 - \alpha$ degrees), they will be equal. Apply Theorem 2.10.

2.6. QUADRILATERALS

Theorem 2.13. *Two quadrilaterals $AB\Gamma\Delta$ and $A'B'\Gamma'\Delta'$, which have equal corresponding sides $|AB| = |A'B'|$, $|B\Gamma| = |B'\Gamma'|$, etc. and two corresponding diagonals equal, for example, $|A\Gamma| = |A'\Gamma'|$, are congruent and conversely.*

Proof. From the given equalities, applying the SSS-criterion, follows the congruence of the triangles $AB\Gamma$, $A'B'\Gamma'$ as well as the congruence of triangles $A\Delta\Gamma$, $A'\Delta'\Gamma'$. From these congruences the equality of corresponding angles of the two quadrilaterals follows directly. The converse is proved similarly.

Exercise 2.58. Show that two quadrilaterals $AB\Gamma\Delta$ and $A'B'\Gamma'\Delta'$, which have equal corresponding sides $|AB| = |A'B'|$, $|B\Gamma| = |B'\Gamma'|$ etc. and two corresponding angles equal, for example $\widehat{AB\Gamma} = \widehat{A'B'\Gamma'}$, are congruent and conversely.

Remark 2.1. The last theorem and exercise are related to the fact, that given the lengths a, b, c and d, if there exists a quadrilateral with these side lengths, then there exist in general infinitely many other quadrilaterals $AB\Gamma\Delta$ having sides equal to these lengths (See Figure 2.33-I). For the determination of one

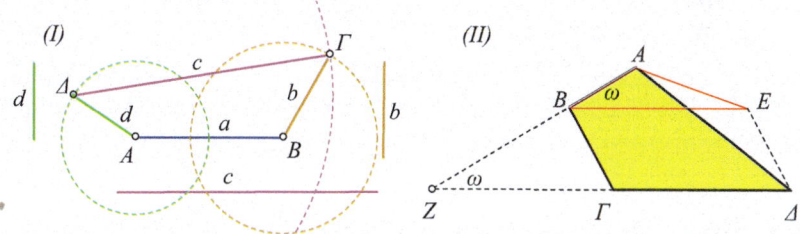

Fig. 2.33: Given side lengths   Quadrilateral construction

of these quadrilaterals it suffices to know one more element for it, like for example the angle $\widehat{BA\Delta}$ and the fact that it is convex. Instead of the angle it also suffices to know one diagonal, for example $B\Delta$, which then determines exactly the two triangles which result after dividing the quadrilateral into two triangles. The distinction between convex or not is again necessary.

Exercise 2.59. Construct a convex quadrilateral $AB\Gamma\Delta$, for which are given the lengths of its sides and the intersection angle ω formed by two opposite sides AB and $\Gamma\Delta$.

Hint: Draw from B a parallel and equal and equal oriented to $\Gamma\Delta$ segment BE (See Figure 2.33-II). Triangle ABE can be constructed, because we know AB, BE (equal to $\Gamma\Delta$) and the angle $\omega = \widehat{ABE} = \widehat{AZ\Delta}$. Therefore AE is of known length, therefore the triangle $AE\Delta$ can be constructed, since we know the lengths of its three sides ($|E\Delta| = |B\Gamma|$). Subsequently the parallelogram $BE\Delta\Gamma$ can be constructed and with this the requested quadrilateral too.

Exercise 2.60. A river has parallel banks at distance δ. On each of the two sides, there are cities A and B. Find the position of the bridge XY, supposed orthogonal to the banks, which makes the sum of distances $|AX|+|XY|+|YB|$ minimal.

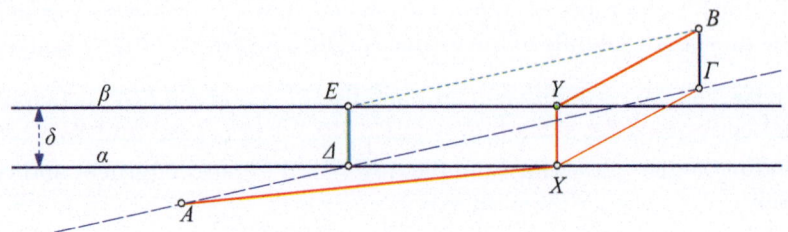

Fig. 2.34: Position of the bridge

Hint: Consider the point Γ at distance δ from B on the orthogonal from B towards the river (See Figure 2.34). The path to minimize $|AX|+|XY|+|YB|$ is always equal to $|AX|+|X\Gamma|+|\Gamma B| = |AX|+|X\Gamma|+\delta$. Therefore it suffices to minimize $|AX|+|X\Gamma|$ (Γ is a fixed point). Note however that this broken line always has length greater than the line segment $A\Gamma$. Therefore point X must coincide with point Δ which is the intersection point of $A\Gamma$ and α.

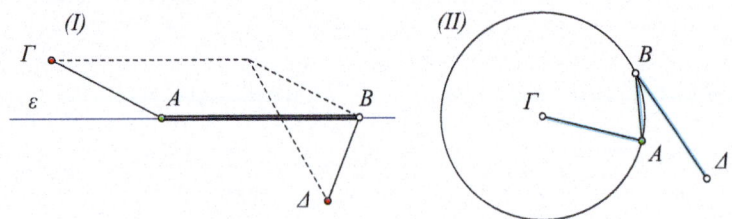

Fig. 2.35: Minimization of $|\Gamma A|+|AB|+|B\Delta|$

Exercise 2.61. The line segment AB, of fixed length δ, is displaced by sliding along the fixed line-carrier ε (See Figure 2.35-I). Fixed points Γ and Δ are given on the two sides of line ε. Find the position of the segment AB on the line ε, for which the length $|\Gamma A|+|AB|+|B\Delta|$ becomes minimal.

Exercise 2.62. An arc AB of constant chord of length δ, is displaced by sliding along the constant circle $\kappa(\Gamma, r)$ carrying it. The fixed point Δ is outside κ. To find the position of the chord AB on κ, such that the length $|\Gamma A|+|AB|+|B\Delta|$ becomes minimal (See Figure 2.35-II).

Exercise 2.63. Construct a parallelogram knowing the length of one of its diagonals and the distances of its opposite sides.

2.6. QUADRILATERALS

Hint: Suppose the parallelogram has been constructed. Project the middle of the given diagonal to the sides of the parallelogram. The projections together with the endpoints and the middle of the diagonal form four right angled triangles, constructible from the given data.

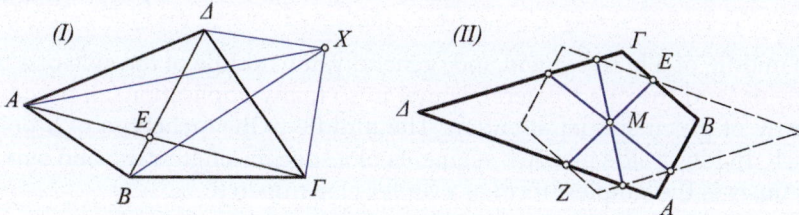

Fig. 2.36: Minimize $|XA|+|XB|+|X\Gamma|+|X\Delta|$ Symmetry application

Exercise 2.64. Given a convex quadrilateral $AB\Gamma\Delta$, show that the sum of distances of a point X, from its vertices, becomes minimal, when it coincides with the intersection point E of its diagonals (See Figure 2.36-I).

Hint: Apply the triangle inequality $|XA|+|X\Gamma| > |A\Gamma|$, $|XB|+|X\Delta| > |B\Delta|$.

Exercise 2.65. Given are the convex quadrilateral $AB\Gamma\Delta$ and a point M in its interior. Construct a line segment EZ, with endpoints on the quadrilateral and middle M (See Figure 2.36-II).

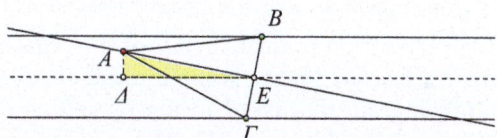

Fig. 2.37: Isosceles between parallels

Exercise 2.66. To construct an isosceles triangle $AB\Gamma$, of which is given the position of the apex A and the other vertices $\{B,\Gamma\}$ are on two parallel lines whose angle with line $B\Gamma$ is known (See Figure 2.37).

Exercise 2.67. Given a side, an adjacent to that side angle and the perimeter of the triangle $AB\Gamma$, is there one or more triangles with these data?

Exercise 2.68. Construct a convex quadrilateral $AB\Gamma\Delta$, for which are given the three sides and the angles adjacent to the fourth side.

Exercise 2.69. Construct a convex quadrilateral $AB\Gamma\Delta$ having three equal sides and whose known are the middles of the equal sides.

2.7 The middles of sides

> He always hurries to the main event and whisks his audience into the middle of things as though they knew already.
>
> *Horace, Ars Poetica*

The middle of a line segment and, generally, the middle of the side of a triangle or/and the quadrilateral, is met in so many applications, it becomes worthy of given special attention. The middle is the archetype of a point which divides a side into two segments of a specific length-ratio and simultaneously is the simplest form of a center of symmetry.

Theorem 2.14. *In a triangle ABΓ, the parallel from the middle M of AB to side BΓ, intersects the third side AΓ at its middle N. Furthermore MN has half the length of BΓ.*

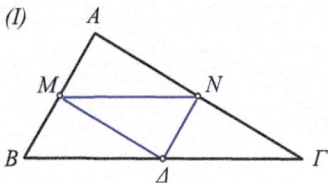

Fig. 2.38: Joining the middles   Parallelogram of middles

Proof. Let N be the intersection point with $A\Gamma$ of the parallel from M to $B\Gamma$ (See Figure 2.38-I). We'll show that N is the middle of $A\Gamma$ using characterizations of the parallelogram discussed in the preceding section. For this, we draw $N\Delta$ parallel to AB, with Δ on $B\Gamma$. $BMN\Delta$ is by construction a parallelogram, therefore $N\Delta$ and BM are equal and parallel. Because also AM and BM are equal, it follows that $AM\Delta N$ is a parallelogram (Theorem 2.12). For the same reason it also follows, that $MN\Gamma\Delta$ is a parallelogram. From the two parallelograms $AM\Delta N$ and $MN\Gamma\Delta$, which have the common side $M\Delta$, follows that AN and $N\Gamma$ are equal. The second statement of the theorem, for half the length, follows again from the parallelograms $MN\Delta B$ and $MN\Gamma\Delta$ which have the side MN in common.

Corollary 2.16. *The line segment MN, which joins the middles of sides AB and AΓ of triangle ABΓ, is parallel to the third side BΓ and has half its length.*

Proposition 2.7. *The middles of the sides of any quadrilateral define a parallelogram.*

Proof. If in quadrilateral $AB\Gamma\Delta$, points E and Θ are the middles of the sides AB and $A\Delta$ respectively (See Figure 2.38-II), then the line $E\Theta$ is parallel to the

2.7. THE MIDDLES OF SIDES

diagonal $B\Delta$ and has half its length. The same happens with line segment ZH which joins the middles of ΓB and $\Gamma\Delta$ respectively. Consequently segments $E\Theta$ and ZH are parallel and equal, therefore they define a parallelogram (Theorem 2.12).

Exercise 2.70. Show that two parallel lines are symmetric relative to any point O belonging to their middle parallel.

Exercise 2.71. Describe all axial- and point-symmetries of the figure which consists of two parallel lines.

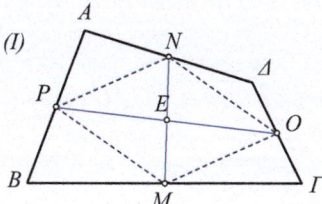 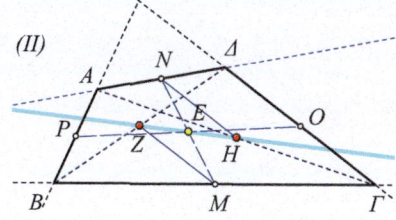

Fig. 2.39: The middles of sides   The Newton line

Exercise 2.72. Show, that in every quadrilateral $AB\Gamma\Delta$, the line segments MN and OP, which join the middles of opposite sides, are bisected by their intersection point E (See Figure 2.39-I).

Hint: MN and OP are the diagonals of the parallelogram of the middles of the sides of the quadrilateral.

Exercise 2.73. Show that, in every quadrilateral $AB\Gamma\Delta$, the intersection point of the line segments MN and OP, which join the middles of opposite sides and the middles H, Z of its diagonals, are contained in the same line and E is the middle of HZ (See Figure 2.39-II).

Hint: In the triangle $B\Gamma\Delta$, the segment MZ joins the middles of sides $B\Gamma$ and $B\Delta$, therefore it is parallel and half of $\Gamma\Delta$. In triangle $A\Gamma\Delta$, the segment HN joins the middles of sides $A\Gamma$ and $A\Delta$, therefore it is parallel and half of $\Delta\Gamma$. We conclude that $HNZM$ is a parallelogram, whose diagonals are bisected.

The line HZ, of the preceding exercise, is defined for every quadrilateral, which is not a parallelogram, and is called **Newton line** (1643-1727) of the quadrilateral. In the case of a parallelogram, points H and Z of the figure coincide and therefore this line cannot be defined.

The two next exercises show that the triangle is completely defined from the position of the middles of its sides. Something similar, however, does not happen with the quadrilateral, in which the middles of its sides define a parallelogram.

Exercise 2.74. Show that, if the positions of the middles *M*, *N* and *O* of the sides of a triangle *ABΓ* are given, then the triangle is completely determined (that is, there exists exactly one triangle with these points as middles of its sides) (See Figure 2.40-I).

Hint: suppose that the requested triangle *ABΓ* has been constructed. By Theorem 2.14, *AΓ* will be double the length of and parallel to *MN*. Similar properties will hold for the other sides of *ABΓ*, therefore it coincides with the triangle which is constructed by drawing parallels from points *O*, *M*, *N* respectively to *MN*, *NO*, *OM*.

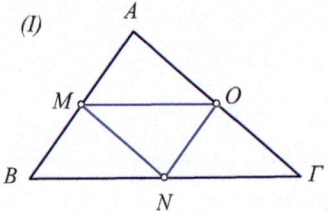
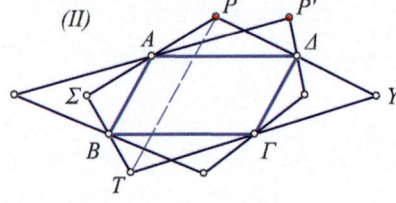

Fig. 2.40: From middles   With same middles

Exercise 2.75. Show that, given the parallelogram *ABΓΔ* and a point *P*, not lying on any of the line-carriers of its sides, there exists exactly one quadrilateral with a vertex at *P* and the side-middles at the parallelogram's vertices.

Hint: Begin with *P*, draw *PA* and double it until *Σ* (See Figure 2.40-II). Then draw *ΣB* and double it until *T*, draw *TΓ* and double it until *Y*. The middle of *YP* is *Δ*. This follows by applying Theorem 2.14 to triangle *PTY* and shows that, for any position of *P*, the resulting construction defines a quadrilateral with the predetermined middles.

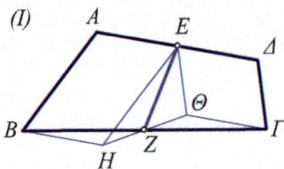
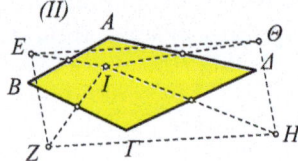

Fig. 2.41: Quadrilateral construction   Parallelogram construction

Exercise 2.76. Construct a convex quadrilateral *ABΓΔ*, for which are given the lengths of its sides, as well as, the length of the line segment *EZ* which joins the middles of the opposite sides *AΔ* and *BΓ* respectively.

Hint: Draw from B a parallel BH, equal and equal oriented to AE (See Figure 2.41-I). Draw from Γ a parallel, equal and equal oriented $\Gamma\Theta$ to ΔE. This defines the parallelograms $AEHB$ and $E\Delta\Gamma\Theta$, therefore $BH\Gamma\Theta$ is a parallelogram with diagonals $H\Theta$ and $B\Gamma$. Therefore the middle of $H\Theta$ is Z (Proposition 2.5). Thus, in the triangle $EH\Theta$, the lengths of its sides EH and $E\Theta$, as well as the length of the median EZ between them are known. According to (Exercise 2.26), triangle $EH\Theta$ can be constructed. Also constructible are the triangles ZHB and $Z\Theta\Gamma$, because their three sides are known. From these one can also construct the parallelograms $HBAE$ and $\Theta\Gamma\Delta E$, which completely determine the quadrilateral.

Exercise 2.77. Given is a convex quadrilateral $AB\Gamma\Delta$ and a point I. Consider the quadrilateral $EZH\Theta$ with vertices the symmetrical points of I relative to the middles of the sides of $AB\Gamma\Delta$. Show that $EZH\Theta$ is a parallelogram. Also show that the parallelograms thus defined, for various positions of I, are all pairwise congruent and have their sides parallel and equal to the diagonals of $AB\Gamma\Delta$ (See Figure 2.41-II).

Exercise 2.78. Construct a parallelogram $AB\Gamma\Delta$, for which are given the positions of the middles for three of its sides.

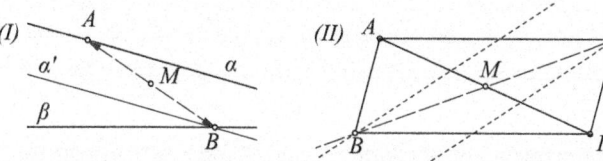

Fig. 2.42: Point symmetry   Parallelogram restricted

Exercise 2.79. Given are the lines α and β and a point M, not lying on them. Find a line segment AB, having its endpoints on the lines and its middle coinciding with M (See Figure 2.42-I).

Exercise 2.80. Construct a parallelogram $AB\Gamma\Delta$, which has vertices A and Γ at given points, vertices B and Δ on given parallels and the diagonal $B\Delta$ passing through a given point Σ (See Figure 2.42-II).

Exercise 2.81. Extend the radius OA of the circle $\kappa(O,\rho)$ towards A to its double until B, and draw tangents of κ from B, with contact points Γ and Δ. Find the angles of the quadrilateral $O\Gamma B\Delta$.

Exercise 2.82. From the middle M, of the side $B\Gamma$ of the triangle $AB\Gamma$, draw an orthogonal line to the bisector of the angle at A. Show that this defines points Δ and E, respectively on AB and $A\Gamma$, such that the lengths of ΔB and $E\Gamma$ are equal to one of $\frac{1}{2}(|AB|+|A\Gamma|)$ and $\frac{1}{2}||AB|-|A\Gamma||$.

2.8 The triangle's medians

> Natural objects should be sought and investigated as they are and not to suit observers, but respectfully as if they were divine beings.
>
> W. Goethe, *Precautions for the Observer* p. 57

The line segments, which join the triangle vertices with the middles of the opposite sides, are called **medians** of the triangle. The two most important properties of medians are expressed in the following theorems. With the help of the first theorem, we can also solve the problem of dividing a line segment into three equal parts. For the more general problem, of dividing a line segment into v parts, we must first learn the theorem of Thales (§ 3.7).

Theorem 2.15. *The three medians of a triangle pass through the same point, which divides each to the ratio 2:1.*

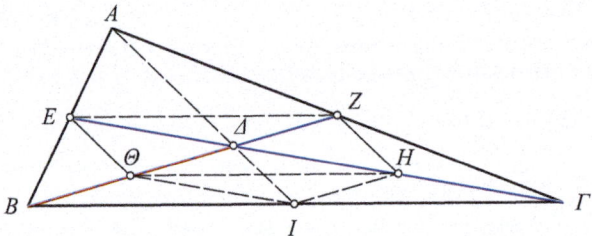

Fig. 2.43: The triangle's center of mass or centroid Δ

Proof. Let BZ and ΓE be two medians of the triangle $AB\Gamma$ (See Figure 2.43). We'll show that the third median passes through their intersection point Δ too. For this, consider the middles Θ and H respectively of $B\Delta$ and $\Gamma\Delta$. By (Theorem 2.14), the line segments ΘE and HZ are parallel to $A\Delta$ and equal to half of it. Therefore they are equal in length and parallel (Corollary 1.62), consequently $EZH\Theta$ is a parallelogram (Theorem 2.12) and its diagonals are bisected at Δ (Proposition 2.5). Draw from H the parallel HI to BZ with I on $B\Gamma$. Applying again the preceding argument we see that HI is parallel to and equal to half of $B\Delta$ and I is the middle of $B\Gamma$. Thus, ΔZHI is also a parallelogram and, consequently, ΔI is on the same line as $A\Delta$ (uniqueness of parallel to ZH from Δ). We conclude therefore, that the third median AI also passes through Δ. Moreover, we saw that Δ, lying on all three medians, divides each to a ratio of 2:1 ($A\Delta$, $B\Delta$, $\Gamma\Delta$ are double of ΔI, ΔZ, ΔE respectively).

The intersection point of the triangle medians is called **center of mass** or **centroid** of the triangle.

Exercise 2.83. Show that if the medians BN and ΓM of the triangle $AB\Gamma$ are equal, then the triangle is isosceles.

2.8. THE TRIANGLE'S MEDIANS

Hint: The intersection point Δ of the medians divides them in ratios 2:1, therefore $\Delta B\Gamma$ is isosceles and the angles $\widehat{\Gamma B\Delta}$ and $\widehat{B\Gamma\Delta}$ are equal (See Figure

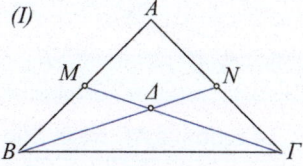 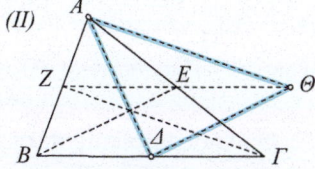

Fig. 2.44: Equal medians   Triangle $A\Delta\Theta$ of medians

2.44-I). Triangles $B\Delta M$ and $\Gamma\Delta N$ are congruent, having (i) the angles at Δ vertical therefore equal, (ii) ΔM and ΔN equal to $1/3$ of equal medians and (iii) ΔB and $\Delta\Gamma$ equal to $2/3$ of equal medians. Then the angles of triangles $MB\Delta$ and $N\Gamma\Delta$ are equal and adding them to the equal angles of the isosceles $\Delta B\Gamma$ we obtain equal angles of $AB\Gamma$ at B and Γ. Note that this exercise expresses the converse of Exercise 1.24, as far as medians are concerned.

Proposition 2.8. *The three triangle medians $A\Delta$, BE, ΓZ, displaced in a parallel manner, so that they become successive, close and form a triangle.*

Proof. $A\Delta$, BE and ΓZ are respective medians of sides $B\Gamma$, ΓA and AB (See Figure 2.44-II). $\Delta\Theta$ is constructed parallel equal and equal oriented to BE. We show that ΘA is also parallel and equal to the third median ΓZ. Indeed, by construction $B\Delta\Theta E$ is a parallelogram, therefore extending $E\Theta$ which passes through the middle E of $A\Gamma$, it will intersect AB at its middle Z (Theorem 2.14). ZE and $E\Theta$ are equal, and EA and $E\Gamma$ are also equal. Therefore in the quadrilateral $\Gamma\Theta AZ$ the diagonals will be bisected, hence it will be a parallelogram (Theorem 2.11).

Triangle $A\Delta\Theta$ of the preceding proposition is called **triangle of medians** of triangle $AB\Gamma$.

Exercise 2.84. Show that the medians of the triangle $A\Delta\Theta$ of the medians of $AB\Gamma$ are parallel and equal to three fourths of the sides of the triangle $AB\Gamma$.

Hint: In the figure 2.44-II point E is the intersection of the medians of $A\Delta\Theta$ and $E\Theta$ is half of $B\Gamma$. Combine these two and complete the proof.

Exercise 2.85. From the vertices of triangle $AB\Gamma$ draw parallels to the opposite sides. Show that this defines a triangle $A'B'\Gamma'$ which has the same angles as $AB\Gamma$ and moreover $AB\Gamma$'s vertices are the middles of its sides.

Hint: Let $A'B'$, $B'\Gamma'$, $\Gamma'A'$ be respectively parallels to AB, $B\Gamma$, ΓA (See Figure 2.45-I). Then $A\Gamma B\Gamma'$ is a parallelogram and the angles at Γ and Γ' are equal.

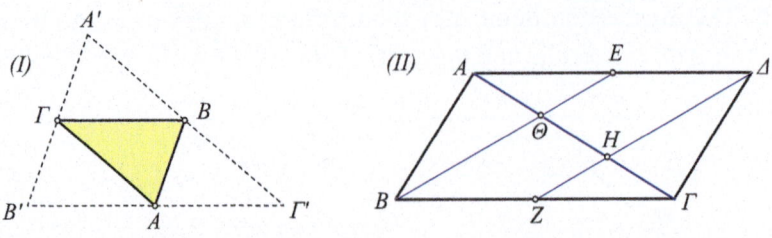

Fig. 2.45: Anticomplementary $A'B'\Gamma'$ Diagonal's $A\Gamma$ trisection

Similarly we prove that the other angles of $A'B'\Gamma'$ are equal to the respective ones of $AB\Gamma$. From the fact that $B\Gamma B'A$ is also a parallelogram, follows that AB' and $A\Gamma'$ are equal to $B\Gamma$.

The triangle $A'B'\Gamma'$, which is mentioned in the preceding exercise, is called **anticomplementary** of the triangle $AB\Gamma$. Next exercise enables us to divide a line segment $A\Gamma$ into three equal parts, by constructing a parallelogram which has $A\Gamma$ as its diagonal (See Figure 2.45-II). Alternatively we could construct a triangle which has $A\Gamma$ as its median and consider the point of intersection of its medians.

Exercise 2.86. Consider the middles E, Z of opposite sides $A\Delta$, $B\Gamma$ of the parallelogram $AB\Gamma\Delta$ and draw BE, ΔZ. Show that these divide the diagonal $A\Gamma$ into three equal segments (See Figure 2.45-II).

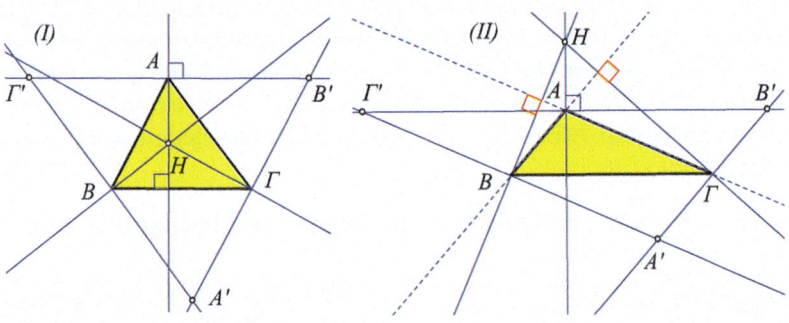

Fig. 2.46: The orthocenter H of $AB\Gamma$ Orthocenter H (obtuse)

Theorem 2.16. *The orthogonal lines from a triangle's vertices to its opposite sides pass through a common point.*

Proof. Consider the anticomplementary $A'B'\Gamma'$ of $AB\Gamma$ (See Figure 2.46-I). The orthogonals from the vertices of $AB\Gamma$ to the opposite sides are medial lines of the sides of $A'B'\Gamma'$, therefore they intersect at a point H (Theorem 2.1).

2.8. THE TRIANGLE'S MEDIANS

Point H, ensured by the preceding proposition, is the intersection point of the three *altitudes* (or their extensions) and is called **orthocenter** of triangle $AB\Gamma$. Note that in a right triangle H coincides with the vertex which carries the right angle and in obtuse triangles the orthocenter belongs to the exterior of the triangle (See Figure 2.46-II).

Exercise 2.87. Show that, if the orthocenter of the triangle $AB\Gamma$ coincides with its center of mass (centroid), then the triangle is equilateral.

Exercise 2.88. Show that, if the orthocenter of the triangle $AB\Gamma$ coincides with its circumcenter, then the triangle is equilateral.

Exercise 2.89. Show that the median to a greater side of a triangle is less than the median to a smaller side.

Exercise 2.90. Show that, if a shape Σ has two different points of symmetry, then its extent is infinite, i.e. for any positive number θ, there are points of Σ at distance greater than θ.

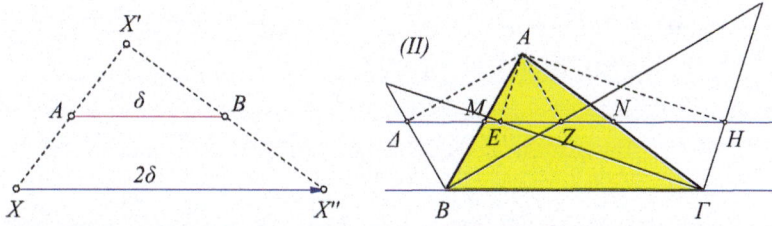

Fig. 2.47: Two point-symmetries   Projecting on bisectors

Hint: Suppose that Σ has two centers of symmetry at A and B (See Figure 2.47-I). Consider an arbitrary point X of Σ and its symmetric X' relative to A, as well as the symmetric X'' of X' relative to B. Because in the created triangle $XX'X''$, the segment AB joins middles of sides, XX'' will have length double that of the length $\delta = |AB|$, the last being fixed and independent of X. Consequently Σ will contain point X'' at distance 2δ from X. Repeating the process with X'', we find a point X'''' at distance 4δ from X, subsequently 6δ from X, and so on. Finally, no matter what positive θ is given, repeating the process ν times, for appropriate ν, we'll find point Y belonging to Σ such that $|XY| = (2\nu)\delta > \theta$.

Exercise 2.91. Show that, if a shape Σ has two parallel axes of symmetry, then its extent is infinite.

Exercise 2.92. Show that the projections of the vertex A of the triangle $AB\Gamma$ on the other angle bisectors, internal and external, belong to the line which passes through the middles M, N of sides AB and $A\Gamma$ (See Figure 2.47-II).

Exercise 2.93. Construct a triangle from its medians $\{\mu_A, \mu_B, \mu_\Gamma\}$ and an isosceles triangle from its altitudes.

Exercise 2.94. Construct an isosceles from the altitude from its apex and the median from a vertex of its base.

Exercise 2.95. The line ε passes through a point Δ of the segment $B\Gamma$. Find a point A on ε, such that for the triangle $AB\Gamma$ the line ε coincides with the bisector at A. Investigate whether there is exactly one, none or infinite many such points.

Exercise 2.96. Given are two intersecting lines $\{OX, OY\}$ and a point A not lying on them. Construct a triangle $AB\Gamma$ which has these lines as bisectors.

Exercise 2.97. Given are two intersecting lines $\{OX, OY\}$, a point E on line OX and a point Δ outside these lines. Construct a triangle $AB\Gamma$, which has the two lines as medians and a side contained in line ΔE.

2.9 The rectangle and the square

> The greatest minds, as they are capable of the highest excellencies, are open likewise to the greatest aberrations; and those who travel very slowly may yet make far greater progress, provided they keep always to the straight road, than those who, while they run, forsake it.
>
> *Descartes, Discource on method, I*

The rectangle, which we so often meet in applications, is one of the more symmetric figures. Among the rectangles the square is a prominent one. It has the most possible symmetries among all quadrilaterals.

Rectangle is called the parallelogram whose angles are all equal (to α). Because the sum of its equal angles is $4\alpha = 360°$, each one must be a right angle ($\alpha = 90°$).

Proposition 2.9. *All the triangles formed by two adjacent sides of the rectangle and the diagonal which joins their endpoints are congruent to each other.*

Proof. There are four such right triangles in the figure ($A\Delta\Gamma$, $\Delta\Gamma B$, ΓBA and $BA\Delta$) (See Figure 2.48-I). The fact, that the triangles, with common side a diagonal, like $AB\Delta$ and $B\Delta\Gamma$, are congruent, has been proved in Theorem 2.7. In rectangles, the other triangles, as well, which have common a side of the rectangle (and not its diagonal) are also congruent, like for example $AB\Delta$ and $AB\Gamma$. Indeed, the two triangles have the side AB in common, sides $A\Delta$ and $B\Gamma$ equal and the contained angles at A and B right, therefore equal. Hence, according to the SAS-criterion, they are congruent.

2.9. THE RECTANGLE AND THE SQUARE

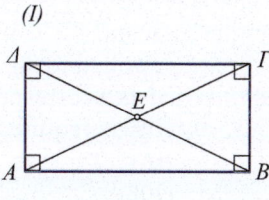
(I)

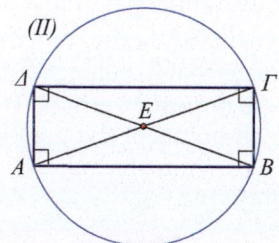
(II)

Fig. 2.48: The rectangle   ... has a circumscribed circle

Theorem 2.17. *A quadrilateral is a rectangle, if and only if its diagonals are equal and they are bisected by their intersection point.*

Proof. From the triangle equalities of the preceding proposition, follows the equality of the diagonals (See Figure 2.48-II). That these are bisected at their middle is a general property of the parallelogram (Proposition 2.5). Conversely, if the diagonals in a quadrilateral are equal and are bisected at their intersection point, then their intersection point E is the center of a circle which contains all the vertices of the quadrilateral. Moreover each diagonal is a diameter of this circle, therefore it will be seen from the opposite vertex under a right angle (Corollary 2.7). Therefore, all the angles of the quadrilateral will be right and it will be a rectangle.

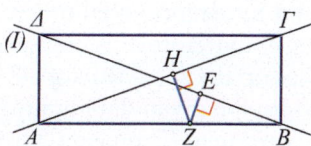
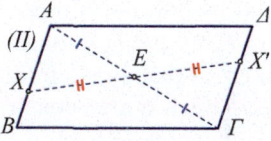

Fig. 2.49: Distances from diagonals   Point symmetry

Exercise 2.98. Show that for every point Z, contained in a side of the rectangle $AB\Gamma\Delta$, the sum of the distances $|ZH| + |ZE|$ from the diagonals is constant (See Figure 2.49-I). Show also the converse property, i.e. if the sum of the distances from the diagonals of a point on a side of the parallelogram, is constant, then the parallelogram is a rectangle.

Exercise 2.99. Show that the intersection point of the diagonals of a parallelogram is a center of symmetry. Conversely, if a quadrangle has a center of symmetry, then it is a parallelogram.

Hint: For any point X of side AB of the parallelogram $AB\Gamma\Delta$, different from its vertices, draw XE, where E is the intersection point of the diagonals, and extend it by doubling it until X' (See Figure 2.49-II). Triangles AXE and $\Gamma EX'$ are congruent (SAS-criterion), consequently angles $\widehat{XAE}$ and $\widehat{X'\Gamma E}$ are equal, which means that X' belongs to $\Gamma\Delta$. The same process applied to the vertices leads to the opposite vertices. This shows that there exists a correspondence between all points of the parallelogram $X \leftrightarrow X'$ exactly as required by the definition of symmetry relative to a point.

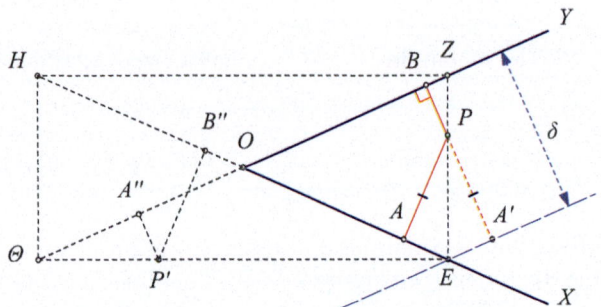

Fig. 2.50: Locus of fixed sum of distances

Exercise 2.100. Given is an angle $\widehat{XOY}$ and a point P, moving in its interior so that the sum of its distances from lines OX and OY is constant: $|PA|+|PB| = \delta$. Determine the geometric locus of point P.

Hint: Extend PB in the direction of P and by length equal to $|PA|$, until A' (See Figure 2.50). BA' has constant length δ, therefore A' is contained in the parallel of OY at distance δ from it and lies on the same side of OY as X (Exercise 1.56). This is a line, constructible from the given data intersects OX at E. The right triangles PEA and PEA' are congruent (Exercise 1.32), therefore PE is the bisector of angle OEA'. Thus, point P is contained in the bisector of angle $\widehat{OEA'}$. Conversely, every point P of this bisector is equidistant from the sides of angle OEA': $|PA| = |PA'|$. By extending PA' until the point B on line OY, we see that the sum of its distances is $|PA|+|PB| = \delta$, therefore it is a point of the locus. The same argument, applied to the other three angles, formed by the two lines, gives the geometric locus of figure 2.50, which is a rectangle $EZH\Theta$ with center point O.

Square is called the rectangle, which, in addition has all its sides equal.

Corollary 2.17. *In a square $AB\Gamma\Delta$, the triangles, which are formed from one of its sides and the intersection point of its diagonals E (for example $EB\Gamma$), are isosceli, right at E and all congruent to each other (See Figure 2.51-I).*

Corollary 2.18. *A parallelogram is a rectangle, if and only if it can be inscribed in a circle, that is, when there exists a circle containing all four of its vertices.*

2.9. THE RECTANGLE AND THE SQUARE

Proof. If it is a rectangle, then (Theorem 2.17) the intersection point of its diagonals is equidistant from its vertices, therefore it can be inscribed in a circle (See Figure 2.51-II). Conversely, if it is a parallelogram and it can be inscribed in a circle then the medial lines of its diagonals will both contain the center O of circle (Corollary 2.3). Consequently this will coincide with the intersection point of its diagonals and the diagonals will be circle diameters.

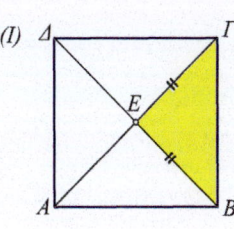 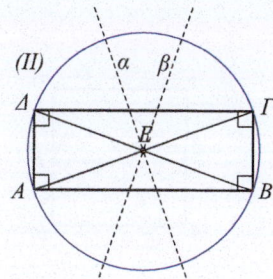

Fig. 2.51: Square   Rectangle characterization

The circle, which is defined in the preceding proposition is called **circumscribed circle** or **circumcircle** of the rectangle and its center, **center** of the rectangle. That this circle is unique, follows from the uniqueness of the circumcircle of a triangle formed by three of its vertices, like e.g. the triangle $AB\Gamma$ (Theorem 2.1).

Exercise 2.101. Show that, in every rectangle, the line which joins the middles of opposite sides is an axis of symmetry. Show that, in a square, its diagonals are also axes of symmetry.

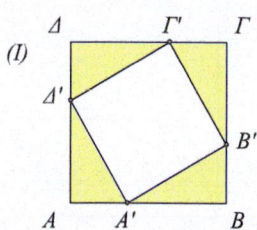 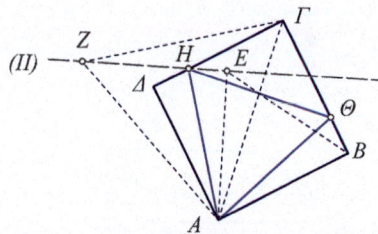

Fig. 2.52:   Square in square   Equilateral in square

Exercise 2.102. Show that, if the square $A'B'\Gamma'\Delta'$ is inscribed in the square $AB\Gamma\Delta$, then it has the same center with it and its sides are hypotenuses of four congruent right triangles (See Figure 2.52-I).

Exercise 2.103. On the side AB, of the square $AB\Gamma\Delta$, we construct an equilateral triangle ABE towards its interior (See Figure 2.52-II). Also on the diagonal $A\Gamma$, we construct an equilateral triangle $A\Gamma Z$ with Z, Δ on the same side of $A\Gamma$. Show that line ZE intersects the side $\Gamma\Delta$ at a point H, which is a vertex of an equilateral triangle $AH\Theta$, inscribed in the square.

Hint: Triangle AEZ is congruent to $AB\Gamma$. Triangle $ZH\Gamma$ is isosceles. Triangle $A\Delta H$ is congruent to $AB\Theta$.

Exercise 2.104. Given are points E and Z on adjacent sides of the square $AB\Gamma\Delta$. Construct a parallelogram, which has all its vertices on the sides of

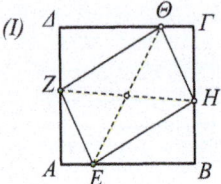
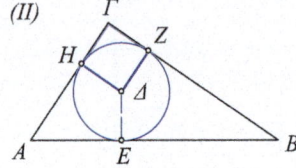

Fig. 2.53: Inscribed parallelogram   Inscribed circle

the square (inscribed in the square) and so that, two of them coincide with E and Z (See Figure 2.53-I). What condition must be satisfied by E and Z, so that this parallelogram is a square?

Exercise 2.105. Show, that the sum of the length of the hypotenuse and the diameter of the inscribed circle of a right triangle is equal to the sum of the lengths of its two vertical sides (See Figure 2.53-II).

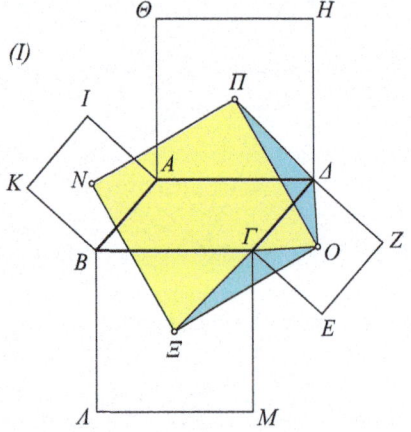
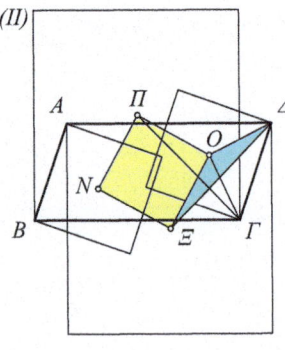

Fig. 2.54: Squares external   ... and internal

Exercise 2.106. On the sides of a parallelogram $AB\Gamma\Delta$ we construct squares towards its exterior. Show that the centers of these squares (intersection

2.10. OTHER KINDS OF QUADRILATERALS

points of their diagonals) define a square $NΞOΠ$ (See Figure 2.54-I). Show the corresponding property for squares constructed towards the interior of $ABΓΔ$ (See Figure 2.54-II).

Exercise 2.107. Given is a circle $κ(O,ρ)$ and a point A not lying on it. For every point X of the circle we extend AX towards X until Y, such that $|AY| = 2|AX|$. To find the geometric locus of Y.

Exercise 2.108. Of the triangle $ABΓ$ is given its base $BΓ$ and the length $μ_B$ of its median from B. To find the geometric locus of its vertex A.

Exercise 2.109. Construct a right-angled triangle of which is given the hypotenuse and the median to one of its orthogonal sides.

Exercise 2.110. Given are two points $\{A,B\}$ and a line $ε$. Construct a rectangle which has two vertices at $\{A,B\}$ and a third vertex on line $ε$. Investigate the number of solutions in dependence of the position of $\{A,B\}$ relative to $ε$.

2.10 Other kinds of quadrilaterals

> The prince has a taste for the arts, and would improve if his mind were not fettered by cold rules and mere technical ideas.
>
> *Goethe, The sorrows of the young Werther*

There are several more, yet, special categories of quadrilaterals which we meet in applications. Currently we'll examine only three. *Trapezia, isosceli trapezia*, which are created by intersecting a triangle with a line parallel to its base, and *rhombi*. **Trapezium** or **trapezoid** is called the quadrilateral, which

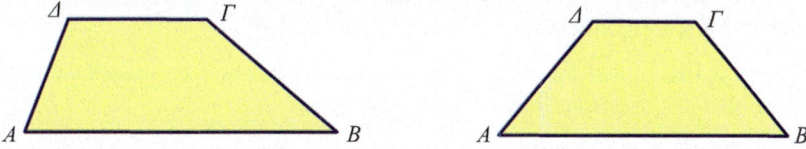

Fig. 2.55: The trapezium and the isosceles trapezium

has (only) two opposite sides parallel (See Figure 2.55). If, additionally, the other (non-parallel) sides are equal, then the trapezium is called **isosceles**.

Exercise 2.111. Show that, in an isosceles trapezium, the angles, adjacent to a parallel side are equal, and conversely: if the adjacent angles to a parallel side are equal then the trapezium is isosceles.

Hint: Consider the isosceles $ABΓΔ$ with the parallel sides AB and $ΓΔ$ (See Figure 2.56-I). Draw from $Γ$ a parallel to $ΔA$ and form the triangle $BΓE$. This triangle is isosceles, because $E\Gamma = A\Delta = B\Gamma$. Therefore angles $\widehat{\Gamma EB}$ and $\widehat{\Gamma BE}$ are equal and consequently angles $\widehat{\Delta AE}$ and $\widehat{\Gamma BE}$ are also equal. Conversely, if $\widehat{\Delta AE}$ and $\widehat{\Gamma BE}$ are equal, then drawing a parallel of $A\Delta$ from Γ and forming the same triangle ΓEB, we'll have $\widehat{\Gamma EB} = \widehat{\Gamma BE}$ and, consequently, the triangle will be isosceles, so $B\Gamma = \Gamma E = \Delta A$ and the trapezium will be isosceles.

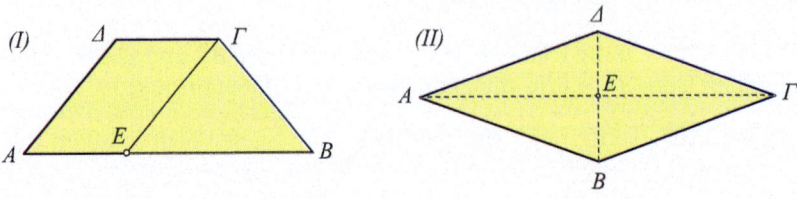

Fig. 2.56: Angles of isosceles   Rhombus

Exercise 2.112. Show that a trapezium is isosceles, if and only if its diagonals are equal.

Hint: According to the preceding exercise, the angles, adjacent to one of its parallel sides, for example side AB, will be equal. Apply therefore the SAS-criterion. The converse is proved the same way.

Exercise 2.113. Construct a trapezium, for which are given the lengths of the four sides.

Hint: Draw a parallel ΓE from one vertex to one of the non parallel sides of the trapezium (See Figure 2.56-I). This defines a triangle ΓEB with known side lengths, therefore constructible. The parallelogram $AE\Gamma\Delta$ also has known side lengths and angles, therefore it is constructible.

Rhombus is called a quadrilateral whose sides are all equal (See Figure 2.56-II). A special case of rhombus is the square in which, besides the sides, all its angles are (right and) also equal to each other.

Theorem 2.18. *A quadrilateral is a rhombus, if and only if its diagonals are orthogonal and are bisected at their intersection point.*

Proof. Indeed, if the quadrilateral $AB\Gamma\Delta$ is a rhombus, then the triangle formed by two of its adjacent sides and the diagonal which joins their endpoints (for example $AB\Gamma$) is isosceles (See Figure 2.56-II). According to Corollary 1.3, the median BE of the isosceles will be orthogonal to its base $A\Gamma$. BE, however, is part of the diagonal $B\Delta$, which also passes through the middle of the other diagonal $A\Gamma$. This proves the property. Conversely, if the diagonals are orthogonal and are bisected at their intersection point E, then, applying the SAS-criterion of congruence, we can easily prove that all the triangles formed by one side and E are congruent to each other. From the congruence of the triangles, follows the equality of the sides of $AB\Gamma\Delta$, and the proof, that it is a rhombus.

2.10. OTHER KINDS OF QUADRILATERALS

Exercise 2.114. Show that, if in a quadrilateral $AB\Gamma\Delta$, with intersection point of diagonals E, the triangles, formed from each side and E, are all isosceli, then the quadrilateral is either a rectangle or an isosceles trapezium.

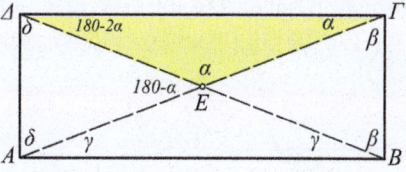

Fig. 2.57: Curious isosceles

Hint: To begin with, we examine what happens, when one leg of these isosceli triangles is a side of the quadrilateral. If the isosceles $E\Delta\Gamma$ had $E\Delta$ and $\Delta\Gamma$ equal with base angles equal to α, then the adjacent isosceles $E\Gamma B$ would not have $E\Gamma$ as a base because the angle $180° - \alpha$ on it would be obtuse (See Figure 2.57). Therefore it would have $B\Gamma$ as a base. For the same reason the isosceles $A\Delta E$ would have $A\Delta$ as a base. If the isosceles

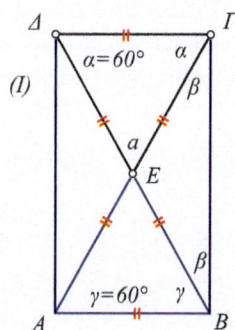
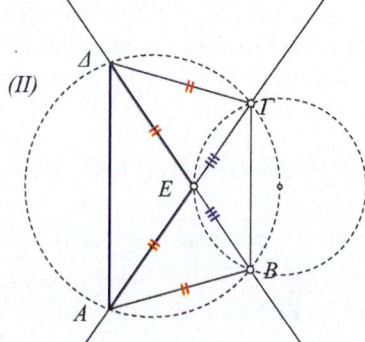

Fig. 2.58: Special rectangle   Special isosceles trapezium

AEB had AB as its base, then $E\Delta$, EA, EB and $E\Gamma$ would be equal, therefore $E\Delta\Gamma$ would be equilateral and from $\alpha = 2\beta = 2\delta$ and $2\beta + 2\gamma = 180°$ results the special rectangle with $\alpha = 60°$, $\beta = 30°$, $\gamma = 60°$ (See Figure 2.58-I). The case of the isosceles EBA having EA as its base leads again to the same special rectangle. The third possibility for the isosceles AEB is for its base to coincide with EB. In this case the triangles $E\Delta\Gamma$ and EBA are congruent (ASA-criterion) and the isosceli $EB\Gamma$ and $E\Delta A$ have respectively equal angles, therefore $B\Gamma$ and $A\Delta$ are parallel and $AB\Gamma\Delta$ is an isosceles trapezium (See Figure 2.58-II). This analysis exhausts all the cases where one of the four isosceli triangles has one of its legs identical with a side of the quadrilateral. If no isosceles has a leg coincident with a quadrilateral side, then all the legs will be equal ($|EA| = |EB| = |E\Gamma| = |E\Delta|$), so the diagonals of $AB\Gamma\Delta$ are equal and are bisected, therefore it is a rectangle (Theorem 2.17).

Exercise 2.115. Show that in the isosceles trapezium of figure 2.58-II, the center of the circumscribed circle of triangle $BE\Gamma$ belongs to the circumscribed circle of $AB\Gamma\Delta$. Examine to what extent the converse holds. That is, if for the isosceles trapezium the preceding property implies, that the trapezium has the form of figure 2.58-II.

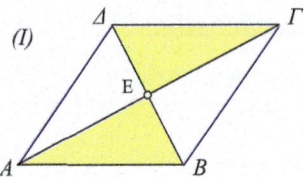 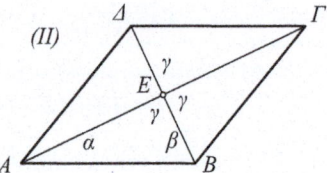

Fig. 2.59: Rhombus characterization   Congruent triangles

Exercise 2.116. Show that, if in a quadrilateral $AB\Gamma\Delta$, with intersection point of diagonals at E, the triangles, which are formed from one of its sides and E, are all pairwise congruent, then the quadrilateral is a rhombus.

Hint: Suppose that one of the triangles, for example EAB, has angles α, β, and γ at vertex E. Then all the others will have the same angles. The question is what are their relative positions. Anyway, at E all the angles must be γ (See Figure 2.59-II). If that's not the case, that is if the angle at E of $BE\Gamma$ was α or β, say α, then it would be an external of EAB and therefore $\alpha = \alpha + \beta$, which implies $\beta = 0$, which is impossible. Therefore all the angles at E are equal and $\gamma = 90°$. This implies that all the sides of the quadrilateral will be pairwise equal, therefore it is a rhombus.

Exercise 2.117. Show that the line segment, which joins the middles of the non parallel sides of a trapezium, is parallel to and equal to half the sum of its parallel sides.

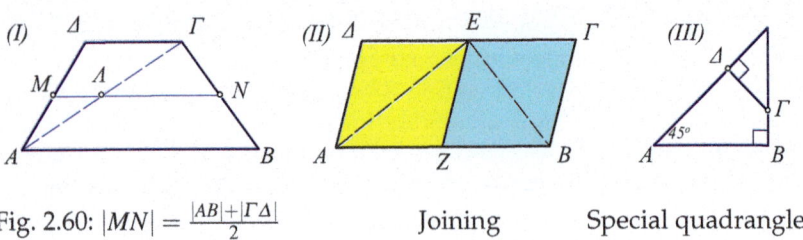

Fig. 2.60: $|MN| = \frac{|AB| + |\Gamma\Delta|}{2}$ Joining   Special quadrangle

Hint: In the trapezium $AB\Gamma\Delta$, consider the segment MN joining the middles of its non parallel sides (See Figure 2.60-I). Considering the middle Λ of the

2.11. POLYGONS, REGULAR POLYGONS

diagonal $A\Gamma$ and applying Corollary 2.16, we see that the three points M, N and Λ are contained in the same line which is the middle parallel (Corollary 1.34) of AB and $\Gamma\Delta$. From Corollary 2.16 we also have that $M\Lambda$ is equal to half the base $\Gamma\Delta$ of triangle $A\Gamma\Delta$ and ΛN is equal to half the base AB of triangle $AB\Gamma$. Adding the lengths of $M\Lambda$ and ΛN, leads to the requested result.

Exercise 2.118. Show that the bisectors of the angles, which are adjacent to a side of a parallelogram, intersect at a point of the opposite side, if and only if the parallelogram results by joining two congruent rhombi, the angle $\widehat{AEB}$ being then a right one (See Figure 2.60-II).

Exercise 2.119. Show that, for a rhombus, there exists a circle, which passes through all of its vertices, if and only if the rhombus is a square.

Exercise 2.120. Construct a convex quadrilateral $AB\Gamma\Delta$, whose angles at A, B, Γ have measure ϕ, 2ϕ, 3ϕ and $\Gamma\Delta$ is half of AB (See Figure 2.60-III).

Exercise 2.121. Show that the diagonals of a rhombus are also axes of symmetry. Also show that the line which joins the middles of the parallel sides in an isosceles trapezium is an axis of symmetry.

2.11 Polygons, regular polygons

> Art does not reproduce the visible; rather, it makes visible.
>
> P. Klee, *Creative Credo sec.1*

A **polygon** is defined by successive line segments $AB, B\Gamma, \Gamma\Delta, ...$, which form a *closed* broken line (§ 1.11). The line segments $AB, B\Gamma, \Gamma\Delta, ...$ are called

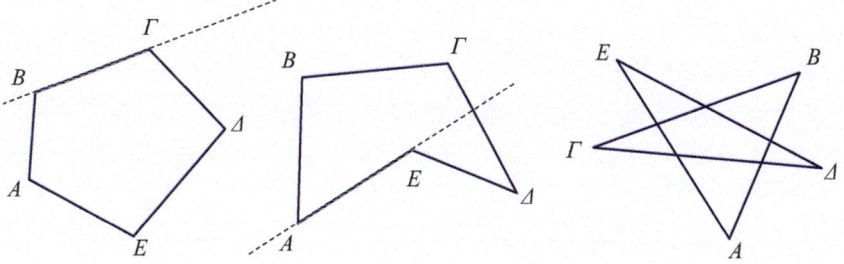

Fig. 2.61: Convex, non-convex and self-intersecting polygon

sides of the polygon, the points $A, B, \Gamma, ...$ are called **vertices** and the angles, formed by two adjacent sides, are called **angles** of the polygon. Here too, as with quadrilaterals, we distinguish between **convex** and non-convex polygons. The convex ones are the those, which have the property, that any of their side, extended to a line, leaves the polygon entirely on one side. Non-convex are called these, which have at least one side, which extended divides the polygon into parts lying on both sides of the extended side. Finally there are also the **self-intersecting** polygons, in which there exist at least two sides which intersect at a point different from the vertices of the polygon. In these lessons we deal exclusively with convex polygons. Two polygons $AB\Gamma\Delta...$ and $A'B'\Gamma'\Delta'...$ are called **congruent**, when they have their respective sides equal and their respective angles also equal. Each line seg-

(I) 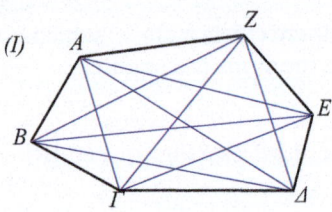 (II)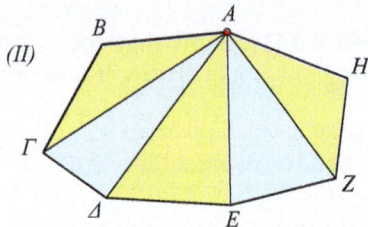

Fig. 2.62: Polygon diagonals   Angles-sum: $(v-2)180°$

ment, whose endpoints are two polygon vertices which are not endpoints of the same polygon side, is called a **diagonal** of the polygon (See Figure 2.62-I). Depending on the number of their sides/angles, we give polygons special names: five-side, six-side, seven-side, etc. If we want to emphasize the angles (polygons have as many sides as angles) we say: pentagon, hexagon, heptagon etc. At present, we'll deal with convex polygons and indeed with some special polygons called **regular**, which are characterized by the property to have all their sides equal and also all their angles equal. Two special cases of such polygons, we have already encountered, are the equilateral triangles and the squares. The sum of the lengths of all the polygon sides is called the **perimeter** of the polygon.

Proposition 2.10. *The sum of the measures of the angles of a convex polygon with v angles is $(v-2)180$ degrees.*

Proof. The proof follows immediately by considering one vertex and connecting it to all its non-adjacent vertices through the polygon's diagonals (See Figure 2.62-II). Since we have omitted exactly two sides and each side along with two diagonals forms a triangle, we get $v-2$ such triangles. The sum of the measures of the angles of these triangles is $(v-2)180°$ and this sum coincides with the sum of the measures of the polygon angles.

Corollary 2.19. *In a regular polygon with v angles, each angle has a measure of $\frac{v-2}{v}180$ degrees.*

2.11. POLYGONS, REGULAR POLYGONS

Theorem 2.19. *Each side AB of a regular polygon, together with the angle bisectors of the angles adjacent to it, forms an isosceles triangle. All these isosceli triangles have the same apex O and are all congruent to each other.*

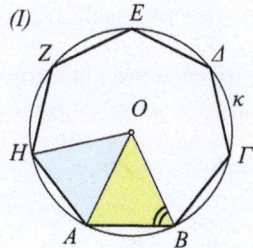 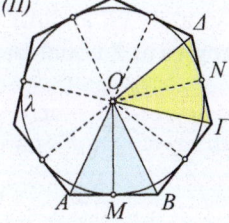

Fig. 2.63: Ccircumscribed circle κ   ... and inscribed circle λ

Proof. The proof follows immediately from the equality of the polygon's angles (See Figure 2.63-I). Drawing the angle bisectors, we define angles which are halves of equal angles therefore also equal to each other. Therefore the triangles constructed this way are isosceli. Applying now the ASA-criterion of congruence to two of them, we see that they are congruent and their apexes coincide and define a point O of the plane.

Point O is called **center** or **circumcenter** of the regular polygon and it is the center of a circle κ passing through all vertices and called **circumscribed** circle or **circumcircle** and its radius **circumradius** of the regular polygon.

Theorem 2.20. *For every regular polygon there is a circle λ, concentric with the circumscribed circle κ, and tangent to the sides of the polygon at their middle (See Figure 2.63-II).*

Proof. For each side AB of the polygon, consider the isosceles ABO. The altitudes, like OM, of all these isosceli are equal and define the radius of a circle, which touches the sides at their middle.

The circle of the preceding proposition is called the **incircle** of the regular polygon and its radius is called the **inradius** of the regular polygon.

Exercise 2.122. The endpoints Γ, Δ of a diameter of a circle are projected, respectively, at points E, Z of a chord of AB. Show that points E, Z divide the chord into segments having the same ratio of lengths (See Figure 2.64-I).

Theorem 2.21. *The square and the regular pentagon are the only regular polygons which have all their diagonals equal.*

Proof. That the aforementioned shapes have equal diagonals, follows by appropriately applying the SAS-criterion of triangle congruence. The converse is more interesting. Indeed, let us consider that the regular polygon has more than four sides and equal diagonals. In such a polygon, consider the quadrilateral $AB\Delta E$, which is formed from three successive sides and one diagonal, as in the figure 2.64-II. According to our hypotheses, triangles $AB\Delta$ and ΔEB are congruent (SSS-criterion) and isosceli. Similarly, triangles ABE and ΔEA are congruent and isosceli. Considering, therefore, the intersection point Z of the diagonals BE and $A\Delta$, we form two triangles AZE and $BZ\Delta$, which have their angles at Z equal, as vertical and additionally they are also isosceli. It

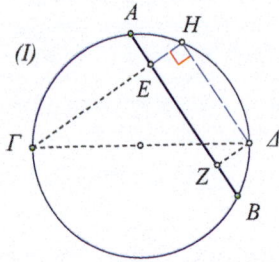
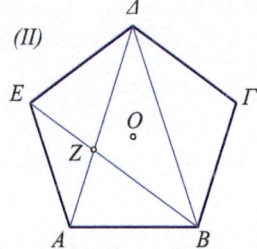

Fig. 2.64:   Chord subdivision   Regular pentagon

follows that the other angles of these triangles are equal, consequently the lines AE and $B\Delta$ are parallel, since their secant EB forms internal and alternate angles equal. This again has as a consequence that BE is a bisector of the angle $\widehat{AB\Delta}$. Therefore, triangle $AB\Delta$ has the noteworthy property of being isosceles and the angle at its apex having measure half that of the angle at its base. Consequently, if the measure of the angle at its vertex is ω, the other two angles will have measure 2ω each. Therefore, the sum of their measures will be $\omega + 2\omega + 2\omega = 5\omega = 180°$. From which follows that $\omega = \frac{180°}{5} = 36°$. All this implies that the angle of the regular polygon at A is $3\omega = 108°$. Thus, in order to find the number of sides of the polygon, it suffices to solve the equation $\frac{N-2}{N} 180° = 108°$ for N. The equation is equivalent to $72N = 360$ and has the solution $N = 5$.

Exercise 2.123. Construct a square having the side length δ.

Exercise 2.124. Given is an equilateral triangle ABH. On side BH we construct another, different from ABH, equilateral $HB\Gamma$, on the side $H\Gamma$ of this equilateral we construct another equilateral $H\Gamma\Delta$, and so on. Show that, after five similar constructions, we return to the initial equilateral and the resulting polygon, which is formed by the sides AB, $B\Gamma$, $\Gamma\Delta$, etc., is a regular hexagon with center H (See Figure 2.65-I).

Exercise 2.125. Given is a circle κ of radius ρ. Show that the following process constructs a regular hexagon inscribed in κ: With center an arbitrary point A of κ, draw the circle of radius ρ, which intersects κ at B. With center at B and radius ρ, draw the circle which intersects κ at Γ. With center at

2.11. POLYGONS, REGULAR POLYGONS

Γ and radius ρ, define the circle which intersects κ at Δ, etc. From the two intersection points, always choose the one in the same direction (counter-clockwise, for example). This way, there are defined a total of six points of κ, which are vertices of a regular hexagon (See Figure 2.65-II)).

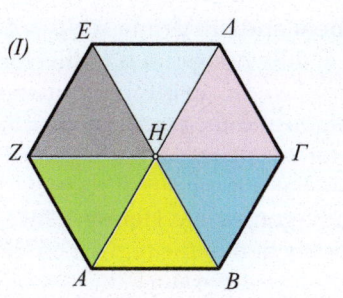 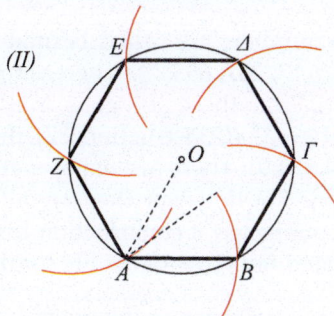

Fig. 2.65: Regular hexagon   Hexagon construction

Remark 2.2. Gauss (1777-1855) stated ([51, p.16]) and Pierre Wantzel (1814-1848) proved in 1837 ([50]) that a regular polygon with ν sides can be constructed, using ruler and compass, if and only if ν is of the form

$$\nu = 2^\mu \cdot \delta_1 \cdot \delta_2 \ldots \delta_\kappa,$$

where the integers $\delta_1, \ldots, \delta_\kappa$ are different and *prime* (in other words they are divisible only by 1 and themselves) and more precisely of the form (when they are different than 1)

$$\delta = 2^\lambda + 1.$$

In the preceding equation for ν, some factors may be missing, for example, some δ's may be 1 or/and the 2^μ may be missing. Next table, constructed according to this criterion, shows the constructible and non-constructible regular polygons for the first 32 numbers.

constructible	3 4 5 6 8 10 12 15 16 17 20 24 25 30 32
non-constructible	7 9 11 13 14 18 19 21 22 23 26 27 28 29 31

Exercise 2.126. For every integer in the first line of the preceding table, determine the integers (ν, μ, δ) which are involved in Gauss' constructibility criterion. Also prove that, each integer in the second line, does not satisfy this criterion.

Exercise 2.127. Show, that every regular polygon has an axis of symmetry, however, only regular polygons with an even number of sides have a center of symmetry. Also show that regular polygons, with ν sides, have ν axes of symmetry.

Remark 2.3. From a theorem by Fermat (1601-1665) ([59, p.1022]), it is proved, that the prime numbers of the form $\delta = 2^\lambda + 1$, must have λ equal to a power of 2 and generally be of the form

$$\delta = 2^{2^\rho} + 1.$$

These numbers are called **Fermat numbers** and are prime for $\rho < 5$ (they are 3, 5, 17, 257, 65337), whereas for $\rho \geq 5$, only non-primes of this form are known. From these polygons, in this lesson, we construct the equilateral triangle ($\nu = 3$), (Construction 2.2), the hexagon and the pentagon ($\nu = 5$, Exercises 4.8, 4.9). There are many methods for constructing the regular 17-gon ([16, p.217]), however they are all somewhat complicated. Richelot (1808-1875) published a construction for the 257-gon, while Hermes (1846-1912) dedicated ten years of his life for the construction of the regular 65337-gon.

More generally, the problem of constructing line segments and angles, using exclusively and only a ruler (without metric subdivisions) and a compass, occupied the Ancient Greeks and its full solution requires knowledge of algebra and specifically field theory ([44, p.117]). There are three famous construction problems, which remained unsloved until the 19th century ([51]): (i) The problem of trisecting an angle, (ii) the problem of doubling the cube (Delian problem) and (iii) the problem of squaring the circle. Despite the fact that these problems cannot be solved, using exclusively a ruler and compass, they can be solved using other geometric methods. Thus, for example, the trisection of an angle and the doubling of the cube can be done using the so called **neusis**, an example of which is given in exercise 5.191 (see also the comments after exercise 5.216).

Exercise 2.128. On the sides of a square and internally construct equilateral triangles ABE, $BZ\Gamma$, $\Gamma \Delta H$, $\Delta A\Theta$ (See Figure 2.66). Show that the middles of their sides $I, K, M, N, O, \Pi, \Sigma, T$ and the middles Λ, Ξ, P, Y respectively of $ZH, H\Theta, \Theta E, EZ$ are vertices of a regular dodecagon ([154, p.137]).

Exercise 2.129. Given is a regular polygon with an even number of sides ν. On its sides we construct alternatively squares and equilateral triangles (tiles) towards the interior of the polygon (See Figure 2.67). Find the least ν for which these tiles are not overlapping.

Exercise 2.130. In the regular decatetragon of Figure 2.67 show that the lines EZ, $\Gamma\Delta$ intersect at an angle ω, double that of $\widehat{HAZ}$. Also show that ω is one third that of external angle $\widehat{IB\Theta}$ of the decatetragon.

Remark 2.4. The problem of constructing a regular polygon with ν sides is equivalent to the problem of constructing an angle of measure $\frac{360°}{\nu}$, which results by joining the center O of the polygon with the endpoints A, B of one of its sides. Therefore, if the regular polygon with ν sides can be constructed, the regular polygon with 2ν sides can be constructed by bisecting

2.11. POLYGONS, REGULAR POLYGONS

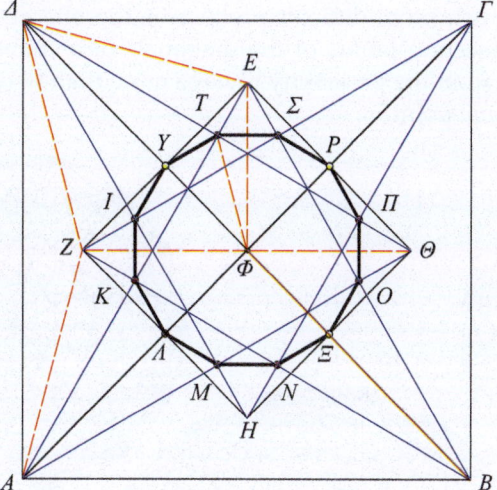

Fig. 2.66: Regular dodecagon

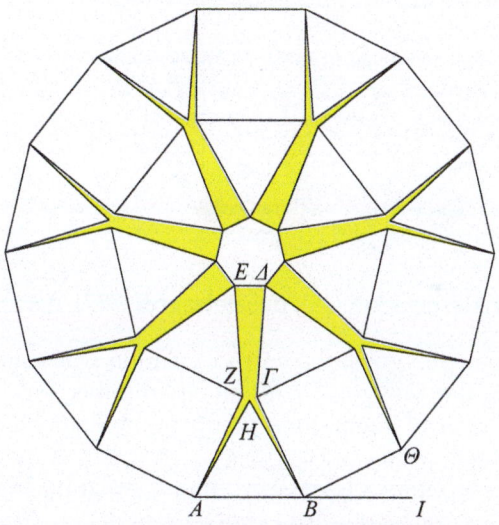

Fig. 2.67: Regular decatetragon

the respective angle $\widehat{AOB}$ of the original polygon. This way, the student can construct successively the regular polygons with 6, 8, 10, 12, 16 etc. sides, beginning with the equilateral, square, pentagon, etc.

Exercise 2.131. Show that the number of diagonals of a convex polygon with v sides is $\frac{v \cdot (v-3)}{2}$.

Remark 2.5. A much more difficult problem is the one of determining the number of intersection points of the diagonals of a convex polygon. The difficulty stems from the possibility to have more than two diagonals intersecting at the same point.

Fig. 2.68: Diagonals: octagon (49), decatetragon (757), decahexagon (1377)

Even for regular polygons, the correct formula was established only in 2006 [55]. For regular polygons, with an *odd* number v of sides, it can be seen, that the number of intersection points of the diagonals is equal to the number of combinations of v objects by four: $C_v = \frac{v \cdot (v-1) \cdot (v-2) \cdot (v-3)}{24}$. This, because, in that case, through every intersection point of the diagonals pass exactly two diagonals. The extremities of these two diagonals define a quadrilateral and, in this way, we see that the number of intersection points is equal to the number of quadrilaterals, which are defined by four vertices of the polygon. The difficult case is the one of regular polygons with an *even* number of sides. In that case there can be intersection points through which pass more than two diagonals, as is seen in figure 2.68, for the regular octagon, decatetragon and decahexagon, with a number of intersection points of diagonals, respectively, 49, 757 and 1377.

2.12. ARCS, CENTRAL ANGLES

Figure 2.69 presents a part of a triantagon (30-gon) and the intersections of its diagonals. Marked are some points, through which pass respectively 5, 6, 7 and 15 diagonals, from a total of 16801 intersection points.

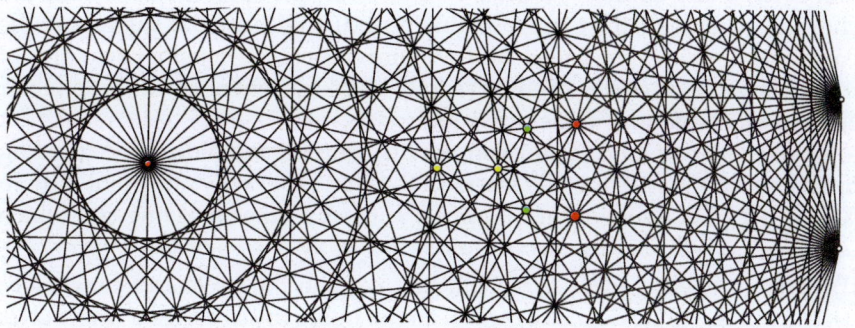

Fig. 2.69: Intersections of diagonals of a triantagon

2.12 Arcs, central angles

> Things are in the saddle, and ride mankind.
>
> R. W. Emerson, Ode Inscribed to W. H. Channing

Two points A and B on the circle define its chord AB, two **arcs** and two angles $\widehat{AOB}$. The first arc $\widehat{A\Gamma B}$ is the part of the circle, which is contained in the

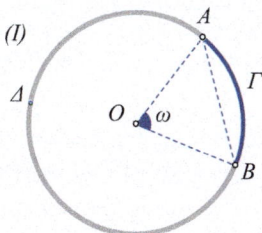

Fig. 2.70: Arcs $\widehat{A\Gamma B}$, $\widehat{A\Delta B}$

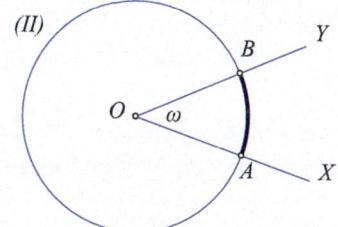

Arcs-angles correspondence

interior of angle $\widehat{AOB}$. The other arc $\widehat{A\Delta B}$ is the remaining part of the circle (See Figure 2.70-I). The two arcs belong to different sides of the line AB and are called **complementary**. The smaller one belongs to the side of the line AB which does not contain the center of the circle. The angle $\omega = \widehat{AOB}$, which contains the arc $\widehat{A\Gamma B}$ in its interior, is called **central** of the arc $\widehat{A\Gamma B}$ which is referred to as **corresponding to the central** angle ω. The same half lines OA and OB define also the arc $\widehat{A\Delta B}$, whose *corresponding central angle* is the non-convex one defined by these half lines. In the special case, where the chord is a diameter, the corresponding central angles are flat angles and the corresponding arcs are called **half circumference**. We consider two arcs of the same circle, or circles of equal radii to be **congruent** or **equal**, when their corresponding central angles are equal.

Consider a half line OX and a circle with center O and radius ρ. According to Axiom 1.8 the half line will intersect the circle at exactly one point A (with $|OA| = \rho$). The same will also happen with any other half line OY, which starts at O. It will also intersect the circle at exactly one point (See Figure 2.70-II). Therefore, each angle $\widehat{XOY}$ defines an arc on the circle and conversely, each arc of the circle defines a central angle. Using the isosceles triangle OAB, defined by the endpoints of the chord, we have immediately the following corollary.

Corollary 2.20. *Two equal chords of the same circle or of congruent circles define two equal or complementary arcs.*

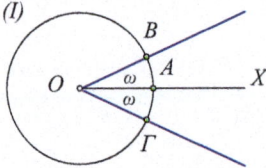
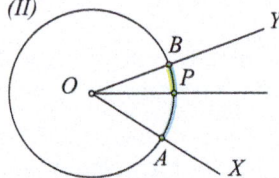

Fig. 2.71: Arcs on both sides   Arc addition

Through the correspondence {arc ↔ corresponding central angle}, notions we defined for angles and properties of angles transfer to circle arcs. For example, axioms 1.10 and 1.11 for the angles of section 1.4 take the following form.

1. Given a half line OX and a circle with center O, which intersects the half line at A, as well as, a number ω with $0 < ω < 180°$, there exist exactly two half lines on each side of OX, which define arcs with corresponding central angle of measure ω (See Figure 2.71-I).
2. For every point P of the circle arc $\widehat{AB}$ we say that the arc $\widehat{AB}$ is the sum of the arcs $\widehat{AP}$ and $\widehat{PB}$ (See Figure 2.71-II).

Theorem 2.22. *The endpoints of two non-intersecting congruent arcs of the same circle define two parallel lines. Conversely, two parallel lines which intersect a circle define two congruent arcs.*

2.12. ARCS, CENTRAL ANGLES

Proof. For the congruent arcs $\widehat{AB}$ and $\widehat{A'B'}$, consider the chord AA' and its middle M (See Figure 2.72-I). Congruence of arcs means equality of their central angles, thus OB and OB' form equal angles with OA and OA' respectively. It follows that OM is a bisector of the isosceles BOB' (or medial line of the diameter BOB', if B, B' are diametrically opposite), therefore it passes through the middle N of BB' and is orthogonal to it, thus AA' and BB' are parallel. For the converse, observe that the isosceli triangles OAA' and OBB', with common vertex the center of the circle and parallel bases, have a common angle bisector and medial line (Corollary 2.1) OM. Therefore the (central) angles $\widehat{AOB}$ and $\widehat{A'OB'}$ are equal.

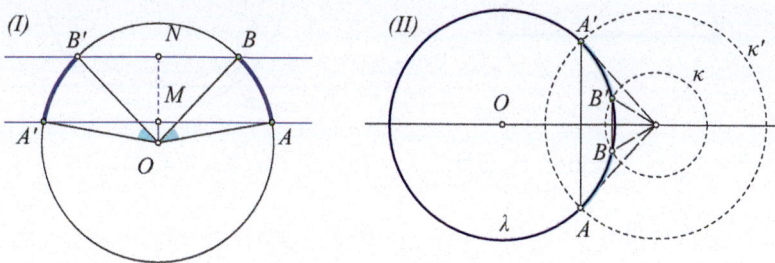

Fig. 2.72: Congruent arcs   Concentric circles

Remark 2.6. The theorem also holds for intersecting arcs. The proof for this is reduced to the preceding case. For example, in the preceding figure 2.72-I, arcs $\widehat{ABB'}$ (with endpoints A, B') and $\widehat{A'B'B}$ (with endpoints A', B) are congruent. The modification of the proof for this case is easy.

Exercise 2.132. Show that two concentric circles κ, κ' excise from a third circle λ congruent arcs (See Figure 2.72-II). When does the converse hold? That is, when, for two congruent arcs of λ, does there exist pair (κ, κ') of concentric circles which excise them from λ? Where are the centers of these concentric circles?

Exercise 2.133. Show, that for a trapezium, there exists a circle, which passes through all of its four vertices (circumscribed), if and only if it is isosceles (See Figure 2.73-I).

Exercise 2.134. Show, that if a circle chord changes its position on the circle κ by retaining its length, then its middle is contained in a circle κ' concentric to κ (See Figure 2.73-II).

Hint: If O is the center of the circle and M is the middle of the chord AB, then the triangles OAM for different positions of the chord, are congruent to each other.

Exercise 2.135. Show the converse of the preceding exercise. That is, if two chords of the same circle have their middles on a concentric circle, then they are equal.

Hint: The segment *OM*, which joins the center with the middle of the chord *AB* (See Figure 2.73-II), is orthogonal to the chord, and the triangles, like *OAM*, for different places of the chord, are mutually congruent.

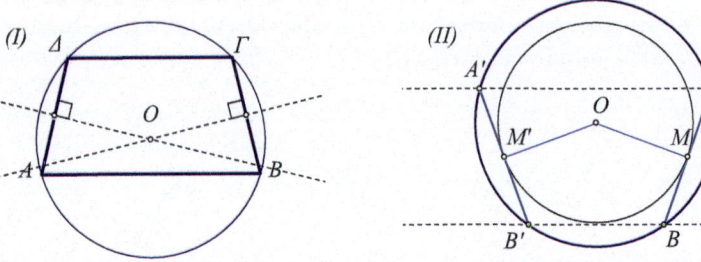

Fig. 2.73: Inscribed in circle   Fixed length chords

Exercise 2.136. Show that the radii to the middles of two arcs, defined by two orthogonal chords of the same circle, form a right angle.

Exercise 2.137. The right angle $\widehat{XAY}$, rotating about its fixed vertex *A*, intersects the sides of another fixed right angle $\widehat{PBT}$ at Γ and Δ. Show that the geometric locus of the middle *M* of the line segment $\Gamma\Delta$, is the medial line of *AB* (See Figure 2.74-I).

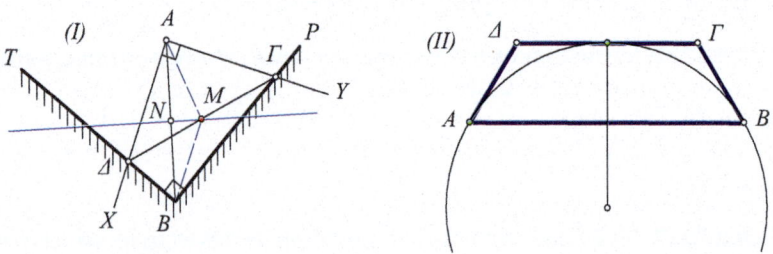

Fig. 2.74: Rotating right angle   Special isosceles

Exercise 2.138. Show that the arc, which is tangent to the two non parallel sides of an isosceles trapezium, at the endpoints of the greater base, is also tangent to its smaller base, if and only if the latter is double the size of its non parallel sides (See Figure 2.74-II).

Exercise 2.139. At the endpoints of an arc $\widehat{AB}$, less than a half circumference, draw tangents intersecting at Γ. Show that the middle *Z* of the altitude $\Gamma\Delta$

2.13. INSCRIBED ANGLES

of the isosceles *ABΓ* is outside the circle which contains the arc (See Figure 2.75-I).

Hint: Apply Exercise 1.39.

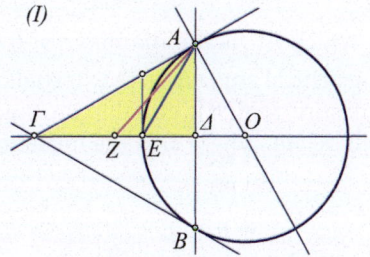

Fig. 2.75: Middle position — Equiperimetric triangles $A'OB'$

Exercise 2.140. Consider the tangents $\{OA, OB\}$ of the circle κ (See Figure 2.75-II). For every point M, of the smaller arc with endpoints A and B, we draw the tangent to κ at M, which intersects OA, OB at A', B' respectively. Show that the triangle $OA'B'$ has a constant perimeter.

2.13 Inscribed angles

> An artist is someone who produces things that people don't need to have but that he-for some reason-thinks it would be a good idea to give them.
>
> *Andy Warhol*

An angle $\widehat{XOY}$ is called **inscribed** in a circle κ, when its vertex is on the circle and its sides intersect the circle at two other points A and B, different

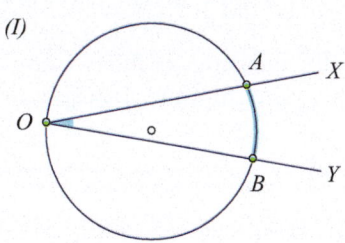
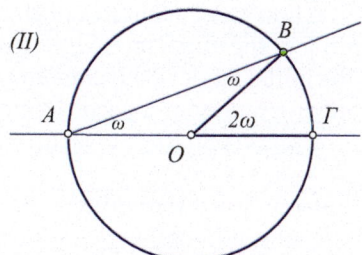

Fig. 2.76: Inscribed angle,   Corresponding central angle

from O (See Figure 2.76-I). The arc AB is called **arc of the inscribed angle**.

Next theorem expresses a fundamental relationship between inscribed and central angles, which, as we say, **subtend the same arc** or **see the same arc**.

Proposition 2.11. *The measure of an inscribed angle, whose one side is a diameter, is half the measure of the corresponding central angle, subtending the same arc.*

Proof. Let $\widehat{BA\Gamma}$ be the inscribed angle, for which $A\Gamma$ is the diameter of the circle (See Figure 2.76-II). The corresponding central angle $\widehat{BO\Gamma}$ is external of the isosceles triangle OAB. Therefore, its measure, which is equal to the sum of the two equal angles of the isosceles, will be 2ω, where ω is the measure of the inscribed angle $\widehat{BA\Gamma}$.

Theorem 2.23. *The measure of the angle inscribed in a circle is half that of the corresponding central angle.*

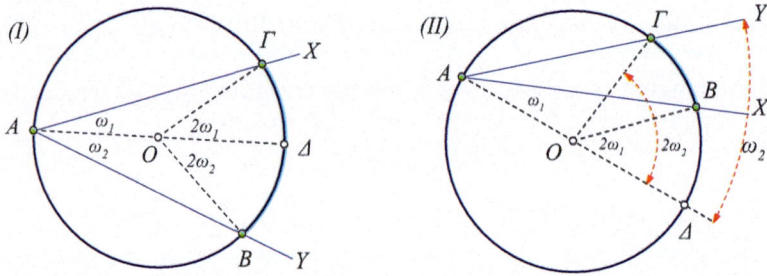

Fig. 2.77: Inscribed angle $\widehat{BA\Gamma}$ and corresponding central angle $\widehat{BO\Gamma}$

Proof. In this theorem we get rid of the restriction of one side being a diameter. We reduce however the proof to the preceding proposition by representing the angle as a sum or difference of two inscribed angles, whose one side is a diameter. When, therefore, no side of the inscribed angle is a diameter, we draw from its vertex A a diameter $A\Delta$.

The first case is that in which, the diameter we draw, is contained in the interior of the initial angle $\widehat{BA\Gamma}$ (See Figure 2.77-I). In this case the measure ω of the initial angle is the sum $\omega = \omega_1 + \omega_2$ of the measures of the two angles. Applying therefore the preceding theorem, we find that the corresponding central angle will have measure $2\omega_1 + 2\omega_2 = 2\omega$.

The second case is that in which, the diameter we draw, is not contained in the interior of the initial angle (See Figure 2.77-II). In this case the measure ω of the initial angle is the difference $\omega = \omega_2 - \omega_1$ of the measures of the two angles. Applying therefore the preceding theorem, we find that the corresponding central angle will have measure $2\omega_2 - 2\omega_1 = 2\omega$.

2.13. INSCRIBED ANGLES

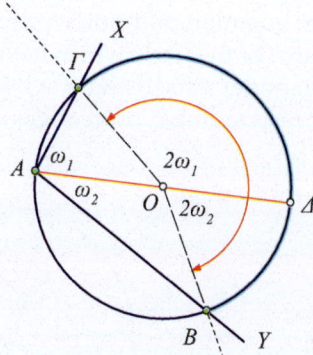

Fig. 2.78: Inscribed obtuse and corresponding central angle (non-convex)

Remark 2.7. There exists a detail in the wording of the proposition which must be noted. When the inscribed angle is obtuse, then the corresponding central has measure greater than 180°, in other words it is non-convex. Its characteristic is that the center of the circle and the arc it subtends are on the same side of the chord $B\Gamma$ (See Figure 2.78).

Corollary 2.21. *If $\widehat{AB\Gamma}$ is an inscribed angle in the circle κ and $\widehat{A\Delta\Gamma}$ labels the corresponding arc of its central angle (the arc "seen by the angle"):*

1. *If the inscribed $\widehat{AB\Gamma}$ is acute then the arc $\widehat{A\Delta\Gamma}$ is less than a half circumference,*
2. *If the inscribed $\widehat{AB\Gamma}$ is right then the arc $\widehat{A\Delta\Gamma}$ is equal to a half circumference,*
3. *If the inscribed $\widehat{AB\Gamma}$ is obtuse then the arc $\widehat{A\Delta\Gamma}$ is greater than a half circumference.*

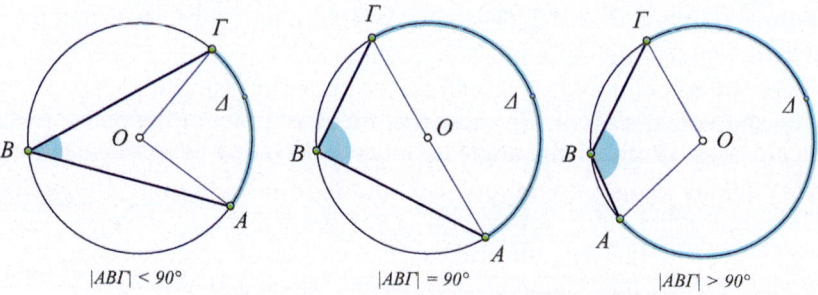

Fig. 2.79: Inscribed and arc which it "sees"

In figure 2.79 the three cases of the preceding corollary are visualized and the arcs $\widehat{A\Delta\Gamma}$ which are *subtended* (or seen) by the corresponding inscribed angle are emphasized. The common and noteworthy characteristic of the circle arcs is that all the points of the circle which belong to the other side of the chord of $A\Gamma$ see this arc under a fixed angle, whose measure is equal to half the measure of the corresponding central angle for this arc (See Figure 2.80).

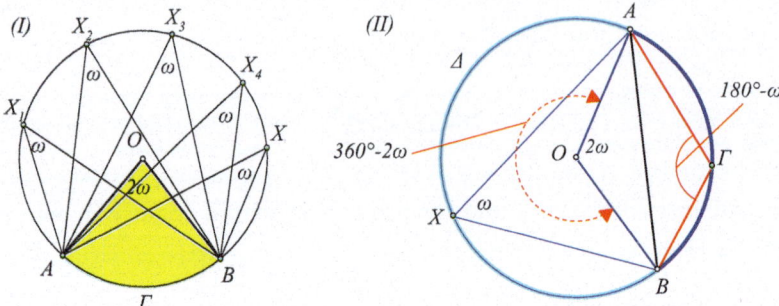

Fig. 2.80:   Equal inscribed angles   Inscribed angles with common chord

Corollary 2.22. *Every chord AB divides the circle into two arcs $\overparen{A\Gamma B}$ and $\overparen{A\Delta B}$, each contained in one of the two half planes, defined by the line AB. These two arcs are complementary. Every point X of the arc $\overparen{A\Delta B}$ sees the other arc $\overparen{A\Gamma B}$ under a fixed angle ω and every point Y of the arc $\overparen{A\Gamma B}$ sees the arc $\overparen{A\Delta B}$ under ω's supplementary $\omega' = 180° - \omega$*

Theorem 2.24. *The geometric locus of points X of the plane, which are on the same side of the line segment AB and see it under given fixed angle ω, is a circle arc which passes through endpoints A and B of the line segment.*

Proof. Let X be a point of the locus i.e. a point which sees the segment AB under angle ω (See Figure 2.80-II), and let κ be the circle which passes through the three points A, B and X (Theorem 2.1). Let point X be contained in the arc $\overparen{A\Delta B}$, with endpoints A and B. According to the preceding corollary, every other point of this arc will see AB under the same angle ω.

To complete the proof, we show that no other point of the plane, except this arc, sees AB under the angle ω. Indeed, if point Y is in the exterior of arc $\overparen{AXB}$ then, joining it with points A and B, forms a triangle AYB, whose, at least one side (why?), for example YB, intersects the arc at some point X (See Figure 2.81-I). Then, the angle ω at X will be external in triangle AXY, therefore greater than the internal opposite, which is θ. Similarly, we show (See Figure 2.81-II) that a point Y, in the interior of the arc, sees AB under an angle θ, which is greater than ω. This completes the proof of the theorem.

Remark 2.8. If we want the geometric locus of *all* points X, which see AB under a fixed angle ω, independent of AB's side, we must repeat the construction of the arc on both sides of AB. Figure 2.82 shows this construction in the

2.13. INSCRIBED ANGLES

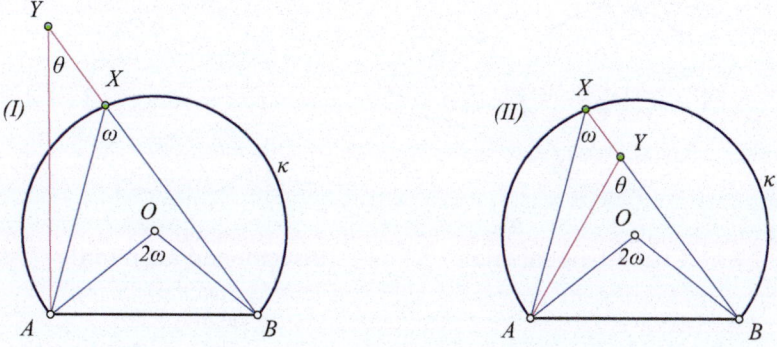

Fig. 2.81: Points Y, which do not see AB under the angle ω

case (i) of an acute angle ω, (ii) of a right angle, where the two arcs are two half circumferences of the same circle and (iii) of an obtuse angle ω.

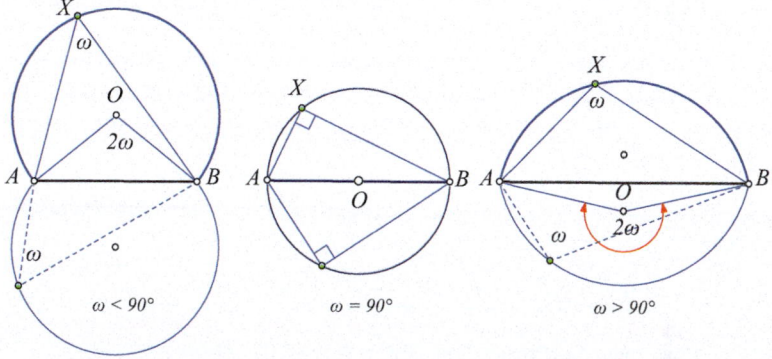

Fig. 2.82: Points which see AB under given angle ω

Theorem 2.25. *The angle ω, formed from a chord AB of a circle and the tangent to the circle at a chord endpoint B, is equal to the inscribed angle, which sees the arc lying between the chord and the tangent. Conversely, if the chord AB and the line BX, define the same angle $\omega = |\widehat{APB}| = |\widehat{ABX}|$, then BX is tangent at B.*

Proof. The angle between chord and tangent at B has its sides orthogonal to the sides of the angle $\widehat{BOM}$, where M is the middle of the chord (See Figure 2.83-I). Both these angles are complementary to angle θ, which is formed by the chord and the radius at B, therefore they are equal. The inscribed angle $\widehat{BPA}$, however, is equal to $\widehat{BOM}$. The converse is proved the same way. If $\omega = \widehat{APB} = \widehat{ABX}$, then $\theta = \widehat{MBO}$ will also be complementary to $\widehat{ABX}$, therefore $\widehat{OBX}$ will be a right angle.

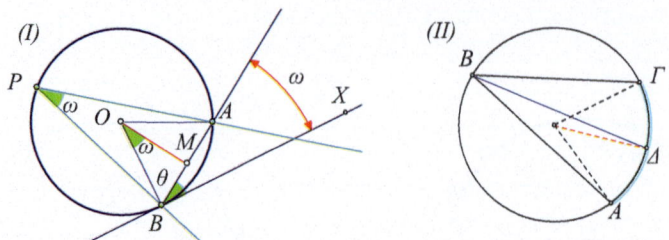

Fig. 2.83: Chord-tangent angle   Bisector and arc middle

Exercise 2.141. If $\widehat{AB\Gamma}$ is an inscribed angle in a circle subtending the arc $\widehat{A\Delta\Gamma}$, then the line $B\Delta$, which is defined from the angle's vertex B and the middle Δ of the arc, is the bisector of the inscribed angle.

Hint: The corresponding central angles of arcs $\widehat{\Gamma\Delta}$ and $\widehat{\Delta A}$ are equal and double these of $\widehat{\Gamma B\Delta}$ and $\widehat{\Delta BA}$ respectively (See Figure 2.83-II).

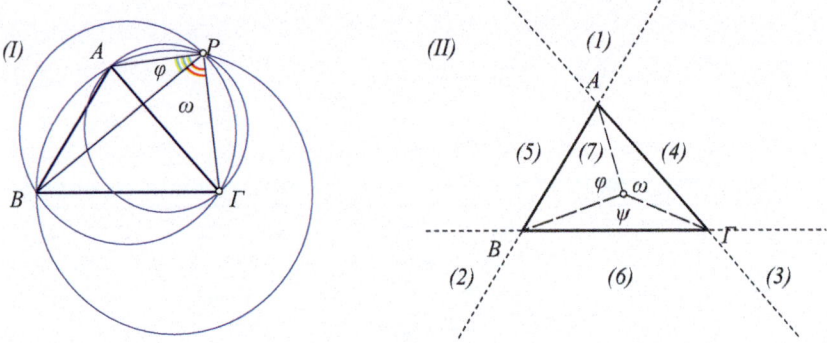

Fig. 2.84: Angles under which P sees the sides of triangle

Exercise 2.142. Show that every point P, in the exterior of a triangle, sees its three sides under angles of measures ω, ϕ, ψ, such that if ψ is the greater, then $\psi = \phi + \omega$ (See Figure 2.84-I). Conversely, circular arcs consisting of points which see the sides of triangle under angles whose measures satisfy the preceding relation, intersect at point P in the exterior of the triangle $AB\Gamma$. Describe the relation between angles ϕ, ψ and ω in each of the seven areas, defined by the sides of a triangle (See Figure 2.84-II).

Exercise 2.143. if H and I are the orthocenter and the incenter of the triangle $AB\Gamma$, determine the angles $\widehat{BH\Gamma}$ (See Figure 2.85-I) and $\widehat{BI\Gamma}$ (See Figure 2.85-II), as a function of the angles of the triangle $AB\Gamma$.

Exercise 2.144. Consider a trapezium $AB\Gamma\Delta$ with parallel sides AB, $\Gamma\Delta$ and E the point of intersection of its diagonals. Show that the circumscribed circles of triangles $\Delta E\Gamma$ and ABE are tangent at E (See Figure 2.86-I).

2.13. INSCRIBED ANGLES

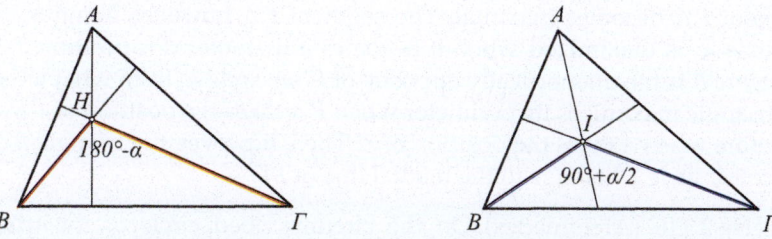

Fig. 2.85: Angles at orthocenter and incenter

Hint: Let EX be the tangent to the circumscribed circle of triangle $\Delta E\Gamma$ at E and EY be the tangent to the circumscribed circle of triangle ABE at E (See Figure 2.86-I). Angle $\widehat{XE\Delta}$ is equal to $\widehat{E\Gamma\Delta}$ (Theorem 2.25), which, in turn, is equal to $\widehat{EAB}$, which, in turn, is equal to $\widehat{BEY}$. Because Δ, E, B are collinear (diagonal of $AB\Gamma\Delta$), the equality of $\widehat{\Delta EX}$ and $\widehat{YEB}$ implies that these two angles are vertical and therefore X, E, Y are collinear.

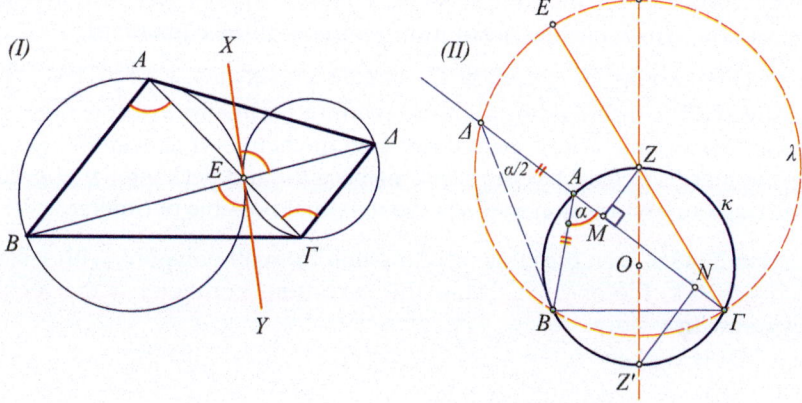

Fig. 2.86: Tangent circles   △$BZ\Gamma$ of maximal perimeter

Exercise 2.145. Given the circle κ, show that from all triangles $AB\Gamma$ which have the base $B\Gamma$ in common and are inscribed in κ, the one with the maximum perimeter is the isosceles (See Figure 2.86-II).

Hint: Extend $A\Gamma$ and consider $A\Delta$ equal to side AB. The triangle $AB\Delta$ is isosceles and its angle at Δ is $\frac{\alpha}{2}$. Consequently, Δ sees $B\Gamma$ under a fixed angle and is therefore contained in the circle λ passing through B and Γ (Theorem 2.24). Observe that the center Z of λ belongs to the medial line of chord $B\Gamma$, as well as, to circle κ, because it sees $B\Gamma$ under angle α (double that at Δ). Returning to the problem, the triangle perimeter is $p = |B\Gamma| + |BA| + |A\Gamma| = |B\Gamma| + |\Gamma\Delta|$ and, because $B\Gamma$ is fixed, p is maximized exactly when the line segment $\Gamma\Delta$ becomes maximal. The segment $\Gamma\Delta$ however, being a chord of circle λ, is maximized when it becomes a diameter (Proposition 2.2). If therefore E is the diametrically opposite of Γ on circle λ, from what we said, the triangle maximizes its perimeter when $\Gamma\Delta$ takes the position of ΓE, and therefore when A takes the position of Z. Then, however, the triangle $BZ\Gamma$ is isosceles, since Z belongs to the medial line of $B\Gamma$.

Exercise 2.146. (Archimedes) On the circumscribed circle of the triangle $AB\Gamma$ consider the middle Z of arc $\widehat{BA\Gamma}$ and project this point to point M of the larger of the two sides AB and $A\Gamma$. Show that M divides the broken line $BA\Gamma$ into two parts of equal length, each of which has length $\frac{1}{2}(b+c)$. The segment AN has also the same length, where N is the projection of the diametrically opposite Z' of Z on the larger of AB, $A\Gamma$.

Hint: In figure 2.86-II, the length of the broken line $BA\Gamma$ is equal to the length of $\Gamma\Delta$, which is a chord of circle λ (of the preceding exercise). Therefore Z, which is the center of λ, is projected at the middle of the chord and consequently $|BA| + |AM| = |M\Gamma|$ ([49, p.1]). For the last claim use Exercise 2.122.

Exercise 2.147. Given the circle κ, show that, from all triangles $AB\Gamma$ inscribed in κ, the one which has the maximum perimeter is the equilateral.

Hint: This property is a simple consequence of Exercise 2.145. Indeed, if triangle $AB\Gamma$ is inscribed in κ, has maximum perimeter p and is not equilateral, for example $|AB| < |A\Gamma|$, then, with the help of that exercise, we can find another (isosceles) having the same base $B\Gamma$ and perimeter $p' > p$, contradicting our assumption that p is the maximal possible perimeter.

Exercise 2.148. Given the circle κ, show that from all polygons with ν sides inscribed in κ, the one which has the maximum perimeter is the regular polygon with ν sides.

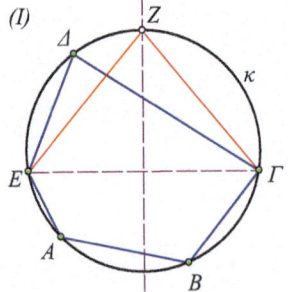
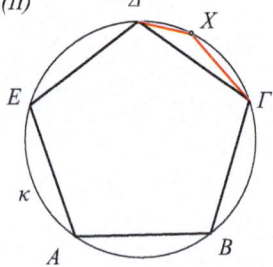

Fig. 2.87: Perimeter maximization   Perimeters of regular polygons

Hint: This exercise is also a simple consequence of Exercise 2.145. Indeed, if the polygon $AB\Gamma\Delta...$ with ν sides is inscribed in κ, has maximum perimeter

p and is not a regular polygon with v sides, then it surely has two consecutive sides unequal for example $\Gamma\Delta$ and ΔE (See Figure 2.87-I). Then, drawing the diagonal ΓE, we form triangle $\Gamma\Delta E$ and according to Exercise 2.145, there is another triangle (isosceles) $\Gamma E Z$ with perimeter greater than $\Gamma E\Delta$. By replacing then the two sides $E\Delta$ and $\Delta\Gamma$ with EZ and $Z\Gamma$, we find a polygon with greater perimeter contradicting the hypothesis, that the initial polygon has maximal perimeter.

Exercise 2.149. Given the circle κ, show that the perimeter of the regular inscribed polygon with $(v+1)$ sides is greater than the perimeter of the regular inscribed polygon with v sides.

Hint: Consider a regular polygon $\Pi = AB\Gamma...$ with v sides inscribed in the circle κ whose perimeter is p. Let X on the circle κ be a point different from the polygon's vertices (See Figure 2.87-II). Suppose that X is contained in the arc defined by the side $\Gamma\Delta$ of the polygon. Replace side $\Gamma\Delta$ with the two sides ΓX and $X\Delta$. The result is a polygon with $v+1$ vertices $\Pi' = AB\Gamma X\Delta...$ and perimeter $p' > p$. According to the preceding exercise, the regular polygon Π'' with $(v+1)$ sides will have perimeter $p'' > p'$. Altogether then, $p'' > p' > p$.

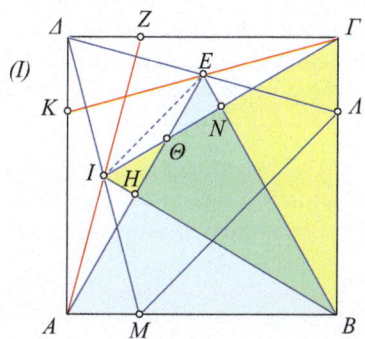 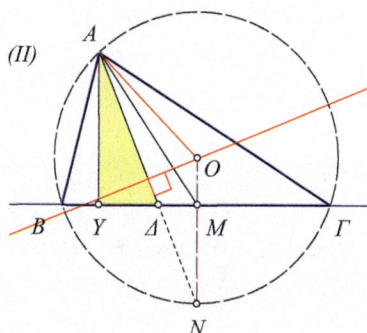

Fig. 2.88: Equilaterals in square   Constr.: altitude+bisector+median

Exercise 2.150. In a square $AB\Gamma\Delta$ we construct on its sides and internally equilaterals ABE and $B\Gamma I$. Show that extending ΔE and ΔI to their intersection with the square creates another equilateral $\Delta\Lambda M$. Investigate figure 2.88-I and find all formed angles in it.

Exercise 2.151. Construct triangle $AB\Gamma$, given the lengths of its altitude $|AY|$, bisector $|A\Delta|$ and median $|AM|$ (See Figure 2.88-II).

Hint: The right triangles $AY\Delta$ and AYM can be constructed directly from the given data. The orthogonal of $Y\Delta$ at M intersects the bisector $A\Delta$ at N, which lies on the circumscribed circle of the requested triangle. It follows that point N can be constructed and the medial line of AN intersects the line

MN at the center O of the circumscribed circle. We draw the circle with center O and radius OA, which passes through A and intersects $Y\Delta$ at points B and Γ defining the required triangle $AB\Gamma$.

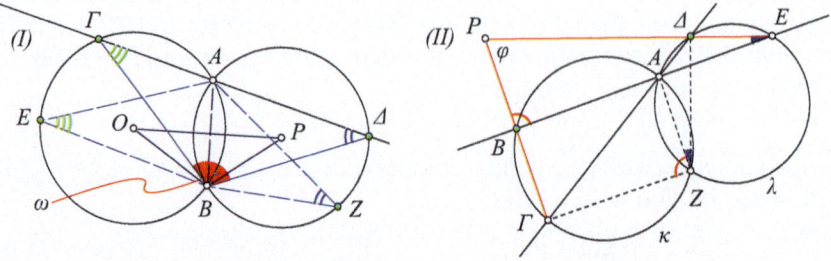

Fig. 2.89: Fixed angles of $B\Gamma\Delta$ — Fixed angle $\phi = 180° - \omega$

Exercise 2.152. Consider two circles intersecting at two points A and B and a line $\Gamma\Delta$ passing through A, and intersecting the circles again at Γ and Δ. Show that the angles of the triangle $B\Gamma\Delta$ are constant and independent of the direction of the line $\Gamma\Delta$. Show also that these angles are respectively equal to the angles of the triangle OBP, where O, P are the centers of the circles (See Figure 2.89-I).

Hint: The angles $\widehat{A\Gamma B}$ and $\widehat{A\Delta B}$ are equal respectively to $\widehat{POB}$ and $\widehat{OPB}$, regardless of the direction of $\Gamma\Delta$.

Exercise 2.153. Through the intersection point A of two circles κ, λ we draw two secants, which define on the circles the chords $B\Gamma$, ΔE. Show that these chords form a constant angle ϕ, which is supplementary to angle ω of the preceding exercise (See Figure 2.89-II).

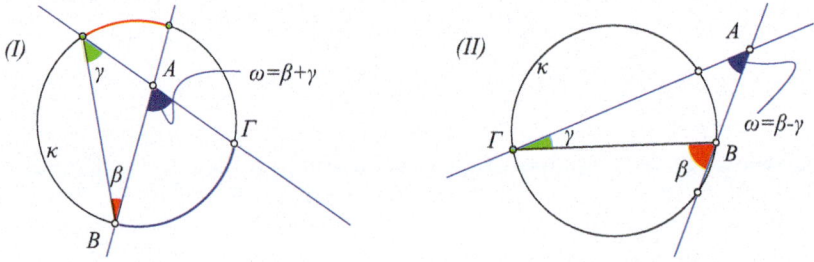

Fig. 2.90: Angle from secants in interior/exterior of circle

Exercise 2.154. Show that the angle $\omega = \widehat{BA\Gamma}$, formed by two secant lines in the interior/exterior of circle κ, is equal to the sum/difference of the two inscribed angles of κ which are defined by the intersection points of the lines with the circle (See Figure 2.90).

2.13. INSCRIBED ANGLES

Exercise 2.155. Given are three points A, B, Γ on the circle κ. Suppose that M, N are the middles of the arcs $\widehat{AB}$ and $\widehat{A\Gamma}$ in the exterior of angle $\widehat{BA\Gamma}$. Show that the chord MN intersects AB and $A\Gamma$ at two points $\{\Delta, E\}$ which are equidistant from A (See Figure 2.91-I).

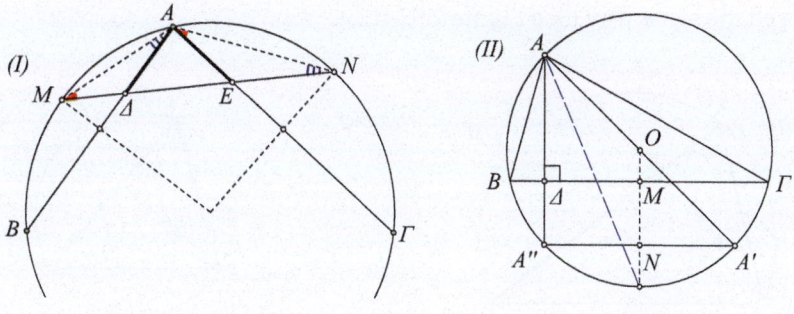

Fig. 2.91:   Isosceles $A\Delta E$ $\qquad$ $A'A''$ parallel of $B\Gamma$

Exercise 2.156. Let O be the circumcenter of the triangle $AB\Gamma$ and point A' be the symmetric of A relative to O and A'' be the intersection point of the altitude $A\Delta$ with the circumscribed circle. Show that $A'A''$ is parallel to $B\Gamma$ and that the angles $\widehat{BA\Gamma}$ and $\widehat{A'AA''}$ have a common angle bisector (See Figure 2.91-II).

Exercise 2.157. Given is a circle κ and two chords of it: $\Gamma\Delta$ fixed and AB variable but of fixed length. Show that the intersection X of the lines $A\Gamma$ and $B\Delta$ describes another circle λ, which passes through the points Γ and Δ of the fixed chord.

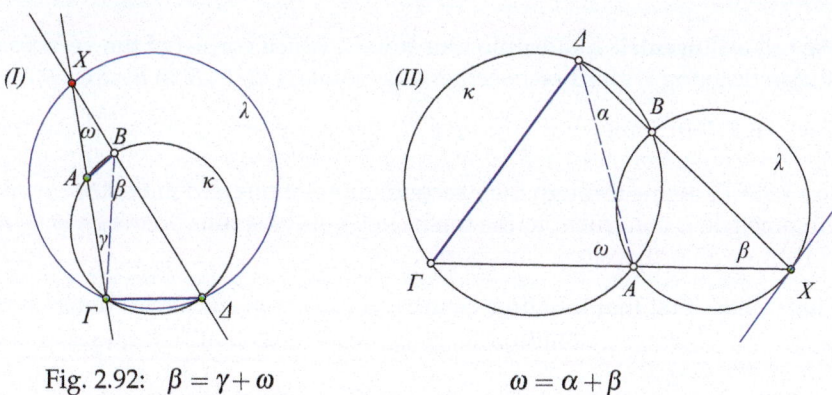

Fig. 2.92: $\beta = \gamma + \omega$ $\qquad$ $\omega = \alpha + \beta$

Hint: Angles β and γ in figure 2.92-I are fixed, as inscribed and subtending arcs of fixed length chords. Angle β is external in the triangle $B\Gamma X$, therefore its measure is equal to the sum of the two opposite internal angles: $\beta = \gamma + \omega$, from which follows that ω is fixed.

Exercise 2.158. Given are two circles κ and λ which intersect at two points A, B. An arbitrary point X of λ defines the second intersections Γ, Δ respectively of XA and XB with circle κ. Show that chord $\Gamma\Delta$ of κ is of fixed length and is parallel to the tangent to λ at X.

Hint: The exercise is a kind of converse for the preceding exercise and is solved the same way (See Figure 2.92-II).

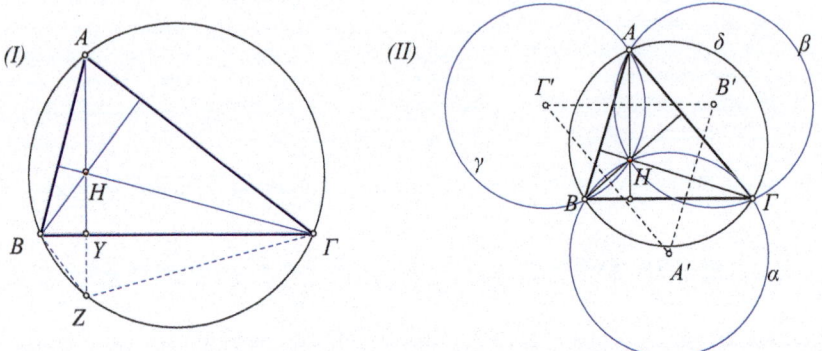

Fig. 2.93: Orthocentric quadruple

Exercise 2.159. Let H be the orthocenter of the triangle $AB\Gamma$. Show that for any triangle τ with vertices three of the points A, B, Γ, H, the remaining fourth point is the orthocenter of τ (See Figure 2.93-I). Also show that the symmetric of H relative to one side belongs to the circumcircle of $AB\Gamma$. Conclude that the circumscribed circles of these triangles are congruent (See Figure 2.93-II).

We call **orthocentric quadruple** four points, which consist of the vertices of a triangle along with its orthocenter, like points A, B, Γ, H in figure 2.93-I.

Exercise 2.160. Somewhat converse to the preceding exercise, show that, three congruent circles $\{\alpha, \beta, \gamma\}$, passing through the same point H (See Figure 2.93-II), define through their second intersections a triangle $AB\Gamma$, whose circumcircle is congruent to the three circles and the four points $\{A, B, \Gamma, H\}$ constitute an orthocentric quadruple.

Hint: Show first that the three centers $\{A', B', \Gamma'\}$ of the circles define a triangle $A'B'\Gamma'$, whose circumcenter is H. Show then that triangles $AB\Gamma$ and $A'B'\Gamma'$ are congruent.

Exercise 2.161. Circle λ is tangent to the circle κ at Γ and to the chord AB of it at Δ (See Figure 2.94-I). Show that line $\Gamma\Delta$ passes through the middle of one of the arcs having endpoints A and B. Also $\Gamma\Delta$ bisects the angle $\widehat{A\Gamma B}$.

2.14. INSCRIPTIBLE OR CYCLIC QUADRILATERALS

Hint: If O and P are respectively the centers of circles κ and λ, define E to be the intersection of κ and the extension of $\Gamma\Delta$. According to Theorem 2.25 angles $\widehat{\Gamma P \Delta}$ and $\widehat{\Gamma O E}$, which are central to the corresponding chords $\Gamma\Delta$ and ΓE, will be double the angle formed by the chord $\Gamma\Delta$ and the tangent ΓX. Therefore, $P\Delta$ and OE will be parallel and, because $P\Delta$ is orthogonal to AB, the same will happen also with OE. However, the orthogonal from the center to a chord passes through the middle of the chord and the middle of the arcs defined by it.

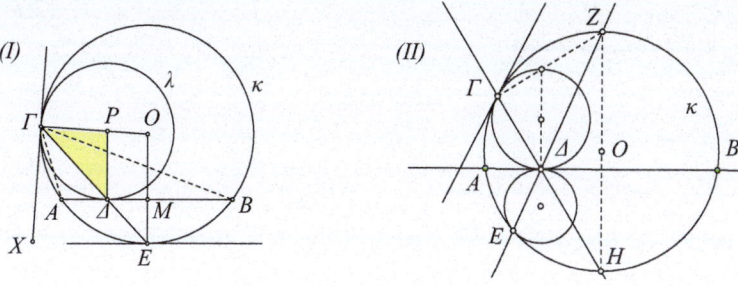

Fig. 2.94: Tangent circles

Exercise 2.162. Two circles are tangent at a point Δ of the chord AB of circle κ and are also tangent to the circle at points Γ, E respectively. Show that $\Gamma\Delta$, $E\Delta$ intersect again circle κ at points H, Z which are diametrically opposite (See Figure 2.94-II).

Exercise 2.163. From a given point A to draw a line ε intersecting from two given and equal circles two equal chords.

2.14 Inscriptible or cyclic quadrilaterals

> ... order and simplification are the first steps toward the mastery of a subject - the actual enemy is the unknown.
>
> Th. Man, *The magic mountain*

Two different points define a line and, for a third point, to be on that line, this is something special. Several theorems have the form of collinearity of three points, which are constructed using a specific recipe from given data. Similarly also, three non collinear points define a circle, and for a fourth point, to be on that circle, this is also something special. Several theorems have the form of the **concyclicity** of four points, which are constructed from the given data using a specific recipe. This game continues as the complexity of the curves we study increases. There are curves defined from exactly v

points and corresponding theorems, which warrant the passing of such a curve through an additional $v + 1$-th point. In this section we'll meet with some characteristics of quadrilaterals, defined from four **concyclic** points.

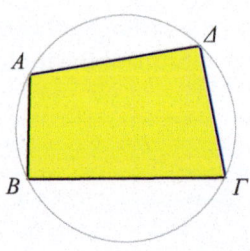 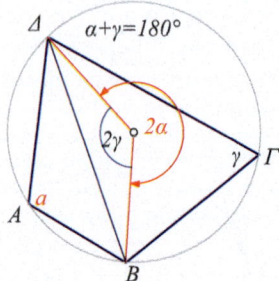

Fig. 2.95: Inscribed quadrilateral ... with opposite angles supplementary

A **cyclic** or **inscribable** or **inscriptible** in a circle quadrilateral is one which can be inscribed in a circle, in other words a quadrilateral for which there is a circle which passes through all four of its vertices (See Figure 2.95). An **inscribed** quadrilateral is also the cyclic quadrilateral which has been drawn inside its circumscribed circle.

Theorem 2.26. *In every convex inscriptible quadrilateral, its opposite angles are supplementary. Conversely, if in a convex quadrilateral, two opposite angles are supplementary, then the quadrilateral is inscriptible (See Figure 2.95).*

Proof. Two opposite angles in an inscriptible quadrilateral, for example $\widehat{A}$ and $\widehat{\Gamma}$ of measures α and γ respectively, see the diagonal $B\Delta$ under supplementary angles, because their corresponding central angles satisfy $2\alpha + 2\gamma = 360°$. Conversely, if the opposite angles of measures α and γ are supplementary, then A belongs to the geometric locus of points which see the segment $B\Delta$ under angle α, which according to Theorem 2.24 is a circle arc and Γ will belong respectively to the complementary arc of the preceding arc relative to the same circle.

Corollary 2.23. *Each angle in an inscriptible convex quadrilateral is equal to its opposite external and conversely, if one angle in a convex quadrilateral is equal to its opposite external then the quadrilateral can be inscribed in a circle (See Figure 2.96-I).*

Proof. The corollary is an immediate consequence of the preceding proposition, because for each quadrilateral angle, its corresponding opposite external is supplementary to the opposite angle.

Theorem 2.27. *In a convex inscriptible quadrilateral, each one of its sides is seen by its other two vertices under equal angles (See Figure 2.96-II). Conversely, if one side of the convex quadrilateral is seen from its two other vertices under equal angles, then the quadrilateral can be inscribed in circle.*

2.14. INSCRIPTIBLE OR CYCLIC QUADRILATERALS

Proof. The proof is again an application of Theorem 2.24.

Remark 2.9. Last theorem holds also for non-convex quadrilaterals. However in the case of non-convex (self-intersecting) inscriptible quadrilaterals opposite angles are equal and not supplementary, as it happens with convex quadrilaterals (See Figure 2.96-III).

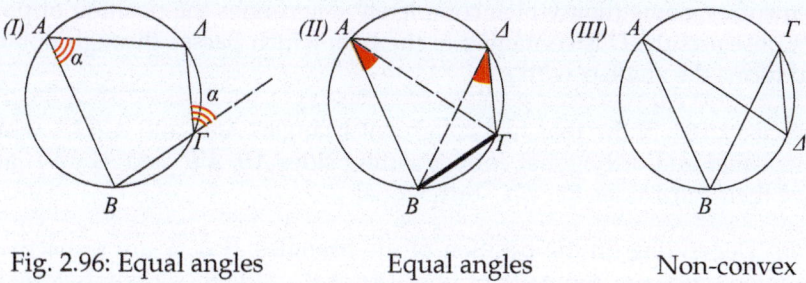

Fig. 2.96: Equal angles　　　　Equal angles　　　　Non-convex

Remark 2.10. It can be proved, that for every quadruple of positive numbers $\{a,b,c,d\}$, such that each of them is less than the sum of the others, there is a cyclic quadrilateral with these side-lengths. The analogous property holds true also for general convex polygons inscribed in the circle. Exercise 5.54 handles the construction of such a quadrilateral. It can be proved ([56]) that, for inscriptible polygons with $n > 4$ sides, the corresponding construction cannot be done with the use of rule and compass.

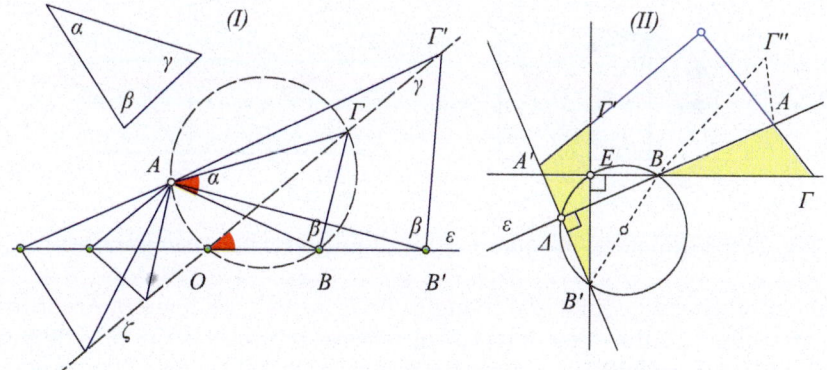

Fig. 2.97: Variable triangle with fixed angles　　　　Rotated triangles

Theorem 2.28. *The triangle $AB\Gamma$ varies in such a way, that the measures of its angles remain fixed, its vertex at A has a fixed position and its vertex B moves along a fixed line ε. Then, its third vertex Γ moves along another fixed line ζ, which forms with ε an angle equal to $\widehat{BA\Gamma}$.*

Proof. Let *ABΓ* be one triangle with the aforementioned properties (See Figure 2.97-I). Consider the circumscribed circle of this triangle and the second intersection point *O* of this circle (different from *B*) with line ε. Because *O* sees *BΓ* under the same angle as *A*, the angle at *O* will be fixed and equal to the triangle angle α (the angles remain fixed, only the dimensions and the position of the triangle change). Because the quadrilateral *AOBΓ* is inscribed in the circle, the angle $\widehat{AOB}$, which is opposite to γ, will be its supplementary, something which completely determines the position of point *O*. Consequently, *Γ* is contained in the line, which passes through point *O* and forms the angle α with ε.

Exercise 2.164. Show that two congruent triangles *ABΓ* and *A'B'Γ'*, which have equal and orthogonal corresponding sides *AB*, *A'B'* and *AΓ*, *A'Γ* also have their third sides *BΓ*, *B'Γ'* orthogonal.

Hint: Depending on the position of the triangles, if Δ, *E* are respectively the intersections of the pairs of lines (*AB,A'B'*), (*BΓ,B'Γ'*), points *B*, *B'*, Δ, *E* form an inscriptible quadrilateral *BB'ΔE*. The proposition is not valid if the triangles are congruent yet have only one side orthogonal to another. In figure 2.97-II this is seen for triangles *A'B'Γ'* and *ABΓ'''*, later being the symmetric of *ABΓ* relative to line ε = *AB*.

Exercise 2.165. Let *P* be a point on the side *AΔ* of the square *ABΓΔ*. Construct an equilateral triangle *PZE*, which has its vertices on the square's sides.

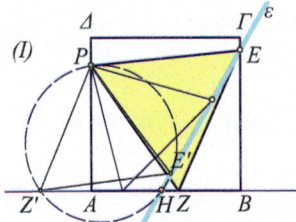 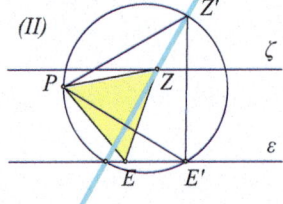

Fig. 2.98: Equilateral in square   Equilateral between parallels

Hint: Consider all equilateral triangles *PZ'E'*, with *Z'* on line *AB*. According to Theorem 2.28, the other vertex *E'* of all these triangles varies on a definite line ε (See Figure 2.98-I). One intersection point *E* of this line and the square, different from the intersection point *H* of ε with *AB*, defines the base of the wanted triangle *PZE*.

Exercise 2.166. Construct an equilateral triangle, which has one vertex at a given point and the other two vertices on given parallel lines (See Figure 2.98-II). Also construct an equilateral triangle which has its three vertices on three given parallel lines.

2.14. INSCRIPTIBLE OR CYCLIC QUADRILATERALS

Exercise 2.167. From one vertex of a regular polygon draw all the diagonals. Show that the angles formed by any two successive diagonals are all equal.

Hint: These are the angles under which the given vertex, which belongs to the circumscribed circle, sees the sides of the polygon (See Figure 2.99-I). Because all sides are equal chords of the circumscribed circle, these angles are all equal between them.

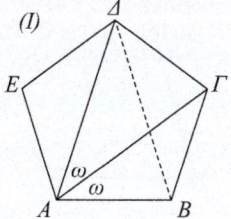 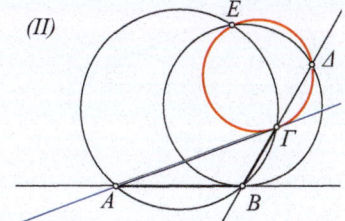

Fig. 2.99: Trapezium of successive sides   Tangent at vertices

Exercise 2.168. Show that three successive sides AB, $B\Gamma$ and $\Gamma\Delta$ of a regular polygon along with the diagonal $A\Delta$ form an isosceles trapezium. Also show that the diagonals $A\Gamma$ and $B\Delta$ of this trapezium bisect the angles adjacent to its base $A\Delta$.

Hint: Angles $\widehat{BA\Gamma}$ and $\widehat{B\Gamma A}$ since $AB\Gamma$ is isosceles (See Figure 2.99-I). From the equality of angles $\widehat{\Delta A\Gamma}$ and $\widehat{A\Gamma B}$ (Exercise 2.167) follows that $A\Delta$ and $B\Gamma$ are parallel.

Exercise 2.169. Given is a triangle $AB\Gamma$ and a point Δ on line $B\Gamma$. Construct the circle which is tangent to AB at B and passes through Δ. Suppose E is the second intersection point of this circle with the circumscribed circle of $AB\Gamma$. Show that the circle $(\Gamma\Delta E)$ is tangent to $A\Gamma$ at Γ.

Hint: $\widehat{E\Delta\Gamma} = \widehat{EBA}$ because of $(E\Delta B)$'s contact to BA at B. Also $\widehat{ABE} = \widehat{A\Gamma E}$ (See Figure 2.99-II). Apply Theorem 2.25 (converse).

Exercise 2.170. Construct a triangle $AB\Gamma$, for which is given the radius ρ of the circumscribed circle and the lengths $v_A = |AY|$ and $\delta_A = |A\Delta|$ of the altitude and angle bisector from A (See Figure 2.100-I).

Exercise 2.171. Construct a *right* triangle, for which are given the traces $\{Y, \Delta, O\}$ of the altitude, of the angle bisector and of the median, all on the line defined by the hypotenuse (See Figure 2.100-I).

Exercise 2.172. Let E be the intersection point of the segments which join the middles of opposite sides of a convex inscriptible quadrilateral $AB\Gamma\Delta$ and point Σ, which is the symmetric of the center O of the circumscribed circle relative to E. Show that the line which joins the middle M of one side with Σ is orthogonal to its opposite side (See Figure 2.100-II).

From the preceding exercise follows that point Σ is contained in the four orthogonals from the middles of the sides to respective opposite sides of a cyclic quadrilateral (See Figure 2.100-II). This point is called **anticenter** of the inscriptible quadrilateral.

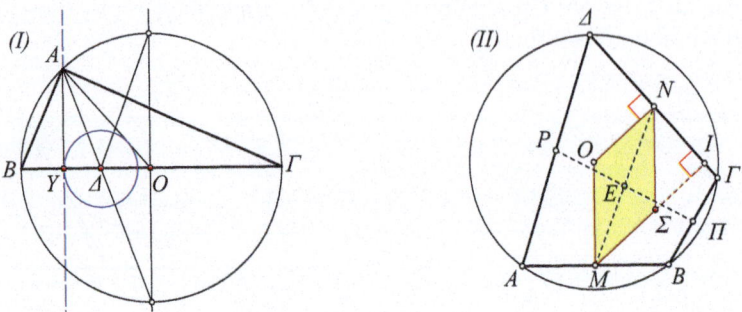

Fig. 2.100: Right triangle from {Y, Δ, O}   Anticenter Σ

Exercise 2.173. Construct a square $AB\Gamma\Delta$, whose sides pass respectively through the vertices of a convex quadrilateral $\Theta IK\Lambda$ (See Figure 2.101-I).

Hint: If Δ sees $K\Lambda$ under a right angle, then the middle E of the semicircle with diameter $K\Lambda$, which does not contain Δ, lies on the diagonal of the square. A similar property holds also for the middle Z of the semicircle with diameter ΘI. Points E, Z are constructible from the given data and determine the diagonal $B\Delta$ of the requested square.

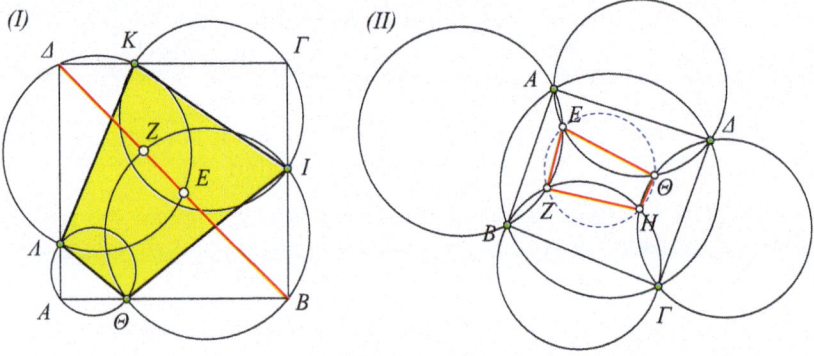

Fig. 2.101: Circumscribed square   Second points of intersection of circles

Exercise 2.174. Show that the second intersection points of circles, which have for chords the sides of an inscriptible quadrilateral $AB\Gamma\Delta$ define also an inscriptible quadrilateral $EZH\Theta$ (See Figure 2.101-II).

2.14. INSCRIPTIBLE OR CYCLIC QUADRILATERALS

Exercise 2.175. Show that triangle ΔEZ, with vertices the traces of the altitudes of the triangle $AB\Gamma$, has the corresponding altitudes as angle bisectors.

Hint: The quadrilateral $B\Delta HZ$ has the opposite lying angles at Δ and Z right (See Figure 2.102-I), therefore it is inscriptible in a circle. It follows that angles $\widehat{ZBH}$ and $\widehat{Z\Delta H}$ are equal. Similarly the angles $\widehat{E\Delta H}$ and $\widehat{E\Gamma H}$ are equal. However $\widehat{E\Gamma H}$ and $\widehat{ZBH}$ are also equal because ΓBZE can also be inscribed, since E and Z see $B\Gamma$ under a right angle.

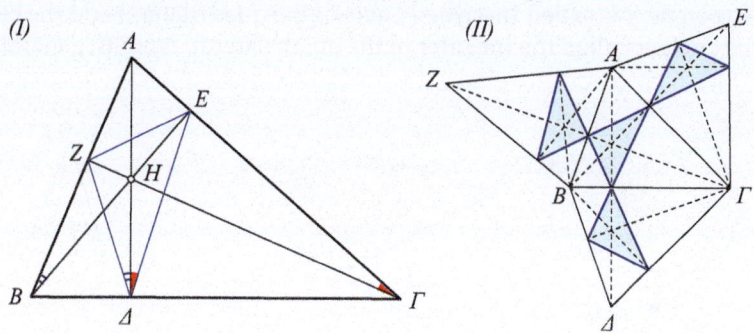

Fig. 2.102: ΔEZ : orthic of $AB\Gamma$ Reflections of the orthic

The triangle ΔEZ defined by the altitude traces of $AB\Gamma$ (See Figure 2.102-I), is called **orthic triangle** of $AB\Gamma$. It can be proved (Theorem 5.24) that for acute-angled triangle ABC, this triangle is of minimal perimeter among all triangles which can be inscribed in $AB\Gamma$, i.e. triangles which have their vertices on the sides of $AB\Gamma$. The orthic triangle of an acute triangle $AB\Gamma$ is also the unique trajectory of a billiard ball which bounces exactly once against each side of a triangular billiard which has the shape of $AB\Gamma$ [43, II, p.213].

Exercise 2.176. On the sides of triangle $AB\Gamma$ we construct congruent to it triangles with vertices Δ, E, Z, the symmetric points of triangle $AB\Gamma$ relative to its opposite sides. Next we construct the orthic triangles of all four triangles (See Figure 2.102-II). Show that the sides of the orthic triangles, taken by three, belong to the same line.

Exercise 2.177. Show that the angle bisectors of pairs of opposite sides in a convex inscriptible quadrilateral intersect orthogonally.

Exercise 2.178. Show that if a triangle $AB\Gamma$ has the property that $\widehat{YAM} = \beta - \gamma$, where $\{AY, AM\}$ are the altitude and the median from A, then it is a right-angled one.

Hint: Show that the bisector $A\Delta$ of α is also bisector of angle $\widehat{YAM}$. Show that triangle AXM is isosceles, where X the intersection point of the circumcircle with the bisector $A\Delta$.

2.15 Circumscribed quadrilaterals

> To appreciate, you will prefer somebody specific.
> When you appreciate everyone, you appreciate nobody.
>
> *Molière, the Misanthrope*

Circumscribed or **circumscriptible** is called a quadrilateral for which there exists a circle κ, such that the sides of the quadrilateral are tangent segments to κ. The circle κ is called **inscribed** circle of the quadrilateral and the center O of the circle is called the **incenter** of the quadrilateral. As with general and

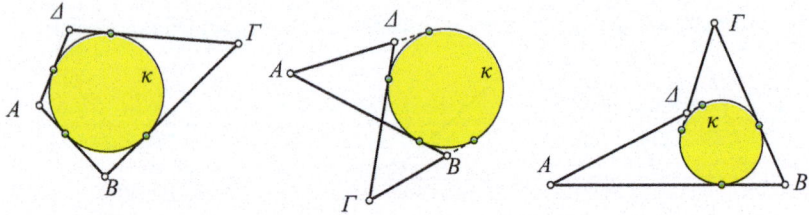

Fig. 2.103: Circumscribed quadrilaterals

inscriptible quadrilaterals, so with circumscribed ones, there are three kinds: the convex, the non-convex and the self-intersecting (See Figure 2.103). In these lessons we'll restrict ourselves to the study of the convex ones, for which next theorem provides a characterization.

Theorem 2.29. *A convex quadrilateral is circumscriptible, if and only if the bisectors of its angles pass through a common point O, which is the center of the inscribed circle (See Figure 2.104-I).*

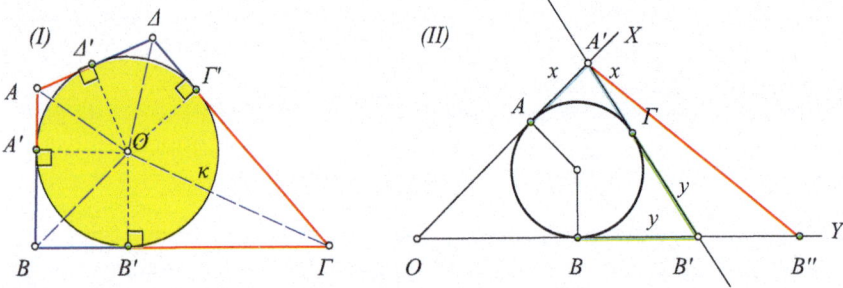

Fig. 2.104: Circumscribed quadrilateral   Characterizing tangent $A'B'$

Proof. If the quadrilateral is circumscriptible, then the center O of the inscribed circle κ has distance from its sides equal to the radius r of the circle,

2.15. CIRCUMSCRIBED QUADRILATERALS

therefore it is found on the bisectors of all its angles. Conversely, if all the bisectors intersect at point O, then all distances of O from the sides of the quadrilateral are equal to r and the sides are tangent to the circle $\kappa(O,r)$.

Next proposition supports the proof of another characteristic property of circumscribed quadrilaterals.

Proposition 2.12. *A circle is tangent to an angle $\widehat{XOY}$ at the points A of OX and B of OY. Then, for two other points A', B' on the half lines OX and OY respectively, the line segment $A'B'$ is also tangent to the circle, if and only if $|A'B'| = |AA'| + |BB'|$ (See Figure 2.104-II).*

Proof. The fact that the above relation for tangents is satisfied follows easily (Corollary 2.14). Conversely, let us suppose that the relation $|A'B''| = |AA'| + |BB''|$ holds, yet $A'B''$ is not a tangent. We'll show that this is contradictory. Indeed, we draw then the tangent $A'B'$ and let us suppose that B' is between B and B''. From the hypothesis we then have $|A'B''| = |AA'| + |BB''| = |AA'| + |BB'| + |B'B''|$. From the first part of the proof however, we have that $|AA'| + |BB'| = |A'B'|$, from which, by substituting it into the preceding equation, we get $|A'B''| = |A'B'| + |B'B''| > |A'B''|$, because of the triangle inequality. A similar contradiction results if we suppose that B'' is between B and B'.

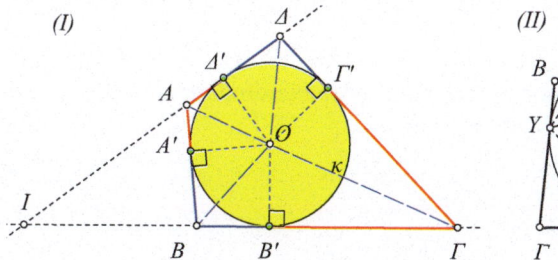
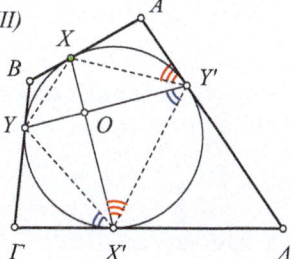

Fig. 2.105: Characterizing a circumscribed   Bicentric quadrilateral

Theorem 2.30. *A convex quadrilateral $AB\Gamma\Delta$ is circumscriptible to a circle, if and only if the sum of the lengths of two opposite sides is equal to the corresponding sum of the two other sides (See Figure 2.105-I).*

Proof. If the quadrilateral is circumscriptible to a circle, then the theorem's property follows directly from Corollary 2.14. Conversely, if the property holds, we consider the circle $\kappa(O,r)$, which is tangent to the quadrilateral's three sides ΔA, AB, $B\Gamma$ and has center O the point of intersection of the bisectors at A and B. Then (See Figure 2.105-I), from the hypothesis $|A\Delta| + |B\Gamma| = |AB| + |\Gamma\Delta|$ and from $|A\Delta'| = |AA'|$, $|BB'| = |BA'|$ therefore also $|AB| = |A\Delta'| + |BB'|$, follows

$$\begin{aligned}|\Delta\Gamma| &= |A\Delta|+|B\Gamma|-|AB|\\ &= |A\Delta|+|B\Gamma|-(|A\Delta'|+|BB'|)\\ &= |A\Delta|-|A\Delta'|+|B\Gamma|-|BB'|\\ &= |\Delta\Delta'|+|B'\Gamma|.\end{aligned}$$

Consequently, by applying the preceding proposition, we conclude that $\Gamma\Delta$ is as well a tangent to the circle κ ([48, p.24]).

Exercise 2.179. The vertex O of the right angle $\widehat{XOY}$ is an interior point of a circle, and its sides intersect the circle at points X, X', Y, Y'. Drawing tangents at these points forms the circumscribed to the circle quadrilateral $AB\Gamma\Delta$ (See Figure 2.105-II). Show that this is a cyclic quadrilateral. Show also the inverse property: if the sides of a cyclic quadrilateral $AB\Gamma\Delta$ are tangent to another circle, then the lines $\{XX', YY'\}$ joining opposite contact points are orthogonal.

Quadrilaterals like $AB\Gamma\Delta$ of the preceding exercise, which are simultaneously circumscriptible to a circle κ and inscriptible in another circle κ', are called **bicentric**.

Exercise 2.180. Circle κ' is tangent to the circumcircle κ of triangle $AB\Gamma$ at the vertex A and intersects the opposite side at two points $\{\Delta, E\}$. Show that angles $\widehat{BA\Delta} = \widehat{EA\Gamma}$.

Exercise 2.181. Extend the sides $\{AB, A\Gamma\}$ of the triangle $AB\Gamma$ to $\{AB', A\Gamma'\}$, so that $B'\Gamma'$ is parallel to the tangent at A of the circumcircle of the triangle. Show that $B\Gamma\Gamma'B'$ is a cyclic quadrilateral.

2.16 Geometric loci

> She did her work with the thoroughness of a mind that reveres details and never quite understands them.
>
> Sinclair Lewis, Babbitt, ch. 18

We already encountered the notion of **geometric loci** in the course of the lesson (Corollary 1.7 and subsequent remark, Corollary 1.36, Exercise 2.6). It refers to sets of points, which are characterized by some property. One of the simplest kinds is the circle $\kappa(O,\rho)$. It is characterized as the geometric locus of points, which are at distance ρ from the fixed point O. Similarly the medial line of a line segment AB: it is the geometric locus of points, which are equidistant from the endpoints A and B of the segment. The simplest geometric loci end up being lines and circles. There also exist, however, simple properties which lead to new curves (for example conic sections II-§ 6.1). In order to find a geometric locus we often use some special points, which we recognize as belonging to it or are related directly to it. Next exercise gives one such example.

Exercise 2.182. Given a circle $\kappa(O,\rho)$ and a line segment AB, from every point Γ of κ we draw a line segment $\Gamma\Delta$ parallel equal and equal oriented to AB. Find the geometric locus of point Δ (See Figure 2.106).

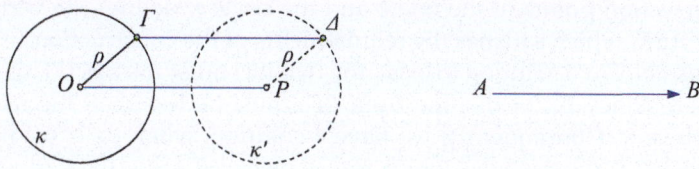

Fig. 2.106: Parallel translation of a circle

Hint: Let P be the point such that OP is parallel, equal and equal oriented to AB. Then $OP\Delta\Gamma$ is a parallelogram, therefore its length $|P\Delta| = \rho$ is a fixed constant and equal to the radius of the given circle. Consequently, the points of the requested locus are contained in the circle $\kappa'(P,\rho)$. Conversely, we show that every point of the circle $\kappa'(P,\rho)$ is a point of the locus. For this, starting from point Δ of κ', we find the point Γ on κ, so that $\Gamma\Delta$ is parallel, equal and equal oriented to AB. This shows that Δ is a point of the locus and completes the proof of coincidence of the circle $\kappa'(P,\rho)$ with the requested geometric locus.

An important application of geometric loci is their use to constructions. Therein we seek a point, which we prove that it is contained in two geometric loci *simultaneously*. The consequence is that the requested point coincides with an intersection point of two geometric loci. The next construction gives an example of this method.

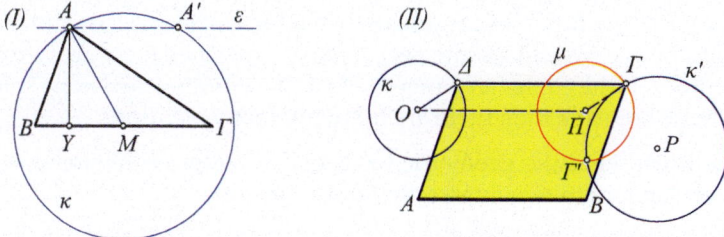

Fig. 2.107: Triangle from $\{a, \upsilon_A, \mu_A\}$ Paralleogram construction

Construction 2.11 *Construct a triangle $AB\Gamma$ from the side $a = |B\Gamma|$, the altitude $\upsilon_A = |AY|$ and the median $\mu_A = |AM|$ (See Figure 2.107-I).*

Construction: Analysis: suppose that the requested triangle has been constructed. The vertex A will be at distance v_A from the base $B\Gamma$, therefore it will be a point on one of the parallel lines ε to $B\Gamma$ at distance v_A from it. Also vertex A will be at distance μ_A from the middle M of $B\Gamma$, therefore it will be a point of the circle $\kappa(M, \mu_A)$. Consequently A will be an intersection point of these two geometric loci.

Synthesis: We construct a line segment $B\Gamma$ of length a and we draw a line ε, parallel to it at distance v_A. We also construct the circle $\kappa(M, \mu_A)$. One of the intersection points of the line ε and the circle κ defines the vertex A of triangle $AB\Gamma$, which satisfies the requirements of the construction.

Discussion: For a solution to exist, the two loci must intersect. This means, that the distance v_A, of the line from $B\Gamma$, must be less than the radius μ_A of the circle κ. Consequently, we have a solution, when $\mu_A \geq v_A$. It is easily proved that the triangles which result, by considering the various intersection points of the two loci, are congruent. Consequently, there is, up to congruence, exactly one triangle, constructible from the given data.

Exercise 2.183. Given is the position and size of a line segment AB and two circles $\kappa(O, \rho)$ and $\kappa'(P, \rho')$. Construct a parallelogram $AB\Gamma\Delta$, whose one side is AB and its vertices Δ and Γ are points of circles κ and κ' respectively.

Hint: Analysis: Suppose that the requested parallelogram $AB\Gamma\Delta$ has been constructed (See Figure 2.107-II). Its side $\Delta\Gamma$ will be parallel, equal and equal oriented to AB, therefore the endpoint Γ will be contained in a known geometric locus defined by the circle $\mu(\Pi, \rho)$ (Exercise 2.182), so that $O\Pi$ is parallel, equal and equal oriented to AB. Γ therefore will be an intersection point of the given circle $\kappa'(P, \rho')$ and of the circle $\mu(\Pi, \rho)$, which can be constructed from the given data.

Synthesis: We construct the circle $\mu(\Pi, \rho)$, as mentioned in the analysis. Let Γ be one point of intersection of the circles $\mu(\Pi, \rho)$ and $\kappa'(P, \rho')$. We locate the point Δ on the circle κ, so that $O\Pi\Gamma\Delta$ is a parallelogram. Then $AB\Gamma\Delta$ is also a parallelogram, which satisfies the requirements of the construction.

Discussion: A solution exists exactly when the circles $\mu(\Pi, \rho)$ and $\kappa'(P, \rho')$ intersect. If these two circles intersect at two points, then there are two different solutions. If the circles are tangent, then there exists exactly one solution. If the circles coincide then there are infinitely many solutions.

Exercise 2.184. Find the geometric locus of the centers K of circles $\kappa(K, \rho)$, which are tangent to two intersecting lines ε and ε'.

Exercise 2.185. Point M varies on a circle with diameter AB. The line AM intersects the medial line of AB at Γ. Find the geometric locus of the circumcenter K of triangle $OM\Gamma$ as well as the geometric locus of the orthocenter H of this triangle (See Figure 2.108-I).

Hint: The circumcircle of triangle $OM\Gamma$ passes through B. This follows from the fact that the angles $\widehat{ABM}$ and $\widehat{A\Gamma O}$ are equal, as complementary to angle

2.16. GEOMETRIC LOCI

$\widehat{BA\Gamma}$. Therefore the center K of circle $(OM\Gamma)$ will be found on the medial line of the segment OB. The orthocenter H of $OM\Gamma$ coincides with the intersection of the orthogonal from M to $O\Gamma$ and the orthogonal from O to $A\Gamma$. Therefore in quadrilateral $OBMH$ opposite sides are parallel and it is a parallelogram. This implies that MH is parallel, equal and equal oriented to BO, therefore the geometric locus of H is the circle with center A and radius equal to $\frac{|AB|}{2}$ (Exercise 2.182).

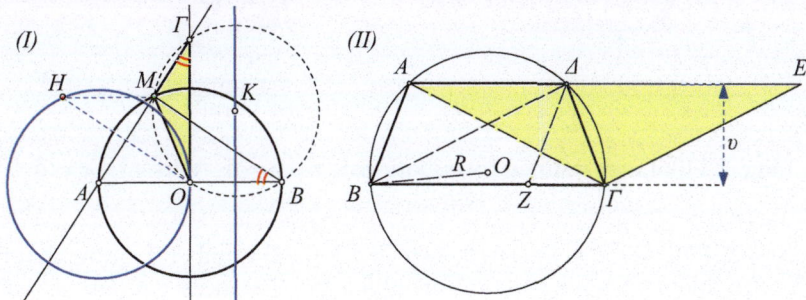

Fig. 2.108: Circumcenter/orthocenter locus Trapezium construction

Exercise 2.186. Construct a trapezium inscribed in a given circle $\kappa(O,R)$, having given altitude υ and given sum/difference of bases (See Figure 2.108-II).

Exercise 2.187. On the sides OX, OY of the angle $\widehat{XOY}$ are defined variable points A and B respectively, such that $|OA|+|OB| = \lambda$ is fixed. Find the geometric locus of the middle I of the line segment AB (See Figure 2.109).

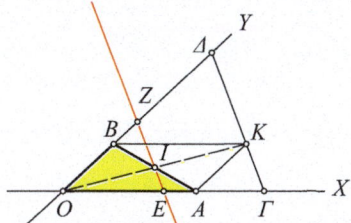

Fig. 2.109: Geometric locus with fixed $|OA|+|OB|$

Hint: Consider the symmetric K of O relative to the middle I of AB. In the quadrilateral $OAKB$ the diagonals AB and OK are bisected at I, therefore it is a parallelogram. Define on OX, OY respectively the points Γ and Δ, so that $|O\Gamma|=|O\Delta|=\lambda$. It follows then that the triangles ΓAK and $KB\Delta$ are isosceli and the points Δ, K and Γ are collinear on a line which can be constructed from the given data. The conclusion is that the middle I of AB, which is also the middle of OK, is contained in line EZ which joins the middles E, Z of the sides of the isosceles $\Gamma O\Delta$.

Exercise 2.188. Construct a trapezium for which are given the lengths of three sides and the angle of its non-parallel sides.

Fig. 2.110: Lines defined by equations: $x - y = \delta$ and $x + y = \delta$

Exercise 2.189. Given is the angle $\widehat{XOY}$ and a fixed number δ. Show that the geometric locus of the points P, for which the distances $x = |PA|$, $y = |PB|$ from OX and OY respectively, satisfy the equations (See Figure 2.110)

$$x - y = \delta, \quad x + y = \delta,$$

are respectively parts of two lines, which are parallel to the bisectors of the angle $\widehat{XOY}$. Find the intersection points of these lines with the sides (or their extensions) of angle $\widehat{XOY}$. What are the positions of these lines when $\delta = 0$?

Exercise 2.190. Construct a point X, which is at distance δ from a given line ε and is also equidistant from two other given lines $\{\varepsilon', \varepsilon''\}$. The same problem to solve when ε is a circle. In both cases investigate the conditions for the existence of solution.

2.17 Comments and exercises for the chapter

> tam diu discendum est, quum diu nescias, et, si proverbio credimus, quam diu vivas: you must continue learning as long as you do not know, and, if we believe the proverb, as long as you live.
>
> *Seneca*

In the following, as usual, (α, β, γ), (a, b, c), $(\mu_A, \mu_B, \mu_\Gamma)$, $(\upsilon_A, \upsilon_B, \upsilon_\Gamma)$ and $(\delta_A, \delta_B, \delta_\Gamma)$ denote respectively the measures/lengths of angles, sides, medians, altitudes and bisectors of the triangle $AB\Gamma$. One problem which comes up naturally in the current stage of the study is the one of *constructing triangles from three of their elements*. The following lists give all the possible problems which may result by combining three of the preceding elements [18].

2.17. COMMENTS AND EXERCISES FOR THE CHAPTER

A (*) means that the construction is not possible with exclusive use of ruler and compass.

A. Constructions containing at least two sides
1. (a,b,c) basic from three sides.
2. $(a,b,\alpha),(a,b,\gamma)$ basic from two sides and one angle.
3. $(a,b,\mu_A),(a,b,\mu_\Gamma)$ two sides, one median.
4. $(a,b,\upsilon_A),(a,b,\upsilon_\Gamma)$ two sides, one altitude.
5. $(a,b,\delta_A)^*,(a,b,\delta_\Gamma)$ two sides, one bisector.

B. Constructions containing exactly one side
1. (a,β,γ) basic from side+angles.
2. $(a,\alpha,\mu_A),(a,\alpha,\mu_B),(a,\beta,\mu_A),(a,\beta,\mu_B),(a,\beta,\mu_\Gamma)$ side+ angle+ median.
3. $(a,\alpha,\upsilon_A),(a,\alpha,\upsilon_B),(a,\beta,\upsilon_A),(a,\beta,\upsilon_B),(a,\beta,\upsilon_\Gamma)$ side+ angle+ altitude.
4. $(a,\alpha,\delta_A),(a,\alpha,\delta_B)^*,(a,\beta,\delta_A)^*,(a,\beta,\delta_B),(a,\beta,\delta_\Gamma)$ side+angle+ bisector.
5. $(a,\mu_A,\mu_B),(a,\mu_B,\mu_\Gamma)$ side + two medians.
6. $(a,\mu_A,\upsilon_A),(a,\mu_A,\upsilon_B),(a,\mu_B,\upsilon_A),(a,\mu_B,\upsilon_B),(a,\mu_B,\upsilon_\Gamma)$ side+ med.+ alt.
7. $(a,\upsilon_A,\upsilon_B),(a,\upsilon_B,\upsilon_\Gamma)$ side + two altitudes.
8. $(a,\mu_A,\delta_A),(a,\mu_A,\delta_B)^*,(a,\mu_B,\delta_A)^*,(a,\mu_B,\delta_B)^*,(a,\mu_B,\delta_\Gamma)^*$ side+ med+bis.
9. $(a,\upsilon_A,\delta_A),(a,\upsilon_A,\delta_B)^*,(a,\upsilon_B,\delta_A)^*,(a,\upsilon_B,\delta_B),(a,\upsilon_B,\delta_\Gamma)$ side+ alt.+bis.
10. $(a,\delta_A,\delta_B)^*,(a,\delta_B,\delta_\Gamma)^*$ side + two bisectors.

C. Constructions containing at least two angles
1. (α,β,γ) from three angles (similarity).
2. $(\alpha,\beta,\mu_A),(\alpha,\beta,\mu_\Gamma)$ two angles, one median.
3. $(\alpha,\beta,\upsilon_A),(\alpha,\beta,\upsilon_\Gamma)$ two angles, one altitude.
4. $(\alpha,\beta,\delta_A),(\alpha,\beta,\delta_\Gamma)$ two angles, one bisector.

D. Constructions containing exactly one angle and secondary elements
1. $(\alpha,\mu_A,\mu_B),(\alpha,\mu_B,\mu_\Gamma)$ angle + two medians.
2. $(\alpha,\mu_A,\upsilon_A),(\alpha,\mu_A,\upsilon_B),(\alpha,\mu_B,\upsilon_A),(\alpha,\mu_B,\upsilon_B),(\alpha,\mu_B,\upsilon_\Gamma)$ ang.+med.+alt.
3. $(\alpha,\mu_A,\delta_A),(\alpha,\mu_A,\delta_B)^*,(\alpha,\mu_B,\delta_A)^*,(\alpha,\mu_B,\delta_B)^*,(\alpha,\mu_B,\delta_\Gamma)^*$ ang.+med.+bis.
4. $(\alpha,\upsilon_A,\upsilon_B),(\alpha,\upsilon_B,\upsilon_\Gamma)$ angle + two altitudes.
5. $(\alpha,\upsilon_A,\delta_A),(\alpha,\upsilon_A,\delta_B)^*,(\alpha,\upsilon_B,\delta_A),(\alpha,\upsilon_B,\delta_B),(\alpha,\upsilon_B,\delta_\Gamma)^*$ ang.+ alt.+ bis.
6. $(\alpha,\delta_A,\delta_B)^*,(\alpha,\delta_B,\delta_\Gamma)^*$ angle+ two bisectors.

E. Constructions from only secondary elements
1. (μ_A,μ_B,μ_Γ) three medians.
2. $(\mu_A,\mu_B,\upsilon_A),(\mu_A,\mu_B,\upsilon_\Gamma),(\mu_A,\mu_B,\delta_A)^*,(\mu_A,\mu_B,\delta_\Gamma)^*$ two medians
3. $(\mu_A,\upsilon_A,\upsilon_B),(\mu_A,\upsilon_B,\upsilon_\Gamma)$ median+ 2 altitudes.
4. $(\mu_A,\upsilon_A,\delta_A),(\mu_A,\upsilon_A,\delta_B)^*,(\mu_A,\upsilon_B,\delta_A)^*,(\mu_A,\upsilon_B,\delta_B)^*,(\mu_A,\upsilon_B,\delta_\Gamma)^*$.
5. $(\mu_A,\delta_A,\delta_B)^*,(\mu_A,\delta_B,\delta_\Gamma)^*$ median+two bisectors.
6. $(\upsilon_A,\upsilon_B,\upsilon_\Gamma)$ three altitudes.
7. $(\upsilon_A,\upsilon_B,\delta_A)^*,(\upsilon_A,\upsilon_B,\delta_\Gamma)$ two altitudes+ bisector.
8. $(\upsilon_A,\delta_A,\delta_B)^*,(\upsilon_A,\delta_B,\delta_\Gamma)^*$ altitude+two bisectors.
9. $(\delta_A,\delta_B,\delta_\Gamma)^*$ three bisectors.

In these listings, the constructions characterized *basic* follow directly from the axioms and have already been examined. Handled also have been slightly more complex constructions, like (a, b, μ_Γ) (Exercise 2.26), (a, μ_A, v_A) (Construction 2.11) and (μ_A, v_A, δ_A) (Construction 2.151). All the constructions of list (A), with the exception of the last two, require only knowledge which has been acquired so far. Also easy and accessible, using the acquired knowledge so far, are the constructions of list (C). From the rest of those which are constructible with ruler and compass, the difficulty varies from very simple (for example (a, β, μ_A)) to difficult $((\alpha, \mu_A, v_A))$. The knowledge needed to deal with them is contained in the first four chapters of the book. As we proceed in them, it is beneficial and useful for the learner to return to these lists and to try his powers by selecting some constructions, either randomly, or redoing those, where he has failed in the past.

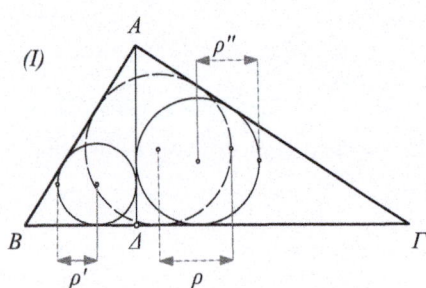
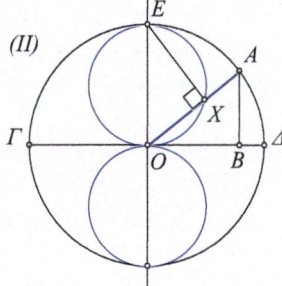

Fig. 2.111: $\rho + \rho' + \rho'' = |A\Delta|$ Point X with $|OX| = |AB|$

Exercise 2.191. A right triangle $AB\Gamma$ is divided by its altitude $A\Delta$ into two other right triangles (See Figure 2.111-I). Show that the sum of the radii of the inscribed circles of the three right triangles equals the length $|A\Delta|$.

Exercise 2.192. A variable point A, of the circle $O(\rho)$, is projected at point B of a fixed diameter $\Gamma\Delta$ (See Figure 2.111-II). Find the geometric locus of the point X of the radius OA for which $|OX| = |AB|$.

Exercise 2.193. Construct a quadrilateral for which are given the lengths of the sides and whose one diagonal bisects one of the adjacent to it angles.

Hint: (See Figure 2.112-I) If $A\Gamma$ bisects the angle at Γ, consider the symmetric E of B relative to $A\Gamma$. Triangle $AE\Delta$ can be constructed. The problem has infinitely many solutions if the points E, Δ coincide.

Exercise 2.194. Given is a convex quadrilateral $AB\Gamma\Delta$. Construct non intersecting circles passing through the pairs of points (A, Γ) and (B, Δ) respectively (See Figure 2.112-II) ([46, p.58]).

Exercise 2.195. To find the geometric locus of the orthocenter and the incenter of a triangle $AB\Gamma$, whose base $B\Gamma$ remains fixed and its angle α has constant measure.

2.17. COMMENTS AND EXERCISES FOR THE CHAPTER

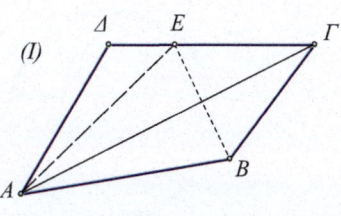
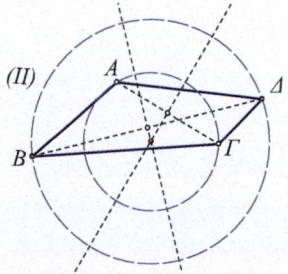

Fig. 2.112: Quadrilateral Construction   Non intersecting circles

Exercise 2.196. Enumerate the vertices of a convex polygon with even number of sides inscribed in a circle. Show that the sum of angles of vertices with even numbers is equal to the sum of angles of vertices with odd numbers.

Exercise 2.197. In the rectangle $AB\Gamma\Delta$ we draw two parallels to one of its diagonals $B\Delta$, at equal distance δ from it. Show that this forms an inscribed parallelogram $EZH\Theta$ with perimeter independent of δ and equal to $2|B\Delta|$ (See Figure 2.113-I).

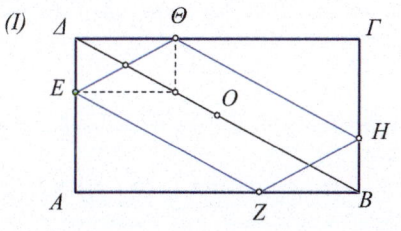
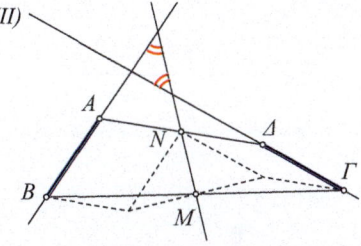

Fig. 2.113: Inscribed parallelogram   Sides $|AB| = |\Delta\Gamma|$

Exercise 2.198. A quadrilateral $AB\Gamma\Delta$ has two opposite sides equal. Show that the line MN of the middles of the unequal sides forms equal angles with the equal sides (See Figure 2.113-II).

Exercise 2.199. Show that the orthogonal from a vertex Γ of a regular polygon to one of its diagonals AB passes through another vertex of the polygon, if the polygon has an even number of sides (See Figure 2.114-I). If the polygon has an odd number of sides then the orthogonal passes through the middle E of the arc of its circumscribed circle defined from one of its sides ZH (See Figure 2.114-II).

Hint: For an even number $2n$, of sides, holds (See Figure 2.114-I):

$$\widehat{\Delta\Gamma B} = 90° - \widehat{\Delta B\Gamma} = 90° - \frac{k}{n} \cdot 90° = (n-k) \cdot \frac{90°}{n},$$

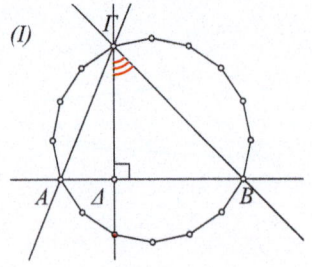 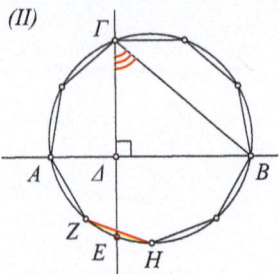

Fig. 2.114: Orthogonals to diagonals of regular polygons

where k is an integer. The proof for an odd number of sides is similar.

Exercise 2.200. Construct a right triangle from the length of the hypotenuse $a = |B\Gamma|$ and the difference of the angles $|\beta - \gamma|$.

Hint: Draw $|BA'| = |A\Gamma|$ (See Figure 2.115-I). The isosceles trapezium $AA'\Gamma B$ is constructible from the given data.

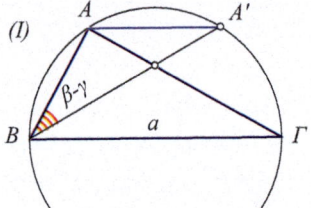 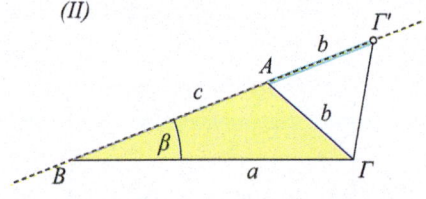

Fig. 2.115: Right from $\{a, |\beta - \gamma|\}$ — Triangle from $\{a, b+c, \beta\}$

Exercise 2.201. Construct a triangle from the lengths $a = |B\Gamma|$, $b+c = |A\Gamma| + |AB|$ and the angle at B.

Hint: In the extension of BA towards A consider Γ', so that $|A\Gamma'| = |A\Gamma|$ (See Figure 2.115-II). The triangle $B\Gamma\Gamma'$ is constructible.

Exercise 2.202. Show that in a triangle $AB\Gamma$, whose angle at A is known, the lengths $b+c$ and the sum of the altitudes $\upsilon_B + \upsilon_\Gamma$ are determined the one from the other.

Hint: From Γ draw a parallel $\Gamma\Gamma''$ to the side BA of length υ_Γ and determine point Γ'' (See Figure 2.116-I). From Γ'' draw a parallel to $A\Gamma$, intersecting BA at A'. Show that $A'\Gamma$ is a bisector of the angle at A' and therefore $b = |A\Gamma| = |AA'|$ and $|BA'| = b+c$. The two lengths to be related are sides of the right triangle $B\Delta A'$ with known angles.

Exercise 2.203. Show that in triangle $AB\Gamma$, whose angle at A is known, the lengths $|b-c|$ and $|\upsilon_B - \upsilon_\Gamma|$ are determined the one from the other (See Figure 2.116-II).

2.17. COMMENTS AND EXERCISES FOR THE CHAPTER

Exercise 2.204. Construct a triangle from its elements $(a = |B\Gamma|, \alpha, v_B + v_\Gamma)$, as well as from the elements $(\alpha, |a-b|, v_B + v_\Gamma)$.

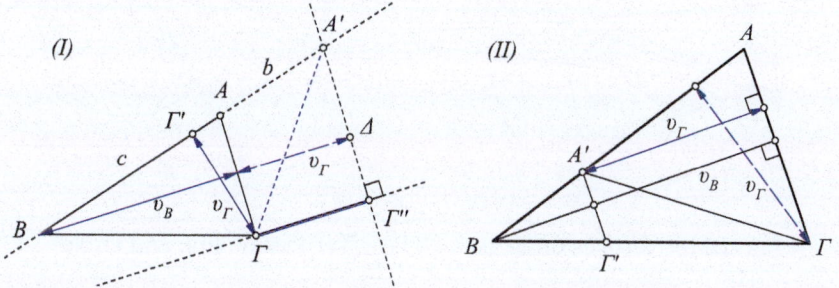

Fig. 2.116: Relation of $b+c$ to $v_B + v_\Gamma$ Relation of $|b-c|$ to $v_B - v_\Gamma$

Exercise 2.205. Construct a triangle from its elements $(a = |B\Gamma|, \gamma, v_B + v_\Gamma)$, as well as from $(a = |B\Gamma|, \gamma, v_B - v_\Gamma)$.

Exercise 2.206. Given two equal line segments AB, $A\Gamma$, find a point Δ on a given circle κ, so that the angles $\widehat{A\Delta B}$ and $\widehat{B\Delta\Gamma}$ are equal. Investigate when the problem has a solution (See Figure 2.117-I).

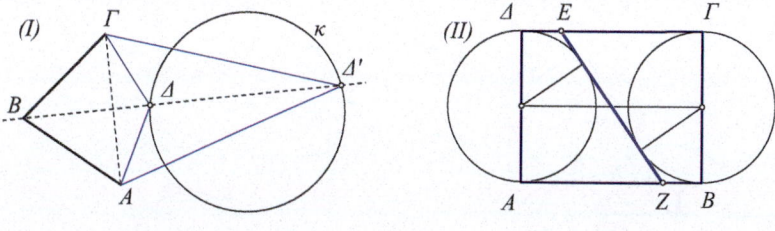

Fig. 2.117: Viewed under equal angles $|EZ| = |AB|$

Exercise 2.207. Construct two circles with diameter, respectively, the opposite sides $A\Delta$, $B\Gamma$ of rectangle $AB\Gamma\Delta$. Show that the internal common tangent to the two circles (when it exists), defines segment EZ equal to the side AB of the rectangle (See Figure 2.117-II).

Exercise 2.208. Construct a triangle $AB\Gamma$ with given angles, whose vertices A, B, Γ are contained respectively in three lines α, β, γ.

Hint: Consider an arbitrary point Γ of γ and triangles $\Gamma A'B'$ with angles equal respectively to the given and the vertex A' on line α (See Figure 2.118-I). Then (Theorem 2.28) vertex B' is contained always in a fixed line ε. Consider the intersection point B of ε and of β and define the triangle $AB\Gamma$.

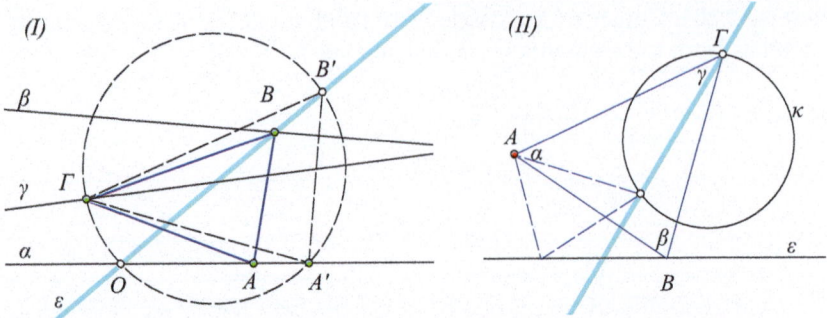

Fig. 2.118: Vertices on lines   Vertices on line and circle

Exercise 2.209. Construct a triangle $AB\Gamma$ with given angles, whose vertex at A is a fixed point, the vertex at B is contained in a line ε and the vertex Γ is contained in a circle κ (See Figure 2.118-II).

Exercise 2.210. The angle $\widehat{XOY}$ of fixed measure ω and with vertex the fixed point O rotates about its vertex O, which is a point inside the fixed circle κ. Its sides intersect the circle at points $\{X,Y\}$ and point Y' is on the opposite of half line OX at distanc $|OY'| = |OY|$. Find the geometric locus of Y'.

Hint: Circle congruent to κ.

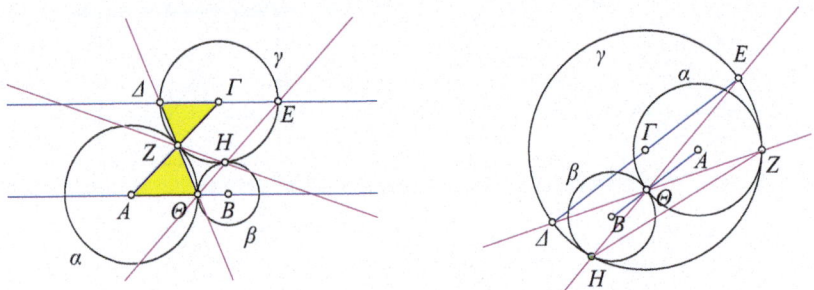

Fig. 2.119: Point of contact between three circles

Exercise 2.211. Given are three circles $\alpha(A)$, $\beta(B)$, $\gamma(\Gamma)$ tangent at three different points Θ, H, Z (See Figure 2.119). Show that the second intersection points E, Δ of lines respectively $H\Theta, Z\Theta$, define on the circle γ diametrically opposite points Δ, E.

Next exercise ([45], [47, p.43]) was proposed by Sylvester (1814-1897) in 1893 and was solved after 40 years!

Exercise 2.212. Given are v points with the property: Every line that joins two of them contains also a third of them. Show that all v points are contained in a line.

Hint: Suppose the conclusion does not hold. Then there exist points A, B, Γ so that the distance $|A\Delta|$ of A from $B\Gamma$ is minimal for all triplets formed by selecting from the ν points (See Figure 2.120-I). Suppose that $B\Gamma$ contains one more point, E. Then at least two out of the three points $\{B, \Gamma, E\}$ will be contained in $B\Gamma$ and on the same side of Δ. Let us suppose that points B, Γ are contained on the same side of Δ and in the order of $\{B..\Gamma..\Delta\}$. Then the distance of Γ from AB will be less than $|A\Delta|$, something which contradicts the hypothesis.

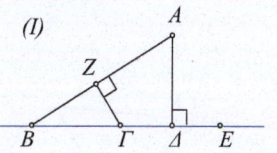

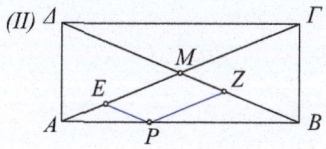

Fig. 2.120: ν collinear points $|PE| + |PZ| = |MA|$

Exercise 2.213. Project a point P, lying on the perimeter of a rectangle, onto the diagonals by parallels to the other diagonals, to the points E and Z and show that $|PE| + |PZ|$ is constant. Conversely, if point P is projected in parallel to the sides of angle $\widehat{AMB}$ at their points E, Z in such a way that $|PE| + |PZ| = \kappa$, where κ is a constant, then point P is contained in the side of a rectangle (See Figure 2.120-II).

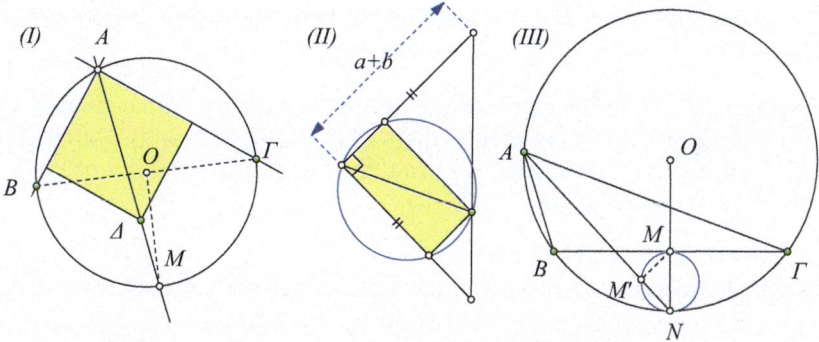

Fig. 2.121: Hints through the corresponding figures

Exercise 2.214. Construct a square, whose two adjacent sides pass through given points B, Γ and the vertex contained in the other two adjacent sides is a given point Δ (See Figure 2.121-I).

Exercise 2.215. In a given circle inscribe a rectangle with a given perimeter (See Figure 2.121-II).

Exercise 2.216. A triangle's base $B\Gamma$ position and length, and the measure of angle $\alpha = \widehat{BA\Gamma}$ remain constant. Find the geometric locus of the projection of the middle M of $B\Gamma$ on the bisector from A (See Figure 2.121-III).

Exercise 2.217. Given a triangle $AB\Gamma$, find the geometric locus of the points P of the plane, for which the lines orthogonal to $PA, PB, P\Gamma$ at A, B, Γ respectively, pass through a common point.

Hint: The circumscribed circle of $AB\Gamma$.

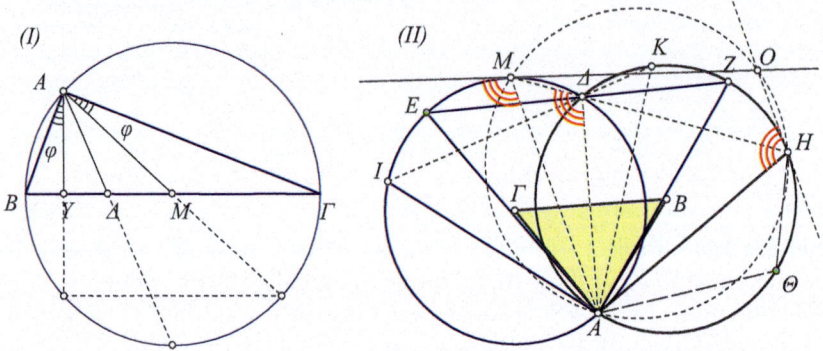

Fig. 2.122: Right-angled characteristic   Angles from variable chord

Exercise 2.218. Show that, besides isosceli, the right triangle is characterized by the fact that its altitude and median from the right angle are equally inclined to the corresponding bisector (See Figure 2.122-I).

Exercise 2.219. From the common point Δ of two circles with centers B and Γ we draw a line, which intersects them at E and Z. Also on the other common point A of the two circles we draw the tangents AI and AH of the two circles. Show that (See Figure 2.122-II):

1. Angle $\widehat{IAH}$ is supplementary to $\widehat{EAZ}$.
2. If M, K are the intersection points of the circles with $\Delta H, I\Delta$ respectively and MO, HO are the tangents at M, H respectively, then $OMAH$ is inscriptible.
3. Angle $\widehat{I\Delta H}$ has measure $|\widehat{I\Delta H}| = 2|\widehat{EAZ}|$.

Hint: See exercise 2.152 and theorem 2.25. For (3): $|\widehat{I\Delta H}| = |\widehat{I\Delta A}| + |\widehat{A\Delta H}| = \frac{1}{2}(|\widehat{I\Gamma A}| + |\widehat{AB\Theta}|)$ etc.

Exercise 2.220. Construct an isosceles triangle $AB\Gamma$ with base AB, for which is given the length of the altitude AE and the length of its median AN.

Hint: The right triangle AEN can be constructed from the given data (See Figure 2.123-I). The intersection point of the medians $Δ$ is a known point of AN and $|ΔA| = |ΔB|$. Consequently, point B is found as an intersection of the circle with radius $ΔA$ with the line EN.

Exercise 2.221. Construct a quadrilateral $ABΓΔ$, for which are given the lengths of its diagonals and the measures of its angles.

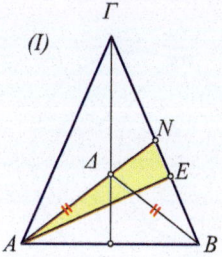
(I)
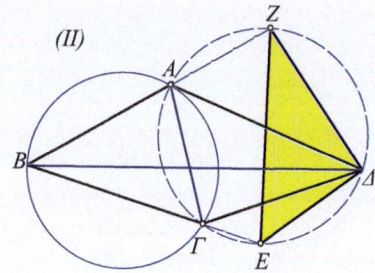
(II)

Fig. 2.123: Isosceles construction, Quadrilateral construction

Hint: Circles $(AΓΔ)$ and $(ABΓ)$ can be constructed, because points $Δ$ and B see under given angles the given diagonal $AΓ$ (See Figure 2.123-II). Extending BA and $BΓ$, we define the points Z and E of the circle $(AΓΔ)$. On the constructed circle $(AΓΔ)$ the known angles $\widehat{ZAΔ}$ and $\widehat{EΓΔ}$ define constructible lengths $|ΔZ|, |ΔE|$. Therefore the triangle $ΔEZ$ can be constructed. B is determined as an intersection of the circle which sees EZ under the given angle $\widehat{B}$ and the circle $Δ(ΔB)$.

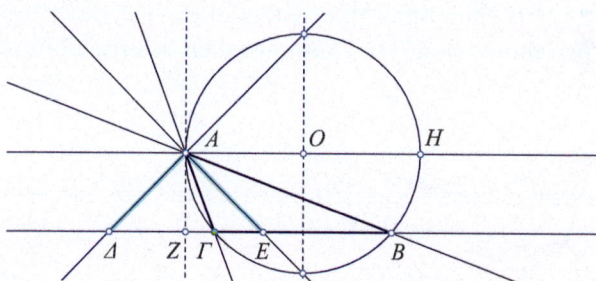

Fig. 2.124: A triangle with equal external/internal bisectors at A

Exercise 2.222. The circumcircle of the triangle $ABΓ$ has a diameter AH parallel to its base $BΓ$. Show that the bisectors external/internal of $\widehat{A}$ are equal. Formulate and show also the converse (See Figure 2.124).

Exercise 2.223. From a given point E inside the square $AB\Gamma\Delta$ we draw a line intersecting it at points Z, H (See Figure 2.125-I). Show that the circles (ΔZE), (EHB) intersect a second time at a point Θ lying on a diagonal of the square.

Exercise 2.224. Given two circles κ, λ intersecting at points E, Z (See Figure 2.125-II), to draw a line EB through E, such that the segment $B\Gamma$, which is defined on it from its second intersection points with the circles has given length δ.

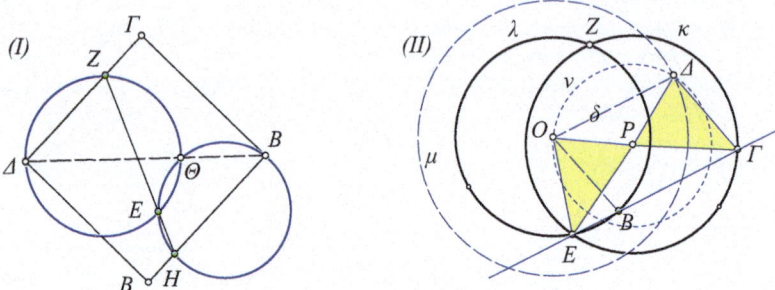

Fig. 2.125: Intersection on the diagonal $B\Gamma$ between two circles

Hint: If $\kappa(P)$, $\lambda(O)$ the circles, consider the variable line through E, which intersects them respectively at Γ, B. Consider also the parallelogram $OB\Gamma\Delta$. Its vertex Δ is contained in a circle ν and the wanted line results when Δ coincides with one from the intersection points $\{I, K\}$ of the circle ν with circle $\mu(O, \delta)$.

Exercise 2.225. Consider the four triangles which are defined from the sides of the parallelogram $AB\Gamma\Delta$ and the intersection point of its diagonals, E (See Figure 2.126-I). Show that the centers of the inscribed circles of these triangles form a rhombus $ZH\Theta I$. Examine when this rhombus is a square.

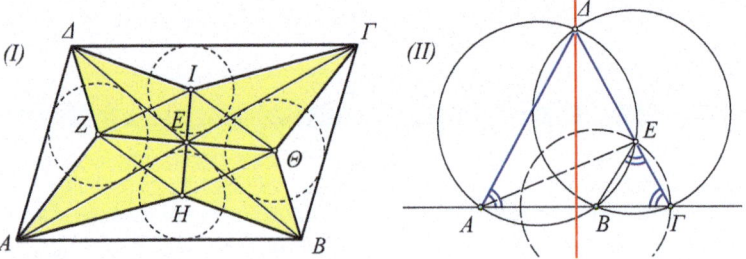

Fig. 2.126: Parallelogram and rhombus   Locus of Δ for equal circles

Exercise 2.226. $AB, B\Gamma$ are successive line segments of the same line. We construct congruent circles which have, respectively, $AB, B\Gamma$ as chords. Find the geometric locus of their other intersection point Δ (See Figure 2.126-II).

2.17. COMMENTS AND EXERCISES FOR THE CHAPTER

Exercise 2.227. Construct a triangle $AB\Gamma$, which has medials of its sides three given lines α, β, γ, passing through a common point O.

Exercise 2.228. The circles $\{\kappa_1, \kappa_2\}$ intersect at points $\{A, B\}$, through which pass correspondingly lines $\{\alpha, \beta\}$. These intersect the circles a second time at points $\{A_1, A_2\}$ the line α and at points $\{B_1, B_2\}$ the line β. Show that the lines $\{A_1B_1, A_2B_2\}$ are parallel. What happens when $\{A_1, B_1\}$ coincide (Exercise 2.158)?

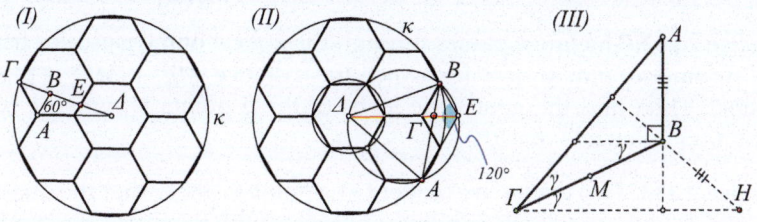

Fig. 2.127: Seven congruent regular hexagons   Triangle construction

Exercise 2.229. In a given circle κ inscribe seven congruent regular hexagons (from Pappus' Synagogue [53, p.1097]) (See Figure 2.127).

Hint: In figure 2.127-I, point A is constructible because it sees $B\Delta$ under the angle of $60°$ and $|\Delta B|/|B\Gamma| = 2$ (proof of Pappus). Similar proof in figure 2.127-II, in which $|A\Gamma|/|\Gamma B| = 2$.

Exercise 2.230. Given is a line ε, a point M of it and a point H outside this line. To construct a triangle $AB\Gamma$ whose base $B\Gamma$ is contained in ε, has its middle at M, its orthocenter at H and the angles at the base satisfy the relation $\beta - \gamma = 90°$ (See Figure 2.127-III).

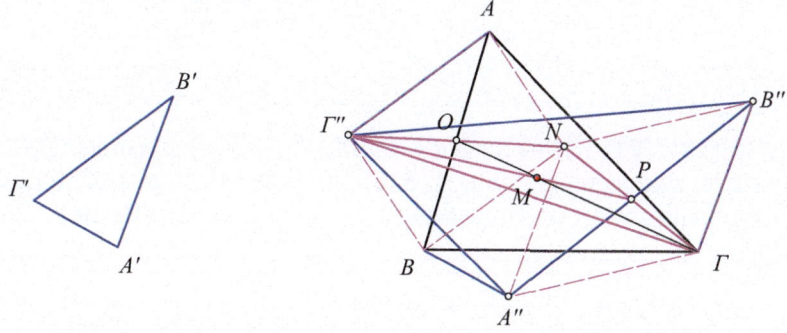

Fig. 2.128: Triangles with common centroid

Exercise 2.231. Given two triangles $AB\Gamma$ and $A'B'\Gamma'$, we construct a third triangle $A''B''\Gamma''$ as follows (See Figure 2.128): From the vertex Γ we draw the segment $\Gamma B''$ parallel and equal to $A'B'$, from A we draw the segment $A\Gamma''$ parallel and equal to $B'\Gamma'$ and from B we draw the segment BA'' parallel and equal to $\Gamma'A'$. Show that the triangle $A''B''\Gamma''$ has the same centroid M as $AB\Gamma$.

Hint: From A'' draw the segment $A''N$ parallel and equal to $\Gamma B''$. The triangle $BA''N$ is congruent to $\Gamma'A'B'$, therefore BN is parallel and equal to $\Gamma''A$. The center of mass of $A''B''\Gamma''$ and $AB\Gamma$ coincides with that of $\Gamma''\Gamma N$.

Exercise 2.232. The line ε passes through the apex A of the isosceles triangle $AB\Gamma$. To find a point M on ε for which the distance $|MB| + |M\Gamma|$ is the least possible. Then find the geometric locus of these points M, when ε rotates about A (circumcircle of $AB\Gamma$).

Exercise 2.233. Two circles are tangent at a point E. Show that the quadrilateral $AB\Gamma\Delta$, with vertices their points of contact with their common tangents, is circumscriptible to a circle with center E (See Figure 2.129-I).

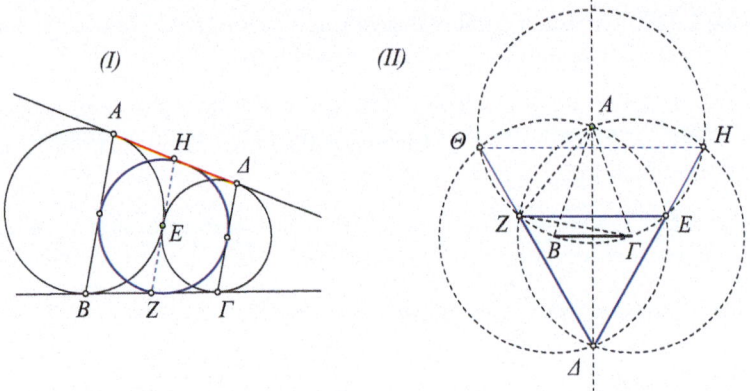

Fig. 2.129: Quadrilateral by tangent circles   Three congruent circles

Exercise 2.234. With radius the leg $r = |AB|$ of an isosceles triangle $AB\Gamma$ we draw circles with centers points A, B, Γ. Show that the intersection points of these circles form two equilateral triangles ΔEZ and $\Delta H\Theta$ (See Figure 2.129-II).

Hint: Because of symmetry $ZE\Delta$ is isosceles, $|\Delta Z| = |\Delta E|$. Also show that the triangles $E\Gamma Z$ and $E\Gamma\Delta$ are congruent and the angle $\widehat{\Gamma EZ}$ is $30°$.

Exercise 2.235. Construct a quadrilateral for which are given the measures of two opposite of its angles, the lengths of its diagonals and the measure of their angle.

2.17. COMMENTS AND EXERCISES FOR THE CHAPTER

Exercise 2.236. Given the triangle $AB\Gamma$, points A', B' and Γ' are defined as the intersections of its bisectors with the circumcircle. Show that the triangle $A'B'\Gamma'$ has the bisectors of $AB\Gamma$ for altitudes (See Figure 2.130-I).

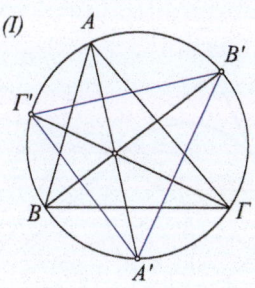
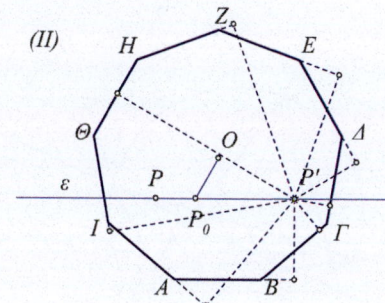

Fig. 2.130: Bisector traces   Sum of distances

Exercise 2.237. Show that for an arbitrary point P in the interior of a regular polygon the sum of its distances from the sides of the polygon is constant.

Hint: Show first, using exercise 2.44, that this sum of distances does not change if the point P moves on a line ε, which is parallel to a side of the polygon (See Figure 2.130-II). Then let the point P take the position P_0 of the projection on ε of the center O of the polygon, parallel to another side of the polygon. Then, apply the same remark for the line P_0O, thus, showing this sum to be equal to the sum of distances of the center O from the sides of the polygon (alternatively, see exercise 3.184).

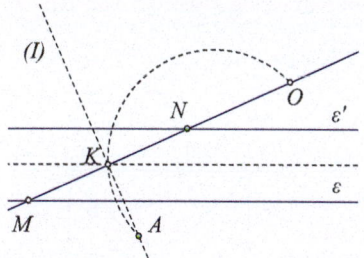
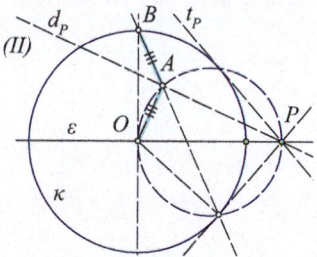

Fig. 2.131: Secant of parallels   Geometric locus

Exercise 2.238. From the point O outside the two parallels $\{\varepsilon,\varepsilon'\}$ we draw lines intersecting them at points correspondingly $\{M,N\}$. To find the secant for which the distances $\{|AM|,|AN|\}$ from a fixed point are equal (See Figure 2.131-I).

Exercise 2.239. Line ε passes through the center of the circle κ(O). For every point P of ε outside the circle, we draw the tangent t_P of κ from P always on the same side of ε. To find the locus of the projections A of the center O of κ on the bisector d_P of the angle of lines $\{ε, t_P\}$ (See Figure 2.131-II).

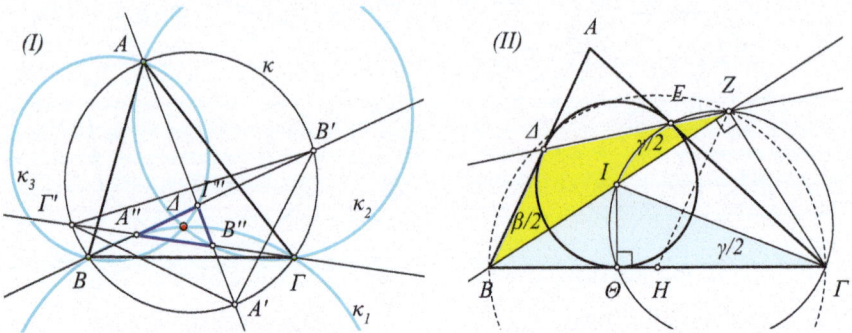

Fig. 2.132: Fixed angles   Intersection on circle

Exercise 2.240. Triangles $\{ABΓ, A'B'Γ'\}$ have a common circumcircle κ. The first remains fixed and the second is rotating inside the circle without to change its shape. Show that the lines $\{AA', BB', ΓΓ'\}$ are side-lines of a triangle $A''B''Γ''$ with constant angles and its vertices are moving on three circles $\{κ_1, κ_2, κ_3\}$ which pass through a common point Δ (See Figure 2.132-I).

Hint: Apply exercise 2.157.

Exercise 2.241. In the triangle ABΓ the bisector from B and line ΔE, joining the contact point of $\{AB, AΓ\}$ with the inscribed circle, intersect at point Z. Show that Z is on the circle with diameter BΓ (See Figure 2.132-II).

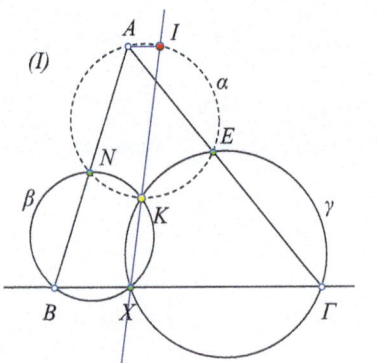 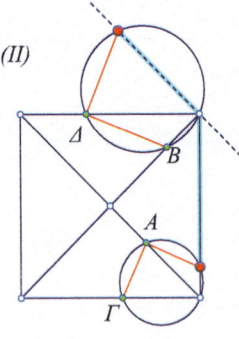

Fig. 2.133: Geometric locus of I   Square from $\{A, B, Γ, Δ\}$

2.17. COMMENTS AND EXERCISES FOR THE CHAPTER

Exercise 2.242. Points $\{N,E\}$ on the sides $\{AB,A\Gamma\}$ of the triangle $AB\Gamma$ are fixed and X moves on the line $B\Gamma$. Show that the common chord XK of the circles $\{\beta = (BXN), \gamma = (X\Gamma E)\}$ passes through a fixed point I, lying on the parallel to $B\Gamma$ from A (See Figure 2.133-I).

Hint: Point K moves on the constant circle $\alpha = (NAE)$.

Exercise 2.243. Construct a square for which we know the positions of two points $\{A,B\}$ respectively on its diagonals, and two points $\{\Gamma,\Delta\}$ on two opposite sides (See Figure 2.133-II).

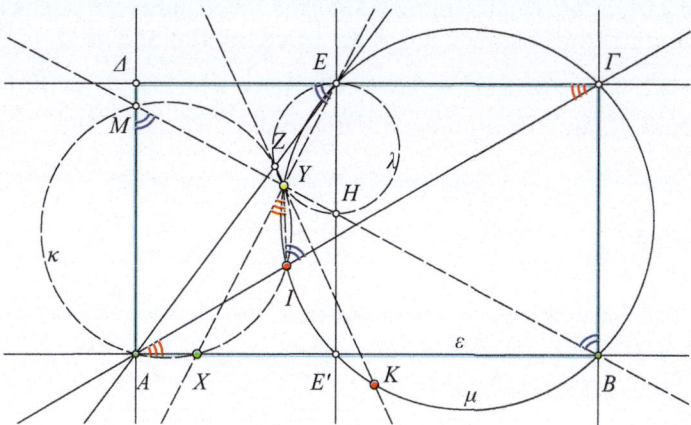

Fig. 2.134: Variable circles passing through two fixed points

Exercise 2.244. Given is a rectangle $AB\Gamma\Delta$, and a fixed point E of the side $\Gamma\Delta$ (See Figure 2.134). For every point X of the line AB we define the orthogonal projection Y of B on line EX, the circle $\kappa = (AXY)$ and the projection E' of E on AB. Show that

1. All the circles κ pass through a fixed point I of the diagonal $A\Gamma$.
2. If H is the intersection point of the lines $\{EE', BY\}$ and Z is the second intersection point of the circles $\{\kappa, \lambda = (EYH)\}$, then line ZY passes through a fixed point K of the circle $\mu = (B\Gamma EE')$.

Hint: For (1) define the intersection point M of the lines $\{A\Delta, BY\}$ and see that M is contained in $\kappa = (AXY)$. Point Y is also contained in the circle $\mu = (B\Gamma EE')$. If I denotes the second intersection point of the circles $\{\kappa, \mu\}$, show that I is contained in $A\Gamma$.

For (2) observe that $\widehat{IYK} = \widehat{\Gamma AE}$, which is a fixed angle.

Exercise 2.245. To inscribe in a given circle a triangle $AB\Gamma$ for which is given the position of A and the measures of the angles $\{\widehat{B}, \widehat{\Gamma}\}$.

Exercise 2.246. In the triangle $AB\Gamma$, from a point X on its base $B\Gamma$ draw parallels to its sides $\{AB, A\Gamma\}$ intersecting the other sides respectively at $\{B_X, \Gamma_X\}$ and define $d_X = |XB_X| + |X\Gamma_X|$. Show that, if for two points $\{X, Y\}$ of the base $B\Gamma$: $d_X = d_Y$, then the triangle is isosceles.

Exercise 2.247. Construct a triangle $AB\Gamma$, for which are given the bisector $A\Delta$ and the radii of the circumcircles of the triangles $AB\Delta$ and $A\Gamma\Delta$.

Exercise 2.248. Given are two circles intersecting at points $\{A, B\}$. To draw from A a line intersecting the circles at second points $\{\Gamma, \Delta\}$, such that A is the middle of the segment $\Gamma\Delta$.

Exercise 2.249. Construct a triangle $AB\Gamma$, for which are given the median $A\Delta$ and the radii of the circumcircles of the triangles $AB\Delta$ and $A\Gamma\Delta$.

Exercise 2.250. Construct a rectangle $AB\Gamma\Delta$, for which is given the length of its diagonal $A\Gamma$ and the difference of the angles $\widehat{\Gamma AB}$ and $\widehat{\Gamma A\Delta}$.

References

41. A. Adler (1906) Theorie der Geometrischen Konstruktionen. Goeschensche Verlagshandlung, Leipzig
42. B. Argunov, M. Balk (1957) Geometritseskie Postroenia na ploskosti (Russian). Gasudarstvenoe Isdatelstvo, Moscva
43. M. Berger (1987) Geometry vols I, II. Springer, Heidelberg
44. R. Courant (1996) What is Mathematics. Oxford University Press, Oxford
45. H. Coxeter, 1948. A Problem of Collinear Points, *American Mathematical Monthly*, 55:26-28
46. H. Coxeter, (1967) CUPM Geometry Conference. Math. Assoc. Amer. Washington DC
47. A. Engel (1998) Problem-Solving Strategies. Springer, Berlin
48. A. Fetisov (1977) Proof In Geometry. Mir Publishers, Moscow
49. R. Honsberger (1995) Episodes in Nineteenth and Twentieth Century Euclidean Geometry. The Mathematical Association of America, Washington
50. N. Kazarinoff, 1968. On who first proved the impossibility... ruler and compass alone, *American Mathematical Monthly*, 75:647
51. F. Klein (1901) Famous Problems of Elementary Geometry. Ginn and Company, Boston
52. G. Martin (1998) Geometric Constructions. Springer, Heidelberg
53. Pappus (1876) Synagoge. Weidmann, Berlin
54. J. Petersen (1901) Constructions Geometriques. Gauthier-Villars, Paris
55. B. Poonen, 2006. The number of intersection points made by the diagonals of a regular polygon, *SIAM J. Discrete Mathematics*, 11:135-156
56. P. Schreiber, 1993. On the Existence and Constructibility of Inscribed Polygons, *Beitraege zur Algebra und Geometrie*, 34:195-199
57. H. Scudder, 1928. How to trisect an angle with a carpenter's square, *American Mathematical Monthly*, 35:250-251
58. J. Steiner (1971) Gesammelte Werke vol. I, II. Chelsea Publishing Company, New York
59. E. Weisstein (1971) CRC concise encyclopedia of mathematics. Chapman and Hall/CRC, Boca Raton
60. D. Wells (1991) Dictionary of Curious and Interesting Geometry. Penguin Books

Chapter 3
Areas, Thales, Pythagoras, Pappus

3.1 Area of polygons

> That I cannot know any substance, as I can have no ideas but of their qualities, and that a thousand qualities of a thing cannot reveal the intimate nature of this thing, which may possess a hundred thousand other qualities that I am unaware of.
>
> Voltaire, *The ignorant Philosopher p. 44*

After segment lengths and angle measures the third magnitude we measure in plane Euclidean Geometry is the **area** of polygons. The area of a convex polygon Π is a positive number $\varepsilon(\Pi)$, of which we require the following properties.

Property 3.1 *Two congruent polygons have equal areas.*

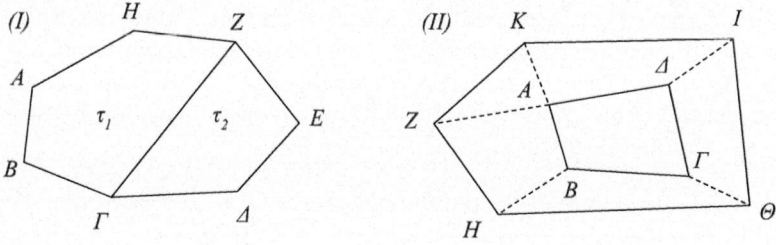

Fig. 3.1: $\varepsilon(\tau_1) + \varepsilon(\tau_2) = \varepsilon(AB\Gamma\Delta EZH)$ $\qquad \varepsilon(AB\Gamma\Delta) < \varepsilon(ZH\Theta IK)$

Property 3.2 *A polygon Π, which is composed from other, not overlapping, finite in number polygons Π', Π'', ... , has area the sum of the areas of the polygons*

$$\varepsilon(\Pi) = \varepsilon(\Pi') + \varepsilon(\Pi'') + ...$$

In figure 3.1-I the two polygons $\tau_1 = AB\Gamma ZH$ and $\tau_2 = \Gamma\Delta EZ$ have the side ΓZ in common and form the new polygon $AB\Gamma\Delta EZH$.

Property 3.3 *A polygon Π contained in another one Π' has less area: $\varepsilon(\Pi) < \varepsilon(\Pi')$ (See Figure 3.1-II).*

Property 3.4 *The area of a unit square (square with side of unit length) is one.*

Remark 3.1. We'll see below that these four accepted area properties, completely determine its form, and for rectangles lead to the well known expression for area, as the product of lengths of its sides. Its existence, therefore, is a consequence of the axioms. For this reason I speak about the *"properties"* and not about the *"axioms"* of area. I could speak about *"axioms"* satisfied by the area-function, which, however do not belong to the *"axioms of geometry"*.

Remark 3.2. Property 3.2 could be reduced inductively to the corresponding additivity of areas for polygons which are composed of only two other polygons. Also the property 3.3 is a consequence of property 3.2 and could be proved using the other properties. However the proof contains several subtle points, which in the first stages of acquaintance with geometry it is not necessary for them to be analyzed further.

Remark 3.3. It is noteworthy to observe the apparent similarities in measurements of lengths and areas. They are special cases of **measures**, i.e. mechanisms of measuring length and more generally, content (area, volume), which are particular cases of the so called, *Jordan measures* and more generally *Lebesque measures* ([80, p.254]). These measures, beginning with measurements of very simple shapes, extend and allow the measurement of more complex shapes than these which concern us in this lesson. Special cases of *measures* are also the volume of polyhedrons which we will meet in solid geometry (II-§ 5.5) as well as the area of spherical polygons (II-§ 5.3). In all cases the mathematical structure of the measurement mechanism is abstractly the same. We require of it certain properties, which could be called *"measurement axioms"*, and we prove that these requirements imply the existence of a unique mechanism which satisfies them.

Exercise 3.1. Show that every parallelogram is divided by one of its diagonals into two triangles of equal area. Show more generally, that every line which passes through the center of a parallelogram divides it into two triangles of equal area.

Hint: The two triangles defined by the diagonal are congruent, therefore they have equal areas.

Exercise 3.2. Show that the two diagonals of a parallelogram divide it into four triangles of equal area.

3.2. THE AREA OF THE RECTANGLE

Hint: Triangles AOB and $\Delta O\Gamma$, where O is the center of the parallelogram $AB\Gamma\Delta$ (point of intersection of the diagonals), are congruent, therefore have equal areas. Consider the symmetric points M, N of O relative to the middles of $B\Gamma$ and $\Delta\Gamma$ respectively (See Figure 3.2). The parallelograms $BO\Gamma M$ and

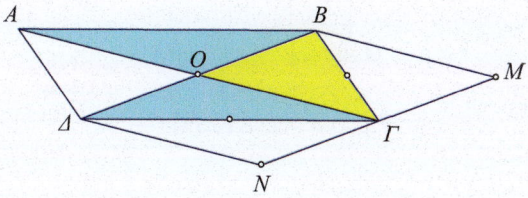

Fig. 3.2: Triangles of equal area

$O\Delta N\Gamma$ are congruent, therefore their halves according to area (according to the preceding exercise), which are $BO\Gamma$ and $\Delta O\Gamma$ will have equal areas.

Exercise 3.3. Show that the median $A\Delta$ of a triangle $AB\Gamma$ divides it into two triangles of equal area.

Hint: Combination of the two preceding exercises (see also Corollary 3.7).

3.2 The area of the rectangle

> The question of what properties, such as angle or area, are reproduced on a map without distortion is of prime interest to mathematicians. The question extends far beyond the confines of geometry, for all mathematics can be considered broadly as a study of maps and mapping.
>
> G.A. Boehm, *The New World of Mathematics, p. 124*

The key for the calculation of areas of polygons is the area of the rectangle. In this section we begin from the area of rectangles whose side lengths are integers and stepping by gradually, with the help of the properties of areas, we arrive at the fundamental expression of the area $\varepsilon = ab$, as the product of the lengths of its sides, for any values of a and b, integer or not.

Lemma 3.1. *We divide the two opposite sides of the unit square (having side of length 1) into μ equal parts and the other two opposite sides into ν equal parts and we draw parallels which join opposite lying points. This forms $\mu\nu$ parallelograms, each of them having area $\frac{1}{\mu\nu}$.*

Proof. Obviously there are formed $\mu\nu$ congruent rectangles which according to the property 3.1 will have the same area E (See Figure 3.3). According to

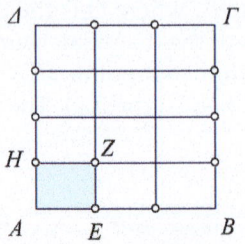

Fig. 3.3: $\varepsilon(AEZH) = \frac{1}{\mu\nu}$

property 3.2 the area of the square which is 1 (Property 3.4) will be the sum of the areas $1 = \mu\nu E$.

Lemma 3.2. *The area ε of a rectangle $AB\Gamma\Delta$, with sides AB and $A\Delta$, whose lengths are rational numbers, is equal to the product of these lengths $\varepsilon = |AB||A\Delta|$.*

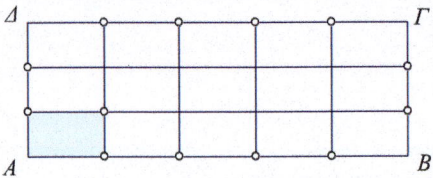

Fig. 3.4: $|AB| = \frac{\alpha}{\beta}, |A\Delta| = \frac{\gamma}{\delta} \Rightarrow \varepsilon(AB\Gamma\Delta) = |AB||A\Delta|$

Proof. We suppose that in the rationals $|AB| = \frac{\alpha}{\beta}, |A\Delta| = \frac{\gamma}{\delta}$ the α, β, γ and δ are positive integers and we divide the unit square into $\beta\delta$ congruent rectangles as in the preceding lemma, according to which each has area $\varepsilon' = \frac{1}{\beta\delta}$ (See Figure 3.4). By the hypothesis, the side AB of the rectangle can be divided into α number segments of length $\frac{1}{\beta}$ and the side $A\Delta$ can be divided into γ number segments of length $\frac{1}{\delta}$. Drawing parallels from the points of division of these sides we therefore form a number of $\alpha\gamma$ congruent rectangles, each of which has area $\frac{1}{\beta\delta}$. The sum of the areas of these rectangles is $\alpha\gamma\frac{1}{\beta\delta}$ and is equal, according to the Property 3.2, to the area of the rectangle.

Lemma 3.3. *For every positive number θ and every natural number ν there exists another natural number μ (or zero), such that the following inequality holds*

$$\left|\theta - \frac{\mu}{\nu}\right| \leq \frac{1}{\nu}.$$

3.2. THE AREA OF THE RECTANGLE

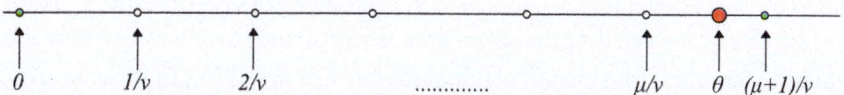

Fig. 3.5: Archimedean axiom

Proof. In essence the lemma coincides with the so called **Archimedean axiom** (Archimedes 287-212 B.C.) for line segments of one line and is pictured in figure 3.5. For the proof, starting at the beginning of a segment of length θ, we place successive intervals of length $\frac{1}{\nu}$. According to the Archimedean axiom, there is a first critical μ, such that $(\mu+1)\frac{1}{\nu}$ exceeds the length θ, in other words θ finds itself in the interval (of length $\frac{1}{\nu}$) between $\frac{\mu}{\nu}$ and $\frac{\mu+1}{\nu}$.

Theorem 3.1. *The area ε of the rectangle $AB\Gamma\Delta$, with sides AB and $A\Delta$, is equal to the product of the lengths of the sides $\varepsilon = |AB||A\Delta|$.*

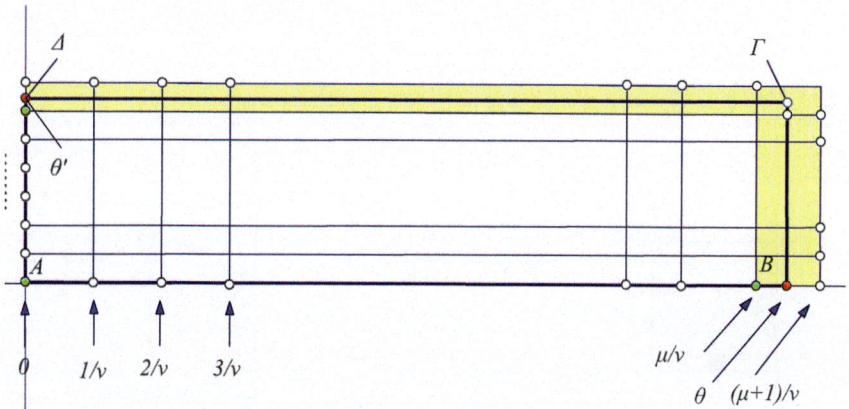

Fig. 3.6: Rectangle approximation

Proof. For the proof we consider the lengths of the sides $\theta = |AB|$ and $\theta' = |A\Delta|$ and a (fairly large) ν. According to the preceding lemma, there exist two corresponding integers μ and μ' satisfying

$$\left|\theta - \frac{\mu}{\nu}\right| \leq \frac{1}{\nu} \quad \text{and} \quad \left|\theta' - \frac{\mu'}{\nu}\right| \leq \frac{1}{\nu}.$$

This way, the area ε of the rectangle (See Figure 3.6) is greater than

$$\frac{\mu}{\nu}\frac{\mu'}{\nu} \leq \varepsilon,$$

since it contains the rectangle with sides of rational length $\frac{\mu}{\nu}$ and $\frac{\mu'}{\nu}$. Thinking similarly, we see that the same area is simultaneously less than the area of the rectangle with sides of rational length $\frac{\mu+1}{\nu}$ and $\frac{\mu'+1}{\nu}$, in other words

$$\varepsilon \leq \frac{\mu+1}{\nu}\frac{\mu'+1}{\nu}.$$

In total therefore we see that the numbers $\theta\theta'$ and ε satisfy the same inequalities

$$\frac{\mu}{\nu}\frac{\mu'}{\nu} \leq \theta\theta' \leq \frac{\mu+1}{\nu}\frac{\mu'+1}{\nu} \quad \text{and} \quad \frac{\mu}{\nu}\frac{\mu'}{\nu} \leq \varepsilon \leq \frac{\mu+1}{\nu}\frac{\mu'+1}{\nu}.$$

Therefore their difference will satisfy the inequality

$$|\varepsilon - \theta\theta'| \leq \frac{\mu+\mu'+1}{\nu^2} = \frac{1}{\nu}\frac{\mu}{\nu} + \frac{1}{\nu}\frac{\mu'}{\nu} + \frac{1}{\nu^2} \leq \frac{1}{\nu}\theta + \frac{1}{\nu}\theta' + \frac{1}{\nu^2} \leq \frac{\theta+\theta'+1}{\nu}.$$

Because the right quantity can become arbitrarily small, provided we choose a big ν, it follows that the (fixed) quantity on the left cannot be strictly positive.

Corollary 3.1. *The area of a square with side length α is equal to α^2.*

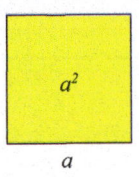
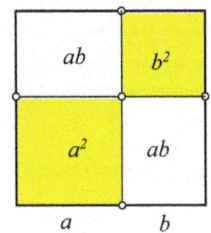

Fig. 3.7: Area of square   Square of sum geometrically

Exercise 3.4. Given the positive numbers a and b, consider the square with side length $a+b$ (and $a-b$) (See Figure 3.7) and show, the identity $(a+b)^2 = a^2 + b^2 + 2ab$.

Exercise 3.5. Similarly to exercise 3.4, show geometrically the identity $(a-b)^2 = a^2 + b^2 - 2ab$.

Exercise 3.6. If E, Z, H, Θ are the side middles of the square $AB\Gamma\Delta$, show that $\Gamma E, \Delta Z, AH$ and $B\Theta$ define the square $IK\Lambda M$, which has area $1/5$ that of $AB\Gamma\Delta$ (See Figure 3.8-I). Compute analogously the area of the central square

3.3. AREA OF PARALLELOGRAM AND TRIANGLE

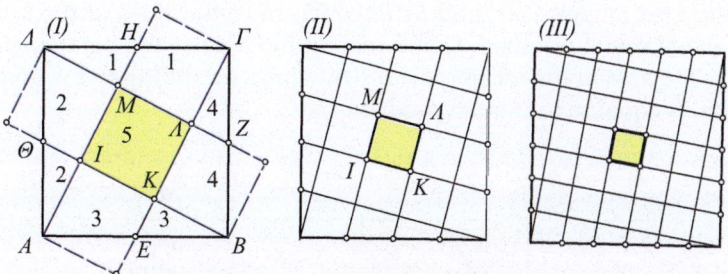

Fig. 3.8: Square division

$IK\Lambda M$, formed by dividing each side of a square in four equal parts (See Figure 3.8-II). Generalize by dividing in 2ν parts (See Figure 3.8-III).

3.3 Area of parallelogram and triangle

> Order is not pressure which is imposed on society from without, but an equilibrium which is set up from within.
>
> *Jose Ortega y Gasset, Mirabeau and Politics*

Knowing the formula for the area of *rectangles*, we can easily find the formula for the area of general parallelograms and, from this, find the area of the triangles, which are nothing but halves (by diagonal sections) of parallelograms.

Theorem 3.2. *The area of a parallelogram $AB\Gamma\Delta$ is equal to the product of the measure of one of its sides times its distance from its opposite parallel side.*

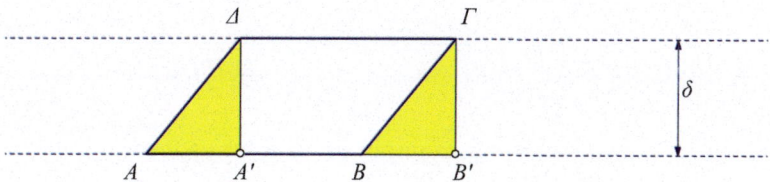

Fig. 3.9: $\varepsilon(AB\Gamma\Delta) = \varepsilon(A'B'\Gamma\Delta) = |AB|\delta$

Proof. Let A' and B' be the projections of vertices Δ and Γ on side AB (See Figure 3.9). Triangles $AA'\Delta$ and $BB'\Gamma$ are congruent, according to the SAS-criterion, having sides $A\Delta$ and $B\Gamma$ equal, sides $\Delta A'$ and $\Delta B'$ also equal and the contained angles $\widehat{A\Delta A'}$ and $\widehat{B\Gamma B'}$ equal, since the sides of the latter are parallel and equal oriented. Consequently the area of the parallelogram is equal to the area of the rectangle $A'B'\Gamma\Delta$, which, according to the preceding theorem, is equal to the stated product.

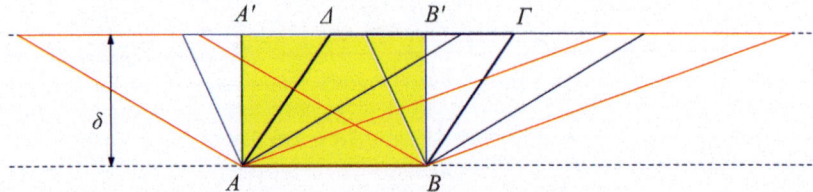

Fig. 3.10: Same base same altitude

From the theorem follows that all parallelograms with fixed base AB and opposite side $\Gamma\Delta$, sliding on a parallel of line AB (See Figure 3.10), have fixed area $\varepsilon = |AB|\delta$, where δ is the distance between the parallels AB and $\Gamma\Delta$. A special case is the rectangle $ABB'A'$, whose area is exactly the product of the lengths of its sides. Observe, that, although all of these parallelograms have the same area, their perimeter is variable and can become as big as we like. Exercise 3.9 characterizes this rectangle as the one of minimal perimeter among all parallelograms with the same base and the same altitude.

Corollary 3.2. *The area of a triangle $AB\Gamma$ is equal to half the product of one of its sides times its altitude falling on this side.*

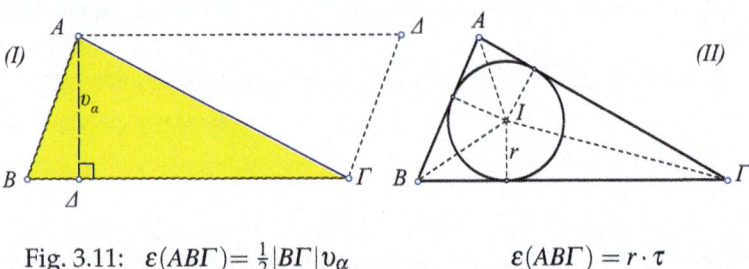

Fig. 3.11: $\varepsilon(AB\Gamma) = \frac{1}{2}|B\Gamma|v_\alpha$ $\qquad\qquad \varepsilon(AB\Gamma) = r \cdot \tau$

Proof. Let us select a side, for example $B\Gamma$ and let us form parallelogram $AB\Gamma\Delta$ by drawing parallels from Γ to AB and from A to $B\Gamma$ (See Figure 3.11-I). This parallelogram has $A\Gamma$ as a diagonal and the two triangles formed: $AB\Gamma$ and $A\Gamma\Delta$ are congruent. According to the preceding theorem, it must hold then $2\varepsilon(AB\Gamma) = |AB|v_\alpha$.

Corollary 3.3. *The area of a right triangle is equal to half the product of its two orthogonal sides.*

3.3. AREA OF PARALLELOGRAM AND TRIANGLE

Corollary 3.4. *The area of a triangle is equal to the product $r \cdot \tau$ of the radius r of its inscribed circle and its half perimeter τ.*

Proof. If I is the center of the inscribed circle (See Figure 3.11-II), the area is
$\varepsilon(AB\Gamma) = \varepsilon(IB\Gamma) + \varepsilon(I\Gamma A) + \varepsilon(IAB) = \frac{1}{2}(r \cdot |AB|) + \frac{1}{2}(r \cdot |B\Gamma|) + \frac{1}{2}(r \cdot |\Gamma A|) = r \cdot (\frac{1}{2}(|AB| + |B\Gamma| + |\Gamma A|)) = r \cdot \tau$.

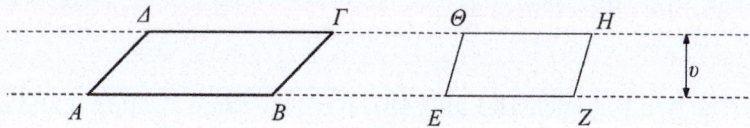

Fig. 3.12: $\quad \dfrac{\varepsilon(AB\Gamma\Delta)}{\varepsilon(EZH\Theta)} = \dfrac{|AB|}{|EZ|}$

Corollary 3.5. *The ratio of the areas of two parallelograms $AB\Gamma\Delta$ and $EZH\Theta$ which have the same altitude towards their bases AB and EZ is equal to the ratio of the lengths of these bases $\dfrac{|AB|}{|EZ|}$.*

Proof. This follows immediately from the fact, that the area is equal to the product of the lengths of the bases and of the corresponding altitude: $\varepsilon(AB\Gamma\Delta) = |AB| \cdot v$ and $\varepsilon(EZH\Theta) = |EZ| \cdot v$ (See Figure 3.12). The aforementioned relation follows by dividing the preceding equations by parts.

Corollary 3.6. *The ratio of the areas of two triangles $AB\Gamma$ and EZH which have the same altitude towards the bases AB and EZ is equal to the ratio of the lengths of these bases $\dfrac{|AB|}{|EZ|}$.*

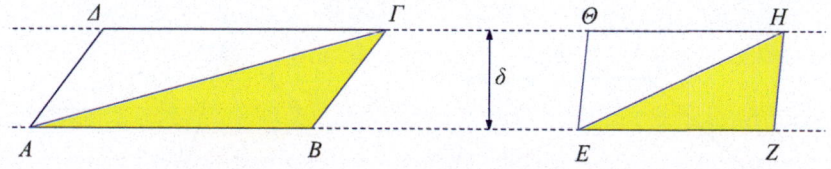

Fig. 3.13: $\quad \dfrac{\varepsilon(AB\Gamma)}{\varepsilon(EZH)} = \dfrac{|AB|}{|EZ|}$

Proof. Follows from the preceding corollary and the fact that the triangle has area half that of the corresponding parallelogram (See Figure 3.13).

Corollary 3.7. *The median $A\Delta$ of the triangle $AB\Gamma$ divides it into two triangles $AB\Delta$ and $A\Delta\Gamma$ of equal area. Conversely, if a line through A divides the triangle into two other triangles of equal area, then this line coincides with the median.*

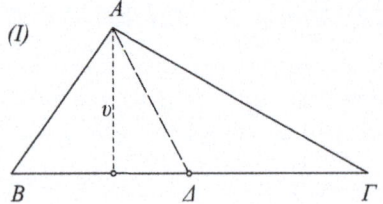 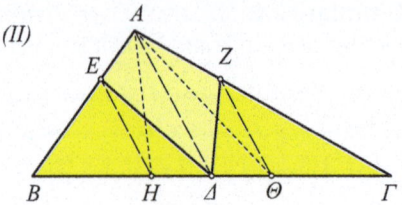

Fig. 3.14: Bisection of area   Trisection of area

Proof. The two triangles $AB\Delta$ and $A\Delta\Gamma$ have the same altitude v and equal bases $|B\Delta| = |\Delta\Gamma|$ (See Figure 3.14-I) (Exercise 3.3 asks for a proof that relies directly on properties of area). The converse follows also directly from Corollary 3.6.

Exercise 3.7. From the middle Δ of base $B\Gamma$ of triangle $AB\Gamma$ draw two lines which partition the triangle into three equal parts.

Hint: Divide the base into three equal parts with the points H, Θ (See Figure 3.14-II). Draw parallels $HE, \Theta Z$ to the median $A\Delta$. This defines two triangles of equal area $BE\Delta, \Delta Z\Gamma$ and the quadrilateral $E\Delta Z A$ which has equal area to the triangles. $BE\Delta$ for example, has area $\frac{1}{3}$ that of triangle $AB\Gamma$. For this, notice that triangles ABH and $A\Theta\Gamma$ have equal bases $BH, \Theta\Gamma$ and the same altitude from A. Hence they have also the same area. But triangles AEH and $EH\Delta$ have also the same bases and corresponding altitude to them, hence also the same area. Thus,

$$\varepsilon(EB\Delta) = \varepsilon(EBH) + \varepsilon(EH\Delta) = \varepsilon(EBH) + \varepsilon(EHA) = \varepsilon(ABH).$$

Analogously show that $\varepsilon(Z\Delta\Gamma) = \varepsilon(A\Theta\Gamma)$ etc. ([63, p.132]).

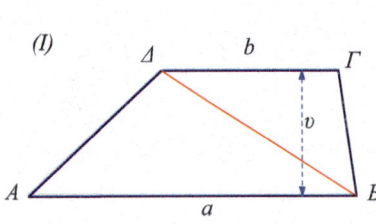 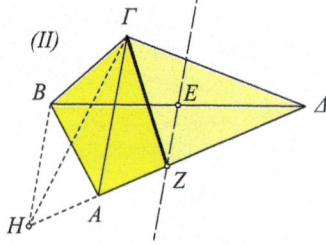

Fig. 3.15: $\varepsilon(AB\Gamma\Delta) = \frac{1}{2}(a+b) \cdot v$ Equiareal partitioning

Corollary 3.8. *The area of the trapezium $AB\Gamma\Delta$ is equal to $\frac{1}{2}(a+b) \cdot v$, where a and b are the lengths of its parallel sides and v is the distance between them.*

Proof. Using the diagonal $B\Delta$, divide the trapezium into two triangles and apply Corollary-3.2, adding the area of the two triangles (See Figure 3.15-I).

3.3. AREA OF PARALLELOGRAM AND TRIANGLE

Exercise 3.8. From the vertex Γ of the quadrilateral $AB\Gamma\Delta$ draw a line partitioning it into two parts of equal area.

Hint: If E is the middle of the diagonal $B\Delta$, draw a parallel EZ to the other diagonal ΓA (See Figure 3.15-II). ΓZ is the desired line. ΓBAZ has the same area as triangle ΓHZ and ΓZ is a median of the triangle $\Gamma H\Delta$.

Corollary 3.9. *From all triangles $AB\Gamma$ with the same base $B\Gamma$ and the same area (equivalently the same altitude v to $B\Gamma$), the isosceles has the least perimeter.*

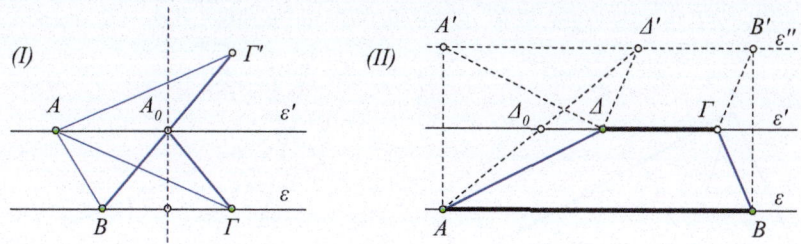

Fig. 3.16: Perimeter minimization

Proof. For the triangles $AB\Gamma$, whose base $B\Gamma$ is fixed and the area $\varepsilon(AB\Gamma) = \frac{1}{2}|B\Gamma| \cdot v$ is also fixed, the altitude v will be fixed, therefore their vertex A will be contained in a line ε', parallel to the base $B\Gamma$ (See Figure 3.16-I). Consider the symmetric Γ' of Γ relative to ε'. The perimeter $|B\Gamma| + |BA| + |A\Gamma|$, because of the constancy of $|B\Gamma|$, will become minimal when the sum $|BA| + |A\Gamma| = |BA| + |A\Gamma'|$ does. However, from the triangle inequality, the last sum is always greater than the length $|B\Gamma'|$. $B\Gamma'$ intersects ε' at a point A_0, lying on the medial line of $B\Gamma$. This defines the isosceles triangle $A_0 B\Gamma$ of least perimeter.

Corollary 3.10. *From all triangles with the same area E, the equilateral with this area has the least perimeter.*

Proof. Let $AB\Gamma$ be the triangle of area E and of least perimeter L. If two of its sides, for example $AB, A\Gamma$, were unequal, then, according to the preceding corollary, there would exist an isosceles with the same basis and area but a perimeter $L' < L$. This would mean that $AB\Gamma$ is not of least perimeter, thereby contradicting the hypothesis.

Exercise 3.9. Show that of all the parallelograms, which have a fixed base and whose opposite base moves on a parallel to the base, the rectangle has the least perimeter.

Exercise 3.10. Two parallelograms, which have the same area and two sides respectively equal have the sides, opposite to their equal sides, at equal distance.

Exercise 3.11. From all trapezia $AB\Gamma\Delta$, of fixed area, having parallel bases AB, $\Gamma\Delta$ of fixed length, the isosceles trapezium has the least perimeter.

Hint: Consider the symmetric A', B' of A, B relative to the line $\Gamma\Delta$ and on it Δ' so that $|B'\Delta'| = |\Gamma\Delta|$ (See Figure 3.16-II). Show that the perimeter is minimized when the sum $|A\Delta| + |\Delta\Delta'|$ is.

Theorem 3.3. *The bisector $A\Delta$ of the triangle $AB\Gamma$ intersects side $B\Gamma$ into two segments $B\Delta$ and $\Delta\Gamma$, whose ratio is the same as the ratio of the sides adjacent to them:* $\frac{|\Delta B|}{|\Delta \Gamma|} = \frac{|AB|}{|A\Gamma|}$. *And conversely if point Δ of the base $B\Gamma$ satisfies the preceding equality, then it is the trace on $B\Gamma$ of the bisector of the angle at A.*

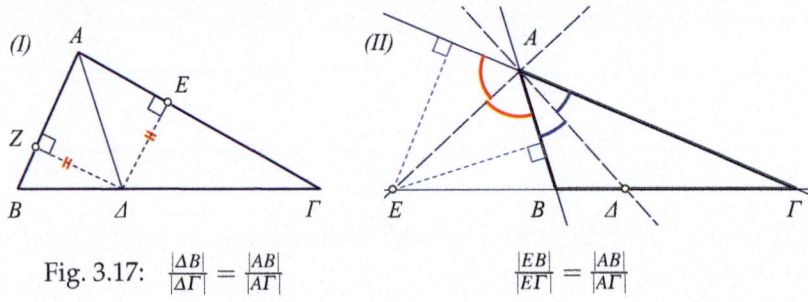

Fig. 3.17: $\quad \frac{|\Delta B|}{|\Delta \Gamma|} = \frac{|AB|}{|A\Gamma|} \qquad\qquad \frac{|EB|}{|E\Gamma|} = \frac{|AB|}{|A\Gamma|}$

Proof. If the triangle is isosceles with base $B\Gamma$, then the result follows from the fact that the bisector of the vertex coincides with the medial line of the base (Corollary 1.3). The proof in the general case follows easily by calculating the areas of triangles $AB\Delta$ and $A\Delta\Gamma$ using two different ways (See Figure 3.17-I). The first time we consider as their bases $B\Delta$ and $\Delta\Gamma$. Because their altitude from A is the same, we have (Corollary 3.6) that the ratio of their areas is equal to $\frac{|B\Delta|}{|\Delta\Gamma|}$. The second time we consider as their bases AB and $A\Gamma$ respectively. Again their altitudes from Δ are equal (Corollary 1.35), therefore (Corollary 3.6) follows that the ratio of their areas will be equal to $\frac{|AB|}{|A\Gamma|}$. For the converse consider that Δ satisfies the relation and draw the bisector $A\Delta'$. Point Δ' on $B\Gamma$ will also satisfy the equality, according to the preceding part of the theorem. It follows that $\frac{|\Delta B|}{|\Delta\Gamma|} = \frac{|\Delta'B|}{|\Delta'\Gamma|}$ and consequently points Δ and Δ' are coincident.

Exercise 3.12. Show that the same conclusion as that of the preceding corollary holds also for the external bisector of a triangle, provided the latter intersects the opposite base. In other words, that the external bisector AE of triangle $AB\Gamma$ intersects the opposite side $B\Gamma$ into segments which satisfy $\frac{|EB|}{|E\Gamma|} = \frac{|AB|}{|A\Gamma|}$. And conversely, if point E of the base $B\Gamma$ satisfies the preceding relation and is external to $B\Gamma$, then it is the trace of the external bisector.

Hint: As previously, calculate the areas of the triangles ABE and $A\Gamma E$ and their ratio using two ways (See Figure 3.17-II). First considering as their

3.3. AREA OF PARALLELOGRAM AND TRIANGLE

bases BE and ΓE and then considering as their bases AB and $A\Gamma$ towards which the altitudes from E are equal (Corollary 1.35). Note that the external bisector at A intersects the opposite base only then, when the triangle *isn't* isosceles with apex A. The converse is proved exactly as in the preceding theorem.

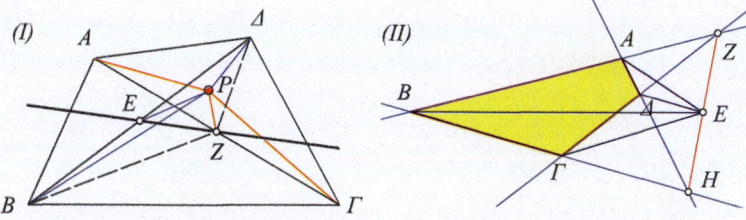

Fig. 3.18: Newton line   Newton's theorem

Exercise 3.13. Given a non-parallelogramic convex quadrilateral $AB\Gamma\Delta$, show that the points P for which hold relations of areas of the form $|\varepsilon(PAB) \pm \varepsilon(P\Gamma\Delta)| = |\varepsilon(PB\Gamma) \pm \varepsilon(PA\Delta)|$ are the points of the Newton line (which joins the middles E, Z of the diagonals) of the quadrilateral.

Hint: Let us suppose that $AB\Gamma\Delta$ is convex and point P is in its interior (See Figure 3.18-I). The difference of areas $\varepsilon(PB\Gamma)$-$\varepsilon(PAB)$ is written with the help of the quadrilaterals $PABZ$ and $PZB\Gamma$, where Z is the middle of the diagonal $A\Gamma$. These two quadrilaterals have equal areas (Corollary 3.7) and holds:

$$\varepsilon(PB\Gamma)\text{-}\varepsilon(PAB) = [\varepsilon(P\Gamma BZ)+\varepsilon(PBZ)]\text{-}[\varepsilon(PABZ)\text{-}\varepsilon(PBZ)] = 2\varepsilon(PBZ).$$

Similarly, because $\varepsilon(PAZ)=\varepsilon(PZ\Gamma)$ (Corollary 3.7),

$$\begin{aligned}\varepsilon(P\Gamma\Delta) - \varepsilon(PA\Delta) &= [\varepsilon(P\Gamma\Delta) + \varepsilon(PZ\Gamma)] - [\varepsilon(PA\Delta) + \varepsilon(PAZ)] \\ &= [\varepsilon(Z\Delta\Gamma) + \varepsilon(PZ\Delta)] - [\varepsilon(ZA\Delta) - \varepsilon(PZ\Delta)] \\ &= 2\varepsilon(PZ\Delta).\end{aligned}$$

It follows then that the relation $\varepsilon(PAB)+\varepsilon(P\Gamma\Delta) = \varepsilon(PB\Gamma)+\varepsilon(PA\Delta)$ is equivalent to

$$\begin{aligned}0 = \varepsilon(PBZ) - \varepsilon(PZ\Delta) &= [\varepsilon(PZB) + \varepsilon(PBE)] - [\varepsilon(PZ\Delta) + \varepsilon(PE\Delta)] \\ &= [\varepsilon(ZBE)+\varepsilon(PZE)] - [\varepsilon(ZE\Delta) - \varepsilon(PZE)] = -2\varepsilon(PZE).\end{aligned}$$

However $\varepsilon(PZE)=0$ means that point P is on the line of E and Z. In the case where point P is outside the quadrilateral we consider the difference of areas and the proof is similar.

Exercise 3.14. Suppose that the opposite sides of the quadrilateral $AB\Gamma\Delta$ intersect at points H and Z. Show that the middle E of ZH is on the Newton line of the quadrilateral.

Hint: Use the preceding exercise. In the case of the quadrilateral shown, it suffices to prove that $|\varepsilon(B\Gamma E) - \varepsilon(A\Delta E)| = |\varepsilon(ABE) - \varepsilon(\Delta\Gamma E)|$ or equivalently $\varepsilon(B\Gamma E) + \varepsilon(\Delta\Gamma E) = \varepsilon(ABE) + \varepsilon(A\Delta E)$ (See Figure 3.18-II). This equality results by expressing the areas through the triangles AHZ, ΓHZ, ΔHZ and BHZ, for example, $\varepsilon(A\Delta E) = \varepsilon(AEH) - \varepsilon(\Delta EH) = \varepsilon(AZH)/2 - \varepsilon(\Delta ZH)/2 = \varepsilon(A\Delta Z)/2$ etc.

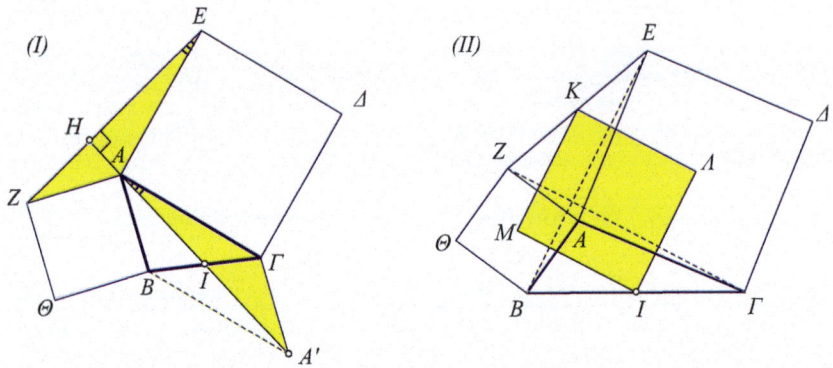

Fig. 3.19: Flank AZE of triangle $AB\Gamma$ Square $K\Lambda IM$ of middles

Exercise 3.15. On the sides AB, $A\Gamma$ of the triangle $AB\Gamma$ we construct squares $AB\Theta Z$ and $A\Gamma\Delta E$ respectively. Show that the median AI of $AB\Gamma$ is orthogonal to the side EZ of AEZ. Also show that the triangle AEZ has the same area with $AB\Gamma$.

Hint: Extend the median AI of $AB\Gamma$ to its double at A' (See Figure 3.19-I). The triangle AZE (called **flank** of the triangle $AB\Gamma$) is congruent to $\Gamma A'A$, because AB and $\Gamma A'$ are equal and parallel and AB is equal to AZ. Also AE and $A\Gamma$ are equal. Finally the angles of the triangles at A and Γ are equal because both are supplementary to $\widehat{BA\Gamma}$. If AH is the altitude of AZE towards ZE, then $|\widehat{HAE}| + |\widehat{\Gamma AA'}| = |\widehat{HAE}| + |\widehat{HEA}| = 90°$. It follows that the sum of angles around A is $|\widehat{HAE}| + |\widehat{EA\Gamma}| + |\widehat{\Gamma AA'}| = 180°$ and the points H, A and A' are collinear. The equality of areas follows easily from the congruence of these triangles.

Exercise 3.16. In the figure of the preceding exercise consider also the middle K of EZ, as well as, the centers Λ and M of the squares $A\Gamma\Delta E$ and $AB\Theta Z$. Show that $I\Lambda KM$ is a square (See Figure 3.19-II).

Hint: The triangles ABE, $AZ\Gamma$ are congruent and have their two sides respectively orthogonal (Exercise 2.164). The exercise is also a special case of theorem 5.27.

Exercise 3.17. Given are three lines through the same point: OX, OY and OZ. Construct a triangle, which has these lines as bisectors.

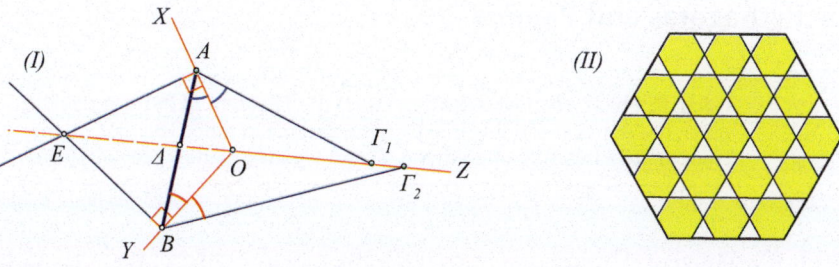

Fig. 3.20: Triangle from bisector lines   Hexagon tiling

Hint: From an arbitrary point A of OX draw an orthogonal to it, intersecting OZ at E (See Figure 3.20-I). Project E onto OY at B. Finally define the symmetric lines of AB: $A\Gamma_1$, relative to OX and $B\Gamma_2$, relative to OB, where Γ_1, Γ_2 are points of OZ. It suffices to show that the two points Γ_1, Γ_2 are coincident. Then the triangle $AB\Gamma$, with $\Gamma = \Gamma_1 = \Gamma_2$ has the desired properties (Theorem 2.3).

The coincidence, however, of Γ_1, Γ_2 follows directly from the fact that, by construction, the triangles $A\Delta\Gamma_1$ and $B\Delta\Gamma_2$ have points O and E as traces of their bisectors (internal/external) at their angles $\widehat{A}$ and $\widehat{B}$ respectively. Consequently (Theorem 3.3, Exercise 3.12, Exercise 1.77) the following relations will hold

$$\frac{|\Delta E|}{|\Delta O|} = \frac{|\Gamma_1 E|}{|\Gamma_1 O|}, \qquad \frac{|\Delta E|}{|\Delta O|} = \frac{|\Gamma_2 E|}{|\Gamma_2 O|}.$$

The left sides on the two equalities are the same, therefore their right sides are equal. Consequently Γ_1, Γ_2 have the same ratio relative to the points E, O, therefore they are coincident (Corollary 1.42).

Exercise 3.18. A regular hexagon with side length δ is tiled with tiles, whose shape is also a regular hexagon of side $\frac{\delta}{5}$. Find the area of the part of the hexagon which is not covered by the tiles (See Figure 3.20-II).

Hint: Analyze each tile into 6 equilaterals congruent to the small triangles which are not covered (See Figure 3.20-II). $\frac{6}{25}$ of the area of the hexagon is the sum of the areas of the triangles which are not covered by the tiles. A small challenge: The calculation of the analogous area for a tiling with tiles of side length $\frac{\delta}{\mu}$, where μ is arbitrary even number.

Exercise 3.19. Show that the maximal, with respect to area, equilateral triangle that can be inscribed inside a rectangle must have a side identical with part of a side of the rectangle.

Exercise 3.20. Show that the maximal with respect to area isosceles triangle that can be inscribed into a rectangle must have a side identical with a side of the rectangle. Find the shapes of these isosceli for a non-square rectangle.

3.4 Pythagoras and Pappus

> Geometry has two great treasures: one is the Theorem of Pythagoras; the other, the division of a line into extreme and mean ratio. The first we may compare to a measure of gold; the second we may name a precious jewel.
>
> *Johannes Kepler*

The importance of the theorem of Pythagoras (apprx. 570-475 B.C.) is made evident by the number of proofs that have been given in various times. The book by Loomis [81] contains 370 proofs of this theorem, the first of which, considered the simplest of all (see § 3.7), is due to Legendre. This theorem, which is proved equivalent to the axiom of parallels (Theorem 3.7), is, one way or another, the basis for all formulas of euclidean geometry.

Theorem 3.4 (Pythagoras). *In every right triangle the square of the hypotenuse is equal to the sum of the squares of its two orthogonal sides.*

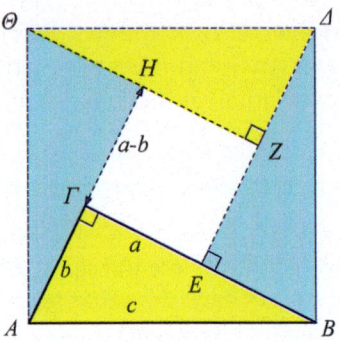

Fig. 3.21: $c^2 = a^2 + b^2$

Proof. Let us suppose that the length of the hypotenuse is $c = |AB|$ and the lengths of the two orthogonal sides are $b = |A\Gamma| \leq a = |B\Gamma|$. Because its angles $\alpha = \widehat{\Gamma AB}$ and $\beta = \widehat{BA\Gamma}$ are complementary, the triangle can be placed on the four sides of the square with side AB, as in figure 3.21. This then forms another square at the center with side length $a - b$. Writing the area of the big square as a sum of the partial areas we have:

$$c^2 = \varepsilon(AB\Delta\Theta) = 4\varepsilon(AB\Gamma) + \varepsilon(\Gamma EZH).$$

Keeping in mind that

$$\varepsilon(AB\Gamma) = \frac{1}{2}ab, \text{ and } \varepsilon(\Gamma EZH) = (a-b)^2 = a^2 + b^2 - 2ab$$

3.4. PYTHAGORAS AND PAPPUS

and substituting the equality of areas we get the wanted equality:
$$c^2 = a^2 + b^2.$$

Theorem 3.5 (converse to Pythagoras' theorem). *If the lengths a, b and c of the sides of a triangle satisfy the equation $a^2 + b^2 = c^2$, then the triangle is a right one with a hypotenuse of length c and orthogonal sides of lengths a and b.*

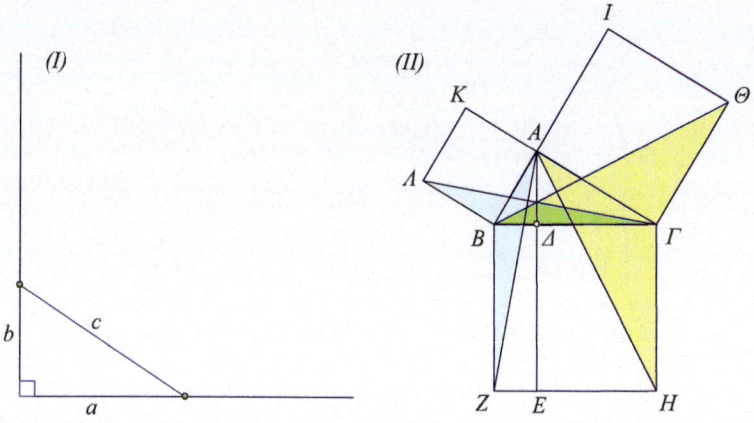

Fig. 3.22: Pythagoras' converse   Proof by Euclid

Proof. On the sides of a right angle we place line segments with lengths, respectively, a and b (See Figure 3.22-I). This forms a right triangle, which according to Pythagoras has hypotenuse of length $a^2 + b^2 = c^2$. The given triangle and the one we constructed have therefore corresponding sides equal, therefore, according to the SSS-criterion, they will be congruent and the given triangle will be right with the right angle opposite to c.

Exercise 3.21. (The proof by Euclid) Prove Pythagoras' theorem by showing, that the area of the square on the hypotenuse is the sum of the areas of the two squares on the orthogonal sides of the triangle.

Hint: In figure 3.22-II, it is to prove that $\varepsilon(B\Gamma HZ)=\varepsilon(A\Gamma\Theta I) + \varepsilon(BAK\Lambda)$. For this, draw the altitude from the right angle $\widehat{A}$ and extend it until the intersection point E with the opposite side ZH of the square $B\Gamma HZ$. The area of this square is equal to the sum of the areas of the two rectangles: $\Delta EH\Gamma$ and ΔEZB. Show that the areas of these rectangles are equal, respectively, to the areas of the squares $A\Gamma\Theta I$ and $BAK\Lambda$. For this, notice that the triangles $A\Gamma H$ and ABZ have area equal to half that of these rectangles (same base and altitude). Notice also that these triangles are congruent respectively with $B\Gamma\Theta$ and $B\Gamma\Lambda$ (SAS-criterion), which in turn have area half that of squares, respectively, $A\Gamma\Theta I$ and $BAK\Lambda$ (same base and same altitude).

Exercise 3.22. In figure 3.22-II show that the lines $\Gamma\Lambda$ and $B\Theta$ intersect on $A\Delta$.

Remark 3.4. Later (Theorem 3.23) we'll see that Pythagoras' theorem can be generalized into a figure which resembles the preceding and in which the square is replaced by a polygon.

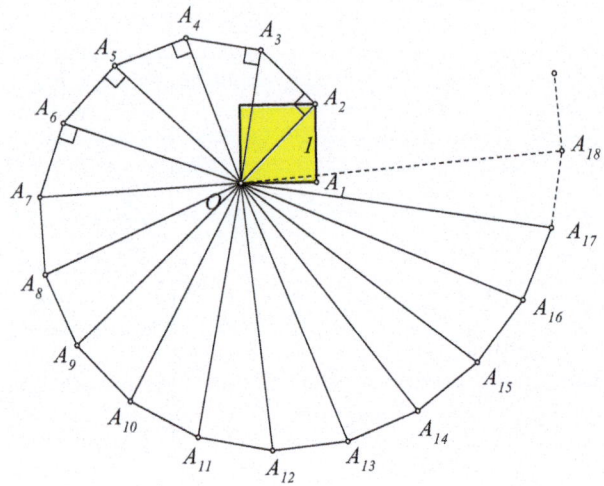

Fig. 3.23: $|OA_2| = \sqrt{2}$, $|OA_3| = \sqrt{3}$, $|OA_4| = \sqrt{4}$, $|OA_5| = \sqrt{5}$,...

Corollary 3.11. *The square roots of the integers $\sqrt{2}$, $\sqrt{3}$, ... can be constructed using the ruler and compass.*

Proof. We begin with the diagonal OA_2 of the unit square which has length $|OA_2| = \sqrt{1+1} = \sqrt{2}$, and using it we construct the right triangle OA_2A_3 with orthogonal sides OA_2 and A_2A_3 with $|A_2A_3| = 1$ (See Figure 3.23). Its hypotenuse is $|OA_3|^2 = |OA_2|^2 + 1^2 = 2 + 1 \Rightarrow |OA_3| = \sqrt{3}$. We repeat the same process using each time the side we constructed ($\sqrt{k}$) and one side of length 1 as orthogonals in order to define the next root $\sqrt{k+1}$.

Remark 3.5. The set of points $\{A_1, A_2, A_3, ...\}$ of figure 3.23 forms the so called **spiral of Theodorus of Kyrenia** (470-399 B.C.). Theodorus is known for his proofs, that the roots of positive integers (which are not perfect squares and are less than 17) are irrational numbers. For his proofs he used the so called *anthyphairesis*, through which the roots of the integers are approximated systematically (see § 1.8). One problem which concerns the spiral of Theodorus, and remains still open, is to find the proper criteria which will determine the best interpolation curve for these points, in other words, to find the best curve which passes through all the points of the spiral. Up to now it seems to be dominant a curve defined in 1993 by Philip Davis, called *curve of Theodorus* ([68], [73]).

3.4. PYTHAGORAS AND PAPPUS

Exercise 3.23. Show that the integers 3, 4 and 5 are lengths of sides of a right triangle, whose radius of the inscribed circle is 1.

Exercise 3.24. Show that for every pair of integers (μ, ν) the integers $a = |\mu^2 - \nu^2|$, $b = 2\mu\nu$, $c = \mu^2 + \nu^2$, provided none is zero, express lengths of sides of a right triangle.

Exercise 3.25. Show that for every positive number a there exists a right triangle with side lengths a, $\frac{a}{2}$, $a\frac{\sqrt{3}}{2}$. What is the connection of such a triangle with the equilateral which has side length a?

Exercise 3.26. Given is a point A in the interior of the circle $\kappa(O, \rho)$. Construct a chord $B\Gamma$ of κ, passing through A and having given length λ.

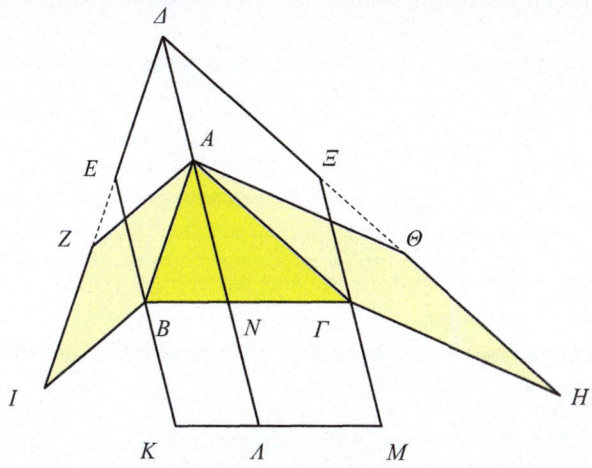

Fig. 3.24: The theorem of Pappus

Theorem 3.6 (Pappus (290-350)). *On the sides AB and $A\Gamma$ of triangle $AB\Gamma$, construct arbitrary parallelograms $ABIZ$ and $A\Gamma H\Theta$. Extend their sides IZ and ΘH until their intersection point Δ. On the third side $B\Gamma$ of the triangle $AB\Gamma$, construct a parallelogram $B\Gamma MK$ such that BK is parallel and equal to ΔA. Then the sum of areas of the parallelograms $ABIZ$ and $A\Gamma H\Theta$ is equal to the area of the parallelogram $B\Gamma MK$.*

Proof. Extend ΔA so that it divides the parallelogram $B\Gamma MK$ to $BN\Lambda K$ and $N\Lambda M\Gamma$ (See Figure 3.24). Obviously $\varepsilon(B\Gamma MK) = \varepsilon(BN\Lambda K) + \varepsilon(N\Lambda M\Gamma)$. However these two parallelograms have areas respectively equal to $\varepsilon(BN\Lambda K) = \varepsilon(BE\Delta A)$ and $\varepsilon(N\Lambda M\Gamma) = \varepsilon(A\Gamma\Xi\Delta)$ because they have equal bases and they are between the same parallels respectively. For the same reason however, equal are also the areas $\varepsilon(IBAZ) = \varepsilon(BE\Delta A)$ and $\varepsilon(A\Gamma H\Theta) = \varepsilon(A\Gamma\Xi\Delta)$.

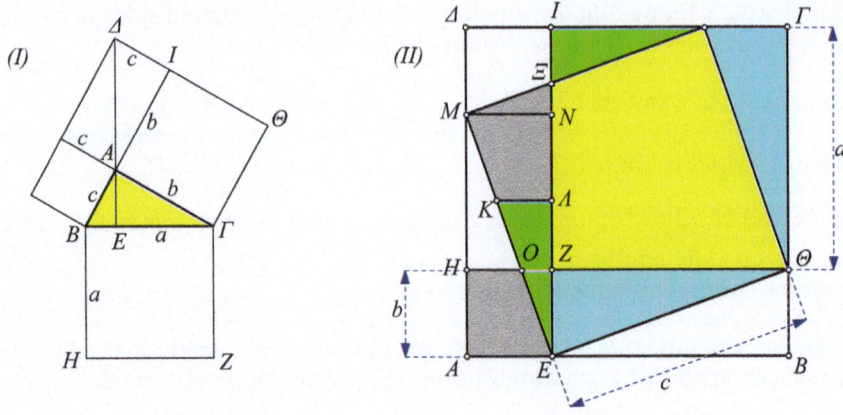

Fig. 3.25: Pythagoras from Pappus   Rejoining squares

Exercise 3.27. Show that the theorem of Pappus generalizes the theorem of Pythagoras. In other words, when $AB\Gamma$ is a right triangle, then $A\Delta$ is orthogonal and equal to $B\Gamma$ and the sum of the areas of the squares on the orthogonal sides is equal to the area of the square on the hypotenuse.

Hint: Proof by the figure (See Figure 3.25-I). $AI\Delta$ is in this case a right triangle congruent to $AB\Gamma$, therefore $B\Gamma$ and $A\Delta$ will be equal. They are also orthogonal because the angles $\widehat{\Delta AI}$ and $\widehat{EAB}$, where AE the altitude from A, are equal and have their sides AI, AB on the same line, therefore they are orthogonal and points Δ, A, E are collinear.

Exercise 3.28. Given two squares with side lengths a, b, show that they can be partitioned into polygons which rearranged in a different order form a new square.

Hint: Using figure 3.25-II. The exercise gives another proof of the Pythagorean theorem (from Thabit Ibn Qurra [65, p.5]).

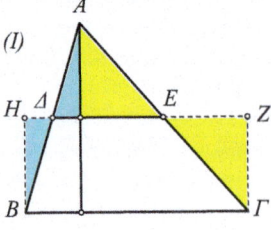 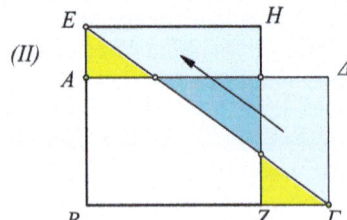

Fig. 3.26: Triangle to rectangle   Rectangle to square

3.4. PYTHAGORAS AND PAPPUS

Exercise 3.29. Show that every triangle can be partitioned into polygons, which rearranged form a rectangle. Show also that every rectangle can be partitioned into polygons, which rearranged form a square.

Hint: For the triangle using Figure 3.26-I. For the rectangle using Figure 3.26-II. In this second figure, $|BE| = \sqrt{|AB||B\Gamma|}$.

Exercise 3.30. Show that every convex polygon of area ε can be partitioned into polygons, which rearranged in different order form a square of the same area.

Hint: If the polygon has v sides, divide the polygon into $v - 2$ triangles using its diagonals from one vertex. Every triangle can be reconstituted into a square of equal area (Exercise 3.29). The result is $v - 2$ squares, which, by applying successively $v - 3$ times exercise 3.28, form, finally, a square.

Fig. 3.27: Partitioning a quadrilateral into parts and rejoining to a square

In figure 3.27 this process is shown for a convex quadrilateral. More about this interesting subject of polygon partitioning in the exercises of the chapter ([16, p.229]).

Proposition 3.1. *For every triangle XAB holds the relation*

$$||XA|^2 - |XB|^2| = 2|AB||X'M|, \tag{3.1}$$

where X' is the projection of X on AB and M the middle of AB.

Proof. Triangles AXX' and BXX' are right at X' (See Figure 3.28-I), therefore $|XA|^2 = |XX'|^2 + |X'A|^2$ and $|XB|^2 = |XX'|^2 + |X'B|^2$. The equality to show follows by subtracting pairwise the equalities, writing $|AX'| = ||AM| - |MX'||$, $|BX'| = |BM| + |MX'|$ and performing calculations.

Corollary 3.12. *Given is a line segment AB. Show that the geometric locus of points X, for which the difference of squares of distances is fixed: $|XA|^2 - |XB|^2 = \kappa$, is a line orthogonal to AB.*

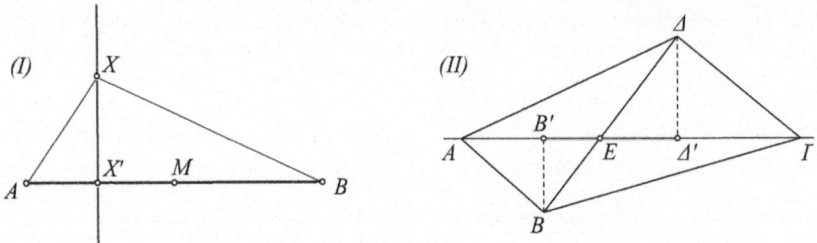

Fig. 3.28: $||XA|^2 - |XB|^2| = 2|AB||X'M|$ Difference of squares

Proof. According to formula (3.1) the position of the projection X' of X on AB will be fixed.

Exercise 3.31. For every quadrilateral $AB\Gamma\Delta$ holds the relation

$$|(|AB|^2 + |\Gamma\Delta|^2) - (|B\Gamma|^2 + |\Delta A|^2)| = 2|A\Gamma||B'\Delta'|,$$

where B' and Δ' are respectively the projections of B, Δ onto the diagonal $A\Gamma$ (See Figure 3.28-II).

Hint: From Pythagoras' theorem and assuming that $|A\Delta| \geq |\Gamma\Delta|$, $|B\Gamma| \geq |BA|$ follow (See also Corollary 3.12) the relations

$$|\Delta A|^2 - |\Delta\Gamma|^2 = |A\Delta'|^2 - |\Delta'\Gamma|^2 = (|A\Delta'| + |\Delta'\Gamma|)(|A\Delta'| - |\Delta'\Gamma|)$$
$$= |A\Gamma|(|A\Delta'| - |\Delta'\Gamma|),$$
$$|B\Gamma|^2 - |BA|^2 = |B'\Gamma|^2 - |AB'|^2 = (|B'\Gamma| + |AB'|)(|B'\Gamma| - |AB'|)$$
$$= |A\Gamma|(|B'\Gamma| - |AB'|) \Rightarrow$$
$$(|\Delta A|^2 + |B\Gamma|^2) - (|\Delta\Gamma|^2 + |BA|^2) = |A\Gamma|(|A\Delta'| - |\Delta'\Gamma| + (|B'\Gamma| - |AB'|))$$
$$= |A\Gamma|((|A\Delta'| - |AB'|) + (|B'\Gamma| - |\Delta'\Gamma|))$$
$$= |A\Gamma|(2|B'\Delta'|).$$

The absolute values guarantee that the expression is valid also in the case where the left side is negative.

Exercise 3.32. Show that if $\{a, b, c\}$ are respectively the lengths of the orthogonal sides and of the hypotenuse of a right triangle and v is the altitude against the hypotenuse, then

$$\frac{1}{v^2} = \frac{1}{a^2} + \frac{1}{b^2}.$$

Exercise 3.33. In the right at Γ triangle $AB\Gamma$ with orthogonal sides of length a, b, we inscribe the square $\Gamma\Delta EZ$, with its other vertices on the sides of the

3.4. PYTHAGORAS AND PAPPUS

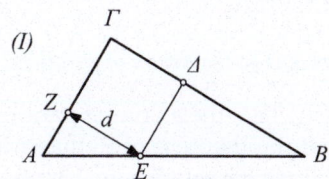
(I)

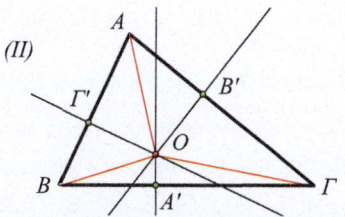
(II)

Fig. 3.29: Square in right triangle   Orthogonals through point

triangle (See Figure 3.29-I). Show that the side d of the square satisfies the relation
$$\frac{1}{d} = \frac{1}{a} + \frac{1}{b}.$$

Exercise 3.34. Consider the points $\{A', B', \Gamma'\}$ respectively on the sides $B\Gamma$, ΓA, and AB of triangle $AB\Gamma$ (See Figure 3.29-II). Show that the orthogonals to these sides at the corresponding points pass through a common point O, if and only if the following relation holds ([65, p.15])
$$|BA'|^2 + |\Gamma B'|^2 + |A\Gamma'|^2 = |A'\Gamma|^2 + |B'A|^2 + |\Gamma'B|^2.$$

Exercise 3.35. Show that, if the sum of squares of two opposite sides of a quadrilateral is equal to the sum of squares of the two other opposite sides, then: (i) the quadrilateral has orthogonal diagonals and (ii) the two line segments which join the middles of opposite sides are equal.

Quadrilaterals with orthogonal diagonals are called **orthodiagonal** (see also Exercise 4.177).

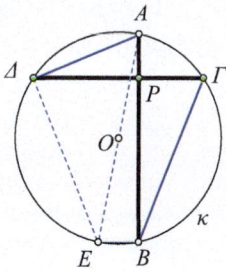
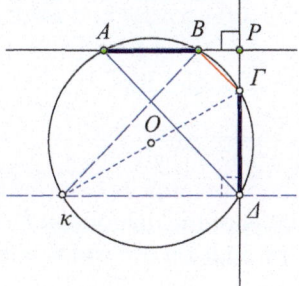

Fig. 3.30: Orthogonal chords

Exercise 3.36. Show that, if the chords AB, $\Gamma\Delta$ of the circle $\kappa(O, \rho)$ intersect orthogonally in the interior/exterior of κ (see Figure 3.30), at a point P, then

$$|PA|^2 + |PB|^2 + |P\Gamma|^2 + |P\Delta|^2 = |A\Delta|^2 + |B\Gamma|^2 = (2\rho)^2.$$

Theorem 3.7. *The theorem of Pythagoras is equivalent to the axiom of parallels.*

Proof. ([84, p.370]) We must show two things: (i) That the axiom of parallels implies the theorem of Pythagoras and (ii) that the theorem of Pythagoras, i.e. that every right triangle with orthogonal sides a,b satisfies $a^2 + b^2 = c^2$, implies the axiom of parallels. In both cases we suppose also the power of all the other axioms of euclidean geometry. (i) is exactly theorem 3.4.

We prove (ii) by finding a triangle, which has sum of angles equal to 180°. This, according to the theorem 1.10 of Legendre, in combination with the theorem 1.13, implies (ii). Indeed, consider the isosceles triangle $AB\Gamma$, right-angled at B and having a, b respectively the lengths of the orthogonal side AB and the hypotenuse $A\Gamma$ (See Figure 3.31). By assumption $a^2 + a^2 = 2a^2 =$

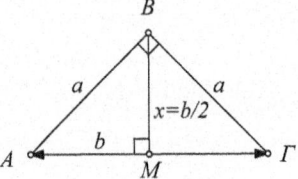

Fig. 3.31: Triangle with sum of angles 180°

$b^2 \Leftrightarrow b = \sqrt{2}a$. Then the median BM, of length x, from the vertex of the right angle is orthogonal to the base of the isosceles (Corollary 1.3) and defines two congruent right triangles. According to the hypothesis we have $x^2 + (b/2)^2 = a^2 \Rightarrow x = b/2$. This means that each one of the triangles BMA, $BM\Gamma$ is isosceles, consequently the angles at their bases will be equal. However their two angles $\widehat{ABM}$ and $\widehat{MB\Gamma}$ at B sum up to 90°, hence also the, equal to them, angles $\widehat{A}$ and $\widehat{\Gamma}$ will add up to 90°, which implies that the triangle $AB\Gamma$ has sum of angles 180°.

Exercise 3.37. (Theorem of Van Aubel) On the sides of a given quadrilateral $AB\Gamma\Delta$ and towards its exterior we construct squares with centers (intersection points of the diagonals) K, M, I, Λ respectively. Show that the line segments IK and ΛM are equal and orthogonal to each other.

Hint: ([94]) Suppose N is the middle of the diagonal $A\Gamma$ (See Figure 3.32-I). Construct the squares $NKPM$, $NIO\Lambda$ (Exercise 3.16). The resulting triangles $N\Lambda M$ and NIK are congruent and have their two sides mutually orthogonal (Exercise 2.164).

Exercise 3.38. In the quadrilateral $K\Lambda IM$ of the preceding exercise we construct anew squares on the sides towards the exterior with centers $OT P\Sigma$.

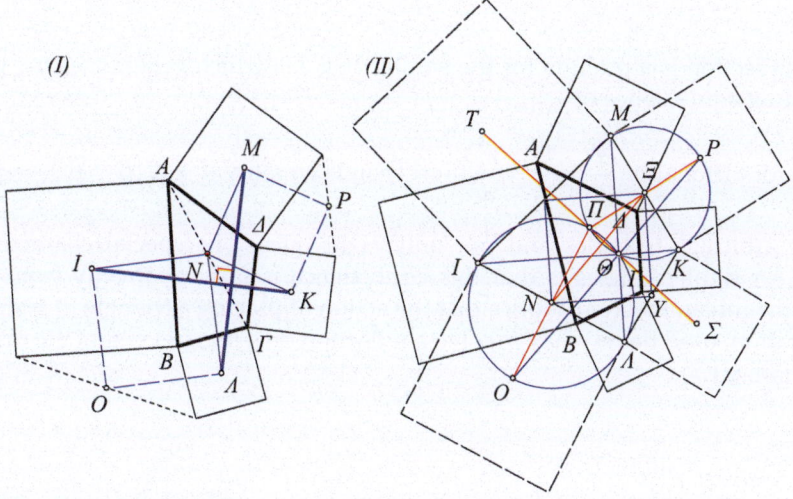

Fig. 3.32: $|IK| = |M\Lambda|$ $\quad |T\Sigma| = |OP|$, $\Theta \in IK \cap M\Lambda \cap OP \cap T\Sigma$

Show that OP and $T\Sigma$ are equal and mutually orthogonal, pass through the intersection point Θ of IK, $M\Lambda$, form with them angle $45°$ and for the ratio of the lengths holds $\frac{|T\Sigma|}{|IK|} = \sqrt{2}$.

Hint: Consider the circles with diameter $I\Lambda$, MK (See Figure 3.32-II). Show that they pass through point Θ and that $N\Xi$ is a hypotenuse of the isosceles triangle $YN\Xi$. A small challenge: show that the other point of intersection Π of the two circles is the middle of $T\Sigma$.

3.5 Similar right triangles

> Professors have a weakness for analogies. So here's one: A gas, any gas, is similar to a crowd of flies. The analogy is dangerous, but we can learn from the dangers.
>
> B.L. Silver, *The Ascent of Science, I*

Similar are called two right triangles which, besides the right angle, have two more corresponding angles equal (therefore all their angles correspondingly equal). In this section we see that this relation between right triangles is equivalent to the relation of **proportionality of their sides**. This again means that the three lengths $\{a,b,c\}$ of the sides of the first triangle and the three lengths $\{a',b',c'\}$ of the corresponding sides of the other triangle satisfy the relation:

$$\frac{a'}{a} = \frac{b'}{b} = \frac{c'}{c}.$$

This section is a prelude for the similarity § 3.9, which generalizes the preceding equivalence of

$$\left(\text{proportionality:}\ \frac{a'}{a} = \frac{b'}{b} = \frac{c'}{c}\right) \Leftrightarrow (\text{Angle equality:}\alpha = \alpha', \beta = \beta', \gamma = \gamma'),$$

for arbitrary triangles and not only right ones. The special characteristic for right triangles is that this equivalence is proved directly from the Pythagorean theorem. Notice that the action of **placing** a triangle in another location, used below, may involve a reflection on a line (reversing the orientation: § 1.7).

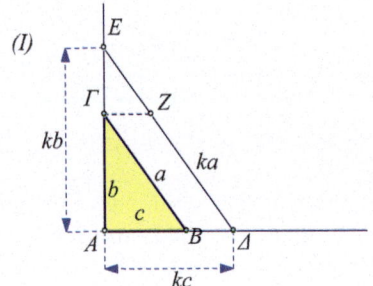

Fig. 3.33: Similar right triangles   Squares inside triangle

Theorem 3.8. *Consider a right angled triangle $AB\Gamma$ with right angle at A, side lengths $a = |B\Gamma|$, $b = |\Gamma A|$, $c = |AB|$ and a positive number k. On the orthogonal half lines AB and $A\Gamma$, consider respectively points Δ and E, such that $|A\Delta| = kc$ and $|AE| = kb$. Then:*

1. *The formed triangle $A\Delta E$ is right angled.*
2. *Its hypotenuse ΔE is parallel to $B\Gamma$.*
3. *Triangles $AB\Gamma$ and $A\Delta E$ have respectively equal angles.*

Proof. The first conclusion follows from the construction (alternatively, the fact that $a^2 = b^2 + c^2$ implies $(ka)^2 = (kb)^2 + (kc)^2$). For the second conclusion, suppose first that $k > 1$ and analyze the area of the triangle $A\Delta E$, which is $(k^2 bc)/2$, as a sum of areas of a triangle and a trapezium (See Figure 3.33-I). To this, draw a parallel from Γ, which intersects ΔE at Z. The area of the resulting trapezium is (Proposition 3.8) $\varepsilon(A\Delta Z\Gamma) = \frac{1}{2}(kc + z)b$, where $z = |\Gamma Z|$. The area of the triangle is $\varepsilon(\Gamma Z E) = \frac{1}{2}z(k-1)b$. The equation between areas gives:

$$\varepsilon(A\Delta E) = \varepsilon(A\Delta Z\Gamma) + \varepsilon(\Gamma Z E),$$

which is equivalent to

3.5. SIMILAR RIGHT TRIANGLES

$$\frac{1}{2}k^2 bc = \frac{1}{2}(kc+z)b + \frac{1}{2}z(k-1)b \quad \Leftrightarrow \quad z = (k-1)c.$$

Because of $|B\Delta| = (k-1)c$, the preceding equality shows that $B\Delta Z\Gamma$ is a parallelogram. From this follows that the two triangles $AB\Gamma$ and $A\Delta E$ have respectively equal angles. The case $k < 1$ can be handled analogously or reduced to the preceding one.

Theorem 3.9. *(Converse of preceding)* Consider the triangle $AB\Gamma$, with right angle at A and side lengths $a = |B\Gamma|$, $b = |\Gamma A|$, $c = |AB|$ and a positive number k. On the orthogonal side AB consider the point Δ, such that $|A\Delta| = kc$ and draw a parallel to $B\Gamma$ from Δ which intersects the other orthogonal at E. Then $|AE| = kb$ and the triangles $AB\Gamma$ and $A\Delta E$ have proportional sides of lengths a, b, c and ka, kb, kc respectively.

Proof. We again use figure 3.33-I, this time with different assumptions. Suppose that $|AE| = k'b$. We will show that $k' = k$. Again, assuming $k > 1$ and performing the analysis of $\varepsilon(A\Delta E) = \varepsilon(A\Delta Z\Gamma) + \varepsilon(\Gamma Z E)$, we get, this time, the equation:

$$\frac{1}{2}kk'bc = \frac{1}{2}((k-1)c+kc)b + \frac{1}{2}(k-1)c(k'-1)b \quad \Leftrightarrow \quad k' = k.$$

From this follows also that $|\Delta E|^2 = (kc)^2 + (kb)^2 = k^2(b^2+c^2) = (ka)^2$ and consequently the stated proportionality of sides. The case $k < 1$ is handled similarly.

Proposition 3.2. *Two right triangles, which have the same angles, have their sides respectively proportional. Conversely, two right triangles, which have their orthogonal sides proportional, have respectively equal angles.*

Proof. For the direct assertion apply Theorem 3.9 placing the two triangles as in the last figure. For the converse apply Theorem 3.8 again placing the triangles in the same way.

Corollary 3.13. *Two right triangles are similar, if and only if they have the same ratio between respective orthogonal sides or when they have the same ratio respectively of one orthogonal side to the hypotenuse.*

Proof. Suppose that the right triangles $AB\Gamma$ and $A'B'\Gamma'$ have the same angles at vertices with same letters and a right angle at A. According to the preceding proposition they are similar, if and only if $\frac{|AB|}{|A'B'|} = \frac{|A\Gamma|}{|A'\Gamma'|}$. The last one, however, is equivalent to $\frac{|AB|}{|A\Gamma|} = \frac{|A'B'|}{|A'\Gamma'|}$. This shows the first claim. The second reduces to the first. Indeed if we suppose that $\frac{|AB|}{|B\Gamma|} = \frac{|A'B'|}{|B'\Gamma'|}$, this one implies

$$\frac{|AB|}{|A\Gamma|} = \frac{|AB|}{\sqrt{|B\Gamma|^2 - |AB|^2}} = \frac{|A'B'|}{\sqrt{|B'\Gamma'|^2 - |A'B'|^2}} = \frac{|A'B'|}{|A'\Gamma'|}.$$

Exercise 3.39. How many quadratic tiles of side δ can be placed in the right angled triangle with orthogonal sides $\{a,b\}$ in the arrangement of the figure 3.33-II ?

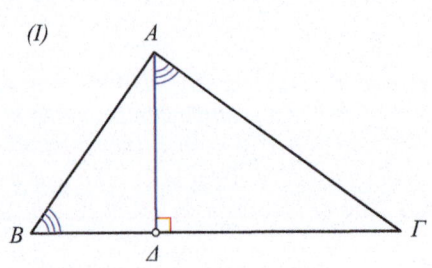
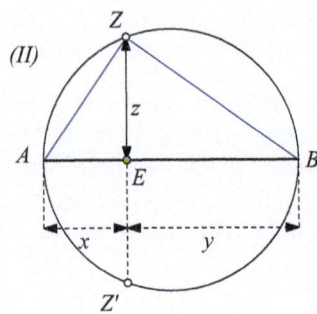

Fig. 3.34: $|A\Delta|^2 = |B\Delta||\Gamma\Delta|$ Mean proportional $z^2 = xy$

Proposition 3.3. *Let Δ be the projection of the vertex A of a right triangle on the hypotenuse $B\Gamma$. Then the following formulas are valid*

$$|A\Delta|^2 = |B\Delta||\Gamma\Delta|, \quad |A\Gamma|^2 = |\Gamma\Delta||\Gamma B|, \quad |AB|^2 = |B\Delta||B\Gamma|.$$

Proof. The right triangles $A\Gamma\Delta$ and $A\Delta B$ have the same angles (See Figure 3.34-I), therefore they will have proportional sides: $\frac{|A\Delta|}{|\Gamma\Delta|} = \frac{|B\Delta|}{|A\Delta|}$, from which the first relation results. Similarly follows the rest by comparing the triangles $AB\Delta$ and $A\Gamma\Delta$ with $AB\Gamma$.

Exercise 3.40. Let $AB\Gamma$ be a right at A triangle and altitude $A\Delta$. Show that $\frac{|AB|^2}{|A\Gamma|^2} = \frac{|B\Delta|}{|\Gamma\Delta|}$.

For three numbers x, y and z, which satisfy the relation $z^2 = xy$, we say that z is the **mean proportional** or **geometric mean** of x and y. The preceding proposition shows one way to construct geometrically the mean proportional of two numbers.

Construction 3.1 *Given the line segments of lengths x and y, construct (using ruler and compass) a line segment having length z, such that $z^2 = xy$.*

Construction: Place the two segments on a line, so that they become successive: AE ($|AE| = x$) and EB ($|EB| = y$) (See Figure 3.34-II). Draw the circle with diameter AB and then draw the orthogonal to AB at E, intersecting the circle at points Z and Z'. The line segments EZ, EZ' have the required length z. Indeed, each one of these points defines a right triangle (Corollary 2.7), whose corresponding segments are the altitudes under the right angle. The conclusion follows by applying the preceding proposition.

3.5. SIMILAR RIGHT TRIANGLES

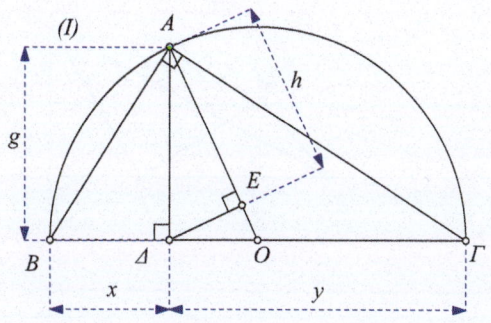
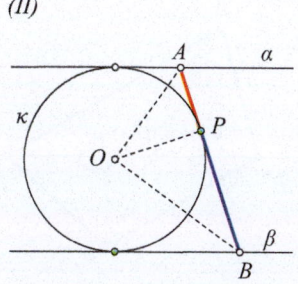

Fig. 3.35: $(h =)\frac{2xy}{x+y} \leq (g =)\sqrt{xy} \leq \frac{x+y}{2}$

Fixed product $|PA||PB|$

For two positive numbers x, y, besides the geometric mean $g = \sqrt{xy}$, we also define the **arithmetic mean** $s = \frac{x+y}{2}$, as well as, the **harmonic mean** $h = \frac{2xy}{x+y}$. Figure 3.35-I, which contains several right triangles, also gives a suggestion for the solution of the next exercise.

Exercise 3.41. Given two positive numbers x, y, the harmonic, the geometric and the arithmetic means satisfy the inequalities

$$\frac{2xy}{x+y} = \frac{2}{\frac{1}{x}+\frac{1}{y}} \leq \sqrt{xy} \leq \frac{x+y}{2}.$$

Equality occurs, if and only if $x = y$.

Exercise 3.42. Show that the product $|PA||PB|$, of the segments on the tangent to circle $\kappa(O, \rho)$ at P, which are excised by two parallel tangents α, β of the circle, is constant (See Figure 3.35-II).

Theorem 3.10. *Two triangles $AB\Gamma$ and $AB\Delta$ have the side $B\Gamma$ in common. Then, the ratio of their areas is equal to the ratio of segments $\frac{|H\Delta|}{|HA|}$, where H is the intersection point of line $B\Gamma$ with $A\Delta$ (See Figure 3.36).*

Proof. The two last figures correspond to the cases where the triangles are on different sides of $B\Gamma$ or on the same side. The proof is the same in both cases. Let ΔE and AZ be the altitudes from the corresponding vertices. Then the ratio of areas is

$$\frac{\varepsilon(B\Gamma\Delta)}{\varepsilon(B\Gamma A)} = \frac{\frac{1}{2}|B\Gamma||\Delta E|}{\frac{1}{2}|B\Gamma||AZ|} = \frac{|\Delta E|}{|AZ|}.$$

The right angled triangles $H\Delta E$ and HAZ are similar, consequently the last ratio is exactly equal to $\frac{|H\Delta|}{|HA|}$.

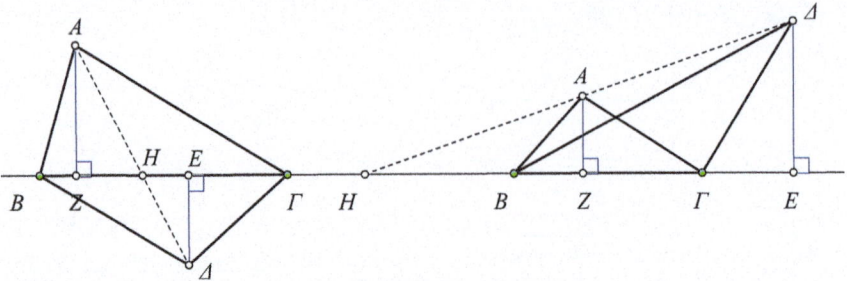

Fig. 3.36: Ratio of areas

Remark 3.6. The last theorem has interesting applications ([74]). Among other things it represents the core of a method of mechanical proof of theorems, which produces, in contrast to other methods, relatively short and readable proofs of theorems in Euclidean Geometry ([66]).

Exercise 3.43. Construct an isosceles triangle $AB\Gamma$, for which is given its circumscribed circle and the ratio of its altitude to its base $\kappa = \frac{|\Gamma M|}{|AB|}$.

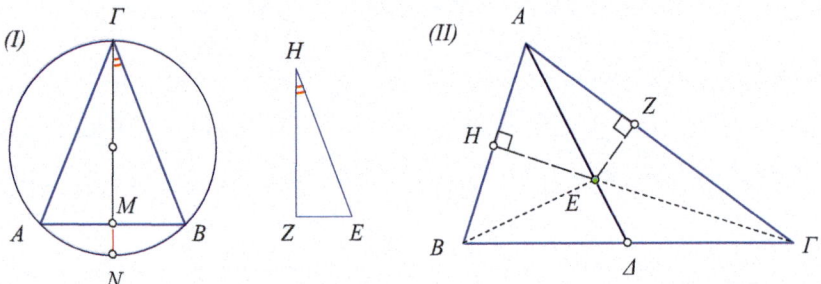

Fig. 3.37: Isosceles construction   Median's points E

Hint: For the right angled triangle ΓMB (See Figure 3.37-I) the ratio of its orthogonal sides is then known.

$$\frac{|\Gamma M|}{|MB|} = \frac{|\Gamma M|}{(|AB|/2)} = 2\kappa.$$

Consequently a similar right triangle HZE to ΓMB can be constructed, with the same ratio of orthogonals, for example, by taking on a right angle $|ZH| = 2\kappa$ and $|ZE| = 1$. Next, on a diameter ΓN of the given circle, we form angle

3.5. SIMILAR RIGHT TRIANGLES

$\widehat{N\Gamma B}$ equal to $\widehat{ZHE}$ and we consider the symmetric A of B relative to the diameter.

Exercise 3.44. Show that the points E of the median $A\Delta$ of triangle $AB\Gamma$ are characterized by the fact, that their distances from the sides AB, $A\Gamma$ are inversely proportional to the lengths of these sides.

Hint: Triangles $B\Delta E$, $\Delta E\Gamma$ have equal area, therefore also BEA, $E\Gamma A$ (See Figure 3.37-II). It follows that $|EH||AB| = |EZ||A\Gamma|$. For the converse, begin from the last relation, which is equivalent to the equality of areas of BEA, $E\Gamma A$. Apply Theorem 3.10 for the ratio of areas and conclude that Δ is the middle of $B\Gamma$.

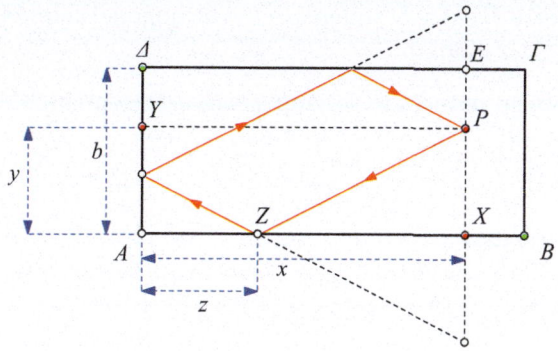

Fig. 3.38: Closed billiard trajectory

Exercise 3.45. Given is a rectangle $AB\Gamma\Delta$ of dimensions $a = |AB|$, $b = |B\Gamma|$ (See Figure 3.37). A billiard ball moves starting from point P and, after being successively reflected on the walls AB, $A\Delta$, $A\Gamma$, returns to its initial position P. Show that if $|AX| = x$, $|AY| = y$ and $|AZ| = z$ are the distances from A of the projections of P on AB, $A\Delta$ and Z is the position of the first bounce on side AB, then the relation holds $z \cdot b = x \cdot (b - y)$ and the sides of the closed parallelogram of the trajectory are parallel to the diagonals of $A\Delta EX$, where E is the projection of P on $\Delta\Gamma$ (see also the similar exercise II-2.83).

Exercise 3.46. On the extensions of sides AB and $A\Gamma$ of the triangle $AB\Gamma$ are taken equal segments BB' and $\Gamma\Gamma'$. Show that the geometric locus of the middles M of segments $B'\Gamma'$ is a line parallel to the bisector of angle $\widehat{A}$ passing through the middle N of $B\Gamma$.

Hint: Draw a parallel and equal $|NX| = |BB'|$, $|NY| = |\Gamma\Gamma'|$ (See Figure 3.39-I). Triangle XNY is isosceles and NM is a bisector of its apex angle (see also Exercise 3.95).

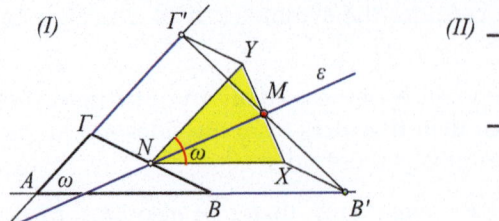 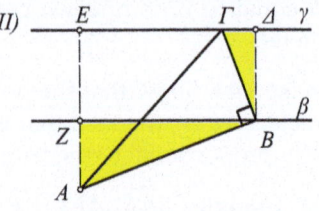

| Fig. 3.39: Line of middles | Vertices on parallels |

Exercise 3.47. Construct a right triangle $AB\Gamma$, of which the vertex of one acute angle $\widehat{A}$ is a given point, the vertices B and Γ lie on given parallels β and γ and the difference of squares of the orthogonal sides $|AB|^2 - |B\Gamma|^2$ is equal to a given constant δ.

Hint: The problem reduces to solving a *biquadratic* equation. Indeed, if we set $x = |AB|$, $y = |B\Gamma|$, then, by assumption, we have $x^2 - y^2 = \delta$ (∗). Also the right triangles BAZ, $B\Gamma\Delta$ (See Figure 3.39-II) have respective sides orthogonal, therefore the angles $\widehat{ABZ}$, $\widehat{\Gamma B\Delta}$ are equal, therefore the triangles are similar. It follows that

$$\frac{|AB|}{|BZ|} = \frac{|B\Gamma|}{|B\Delta|} \Rightarrow \frac{|AB|}{\sqrt{|AB|^2 - |AZ|^2}} = \frac{|B\Gamma|}{|B\Delta|} \Rightarrow \frac{x}{\sqrt{x^2 - \kappa^2}} = \frac{y}{\lambda},$$

where $\kappa = |AZ|$, $\lambda = |B\Delta|$, fixed constants. The last, in combination with (∗), determines completely the lengths x, y and consequently the right triangle.

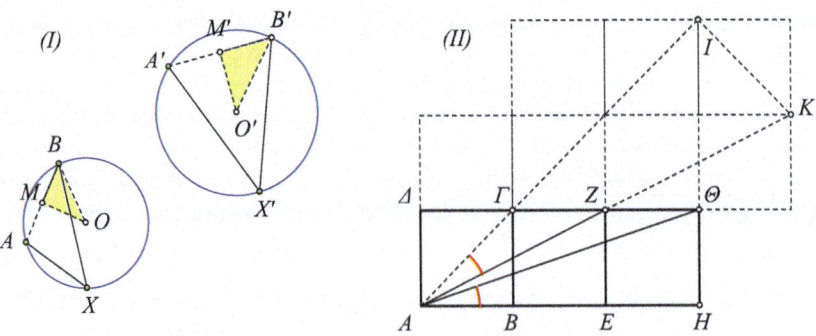

Fig. 3.40: Similar right triangles

3.6. THE TRIGONOMETRIC FUNCTIONS

Exercise 3.48. Show that if the points X, X' see respectively the line segments $AB, A'B'$ under equal or supplementary angles, then the ratio $\frac{|AB|}{|A'B'|}$ is equal to the ratio of the diameters $\frac{d}{d'}$ of the circles $(AXB), (A'X'B')$ (See Figure 3.40-I).

Exercise 3.49. Given are three congruent squares $AB\Gamma\Delta, BEZ\Gamma, EH\Theta Z$, placed successively as in figure 3.40-II. Show that the angles $\widehat{ZA\Gamma}$ and $\widehat{HA\Theta}$ are equal.

Hint: Construct the additional squares as in figure 3.40-II. Show that triangles $AH\Theta$ and AIK are similar right triangles.

3.6 The trigonometric functions

> And those who seek for the best kind of song and music ought not to seek for that which is pleasant, but for that which is true.
>
> *Plato, Laws book II*

The just proved equivalence of similarity and proportionality of sides for right triangles leads directly to the definition of the **trigonometric functions** of an angle. Currently we limit ourselves to angles which appear in right triangles, in other words acute angles. Indeed, let $\widehat{XOY}$ be an acute angle

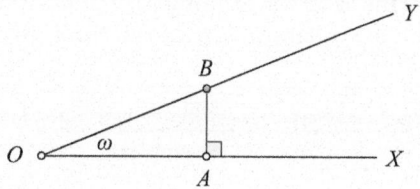

Fig. 3.41: $\frac{|AB|}{|OB|}$ independent of A's position

and B be a point of the side OY, which we project to A on OX, so that the right triangle OAB is created. According to the conclusions of the preceding section, the ratios

$$s = \frac{|AB|}{|OB|}, \quad c = \frac{|OA|}{|OB|},$$

and therefore also the ratios

$$t = \frac{|AB|}{|OA|} = \frac{s}{c}, \quad t' = \frac{|OA|}{|AB|} = \frac{1}{t}$$

do not depend on the particular position of B on the half line OY, since the resulting triangles have proportional sides. Consequently these quantities depend exclusively on the measure ω of the angle $\widehat{XOY}$ and vary only when this angle changes. Therefore there are defined corresponding functions of the measure ω of the angle:

$$\sin(\omega) = s, \quad \cos(\omega) = c, \quad \tan(\omega) = t, \quad \cot(\omega) = t',$$

which are called respectively **sine, cosine, tangent** and **cotangent** of the angle ω. These functions are not mutually independent but are connected with some relations. In the last two this is obvious from their very definition:

$$\tan(\omega) = \frac{1}{\cot(\omega)} = \frac{\sin(\omega)}{\cos(\omega)}.$$

But also for the first two it follows from the theorem of Pythagoras (See Figure 3.41) that

$$|AB|^2 + |OA|^2 = |OB|^2 \implies \left(\frac{|AB|}{|OB|}\right)^2 + \left(\frac{|OA|}{|OB|}\right)^2 = 1$$

in other words

$$\sin^2(\omega) + \cos^2(\omega) = 1.$$

Essentially therefore, the independent function is one and we can choose anyone from the four we want and express the rest of the functions in terms of the one we chose. This way for example, if I choose $\sin(\omega)$ as the basic trigonometric function, the others are expressed through this one:

$$\cos(\omega) = \sqrt{1 - \sin^2(\omega)}, \quad \tan(\omega) = \frac{\sin(\omega)}{\sqrt{1 - \sin^2(\omega)}}, \quad \cot(\omega) = \frac{\sqrt{1 - \sin^2(\omega)}}{\sin(\omega)}.$$

Exercise 3.50. Show that for every angle with measure $0° < \omega < 90°$ the trigonometric functions satisfy

$$|\sin(\omega)| < 1 \quad \text{and} \quad |\cos(\omega)| < 1,$$

as well as

$$\sin(90° - \omega) = \cos(\omega), \quad \cos(90° - \omega) = \sin(\omega),$$
$$\tan(90° - \omega) = \cot(\omega), \quad \cot(90° - \omega) = \tan(\omega).$$

Exercise 3.51. Show that the trigonometric functions are expressed using $\cos(\omega)$ and the formulas:

3.6. THE TRIGONOMETRIC FUNCTIONS

$$\sin(\omega) = \sqrt{1-\cos^2(\omega)}, \quad \tan(\omega) = \frac{\sqrt{1-\cos^2(\omega)}}{\cos(\omega)}, \quad \cot(\omega) = \frac{\cos(\omega)}{\sqrt{1-\cos^2(\omega)}}.$$

Exercise 3.52. Show that the trigonometric functions are expressed using $\cot(\omega)$ and the formulas:

$$\sin(\omega) = \frac{\tan(\omega)}{\sqrt{1+\tan^2(\omega)}}, \quad \cos(\omega) = \frac{1}{\sqrt{1+\tan^2(\omega)}}, \quad \cot(\omega) = \frac{1}{\tan(\omega)}.$$

Exercise 3.53. Show that the trigonometric functions are expressed using $\tan(\omega)$ and the formulas:

$$\sin(\omega) = \frac{1}{\sqrt{1+\cot^2(\omega)}}, \quad \cos(\omega) = \frac{\cot(\omega)}{\sqrt{1+\cot^2(\omega)}}, \quad \tan(\omega) = \frac{1}{\cot(\omega)}.$$

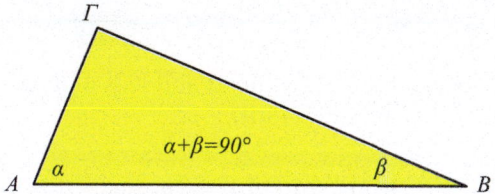

Fig. 3.42: Complementary angles

Theorem 3.11. *In every right triangle $AB\Gamma$, with acute angles $\alpha = \widehat{BA\Gamma}$ and $\beta = \widehat{AB\Gamma} = 90° - \alpha$, the following relations are valid (See Figure ??):*

1. $|B\Gamma| = |AB|\sin(\alpha)$: *the opposite orthogonal is equal to the hypotenuse times the sine of the angle.*
2. $|A\Gamma| = |AB|\cos(\alpha)$: *the adjacent orthogonal is equal to the hypotenuse times the cosine of the angle.*
3. $\tan(\alpha) = \frac{|B\Gamma|}{|A\Gamma|}$: *the tangent is equal to the ratio of the opposite over the adjacent orthogonal.*
4. $\cot(\alpha) = \frac{|A\Gamma|}{|B\Gamma|}$: *the cotangent is equal to the ratio of the adjacent over the opposite orthogonal.*
5. $\sin(\alpha) = \cos(\beta), \quad \cos(\alpha) = \sin(\beta), \quad \tan(\alpha) = \cot(\beta), \quad \cot(\alpha) = \tan(\beta).$

In every triangle $AB\Gamma$, the following holds for area:

$$\varepsilon(AB\Gamma) = \frac{1}{2}|AB||A\Gamma|\sin(\alpha) = \frac{1}{2}|B\Gamma||BA|\sin(\beta) = \frac{1}{2}|\Gamma A||\Gamma B|\sin(\gamma).$$

Proof. All properties are direct consequences of the definitions. The first four are put forward for their memorization and only. For the last, read the proof from the figure 3.43-I.

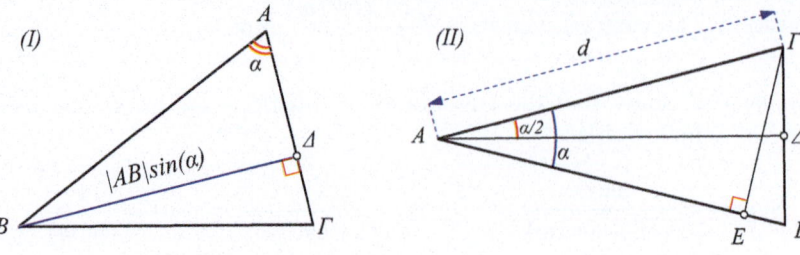

Fig. 3.43: $\varepsilon(AB\Gamma) = \frac{1}{2}|AB||A\Gamma|\sin(\alpha)$ Double angle

Exercise 3.54. Show that for an acute angle α the following formulas hold:

$$\sin(\alpha) = 2\sin\left(\tfrac{\alpha}{2}\right)\cdot\cos\left(\tfrac{\alpha}{2}\right), \quad \cos(\alpha) = \cos^2\left(\tfrac{\alpha}{2}\right) - \sin^2\left(\tfrac{\alpha}{2}\right),$$

$$\tan(\alpha) = \frac{2\tan\left(\tfrac{\alpha}{2}\right)}{1-\tan^2\left(\tfrac{\alpha}{2}\right)}, \quad \cot(\alpha) = \frac{\cot^2\left(\tfrac{\alpha}{2}\right)-1}{\cot\left(\tfrac{\alpha}{2}\right)}.$$

Hint: Consider an isosceles with vertex α and legs of length d and calculate its area in two ways (See Figure 3.43-II).

$$\varepsilon(AB\Gamma) = \frac{1}{2}|B\Gamma|\cdot|A\Delta| = |B\Delta|\cdot|A\Delta| = \left(d\cdot\sin\left(\tfrac{\alpha}{2}\right)\right)\left(d\cdot\cos\left(\tfrac{\alpha}{2}\right)\right)$$
$$= d^2\cdot\sin\left(\tfrac{\alpha}{2}\right)\cdot\cos\left(\tfrac{\alpha}{2}\right).$$
$$\varepsilon(AB\Gamma) = \frac{1}{2}|AB|\cdot|\Gamma E| = \frac{1}{2}d\cdot(d\cdot\sin(\alpha)) = \frac{1}{2}d^2\cdot\sin(\alpha).$$

The first equality follows by equating the two expressions for the same area. The second equality follows from the first and the basic properties of the trigonometric functions:

$$\cos(\alpha) = \sqrt{1-\sin^2(\alpha)} = \sqrt{1-\left(2\sin\left(\tfrac{\alpha}{2}\right)\cdot\cos\left(\tfrac{\alpha}{2}\right)\right)^2}$$
$$= \sqrt{1-\left(4\sin^2\left(\tfrac{\alpha}{2}\right)\cdot\cos^2\left(\tfrac{\alpha}{2}\right)\right)}$$
$$= \sqrt{1-\left(4\sin^2\left(\tfrac{\alpha}{2}\right)\cdot\left(1-\sin^2\left(\tfrac{\alpha}{2}\right)\right)\right)}$$
$$= \sqrt{1-4\sin^2\left(\tfrac{\alpha}{2}\right)+4\sin^4\left(\tfrac{\alpha}{2}\right)} = \sqrt{\left(1-2\sin^2\left(\tfrac{\alpha}{2}\right)\right)^2}$$
$$= 1-2\sin^2\left(\tfrac{\alpha}{2}\right) = \left(1-\sin^2\left(\tfrac{\alpha}{2}\right)\right)-\sin^2\left(\tfrac{\alpha}{2}\right) = \cos^2\left(\tfrac{\alpha}{2}\right)-\sin^2\left(\tfrac{\alpha}{2}\right).$$

The remaining equalities follow directly from the preceding two.

Exercise 3.55. Show that for an acute angle α the following formulas hold:

3.6. THE TRIGONOMETRIC FUNCTIONS

$$\sin^2\left(\frac{\alpha}{2}\right) = \frac{1-\cos(\alpha)}{2} \;,\; \cos^2\left(\frac{\alpha}{2}\right) = \frac{1+\cos(\alpha)}{2},$$
$$\tan\left(\frac{\alpha}{2}\right) = \frac{\sin(\alpha)}{1+\cos(\alpha)} \;,\; \cot\left(\frac{\alpha}{2}\right) = \frac{1+\cos(\alpha)}{\sin(\alpha)}.$$

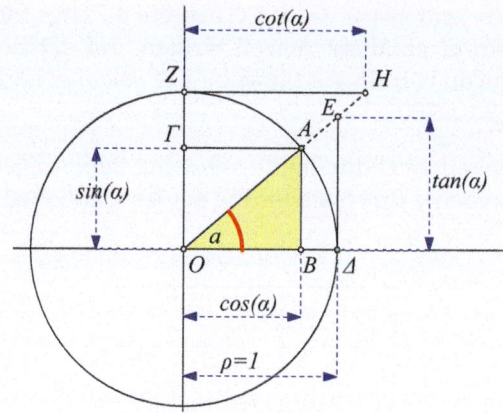

Fig. 3.44: Trigonometric circle

Remark 3.7. By their very definition the trigonometric functions are intimately related to the unit circle (circle with radius equal to the measuring unit). To see this, draw from the center of the unit circle two mutually orthogonal lines $O\Delta$ and OZ (See Figure 3.44). Then every point A of the arc $\overset{\frown}{\Delta Z}$ defines a central angle $\alpha = \overset{\frown}{\Delta O A}$ as well as the right triangle OAB, where B is the projection of A on $O\Delta$. The numbers $\cos(\alpha), \sin(\alpha), \tan(\alpha), \cot(\alpha)$ are expressed then as *lengths* of the sides of the triangle OAB and its similar ones $O\Delta E$ and OZH, which result by intersecting OA with the tangents at Δ and Z respectively.

$$\cos(\alpha) = |OB|, \sin(\alpha) = |AB|, \tan(\alpha) = |\Delta E|, \cot(\alpha) = |ZH|.$$

According to their definition as ratios, for the sine e.g. $\sin(\alpha) = \frac{|AB|}{|OA|} = |AB|$, because $|OA| = 1$. Similarly, all the other trigonometric numbers result from the ratios with which they are defined and the fact, that in this specific form the denominator is 1. I will not expand further this subject. I am just touching on here as a prelude and first contact with the so called **trigonometric circle**, which is exactly the unit circle we saw and the description of its points as pairs of numbers

$$(x = \cos(\alpha), y = \sin(\alpha)).$$

Remark 3.8. The trigonometric functions sin(x), cos(x), ... together with the elementary functions x, x^2, x^3, ... of powers of x are met in almost all applications of mathematics. They therefore are of special interest and, as we saw from their definition, their existence is a direct consequence of the Pythagorean theorem. As it happens with the basic metric relations between lengths and angles we just saw for right triangles, more generally also all relations between lengths and angles in figures of Euclidean geometry (all of the trigonometry) eventually reduce to the Pythagorean theorem. Its power seals this geometry and as we proved (Theorem 3.7) this theorem is equivalent to the axiom of parallels. A well written and detailed study of the trigonometric functions from the viewpoint of calculus is contained in the book by Spivak [90, p.300].

Exercise 3.56. Show the validity of the following table, which gives the values of the trigonometric functions for the angles mentioned in the first column.

angle	sin	cos	tan	cot
0	0	1	0	$\pm\infty$
30°	$\frac{1}{2}$	$\frac{\sqrt{3}}{2}$	$\frac{\sqrt{3}}{3}$	$\sqrt{3}$
45°	$\frac{\sqrt{2}}{2}$	$\frac{\sqrt{2}}{2}$	1	1
60°	$\frac{\sqrt{3}}{2}$	$\frac{1}{2}$	$\sqrt{3}$	$\frac{\sqrt{3}}{3}$
90°	1	0	$\pm\infty$	0

Exercise 3.57. Show that the trigonometric functions sin(x), tan(x) are increasing and the functions cos(x), cot(x) are decreasing in the interval $0° \leq x \leq 90°$

Hint: To begin with, a function $f(x)$ is called *increasing*, when $x < x'$ implies $f(x) < f(x')$. $f(x)$ is called respectively *decreasing*, when $x < x'$ implies $f(x) > f(x')$. The property of the exercise follows directly from the representation of sin(x), cos(x), ... etc. as line segments in the unit circle (See Figure 3.44). As the angle α increases, the line segments which represent sin(α), tan(α) themselves increase, while the line segments which represent cos(α), cot(α) decrease.

Exercise 3.58. Show that for every regular polygon with v sides the ratios of its side δ, respectively of the radius σ of its inscribed circle, over the radius ρ of its circumscribed circle are equal to (See Figure 3.45-I):

$$\frac{\delta}{\rho} = 2\sin\left(\frac{180°}{v}\right) \quad \text{and} \quad \frac{\sigma}{\rho} = \cos\left(\frac{180°}{v}\right).$$

(I) 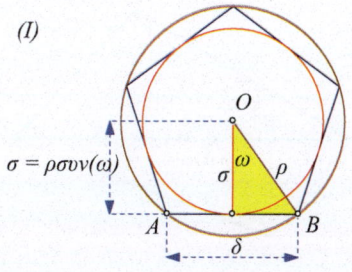   (II)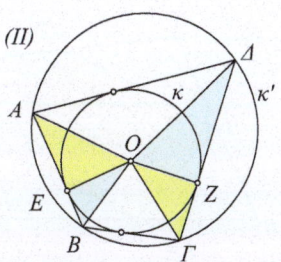

Fig. 3.45: Circumscribed/inscribed radii   Bicentric quadrilateral

Exercise 3.59. Show that the circumscriptible in a circle, convex quadrilateral $AB\Gamma\Delta$, is simultaneously also inscriptible, if and only if the contact points E, Z of AB and $\Gamma\Delta$ satisfy

$$\frac{|EA|}{|EB|} = \frac{|Z\Delta|}{|Z\Gamma|}.$$

Hint: If the quadrilateral is inscribed in circle κ' and its inscribed circle κ has radius ρ, then the angles at A and Γ are supplementary and the right triangles AEO and $OZ\Gamma$ are similar (See Figure 3.45-II). It follows that $\frac{\rho}{|AE|} = \frac{|Z\Gamma|}{\rho} \Rightarrow |Z\Gamma||AE| = \rho^2$. Similarly we show that $|EB||Z\Delta| = \rho^2$, which, in combination with the preceding, gives the requested relation. For the converse we show that if the quadrilateral is not inscriptible, then the equality does not hold. Indeed, if $AB\Gamma\Delta$ is not inscriptible, then it will have a pair of opposite angles, for example $\widehat{A}$, $\widehat{\Gamma}$, with sum of measures respectively $\alpha + \gamma < 180°$ and the other pair $\beta + \delta > 180°$. Then $\frac{\alpha}{2} < 90° - \frac{\gamma}{2}$ and consequently (Exercise 3.57) $\frac{\rho}{|AE|} = \tan\left(\frac{\alpha}{2}\right) < \tan\left(90° - \frac{\gamma}{2}\right) = \frac{|Z\Gamma|}{\rho} \Rightarrow \rho^2 < |EA||Z\Gamma|$. Similarly, from the other inequality follows $\frac{\rho}{|Z\Delta|} = \tan\left(\frac{\delta}{2}\right) > \tan\left(90° - \frac{\beta}{2}\right) = \frac{|EB|}{\rho} \Rightarrow \rho^2 > |EB||Z\Delta|$. From the two inequalities $|EA||Z\Gamma| > \rho^2 > |EB||Z\Delta| \Rightarrow \frac{|EA|}{|EB|} > \frac{|Z\Delta|}{|Z\Gamma|}$.

A quadrilateral, like the one of the preceding exercise, which is simultaneously circumscriptible in circle κ and inscriptible in another circle κ', is called **bicentric** (see also exercise 2.179). This simple characterization of the bicentric quadrilateral is relatively recent (Sinefakopoulos 2003 [89]).

Exercise 3.60. Let $AB\Gamma\Delta$ be a convex quadrilateral with E, Z, H, Θ the middles of its sides. Show that the area of the parallelogram $EZH\Theta$ is half that of the quadrilateral and equals

$$\varepsilon(EZH\Theta) = \frac{1}{4}|A\Gamma||B\Delta|\sin(\omega),$$

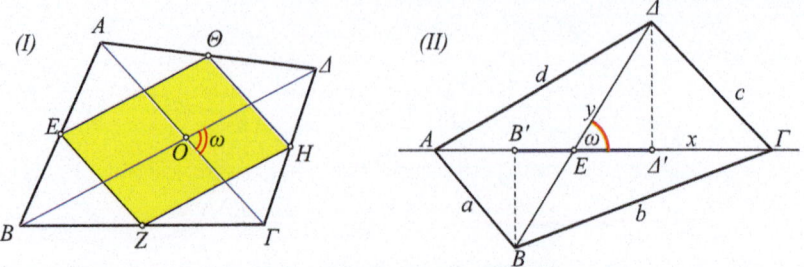

Fig. 3.46: Quadrilateral area

where ω is the acute or right angle between the diagonals of the quadrilateral (See Figure 3.46-I).

Exercise 3.61. Consider a convex quadrilateral $AB\Gamma\Delta$ with side lengths $a = |AB|$, $b = |B\Gamma|$, $c = |\Gamma\Delta|$, $d = |\Delta A|$ and diagonal lengths $x = |A\Gamma|$, $y = |B\Delta|$. Show that
$$\varepsilon(AB\Gamma\Delta) = \frac{1}{4}\sqrt{4x^2y^2 - (a^2 + c^2 - b^2 - d^2)^2}.$$

Hint: If ω is the measure of the angle between the diagonals of the quadrilateral, then, according to Exercise 3.60, the area $\varepsilon = \varepsilon(AB\Gamma\Delta)$ is equal to

$$\varepsilon^2 = \frac{1}{4}x^2y^2\sin(\omega)^2 = \frac{1}{4}x^2y^2(1 - \cos^2(\omega))$$
$$= \frac{1}{4}(x^2y^2 - (xy\cos(\omega))^2).$$

However Exercise 3.31, the last product is equal to
$$xy\cos(\omega) = |A\Gamma||B'\Delta'| = \frac{1}{2}|a^2 + c^2 - b^2 - d^2|,$$

where B' and Δ' are the projections of B and Δ respectively on $A\Gamma$ (See Figure 3.46-II). The formula follows by replacing the last expression into the preceding one.

Remark 3.9. The so called **addition formulas** of the trigonometric functions ($\cos(\alpha + \beta)$, $\sin(\alpha + \beta)$, etc.) are examined in § 3.11.

Exercise 3.62. Express the trigonometric functions of an acute angle inscribed in a circle in terms of the diameter and the chords defined by the angle.

Exercise 3.63. Given two points A and B and a circle $\kappa(O, \rho)$, such that A, B and O are not collinear, draw two parallels AX and BY, so that they excise from the circle two equal chords. Investigate when the problem has a solution.

3.7 The theorem of Thales

> Although it is true that it is the goal of science to discover rules which permit the association and foretelling of facts, this is not its only aim. It also seeks to reduce the connections discovered to the smallest possible number of mutually independent conceptual elements.
>
> A. Einstein, *Science and Religion p. 49*

The theorem of Thales (624-547 B.C.), which, along with the theorem of Pythagoras, are the two most important theorems of euclidean geometry, examines the relation of lengths of line segments *OA* and *OB*, which are defined on the sides of an angle $\widehat{XOY}$ by a line which is displaced but remains parallel to a fixed line ε. The theorem says that there exists a constant κ, such

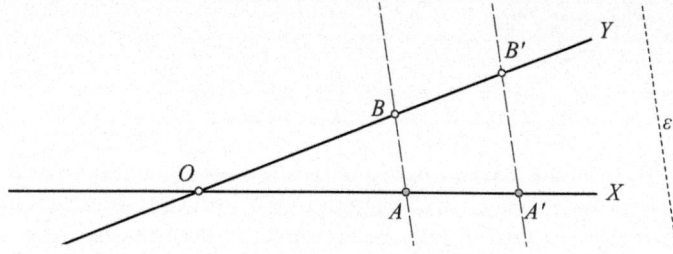

Fig. 3.47: $|OB| = \kappa \cdot |OA|$

that $|OB| = \kappa \cdot |OA|$ for every position of *A* on *OX* (See Figure 3.47). In the preceding section we examined a special case of this phenomenon, according to which line ε is orthogonal to *OY*. Then the corresponding κ defined exactly $\cos(\widehat{XOY})$.

Theorem 3.12. *Given is an angle $\widehat{XOY}$ and a line ε intersecting both sides of the angle. From the points A of OX we draw parallels to ε intersecting OY at B. Then the ratio*

$$\kappa = \frac{|OB|}{|OA|}$$

is fixed and independent of the position of A on OX. Conversely, if for two points A, A' of OX and two points B, B' of OY the ratios

$$\frac{|OB|}{|OA|} = \frac{|OB'|}{|OA'|}$$

are equal, then AB and A'B' are parallel.

Proof. For the first claim it suffices, for a second line $A'B'$ parallel to ε, and consequently also parallel to *AB* (See Figure 3.48-I), to show the equality of

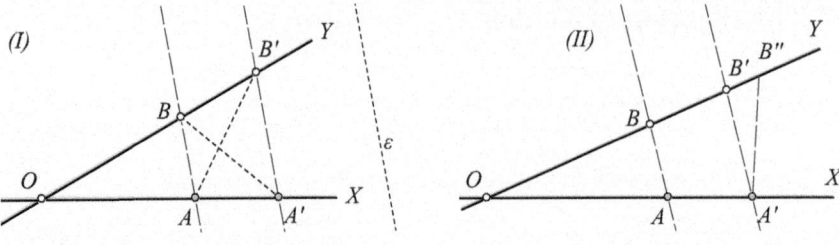

Fig. 3.48: Thales, the proof ...and the converse

ratios
$$\frac{|OB|}{|OA|} = \frac{|OB'|}{|OA'|} \Leftrightarrow \frac{|OA'|}{|OA|} = \frac{|OB'|}{|OB|}.$$

In the last equality, the two ratios are respectively equal to the ratios of areas of the triangles (Corollary 3.6):

$$\frac{|OA'|}{|OA|} = \frac{\varepsilon(OBA')}{\varepsilon(OBA)}, \quad \frac{|OB'|}{|OB|} = \frac{\varepsilon(OAB')}{\varepsilon(OAB)}.$$

Finally, the ratios of areas are equal, as the denominators are equal and the numerators also equal, since the two triangles OAB' and OBA' have common part OAB (figure 3.48) and differ only with respect to the triangles ABA' and ABB' which have equal areas (Corollary 3.6).

For the converse we use the proved first part (See Figure 3.48-II). Suppose therefore that $\frac{|OB|}{|OA|} = \frac{|OB'|}{|OA'|}$ and let $A'B''$ be the parallel from A' to AB. It suffices to show that $B' = B''$. Indeed, according to the first part, $\frac{|OB|}{|OA|} = \frac{|OB''|}{|OA'|}$ will be valid, which combined with the preceding equality gives $\frac{|OB'|}{|OA'|} = \frac{|OB''|}{|OA'|}$, from which follows $|OB'| = |OB''|$ and consequently the desired identity $B' = B''$.

Corollary 3.14. *With the assumptions of the theorem, for every line segment AA' on OX and the corresponding segment BB' on OY the ratio $\frac{|BB'|}{|AA'|}$ will be fixed ($= \kappa$) and independent of the position of A, A' on OX (See Figure 3.48-I).*

Proof. Let us suppose that A is between O and A', therefore (Axiom 1.7)

$$|OA'| = |OA| + |AA'|,$$

and because of the fact that AB and $A'B'$ are parallel, B will also be between O and B' consequently will also hold

$$|OB'| = |OB| + |BB'|.$$

According to Thales, there exists κ such that $|OB| = \kappa \cdot |OA|$ for every A on OX and its corresponding B on OY. Multiplying therefore with κ the first

3.7. THE THEOREM OF THALES

relation we have
$$\kappa \cdot |OA'| = \kappa \cdot |OA| + \kappa \cdot |AA'|,$$
which means that $|OB'| = |OB| + k|AA'|$ and comparing with $|OB'| = |OB| + |BB'|$ follows $|BB'| = \kappa \cdot |AA'|$.

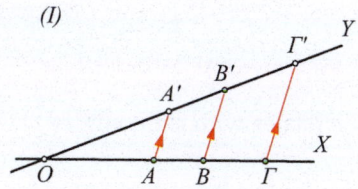 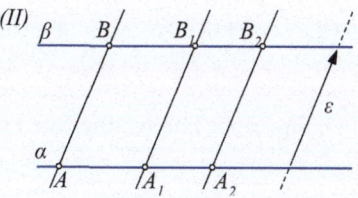

Fig. 3.49: Ratio transfer   Generalized of property

Corollary 3.15. *Let A, B, Γ be points on the line OX and A', B', Γ' points on the line OY, such that AA', BB' and $\Gamma\Gamma'$ are parallel lines. Then*
$$\frac{|AB|}{|B\Gamma|} = \frac{|A'B'|}{|B'\Gamma'|}.$$

Conversely, if this relation holds and two out of the three segments $\{AA', BB', CC'\}$ are parallel, then all three segments are parallel.

Proof. The relation (See Figure 3.49-I) is equivalent to $\frac{|AB|}{|A'B'|} = \frac{|B\Gamma|}{|B'\Gamma'|}$, which is valid according to the preceding corollary. The converse is proved using the already proved part of the corollary, exactly as with the converse of the theorem of Thales (see proposition 3.4 below).

Remark 3.10. The formulation of the theorem of Thales, which is expressed with the two corollaries, generalizes the property concerning two parallel lines α and β intersected with other lines parallel to a fixed line ε (See Figure 3.49-II). From the equality of opposite sides of a parallelogram follows then that $\frac{|BB_1|}{|AA_1|} = 1$. In other words in the case where the α, β are parallels the conclusion of the corollary still holds with constant $\kappa = 1$. In the case where the α, β intersect, the difference lies in the value of the fixed constant, and we have in general (see Exercise 3.70) $\kappa \neq 1$.

Corollary 3.16. *Let the points A, A' be on side OX and B, B' on side OY of the angle $\widehat{XOY}$, so that AB and $A'B'$ are parallel. Then holds $\frac{|OA'|}{|OA|} = \frac{|A'B'|}{|AB|}$.*

Proof. Draw AE parallel to OB and intersecting $A'B'$ at E (See Figure 3.50-I). According to Corollary 3.14, applied to the angle $\widehat{OA'B'}$ and on the two parallels AE, BB':
$$\frac{|A'O|}{|A'B'|} = \frac{|AO|}{|EB'|}.$$

However $|AB| = |EB'|$, therefore $\frac{|OA|}{|AB|} = \frac{|OA'|}{|A'B'|}$.

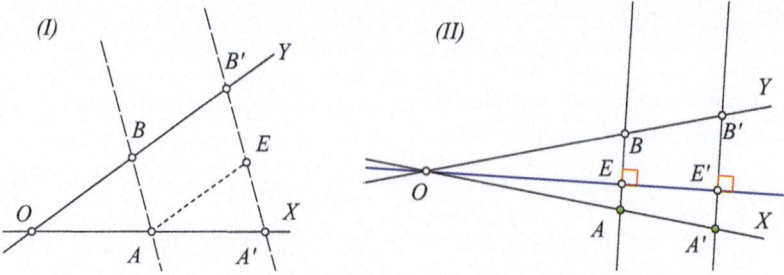

Fig. 3.50: Thales, the third side   Thales with right angles

Exercise 3.64. Give a proof of Thales' theorem using the similarity of two right triangles.

Hint: The exercise is related to Proposition 3.2. Draw the orthogonal from O to ε and define the altitudes OE and OE' of the triangles OAB and $OA'B'$ (See Figure 3.50-II). The created right triangles are similar: OAE to $OA'E'$ and OBE to $OB'E'$. According to the aforementioned proposition, they will have proportional sides: $\frac{|OA|}{|OA'|} = \frac{|OE|}{|OE'|}$ as well as $\frac{|OB|}{|OB'|} = \frac{|OE|}{|OE'|}$, therefore finally $\frac{|OB|}{|OB'|} = \frac{|OA|}{|OA'|}$.

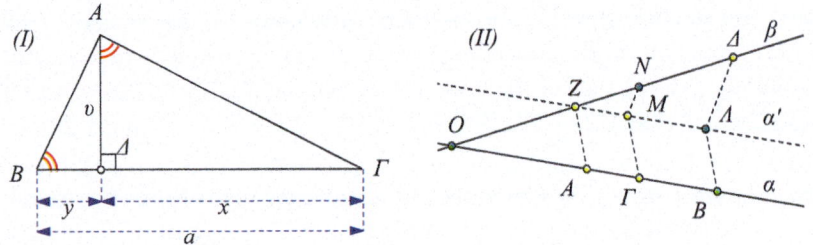

Fig. 3.51: Pythagoras from Thales   Ratio transfer

Remark 3.11. The proof we gave for the theorem of Thales does not use the theorem of Pythagoras. The proof of the exercise however uses it, therefore, one may say that the theorem of Thales is a consequence of the theorem of Pythagoras. The converse also holds. One can show that the theorem of Pythagoras follows from the theorem of Thales. First one must observe that by Drawing the altitude $A\Delta$ ($\upsilon = |A\Delta|$, $x = |\Gamma\Delta|$, $y = |\Delta B|$) from the right angle of the right triangle forms the right triangles $A\Gamma\Delta$ and $A\Delta B$ which have the same angles with the initial triangle (See Figure 3.51-I). Placing these triangles so their equal angles become coincident and their opposite sides become parallel the following relations result according to Thales:

3.7. THE THEOREM OF THALES

$$\frac{c}{y} = \frac{b}{v} = \frac{a}{c}, \text{ and } \frac{v}{x} = \frac{y}{v} \Rightarrow v^2 = xy.$$

From these, by squaring, results:

$$c^2 = \frac{a^2}{c^2}y^2, \quad b^2 = \frac{a^2}{c^2}v^2,$$

$$c^2 + b^2 = \frac{a^2}{c^2}(y^2 + v^2) = \frac{a^2}{c^2}(y^2 + xy) = \frac{a^2}{c^2}y(y+x) = \frac{a^2}{c^2}y \cdot a = a^2.$$

Exercise 3.65. Complete all the details which were omitted in the preceding remark.

Exercise 3.66. Given are two lines α and β intersecting at O. Given also are three points A, B, Γ on α and two points Z, Δ on β (See Figure 3.51-II). Find a point N on β such that $\frac{|\Gamma A|}{|\Gamma B|} = \frac{|NZ|}{|N\Delta|}$.

Hint: Draw from Z the parallel α' to α and from Γ and B parallels to AZ, intersecting the parallel α' at M and Λ respectively. Draw $\Lambda\Delta$ and from M the parallel to $\Lambda\Delta$, intersecting β at N. This is the wanted point. Because of parallels ΓM and $B\Lambda$: $\frac{|\Gamma A|}{|\Gamma B|} = \frac{|MZ|}{|M\Lambda|}$. Also because of the parallels MN and $\Lambda\Delta$: $\frac{|MZ|}{|M\Lambda|} = \frac{|NZ|}{|N\Delta|}$.

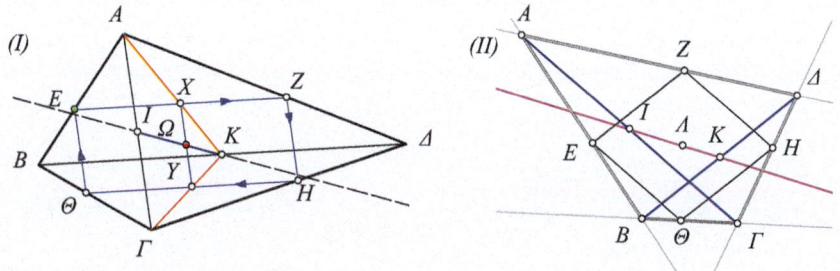

Fig. 3.52: Parallelograms in quadrilateral... and a rhombus: $\frac{|A I|}{|A K|} = \frac{|A\Gamma|}{|B\Delta|}$

Exercise 3.67. From an arbitrary point E on the side AB of the convex quadrilateral $AB\Gamma\Delta$, which is not a parallelogram, we draw the parallel EZ to the diagonal $B\Delta$ (See Figure 3.52-I). From Z we draw the parallel ZH to the diagonal $A\Gamma$ and from H the parallel $H\Theta$ to $B\Delta$. Show that $EZH\Theta$ is a parallelogram. Also show that the point of intersection Ω of the diagonals of $EZH\Theta$ lies on the Newton line of the quadrilateral.

Hint: Let $\frac{|EA|}{|EB|} = \kappa$ (See Figure 3.52-I). Then also $\frac{|AZ|}{|Z\Delta|} = \kappa$ (Corollary 3.15). Similarly also $\frac{|\Gamma H|}{|H\Delta|} = \kappa$ and $\frac{|\Gamma\Theta|}{|\Theta B|} = \kappa$ and finally $\frac{|EA|}{|EB|} = \frac{|\Theta\Gamma|}{|\Theta B|}$, which implies that $A\Gamma$ and ΘE are parallel. The second part of the exercise follows from the fact that the intersection point of the diagonals of the parallelogram $EZH\Theta$ coincides with the middle of the segment XY, where X, Y are the middles of EZ and ΘH. AX and ΓY intersect at the middle of $B\Delta$ and $K\Omega$ passes through the middle I of $A\Gamma$. Complete the proof of these claims and with them the proof of the exercise.

Exercise 3.68. Show that in every convex quadrilateral $AB\Gamma\Delta$, which is not a parallelogram, Newton's line intersects each pair of opposite sides into proportional parts.

Hint: From the preceding exercise follows that the Newton line IK passes through the center of the parallelogram $EZH\Theta$ (See Figure 3.52-I). It follows that if E belongs to IK, then H also belongs to IK etc.

Exercise 3.69. Show that every convex quadrilateral $AB\Gamma\Delta$ with diagonal middles I, K has exactly one rhombus inscribed with sides parallel to the quadrilateral's diagonals. The intersection point Λ of the diagonals of the rhombus belongs to the Newton line of $AB\Gamma\Delta$ and divides the interval IK into a ratio equal to the ratio of the quadrilateral's diagonals (See Figure 3.52-II).

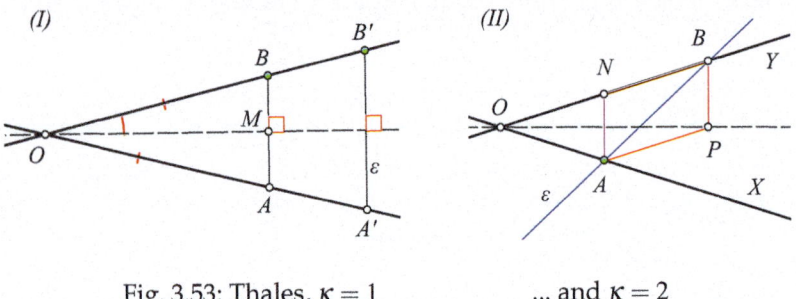

Fig. 3.53: Thales, $\kappa = 1$ ... and $\kappa = 2$

Exercise 3.70. Given is an angle $\widehat{XOY}$. Find a line ε such that the parallels to ε excise on the angle sides respective segments AA' and BB' (See Figure 3.53-I), such that $\frac{|BB'|}{|AA'|} = 1$.

Exercise 3.71. Given is an angle $\widehat{XOY}$. Find a line ε, such that the parallels to ε excise on the angle sides respective segments AA' and BB' such that $\frac{|BB'|}{|AA'|} = 2$ (See Figure 3.53-II).

Hint: Take an arbitrary point A on OX and then a point B on OY, such that $|OB| = 2|OA|$. Line AB is the requested one. Another construction for a parallel to the preceding line AB is the following. From point A of side OX draw a parallel to the other side which intersects the bisector of $\widehat{XOY}$ at P. From P draw orthogonal to the bisector which intersects OY at B. Line AB is the requested one ($|OB| = 2|OA|$).

3.8. PENCILS OF LINES

Exercise 3.72. Next paradox follows from the sloppy usage of figure 3.54, in which line *EH* is drawn thick in order to mislead the reader ([85, p.14]). Here is the puzzle: A square $AB\Gamma\Delta$ is divided into $8 \times 8 = 64$ congruent squares. Next, drawing the lines of the left figure, we divide it into two congruent quadrilaterals and two congruent triangles. We move these parts and place them as in the figure to the right, so that a rectangle with $5 \times 13 = 65$ little squares results. Because the total area must be the same, irrespective of the placing of the parts, it holds 64=65. How does Thales restores here the truth?

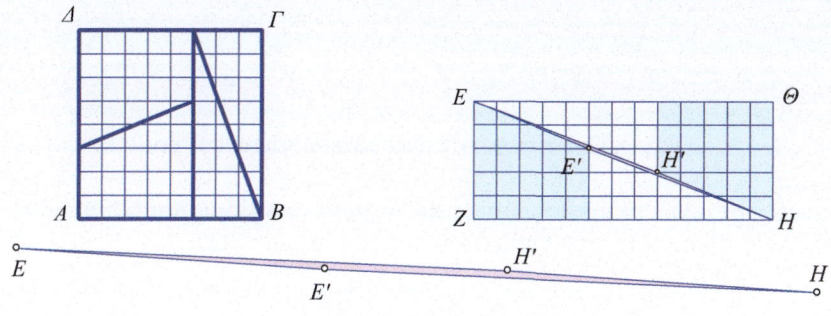

Fig. 3.54: $64 = 65$!!

Hint: The "line" *EH*, in reality is a very thin parallelogram of area equal to that of the little square. A magnification of this parallelogram can be seen in the bottom part of the figure.

3.8 Pencils of lines

> A picture of reality drawn in a few sharp lines cannot be expected to be adequate to the variety of all its shades. Yet even so the draftsman must have the courage to draw the lines firm.
>
> H. Weyl, *Philosophy of Mathematics, p. 274*

The theorem of Thales is connected with the so called line **pencils**. These consist of all the lines that pass through a given point *O*, called the **center** of the pencil. We also call **parallel** pencil the set of lines, which are parallel to a specific line, called the **direction** of the pencil (See Figure 3.55). We often use the term *pencil* for *some* of the lines which either pass through a point *O* or are parallel to a given line (and not *all* lines with this property). The first version of the theorem talks about the segments which are defined on every line which intersects a parallel pencil.

Theorem 3.13. *A pencil of parallel lines, which intersects lines* $\{\varepsilon, \varepsilon_1, \varepsilon_2, ...\}$, *defines segments on them:* $\{AB, B\Gamma, \Gamma\Delta, ...\}$, $\{A_1B_1, B_1\Gamma_1, \Gamma_1\Delta_1, ...\}$, *such that any* two of the $\{|AB|, |B\Gamma|, |\Gamma\Delta|, ...\}$ have the same ratio with the corresponding two from the $\{|A_1B_1|, |B_1\Gamma_1|, |\Gamma_1\Delta_1|, ...\}$.

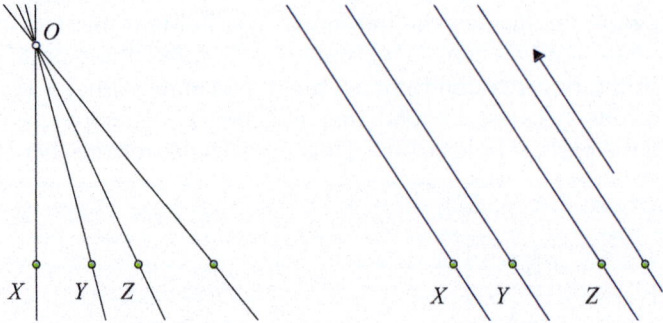

Fig. 3.55: Point pencil and parallel pencil of lines

Proof. The proof of the equality of the respective ratios follows immediately from Corollary 3.15 (See Figure 3.56), according to which

$$\frac{|A_1B_1|}{|AB|} = \frac{|B_1\Gamma_1|}{|B\Gamma|} = \frac{|\Gamma_1\Delta_1|}{|\Gamma\Delta|} = ...,$$

which is a series of equalities equivalent to the mentioned ones.

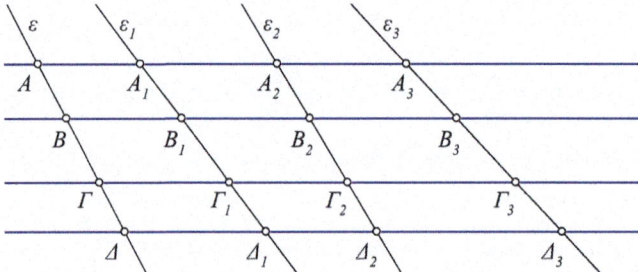

Fig. 3.56: Segments defined on lines intersecting a pencil of parallels

Remark 3.12. Last proposition underlines the fact that the ratios between segments $AB, B\Gamma, \Gamma\Delta, ...$, is a characteristic of the parallel pencil and not of the special line which they intersect. These ratios, for example, are equal to the ratios of the segments defined on a line, which is orthogonal to the lines of the parallel pencil (See Figure 3.57). The lengths of these segments define the distances $a, b, c, ...$ of the parallels in pairs. The ratios therefore of the segments which are excised on other lines are equal to the respective ratios of parallel distances.

$$\frac{|AB|}{|B\Gamma|} = \frac{a}{b}, \quad \frac{|B\Gamma|}{|\Gamma\Delta|} = \frac{b}{c}, \quad ...$$

3.8. PENCILS OF LINES

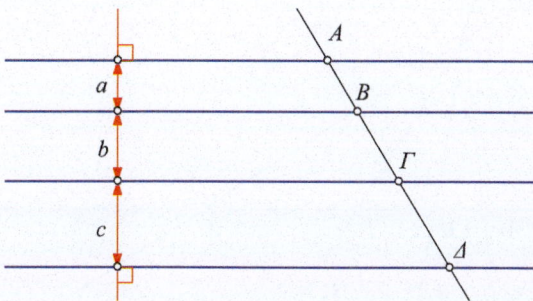

Fig. 3.57: Ratios of distances between parallels

Construction 3.2 *Divide a line segment AB into ν equal parts (See Figure 3.58).*

Construction: At the end A of the line segment draw an arbitrary line segment AE_1 and subsequently define on the line AE_1 the equidistant points $\{A = E_0, E_1, E_2, E_3, ..., E_{\nu-1}, E_\nu = \Gamma\}$, dividing $A\Gamma$ into ν equal parts. Draw then ΓB and subsequently the parallels to it $\{E_1 A_1, E_2 A_2, ...\}$, which define on AB the points $A_1, A_2, ..., A_\nu = B$. According to the last proposition, the ratios $\frac{|AA_1|}{|A_1 A_2|}$, $\frac{|A_1 A_2|}{|A_2 A_3|}$, ... will be respectively equal to the ratios $\frac{|AE_1|}{|E_1 E_2|}$, $\frac{|E_1 E_2|}{|E_2 E_3|}$, ... which by construction are all equal to one.

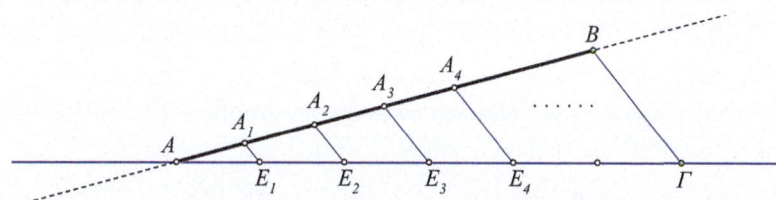

Fig. 3.58: Division into ν equal parts

The second version of Thales' theorem deals with segments which are defined on the lines of a parallel pencil when latter are intersected by the lines of a point pencil.

Theorem 3.14. *A point pencil intersecting the parallel lines $\{\varepsilon, \varepsilon_1, \varepsilon_2, ...\}$, defines line segments on each of them: $\{AB, B\Gamma, \Gamma\Delta, ...\}$, $\{A_1 B_1, B_1 \Gamma_1, \Gamma_1 \Delta_1, ...\}$, such that the ratios of any two of the one set are equal to the ratios of the corresponding segments from the other set (See Figure 3.59).*

Proof. The proof results directly from Corollary 3.16. According to it, we have equality for the ratios $\frac{|AB|}{|A_1 B_1|} = \frac{|OA|}{|OA_1|}$ and similarly for the other ratios, for example $\frac{|B\Gamma|}{|B_1 \Gamma_1|} = \frac{|OB|}{|OB_1|}$. But, by Thales the following ratios are also equal $\frac{|OA|}{|OA_1|} = \frac{|OB|}{|OB_1|}$, implying that $\frac{|AB|}{|A_1 B_1|} = \frac{|B\Gamma|}{|B_1 \Gamma_1|}$, which is equivalent to $\frac{|AB|}{|B\Gamma|} = \frac{|A_1 B_1|}{|B_1 \Gamma_1|}$, which is one of the requested equalities of ratios. The others are proved in exactly the same way.

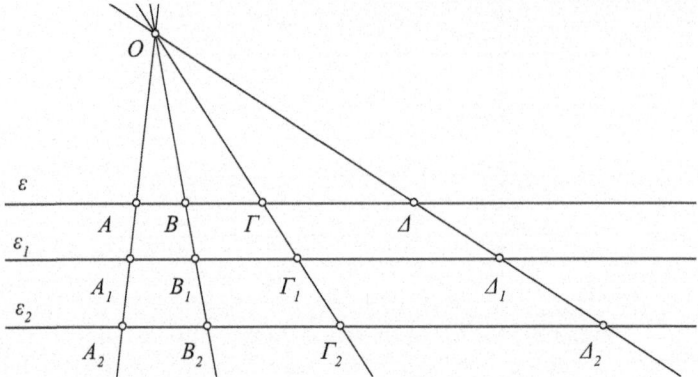

Fig. 3.59: Point pencil incident on parallels

The last two propositions have also converses, frequently used when we want to show that three lines are parallel or that they pass through a common point.

Proposition 3.4. *If three lines are intersected by two other lines ε and ε_1 at points $\{A, B, \Gamma\}$, respectively at $\{A_1, B_1, \Gamma_1\}$ and make equal ratios $\frac{|AB|}{|B\Gamma|} = \frac{|A_1 B_1|}{|B_1 \Gamma_1|}$ and the first two lines AA_1 and BB_1 are parallel, then the third line $\Gamma\Gamma_1$ is also parallel to the first two (See Figure 3.60).*

Proof. From Γ draw the parallel to AA_1 and BB_1 which intersects ε_1 at point Γ'. By Thales $\frac{|AB|}{|B\Gamma|} = \frac{|A_1 B_1|}{|B_1 \Gamma'|}$. However also by assumption $\frac{|AB|}{|B\Gamma|} = \frac{|A_1 B_1|}{|B_1 \Gamma_1|}$, therefore $|B_1 \Gamma'| = |B_1 \Gamma_1|$, which means that Γ' and Γ_1 are coincident.

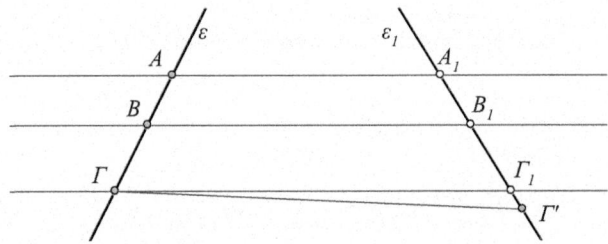

Fig. 3.60: Criterion for three lines to be parallel

Proposition 3.5. *If two parallel lines ε and ε_1 intersected by three others at points $\{A, B, \Gamma\}$ and $\{A_1, B_1, \Gamma_1\}$ respectively, make equal ratios $\frac{|AB|}{|B\Gamma|} = \frac{|A_1 B_1|}{|B_1 \Gamma_1|}$, then the three lines are either parallel or they pass through the same point O.*

Proof. If two of the lines are parallel, for example AA_1 and BB_1, then, applying the preceding criterion, we see that the third line $\Gamma\Gamma_1$ will also be parallel

3.8. PENCILS OF LINES

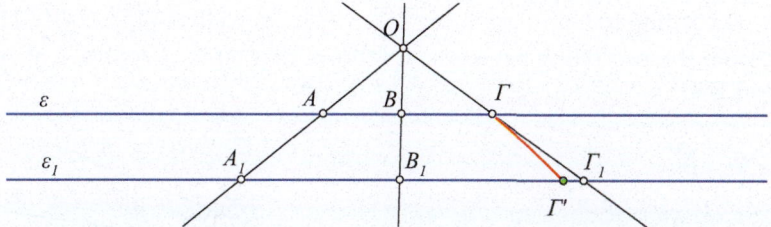

Fig. 3.61: Criterion of three lines to pass through a point

to the other two. Suppose therefore that the first two lines are not parallel and let O be their point of intersection (See Figure 3.61). We will show that the third line also passes through the same point. The proof is the same with the preceding one. We draw $O\Gamma$ and consider the intersection point Γ' of this line with ε_1. According to Theorem 3.14, we have $\frac{|AB|}{|B\Gamma|} = \frac{|A_1 B_1|}{|B_1 \Gamma'|}$. But also by assumption $\frac{|AB|}{|B\Gamma|} = \frac{|A_1 B_1|}{|B_1 \Gamma_1|}$, therefore $|B_1 \Gamma'| = |B_1 \Gamma_1|$, which means that Γ' and Γ_1 are coincident.

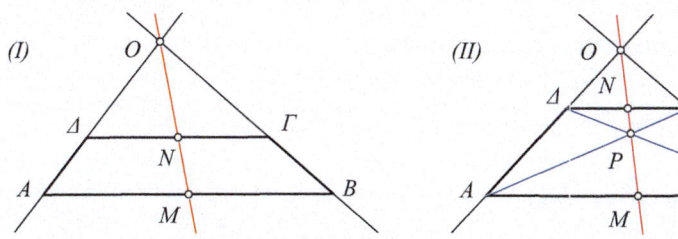

Fig. 3.62: Property of the trapezium   ...and its diagonals

Exercise 3.73. Show that the line MN, which joins the middles of the parallel sides of a trapezium $AB\Gamma\Delta$, passes through the intersection point O of its two non-parallel sides.

Hint: Apply the last criterion for the ratios $\frac{|\Delta N|}{|N\Gamma|} = \frac{|AM|}{|MB|} = 1$ (See Figure 3.62-I).

Exercise 3.74. Show that the line MN, which joins the middles of the parallel sides of trapezium $AB\Gamma\Delta$, passes through the intersection point P of its diagonals.

Hint: Almost the same with the preceding one (See Figure 3.62-II): Apply the last criterion for the ratios $\frac{|\Delta N|}{|N\Gamma|} = \frac{|BM|}{|MA|} = 1$. Which is the subtle difference from the preceding exercise?

Exercise 3.75. Show in figure 3.62-II that the ratios $\frac{|PN|}{|PM|}$ and $\frac{|ON|}{|OM|}$ are equal, therefore (M,N,O,P) form a harmonic quadruple.

Hint: According to Corollary 3.16: $\frac{|PN|}{|PM|} = \frac{|N\Delta|}{|MB|}$, as well as $\frac{|ON|}{|OM|} = \frac{|N\Delta|}{|MA|}$. However, from the preceding exercises, it follows that $|MA| = |MB|$ (see also Exercise 3.164).

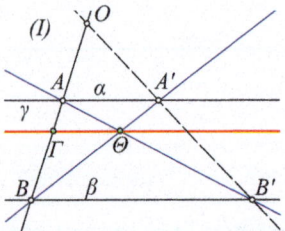
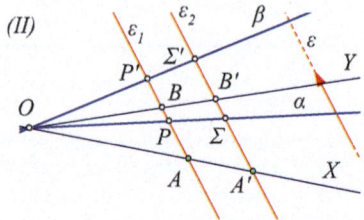

Fig. 3.63: Criterion through O constant ratio $\frac{|PA|}{|PB|} = \kappa$

Exercise 3.76. Let A, B be points lying respectively on the parallel lines α, β. Let also A', B' be variable points lying respectively, on α, β (See Figure 3.63-I). Show that the line $A'B'$ passes through a fixed point O of AB, if and only if the intersection point Θ of AB' and $A'B$ is contained in a line γ parallel to α, β and passing through the harmonic conjugate Γ of O relative to A, B.

Hint: Application of the preceding exercise.

Proposition 3.6. *Consider an angle $\widehat{XOY}$, a line ε and a positive number κ. There exist exactly two lines α and β, such that for every point P of these lines we have $\frac{|PA|}{|PB|} = \kappa$, where A and B are respectively the intersections with OX and OY of the parallel to ε from P (See Figure 3.63-II).*

Proof. Consider a parallel ε_1 to ε, which intersects OX, OY at A, B respectively. By Corollary 1.42 there exist exactly two points P and P' on ε_1 with the property $\frac{|PA|}{|PB|} = \frac{|P'A|}{|P'B|} = \kappa$. The lines $\alpha = OP$ and $\beta = OP'$ are the requested ones. Indeed, by Theorem 3.14 every other parallel ε_2 of ε will also intersect the four lines OX, OY, α and β at points A', B', Σ, Σ' with the same property: $\frac{|\Sigma A'|}{|\Sigma B'|} = \frac{|\Sigma' A'|}{|\Sigma' B'|} = \kappa$. Conversely, if on a parallel ε_2 the points A', B', Σ, Σ' define ratios such that the preceding relation holds, then by Proposition 3.5, OP and $O\Sigma$ will coincide and similarly OP' and $O\Sigma'$ will coincide as well.

Remark 3.13. We'll call later lines α and β of the preceding proposition (§ 5.15), **Harmonic conjugate** relative to OX, OY. Preserving the positions of the other lines and changing only the direction of ε we easily find that, on lines parallel to ε, there are defined (by α, β) again points P, P' with the same ratio $\frac{|PA|}{|PB|} = \frac{|P'A|}{|P'B|} = \kappa'$. In general $\kappa' \neq \kappa$.

Proposition 3.7. *For every angle $\widehat{XOY}$ and positive number κ, there exist exactly two lines α, β whose points P have ratio of distances from the sides of the angle equal to κ: $\frac{|PA|}{|PB|} = \kappa$ (See Figure 3.64-I).*

3.8. PENCILS OF LINES

Proof. Suppose first that the angle $\widehat{XOY}$ is not right and consider a point P with this property and extend PB until it intersects OX at Γ (See Figure 3.64-I). It holds $|P\Gamma| = |PA|/\cos(\omega)$, where $\omega = \widehat{XOY}$. Consequently, if $\frac{|PA|}{|PB|} = \kappa$, then also the ratio $\kappa' = \frac{|P\Gamma|}{|PB|} = \kappa/\cos(\omega)$ will be fixed. Therefore the proposition is a consequence of Proposition 3.6, since line PB is orthogonal to OY and the ratio κ' is fixed. In the case where the angle $\widehat{XOY}$ is right (See Figure 3.64-II), the proof is easier and leads to two lines α and β, which have the sides of $\widehat{XOY}$ as bisectors of their angle.

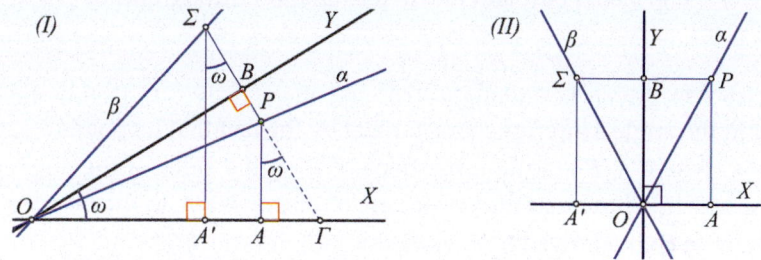

Fig. 3.64: Line(s) carrying points with constant ratios $\frac{|PA|}{|PB|}$

Exercise 3.77. Prove the preceding exercise in the case where $\widehat{XOY}$ is right.

Hint: The right triangles OAP and $OA'\Sigma$ are similar (equal actually) with fixed ratio of orthogonal sides $|PA|/|AO| = \kappa$.

Remark 3.14. The two last propositions give two equivalent ways of determining lines, like α, β, which pass through the intersection point O of two othe given lines OX, OY. Points P of such a line are determined through the ratio $\frac{|PA|}{|PB|}$ of their distances from OX, OY or through the ratio they define on parallels to a fixed line ε, together with the information whether they belong to the interior of the angle $\widehat{XOY}$ or its exterior.

Exercise 3.78. Point P of the side AB of the triangle $AB\Gamma$ is projected orthogonally to points A', B', and parallel to the sides $B\Gamma, A\Gamma$ to points A'', B'' of the other sides (See Figure 3.65-I). Show that the distances $x = |PA|, y = |PB|$, $x' = |PA'|, y' = |PB'|, x'' = |PA''|, y'' = |PB''|$ satisfy the following relations:

$$\frac{x}{y} = \lambda, \quad \frac{x'}{y'} = \lambda \frac{\sin(\widehat{A})}{\sin(\widehat{B})} = \frac{x''}{y''}.$$

Show also that these ratios are equal, if and only if the triangle is an isosceles with base AB.

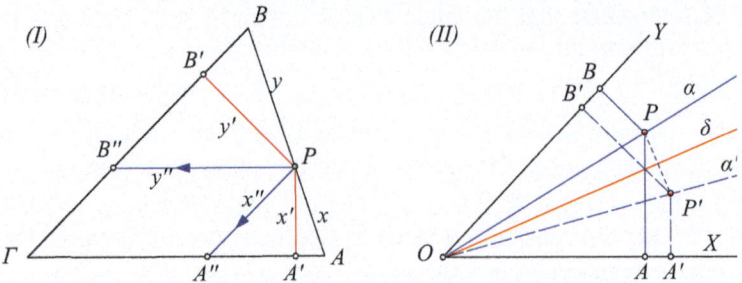

Fig. 3.65: Point and line determination through ratios

Exercise 3.79. Given is the angle $\widehat{XOY}$ and the ratio $\kappa > 0$. Define the half line α in the angle's interior characterized by the fact that its points P satisfy $\frac{|PA|}{|PB|} = \kappa$, where A, B are the projections of P on the sides of the angle (See Figure 3.65-II). Show that the symmetric α' of α, relative to the bisector δ of the angle, is characterized by the corresponding ratio $\frac{|P'A'|}{|P'B'|} = \frac{1}{\kappa}$.

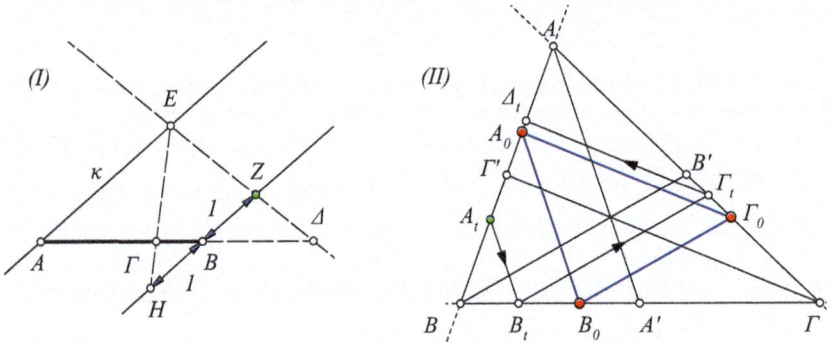

Fig. 3.66: Divide into ratio κ \qquad Project parallel to medians

Exercise 3.80. To find a point Γ on line segment AB, such that $\frac{|\Gamma A|}{|\Gamma B|} = \kappa$, where κ is a given positive number. To find also a point Δ on the extension of AB ,such that $\frac{|\Delta A|}{|\Delta B|} = \kappa$. Notice that $\{\Gamma, \Delta\}$ are harmonic conjugate to $\{A, B\}$.

Hint: From the endpoints of the line segment AB draw two parallel lines and define also a segment AE of length κ on the parallel through A (See Figure 3.66-I), while on the other parallel, the one passing through B, define segments of length 1. Segment BZ equal oriented and BH opposite oriented to AE. The intersection of EH with AB defines Γ and the intersection of EZ with the extension of AB defines point Δ.

3.8. PENCILS OF LINES

Exercise 3.81. $\{AA', BB', \Gamma\Gamma'\}$ are the medians of the triangle $AB\Gamma$. A point A_t on side AB is projected parallel to AA' to point B_t on side $B\Gamma$. This is projected parallel to BB' on side ΓA to Γ_t. Finally this is projected parallel to $\Gamma\Gamma'$ to point Δ_t on side AB. Show that point A_0 on side AB, satisfying $|AA_0|/|A_0B| = 0.5$, is the unique point on this side, such that $A_t = \Delta_t$ (See Figure 3.66-II).

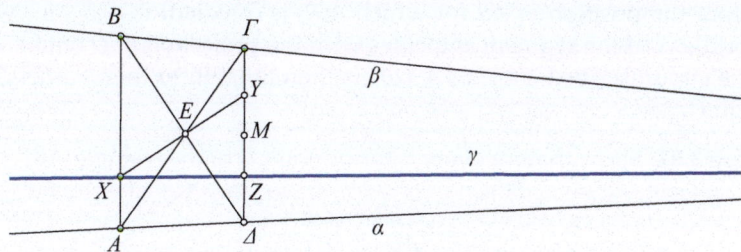

Fig. 3.67: Towards the intersection point of α and β

Exercise 3.82. Let α, β be two lines intersecting at a point P lying far away outside the drawing sheet, and X a given point not lying on these lines. Find a third line γ through X, which passes through P.

Hint: draw from X an arbitrary line intersecting α and β respectively at A and B. Also draw a parallel $\Delta\Gamma$ to AB and find the intersection point E of the diagonals of the trapezium $AB\Gamma\Delta$. Let Y be the intersection point of EX with $\Gamma\Delta$ (See Figure 3.67). According to Theorem 3.14 $\frac{|AX|}{|XB|} = \frac{|\Gamma Y|}{|Y\Delta|}$. Consider the symmetric Z of Y relative to the middle M of $\Gamma\Delta$ ($|YM| = |MZ|$). Then $\frac{|AX|}{|XB|} = \frac{|\Delta Z|}{|Z\Gamma|}$ and XZ will pass through the intersection point P of α and β (Proposition 3.5).

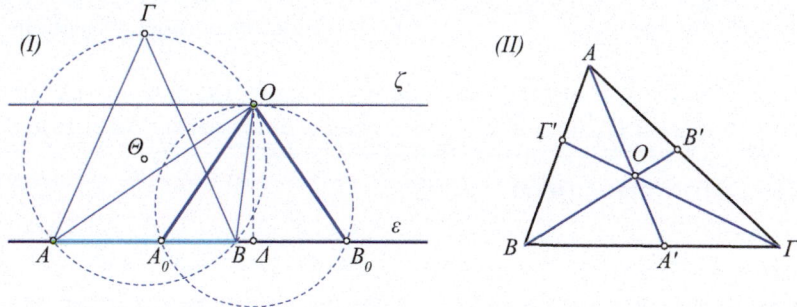

Fig. 3.68: Maximal $\widehat{AOB}$ for $|AB|$ fixed   Coincidence at a O

Exercise 3.83. The line segment AB has fixed length and slides on the line ε. From a point O not lying on ε we form the angle $\widehat{AOB}$. Find the position of the segment AB, for which the angle $\widehat{AOB}$ is maximal.

Hint: Consider the circumscribed circle of the triangle AOB and the isosceles $AB\Gamma$ which is inscribed in this circle (See Figure 3.68-I). Such an isosceles results for each position of AB and the bases of all these isosceli have the same length $|AB|$, while their apical angle is equal to $\widehat{AOB}$. According to Exercise 1.34, the greater apical angle will correspond to the lesser altitude of the isosceles, which is always greater than the distance $O\Delta$ of O from ε. Therefore the position of AB for which AOB is maximal is A_0B_0, for which the middle Δ of the segment coincides with the projection of O onto ε. Note that the circumscribed circle of A_0OB_0 is then tangent to the parallel ζ to ε through O.

Exercise 3.84. Show that for every interior point O of the triangle $AB\Gamma$ holds

$$\frac{|OA'|}{|AA'|} + \frac{|OB'|}{|BB'|} + \frac{|O\Gamma'|}{|\Gamma\Gamma'|} = 1,$$

where A', B', Γ' are the points of intersection of $OA, OB, O\Gamma$ with the opposite sides.

Hint: $\frac{|OA'|}{|AA'|} = \frac{\varepsilon(OB\Gamma)}{\varepsilon(AB\Gamma)}$ (See Figure 3.68-II), etc. Generalization in Exercise 5.117.

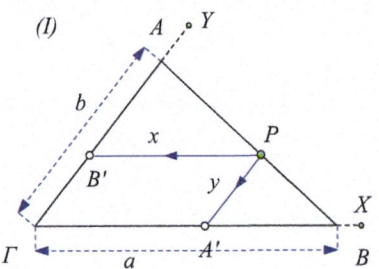
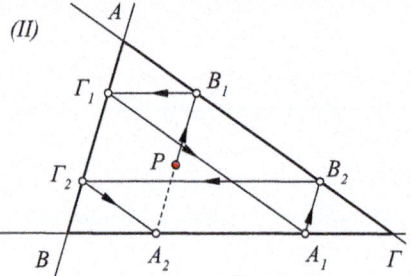

Fig. 3.69: Line equation   Recurring after parallel projections

Exercise 3.85. Point P of the side AB of the triangle $AB\Gamma$ is projected to its sides at points A', B', such that PA' and PB' are respectively parallel to the sides $A\Gamma$ and $B\Gamma$. Show that the lengths $x = |PB'|$, $y = |PA'|$ for $a = |B\Gamma|$, $b = |A\Gamma|$, satisfy the equation

$$\frac{x}{a} + \frac{y}{b} = 1.$$

Hint: $\frac{x}{a} = \frac{|AP|}{|AB|}$, $\frac{y}{b} = \frac{|PB|}{|AB|}$ and $|AP| + |PB| = |AB|$ (See Figure 3.69-I).

Remark 3.15. The preceding relation between pairs (x,y) and (a,b) has also an interesting different interpretation. Given the lines ΓY, ΓX, point P is completely determined by (x,y). Also the line AB is completely determined from (a,b). Now if we consider $P(x,y)$ fixed and the line AB variable, the

3.9. SIMILAR TRIANGLES

property of the line, to pass from *P*, is equivalent to the preceding relation. This leads to the expression of the next criterion of a variable line passing through a fixed point:

The variable points $A(b)$, $B(a)$ of lines ΓY, ΓX define a variable line AB passing through the fixed point $P(x,y)$, if and only if $xb + ya = ab$.

This again is equivalent to the fact that the product $(a-x)(b-y)$ is fixed and equal to xy.

Exercise 3.86. Point *P* in the interior of triangle $AB\Gamma$ is projected in parallel to and successively on its sides at points $\{B_1, \Gamma_1, A_1, B_2, \Gamma_2, A_2\}$ (See Figure 3.69-II). Show that the points of impact of the created trajectory define equal segments on the sides $|BA_2| = |A_1\Gamma|$, $|\Gamma B_2| = |B_1 A|$, $|A\Gamma_1| = |\Gamma_2 B|$. Conclude that the triangles $\{AB_1\Gamma_1, \Gamma_2 BA_2, B_2 A_1 \Gamma\}$ are congruent and the points P, B_1, A_2 are collinear.

3.9 Similar triangles

> To present a scientific subject in an attractive and stimulating manner is an artistic task, similar to that of a novelist or even a dramatic writer. The same holds for writing textbooks.
>
> M. Born, *My Life and My Views*, p. 48

We met already the concept of **similar** triangles (for right triangles) in § 3.5 referring to *triangles which have respective angles equal*. The existence of similar, but not congruent triangles, is characteristic of Euclidean geometry and it is proved to be equivalent to the axiom of parallels. Next proposition shows a necessary and sufficient condition for two triangles to be similar. Notice that the action of **placing** a triangle in another location, used below, may involve a reflection on a line (reversing the orientation: § 1.7).

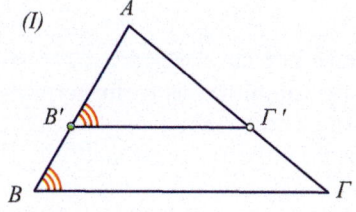
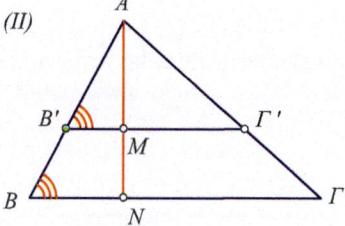

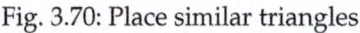

Fig. 3.70: Place similar triangles   Reduction to right triangles

Proposition 3.8. *Two are similar, if and only if they can be placed in such a way, that two of their respective angles coincide and the, opposite to these angles, sides are either parallel or they coincide.*

Proof. Suppose first that the two triangles $AB\Gamma$ and $A'B'\Gamma'$ can be placed so, that their equal angles at A and A' coincide and their sides AB and $A'B'$ also coincide, and Γ and Γ' are on the same side of AB (See Figure 3.70-I). Then the equality of angles at B and B', implies that either B coincides with B', in which case the two triangles are congruent (ASA-criterion) and consequently $B\Gamma$ and $B'\Gamma'$ are coincident, or that $B\Gamma$ and $B'\Gamma'$ are parallel.

Conversely, if the two triangles are similar and equally oriented then, placed in the way described above, will either have coincident $B\Gamma$ and $B'\Gamma'$ or these will be parallel, because of the equal angles formed with AB.

Theorem 3.15. *Two triangles are similar, if and only if their respective sides are proportional.*

Proof. This proposition is also a generalization of the analogue we saw (Proposition 3.2) for right triangles. If the triangles are similar, then we place them as in the preceding proposition (See Figure 3.70-I). Applying next Theorem 3.12 and Proposition 3.16, we find that the sides are proportional, in other words they satisfy the relation:

$$\frac{|A'B'|}{|AB|} = \frac{|B'\Gamma'|}{|B\Gamma|} = \frac{|\Gamma'A'|}{|\Gamma A|}.$$

Suppose now that the preceding relations hold and the common ratio of the fractions is κ. On side AB and towards B we take the point B'', such that $|AB''| = \kappa \cdot |AB|$. On side $A\Gamma$ and towards Γ we take the point Γ'', such that $|A\Gamma''| = \kappa \cdot |A\Gamma|$. This defines the triangle $AB''\Gamma''$, whose third side $B''\Gamma''$ is parallel to $B\Gamma$ (Theorem 3.12) and satisfies $|B''\Gamma''| = \kappa \cdot |B\Gamma|$ (Corollary 3.2). Triangles $AB\Gamma$ and $AB''\Gamma''$ have the same angles and the second one is congruent to $A'B'\Gamma'$ according to the SSS-criterion, since the two triangles have equal sides.

Exercise 3.87. Give a proof of the preceding proposition using the similarity of right triangles.

Hint: The exercise is related to Corollary 3.2. If the triangles $AB\Gamma$ and $A'B'\Gamma'$ are similar, then similar are also the pairs of right triangles $(ABN, AB'M)$ and $(AN\Gamma, AM\Gamma')$, which are created by drawing the altitudes from vertices A and A' (See Figure 3.70-II). From the similarity of the right triangles, follows the proportionality of their sides and from this, the proportionality of the sides of the initial triangles. The converse, in other words, when the sides are proportional, then the triangles are similar, is proved in the way of the preceding proposition. We need first to show, that the triangle altitudes are proportional to the sides as well.

Exercise 3.88. Show that two isosceli triangles are similar, if and only if they have equal angles at their apex. Also when they have the same ratio of base length to corresponding altitude.

3.9. SIMILAR TRIANGLES

Theorem 3.16. *Two triangles, which have one angle equal or supplementary, have ratio of areas equal to the ratio of the products of the lengths of the sides which contain these angles.*

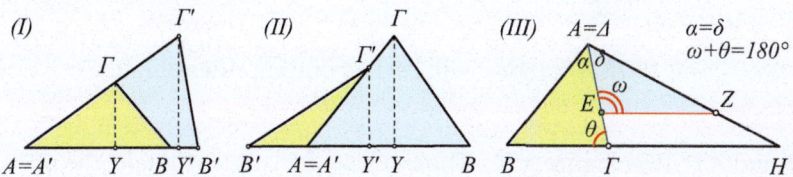

Fig. 3.71: Triangles with one angle equal or/and supplementary

Proof. Assuming that the two triangles $AB\Gamma$, $A'B'\Gamma'$ have their angles at A, A' equal or supplementary (figures 3.71-I, II), the relation follows from the expression for area $2\varepsilon(AB\Gamma) = |AB||A\Gamma|\sin(\alpha)$, applied to the two triangles and dividing in pairs.

Corollary 3.17. *Two triangles, which have respectively two angles equal and two supplementary, have ratio of sides opposite to the equal angles equal to the ratio of sides opposite to the supplementary angles.*

Proof. Apply two times the preceding theorem to triangles $AB\Gamma$, ΔEZ, which have equal angles at A, Δ and supplementary at Γ, E (See Figure 3.71-III):

$$\frac{\varepsilon(AB\Gamma)}{\varepsilon(\Delta EZ)} = \frac{|AB||A\Gamma|}{|\Delta E||\Delta Z|} = \frac{|B\Gamma||A\Gamma|}{|\Delta E||EZ|} \Rightarrow \frac{|B\Gamma|}{|EZ|} = \frac{|AB|}{|\Delta Z|}.$$

Exercise 3.89. Show that the similarity relation between two triangles is transitive. That is, if $AB\Gamma$ is similar to $A'B'\Gamma'$ and this is similar to $A''B''\Gamma''$, then $AB\Gamma$ is also similar to $A''B''\Gamma''$.

Proposition 3.9. *Two similar triangles have respectively proportional altitudes and the ratio of their areas is the square of the ratio of their sides.*

Proof. Referring to Proposition 3.8 and its figure, according to the theorem of Thales, the ratio of altitudes will be $\frac{|AM|}{|AN|} = \frac{|AB'|}{|AB|} = \kappa$, therefore the corresponding areas will be

$$\varepsilon(AB'\Gamma') = \frac{1}{2}|B'\Gamma'||AM| = \frac{1}{2}(\kappa|B\Gamma|)(\kappa|AN|) = \kappa^2\frac{1}{2}|B\Gamma||AN| = \kappa^2\varepsilon(AB\Gamma).$$

Proposition 3.10. *Two triangles $AB\Gamma$ and $A'B'\Gamma'$, which have their corresponding sides parallel are similar (See Figure 3.72).*

Proof. Translate triangle $A'B'\Gamma'$ and place it in such a way, that the vertices A and A' coincide and the lines of their sides AB, $A\Gamma$ coincide respectively with

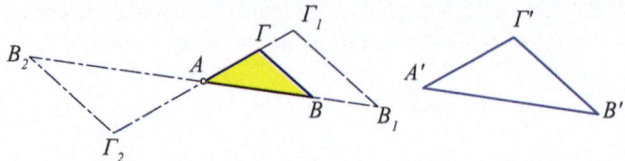

Fig. 3.72: Triangles with corresponding sides parallel

$A'B'$ and $A'\Gamma'$ (See Figure 3.72). The translated triangle will take the position $AB_1\Gamma_1$ or $AB_2\Gamma_2$, with its third side parallel to $B\Gamma$. Therefore, it will be similar to $AB\Gamma$, while it is also congruent to the initial $A'B'\Gamma'$.

Two triangles, which have their sides parallel, are called **homothetic**. According to the preceding proposition, two homothetic triangles are also similar. Next proposition shows a characteristic of two homothetic triangles. Distinguished is the case, where the homothetic triangles are, not only similar, but also congruent.

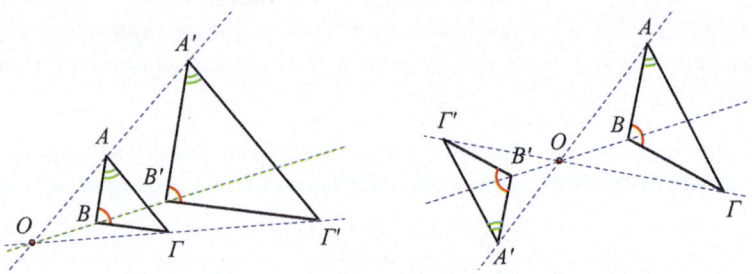

Fig. 3.73: Homothetic triangles

Theorem 3.17. *For two homothetic triangles $AB\Gamma$ and $A'B'\Gamma'$, the lines AA', BB' and $\Gamma\Gamma'$, which join the vertices with the corresponding equal angles, either pass through a common point or are parallel and the triangles are congruent.*

Proof. (See Figure 3.73) Let O be the intersection point of AA' and BB'. We will show that $\Gamma\Gamma'$ also passes through point O. According to Thales, we have equal ratios $\frac{|AB|}{|A'B'|} = \frac{|OA|}{|OA'|} = \frac{|OB|}{|OB'|} = \kappa$. Consider therefore on $O\Gamma$ point Γ'' with $\frac{|O\Gamma|}{|O\Gamma''|} = \kappa$. The created triangle $A'B'\Gamma''$ has sides proportional to those of $AB\Gamma$, therefore it is similar to it and consequently has the same angles. It follows, that $A'B'\Gamma'$ and $A'B'\Gamma''$ have $A'B'$ in common and same angles at A' and B', therefore they coincide and $\Gamma' = \Gamma''$, in other words, $O\Gamma$ passes through Γ' too.

This reasoning shows also that, if the two lines AA' and BB' do not intersect, that is if they are parallel, then the third line will also be necessarily parallel to them and $ABB'A'$, $B\Gamma\Gamma'B'$ and $A\Gamma\Gamma'A'$ will be parallelograms, therefore the triangles will have corresponding sides equal.

3.9. SIMILAR TRIANGLES

Point O, guaranteed by the preceding proposition for two homothetic and not-congruent triangles, is called **homothety center** of the two triangles. The ratio κ of two corresponding sides of the homothetic triangles is called **homothety ratio** of the two triangles. Two corresponding points like A and A' are called **homologous**. In the case where the homothety center O is *between* the homologous points we say that the triangles are **antihomothetic**.

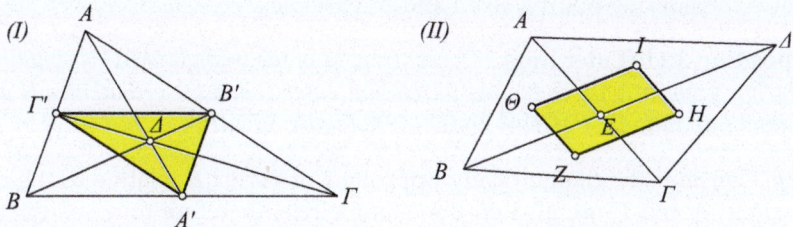

Fig. 3.74: Complementary of $AB\Gamma$ Parallelogram of centroids

The most famous pair of (anti)homothetic triangles and of corresponding homothety center is that of a triangle $AB\Gamma$ and of the triangle $A'B'\Gamma'$ of the middles of its sides, which is often mentioned as the **complementary** triangle of $AB\Gamma$ (See Figure 3.74-I). The homothety center in this case coincides with the intersection point of the medians Δ (centroid) of triangle $AB\Gamma$ and the homothety ratio of $A'B'\Gamma'$ over $AB\Gamma$ is $\frac{1}{2}$. The last proposition gives another proof, different from the one we saw in § 2.8, for the fact that the three triangle medians pass through the same point.

Exercise 3.90. The diagonals $A\Gamma$ and $B\Delta$ of the quadrilateral $AB\Gamma\Delta$ define, through their intersection point E, four triangles $EAB, EB\Gamma,...$ etc. (See Figure 3.74-II). Show that the centroids of these four triangles form a parallelogram whose area is $2/9$'s that of the area of $AB\Gamma\Delta$.

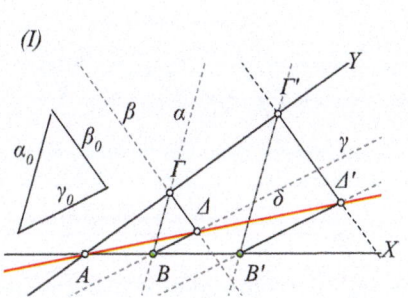 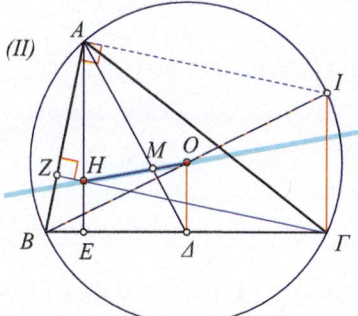

Fig. 3.75: Locus of vertices Δ Euler line HO

Corollary 3.18. *Given is the angle $\widehat{XAY}$ and lines α_0, β_0, γ_0. Let the parallel α of α_0 intersect AX, AY, respectively, at points B, Γ, from which we draw, respectively, parallels γ, β to γ_0, β_0. The geometric locus of the intersection point Δ of β, γ is a line δ passing through A (See Figure 3.75-I).*

Proof. The created triangles $B\Gamma\Delta$, $B'\Gamma'\Delta'$ are homothetic, therefore, according to the preceding theorem, the line $\Delta\Delta'$ passes through A. The conclusion follows considering point Δ fixed and Δ' variable.

Proposition 3.11 (Euler line). *The centroid M of the triangle $AB\Gamma$ is contained in the line segment with endpoints the circumcenter O and the orthocenter H and divides it into ratio 1:2 ($|MH| = 2|MO|$) (See Figure 3.75-II).*

Proof. Consider the diametrically opposite I of vertex B relative to the circumcircle. A and Γ are also on the circumcircle, therefore they see the diameter BI under a right angle. It follows easily that $AH\Gamma I$ is a parallelogram, hence $|AH| = |I\Gamma|$. However, if Δ is the middle of $B\Gamma$, $O\Delta$ joins side middles of the triangle $B\Gamma I$, therefore $|I\Gamma| = 2|O\Delta|$ and consequently AH is double of and parallel to $O\Delta$. Let M be the intersection point of the median $A\Delta$ with OH. Triangles AHM and ΔOM are similar, having corresponding angles equal. Consequently, their sides will be proportional and, because $|AH| = 2|O\Delta|$, the same will be valid also for the other corresponding sides, in other words:

$$|AM| = 2|M\Delta| \text{ and } |HM| = 2|MO|.$$

Line OH is called **Euler line** of the triangle. Besides the circumcenter O and the orthocenter H, this line contains also many other interesting points related to the triangle (Theorem 5.3).

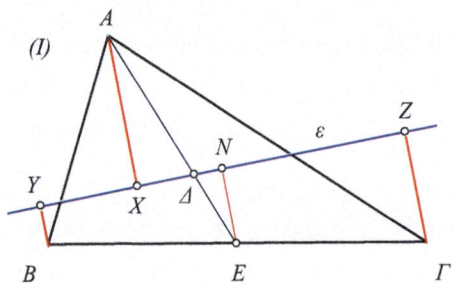
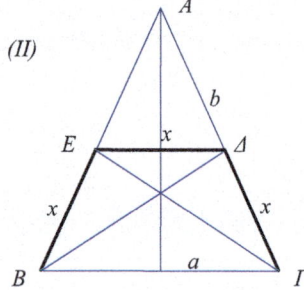

Fig. 3.76: Property of centroid   Isosceles trapezium

Exercise 3.91. Let ε be a line passing through the centroid Δ of triangle $AB\Gamma$. If the points X, Y and Z of ε are the projections of A, B and Γ respectively, show that the greater of the line segments AX, BY and ΓZ is equal to the sum of the other two (See Figure 3.76-I).

3.9. SIMILAR TRIANGLES

Hint: Suppose AX is greater from $\{BY, \Gamma Z\}$. The segment EN, which joins the middles of $B\Gamma$ and YZ is, according to Exercise 2.117, equal to the half sum of BY and ΓZ. However, the triangles ΔAX and ΔEN are (anti)homothetic with ratio 2. Consequently $|AX| = 2|EN| = |BY| + |\Gamma Z|$ ([87, p.5]).

Exercise 3.92. Let $AB\Gamma$ be an isosceles triangle. Find a parallel ΔE to the base $B\Gamma$, such that the isosceles trapezium has equal sides $|\Gamma\Delta| = |\Delta E| = |EB| = x$. Show that $x = \frac{ab}{a+b}$, where $a = |B\Gamma|$ and $b = |AB|$ (See Figure 3.76-II).

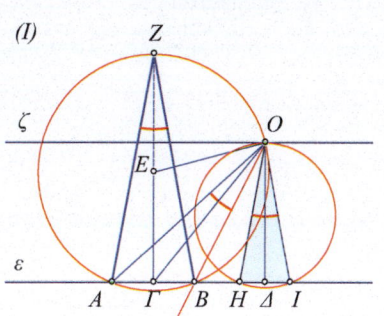
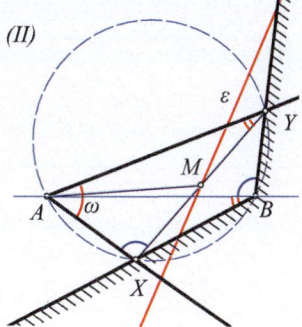

Fig. 3.77: Minimal segment AB — Rotating angle

Exercise 3.93. The angle $\widehat{AOB}$, of fixed measure ω, rotates about the fixed point O excising on line ε the segment AB. Find the position of $\widehat{AOB}$ for which the length $|AB|$ becomes minimal (See Figure 3.77-I).

Hint: Consider the circumscribed circle of the triangle AOB and its intersection point Z with the medial line of AB on the side of ε, on which also lies O. Triangle ABZ is isosceles with apical angle of measure ω. Consequently, each position of the angle $\widehat{AOB}$ creates such an isosceles with base AB. All these isosceli moreover are similar, since they have the same apical angle ω. The altitude $Z\Gamma$ of these isosceli is always greater than the distance $O\Delta$ of O from ε. Consequently the minimal AB will be obtained at the position HI, for which the altitude of the isosceles coincides with the distance $O\Delta$ of O from ε. Note that, in this position, its circumscribed circle OHI is tangent to the parallel ζ of ε from O (See the analogue for circle Exercise 3.115).

Exercise 3.94. The angle $\widehat{XAY}$, of fixed measure ω, rotates about its vertex A and intersects the fixed angle $\widehat{XBY}$ at points X and Y. Show that, if the angle $\eta = \widehat{XBY}$ is supplementary to ω, then the geometric locus of the middle M of the line segment XY is a line (See Figure 3.77-II).

Hint: If $\omega + \eta = 180°$, then the quadrilateral $AXBY$ is inscriptible in a circle. From this follows, that the triangle AXY has fixed angles and, consequently, AXM which is formed from the middle of XY, will also have fixed angles.

It follows, that triangle *AXM* remains similar to itself with fixed *A* and *X* moving on fixed line (side of the fixed angle). Therefore, its third vertex *M* will be moving also on a fixed line ε (Theorem 2.28) (see also Exercise 2.137).

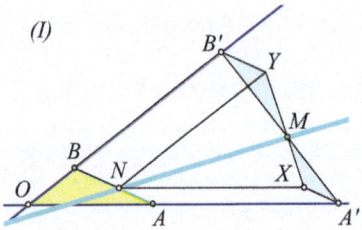

Fig. 3.78: Locus of point *M*   Locus of point *N*

Exercise 3.95. On the extensions of sides *OA*, *OB* of the triangle *OAB* are taken respectively equal segments *AA'*, *BB'*. Find the geometric locus of point *M* of *A'B'*, which divides this segment into the fixed ratio: $\frac{|MA'|}{|MB'|} = \kappa$ (See Figure 3.78-I).

Hint: Suppose *N* is the point of segment *AB* for which $\frac{|NA|}{|NB|} = \kappa$. From *N* draw parallel and equal segments $|NX| = |AA'|$, $|NY| = |BB'|$. The intersection point *M* of *XY* with *A'B'* forms similar triangles *XMA'* and *YMB'*, with $\frac{|MA'|}{|MB'|} = \frac{|XA'|}{|YB'|} = \kappa$. However the created isosceli triangles *XNY*, for the different positions of *A'*, *B'* are all similar to each other and *M* on *XY* divides their base *XY* into segments of ratio κ. Therefore *M* is contained in a fixed line passing through *N* (Proposition 3.6) (The exercise generalizes Exercise 3.46).

Exercise 3.96. Let $\{M, K\}$ be points respectively on segments $\{AB, \Gamma\Delta\}$ dividing them in ratio $\frac{|MA|}{|MB|} = \frac{|K\Gamma|}{|K\Delta|} = \kappa$ (See Figure 3.78-II). Let also $\{E, Z\}$ be points respectively on lines $\{A\Gamma, B\Delta\}$ dividing the segments in ratio $\frac{|EA|}{|E\Gamma|} = \frac{|ZB|}{|Z\Delta|} = \lambda$. Show that point *N* of *EZ* for which $\frac{|NE|}{|NZ|} = \kappa$ is contained in line *MK*.

Hint: Draw parallels and equals $E\Theta$, *EH* from *E*, respectively to *AB*, $\Gamma\Delta$. ΘH and $B\Delta$ intersect at *Z*. Similarly, the parallels of $A\Gamma$ from *M*, *K* intersect $E\Theta$, *EH* at respective points Λ, *I*. Lines *MK*, ΛI, *EZ* intersect at *N*.

Theorem 3.18. *Suppose that the lengths $x = |OA|$, $y = |OB|$ of the projections of point P, parallel to the sides of angle XOY (See Figure 3.79-I), satisfy the equation $ax + by = c$, for constants a, b, c. Then P is contained in the line ε which intersects OX, OY respectively at points A_0, B_0 with $|OA_0| = \frac{c}{a}$ and $|OB_0| = \frac{c}{b}$ respectively. Conversely, if P is contained in this line ε, then the respective x, y will satisfy this equation.*

3.9. SIMILAR TRIANGLES

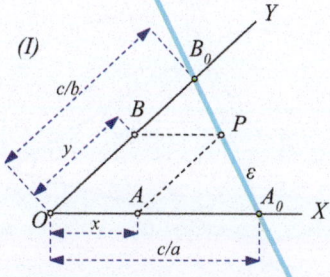
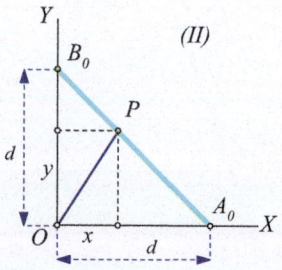

Fig. 3.79: Line equation — Maximization of area

Proof. The proof relies on the similarity of triangles PBB_0 and PA_0A, which is equivalent to the equation. Indeed the two triangles have the same angles at A and B and proportional sides

$$\frac{\frac{c}{b}-y}{x} = \frac{y}{\frac{c}{a}-x} \quad \Leftrightarrow \quad ax+by+c=0.$$

If, therefore, the equation is satisfied, then the triangles will be similar and the sum of the angles at P will be $180°$ and conversely.

Exercise 3.97. The positive numbers x, y have fixed sum $x+y=d$. Show that their product xy becomes maximal, when $x=y$.

Hint: On the sides of a right angle $\widehat{XOY}$ define equal segments $|OA_0| = |OB_0| = d$. According to Exercise 3.85, point P will lie on A_0B_0 (See Figure 3.79-II). Also $xy = \frac{1}{2}((x+y)^2 - (x^2+y^2)) = \frac{1}{2}(d^2 - (x^2+y^2))$. Therefore xy will become maximal when x^2+y^2 becomes minimal. This however is the length $|OP|^2$, which becomes minimal when P is the middle of A_0B_0.

Exercise 3.98. The positive numbers x, y satisfy the equation $ax+by=c$, where a, b, c are positive constants. Show that their product xy becomes maximal, when $ax = by$.

Hint: Define $x' = ax$, $y' = by$. These satisfy $x'+y' = c$ and $x'y'$ is maximized exactly when xy is. Apply Exercise 3.97 on x', y'.

Exercise 3.99. Point P is moving along side $B\Gamma$ of the triangle $AB\Gamma$. Drawing parallels to the other sides, creates a parallelogram $AB'P\Gamma'$ (See Figure 3.80-I). Show that the area of this parallelogram is maximized when P coincides with the middle M of $B\Gamma$.

Hint: According to Exercise 3.98, the lengths of the sides of the parallelogram x, y will satisfy an equation $ax+by = c$. Also the area will equal

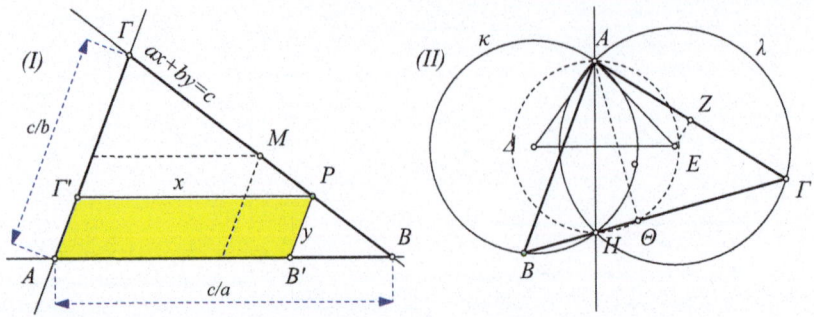

Fig. 3.80: Maximum area of $AB'P\Gamma'$ Maximum area of $AB\Gamma$

$\varepsilon(AB'P\Gamma') = xy\sin(\widehat{A})$ and will be maximized when the product xy is. According to Exercise 3.98 this happens when $ax = by$. Then, from the similar triangles $PB'B$ and $\Gamma\Gamma'P$, we will have

$$\frac{|P\Gamma|}{|PB|} = \frac{x}{\frac{c}{a}-x} = \frac{ax}{c-ax} = \frac{by}{c-ax} = 1.$$

Exercise 3.100. Circles $\{\kappa(\Delta), \lambda(E)\}$ intersect at two different points $\{A, H\}$ and the triangle $AB\Gamma$ has its vertices $\{B, \Gamma\}$ respectively on $\{\kappa, \lambda\}$ and H is on $B\Gamma$ (See Figure 3.80-II). Show that

1. Triangles $\{AB\Gamma, A\Delta E\}$ are similar, with similarity-ratio $k \leq 2$.
2. The maximal in area triangle $AB\Gamma$ is obtained when $B\Gamma$ is parallel to ΔE.
3. Locate a point B on κ, such that the corresponding $AB\Gamma$ is equal to $A\Delta E$.
4. Determine the locus of the circumcenters, centroids, orthocenters, incenters of the triangles $AB\Gamma$ (see Theorem 3.21).

3.10 Similar polygons

> "Knowledge", comprehension, the paradoxical production of problems and questions which introduce difficulties and contradictions in the *natural* course of our intellectual life, are *forms of pain*, exploitable and advanced ...
>
> Paul Valery, Thoughts and Aphorisms

The relation of similarity, which we initially examined for right triangles § 3.5 and we just generalized for all triangles, can be generalized even further to include any shape of the plane or space. Here I will confine myself to the definition for general polygons. Two polygons, with the same number of

3.10. SIMILAR POLYGONS

sides (and consequently angles), are called **similar** when: (i) they have corresponding angles equal and (ii) they have corresponding sides proportional. Here the word **corresponding**, like for triangles, is critical. It means, that we can describe the polygons with the same letters $AB\Gamma\Delta...$ and $A'B'\Gamma'\Delta'$ ... and the angles at the vertices, which correspond to same letters, are equal and the adjacent to them sides proportional ($\frac{|AB|}{|A'B'|}=\frac{|B\Gamma|}{|B'\Gamma'|}=\frac{|\Gamma\Delta|}{|\Gamma'\Delta'|}=...=\kappa$). The two sides which are involved in each of the preceding ratios are mentioned as **homologous**. The constant κ is called **similarity ratio** of the polygon $AB\Gamma\Delta...$ to the polygon $A'B'\Gamma'\Delta'$ ...

Note that, in the case of triangles, the equality of angles implies the proportionality of sides and conversely (Proposition 3.15), the proportionality of sides, implies the equality of angles. This does not hold more generally, not even for quadrilaterals. For example, the square and a rectangle, which is not a square, have all angles equal (all right) but no proportional sides. A square and a non-square rhombus, with the same side length, have proportional sides (in fact equal: $\kappa = 1$) but their angles are not correspondingly equal.

On the other side, all the triangles which have two angles respectively equal are similar. All the right triangles which have one of their acute angles respectively equal are similar. All the squares are similar. All the rhombi, which have respectively equal one of their angles are similar.

Notice that the action of **placing** a polygon in another location, used below, may involve a reflection on a line (reversing the orientation: § 1.7).

Exercise 3.101. Show that two rectangles are similar, if and only if they have the same ratio of orthogonal sides.

Hint: If $AB\Gamma\Delta$ and $A'B'\Gamma'\Delta'$ are similar, then by definition, they will have $\frac{|AB|}{|A'B'|}=\frac{|B\Gamma|}{|B'\Gamma'|}$, which is equivalent to $\frac{|AB|}{|B\Gamma|}=\frac{|A'B'|}{|B'\Gamma'|}$. Conversely, if this relation holds, since the angles are also respectively equal (all right), the definition of similarity is satisfied.

Exercise 3.102. Show that two parallelograms are similar, if and only if they can be placed in such a way, as to have a common angle and the opposite to it diagonals be coincident or parallel.

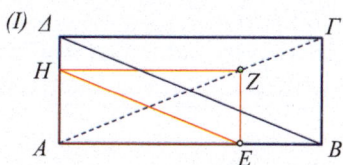

Fig. 3.81: Similar rectangles

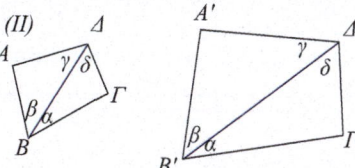

Similar quadrilaterals

Hint: The condition is equivalent to the fact that the created triangles from the two successive sides of the parallelogram and the diagonal which joins their non-common endpoints are similar (See Figure 3.81-I).

Exercise 3.103. Show that two quadrilaterals are similar, if and only if they have the angles of their sides with one diagonal respectively equal (See Figure 3.81-II).

Hint: If the quadrilaterals $ABΓΔ$, $A'B'Γ'Δ'$ are similar and the angles at A, A' equal and the adjacent to them sides proportional, then the triangles $ABΔ$ and $A'B'Δ'$ are similar, therefore also the angles of the triangles adjacent to the diagonal $BΔ$ and $B'Δ'$ are respectively equal. It follows also that the triangles $ΔΓB$ and $Δ'Γ'B'$ are similar and the corresponding angles, the adjacent to their diagonals are respectively equal. Conversely if angles $α$, $β$, $γ$, $δ$ adjacent to the diagonal $BΔ$ are respectively equal to the ones adjacent to the diagonal $B'Δ'$, then the triangles $ABΔ$, $A'B'Δ'$ are similar and the triangles $ΔΓB$ and $Δ'Γ'B'$ are similar. Consequently the sides of these triangles will be proportional and in fact under the same ratio, which is defined by the fraction $\frac{|BΔ|}{|B'Δ'|}$, therefore the corresponding sides of the quadrilaterals are proportional.

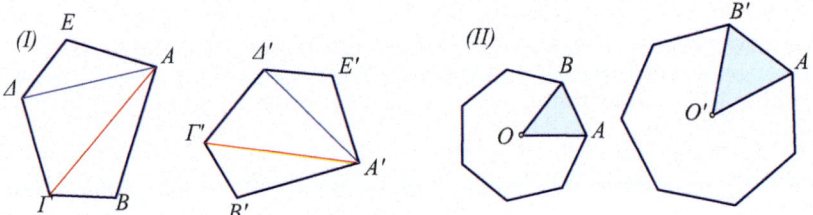

Fig. 3.82: Similar polygons

Exercise 3.104. Show that two polygons with the same number of sides are similar, if and only if they have two corresponding vertices A, A' such that the diagonals passing through them divide the polygons into corresponding triangles which are similar (See Figure 3.82-I).

Hint: For those, who are not familiar with a proof by induction, it suffices to show the proposition for pentagons. In essence the proof is the same with that of the preceding exercise.

Exercise 3.105. Show that two regular polygons, with the same number of sides, are similar (See Figure 3.82-II).

Hint: As we saw in Corollary 2.19, the angles of a regular polygon with $ν$ sides are all equal to $\frac{ν-2}{ν}180°$. Consequently two regular polygons with $ν$

3.10. SIMILAR POLYGONS

sides have respectively equal angles. If the ratio of two of their sides is κ, then, because all the sides of a regular polygon are equal, the ratio of any other two corresponding sides will be κ as well. The definition of similarity between the two polygons is therefore satisfied.

Two similar polygons $AB\Gamma...$ and $A'B'\Gamma'...$, which have corresponding sides parallel, are called **homothetic**.

Theorem 3.19. *For two homothetic polygons $AB\Gamma...$ and $A'B'\Gamma'...$ the lines AA', BB', $\Gamma\Gamma'$, ..., which join corresponding vertices either pass through a common point O or they are parallel and the two polygons are congruent.*

Proof. The same proof as that of theorem 3.17. Point O, when it exists, is called **homothety center** and the length ratio of corresponding sides **homothety ratio** of the two polygons.

Remark 3.16. As it is seen also from the examples, similar shapes are these, which have the same ... "shape". One can say that two similar shapes are essentially the same "shape", which we see from greater or lesser distance. Another way of consideration is the change of **scale**. If we measured with meters and, for example, a triangle with sides 3, 4, 5, meant 3 meters, 4 meters and 5 meters, now we measure with another measure, for example centimeters, and saying: construct a triangle with sides 3, 4, 5, we mean lengths of 3 centimeters, 4 centimeters and 5 centimeters respectively. The first triangle is certainly 100 times greater than the second (relative to lengths of sides) and could conceivably correspond to a real small land site, while the second could be considered the drawing of this land site on paper.

The ability we have in Euclidean geometry, to draw under scale, is characteristic of this geometry and can be proved equivalent to the axiom of parallels. In other words, if the axioms about lines, angles and triangles, which we accepted in sections 1.2-1.6 hold and in our plane are found two similar (but not congruent) triangles, then the geometry of our plane is the Euclidean one and, from one point not lying on a line, we can draw only one parallel to it. The converse also holds, if we prove that in our plane (with the axioms of sections 1.2-1.6 holding and) there are no similar triangles, then from a point not lying on a line there can be drawn more than one parallels to it and we are inside a non-Euclidean plane. In such a plane, it can be proved, that two triangles which have corresponding angles equal are congruent (they also have their sides respectively equal (§ 1.13, § 1.14)).

Remark 3.17. In two similar polygons $AB\Gamma$ ... and $A'B'\Gamma'$... the equality of angles and the proportionality of the sides passes also to shapes which result from these using some well defined process. For example, the diagonals $A\Gamma$ and $A'\Gamma'$ (See Figure 3.83) are also proportional with ratio the same ratio of similarity of the polygons. Also the altitudes AX and AX' to $A\Gamma$ and $A'\Gamma'$ are also proportional with the same ratio and form respectively equal angles with the sides of the polygons.

Fig. 3.83: Similar polygons II

Given two similar polygons $p = AB\Gamma\Delta...$ and $p' = A'B'\Gamma'\Delta'...$, we say that the points M and M' have respectively the **same relative position** with respect to the polygons, when all the triangles which are formed by connecting M and M' with corresponding vertices of the polygons are respectively similar. In figure 3.84 points M and M' have the same relative position with respect to the two similar polygons.

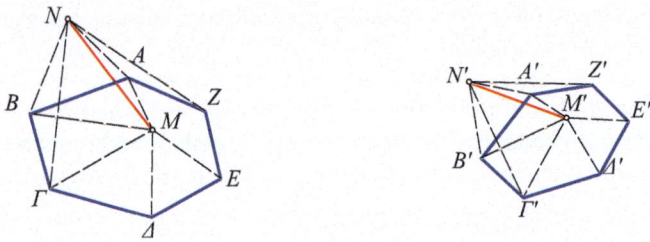

Fig. 3.84: M and M' have the same relative position

The ratios $\frac{|MA|}{|M'A'|} = \frac{|MB|}{|M'B'|} = ...$ are all equal to the similarity ratio of the two polygons. It follows directly from the definition, that the distances between two points with the same relative positions have themselves ratio $\frac{|MN|}{|M'N'|}$ equal to the ratio of similarity of the two polygons. More generally, it can be proved easily, that if the points $X, Y, Z, ...$ and $X', Y', Z', ...$ have respectively the same relative position with respect to the similar polygons p and p', then the polygons $XYZ...$ and $X'Y'Z'...$ are similar and their similarity ratio is equal to the similarity ratio of p, p'. Some special points in p, p' which have the same relative position, are corresponding vertices, the middles of corresponding sides, etc.

Remark 3.18. If we had the knowledge of II-§ 2.7, we would know that two similar polygons p, p' define a similarity f which maps one to the other: $f(p) = p'$ and we would say that X, X' have the same relative position with respect to p, p', exactly when they correspond under f: $f(X) = X'$.

3.10. SIMILAR POLYGONS

Often in applications we consider polygons $p = AB\Gamma\Delta...$ which vary, remain however similar to a fixed polygon. We say then that the polygon varies **by similarity**. Points which are co-varied with such a polygon, retaining however their relative position with respect to the similar polygons, we say that they remain **similarly invariant**. In figure 3.84 polygon $p' = A'B'\Gamma'...$ results by similarity from $p = AB\Gamma...$ and, following this change, points M, N remain similarly invariant relative to the changing polygon, taking the same relative positions M', N' with respect to the similar polygon p'. Such a variation by similarity we met already in Theorem 2.28, which can be re-expressed as follows:

If the triangle $AB\Gamma$ varies by similarity, so that one of its vertices remains fixed, while another moves on a line, then the third vertex as well moves on a line.

The next two theorems generalize this property.

Theorem 3.20. *Suppose that the polygon $p = AB\Gamma\Delta...$ varies by similarity and in such a way, that point M of its plane retains its position fixed, not only its relative position with respect to p, but also its absolute position on the plane. Suppose further that, under this variation, the similarly invariant point X of the polygon moves on a line ε_X, then every other similarly invariant point Y of p will move on a line ε_Y (See Figure 3.85).*

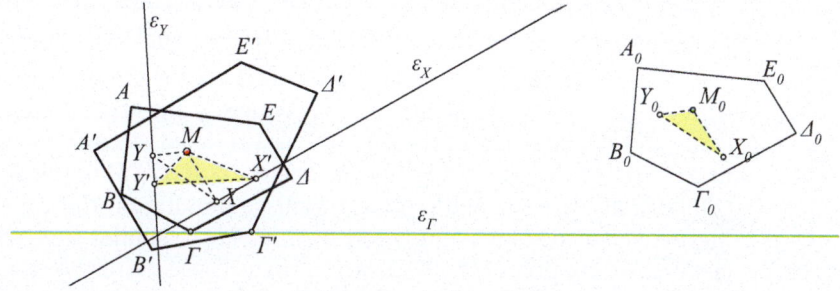

Fig. 3.85: Variation by similarity with M's position absolutely fixed

Proof. The proof follows directly by applying Theorem 2.28 to triangle MXY, which, according to the preceding remarks, varies but remains similar to itself. Let us note that the polygon's vertices are special points which remain similarly invariant, therefore they too will describe lines. In figure 3.85 the line ε_Γ is shown, which the vertex Γ of the polygon slides on. In the same figure can also be seen a similar polygon $p_0 = A_0 B_0 \Gamma_0...$ to p, on which the corresponding points M_0, X_0, Y_0 can be distinguished. These points have in p_0 the same relative position with that of M, X, Y in p.

Theorem 3.21. *Suppose that the polygon $p = AB\Gamma\Delta...$ varies by similarity and in such a way, that the point of its plane M retains its position fixed, not only its relative position with respect to p, but also its absolute position on the plane. Suppose also that, during this variation, the similarly invariant point X of the polygon moves* on a circle κ_X, then every other similarly invariant point Y of p will be moving on a circle κ_Y (See Figure 3.86).

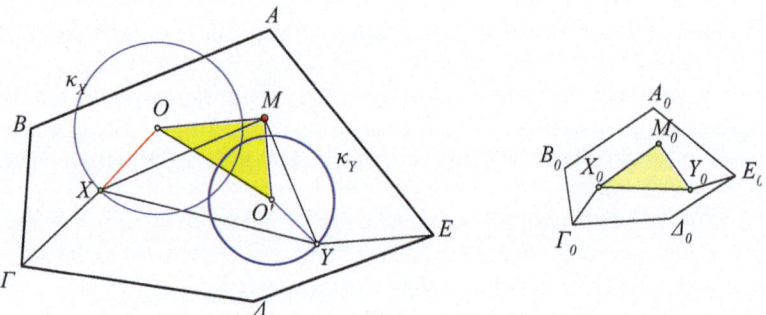

Fig. 3.86: Variation by similarity with the position of M absolutely fixed II

Proof. During the variation of the polygon p, the triangle MXY, whose vertices remain similarly invariant, will remain similar to itself. The angles of this triangle will remain fixed, point M will remain fixed and X will move on a fixed circle $\kappa_X(O, \rho)$ (See Figure 3.86). We consider the triangle MOX and we construct its similar $MO'Y$, such that the angle $\widehat{YMO'}$ is equal to $\widehat{XMO}$ and $\widehat{O'YM}$ is equal to $\widehat{OXM}$. Triangles MOO' and MXY have then their sides at M proportional and their angles at M equal. Consequently the triangles are similar. Because OM is fixed, it follows that MO' is also fixed, consequently point O' will be fixed. From the similarity of triangles OXM, $O'YM$ it follows that $\rho' = |O'Y|$ will also be fixed, therefore Y will be moving on the circle $\kappa_Y(O', \rho')$. Figure 3.86 also shows a polygon $p_0 = A_0 B_0 \Gamma_0 ...$, similar to p, on which the corresponding points M_0, X_0, Y_0 are shown. These points have on p_0 the same relative position with that of M, X, Y on p.

Theorem 3.22. *The ratio of areas of two similar polygons Π and Π' is*

$$\frac{\varepsilon(\Pi)}{\varepsilon(\Pi')} = \kappa^2,$$

where κ is the similarity ratio of the two polygons.

Proof. From one vertex in Π and its corresponding in Π' draw the diagonals and divide Π and respectively Π' into triangles respectively similar to the preceding Exercise 3.104. The areas of the polygons are written as a sum of the areas of these triangles $\varepsilon(\Pi) = \varepsilon(t_1) + \varepsilon(t_2) + ...$ and respectively $\varepsilon(\Pi') = \varepsilon(t'_1) + \varepsilon(t'_2) + ...$. The corresponding triangles are similar therefore (Proposition 3.9) $\varepsilon(t'_1) = \kappa^2 \varepsilon(t_1)$, $\varepsilon(t'_2) = \kappa^2 \varepsilon(t_2), ...$ and the assertion follows using these relations in the preceding equations for areas:

3.10. SIMILAR POLYGONS

$$\begin{aligned}\varepsilon(\Pi') &= \varepsilon(t_1') + \varepsilon(t_2') + ... \\ &= \kappa^2 \varepsilon(t_1) + \kappa^2 \varepsilon(t_2) + ... \\ &= \kappa^2 \cdot (\varepsilon(t_1) + \varepsilon(t_2) + ...) \\ &= \kappa^2 \varepsilon(\Pi).\end{aligned}$$

Remark 3.19. The preceding relation, between areas of similar polygons, leads to another form of the Pythagorean theorem in which, instead of squares on the sides of the right triangle, we construct polygons similar to a given polygon.

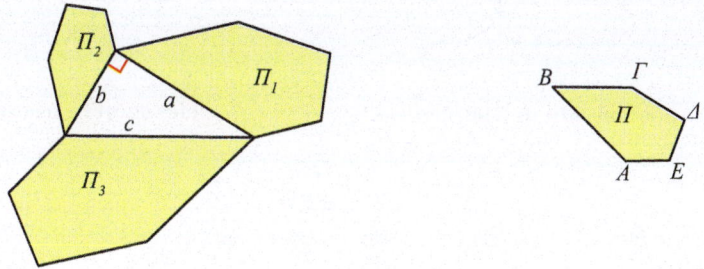

Fig. 3.87: Generalized theorem of Pythagoras

Theorem 3.23. *Given is a polygon* $\Pi = AB\Gamma\Delta...$ *and a right triangle. On the sides of the right triangle are constructed polygons* Π_1, Π_2, Π_3 *similar to* Π, *such that the sides of the right triangle a, b, c are homologous to the side AB of* Π *(See Figure 3.87). Then the sum of the areas of the polygons on the orthogonal sides is equal to the area of the polygon on the hypotenuse*

$$\varepsilon(\Pi_1) + \varepsilon(\Pi_2) = \varepsilon(\Pi_3).$$

Proof. Let $|AB| = d$ be the length of the side AB of Π. According to Theorem 3.22, the ratios of areas will be

$$\frac{\varepsilon(\Pi_1)}{\varepsilon(\Pi)} = \frac{a^2}{d^2}, \quad \frac{\varepsilon(\Pi_2)}{\varepsilon(\Pi)} = \frac{b^2}{d^2}, \quad \frac{\varepsilon(\Pi_3)}{\varepsilon(\Pi)} = \frac{c^2}{d^2}.$$

The claim follows directly by solving for a^2, b^2, c^2 the preceding relations and replacing in the theorem of Pythagoras: $a^2 + b^2 = c^2$.

Exercise 3.106. Construct a square ΔEZH, inscribed in the triangle $AB\Gamma$ with its side ΔE on $B\Gamma$.

Hint: Suppose that the square has been constructed (See Figure 3.88-I). We extend $A\Delta$ and AE until they meet the orthogonals of $B\Gamma$ at B and Γ, respectively, at points K and I. $(AH\Delta, ABK)$ and $(AZE, A\Gamma I)$ are pairs of similar triangles, consequently

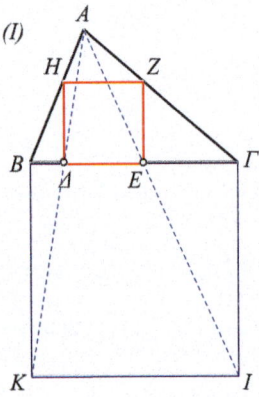
(I) Fig. 3.88: Square in triangle

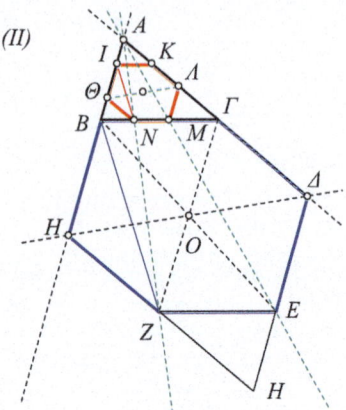
(II) Hexagon in triangle

$$\frac{|H\Delta|}{|BK|} = \frac{|AH|}{|AB|}, \quad \frac{|ZE|}{|\Gamma I|} = \frac{|AZ|}{|A\Gamma|}.$$

Because HZ and $B\Gamma$ are parallel, the ratios on the right sides of the equalities are equal, therefore the ratios on the left sides will be equal. This implies that $|BK| = |\Gamma I|$, therefore $B\Gamma IK$ is a rectangle and consequently the triangles $A\Delta E$ and AKI are similar. Because $\frac{|\Delta E|}{|KI|} = \frac{|A\Delta|}{|AK|} = \frac{|AH|}{|AB|}$, it follows that KI and BK are equal, therefore $B\Gamma IK$ is a square. This square can be constructed directly and by drawing AK and AI we define ΔE on $B\Gamma$ and from this ΔEZH, which is proved being a square with similar reasoning (see the related § 5.23).

Exercise 3.107. Construct an equilateral hexagon with three sides on the sides of triangle $AB\Gamma$ and three parallel to them.

Hint: Let $\Theta IK\Lambda MN$ be the requested hexagon (See Figure 3.88-II). Construct the similar to it on side $B\Gamma$. For this, define equal segments on the extensions of AB and $A\Gamma$ respectively: $|BH| = |B\Gamma| = |\Gamma\Delta|$ and consider the middle O of $H\Delta$. Next consider the symmetric E, Z relative to O of B, Γ respectively. The first hexagon is equilateral by assumption and the second by construction. Also the two hexagons have corresponding sides parallel, hence they are also homothetic (Theorem 3.19). The three points A, N and Z are collinear, as well as the three points A, M and E. Triangles $I\Theta N$ and BHZ are isosceli with equal apical angles Θ and H. Therefore IN and BZ are parallel and their ratio is

$$\frac{|IN|}{|BZ|} = \frac{|\Theta I|}{|HB|} = \frac{|IK|}{|B\Gamma|}.$$

However AIK and $AB\Gamma$ are similar, consequently

$$\frac{|IK|}{|B\Gamma|} = \frac{|AI|}{|AB|} = \kappa.$$

3.10. SIMILAR POLYGONS

The two hexagons, $\Theta IK\Lambda MN$ and $HB\Gamma\Delta EZ$ are homothetic relative to A with homothetic ratio κ. The equilateral hexagon $HB\Gamma\Delta EZ$ can be constructed directly and, by drawing AZ and AE, we define the points N and M on $B\Gamma$. From there on the construction of $\Theta IK\Lambda MN$ is easy, through parallels to the sides of the triangle $AB\Gamma$.

Exercise 3.108. Show that the equilateral hexagon of the preceding exercise is unique, consequently the three hexagons which result from the preceding construction, but starting from a different triangle side, coincide.

Hint: From the preceding exercise, it follows that for every hexagon like $\Theta IK\Lambda MN$ (See Figure 3.88-II), its vertex N is contained in a line AZ which is dependent only from the triangle $AB\Gamma$. This line intersects $B\Gamma$ at exactly one point, N, and consequently there is only one triangle $I\Theta N$ defined and similar to BHZ with its vertex N on $B\Gamma$. From the uniqueness of N follows that of Θ, next of I etc. The second part of the exercise is a direct logical consequence of its first part.

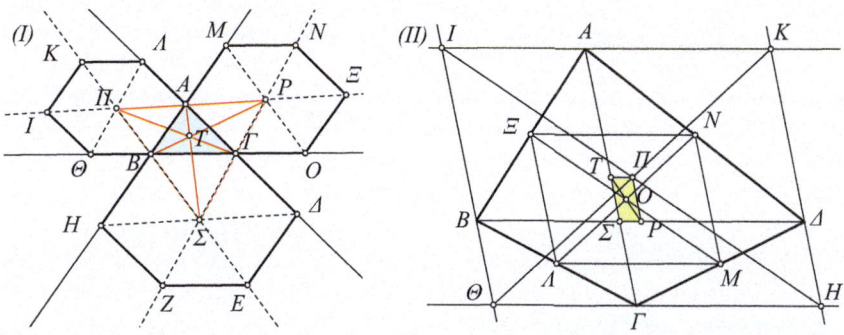

Fig. 3.89: Escribed hexagons   Circumscribed parallelogram

Exercise 3.109. Construct the three escribed equilateral hexagons on the sides of triangle $AB\Gamma$ (See Figure 3.89-I). Show that the lines which join the centers of symmetry Π, P and Σ of these hexagons pass through corresponding vertices of the triangle $AB\Gamma$.

Hint: Show first that the halves of two such polygons, like for example the quadrilaterals $\Gamma\Delta HB$ and $\Gamma A\Xi O$ are similar.

Exercise 3.110. From the vertices of the convex quadrilateral $AB\Gamma\Delta$ draw parallels to the diagonals, which do not contain them. This forms a parallelogram $H\Theta IK$ (See Figure 3.89-II). Show that the point Π of the intersection of its diagonals, the intersection point Σ of the diagonals of $AB\Gamma\Delta$ as well as the middles of its diagonals, define the vertices of another parallelogram. Also show that the parallelogram $\Lambda MN\Xi$ of the middles of the sides of $AB\Gamma\Delta$ is homothetic to $H\Theta IK$ and find the ratio and the center of the homothety.

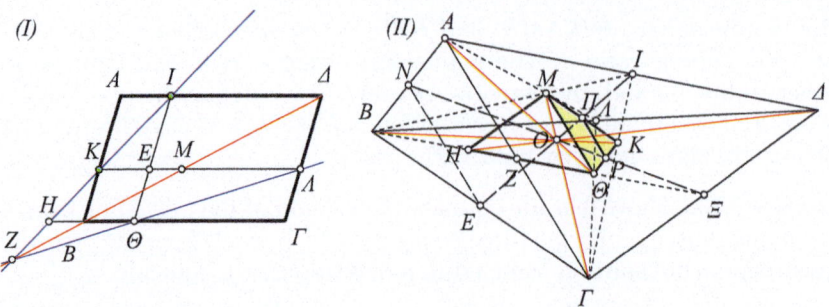

Fig. 3.90: Parallelogram secants   Homothetic to quadrilateral

Exercise 3.111. From a point E in the interior of a parallelogram, draw parallels to the sides ΘI and $K\Lambda$. Show that lines KI and $\Theta \Lambda$ intersect on one of its diagonals (See Figure 3.90-I).

Hint: Let Z be the intersection point of $\Theta \Lambda$ and $B\Delta$ and Z' be the intersection point of KI and $B\Delta$. We show that Z and Z' coincide. For this it suffices to show that the ratios $\frac{|ZB|}{|ZM|} = \frac{|\Theta B|}{|\Lambda M|}$ and $\frac{|Z'B|}{|Z'M|} = \frac{|HB|}{|KM|}$ are equal. However $\frac{|\Theta B|}{|HB|} = \frac{|AI|}{|BH|} = \frac{|AK|}{|KB|}$ and $\frac{|KM|}{|M\Lambda|} = \frac{|KB|}{|\Lambda\Delta|} = \frac{|KB|}{|KA|}$.

Exercise 3.112. In the convex quadrilateral $AB\Gamma\Delta$ we define the centroids H, Θ, K, M of the triangles $AB\Gamma, B\Gamma\Delta, \Gamma\Delta A, \Delta AB$, respectively. Show that $H\Theta KM$ is homothetic to $AB\Gamma\Delta$ with homothety ratio $\frac{1}{3}$ and homothety center the common middle O of the line segments which join the middles of its opposite sides (See Figure 3.90-II).

Hint: First show that the triangles like BNE and $KP\Pi$ are homothetic relative to the homothety with center O and ratio 3. Points P, Π are respectively the middles of $K\Theta$ and KM.

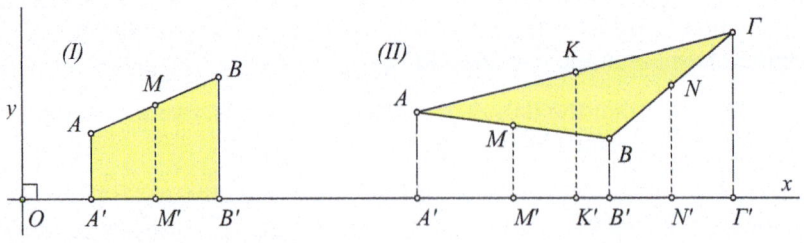

Fig. 3.91: Trapezium's area   Triangle's area

Exercise 3.113. Given the right angle $\widehat{xOy}$, we consider points $\{A, B, \Gamma, \ldots\}$ in its interior. Every point A inside this angle is projected to A' on Ox and

3.10. SIMILAR POLYGONS

defines the numbers $(x = |OA'|, y = |A'A|)$, which, in order to remember their relation to A, we write $A(x,y)$. Show that:

1. For two points $\{A(x,y), B(x',y')\}$, the area of the trapezium $ABB'A'$ is expressed by the formula $\frac{1}{2}|(x'-x)(y'+y)|$ (See Figure 3.91-I).
2. For three points $\{A(x,y), B(x',y'), \Gamma(x'',y'')\}$ (See Figure 3.91-II), the area of the corresponding triangle is equal to the absolute value of the expression
$$\frac{1}{2}((x'y - xy') + (x''y' - x'y'') + (xy'' - x''y)).$$
3. Show that the sign of this expression is positive for positive oriented triangles $AB\Gamma$ (whose order $A \to B \to \Gamma$ is reverse to the clock) and negative for negative oriented triangles.
4. For n points $\{A_1(x_1,y_1), A_2(x_2,y_2), \ldots A_n(x_n,y_n)\}$, which are vertices of a convex polygon, the area of the polygon $A_1A_2\ldots A_n$ is equal to the absolute value of the expression
$$\frac{1}{2}((x_2y_1 - x_1y_2) + (x_3y_2 - x_2y_3) + \cdots + (x_1y_n - x_ny_1)).$$

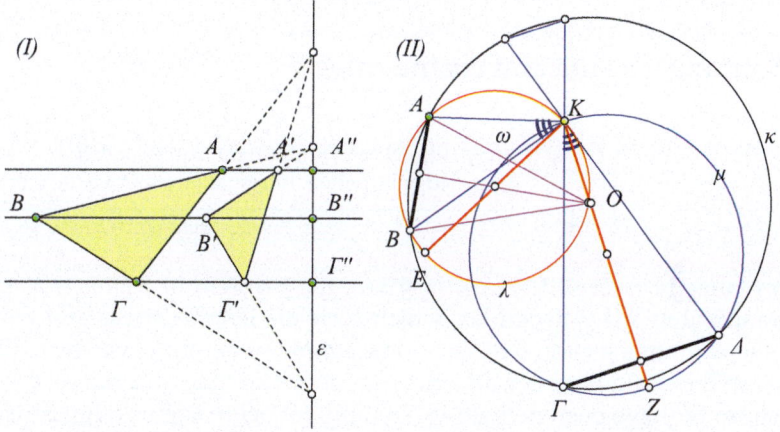

Fig. 3.92: Given ratio of areas   Chord maximization

Exercise 3.114. The vertices of the polygon $AB\Gamma\Delta\ldots$ are projected orthogonally to a fixed line ε, to corresponding points $A'', B''', \Gamma''', \ldots$ On the segments $AA'', BB''', \Gamma\Gamma', \ldots$ we consider corresponding points $A', B', \Gamma', \ldots$ such that the ratio $\frac{|A'A|}{|A'A''|} = \frac{|B'B|}{|B'B'''|} = \frac{|\Gamma'\Gamma|}{|\Gamma'\Gamma'''|}\ldots = \kappa$ is constant. Show that the ratio of the areas is $\frac{\varepsilon(AB\Gamma\Delta\ldots)}{\varepsilon(A'B'\Gamma'\Delta'\ldots)} = 1 + \kappa$ (See Figure 3.92-I).

Hint: Apply the formula of the preceding exercise, for the area of a polygon.

Exercise 3.115. An angle of fixed measure ω rotates about a given point K in the interior of the circle $\kappa(O)$. Find its positions for which the chord AB of the circle, excised by its sides, has respectively minimum/maximum possible length.

Hint: Show that the diameter KE of the circumcircle of the triangle AKB is smaller than the diameter KZ of the circumcircle of the triangle $K\Gamma\Delta$, which is symmetric relative to OK and has an angle of measure ω at K (See Figure 3.92-II).

Exercise 3.116. From the vertices of the triangle $AB\Gamma$ we draw parallels to a given line ε, which intersect its opposite sides at points A', B', Γ'. Show that the area of $A'B'\Gamma'$ is double that of $AB\Gamma$.

Exercise 3.117. Show that a rectangle $AB\Gamma\Delta$ is a square, if and only if in it can be inscribed a rectangle $A'B'\Gamma'\Delta'$ similar to $AB\Gamma\Delta$ ([69]).

Exercise 3.118. Show that the quadrilateral formed by the inner bisectors of the angles of an arbitrary convex quadrilateral is cyclic. Show the same property for the quadrilateral formed by the external bisectors.

3.11 Triangle's sine and cosine rules

> Imagination is more important than knowledge.
> Knowledge is limited. Imagination encircles the world.
>
> A. Einstein, *Interview October 26, 1929*

As we already noticed (remark in § 3.6), the theorem of Pythagoras produces infinitely many formulas, which correlate lengths, areas and angles in a shape. In this section we'll see some simple such formulas. The expressions we'll examine involve also the trigonometric functions $\sin(\phi)$, $\cos(\phi)$, which, so far have been defined (§ 3.6) only for acute angles (angles which appear in right triangles). Here we need their extensions (which show up in calculus ([90, p.300]), [76, I, p.350]) for angles ϕ in the range $0° \leq \phi \leq 180°$ for which the following relations are valid:

$$\begin{aligned}\sin(0°) &= 0 & \cos(0°) &= 1 \\ \sin(90°) &= 1 & \cos(90°) &= 0 \\ \sin(180°) &= 0 & \cos(180°) &= -1 \\ \sin(\phi) &= \sin(180° - \phi) & \cos(\phi) &= -\cos(180° - \phi)\end{aligned}$$

Proposition 3.12. *The length of the altitude $v_A = |A\Delta|$, from the vertex A of the triangle $AB\Gamma$, can be expressed through the lengths of the sides $b = |A\Gamma|$, $c = |AB|$ (See Figure 3.93):*

$$v_A = b\sin(\gamma) = c\sin(\beta).$$

3.11. TRIANGLE'S SINE AND COSINE RULES

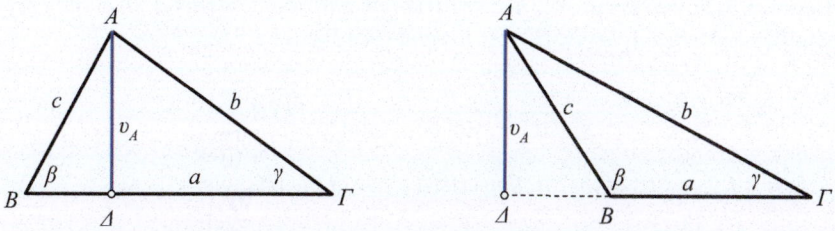

Fig. 3.93: Altitude expressed through sides

Proof. For acute angles, the proof follows directly from the definition of the trigonometric functions (See also Theorem 3.11). For obtuse angles, for which the trace of the altitude falls outside the opposite base, the resulting configuration involves again a right triangle and the following relations hold $v_A = c\sin(180° - \beta) = c\sin(\beta)$, because of the preceding properties.

Corollary 3.19. *Twice the area of a triangle is equal to the product $a \cdot b \cdot \sin(\widehat{\Gamma})$ of the lengths of two of its sides times the sine of the angle defined between them.*

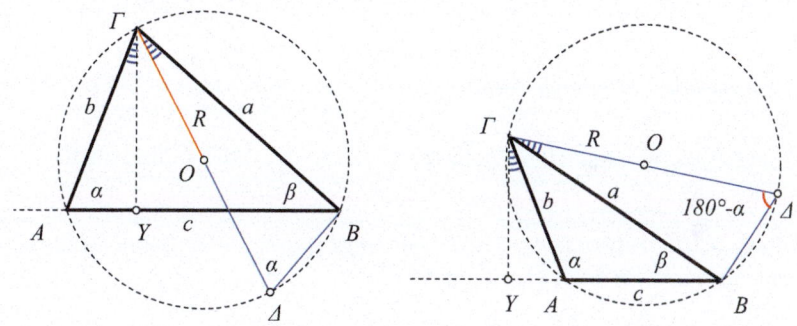

Fig. 3.94: Sine formula or sine rule for triangles

Proposition 3.13 (Sine rule). *In every triangle with side lengths a, b, c and corresponding opposite angles α, β, γ, it is valid:*

$$\frac{a}{\sin(\alpha)} = \frac{b}{\sin(\beta)} = \frac{c}{\sin(\gamma)}.$$

Proof. The last equality follows from the equality of the preceding proposition by dividing it in pairs with the product $\sin(\beta)\sin(\gamma)$. The first equality follows similarly from the corresponding equality for the altitude from Γ: $v_\Gamma = b\sin(\alpha) = a\sin(\beta)$

Theorem 3.24 (Sine rule). *In every triangle with side lengths a, b, c and corresponding opposite angles α, β, γ, it is valid (See Figure 3.94):*

$$\frac{a}{\sin(\alpha)} = \frac{b}{\sin(\beta)} = \frac{c}{\sin(\gamma)} = 2R,$$

where R is the radius of the circumscribed circle of the triangle.

Proof. For the angle α of the triangle, consider the diametrically opposite point Δ of Γ on the circumcircle of triangle. Angle $\widehat{B\Delta\Gamma}$ has measure α or $180° - α$, depending on whether it is acute or obtuse (Theorem 2.24). Also triangle $B\Gamma\Delta$ is right at B, since angle $\widehat{\Delta B\Gamma}$ sees a diameter of the circle (Corollary 2.7). From the trigonometric relations for the right triangle (Theorem 3.11), follows $|\Gamma B| = |\Gamma\Delta| \cdot \sin(\alpha)$, which is the requested relation for the angle α, the proof for the other angles being the same.

Exercise 3.119. Show that in figure 3.94 the angles $\widehat{A\Gamma Y} = \widehat{\Delta\Gamma B}$, where AY is the altitude from Γ.

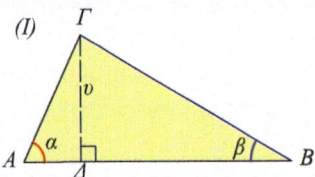
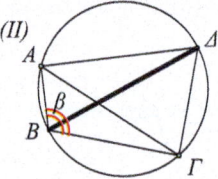

Fig. 3.95: $v = a\sin(\beta) = b\sin(\alpha)$ $|A\Gamma| = |B\Delta|\sin(\beta)$

Corollary 3.20. *In every triangle with side lengths a, b, c, area E and radius of the circumscribed circle R holds (See Figure 3.95-I):*

$$a \cdot b \cdot c = 4 \cdot R \cdot E \Leftrightarrow E = \frac{a \cdot b \cdot c}{4 \cdot R}.$$

Proof. According to the preceding proposition, the product of lengths is $a \cdot b \cdot c = a \cdot (2R\sin(\beta)) \cdot c$. However $v = a \cdot \sin(\beta)$ is the altitude of the triangle on AB of length $|AB| = c$. Therefore $a \cdot b \cdot c = (2E)(2R)$.

Corollary 3.21. *If in the inscriptible in a circle quadrilateral $AB\Gamma\Delta$, the diagonal $B\Delta$ is a circle diameter, then the other diagonal $A\Gamma$ and its opposite angle are connected with the formula $|A\Gamma| = |B\Delta|\sin(\beta)$, where $\beta = \widehat{AB\Gamma}$ (See Figure 3.95-II).*

Corollary 3.22. *In every triangle the product of the lengths of two of its sides is equal to the product of the diameter of its circumcircle times the altitude to its third side ($b \cdot c = 2R \cdot v_A$).*

3.11. TRIANGLE'S SINE AND COSINE RULES

Corollary 3.23. *For every convex quadrilateral $AB\Gamma\Delta$ inscribed in a circle of radius R, its area is equal to*

$$\varepsilon(AB\Gamma\Delta) = 2R^2 \sin(\widehat{A}) \sin(\widehat{B}) \sin(\omega),$$

where ω is the angle between the diagonals of the quadrilateral.

Hint: Application of the sine rule and exercise 3.60.

Theorem 3.25 (Cosine rule). *In every triangle with side lengths a, b, c and corresponding opposite angles α, β, γ, it holds*

$$a^2 = b^2 + c^2 - 2b \cdot c \cdot \cos(\alpha).$$

(I) 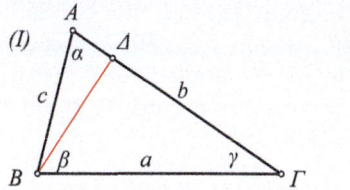   (II)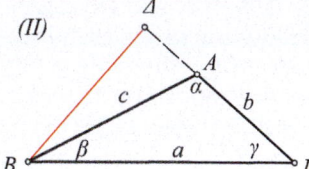

Fig. 3.96: Theorem of cosines

Proof. If the angle α is acute (See Figure 3.96-I), then the altitude $B\Delta$ is contained in it and applying the Pythagorean theorem to the right triangle $A\Delta B$ we have:

$$|\Gamma\Delta| = |\Gamma A| - |\Delta A| \Rightarrow$$
$$|\Gamma\Delta|^2 = |\Gamma A|^2 + |\Delta A|^2 - 2 \cdot |\Gamma A| \cdot |\Delta A| \Rightarrow$$
$$a^2 - |B\Delta|^2 = b^2 + |\Delta A|^2 - 2 \cdot b \cdot |\Delta A| \Rightarrow$$
$$a^2 = b^2 + |\Delta A|^2 + |B\Delta|^2 - 2 \cdot b \cdot |\Delta A| \Rightarrow$$
$$a^2 = b^2 + c^2 - 2 \cdot b \cdot |\Delta A| \Rightarrow$$
$$a^2 = b^2 + c^2 - 2 \cdot b \cdot c \cdot \cos(\alpha).$$

If the angle α is obtuse (See Figure 3.96-II), then the altitude $B\Delta$ is external to angle α and

$$|\Gamma\Delta| = |\Gamma A| + |A\Delta|.$$

Doing the same calculations, as in the preceding case, we find the rule

$$a^2 = b^2 + c^2 + 2 \cdot b \cdot c \cdot \cos(180° - \alpha),$$

which, because of $\cos(180° - \alpha) = -\cos(\alpha)$, gives the requested form of the expression for the cosine rule whether angle α is acute or obtuse.

Remark 3.20. Of course similar formulas are valid for the other angles of the triangle:

$$b^2 = c^2 + a^2 - 2c \cdot a \cdot \cos(\beta), \qquad c^2 = a^2 + b^2 - 2a \cdot b \cdot \cos(\gamma)$$

Remark 3.21. The cosine rule generalizes the theorem of Pythagoras. Indeed, for $\alpha = 90°$, $\cos(\alpha) = 0$ and the rule becomes $a^2 = b^2 + c^2$.

Proposition 3.14. *For two numbers α and β with $0° \leq \alpha + \beta \leq 180°$ the following relations hold*

$$\sin(\alpha + \beta) = \sin(\alpha)\cos(\beta) + \cos(\alpha)\sin(\beta), \qquad (3.2)$$
$$\cos(\alpha + \beta) = \cos(\alpha)\cos(\beta) - \sin(\alpha)\sin(\beta). \qquad (3.3)$$

Proof. The cases $\alpha + \beta = 0°$ or $\alpha + \beta = 180°$ are reduced to the equalities in the beginning of the section. Suppose then that $\alpha + \beta < 180°$ and construct triangle $AB\Gamma$ with a side of arbitrary length $|AB| = c$, and adjacent to it angles α and β (Proposition 1.10). Writing sines and cosines according to Theorem 3.24 and Theorem 3.25 we have:

$$\sin(\alpha)\cos(\beta) + \cos(\alpha)\sin(\beta) = \left(\frac{a}{2R}\right)\left(\frac{a^2 + c^2 - b^2}{2ac}\right) + \left(\frac{b}{2R}\right)\left(\frac{b^2 + c^2 - a^2}{2bc}\right)$$
$$= \frac{1}{4Rc} \cdot (a^2 + c^2 - b^2 + b^2 + c^2 - a^2)$$
$$= \frac{2c^2}{4Rc} = \frac{c}{2R}$$
$$= \sin(\gamma) = \sin(180° - (\alpha + \beta)) = \sin(\alpha + \beta).$$

These equalities prove the first formula. The second is proved using the preceding formula and the basic relations of the trigonometric functions:

$$\cos(\alpha + \beta) = \cos(180° - \gamma) = -\cos(\gamma) = -\frac{a^2 + b^2 - c^2}{2ab}$$
$$= \frac{(2R\sin(\gamma))^2 - (2R\sin(\alpha))^2 - (2R\sin(\beta))^2}{2(2R\sin(\alpha))(2R\sin(\beta))}$$
$$= \frac{\sin^2(\gamma) - \sin^2(\alpha) - \sin^2(\beta)}{2\sin(\alpha)\sin(\beta)}.$$

In the last representation we replace $\sin(\gamma)$ with the formula we showed already

$$\sin^2(\gamma) = \sin^2(180° - (\alpha + \beta)) = \sin^2(\alpha + \beta) = (\sin(\alpha)\cos(\beta) + \sin(\beta)\cos(\alpha))^2,$$

as well as $\sin^2(\alpha)$, $\sin^2(\beta)$ with the formulas

3.11. TRIANGLE'S SINE AND COSINE RULES

$$\sin^2(\alpha) = \sin^2(\alpha) \cdot 1 = \sin(\alpha)^2(\cos^2(\beta) + \sin^2(\beta))$$
$$= \sin^2(\alpha)\cos^2(\beta) + \sin^2(\alpha)\sin^2(\beta),$$
$$\sin^2(\beta) = \sin^2(\beta) \cdot 1 = \sin^2(\beta)(\cos^2(\alpha) + \sin^2(\alpha))$$
$$= \sin^2(\beta)\cos^2(\alpha) + \sin^2(\beta)\sin^2(\alpha).$$

After a simple calculation, it follows that

$$\cos(\alpha+\beta) = \frac{\sin^2(\gamma) - \sin^2(\alpha) - \sin^2(\beta)}{2\sin(\alpha)\sin(\beta)}$$
$$= \frac{2\sin(\alpha)\cos(\alpha)\sin(\beta)\cos(\beta) - 2\sin^2(\alpha)\sin^2(\beta)}{2\sin(\alpha)\sin(\beta)}$$
$$= \cos(\alpha)\cos(\beta) - \sin(\alpha)\sin(\beta).$$

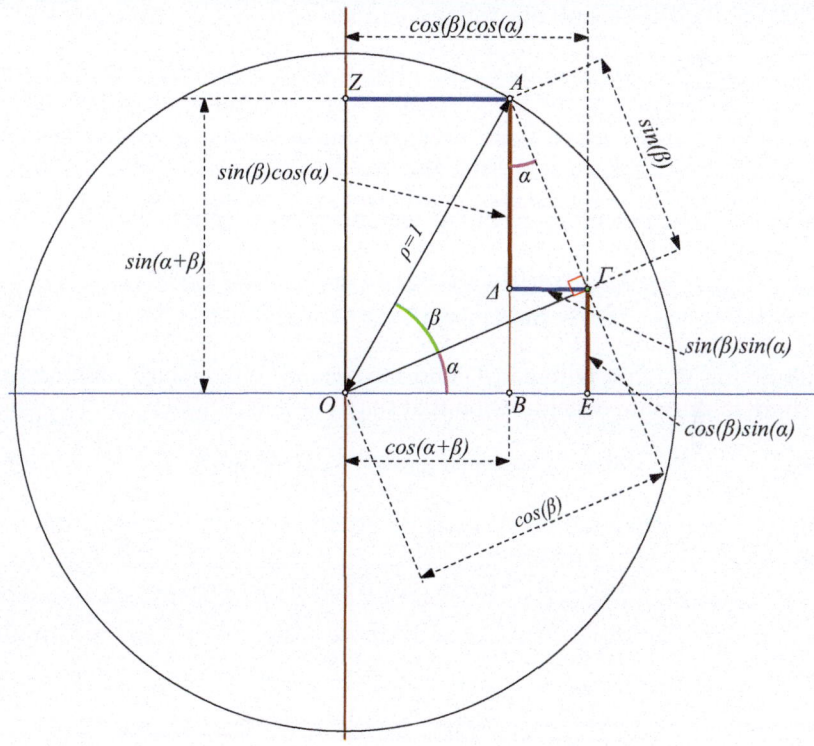

Fig. 3.97: Trigonometric formulas for sum of angles

Remark 3.22. The proof of Proposition 3.14 can also be given geometrically, by expressing the trigonometric numbers as side lengths of right triangles (see remark 3.7 in § 3.6). Figure 3.97 shows a way this can be done, using a circle of radius 1 (trigonometric circle) and forming the angles α and β as central and successive $\alpha = \widehat{EO\Gamma}$, $\beta = \widehat{\Gamma OA}$. In the figure $\sin(\alpha+\beta) = |OZ|$. This length however is $|OZ| = |A\Delta| + |\Gamma E|$. Triangle $A\Gamma\Delta$ is similar to $OE\Gamma$ and $|A\Delta| = |A\Gamma|\cos(\alpha)$. However $|A\Gamma| = \sin(\beta)$ and, by doing substitutions, we find the formula $\sin(\alpha+\beta) = \sin(\alpha)\cos(\beta) + \sin(\beta)\cos(\alpha)$. Similarly, from $|OB| = |OE| - |BE|$ follows also the formula for the cosine $\cos(\alpha+\beta) = \cos(\alpha)\cos(\beta) - \sin(\alpha)\sin(\beta)$.

Exercise 3.120. Show that for two numbers α and β, with $0° \le \alpha + \beta \le 180°$, the following formulas are valid

$$\tan(\alpha+\beta) = \frac{\tan(\alpha)+\tan(\beta)}{1-\tan(\alpha)\tan(\beta)}, \quad \cot(\alpha+\beta) = \frac{\cot(\alpha)\cot(\beta)-1}{\cot(\alpha)+\cot(\beta)}.$$

Hint: For the first formula divide the first of the preceding ones with $\cos(\alpha)\cos(\beta)$. The second one follows similarly.

Remark 3.23. We can get immediately similar formulas for the difference $\alpha - \beta$ of two angles also, using the preceding formulas and writing the angle as

$$\alpha = (\alpha - \beta) + \beta.$$

By applying the preceding formulas we find the two equations:

$$\sin(\alpha) = \cos(\beta)\sin(\alpha-\beta) + \sin(\beta)\cos(\alpha-\beta),$$
$$\cos(\alpha) = -\sin(\beta)\sin(\alpha-\beta) + \cos(\beta)\cos(\alpha-\beta).$$

Multiplying the first with $\cos(\beta)$ and the second with $-\sin(\beta)$ and adding we find, taking into consideration that $\sin(\beta)^2 + \cos(\beta)^2 = 1$:

$$\sin(\alpha-\beta) = \sin(\alpha)\cos(\beta) - \cos(\alpha)\sin(\beta). \tag{3.4}$$

Multiplying the first with $\sin(\beta)$ and the second with $\cos(\beta)$ and adding we find

$$\cos(\alpha-\beta) = \cos(\alpha)\cos(\beta) + \sin(\alpha)\sin(\beta). \tag{3.5}$$

Combining equations 3.2, 3.3, 3.4 and 3.5, we easily prove also the formulas

$$\sin(\alpha)\sin(\beta) = \frac{1}{2}(\cos(\alpha-\beta) - \cos(\alpha+\beta)), \tag{3.6}$$

$$\cos(\alpha)\cos(\beta) = \frac{1}{2}(\cos(\alpha-\beta) + \cos(\alpha+\beta)), \tag{3.7}$$

$$\sin(\alpha)\cos(\beta) = \frac{1}{2}(\sin(\alpha-\beta) + \sin(\alpha+\beta)). \tag{3.8}$$

From these again, follow directly

3.11. TRIANGLE'S SINE AND COSINE RULES

$$\sin(\alpha) + \sin(\beta) = 2\sin\left(\frac{\alpha+\beta}{2}\right)\cos\left(\frac{\alpha-\beta}{2}\right), \tag{3.9}$$

$$\sin(\alpha) - \sin(\beta) = 2\cos\left(\frac{\alpha+\beta}{2}\right)\sin\left(\frac{\alpha-\beta}{2}\right), \tag{3.10}$$

$$\cos(\alpha) + \cos(\beta) = 2\cos\left(\frac{\alpha+\beta}{2}\right)\cos\left(\frac{\alpha-\beta}{2}\right), \tag{3.11}$$

$$\cos(\alpha) - \cos(\beta) = -2\sin\left(\frac{\alpha+\beta}{2}\right)\sin\left(\frac{\alpha-\beta}{2}\right). \tag{3.12}$$

Exercise 3.121. Show that the following formulas are valid

$$\tan(\alpha) \pm \tan(\beta) = \frac{\sin(\alpha \pm \beta)}{\cos(\alpha)\cos(\beta)}, \tag{3.13}$$

$$\cot(\alpha) \pm \cot(\beta) = \pm\frac{\sin(\alpha \pm \beta)}{\sin(\alpha)\sin(\beta)}, \tag{3.14}$$

$$\tan(\alpha) + \cot(\beta) = \frac{\cos(\alpha - \beta)}{\cos(\alpha)\sin(\beta)}, \tag{3.15}$$

$$\cot(\alpha) - \tan(\beta) = \frac{\cos(\alpha + \beta)}{\sin(\alpha)\cos(\beta)}. \tag{3.16}$$

Exercise 3.122. Show that, for three angles α, β, γ the following relations are valid,

$$\sin(\alpha)\sin(\beta)\sin(\gamma) = \frac{1}{4}(-\sin(\alpha+\beta+\gamma)$$
$$+ \sin(\beta+\gamma-\alpha) + \sin(\gamma+\alpha-\beta) + \sin(\alpha+\beta-\gamma)),$$

$$\sin(\alpha)\cos(\beta)\cos(\gamma) = \frac{1}{4}(\sin(\alpha+\beta-\gamma)$$
$$- \sin(\beta+\gamma-\alpha) + \sin(\gamma+\alpha-\beta) + \sin(\alpha+\beta+\gamma)),$$

$$\sin(\alpha)\sin(\beta)\cos(\gamma) = \frac{1}{4}(-\cos(\alpha+\beta-\gamma)$$
$$+ \cos(\beta+\gamma-\alpha) + \cos(\gamma+\alpha-\beta) - \cos(\alpha+\beta+\gamma)),$$

$$\cos(\alpha)\cos(\beta)\cos(\gamma) = \frac{1}{4}(\cos(\alpha+\beta-\gamma)$$
$$+ \cos(\beta+\gamma-\alpha) + \cos(\gamma+\alpha-\beta) + \cos(\alpha+\beta+\gamma)).$$

Exercise 3.123. Show that for three angles satisfying $\alpha + \beta + \gamma = \pi$ we have:

$$4\sin(\alpha)\sin(\beta)\sin(\gamma) = \sin(2\alpha) + \sin(2\beta) + \sin(2\gamma)$$
$$1 + 4\cos(\alpha)\cos(\beta)\cos(\gamma) = \cos(2\alpha) + \cos(2\beta) + \cos(2\gamma).$$

Remark 3.24. The formulas of Proposition 3.14, applied for $\beta = \alpha$, give

$$\sin(2\alpha) = 2\sin(\alpha)\cos(\alpha),$$
$$\cos(2\alpha) = \cos^2(\alpha) - \sin^2(\alpha),$$

which are equivalent to the formulas we proved in Exercise 3.54.

Exercise 3.124. Show that the trigonometric functions for every angle α are expressed through $\tan(\frac{\alpha}{2})$ with the formulas:

$$\sin(\alpha) = \frac{2\tan(\frac{\alpha}{2})}{1+\tan^2(\frac{\alpha}{2})} \; , \; \cos(\alpha) = \frac{1-\tan^2(\frac{\alpha}{2})}{1+\tan^2(\frac{\alpha}{2})},$$

$$\tan(\alpha) = \frac{2\tan(\frac{\alpha}{2})}{1-\tan^2(\frac{\alpha}{2})} \; , \; \cot(\alpha) = \frac{1-\tan^2(\frac{\alpha}{2})}{2\tan(\frac{\alpha}{2})}.$$

Exercise 3.125. Show that for every angle α holds

$$\sin(3\alpha) = 3\sin(\alpha) - 4\sin^3(\alpha) \; , \; \cos(3\alpha) = -3\cos(\alpha) + 4\cos^3(\alpha).$$

Remark 3.25. These equations play a central role in a short proof of the fact that a trisection of an angle with ruler and compass is in general impossible ([64, p.33], [70, p.177], [78, p.54]).

Exercise 3.126. Let E be the intersection point of the diagonals $A\Gamma$ and $B\Delta$ of the inscriptible in a circle quadrilateral $AB\Gamma\Delta$. Show that

$$\frac{|AE|}{|E\Gamma|} = \frac{|AB||A\Delta|}{|\Gamma B||\Gamma\Delta|}.$$

Theorem 3.26. *Of all convex quadrilaterals $AB\Gamma\Delta$, having sides with given lengths $|AB| = a$, $|B\Gamma| = b$, $|\Gamma\Delta| = c$, $|\Delta A| = d$, the one having maximal area is the inscriptible in a circle with these side lengths (See Figure 3.98-I).*

Proof. The area E of the quadrilateral is the sum of the areas of the triangles $AB\Delta$ and $B\Delta\Gamma$. Consequently, using the opposite angles α, γ, we'll have

$$E = \frac{1}{2}b \cdot c \cdot \sin(\gamma) + \frac{1}{2}a \cdot d \cdot \sin(\alpha) \Rightarrow \qquad (3.17)$$
$$b \cdot c \cdot \sin(\gamma) + a \cdot d \cdot \sin(\alpha) = 2E. \qquad (3.18)$$

Applying the cosine rule twice for $B\Delta$ we also have

$$a^2 + d^2 - 2a \cdot d \cdot \cos(\alpha) = b^2 + c^2 - 2b \cdot c \cdot \cos(\gamma) \Rightarrow \qquad (3.19)$$
$$b \cdot c \cdot \cos(\gamma) - a \cdot d \cdot \cos(\alpha) = \frac{1}{2}(b^2 + c^2 - (a^2 + d^2)). \qquad (3.20)$$

Squaring both sides of the equations (3.19) and (3.20) and adding by parts the resulting equations we get

3.11. TRIANGLE'S SINE AND COSINE RULES

$$(b \cdot c)^2 + (a \cdot d)^2 + 2a \cdot b \cdot c \cdot d \cdot (\sin(\alpha)\sin(\gamma) - \cos(\alpha)\cos(\gamma))$$
$$= 4E^2 + \frac{1}{4}(b^2 + c^2 - (a^2 + d^2))^2,$$

which is equivalent to

$$(b \cdot c)^2 + (a \cdot d)^2 - 2a \cdot b \cdot c \cdot d \cdot \cos(\alpha + \gamma) = 4E^2 + \frac{1}{4}(b^2 + c^2 - (a^2 + d^2))^2.$$

Given that a, b, c, d are fixed, E is maximized when the left side is maximized, that is, when $\cos(\alpha + \gamma)$ supposes the maximum possible negative value which is $\cos(\alpha + \gamma) = -1$. This, for the angles of the quadrilateral, means that $\alpha + \gamma = \pi$, or that the quadrilateral is inscriptible, (alternatively [62, p.45], [93]).

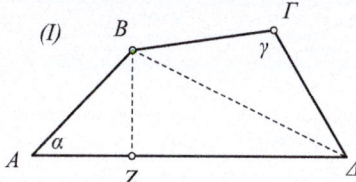
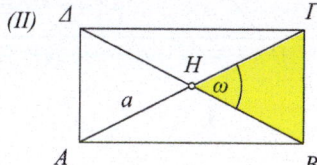

Fig. 3.98: Area maximization $AB\Gamma\Delta$ Diagonal minimization

Remark 3.26. We should notice that the theorem *supposes* the existence of a quadrilateral with given side lengths $\{a,b,c,d\}$, as well as the existence of a convex quadrilateral with the same sides and maximal area. A necessary condition for the existence of these quadrilaterals is, the similar to the triangle inequality requirement, that each of these four numbers is less than the sum of the other three. It can be proved that this condition is also sufficient and when it is satisfied, then there are inscriptible quadrilaterals with the given sides (Exercise 5.54). The general problem of constructibility of an inscriptible polygon with given side-lengths has been relatively recently solved (Pinelis 2005 [86]).

Exercise 3.127. Show that, of all rectangles with fixed area E the square of area E has the diagonal of least length. Equivalently: Given variable positive numbers x, y with $x \cdot y = d^2$, where d constant, $x^2 + y^2$ is minimized when $x = y$.

Hint: Consider a rectangle $AB\Gamma\Delta$ of area E with diagonals of length a and angle between them ω. It holds $E = 4\varepsilon(HB\Gamma)$ (See Figure 3.98-II). Consequently (Corollary 3.19)

$$\frac{E}{4} = \left(\frac{a^2}{4}\right)\sin(\omega) \quad \Rightarrow \quad a^2 = \frac{E}{\sin(\omega)}.$$

Consequently, a becomes minimal when $\omega = 90°$.

3.12 Stewart, medians, bisectors, altitudes

> The rules that describe nature seem to be mathematical....
> Why nature is mathematical is, again, a mystery.
>
> R. Feynmann, *The meaning of it All*, p. 24

Stewart's theorem relates the distance $|A\Delta|$ of a point Δ on the base $B\Gamma$ of a triangle $AB\Gamma$ to the lengths $\{a,b,c\}$ of its sides.

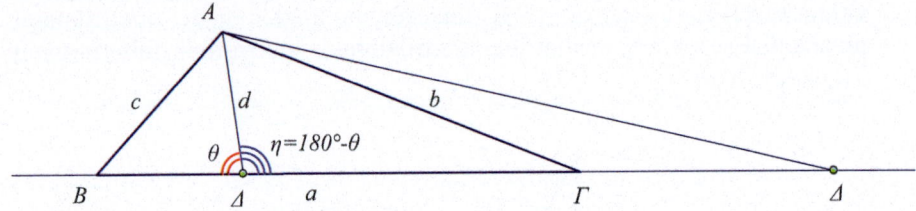

Fig. 3.99: Theorem of Stewart

Lemma 3.4. *Let Δ be a point on the side $B\Gamma$ (or on its extension) of a triangle $AB\Gamma$ and $d = |A\Delta|$. Then the following relation holds (See Figure 3.99)*

$$\frac{c^2 - d^2 - |B\Delta|^2}{b^2 - d^2 - |\Delta\Gamma|^2} = \mp \frac{|B\Delta|}{|\Delta\Gamma|},$$

where the sign is negative for points Δ in the interior of $B\Gamma$ and positive for points in the exterior of $B\Gamma$.

Proof. If Δ is in the interior of $B\Gamma$, then the angles $\theta = \widehat{B\Delta A}$ and $\eta = \widehat{\Gamma\Delta A}$ are supplementary and $\cos(\theta) = -\cos(\eta)$. If Δ is in the exterior of $B\Gamma$ then $\theta = \eta$. Taking into consideration this observation, we apply the cosine rule to the triangles $AB\Delta$ and $A\Delta\Gamma$:

$$c^2 = d^2 + |B\Delta|^2 - 2d|B\Delta|\cos(\theta) \Leftrightarrow c^2 - d^2 - |B\Delta|^2 = -2d|B\Delta|\cos(\theta),$$
$$b^2 = d^2 + |\Delta\Gamma|^2 - 2d|\Delta\Gamma|\cos(\eta) \Leftrightarrow b^2 - d^2 - |\Delta\Gamma|^2 = -2d|\Delta\Gamma|\cos(\eta).$$

The conclusion follows by dividing the equalities in pairs.

Theorem 3.27 (Stewart's formula). *If the point Δ in the interior of side $B\Gamma$ of triangle $AB\Gamma$ divides it into segments $|B\Delta| = \kappa \cdot a$, $|\Delta\Gamma| = \lambda \cdot a$, then it holds*

$$d^2 = b^2\kappa + c^2\lambda - a^2\kappa \cdot \lambda.$$

Proof. To begin with notice, that if we write $|B\Delta|$, $|\Delta\Gamma|$ the way the theorem suggests, then $\kappa + \lambda = 1$ since $|B\Delta| + |\Delta\Gamma| = |B\Gamma| \Leftrightarrow \kappa a + \lambda a = a$. The

3.12. STEWART, MEDIANS, BISECTORS, ALTITUDES

assumption that Δ is inside $B\Gamma$ implies that in the formula of the preceding lemma we must consider a negative sign. Also we must take into account that $|B\Delta| + |\Delta\Gamma| = a$. The relation of the lemma therefore becomes

$$(c^2 - d^2 - |B\Delta|^2)|\Delta\Gamma| + (b^2 - d^2 - |\Delta\Gamma|^2)|B\Delta| = 0 \Leftrightarrow$$
$$c^2|\Delta\Gamma| + b^2|B\Delta| - d^2(|\Delta\Gamma| + |B\Delta|) - |B\Delta||\Delta\Gamma|(|\Delta\Gamma| + |B\Delta|) = 0 \Leftrightarrow$$
$$c^2(a\lambda) + b^2(a\kappa) - d^2 a - a^3 \kappa\lambda = 0$$

The formula results by dividing the last relation with a.

Corollary 3.24. *The length $\mu_A = |A\Delta|$ of the median of triangle $AB\Gamma$ is expressed using the sides through the formula*

$$|A\Delta|^2 = \frac{1}{2}(b^2 + c^2) - \frac{1}{4}a^2 \Leftrightarrow b^2 + c^2 = 2(|A\Delta|^2 + |B\Delta|^2).$$

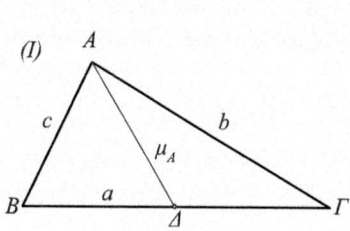

(I)

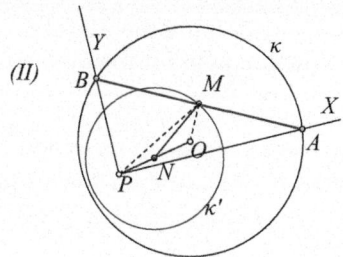

(II)

Fig. 3.100: Median's length   Locus of middle (M) of AB

Proof. Application of the preceding formula with $\kappa = \lambda = \frac{1}{2}$.

Exercise 3.128. Given is a circle $\kappa(O, \rho)$ and a point P in its interior. Let a right angle $\widehat{XPY}$ rotate about P, defining the chord AB (See Figure 3.100-II). Show that the geometric locus of the middle M of the chord AB is the circle κ' with center the middle N of OP and radius ρ' which satisfies $\rho'^2 = \frac{\rho^2}{2} - \frac{OP^2}{4}$.

Corollary 3.25. *The sum of the squares of distances of the centroid M of a triangle from its vertices is equal to*

$$|MA|^2 + |MB|^2 + |M\Gamma|^2 = \frac{4}{9}(\mu_A^2 + \mu_B^2 + \mu_\Gamma^2) = \frac{1}{3}(a^2 + b^2 + c^2).$$

Corollary 3.26. *Using the notation of figure 3.100-I, the following inequality is valid:*

$$|A\Delta| > \frac{1}{2}(b + c - a).$$

Proof. Combination of the corollary 3.24 and of the general inequality $\frac{1}{2}(b^2 + c^2) > \frac{1}{4}(b + c)^2$.

Corollary 3.27. *The sum of the squares of the four sides of a quadrilateral is equal to the sum of the squares of its diagonals plus four times the segment which joins the middles of its diagonals.*

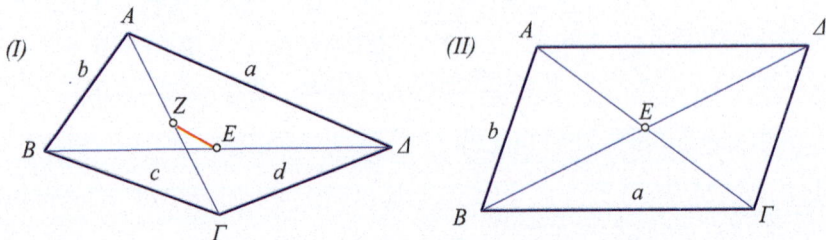

Fig. 3.101: $a^2+b^2+c^2+d^2 = |A\Gamma|^2+|B\Delta|^2+4|EZ|^2$, $2(a^2+b^2) = |A\Gamma|^2+|B\Delta|^2$

Proof. We met figure 3.101-I already in Exercise 2.73. If $AB\Gamma\Delta$ is the quadrilateral and Z, E are the middles of its diagonals $A\Gamma$ ad $B\Delta$, we apply three times the theorem for the median, on triangles $AB\Delta$, $B\Gamma\Delta$ and $A\Gamma E$.

$$|AE|^2 = \frac{1}{2}(|AB|^2 + |A\Delta|^2) - \frac{1}{4}|BD|^2$$

$$|\Gamma E|^2 = \frac{1}{2}(|\Gamma B|^2 + |\Gamma\Delta|^2) - \frac{1}{4}|BD|^2$$

$$|ZE|^2 = \frac{1}{2}(|AE|^2 + |\Gamma E|^2) - \frac{1}{4}|A\Gamma|^2.$$

The conclusion follows by replacing the expressions for $|AE|$, $|\Gamma E|$ of the first two formulas in the third.

Corollary 3.28. *The sum of the squares of the four sides of a parallelogram is equal to the sum of the squares of its diagonals.*

Proof. Follows from the preceding corollary and the fact that in a parallelogram the middles of its diagonals coincide, therefore $|EZ| = 0$ (See Figure 3.101-II).

Exercise 3.129. Show that the length δ_A of the internal bisector of the angle $\widehat{A}$ of the triangle $AB\Gamma$ is

$$\delta_A^2 = b\cdot c\cdot\left(1-\frac{a^2}{(b+c)^2}\right) = b\cdot c - b'\cdot c',$$

where b', c' are the segments to which δ_A divides $B\Gamma$.

Hint: From Theorem 3.3 we know that $\frac{|B\Delta|}{|\Delta\Gamma|} = \frac{c}{b}$ (See Figure 3.102-I). If we set $|B\Delta| = \kappa a$, $|\Delta\Gamma| = \lambda a$ we find $\frac{\kappa}{\lambda} = \frac{c}{b}$ and $a = \kappa a + \lambda a \Rightarrow \kappa + \lambda = 1$. From these it follows immediately that

3.12. STEWART, MEDIANS, BISECTORS, ALTITUDES

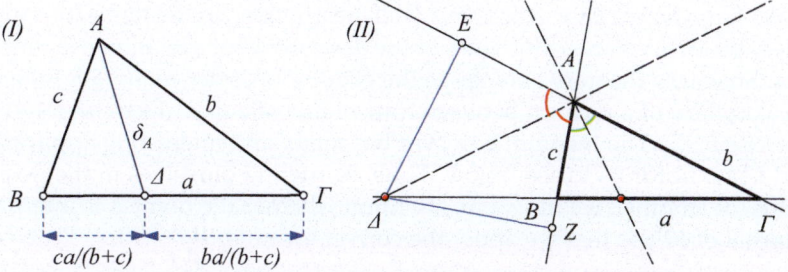

Fig. 3.102: Length of internal and external bisector

$$\kappa = \frac{c}{b+c}, \quad \lambda = \frac{b}{b+c}.$$

The formula follows immediately by replacing these κ, λ in Stewart's formula.

Exercise 3.130. Construct a triangle for which are given the lengths b, c and the bisector δ_A.

Theorem 3.28 (Stewart's formula). *If point Δ on the extension of side $B\Gamma$ of the triangle $AB\Gamma$ divides it into segments such that $|B\Delta| = \kappa \cdot a$, $|\Delta\Gamma| = \lambda \cdot a$, then the following relation is valid*

$$d^2 = c^2 \frac{\lambda}{\lambda - \kappa} - b^2 \frac{\kappa}{\lambda - \kappa} + a^2 \kappa \cdot \lambda.$$

Proof. To begin with, notice that if we write $|B\Delta|$, $|\Delta\Gamma|$ the way the theorem suggests, then $\kappa - \lambda = \pm 1$, since $|B\Delta| - |\Delta\Gamma| = \pm a = a\kappa - a\lambda$. The plus sign concerns the case $\kappa > \lambda$, in other words when Γ and Δ belong to the same half line from those defined by B. Respectively the minus sign concerns the case when Δ, Γ belong to different half lines starting at B. The proof follows the pattern of Theorem 3.27.

Remark 3.27. The difference $\lambda - \kappa$, with which we divide the last formula, is ± 1 and this allows us to express this formula more uniformly. Otherwise, we would have to write two formulas. Taking into account also that in the case of Theorem 3.27 holds $\kappa + \lambda = 1$ we can unify the formulas of Stewart into one, where, depending on the position of Δ, we would have to choose the correct version:

$$d^2 = c^2 \frac{\lambda}{\lambda \pm \kappa} \pm b^2 \frac{\kappa}{\lambda \pm \kappa} \mp a^2 \kappa \lambda.$$

We choose the upper signs when Δ is inside $B\Gamma$ and the lower signs when it is outside.

We have here a case where our (self-)restriction to consider ratios exclusively positive creates difficulties in the description of a relation in a uniform way. Stewart's theorem belongs to the chapter of *signed ratios* (§ 5.14) and it is a corollary of a relation between four points of a line (Exercise 5.100 and Exercise 5.97). The theorem however has many and interesting applications and I saw fit for us to see it now. Thus, restricting ourselves in the positive ratios (of lengths), we must pay attention whether the point Δ is internal or external of $B\Gamma$, so that we apply the correct formula.

Exercise 3.131. Show that the length δ'_A of the external bisector of angle $\widehat{A}$ of a triangle $AB\Gamma$, whose the adjacent sides to A are not equal, is given by the formula

$$\delta'^2_A = b \cdot c \cdot \left(\frac{a^2}{(b-c)^2} - 1 \right) = b' \cdot c' - b \cdot c,$$

where $b' = |\Delta B|$ and $c' = |\Delta \Gamma|$.

Hint: Notice first that $\kappa = \frac{c}{b-c}$, $\lambda = \frac{b}{b-c}$, $\lambda - \kappa = 1$ (See Figure 3.102-II), since $b > c$ and apply Theorem 3.28. The requirement for not equal sides adjacent to the angle $\widehat{A}$ is necessary for the existence of Δ, otherwise $A\Delta$ would be parallel to $B\Gamma$.

Exercise 3.132. Construct a triangle for which are given the lengths b, c and the length of its external bisector δ'_A.

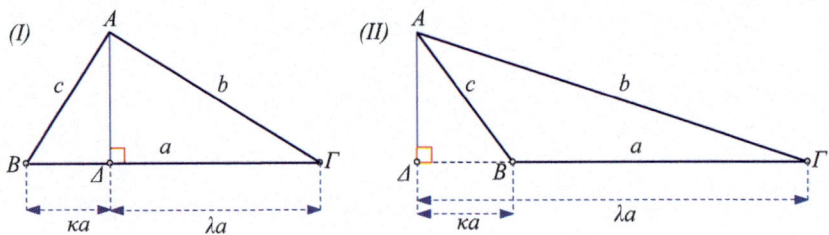

Fig. 3.103: Altitude length $v_A^2 = \frac{1}{4a^2}(a+b+c)(b+c-a)(c+a-b)(a+b-c)$

Proposition 3.15. *The square of the length of the altitude v_A of angle $\widehat{A}$ of the acute-angled triangle $AB\Gamma$ is (See Figure 3.103)*

$$v_A^2 = \frac{1}{4a^2}(a+b+c)(b+c-a)(c+a-b)(a+b-c).$$

Proof. To begin with, we calculate κ and λ, which express the segments $|B\Delta| = \kappa \cdot a$ and $|\Delta\Gamma| = \lambda \cdot a$, in which the altitude $A\Delta$ divides the opposite side $B\Gamma$ of the triangle $AB\Gamma$. As usual $\kappa \cdot a + \lambda \cdot a = a$ gives $\kappa + \lambda = 1$. We get a second equation from the right triangles $AB\Delta$ and $A\Delta\Gamma$:

$$v_A^2 = c^2 - (\kappa a)^2 = b^2 - (\lambda a)^2 \Rightarrow \kappa - \lambda = \frac{c^2 - b^2}{a^2}.$$

3.12. STEWART, MEDIANS, BISECTORS, ALTITUDES

Solving the two equations for κ and λ we find:

$$\kappa = \frac{1}{2}\left(1 + \frac{c^2 - b^2}{a^2}\right), \quad \lambda = \frac{1}{2}\left(1 - \frac{c^2 - b^2}{a^2}\right).$$

Replacing into Stewart's formula (Theorem 3.27) these κ and λ, we find:

$$v_A^2 = b^2\kappa + c^2\lambda - a^2\kappa\lambda$$
$$= \frac{b^2}{2}\left(1 + \frac{c^2 - b^2}{a^2}\right) + \frac{c^2}{2}\left(1 - \frac{c^2 - b^2}{a^2}\right) - \frac{a^2}{4}\left(1 + \frac{c^2 - b^2}{a^2}\right)\left(1 - \frac{c^2 - b^2}{a^2}\right)$$

A little adventure in the calculations leads finally to

$$v_A^2 = \frac{1}{4a^2}(2a^2b^2 + 2b^2c^2 + 2c^2a^2 - a^4 - b^4 - c^4).$$

The contents of the parenthesis can be factored and take the form we desire

$$2a^2b^2 + 2b^2c^2 + 2c^2a^2 - a^4 - b^4 - c^4 = (a+b+c)(b+c-a)(c+a-b)(a+b-c).$$

One sees this easier by passing from right to left:

$$(a+b+c)(b+c-a)(c+a-b)(a+b-c)$$
$$= ((b+c)^2 - a^2)(a^2 - (b-c)^2)$$
$$= (b+c)^2 a^2 - a^4 - (b+c)^2(b-c)^2 + a^2(b-c)^2$$
$$= (b^2 a^2 + 2bca^2 + c^2 a^2) - a^4 - (b^2 - c^2)^2 + (a^2 b^2 - 2a^2 bc + a^2 c^2)$$
$$= (b^2 a^2 + 2bca^2 + c^2 a^2) - a^4 - (b^4 - 2b^2 c^2 + c^4) + (a^2 b^2 - 2a^2 bc + a^2 c^2)$$
$$= 2a^2 b^2 + 2b^2 c^2 + 2c^2 a^2 - a^4 - b^4 - c^4.$$

Exercise 3.133. Show that the formula for the altitude $A\Delta$ of a triangle holds also when its trace Δ is outside of $B\Gamma$, in other words, when the angle at B is obtuse (See Figure 3.103-II).

Hint: Calculate first κ, λ and see that it is

$$\kappa = \frac{1}{2}\left(\frac{b^2 - c^2}{a^2} - 1\right), \quad \lambda = \frac{1}{2}\left(\frac{b^2 - c^2}{a^2} + 1\right).$$

Next, replace these κ, λ in Theorem 3.28 and see that the following expression results again

$$\frac{1}{4a^2}(2a^2b^2 + 2b^2c^2 + 2c^2a^2 - a^4 - b^4 - c^4),$$

which can be factored as in Proposition 3.15.

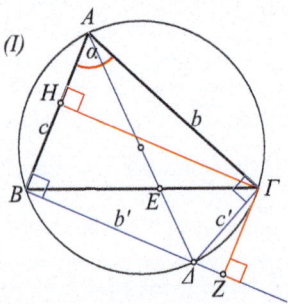 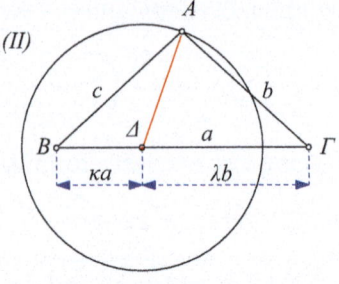

Fig. 3.104: Diametral opposite vertices   Fixed sum $\kappa b^2 + \lambda c^2$

Exercise 3.134. Let Δ be the diametrically opposite of A relative to the circumcircle of triangle $AB\Gamma$. Using figure 3.104-I, show that

$$c = b\cos(\alpha) + c'\sin(\alpha), \quad b = c\cos(\alpha) + b'\sin(\alpha).$$

Exercise 3.135. Given a line segment AB, show that the geometric locus of points X for which the sum $|XA|^2 + |XB|^2 = \kappa$ is fixed, is a circle with center at the middle of AB.

Exercise 3.136. Given are two fixed points B, Γ and two positive numbers κ, λ with $\kappa + \lambda = 1$. Show that the geometric locus of points A, for which the sums $\kappa|A\Gamma|^2 + \lambda|AB|^2$ are fixed and equal to μ^2, is a circle (See Figure 3.104-II).

Hint: The center of the circle is the point Δ of $B\Gamma$ which divides it into segments $|B\Delta| = \kappa a$, $|\Delta\Gamma| = \lambda a$. Apply the formula of Stewart.

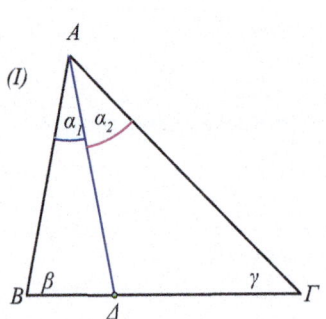 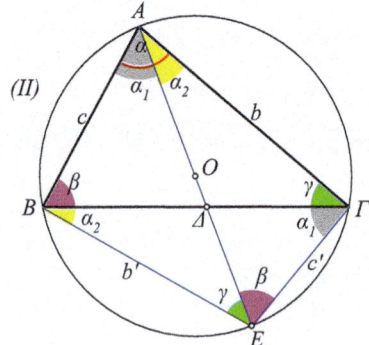

Fig. 3.105: Stewart with angles   Product of tangents

3.12. STEWART, MEDIANS, BISECTORS, ALTITUDES

Exercise 3.137. Show that the point Δ on the base $B\Gamma$ of the acute-angled triangle $AB\Gamma$ defines $A\Delta$ and the angles $\alpha_1 = \widehat{BA\Delta}$, $\alpha_2 = \widehat{\Delta A\Gamma}$, such that the following relation is valid

$$|A\Delta| = \frac{bc}{a} \cdot \frac{\sin(\beta)\sin(\alpha_1) + \sin(\gamma)\sin(\alpha_2)}{\sin(\beta)\sin(\gamma) + \sin(\alpha_1)\sin(\alpha_2)}.$$

Hint: From the sine formula follows that (See Figure 3.105-I):

$$\frac{|A\Delta|}{\sin(\beta)} = \frac{|B\Delta|}{\sin(\alpha_1)} \Rightarrow |B\Delta| = |A\Delta|\frac{\sin(\alpha_1)}{\sin(\beta)},$$

$$\frac{|A\Delta|}{\sin(\gamma)} = \frac{|\Gamma\Delta|}{\sin(\alpha_2)} \Rightarrow |\Gamma\Delta| = |A\Delta|\frac{\sin(\alpha_2)}{\sin(\gamma)}.$$

The conclusion follows by replacing in Stewart's formula (Theorem 3.27) and taking into account the sine rule for the triangle $AB\Gamma$.

Exercise 3.138. Let E be the diametrically opposite of A relative to the circumcenter O of the acute triangle $AB\Gamma$ and Δ be the point at which AE intersects $B\Gamma$. Show that

$$\frac{|A\Delta|}{|E\Delta|} = \tan(\beta)\tan(\gamma), \quad \frac{|B\Delta|}{|\Delta\Gamma|} = \frac{\sin(2\beta)}{\sin(2\gamma)}.$$

Hint: According to Exercise 3.137 the following relations are valid (See Figure 3.105-II):

$$|A\Delta| = \frac{bc}{a} \cdot \frac{\sin(\beta)\sin(\alpha_1) + \sin(\gamma)\sin(\alpha_2)}{\sin(\beta)\sin(\gamma) + \sin(\alpha_1)\sin(\alpha_2)},$$

$$|E\Delta| = \frac{b'c'}{a} \cdot \frac{\sin(\alpha_2)\sin(\gamma) + \sin(\alpha_1)\sin(\beta)}{\sin(\alpha_1)\sin(\alpha_2) + \sin(\beta)\sin(\gamma)}.$$

Dividing the equalities in pairs we get the equation

$$\frac{|A\Delta|}{|E\Delta|} = \frac{bc}{b'c'} = \left(\frac{b}{c'}\right)\left(\frac{c}{b'}\right) = \tan(\beta)\tan(\gamma).$$

For the second one, we use again the preceding exercise, which implies that

$$\frac{|B\Delta|}{|\Delta\Gamma|} = \frac{\frac{\sin(\alpha_1)}{\sin(\beta)}}{\frac{\sin(\alpha_2)}{\sin(\gamma)}} = \frac{\frac{\cos(\gamma)}{\sin(\beta)}}{\frac{\cos(\beta)}{\sin(\gamma)}} = \frac{\sin(2\gamma)}{\sin(2\beta)}.$$

Exercise 3.139 (Steiner-Lehmus theorem). Show that, if in a triangle the two internal bisectors are equal, then the triangle is isosceles.

Hint: Proof by a calculation, through application of the Exercise 3.129 (alternatively Theorem 2.8). If δ_A, δ_B and σ are respectively the bisectors and the perimeter, then (Exercise 3.129):

$$\delta_A^2 - \delta_B^2 = b \cdot c \cdot \left(1 - \frac{a^2}{(b+c)^2}\right) - c \cdot a \cdot \left(1 - \frac{b^2}{(c+a)^2}\right)$$

$$= c(b-a) - cab\left(\frac{a}{(\sigma-a)^2} - \frac{b}{(\sigma-b)^2}\right)$$

$$= c(b-a) - cab(a-b)\frac{\sigma^2 - ab}{(\sigma-a)^2(\sigma-b)^2}$$

$$= c(b-a)\left(1 + ab\frac{\sigma^2 - ab}{(\sigma-a)^2(\sigma-b)^2}\right).$$

The parenthesis is always positive, therefore to zero out the expression, we must have $(b-a) = 0$.

Exercise 3.140. Construct a triangle, for which are given the side $a = |B\Gamma|$ the altitude v_A from A and the sum $|AB|^2 + |A\Gamma|^2$.

Exercise 3.141. Construct a triangle, for which are given the side $a = |B\Gamma|$, the angle $\alpha = \widehat{BA\Gamma}$ and the difference $|AB|^2 - |A\Gamma|^2$.

Exercise 3.142. Show that all the circumscribed parallelograms of a circle are rhombi and that, of these, the circumscribed square has the least area. Also show, that all the inscribed in a circle parallelograms are rectangles and of these, the inscribed square has the maximum area.

Exercise 3.143. Triangle $AB\Gamma$ is very long and its vertex A is outside the drawing sheet, which contains only its basis $B\Gamma$ and parts of its sides $\{BB', \Gamma A''\}$. To draw the line AM carrying the median from A. Similar problem for the altitude AY from A.

Exercise 3.144. The lines $\{\alpha, \beta\}$ intersect outside the drawing sheet. To draw the bisector of their angle.

Exercise 3.145. For the triangle $AB\Gamma$ we have $\{c = |AB| < b = |A\Gamma|\}$ and $\{BB', \Gamma\Gamma'\}$ are the inner bisectors. Show that the line $B'\Gamma'$ intersects $B\Gamma$ towards B.

Exercise 3.146. Construct a triangle from its bisector δ_A the side-length a and the difference of the angles $\beta - \gamma$.

Exercise 3.147. Show that if two altitudes (medians) of the triangle are proportional to the corresponding sides ($v_A/v_B = a/b$, $\mu_A/\mu_B = a/b$), then the triangle is isosceles.

3.13 Antiparallels, symmedians

> Every mathematical book that is worth reading must be read "backwards and forwards", if I may use the expression. I would modify Lagrange's advice a little and say, "Go on, but often return to strengthen your faith." When you come on a hard or dreary passage, pass it over, and come back to it after you have seen its importance or found the need for it further on.
>
> G. Chrystal, Algebra, preface p. viii

Given a triangle $AB\Gamma$, **antiparallels to side** $B\Gamma$, are called the lines XY, which are defined by circles κ passing through vertices B and Γ and intersecting the other sides a second time at the points X and Y. Since, by definition, the

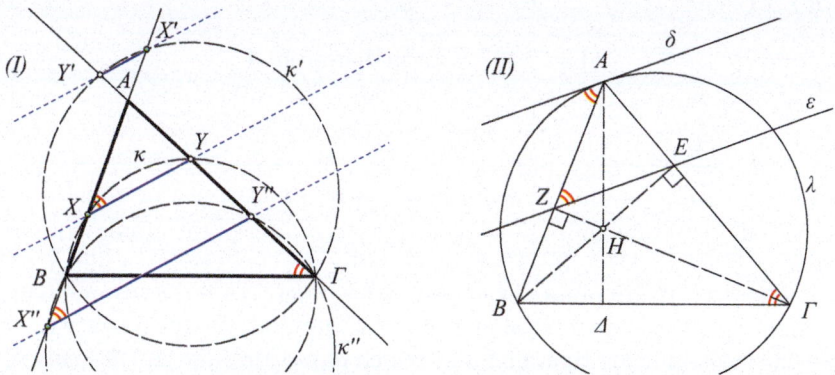

Fig. 3.106: Antiparallels $XY, X'Y', X''Y'''$ and δ, ε to side $B\Gamma$

quadrilaterals $B\Gamma YX$ are inscriptible in a circle, it follows that their angles $\widehat{YXB}$ are supplementary or equal to $\widehat{\Gamma}$, therefore all these lines XY are parallel to each other (See Figure 3.106-I). Two special circles define respectively two special antiparallels. The first is the circumcircle λ and the tangent δ at A. This can be considered as antiparallel, with points X, Y coinciding with vertex A. Indeed, the angle at A formed by the chord AB and the tangent δ is equal to the corresponding inscribed $\widehat{\Gamma}$. The second special circle is the one which has $B\Gamma$ as a diameter. The corresponding antiparallel, which it defines, passes through the traces E, Z of the altitudes (See Figure 3.106-II).

The **symmedian** from A is the line which contains the middles M of all the antiparallels of $B\Gamma$. Next theorem formulates its main properties.

Theorem 3.29. *The middles M of the antiparallels of side $B\Gamma$ are contained in a line passing through the vertex A. This line is symmetric to the median from A, relative to the bisector from A, and is characterized by the fact that its points M have fixed* ratio $\frac{|MM_B|}{|MM_\Gamma|} = \sigma$ of distances from the sides $\{AB, A\Gamma\}$, which is equal to the ratio of these sides $\sigma = \frac{|AB|}{|A\Gamma|}$ (See Figure 3.107-I).

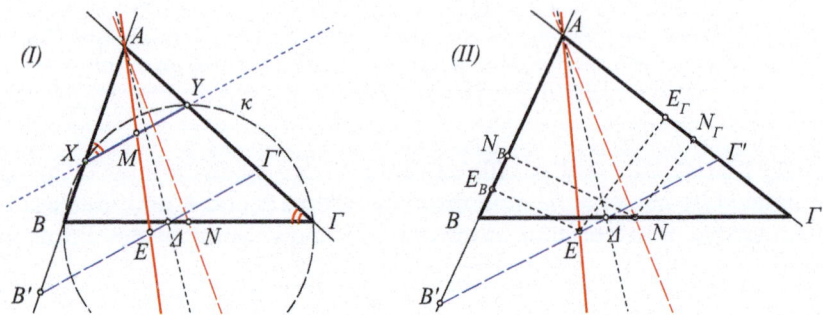

Fig. 3.107: Symmedian from A

Proof. The fact, that the middles M of the parallel segments XY are contained in a line passing through A, is a consequence of Theorem 3.14. Indeed, if M is the middle of one antiparallel XY, then, according to the aforementioned theorem, the line AM will contain the middle of every other parallel $X'Y'$ to XY, where X', Y' are points of AB, $A\Gamma$ respectively. That the symmedian is symmetric to the median relative to the bisector, follows by considering the bisector $A\Delta$ and the antiparallel $B'\Gamma'$ which passes through Δ. We easily see, that the triangles $AB\Gamma$ and $A\Gamma'B'$ are congruent and lie symmetrically with respect to $A\Delta$. It follows that also the median AE of $AB'\Gamma'$, which coincides with the symmedian, is symmetric, relative to $A\Delta$, to the median AN of the triangle $AB\Gamma$.

The last claim of the theorem follows from Exercise 3.79, according to which, if the line which passes through A, like the median AN, is characterized by the ratio τ of the distances of its points from its sides, then its symmetric relative to the bisector will be characterized by the corresponding ratio $\sigma = \frac{1}{\tau}$ (See Figure 3.107-II). However, from the equality of areas of triangles ABN, $AN\Gamma$ we have that

$$|AB||NN_B| = |A\Gamma||NN_\Gamma| \Rightarrow \tau = \frac{|NN_B|}{|NN_\Gamma|} = \frac{|A\Gamma|}{|AB|} \Rightarrow \sigma = \frac{|EE_B|}{|EE_\Gamma|} = \frac{1}{\tau} = \frac{|AB|}{|A\Gamma|}.$$

Exercise 3.148. Show that the antiparallels XY to $B\Gamma$ of the triangle $AB\Gamma$ are the symmetrics, with respect to the bisector from A, of the segments $X'Y'$ which are parallel to $B\Gamma$.

Hint: Consider the point Δ variable on the bisector $A\Delta$ of the angle $\widehat{A}$ and the corresponding triangles $AB\Gamma$ and $A\Gamma'B'$ as in figure 3.107-I. These are con-

3.13. ANTIPARALLELS, SYMMEDIANS

gruent and lie symmetrically with respect to the bisector $A\Delta$, for all positions of Δ on this bisector.

Theorem 3.30. *The three symmedians of the triangle $AB\Gamma$ pass through a common point K. In addition the symmedians pass also through the vertices of the tangential triangle $A'B'\Gamma'$ of $AB\Gamma$.*

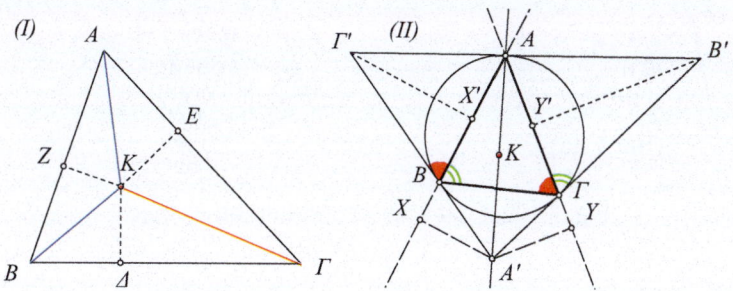

Fig. 3.108: Symmedian point K of triangle $AB\Gamma$

Proof. For the first claim, consider the intersection K of the two symmedians from A and B. We see immediately that the third must pass through K. Indeed, if Δ, E, Z are the orthogonal projections of K to the sides of the triangle (See Figure 3.108-I), then the symmedian AK is characterized by the ratio $\frac{|KZ|}{|KE|} = \frac{|AB|}{|A\Gamma|}$. Respectively the symmedian BK is characterized by the ratio $\frac{|K\Delta|}{|KZ|} = \frac{|B\Gamma|}{|BA|}$. Multiplying the two relations in pairs we have $\frac{|K\Delta|}{|KE|} = \frac{|B\Gamma|}{|A\Gamma|}$, which characterizes the symmedian from Γ. Consequently K is contained also in the symmedian from Γ.

The **tangential triangle** $A'B'\Gamma'$ of triangle $AB\Gamma$ has for sides the tangents to the circumcircle at the vertices of ABC. For the second claim of the theorem we use again the characteristic ratio $\frac{|A'X|}{|A'Y|}$, where $X, Y, X', Y', ...$ are the orthogonal projections of points on the sides of the triangle (See Figure 3.108-II). Given that the two tangents from a point to a circle are equal and that the right triangles, like $A'BX$, $\Gamma'X'B$, $\Gamma'X'A$ are similar, we have

$$\frac{|A'X|}{|A'Y|} = \frac{|A'X|/|A'B|}{|A'Y|/|A'\Gamma|} = \frac{|\Gamma'X'|/|\Gamma'B|}{|B'Y'|/|B'\Gamma|} = \frac{\sin(\widehat{\Gamma})}{\sin(\widehat{B})} = \frac{|AB|}{|A\Gamma|},$$

where the last equality results from the sine rule for triangle $AB\Gamma$.

The intersection point K of the symmedians is called **symmedian center** (or symmedian point) of the triangle $AB\Gamma$. Even though it was discovered relatively recently (1873 by Lemoine), this point has a plethora of noteworthy properties ([49, p.53], [82], [83]), which make it as worthy as other famous points of the triangle, like the orthocenter and the centroid. The following exercises describe some of these properties.

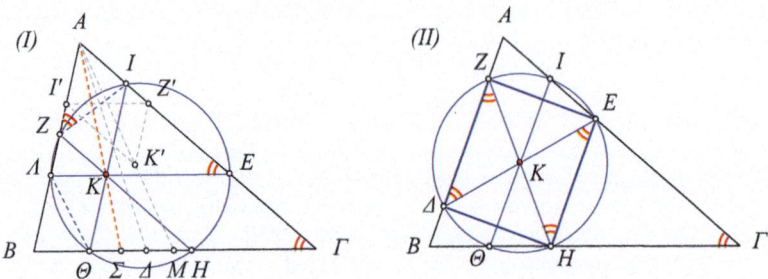

Fig. 3.109: First and second Lemoine circles

Exercise 3.149. Show that the points K of the symmedian $A\Sigma$ from A are also characterized by the property: The parallels KI, KZ respectively to the sides AB, $A\Gamma$ define a parallelogram, whose second diagonal ZI is antiparallel of $B\Gamma$.

Hint: Let K' be the symmetric of K relative to the bisector $A\Delta$ (See Figure 3.109-I). From K' draw parallels $K'I'$, $K'Z'$ respectively to the sides $A\Gamma$, AB. Point K' is on the median of $AB\Gamma$ from A and the parallelograms $K'Z'AI'$ and $KIAZ$ are symmetric relative to $A\Delta$ and congruent. Apply exercise 3.148.

Exercise 3.150. Show that the parallels to the sides of triangle $AB\Gamma$, which are drawn from the symmedian point K, define respective points on its sides $\Lambda, E, Z, H, \Theta, I$ contained in a circle.

Hint: According to Exercise 3.149, ZI is antiparallel of $B\Gamma$ (See Figure 3.109-I). This implies that the quadrilateral ΛEIZ is inscriptible in a circle κ. Show also that the quadrilaterals $IZ\Lambda\Theta$ and $\Lambda\Theta HE$ are inscriptible. Because the three inscriptible quadrilaterals, have, in pairs, three common vertices, it follows that their circumscribed circles coincide.

Exercise 3.151. Show that, in the antiparallels of the sides of the triangle $AB\Gamma$, which are drawn from the symmedian center K, the other sides define respectively equal segments $|\Delta E| = |ZH| = |\Theta I|$, which are diameters of a circle with center K (See Figure 3.109-II).

Hint: The antiparallel ΔE of $B\Gamma$ forms an inscriptible quadrilateral $\Delta E\Gamma B$. It follows that angle $\widehat{E\Delta A}$ is equal to $\widehat{\Gamma}$. Similarly follows that also angle $\widehat{BZH}$ is equal to $\widehat{\Gamma}$. On the other hand, K is the middle of the antiparallels ΔE, ZH, therefore the quadrilateral $HEZ\Delta$ has diagonals bisected by K, therefore it is a parallelogram. From the equality of angles $\widehat{E\Delta Z}$, $\widehat{\Delta ZH}$, follows that

3.13. ANTIPARALLELS, SYMMEDIANS

the parallelogram is a rectangle. This implies that ZH and ΔE are circle diameters of a circle with center K. A similar comparison also with the third antiparallel leads to the conclusion of the exercise.

The circles defined in exercises 3.150 and 3.151 are called respectively **first** and **second Lemoine circles** of the triangle $AB\Gamma$.

Exercise 3.152. Show that the centers P of the rectangles, inscribed in the triangle $AB\Gamma$ and having one side on the line $B\Gamma$, are contained in a line which passes through the middle of $B\Gamma$, the middle of the altitude AY from A and the symmedian center K of the triangle.

Hint: Such a rectangle is ΘHIZ, in figure 3.109-II, with respective center the symmedian center K of the triangle $AB\Gamma$. The general rectangle, which is

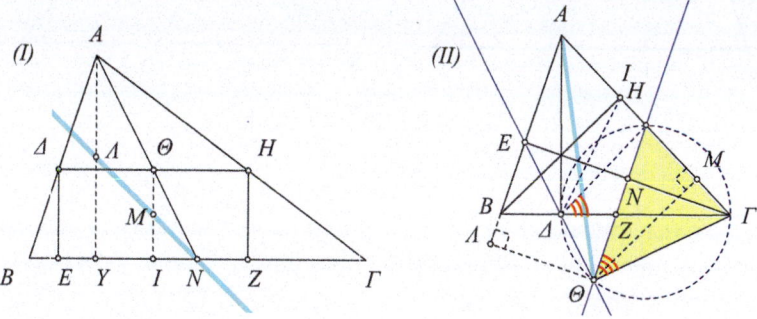

Fig. 3.110: Rectangle centers   Symmedian property

mentioned in the exercise, results from a point Δ moving on side AB, which is projected parallel to $B\Gamma$ to point H of $A\Gamma$ (See Figure 3.110-I). This defines the rectangle ΔEZH, where Z is the projection of H on $B\Gamma$. The center of the rectangle, that is the intersection point of its diagonals is the middle of segment ΘI, where Θ, I are respectively the middles of opposite sides ΔH, EZ of the rectangle. M is contained in the median $N\Lambda$ of triangle AYN from the middle N of $B\Gamma$.

Exercise 3.153. Points E, Δ are respectively the altitude traces of the triangle $AB\Gamma$ from Γ and A. Points H, Z are respectively the middles of $A\Gamma$ and $B\Gamma$. Show that the lines $E\Delta$ and ZH intersect at a point Θ of the symmedian from the vertex A.

Hint: According to Exercise 2.175, the altitudes $A\Delta$, BI, ΓE are bisectors of the orthic triangle ΔIE (See Figure 3.110-II). Also ΔI is antiparallel of HZ and ΔZHI is inscriptible in circle. There results the equality of angles $\widehat{\Theta \Delta \Gamma} = \widehat{\Gamma \Delta I} = \widehat{ZH\Gamma}$, from which follows that $\Gamma \Theta \Delta H$ is inscriptible in a circle. There results the equality of angles $\widehat{\Gamma \Theta H} = \widehat{\Gamma \Delta H} = \widehat{H\Gamma \Delta}$, from which follows that triangles $H\Theta \Gamma$ and $AB\Gamma$ are similar. Then, it follows

$$\frac{|\Theta \Lambda|}{|\Theta M|} = \frac{|\Gamma N|}{|\Theta M|} = \frac{|BI|}{|\Gamma E|} = \frac{|AB|}{|A\Gamma|}.$$

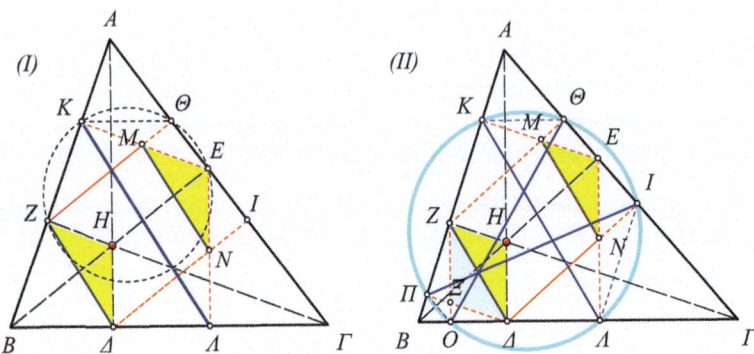

Fig. 3.111: Taylor's circle

Exercise 3.154. The traces Δ, E, Z of the altitudes of the triangle $AB\Gamma$ are projected orthgonally on its other sides, at points Λ, I, Θ, K, Π, O (See Figure 3.111-II). Show that

1. The triangles $HZ\Delta$ and EMN are congruent (See Figure 3.111-I).
2. The lines MN, $Z\Delta$, $K\Lambda$ are mutually parallel and antiparallel to $A\Gamma$.
3. The lines $K\Theta$, $I\Lambda$, $O\Pi$ are respectively parallel to the sides of the triangle.
4. The line segments $K\Lambda$, $I\Pi$, $O\Theta$ are equal.
5. The points O, Λ, I, Θ, K, Π are concyclic.

Hint: (1) The triangles $HZ\Delta$, $EK\Lambda$ have parallel sides, therefore they are similar. The fact that their sides are parallel at vertices H, E is evident. That $Z\Delta$ and $K\Lambda$ are parallel follows from the equalities $\frac{|BZ|}{|BK|} = \frac{|BH|}{|BE|} = \frac{|B\Delta|}{|B\Lambda|}$. (2) $K\Lambda$, MN are parallel to $Z\Delta$, which is antiparallel of $A\Gamma$. (3) The quadrilateral $K\Theta EZ$ is inscriptible, because the angles $\widehat{KEA}$ and $\widehat{AZ\Theta}$ are equal. It follows that $\widehat{A\Theta K} = \widehat{EZA} = \widehat{A\Gamma B}$, where the last equality holds because the quadrilateral $EZB\Gamma$ is inscriptible. (4) The quadrilateral $AK\Lambda\Gamma$ is inscriptible, therefore $\widehat{K\Lambda B} = \widehat{BA\Gamma}$. Similarly $\widehat{\Gamma O\Theta} = \widehat{BA\Gamma}$. It follows that the trapezium $K\Theta\Lambda O$ has equal diagonals. (5) Angles $\widehat{\Pi K\Lambda}$ and $\widehat{\Pi I \Lambda}$ are equal and $K\Theta I\Pi$ is inscriptible etc.

The circle which contains the six points in Exercise 3.154(5) is called **Taylor circle** 1842-1927 of the triangle $AB\Gamma$. This circle, along with the circles of Lemoine (Exercise 3.150, Exercise 3.151) belong to the wider family of **Tucker circles** of the triangle $AB\Gamma$. Later are constructed as circumscribed

3.13. ANTIPARALLELS, SYMMEDIANS

circles of special hexagons, which have sides, alternatively, parallel and antiparallel, respectively, to the sides of the triangle. Next exercise gives the details of their construction and their basic properties [49, p.89].

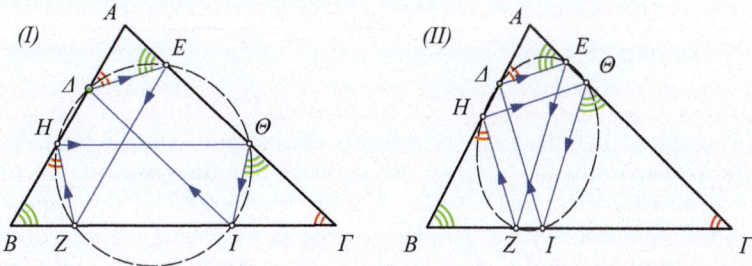

Fig. 3.112: Alternatively antiparallel-parallel and onnly antiparallels

Exercise 3.155. From an arbitrary point Δ of the side AB of the triangle $AB\Gamma$ we draw successively: antiparallel ΔE to $B\Gamma$, parallel EZ to AB, antiparallel ZH to ΓA, parallel $H\Theta$ to $B\Gamma$, antiparallel ΘI to AB. Then $I\Delta$ is parallel to $A\Gamma$ and the resulting hexagon $\Delta EZH\Theta I$ is inscriptible in a circle.

Hint: Because of the antiparallels, the quadrilaterals ΔEZH and $H\Theta IZ$ are inscriptible (See Figure 3.112-I). Also because of the antiparallels, follows that these quadrilaterals are isosceli trapezia. It follows that $|\Delta E| = |HZ| = |I\Theta|$ and $\widehat{\Delta EA} = \widehat{I\Theta\Gamma}$, therefore $\Delta E\Theta I$ is also an isosceles trapezium. This shows that the hexagon $\Delta EZH\Theta I$ has sides, alternatively, parallel and antiparallel to the sides of the triangle. The fact that this hexagon is inscriptible follows from easily provable equality of angles, like for example, $\widehat{\Delta EZ} = \widehat{\Delta IZ}$. Note that a similar proof may be given also in the case in which in the definition we start with a parallel to one side. One such case, for example, is that of the polygon $\Delta I\Theta HZE$ (See Figure 3.112-I), which, essentially, is the same with the preceding, the difference being in the reversal of the orientation.

Figure 3.112-II, shows a polygon resulting under a small variation of the preceding procedure. Instead of having alternating parallels/antiparallels, we use only antiparallels. Specifically, from an arbitrary point Δ of the side AB of the triangle $AB\Gamma$ we draw successively antiparallels ΔE, EZ, ZH, $H\Theta$, ΘI respectively to the sides $B\Gamma$, AB, $A\Gamma$, $B\Gamma$, AB. It can be proved (Exercise II-2.92) that $I\Delta$ is also antiparallel of $A\Gamma$. However, from the vertices of the formed hexagon $\Delta EZH\Theta I$ passes, in general, a conic section (II-§ 6.1) and not a circle. Another relevant application we have already met in Exercise 3.86.

3.14 Comments and exercises for the chapter

> Life does not consist mainly-or even largely- of facts and happenings. It consists mainly of the storm of thoughts that is forever blowing through one's head.
>
> Marc Twain, Autobiography

This section contains some of the more useful formulas, which connect lengths, angles, distances etc, of various elements of simple shapes. In a triangle $AB\Gamma$, as always, (α, β, γ) are symbols for the measures of its angles, respectively, at (A, B, Γ) and a, b, c are the lengths of the opposite to them sides. Also with E we denote the area of the triangle and with 2τ its perimeter. In what follows, to a formula, which includes the sides/angles of a triangle, correspond analogous formulas, which result by replacing the letters through a cyclic permutation: $\alpha \to \beta \to \gamma \to \alpha$ and $A \to B \to \Gamma \to A$ and $a \to b \to c \to a$.

Exercise 3.156. Show that for every triangle hold the formulas

$$\cot(\alpha) = \frac{b^2 + c^2 - a^2}{4E} \quad \Rightarrow \quad \cot(\alpha) + \cot(\beta) + \cot(\gamma) = \frac{a^2 + b^2 + c^2}{4E}.$$

Exercise 3.157. Show that in every triangle, with half perimeter $\tau = \frac{1}{2}(a+b+c)$, the following formulas are valid

$$\sin\left(\frac{\alpha}{2}\right) = \sqrt{\frac{(\tau-b)(\tau-c)}{bc}}, \quad \sin\left(\frac{\beta}{2}\right) = \sqrt{\frac{(\tau-c)(\tau-a)}{ca}},$$

$$\sin\left(\frac{\gamma}{2}\right) = \sqrt{\frac{(\tau-a)(\tau-b)}{ab}}.$$

Hint: Start with the formula for the cosine and show first that

$$2\sin^2\left(\frac{\alpha}{2}\right) = \frac{(a-b+c)(a+b-c)}{2bc}.$$

Exercise 3.158. Show that for every triangle with half perimeter τ the following formulas are valid

$$\cos\left(\frac{\alpha}{2}\right) = \sqrt{\frac{\tau(\tau-a)}{bc}}, \quad \cos\left(\frac{\beta}{2}\right) = \sqrt{\frac{\tau(\tau-b)}{ca}}, \quad \cos\left(\frac{\gamma}{2}\right) = \sqrt{\frac{\tau(\tau-c)}{ab}}.$$

$$\tan\left(\frac{\alpha}{2}\right) = \sqrt{\frac{(\tau-b)(\tau-c)}{\tau(\tau-a)}}, \quad \tan\left(\frac{\beta}{2}\right) = \sqrt{\frac{(\tau-c)(\tau-a)}{\tau(\tau-b)}},$$

$$\tan\left(\frac{\gamma}{2}\right) = \sqrt{\frac{(\tau-a)(\tau-b)}{\tau(\tau-c)}}.$$

Exercise 3.159. Show that for every triangle the following formulas are valid

$$\frac{a-b}{c} = \frac{\sin\left(\frac{1}{2}(\alpha-\beta)\right)}{\cos\left(\frac{1}{2}\gamma\right)}, \quad \frac{a+b}{c} = \frac{\cos\left(\frac{1}{2}(\alpha-\beta)\right)}{\sin\left(\frac{1}{2}\gamma\right)}.$$

Exercise 3.160. Let M be the centroid of a triangle $AB\Gamma$ and X be an arbitrary point. Show that

$$|XA|^2 + |XB|^2 + |X\Gamma|^2 = |MA|^2 + |MB|^2 + |M\Gamma|^2 + 3|XM|^2.$$

Also show the converse, that is, if the point M satisfies the above equation, for every point X of the plane, then it coincides with the centroid of the triangle.

Exercise 3.161. Show that, for a triangle $AB\Gamma$ with circumcircle $\kappa(O,R)$ and centroid M, the following formula is valid

$$|OM|^2 = R^2 - \frac{1}{9}(a^2 + b^2 + c^2).$$

Exercise 3.162. Show that the median μ_A and the bisector δ_A, from A, of triangle $AB\Gamma$, are given, respectively, by the formulas

$$\mu_A = \frac{1}{2}\sqrt{b^2 + c^2 + 2bc\cos(\alpha)}, \qquad \delta_A = \frac{2bc\cos\left(\frac{\alpha}{2}\right)}{b+c}.$$

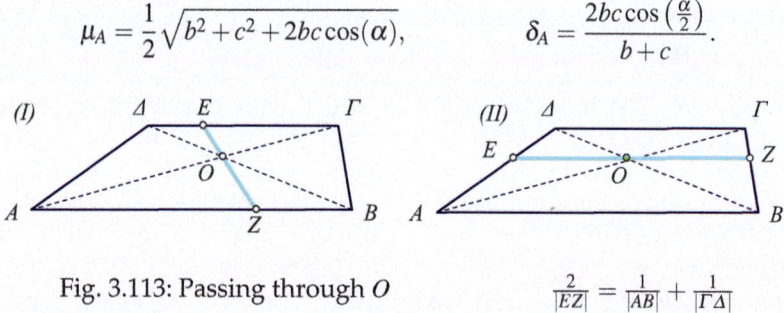

Fig. 3.113: Passing through O $\qquad \frac{2}{|EZ|} = \frac{1}{|AB|} + \frac{1}{|\Gamma\Delta|}$

Exercise 3.163. Given are the points E, Z on the parallel sides of the trapezium $AB\Gamma\Delta$. Show that EZ passes through the intersection point of the diagonals O, if and only if $\frac{|E\Delta|}{|E\Gamma|} = \frac{|ZB|}{|ZA|}$ (See Figure 3.113-I).

Exercise 3.164. The points E, Z on the non-parallel sides of trapezium $AB\Gamma\Delta$, define the segment EZ which is parallel of AB and passes through the intersection point O of the diagonals (See Figure 3.113-II). Show that $|EZ|$ is the harmonic mean of $|AB|$ and $|\Gamma\Delta|$, in other words holds $\frac{2}{|EZ|} = \frac{1}{|AB|} + \frac{1}{|\Gamma\Delta|}$.

Hint: Since $|EO| = |OZ|$, this is equivalent to $\frac{|EO|}{|AB|} + \frac{|OZ|}{|\Gamma\Delta|} = \frac{|\Delta O|}{|AB|} + \frac{|OB|}{|\Delta B|} = 1$.

Exercise 3.165. Construct a triangle for which are given the positions of the traces of its three altitudes.

Exercise 3.166. Show that if the positive numbers x, y have fixed product $x \cdot y = d^2$, then $x^2 + y^2$ is minimal when $x = y$.

Hint: The expression $x^2 + y^2$ represents the area E of a square with side $a = \sqrt{x^2 + y^2}$. E becomes minimal exactly when a becomes minimal. a however is the length of the diagonal of a rectangle with sides x, y. Apply Exercise 3.127.

Exercise 3.167. Let $\{a, b, x, y\}$ be positive numbers. Suppose that x, y change while keeping the product $x \cdot y = d^2$ fixed, then $ax^2 + by^2$ becomes minimal when $x\sqrt{a} = y\sqrt{b}$.

Hint: Define $x' = x\sqrt{a}$, $y' = y\sqrt{b}$ and apply Exercise 3.166 for x', y'.

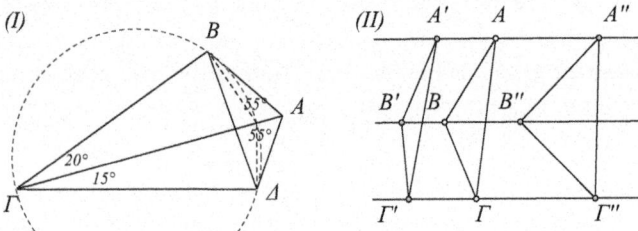

Fig. 3.114: Special quadrilateral   Combining triangles

Exercise 3.168. Quadrilateral $AB\Gamma\Delta$ has angles $\widehat{BA\Gamma} = \widehat{\Gamma A\Delta} = 55°$ and $\widehat{A\Gamma B} = 20°$, $\widehat{A\Gamma\Delta} = 15°$ (See Figure 3.114-I). To find the measure of angle $\widehat{B\Delta A}$.

Exercise 3.169. The triangles $\{A'B'\Gamma', A''B''C''\}$ have their vertices on three parallel lines (See Figure 3.114-II) and points $\{A, B, \Gamma\}$ divide the segments in the same ratio $k = |AA'|/|AA''| = |BB'|/|BB''| = |\Gamma\Gamma'|/|\Gamma\Gamma''|$. Show that the areas of the corresponding triangles satisfy the relation $(k+1)\varepsilon(AB\Gamma) = \varepsilon(A'B'\Gamma') + k\varepsilon(A''B''\Gamma'')$.

Exercise 3.170. The circles α, β are mutually tangent at the point X and are also tangent to line ε at points A, B, respectively (See Figure 3.115-I). Show that the line AX passes through the diametrically opposite point Z of B.

Exercise 3.171. The circles $\lambda(O)$, $\mu(P)$ are tangent at a point A and are also tangent to the circle $\kappa(\Pi)$ at points, respectively E, B (See Figure 3.115-II). Show that, keeping the other circles fixed and varying only circle μ, in such a way that it is always tangent to κ, λ, line AB passes through a fixed point Γ of the center-line of κ, λ. Show that points Γ, E are harmonic conjugate relative to O, Π.

3.14. COMMENTS AND EXERCISES FOR THE CHAPTER

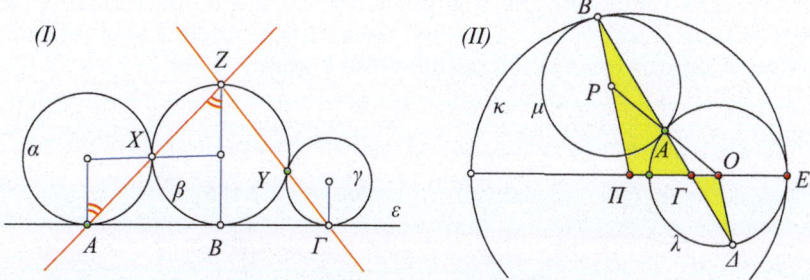

Fig. 3.115: Property of tangent circles

Exercise 3.172. Three circles α, β, γ are mutually tangent at points X, Y and all are tangent to a circle ε at points A, B and Γ. Show that

1. Lines BX, BY intersect circles α, γ, respectively, at points A', Γ' and lines BA, $B\Gamma$ at points A'', Γ'', such that $A'A''$, $\Gamma'\Gamma''$ are diameters of α, γ, parallel to the diameter of β through B (See Figure 3.116-I).
2. Lines AX, ΓY intersect line BO at the same point I, which is harmonic conjugate of B relative to O, O_β, where O, O_β are the centers of ε and β respectively (See Figure 3.116-II).

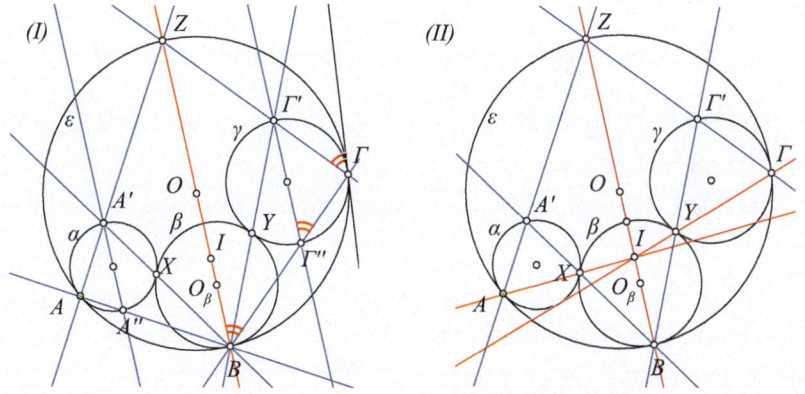

Fig. 3.116: Chain of pairwise tangent circles

Exercise 3.173. Show that there is no point O of the plane of the triangle $AB\Gamma$ with the property: Every line through O divides the triangle into two other polygons with the same area.

Hint: If there were such a point, then OA should be coincident with the median from A, therefore O should be coincident with the centroid of $AB\Gamma$.

Exercise 3.174. Determine the triangular trajectory $AB\Gamma$ of a billiard ball, which starting from a point A in the interior of the circle κ and reflecting twice on it, returns to its initial position (See Figure 3.117-I).

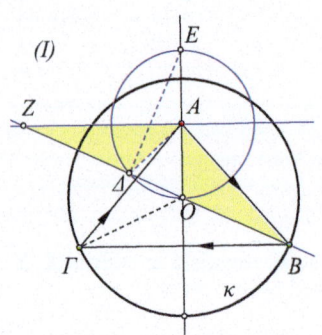
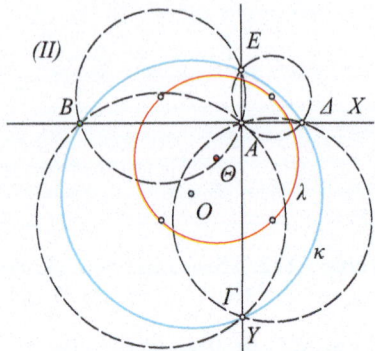

Fig. 3.117: Cyclic billiard   Rotating right angle

Hint: ([9, p.156]) Because, upon reflection at B and Γ, the angle to the radius is preserved, triangle $AB\Gamma$ is seen to be isosceles. The key in the problem is the circle $A(|AO|)$, which intersects again OA at E and OB at Δ. If Z is the intersection of OB with the orthogonal to OA at A, then: (i) the triangles AOB and $A\Delta Z$ are congruent and (ii) the triangles $O\Delta E$, OAZ are similar right triangles. From the congruent triangles follows that $|OZ| - |O\Delta| = |Z\Delta| = |OB|$ and from the similar triangles follows that $|OZ|/|OA| = |OE|/|O\Delta|$ $\Leftrightarrow$ $|O\Delta||OZ| = |OA||OE| = 2|OA|^2$. Thus, $|O\Delta|$ and $|OZ|$ can be determined from these two equations.

Exercise 3.175. The right angle $\widehat{XAY}$ rotates about a fixed point A in the interior of the circle $\kappa(O,\rho)$ and its sides intersect the circle at points B, Γ, Δ, E (See Figure 3.117-II). Show that the centers of the four circles with diameters $B\Gamma$, $\Gamma\Delta$, ΔE, EB are vertices of a rectangle and are contained in a circle λ, whose center Θ is the middle of OA.

Hint: See Exercise 3.128.

Exercise 3.176. Extend sides HZ, ME of a regular octagon $EZH\Theta IK\Lambda M$, so that the right angle $\widehat{NTE}$ is formed (See Figure 3.118-I). Show that $\frac{|Z\Gamma|}{|ZN|} = \sqrt{2}$, where N is the middle of ZH.

Exercise 3.177. Four equal circles $\{\kappa_1, \kappa_2, \kappa_3, \kappa_4\}$ pass through a point M and intersect for a second time at points $\{A,B,\Gamma,\Delta,E,Z\}$ (See Figure 3.118-II). Show that the line segments $\{A\Gamma, B\Delta, EZ\}$ have a common middle. How can be this formulated as a property of the inscriptible quadrilateral ([75, p.239])?

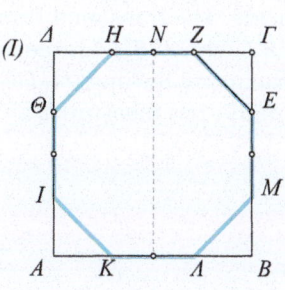
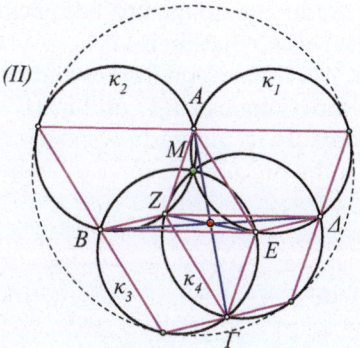

Fig. 3.118: Regular octagon   Four equal circles

Exercise 3.178. The triangles $AB\Gamma$ and $AB'\Gamma'$ are similar and similarly oriented. In the circles $B(|BB'|)$, $\Gamma(|\Gamma\Gamma'|)$ we form equal and similarly oriented central angles $\widehat{B'BB''}$ and $\widehat{\Gamma'\Gamma\Gamma''}$. Show that the triangle $AB''\Gamma''$ is also similar to $AB\Gamma$, $AB'\Gamma'$ (See Figure 3.119-I).

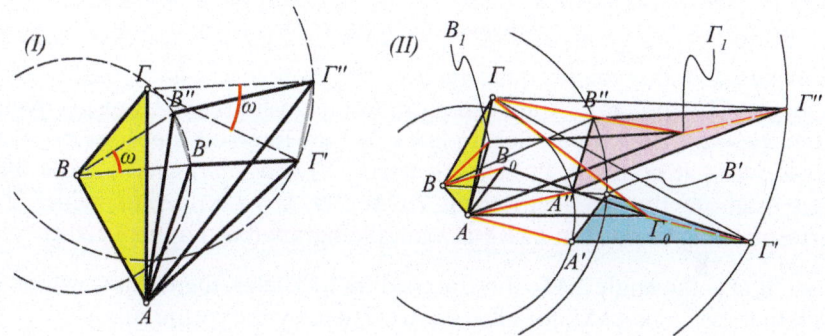

Fig. 3.119: Similar triangles with vertices on circles

Hint: The triangles BAB' and $\Gamma A\Gamma'$ are similar with ratio, say λ. Triangles $BB'B''$ and $\Gamma\Gamma'\Gamma''$ are also similar with the same ratio. It follows that $AB'B''$ and $A\Gamma'\Gamma''$ are similar with the same ratio. From this, it follows that $\widehat{B''A\Gamma''}$ and $\widehat{B'A\Gamma'}$ are equal, as well as the ratios $\frac{|AB''|}{|A\Gamma''|} = \frac{|AB'|}{|A\Gamma'|}$.

Exercise 3.179. The triangles $AB\Gamma$ and $AB'\Gamma'$ are similar and similarly oriented. In the circles $A(|AA'|)$, $B(|BB'|)$ and $\Gamma(|\Gamma\Gamma'|)$ we form equal and similarly oriented central angles $\widehat{A'AA''}$, $\widehat{B'BB''}$ and $\widehat{\Gamma'\Gamma\Gamma''}$ of measure ω. Show that the triangle $A''B''\Gamma''$ is also similar to $AB\Gamma$, $A'B'\Gamma'$ (See Figure 3.119-II).

Hint: Reduction to the preceding exercise. Translate the triangles parallel to themselves so that their vertices A, A', A'' coincide. The translated triangles $AB_0\Gamma_0$, $AB_1\Gamma_1$ have equal sides and parallel to $A'B'\Gamma'$ and $A''B''\Gamma''$ respectively. Then the triangles $\Gamma\Gamma_0\Gamma'$ and $\Gamma\Gamma_1\Gamma''$ are congruent and the angle $\Gamma_0\Gamma\Gamma_1$ also has measure ω. Similarly we show that angle B_0BB_1 has measure ω as well, so that the preceding exercise can be applied.

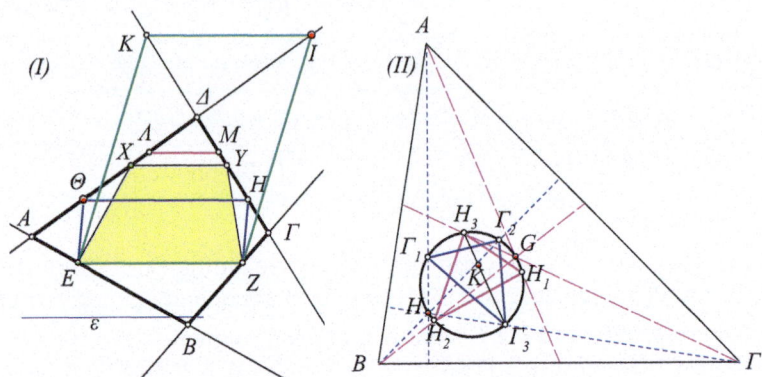

Fig. 3.120: Inscribed trapezia   Orthocenter, centroid

Exercise 3.180. In a convex quadrilateral $AB\Gamma\Delta$ are inscribed trapezia $EZYX$ as follows: The points E, X are taken randomly on sides AB, $A\Delta$ respectively. From these are drawn parallels to line ε, which intersect the sides $B\Gamma$, $\Gamma\Delta$, respectively, at points Z, Y. Leaving fixed EZ, and varying XY, show that there exist exactly two positions of X : $X = \Theta$, $X = I$ on $A\Delta$, for which the corresponding trapezium $EZYX$ is a parallelogram (See Figure 3.120-I).

Hint: If ω is the angle $A\Delta\Gamma$, then, according to Thales, there is a constant κ, such that $|\Delta Y| = \kappa \cdot |\Delta X|$, for all X on $A\Delta$. Then, by the cosine rule,

$$|XY|^2 = |\Delta X|^2 + |\Delta Y|^2 - 2|\Delta X||\Delta Y|\cos(\omega)$$
$$= |\Delta X|^2(1 + \kappa^2 - 2\kappa\cos(\omega)) = |\Delta X|^2|\Lambda M|^2,$$

where Λ is defined so that the square of its length is equal to the preceding parenthesis. From the preceding formula, the equation $|XY|^2 = |EZ|^2$ gives two solutions ΔX for $X = \Theta$, $X = I$, which are the points of $A\Delta$ lying symmetrically relative to Δ.

Exercise 3.181. The orthocenter H of a triangle $AB\Gamma$ is projected on the medians (See Figure 3.120-II), at points H_1, H_2, H_3 and the centroid G is projected on the altitudes, at points Γ_1, Γ_2, Γ_3. Show that all six points lie on circle with diameter HG and triangle $\Gamma_1\Gamma_2\Gamma_3$ is similar to $AB\Gamma$, while $H_1H_2H_3$ is similar to the triangle of the medians. Also show that the lines $\Gamma_1 H_1$, $\Gamma_2 H_2$, $\Gamma_3 H_3$ pass through the symmedian center K of the triangle.

Exercise 3.182. Show that a secant z of the non parallel sides of trapezium, dividing it into two similar convex quadrilaterals is the mean proportional of the parallel sides $x, y : z^2 = xy$ (See Figure 3.121-I). In which cases there are one or two such secants?

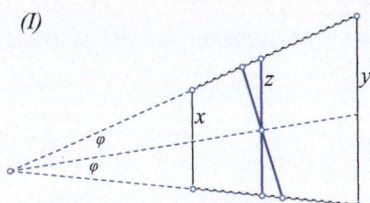
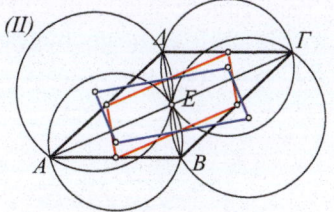

Fig. 3.121: Trapezium division   Similar parallelograms

Exercise 3.183. Given is a parallelogram $AB\Gamma\Delta$. Show that the circumcenters of the triangles, which are formed from its sides and its center E, are vertices of a parallelogram similar to the parallelogram of the middles of its sides (See Figure 3.121-II).

Exercise 3.184. Let P be a point inside the regular ν-gon Π. Using areas, show that the sum of the distances of P from the sides of the polygon is equal to $\nu\rho$, where ρ the radius of the inscribed circle of the polygon (see also exercise 2.237).

Exercise 3.185. The radii AB and $\Delta\Gamma$ of two circles, respectively α and β rotate in such a way as to always form the fixed angle ω (See Figure 3.122-I).

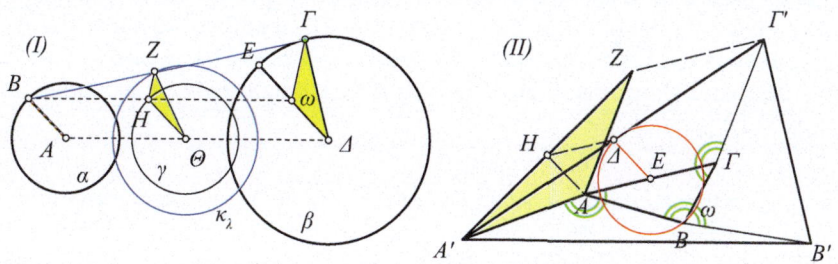

Fig. 3.122: Geometric locus I   Geometric locus II

On $B\Gamma$ define point Z such that the ratio $\frac{|ZB|}{|Z\Gamma|} = \lambda$ remains fixed. Show that the geometric locus of Z is a circle with center a point Θ of $A\Delta$.

Exercise 3.186. From the vertices of the triangle $AB\Gamma$ and with the same orientation we draw equal line segments AA', BB', $\Gamma\Gamma'$, which form, respectively, the same angle ω with sides $A\Gamma$, BA, ΓB (See Figure 3.122-II). Show that for variable angle ω the geometric loci of the middles of the sides of the resulting triangle $A'B'\Gamma'$ are circles.

Exercise 3.187. From the vertices of the triangle $AB\Gamma$ and with the same orientation we draw equal line segments AA', BB', $\Gamma\Gamma'$ which form the angle ω, respectively, with sides $A\Gamma$, BA, ΓB (See Figure 3.123-I). Show that, for variable angle ω the geometric locus of the centroid of the resulting triangle $A'B'\Gamma'$ is a circle.

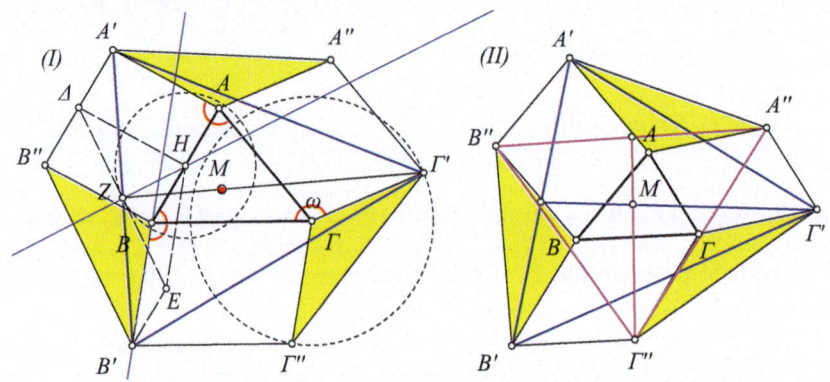

Fig. 3.123: Geometric locus of centroid of a variable triangle

Hint: From Exercise 3.186 we know that the middle Z of $A'B'$ is contained in a fixed circle α. Also, by hypothesis, point Γ' is contained in a fixed circle β. Because the centroid M divides $Z\Gamma'$ into the ratio 1:2, it suffices, according to Exercise 3.185, to show that the radii HZ, $\Gamma\Gamma'$ of circles α, β respectively form a fixed angle. Because the angles are fixed, it suffices to show that the angle between AA' and ZH is constant (See Figure 3.123-I). Construct therefore the parallelograms $A'AH\Delta$ and $HBB'E$. Triangle $H\Delta E$ is isosceles, the angle $\widehat{\Delta HE}$ is fixed etc.

Exercise 3.188. In the figure of the preceding exercise (See Figure 3.123-II), show that the triangles $A'B'\Gamma'$ and $A''B''\Gamma''$ have the same centroid M.

Hint: Apply the conclusion of Exercise 2.231.

Exercise 3.189. From the centers A, B of two mutually external circles κ, λ we draw tangents to them. Show that they excise on the circles chords $\Gamma\Delta$, EZ of equal length (See Figure 3.124-I).

Hint: Show that ΓE is parallel to the centerline AB, through the relation $\frac{|H\Gamma|}{|HA|} = \frac{|HE|}{|HB|}$, which results from the similar triangles of the figure.

Exercise 3.190. Let E be a point on the side AB of the rectangle $AB\Gamma\Delta$. Construct another rectangle, which has one vertex on each side of the given one and one of them coincides with E. How many such rectangles exist? Is there a point E, such that one of the corresponding inscribed rectangles with vertex at E has the same side ratio with that of $AB\Gamma\Delta$ (See Figure 3.124-II) ?

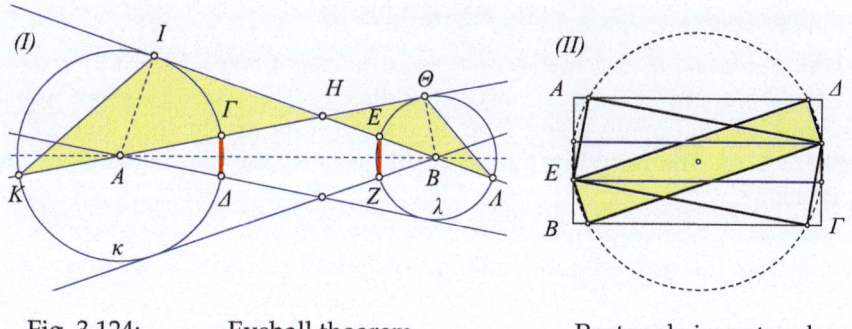

Fig. 3.124:   Eyeball theorem   Rectangle in rectangle

Exercise 3.191. Find points I and K on the sides AB and $A\Gamma$ of triangle $AB\Gamma$ such that $|BI| = |IK| = |K\Gamma|$ (See Figure 3.125-I).

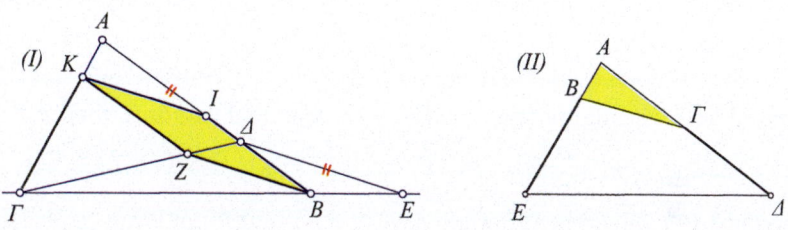

Fig. 3.125: Equal segments   Construction from $\{\widehat{A}, b+a, c+a\}$

Hint: Suppose that the requested points K and I were found. Also suppose that AB is greater than $A\Gamma$ and consider the point Δ on AB, such that $|A\Delta| = |A\Gamma|$. Draw from K a parallel to AB intersecting $\Gamma\Delta$ at Z. Because of the similarity to the isosceles $A\Gamma\Delta$, triangle ΓKZ is an isosceles triangle. It follows that $|KZ| = |K\Gamma| = |IB|$, therefore $KZBI$ is a rhombus. Define the intersection point E of the parallel from Δ to ZB with $B\Gamma$. From the similar triangles $BZ\Gamma$, $E\Delta\Gamma$ follows $\frac{|\Delta E|}{|ZB|} = \frac{|\Gamma\Delta|}{|\Gamma Z|}$. From the similar triangles $A\Gamma\Delta$, $K\Gamma Z$ follows $\frac{|\Delta E|}{|ZB|} = \frac{|\Gamma\Delta|}{|\Gamma Z|}$. From the two equalities follows obviously $|\Delta E| = |\Delta A|$. Thus, point E is constructible from the given data: by constructing first point Δ and writing a circle with center Δ and radius ΔA, which intersects $B\Gamma$ at E (generalization in exercise 4.137).

Relying on the preceding analysis, construct first the points Δ, E and then draw a parallel to $E\Delta$ from B, intersecting $\Gamma\Delta$ at Z. From Z draw a parallel to BA intersecting $A\Gamma$ at K. Consider, finally, the point I on BA, such that $|BI| = |ZK|$. K and I, constructed this way, are the wanted points.

Exercise 3.192. Construct a triangle from angle $\alpha = \widehat{BA\Gamma}$ and the sums of lengths of sides $b+a$, $c+a$ (See Figure 3.125-II).

Hint: The triangle $A\Delta E$ with angle measure α at A and adjacent to it sides $|A\Delta| = b+a$, $|AE| = c+a$ is constructed directly. Then, using the preceding exercise we find points B and Γ on AE and $A\Delta$ respectively, such that $|EB| = |B\Gamma| = |\Gamma\Delta|$. The resulting triangle $AB\Gamma$ satisfies the requirements of the problem.

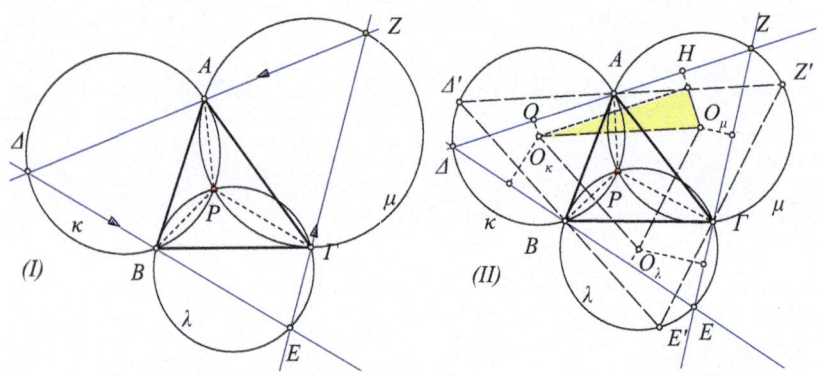

Fig. 3.126: ΔEZ circumscribes $AB\Gamma$ Maximal circumscribed $\Delta' E' Z'$

Exercise 3.193. Given is a triangle $AB\Gamma$ and a point P not lying on its sides. We construct the circles $\kappa = (PAB)$, $\lambda = (PB\Gamma)$, $\mu = (P\Gamma A)$ and consider an arbitrary point Δ on the circle κ and the second intersection point E of ΔB with circle λ. Let Z be the intersection point of line $E\Gamma$ with the circle μ. Show that points Z, A and Δ are collinear and the triangle ΔEZ has always the same angles, regardless of the position of Δ on κ (See Figure 3.126-I).

Hint: The angles at P: $\widehat{APB}, \widehat{BP\Gamma}, \widehat{\Gamma PA}$ have a sum of measure 360 degrees. The quadrilaterals $APB\Delta$ and $BP\Gamma E$ are inscriptible quadrilaterals and the following relations for the angles are valid

$$\widehat{A\Delta B} = 180° - \widehat{APB}, \quad \widehat{BE\Gamma} = 180° - \widehat{BP\Gamma} \Rightarrow$$
$$\widehat{A\Delta B} + \widehat{BE\Gamma} = 360° - (\widehat{BP\Gamma} + \widehat{BP\Gamma}) = \widehat{AP\Gamma}.$$

Thus, if we define Z as the intersection point of $A\Delta$ and ΓE, then by the preceding relations, in triangle ΔEZ we have $\widehat{AZ\Gamma} = 180° - \widehat{A\Delta B} - \widehat{BE\Gamma} =$

3.14. COMMENTS AND EXERCISES FOR THE CHAPTER

$180° - \widehat{AP\Gamma}$, which means that the quadrilateral $AP\Gamma Z$ is inscriptible and point Z is contained in the circle μ.

Exercise 3.194. Given are two triangles $AB\Gamma$ and XYZ. Construct a triangle ΔEZ, circumscribed of $AB\Gamma$ and similar to XYZ (See Figure 3.126-I).

Hint: Suppose the requested triangle ΔEZ has been constructed. Show first that the circles $\kappa = (A\Delta B)$, $\lambda = (B\Gamma E)$ and $\mu = (\Gamma Z A)$ pass through a common point P. Show next that the position of P is determined from the given data. Finally, follow the procedure in the preceding exercise to define ΔEZ.

Exercise 3.195. Let the triangle ΔEZ be circumscribed to the triangle $AB\Gamma$ and O_κ, O_λ, O_μ be the centers of the circles $\kappa = (AB\Delta)$, $\lambda = (\Gamma BE)$ and $\mu = (A\Gamma Z)$. Show that the triangle $\Delta' E' Z'$, with sides parallel to those of the triangle $O_\kappa O_\lambda O_\mu$ and passing through A, B, Γ, is similar to ΔEZ and has perimeter greater than or equal that of triangle ΔEZ (See Figure 3.126-II).

Hint: Project O_κ, O_λ, O_μ onto the sides of triangle ΔEZ. form the right triangle OHO_μ. OH has half the length of ΔZ and, as orthogonal to HO_μ is less than the hypotenuse $O_\kappa O_\mu$ which in turn has length half that of $\Delta' Z'$.

Exercise 3.196. Construct the maximal in perimeter, equilateral triangle ΔEZ which is circumscribed to a given triangle $AB\Gamma$ (See Figure 3.126-II).

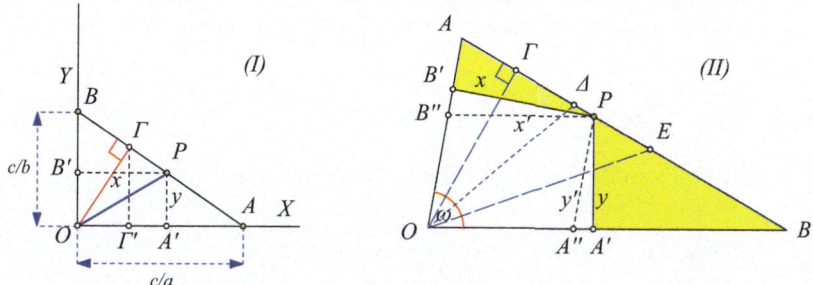

Fig. 3.127: Minimization $x^2 + y^2$ Maximization $\varepsilon(OA'PB')$

Exercise 3.197. The variable positive numbers x, y satisfy the equation $ax + by = c$, where a, b, c are positive constants. Show that expression $x^2 + y^2$ is minimized when $\frac{x}{y} = \frac{a}{b}$.

Hint: On the sides of the right angle $\widehat{XOY}$ define respectively segments of length c/a, c/b (See Figure 3.127-I). According to Theorem 3.18, point P, whose lengths of projections on the sides of angle $x = |PB'|$, $y = |PA'|$ satisfy the equation $ax + by = c$ and are contained in line AB. Expression $x^2 + y^2$ is the square of the length $|OP|$, which is minimized when OP coincides with the altitude $O\Gamma$ of triangle OAB from point O. Then the corresponding triangle $\Gamma' \Gamma' O$ is similar to AOB therefore the expression $\frac{x}{y} = \frac{a}{b}$ follows.

Exercise 3.198. The variable positive numbers x, y satisfy the equation $ax + by = c$, where a, b, c are positive constants. Show that expression $ux^2 + vy^2$, where u, v are positive constants, is minimized when $\frac{x}{y} = \frac{a \cdot v}{b \cdot u}$.

Hint: Define $x' = x\sqrt{u}$, $y' = y\sqrt{v}$. x', y' satisfy the equation $a\frac{x'}{\sqrt{u}} + b\frac{y'}{\sqrt{v}} = c$ and the minimization of expression $ux^2 + vy^2$ is equivalent to the minimization of expression $x'^2 + y'^2$. According to Exercise 3.197 this will happen when $\frac{x'}{y'} = \frac{a/\sqrt{u}}{b/\sqrt{v}}$, which is equivalent to $\frac{x}{y} = \frac{a \cdot v}{b \cdot u}$.

Exercise 3.199. Point P of side AB of an acute angled triangle OAB is projected in points A', B' of the other sides (See Figure 3.127-II). Show that the quadrilateral $OA'PB'$ has maximum area when P is the intersection E of AB with the symmetric OE of the altitude $O\Gamma$ relative to the bisector of angle $\widehat{O}$.

Hint: According to Exercise 3.85, the lengths $x' = |PB''| = x/\sin(\omega)$, $y' = |PA''| = y/\sin(\omega)$, where A'', B'' are the projections of P parallel to the sides OA, OB, satisfy the equation

$$\frac{x'}{a} + \frac{y'}{b} = 1 \Leftrightarrow \frac{x}{a} + \frac{y}{b} = \sin(\omega),$$

with $a = |OB|$, $b = |OA|$. Also, the quadrilateral $OA'PB'$ will have maximum area exactly when the sum of the areas of the two triangles $\delta = \varepsilon(PB'A) + \varepsilon(PA'B)$ becomes minimal. This sum, however, is calculated easily

$$\varepsilon(PB'A) = \frac{x^2}{2\tan(A)}, \quad \varepsilon(PA'B) = \frac{y^2}{2\tan(B)} \Rightarrow \delta = \frac{x^2}{2\tan(A)} + \frac{y^2}{2\tan(B)}.$$

According to Exercise 3.198, δ will be minimized exactly when

$$\frac{x}{y} = \frac{1/(a \cdot \tan(B))}{1/(b \cdot \tan(A))} = \frac{\cos(B)}{\cos(A)}.$$

However the ratio $\frac{x}{y} = \frac{\cos(A)}{\cos(B)} = \kappa$ characterizes the line-carrier $A\Gamma$ of the altitude and the ratio $\frac{1}{\kappa}$ the symmetric of AE relative to the bisector of angle $\widehat{O}$ (Exercise 3.79).

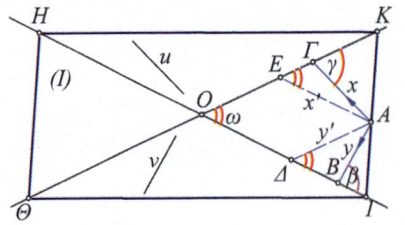
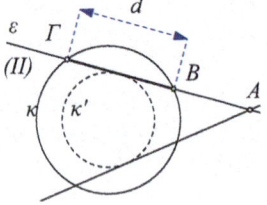

Fig. 3.128: $x + y = \delta$ Excision of given length

3.14. COMMENTS AND EXERCISES FOR THE CHAPTER

Exercise 3.200. Point *A* is projected in parallel to the fixed directions *u*, *v* on the sides of angle $\widehat{BO\Gamma}$ respectively at points Γ and *B*. Show that the geometric locus of *A*, for which the summ $\delta = |AB| + |A\Gamma| = \delta$ is fixed, is a parallelogram with center at the point *O*.

Hint: Using the notation of figure 3.128-I, $x+y=\delta$ is equivalent to $\frac{x'}{a}+\frac{y'}{b}=1$, where $a = \delta \sin(\gamma)/\sin(\omega)$ and $b = \delta \sin(\beta)/\sin(\omega)$. Reduce into Exercise 3.85.

Exercise 3.201. Let *A* be a point not lying on the circle κ. Draw a line through *A*, intersecting κ at points *B*, Γ such that $B\Gamma$ has a given length (See Figure 3.128-II).

Hint: The chords of a circle κ with given fixed length *d* are tangent to a concentric circle κ' of κ.

Exercise 3.202. An isosceles trapezium has fixed basis and circumscribed circle κ. The side parallel to its base is a chord of κ. Find the place of the chord which maximizes the area of the trapezium.

Exercise 3.203. For a given triangle $AB\Gamma$ to find a fourth point Δ so that the quadrilateral $AB\Gamma\Delta$ is simultaneously inscriptible and circumscribable.

Exercise 3.204. Point *P* is projected on the sides and the bisector of the angle $\widehat{XOY}$ at corresponding points $\{A, B, \Delta\}$. Show that $|\Delta A| = |\Delta B|$.

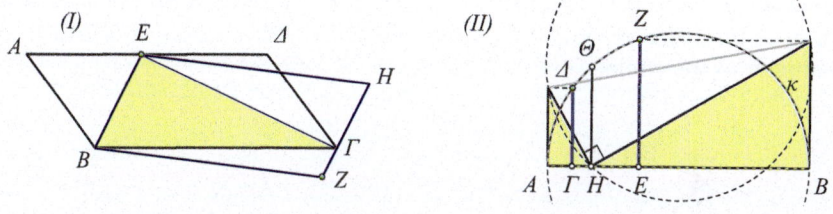

Fig. 3.129: Equal areas   Mean proportional

Exercise 3.205. The two parallelograms of figure 3.129-I have common the vertex *B*, point *E* is on $A\Delta$ and Γ on ZH. Show that they have equal area.

Exercise 3.206. Let $\Gamma\Delta$, EZ be two semi-chords orthogonal to the diameter AB of a half circle (See Figure 3.129-II). Construct a parallel to them chord $H\Theta$, such that $|H\Theta|^2 = |\Gamma\Delta||EZ|$.

Exercise 3.207. Let α, β, γ be three lines passing through a common point *H*. Construct triangles which have these lines as carriers of their altitudes.

Exercise 3.208. Show that the lines $\{A'\Delta', B'E', \Gamma'Z'\}$, which join the middles of sides of the triangle $AB\Gamma$ with the middles of sides of its orthic ΔEZ, intersect at a point Θ, coinciding with the circumcenter of the orthic triangle (See Figure 3.130-I).

Exercise 3.209. Show that the projections of the trace of Δ, of the altitude of triangle $AB\Gamma$ from A, on the other sides and the other altitudes $\{BE, \Gamma Z\}$ are points of the same line ε (See Figure 3.130-II). Show also that ε is parallel to $\varepsilon' = ZE$.

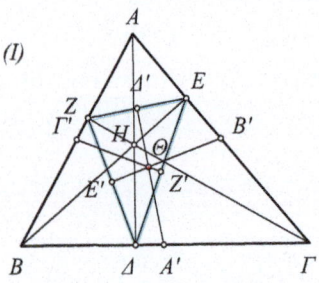
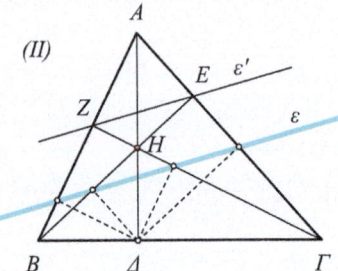

Fig. 3.130: Circumcenter of orthic   Projections to altitudes

Exercise 3.210. To find the geometric locus of points P whose projections on the sides of a fixed angle $\widehat{XOY}$ at corresponding points $\{A, B\}$ define lines AB of a fixed direction.

Exercise 3.211. Angle $\omega = \widehat{XOY}$ of constant measure is turning about the center O of the circle $\kappa(O)$ intersecting it at points $\{A, B\}$. To find the geometric locus of intersection points P of the lines $\{AA', BB'\}$, where $A'B'$ is a fixed diameter of the circle.

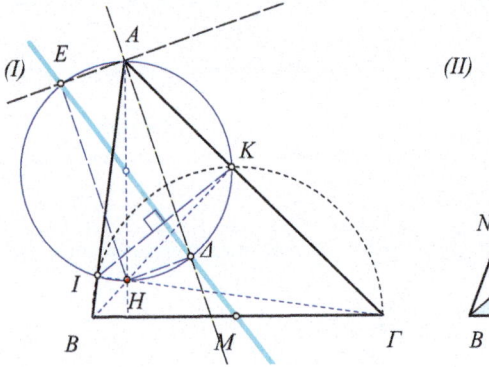
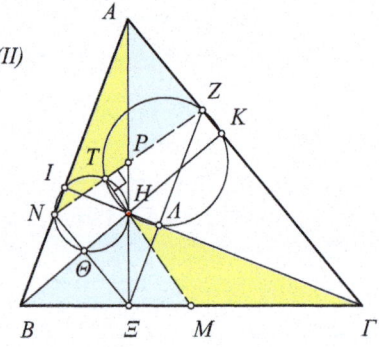

Fig. 3.131: Projections to bisectors   Projections to altitudes

Exercise 3.212. The orthocenter H of triangle $AB\Gamma$ is projected to the bisectors of angle $\widehat{A}$, at points Δ and E (See Figure 3.131-I). Show that the line ΔE passes through the middle M of $B\Gamma$.

Hint: The circle with diameter ΔE passes through A, H, K, I, where K, I are the traces of the altitudes from B and Γ respectively (See Figure 3.131-I). Also the triangle KEI is isosceles, has $E\Delta$ as a bisector on its vertex and IK is a chord of the circle with diameter $B\Gamma$.

Exercise 3.213. If $\{A\Xi, BK, \Gamma I\}$ are the altitudes of triangle $AB\Gamma$ and H its orthocenter, point Ξ is projected to the other altitudes $\{BK, \Gamma I\}$, correspondingly at points $\{\Theta, \Lambda\}$. Show that the second intersection point T of the circles $\{(H,I,N),(H,\Lambda,Z)\}$ is contained in the line HM, where M is the middle of $B\Gamma$ (See Figure 3.131-II).

Hint: Triangles $\{BMH, APZ\}$, as well as $\{\Gamma MH, APN\}$, are similar.

Exercise 3.214. Points $\{A, B\}$ move on parallel lines $\{\alpha, \beta\}$, so that the segment AB has a constant direction. From point P outside the band of the parallels we form the angle $\omega = \widehat{APB}$. To locate the place of AB, for which this angle becomes maximal.

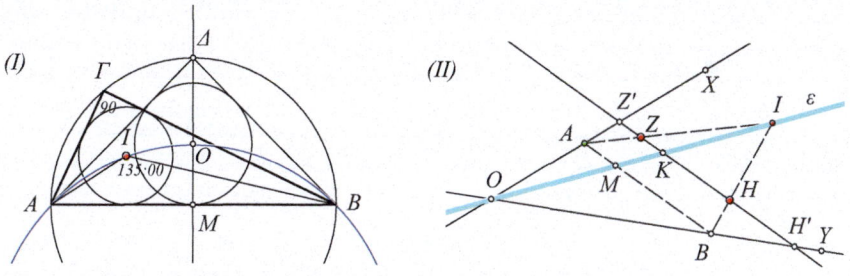

Fig. 3.132: Ratio of radii   Geometric locus

Exercise 3.215. Show that for every right angled triangle the ratio r/R of the radius of the inscribed to that of the circumscribed circle is less than or equal to $\sqrt{2} - 1$ (See Figure 3.132-I).

Exercise 3.216. Show that the relation $\sqrt{2} - 1 \geq r/R$, of the preceding exercise, is the necessary and sufficient condition for the existence of a right angled triangle with circumradius R and inradius r. Then, construct the orthogonal sides of such a triangle for given $\{r, R\}$.

Exercise 3.217. Let OX, OY be two intersecting lines and Z, H be two points. From every point A of OX we draw a parallel to ZH, which intersects OY at B. Show that the geometric locus of the point of intersection I of AZ, BH is a line ε passing through O (See Figure 3.132-II).

Hint: $\frac{|KZ|}{|KH|} = \frac{|MA|}{|MB|}$. Also $\frac{|KZ'|}{|KH'|} = \frac{|MA|}{|MB|}$, therefore $\frac{|KZ|}{|KH|} = \frac{|KZ'|}{|KH'|}$. From this relation follows that K is a fixed point.

Exercise 3.218. Partition an equilateral triangle into polygons and rearrange them into a square (Figure 3.133).

Figure 3.133 is simultaneously also a hint for the last exercise. The method applied is the general one, which is described in exercises 3.28, 3.29, 3.30 and gives a partition into 5 polygons. There is however also another way with only 4, which was discovered in 1902 by Henry E. Dudeney ([16, p.260]).

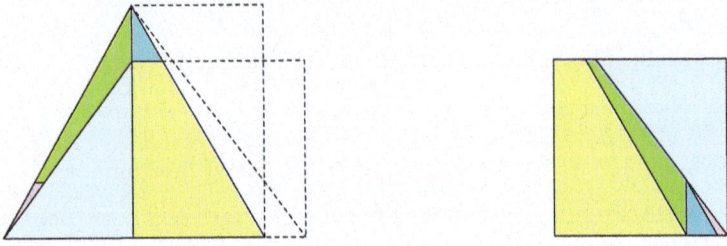

Fig. 3.133: Partition and rearrangement of parts into a square

A characteristic of this partition is that, by connecting the 4 parts by two with joints at vertices Δ, Z, E (See Figure 3.134), we can "unwind" the equilateral triangle and "rewind" it into a square. In figure 3.134 the "unwinded" joint shows in the middle. Choosing the last tile (4) and winding clockwise we get the figure to the left, that of the equilateral. Winding the joint counter-clockwise we get the square to the right ([67, p.142]).

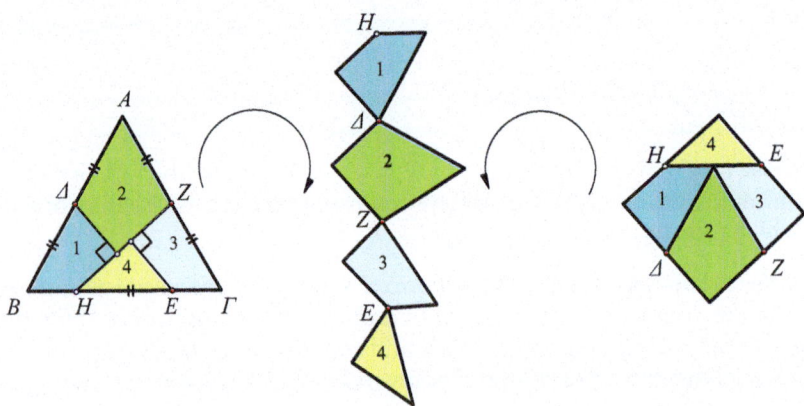

Fig. 3.134: Dudeney's partition of an equilateral

There are many interesting problems connected with polygon partitions, several of which remain open ([79, p.50]). One of these is the determination, for given v, of the least number $\kappa(v)$ of polygons to which a convex v-gon may be partitioned, in such a way as for its parts to reconstitute a square. Corresponding problems may be formulated also for space shapes and the third of the famous 23 problems of Hilbert, which he announced

3.14. COMMENTS AND EXERCISES FOR THE CHAPTER

in 1900, during the second international symposium, in Paris, ([59, p.1378], [91]) is related to this subject. This problem is equivalent to the fact that a regular tetrahedron cannot be partitioned into sub-polyhedrons, which reconstitute a cube ([61]). This property was proved in 1902 by Max Dehn. The preceding exercises show that the corresponding problem on the plane, i.e. the partition of an equilateral triangle to parts which reconstitute a square, is possible. Let us note here the following property:

If the (convex) polygons A, B have the same area, then A can be partitioned into sub-polygons, which rearranged in different order form polygon B.

The proof of this property is a consequence of the exercise 3.30. In fact, by that exercise, both polygons can be partitioned into sub-polygons which reconstitute the same square. We therefore have two tilings of the same square in two different ways. These two tilings define a third, of which the tiles result as intersections of the tiles of the two preceding tilings. With this third tiling we easily see that both the original polygons can be reconstituted. A more detailed examination of the subject also shows that the assumption of convexity is redundant. Despite the theoretical possibility, however, the practical partition and reconstitution of two specific polygons with equal areas is not at all obvious or easy.

Exercise 3.219. Show that the partition of the equilateral suggested by figure 3.134-left, where $|HE| = \frac{|B\Gamma|}{2}$ is symmetric relative to the middle of $B\Gamma$, really gives the square. If, however, the partition relies on the figure, with $|HE| = \frac{|B\Gamma|}{2}$, but with $|BH| \neq |E\Gamma|$, then, after the rearrangement of the parts results a rectangle. Examine also, which are the possible ratios $\frac{a}{b}$, of the bigger to the smaller side, of the parallelograms, which result this way.

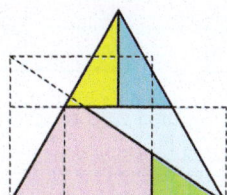 

Fig. 3.135: Partition and rearrangement of parts into a square II

Exercise 3.220. Determine the length of the sides of triangles, which belong to the partition of the equilateral triangle in figure 3.135.

Related to polygon partitions are also the two traditional puzzles: **Archimedes' Stomachion** and **Tangram**. The first is a partition of the square into 14 tiles. These are defined by dividing the square into 12×12 congruent squares and Drawing the lines which show up in figure 3.136-I ([71, p.109]). The figure also shows two rearrangements which recompose the square. It can be proved that there are 536 different ways of rearranging the tiles in such a way as to reconfigure a square ([77, p.56]).

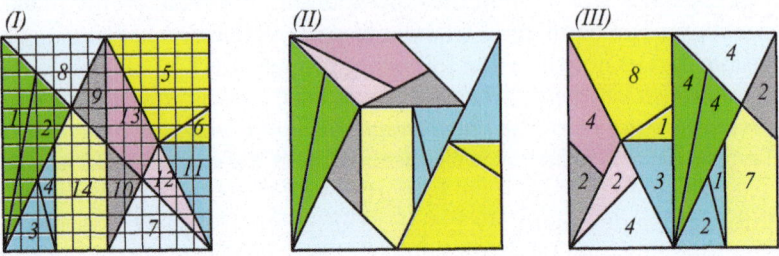

Fig. 3.136: Archimedes' "stomachion" (14 tiles) and two replacements

Exercise 3.221. Show that the areas of the tiles of stomachion are rational multiples of the area of the square. Specifically, show that each tile has area $\frac{x}{48}$ of the area of the square, where x is the tile's label on the right part of figure 3.136-III.

It can also be proved for all possible positions, the tile pairs $(1,2)$, $(3,4)$, $(5,6)$ pass together and always in the same order (See Figure 3.136-I).

Fig. 3.137: The 11 tiles of the *reduced stomachion* and two rearrangements

Glueing together the two tiles of the three preceding pairs, creates the **reduced stomachion** with 11 tiles (See Figure 3.137). Because in the reduced stomachion show up two additional congruent triangular tiles, the number of the different rearrangements of the tiles producing a square, reduces to half the preceding: 268.

Exercise 3.222. Construct the reduced stomachion from paper and use the symmetries you see in it, in order to create as many rearrangements of the tiles as possible, to a square.

The **Tangram** consists of a partition of the square into 7 tiles and was a game in ancient China, with main problem, the rearrangement of the tiles so that a polygon of given outline results ([92]). Relatively recently (1942), for example, it was proved that there are only 13 convex polygons which can be constructed with these tiles. Figure 3.138 shows the initial partition of the square into 7 tiles, as well as, rearrangements of the tiles which form each of these 13 polygons ([95]).

Fig. 3.138: The 13 convex polygons which are constructed using Tangram

Exercise 3.223. For a point X, not lying on the sides of the rectangle $AB\Gamma\Delta$, show that $|XA|^2 + |X\Gamma|^2 = |XB|^2 + |X\Delta|^2$.

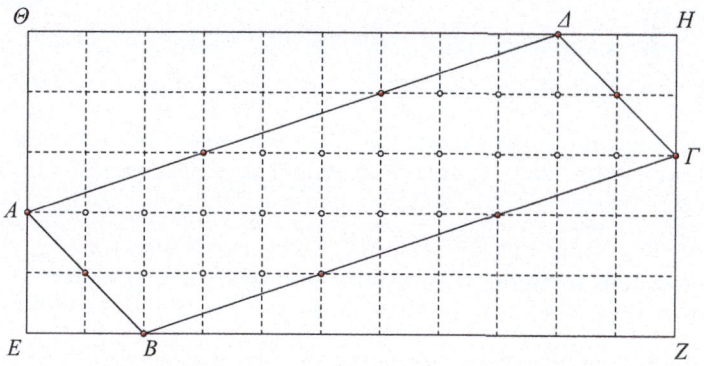

Fig. 3.139: Theorem of Pick for parallelograms

Exercise 3.224 (Pick's theorem for parallelograms). Show, that every parallelogram p with its vertices at nodal points of squared paper by unit-squares and containing in its interior κ nodal points and at its boundary λ nodal points, has area given by the formula (see Figure 3.139)

$$\varepsilon(p) = \kappa + \frac{\lambda}{2} - 1$$

Hint: Three steps. (i) We easily see that the formula holds for parallelograms like *EZHΘ*, with horizontal and vertical sides. (ii) relying on (i), we see again easily that the formula holds for right triangles, like *AEB* with one side horizontal and the other vertical. (iii) Combination of the preceding and the fact that each parallelogram *ABΓΔ* defines another one which encloses it and has horizontal and vertical sides, like *EZHΘ*.

The theorem holds more generally for a polygon with vertices at nodal points ([34, p.277], [72]). In the interesting collection of problems *"for children from 5 to 15 years old"* ([60, p.10]) Arnold proposes the simpler problem of the next exercise without mentioning the formula of Pick. A consequence of this formula is formulated in exercise 3.226.

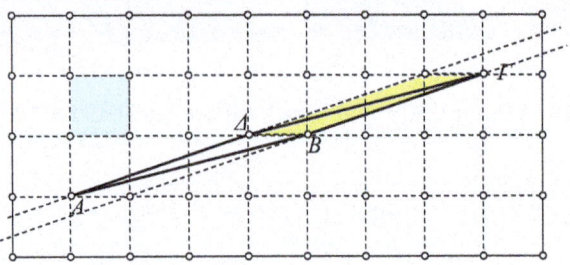

Fig. 3.140: Parallelogram of area 1

Exercise 3.225. Show that on squared paper, a parallelogram with its vertices at nodal points (See Figure 3.140), which does not contain in its interior or on its side another nodal point, has area equal to that of the nodal square.

Exercise 3.226. Show that the formula of Pick holds also for triangles and that there is no equilateral triangle with vertices at nodal points.

Hint: The proof for triangles reduces to that for parallelograms. For the equilateral triangle with vertices at nodal points, the square of the altitude v^2 is calculated easily when it is rational. The area ε, according to Pick, is also rational, therefore also $\frac{\varepsilon}{v^2}$ will be rational, which is however equal to $\frac{2}{\sqrt{3}}$.

Exercise 3.227. Show that the radii of two externally tangent circles and tangent to two lines which intersect at angle α (See Figure 3.141), is given by the formula

$$r' = r \frac{1+\sin\left(\frac{\alpha}{2}\right)}{1-\sin\left(\frac{\alpha}{2}\right)}.$$

Find the formula which connects the radii of the first and the last circle of a chain of ν circles, as in figure 3.141.

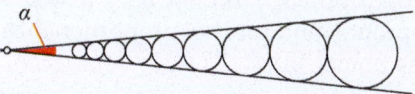

Fig. 3.141: Sequence of tangent circles

Exercise 3.228. Given are three fixed points $\{A_1, A_2, A_3\}$ and positive constants $\{\lambda_1, \lambda_2, \lambda_3, \mu\}$. Show that the geometric locus of points X, for which the sum is

$$\lambda_1 |XA_1|^2 + \lambda_2 |XA_2|^2 + \lambda_3 |XA_3|^2 = \mu,$$

is a circle. Show the corresponding property for four points $\{A_1, A_2, A_3, A_4\}$.

Hint: Use exercise 3.136. Replace $\lambda_1 |XA_1|^2 + \lambda_2 |XA_2|^2$ with a term of the form $\nu |XB|^2$, using a suitable fixed point B and a fixed number ν.

Next two exercises stem respectively from Euler and Fermat. The first one, which is solved easily using the figure, is used by Euler to solve the second ([88, p.10], [96, p.314]). Here, the hint of the second exercise, leads to a simpler solution which doesn't use the first one.

Exercise 3.229. Show that for four collinear points $\{A, P, \Sigma, B\}$ in this order (see Figure 3.142-I), the following relation holds $|AB| \cdot |P\Sigma| + |AP| \cdot |\Sigma B| = |A\Sigma| \cdot |PB|$.

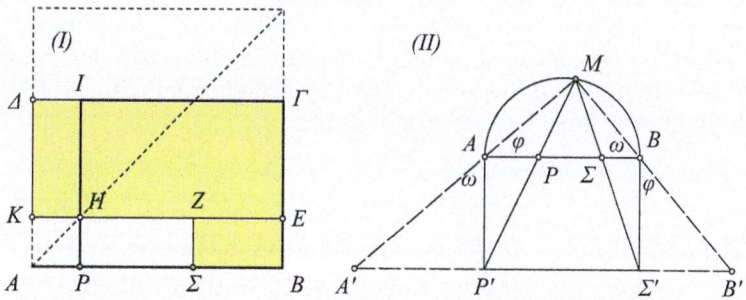

Fig. 3.142: Relation of lengths   Problem of Fermat

Exercise 3.230. The rectangle $AB\Sigma'P'$ has sides $a = |AP'|$, $b = |AB| = a\sqrt{2}$ and point M is found on the half circle with diameter AB (See Figure 3.142-II). Show that the lines $\{MP', M\Sigma'\}$ intersect AB at points $\{P, \Sigma\}$, for which holds $|A\Sigma|^2 + |PB|^2 = |AB|^2$.

Hint: $\{AA', BB'\}$ form a right triangle with altitude equal to $a = |AP'|$. Consequently holds (Proposition 3.3) $|A'P'||\Sigma'B'| = a^2 = b^2/2$ (*). The following relations prove the proposition for the, proportional to the given ones segments defined by the points $\{A', P', \Sigma', B'\}$.

$$\begin{aligned}
|A'\Sigma'|^2 + |P'B'|^2 &= (|A'B'| - |\Sigma'B'|)^2 + (|A'B'| - |A'P'|)^2 \\
&= 2|A'B'|^2 + |\Sigma'B'|^2 + |A'P'|^2 - 2|A'B'|(|\Sigma'B'| + |A'P'|) \\
&= 2|A'B'|^2 + |\Sigma'B'|^2 + |A'P'|^2 - 2|A'B'|(|A'B'| - b) \\
&= 2|A'B'|^2 + (|\Sigma'B'| + |A'P'|)^2 - [2|\Sigma'B'||A'P'|] - 2|A'B'|(|A'B'| - b) \\
&\stackrel{(*)}{=} 2|A'B'|^2 + (|A'B'| - b)^2 - [b^2] - 2|A'B'|(|A'B'| - b) = |A'B'|^2.
\end{aligned}$$

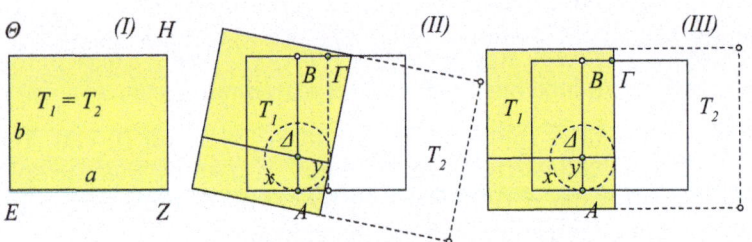

Fig. 3.143: Table that unfolds and doubles its surface

Exercise 3.231. The table $EZH\Theta$ has two rectangular plates $\{T_1, T_2\}$ lying one on top of the other and connected with a join along EZ (See Figure 3.143-I). At a certain point Δ on the table's frame, beneath of T_1, plate T_1 is mounted so that it can turn around Δ (See Figure 3.143-II) and unfolding plate T_2, after a turn by $90°$ makes a rectangle of the double area, sides parallel to the initial rectangle and the same symmetry center (See Figure 3.143-III). At which point of T_1 must be located point Δ?

Hint: $\{x = (2a - b)/4, \; y = b/4\}$.

Exercise 3.232. From a point A exterior to the cirlce $\kappa(O)$, we draw a line intersecting it at points $\{B, \Gamma\}$, and draw also the tangents $\{\tau_B, \tau_\Gamma\}$ to κ at these points. Suppose that these tangents intersect the orthogonal δ of AO at A at points $\{B', \Gamma'\}$. Show that A is the middle of $B'\Gamma'$.

Exercise 3.233. With diameters the sides $\{A'B', B'\Gamma', \Gamma'A'\}$ of the orthic $A'B'\Gamma'$ of the triangle $AB\Gamma$ we draw circles correspondingly $\{\gamma, \alpha, \beta\}$ intersecting triangle's sides: (i) α the sides $\{AB, A\Gamma\}$ at $\{A_1, A_2\}$, (ii) β the sides

$\{BA, B\Gamma\}$ at $\{B_1, B_2\}$, (iii) γ the sides $\{\Gamma A, \Gamma B\}$ at $\{\Gamma_1, \Gamma_2\}$. Show that points $\{A_1, A_2, B_1, B_2, \Gamma_1, \Gamma_2\}$ are concyclic.

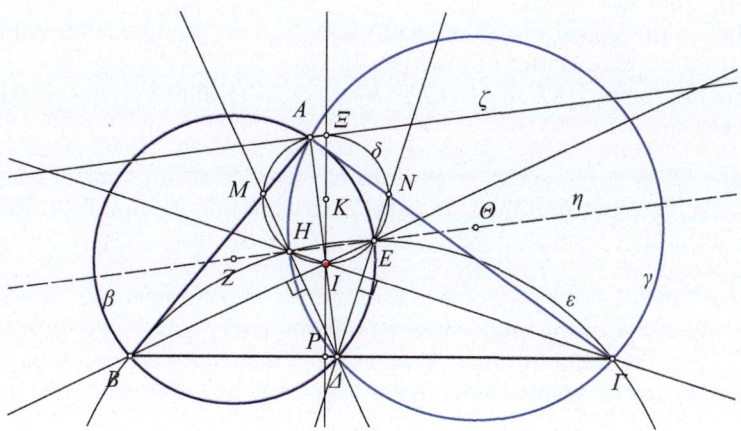

Fig. 3.144: Circles related to bisectors

Exercise 3.234. Consider the circles $\{\beta = (AB\Delta), \gamma = (A\Gamma\Delta)\}$, where Δ is the trace of the bisector $A\Delta$ on side $B\Gamma$ of the triangle $AB\Gamma$ (See Figure 3.144). Prove the following relations:

1. If $\{Z, \Theta\}$ are the centers of $\{\beta, \gamma\}$ and $\{E, H\}$ the intersections of the bisectors $\{BI, \Gamma I\}$ with the circles $\{\beta, \gamma\}$, then the points $\{E, H\}$ are on the line $Z\Theta$, where I is the incenter of $\triangle AB\Gamma$.
2. The points $\{B, \Gamma, E, H\}$ are contained in the circle ε.
3. If $\{M, N\}$ are the intersection points of $\{\Delta H, \Delta E\}$ with the corresponding sides $\{AB, A\Gamma\}$, then the 6 points $\{A, M, H, I, E, N\}$ are contained in a circle δ.
4. The pairs $\{(KI, B\Gamma), (I\Gamma, \Delta E), (IB, \Delta H)\}$ consist of orthogonal lines, where K is the center of the circle δ.

Hint: (1) is obvious, since $\{E, H\}$ will be the middles of the arcs $\{\widehat{AE\Delta}, \widehat{AH\Delta}\}$ of the circles $\{\beta, \gamma\}$.

(2) results immediately by showing that $\widehat{BEH} = \widehat{\Gamma}/2$.

For (3) notice that $\widehat{A}/2 = \widehat{\Delta A\Gamma} = \widehat{BE\Delta} = \widehat{\Delta H\Gamma}$, showing that $AIEN$ and $AMHI$ are cyclic quadrilaterals. Also $\widehat{HA\Delta} = \widehat{H\Gamma\Delta}$ and $\widehat{EA\Delta} = \widehat{EB\Delta}$ imply that $AHIE$ is a cyclic quadrilateral.

For (4), the orthogonality of $\{IK, B\Gamma\}$ follows from the fact that $\eta = Z\Theta$ is parallel to the external bisector ζ at A and angle $\widehat{BAI} = \widehat{A\Xi I}$, where Ξ is the intersection of IK with δ. The two other pairs of orthogonals result by easy angle measurements.

Exercise 3.235. Continuing the preceding exercise prove also the following relations:

1. The quadrilaterals $\{ZHKA, \Theta EKA, ZH\Lambda B, \Gamma\Theta E\Lambda\}$ are cyclic, where Λ is the center of ε.
2. The quadrilateral $\Theta\Lambda ZK$ is also cyclic and its circumcenter coincides with the circumcenter of the triangle $AB\Gamma$.
3. The line pairs $\{(KH, E\Lambda), (KE, \Lambda H), (\Gamma\Theta, BZ)\}$ intersect on the external bisector ζ.

Exercise 3.236. An isosceles trapezium has fixed basis a diameter of its circumscribed circle κ and the side parallel to its base is a chord of κ. Find the place of the chord which maximizes the area of the trapezium.

Exercise 3.237. Let A be a point in the interior of the angle $\widehat{XOY}$ (see Figure 3.145). We think of the angle sides as mirrors and point A as a light source and AB a light beam with B moving on OX. The beam AB is reflected to $B\Gamma$, this to $\Gamma\Delta$, this to ΔE etc. Show that:

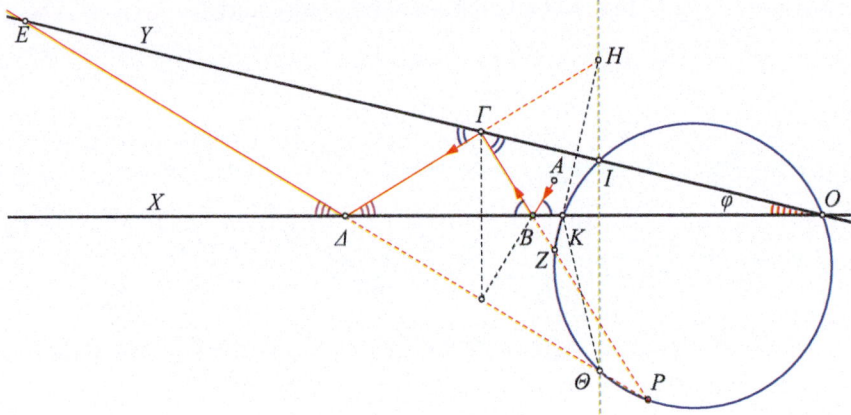

Fig. 3.145: Reflected lines

1. All lines $B\Gamma$ pass through a fixed point Z.
2. All lines $\Gamma\Delta$ pass through a fixed point H.
3. All lines ΔE pass through a fixed point Θ.
4. The intersection point P of lines $B\Gamma$ and ΔE belongs to circle $(OZ\Theta)$,
5. This circle passes through the intersection point I of OY with $H\Theta$.

Hint: Point Z is the symmetric of A with respect to OX. Point H is the symmetric of Z with respect to OY. Point Θ is the symmetric of H with respect to OX. Angle $\widehat{\Theta PZ} = \widehat{ABK} - \widehat{\Gamma\Delta B} = (\widehat{AB\Gamma} - \widehat{B\Gamma O}) + (\widehat{\Delta\Gamma E} - \widehat{E\Delta X}) = 2\phi$. Similarly $\widehat{ZI\Theta} = 2\widehat{ZH\Theta} = 2\phi$ and $\widehat{ZK\Theta} = 2\widehat{ZH\Theta} = 2\phi$.

Exercise 3.238. Let the points $\{A,B\}$ be in the interior of the angle $\widehat{XOY}$. To find the trajectory of a ball (broken line) which starting from A and reflecting alternatively on sides OX, OY and finally again on OX, passes through B.

Hint: Use exercise 3.237

Exercise 3.239. Construct circles $\{\alpha,\beta,\gamma\}$ with diameters the medians AA', BB', and $\Gamma\Gamma'$ of triangle $AB\Gamma$. Show that these circles intersect by pairs at points $\{A_1,A_2,B_1,B_2,\Gamma_1,\Gamma_2\}$ contained in the altitudes of the triangle.

Exercise 3.240. If G is the centroid of the triangle $AB\Gamma$ and X is an arbitrary point of the plane, show that

$$|XA|^2+|XB|^2+|XC|^2 = \frac{1}{3}(|AB|^2+|B\Gamma|^2+|\Gamma A|^2)+3|XG|^2. \tag{3.21}$$

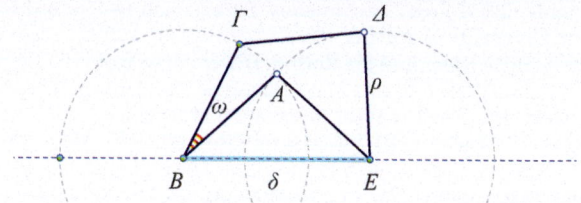

Fig. 3.146: Equilateral non-convex pentagon

Exercise 3.241. Figure 3.146 shows a non-convex and equilateral pentagon $\pi = AB\Gamma \Delta E$ with sides of length ρ. Show that:

1. Its shape, up to isometry, is determined by the parameters $\{\delta, \rho, \omega\}$ shown.
2. For fixed δ there is a maximal $\rho = \rho_0$ such that π is non self-intersecting.
3. For this ρ_0 the vertex A of π_0 lies in $\Gamma\Delta$ for all values of ω.

Exercise 3.242. Review the question of exercise 2.67.

References

60. V. Arnlold (2004) Problems for children from 5 to 15. MCCME, Moscow
61. D. Benko, 2007. A New Approach to Hilbert's Third Problem, *The American Mathematical Monthly*, 114:665-676
62. W. Blaschke (1936) Kreis und Kugel. Walter de Gruyter, Berlin
63. W. Blythe (1900) Geometrical Drawing. Cambridge University Press, Cambridge
64. B. Bold (1982) Famous Problems of Geometry. Dover Publications, Inc., New York
65. O. Bottema (1900) Topics in Elementary Geometry. Springer Verlag, Heidelberg
66. S. Chou, G. Shan, Z. Jing-Zhong (1994) Machine Proofs in Geometry. World Scientific, Singapore
67. D. Darling (2004) The Universal Book of Mathematics. Wiley, New York

68. P. Davis (1993) Spirals, From Theodorus to Chaos. A K Peters, Wellesley Massachusetts
69. D. DeTemple, J. Robertson, 1981. A Billiard Path Characterization of Regular Polygons, *Mathematics Magazine*, 54:73-75
70. H. Doerrie (1965) 100 Great Problems of Elementary Mathematics. Dover Publications, Inc. New York
71. E. Fourray (1900) Curiosities Geometriques. Vuibert editeurs, Paris
72. R. Gaskell, M. Klamkin, P. Watson, 1976. Triangulations and Pick's Theorem, *Mathematics Magazine*, 49:35-37
73. D. Gronau, 2004. The Spiral of Theodorus, *The American Mathematical Monthly*, 111:230-237
74. B. Grunbaum, G. Shephard, 1995. Ceva, Menelaus, and the Area Principle, *Mathematics Magazine*, 68:254-268
75. V. Gusev, V. Litvinenko, A. Mondkovich (1988) Solving Problems in Geometr. Mir Publishers, Moscow
76. W. Kaplan (2007) Calculus and Linear Algebra, vol. I, II. Sholarly Publishing Office, Ann Arbor
77. K. Katsigiannis (2010) From Archimedes' Stomachion to Pick's theorem Diploma Dissertation. University of Athens
78. N. Kazarinoff (1970) Ruler and the Round. Dover, New York
79. V. Klee, S. Wagon (1991) Old and New Unsolved Problems in Plane Geometry and Number Theory. The Mathematical Association of America, Washington
80. A. Kolmogorov, S. Fomin (1970) Introductory Real Analysis. Dover, New York
81. E. Loomis (1968) The Pythagorean Proposition. The National Council of Teachers of Mathematics, Washington
82. J. Mackay, 1895. Symmedians of a Triangle and their concomitant circles, *Proceedings of the Edinburgh Mathematical Societ*, 14:37-103
83. J. Mackay, 1895. Early history of the symmedian point, *Proceedings of the Edinburgh Mathematical Societ*, 11:92-103
84. M. O'Leary (2010) Revolutions of Geometry. Wiley, New York
85. M. Petkovic (2009) Famous Puzzles. American Mathematical Society, Providence
86. I. Pinelis, 2005. Cyclic polygons with given edge lengths, *Journal of Geometry*, 82:156-171
87. A. Posamentier (1988) Challenging Problems in Geometry. Dover, New York
88. E. Sandifer (2015) How Euler Did Even More.Mathematical Association of America, New York
89. A. Sinefakopoulos, 2001. Circumscribing an Inscribed Quadrilateral, *The American Mathematical Monthly*, 108:378
90. M. Spivak (1994) Calculus, Third Edition. Publish or Perish, Houston
91. R. Thiele, 2003. Hilbert's Twenty-Fourth Problem, *The American Mathematical Monthly*, 110:1-24
92. X. Tian, 2012. The Art and Mathematics of Tangrams, *Bridges, Mathematics, Music, Art, Architecture, Culture*, 1:553-556
93. A. Varverakis, 2001. A Maximal Property of Cyclic Quadrilaterals, *Forum Geometricorum*, 5:63-64
94. M. Villiers, 1998. Dual Generalizations of Van Aubel's theorem, *The Mathematical Gazette*, 11:405-412
95. F. Wang, Ch. Hsiung, 1942. A theorem on the Tangram, *The Mathematical Gazette*, 49:596-599
96. H. White, 2007. The Geometry of Leonhard Euler, Life Work and Legacy, *Studies In the history and philosophy of mathematics*, 5:303-323

Chapter 4
The power of the circle

4.1 Power with respect to a circle

> Words being arbitrary must owe their power to association, and have the influence, and that only, which custom has given them. Language is the dress of thought.
>
> S. Johnson, Lives of the English Poets, "Cowley"

An important and with many applications quantity, depending on a circle and a point is that of the **power of point with respect to a circle**. This quantity resides on the two next propositions, which correspond to theorems 35, 36 and 37 of the third book of Euclid's elements ([21, II, p.73]). Extended use of it appears for the first time in the work of Louis Gaultier (1803, [100]), who introduced the notion of radical axis (and radical plane of two spheres). The full development and its systematic use though, is due to Jacob Steiner [58, I, pp.17-76], who introduced the word *power*.

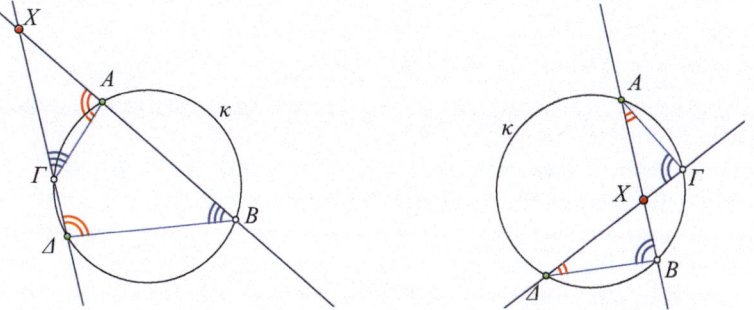

Fig. 4.1: Power of point X relative to a circle: $\pm|XA||XB|$

Theorem 4.1. *Given a circle κ and a point X, suppose that a line through X intersects the circle at points A and B. Then the product $|XA||XB|$ is independent of the direction of the line and depends only on the position of the point relative to the circle (See Figure 4.1).*

Proof. Consider two different lines which pass through point X and intersect the circle κ at points A, B and Γ, Δ respectively. The triangles $XA\Gamma$ and $X\Delta B$ are similar (Corollary 2.23). Consequently they have proportional sides:

$$\frac{|XA|}{|X\Gamma|} = \frac{|X\Delta|}{|XB|} \Rightarrow |XA||XB| = |X\Gamma||X\Delta|.$$

The number $p(X) = |XA||XB|$ if the point is external, $p(X) = -|XA||XB|$ if the point is internal and $p(X) = 0$ if X lies on the circle, is called **power of the point X relative to the circle** κ. Often we'll use also the symbols $p(X, \kappa)$ or/and $p_\kappa(X)$. In the case where X is an external point we have two special positions of the line: the tangents from X. We may consider this as a limiting case, in which A and B coincide. Yet again similar triangles $XA\Gamma$ and $X\Delta A$

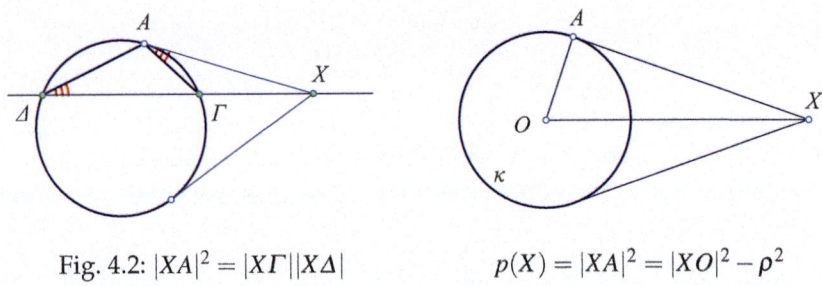

Fig. 4.2: $|XA|^2 = |X\Gamma||X\Delta|$ $\qquad\qquad p(X) = |XA|^2 = |XO|^2 - \rho^2$

are formed (See Figure 4.2-I), because of the fact that the inscribed angle $\widehat{X\Delta A}$ is equal to $\widehat{\Gamma AX}$ which is formed from the chord ΓA and the tangent at its end A (Theorem 2.25). Therefore it follows that $\frac{|XA|}{|X\Gamma|} = \frac{|X\Delta|}{|XA|}$, which implies $p(X) = |X\Gamma||X\Delta| = |XA|^2$. We proved then the following corollary.

Corollary 4.1. *For every point X, external to the circle κ, the power $p(X)$ is equal to the square of the tangent to κ from X.*

A third expression for the power $p(X)$ is found by drawing XO, which joins the point X with the center O of the circle κ. If A is the point of contact of the tangent XA from X, then from the right triangle OAX we have $p(X) = |XA|^2 = |XO|^2 - \rho^2$, where ρ is the radius of the circle (See Figure 4.2-II). For points X internal to the circle there are no tangents to it. However the preceding expression, this time with opposite sign, again gives the power of the point. This is seen by drawing the chord of the circle which passes through X and is orthogonal to XO (See Figure 4.3-I). The definition of the power of X and the right triangle OXE lead to: $-p(X) = |XA||XB| = |XE|^2 = \rho^2 - |OX|^2$. Taking into account the fact that for points on the circle holds $p(X) = \rho^2 - |XO|^2 = 0$, we have the next proposition.

4.1. POWER WITH RESPECT TO A CIRCLE

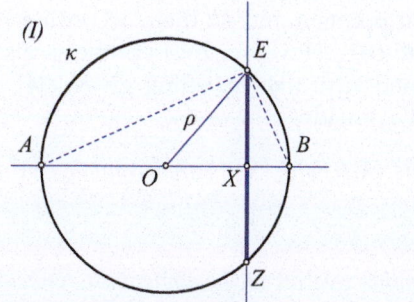
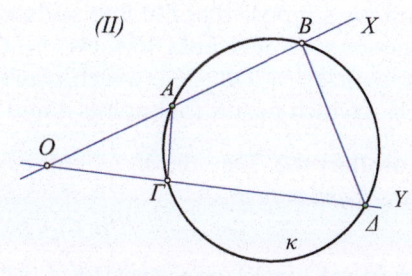

Fig. 4.3: $p(X) = |XO|^2 - \rho^2$ Concyclicity criterion

Theorem 4.2. *Given a circle $\kappa(O,\rho)$ of center O and radius ρ, the power of any point X on the plane relative to the circle κ is given by the formula*

$$p(X) = |OX|^2 - \rho^2.$$

Theorem 4.3. *Given two intersecting lines OX and OY the points A, B on OX and Γ, Δ on OY are concyclic, if and only if $|OA||OB| = |O\Gamma||O\Delta|$.*

Proof. As we saw, when points A, B, Γ and Δ are on the same circle (concyclic) (See Figure 4.3-II), then the equation holds. Conversely, suppose that the equation holds and κ is the circle which passes through the three points: A, B and Γ (Theorem 2.1). Let Δ' be the second intersection point of this circle with OY. Then, according to Theorem 4.1 we have $|OA||OB| = |O\Gamma||O\Delta'|$. But from hypothesis we also have $|OA||OB| = |O\Gamma||O\Delta|$ and therefore $|O\Delta'| = |O\Delta|$ and points Δ and Δ' coincide.

Theorem 4.4. *Given are three non collinear points, A on the line OX and B, Γ on line OY. The circle $(AB\Gamma)$ is tangent to OX at A, if and only if*

$$|OA|^2 = |OB||O\Gamma|.$$

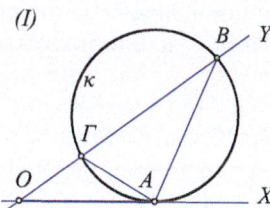
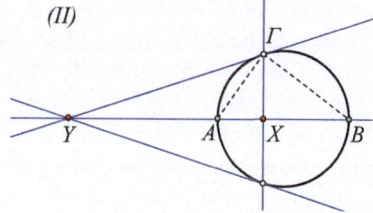

Fig. 4.4: $\kappa = (AB\Gamma)$ tangent to OX Harmonic quadruple

Proof. If OA is tangent to the circle $\kappa = (AB\Gamma)$ at A (See Figure 4.4-I), then the relation holds according to Corollary 4.1. Conversely, if the relation holds and we suppose that *OX* intersects κ at points *A* and *A'*, then, according to Theorem 4.3 will hold $|OA||OA'| = |OB||O\Gamma|$. However by hypothesis also holds $|OA|^2 = |OB||O\Gamma|$, which combined with the preceding gives $|OA| = |OA'|$, which means that points *A* and *A'* coincide.

Corollary 4.2. *The pairs of collinear points (A,B) and (X,Y) are harmonic conjugate, if and only if*
$$p(Y) + p(X) = |XY|^2,$$
where $p(X)$, $p(Y)$ are the powers of the points relative to the circle with diameter AB (See Figure 4.4-II).

Proof. It follows from Corollary 1.44 that
$$p(Y) = |YA||YB| = |Y\Gamma|^2, \quad -p(X) = |XA||XB| = |X\Gamma|^2,$$
where $Y\Gamma$ is the tangent to the circle from *Y*, by hypothesis lying external to the circle.

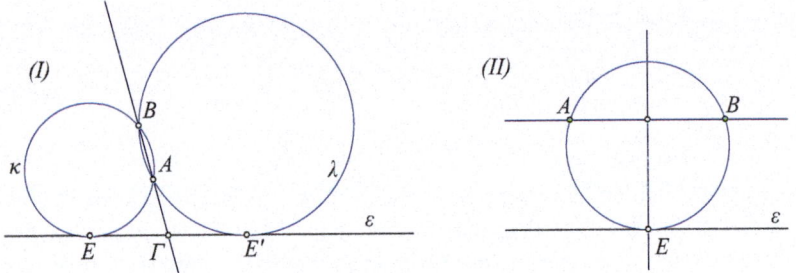

Fig. 4.5: Circle through *A*, *B* tangent to line ε

Exercise 4.1. Construct a circle κ, such that it passes through two given points *A* and *B* and is tangent to a given line ε.

Hint: Suppose that the requested circle has been constructed. Let further Γ be the point of intersection of *AB* with the given line ε and *E* be the point of contact of the circle with ε (See Figure 4.5-I). Point Γ is determined from the given data of the problem and its power relative to the circle will be $p(\Gamma) = |\Gamma A||\Gamma B| = |\Gamma E|^2$. Therefore the length $|\Gamma E|$ and consequently the position of *E* is also determined from the given data. Consequently one more point (*E*) of the circle is determined. In general there are two solutions which correspond to two positions of *E* which are symmetric relative to Γ. Note that the problem doesn't have a solution when the points *A* and *B* belong to different sides of ε.

Also note, that in the case where Γ doesn't exist, in other words, when the line *AB* is parallel to ε (See Figure 4.5-II), the problem has only one solution and point *E* is determined by intersecting line ε with the medial line of *AB*.

4.1. POWER WITH RESPECT TO A CIRCLE

Exercise 4.2. Given is a circle κ and a constant δ. Find the geometric locus of points X for which the power $p(X)$ relative to the circle κ is equal to δ.

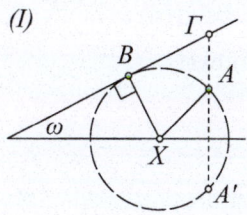

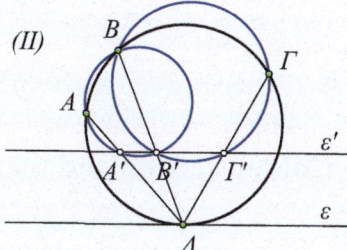

Fig. 4.6: Point $X : |XA| = |XB|$ — Collinear points A', B', Γ'

Exercise 4.3. Given is a convex angle ω and a point A in its interior. Find a point X on one side of the angle which is equidistant from the point A and its other side (See Figure 4.6-I).

Exercise 4.4. From a point Δ on the circle κ are drawn three chords ΔA, ΔB, $\Delta \Gamma$, as well as a parallel ε' of the tangent to Δ, which intersects the chords at A', B' and Γ' respectively. Show that $|\Delta A||\Delta A'| = |\Delta B||\Delta B'| = |\Delta \Gamma||\Delta \Gamma'|$ (See Figure 4.6-II).

Exercise 4.5. The median AM of a given triangle $AB\Gamma$ intersects the circumcircle of the triangle at Δ. Show that

$$|AM||A\Delta| = \frac{1}{2}(|AB|^2 + |A\Gamma|^2).$$

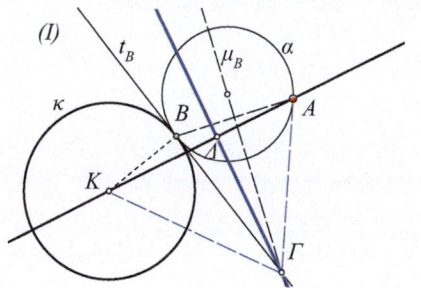

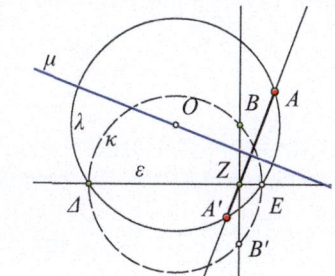

Fig. 4.7: Locus of point Δ — Locus of centers of circles

Exercise 4.6. The variable circle α is tangent to circle $\kappa(K)$ at B and passes through the fixed point A. To find the locus of intersections Γ of the medial line μ_B of AB with the tangent t_B to κ at B (See Figure 4.7-I).

Exercise 4.7. The variable points $\{\Delta, E\}$ of line ε have constant product of distances from the point Z of ε and we consider circles λ, passing through $\{\Delta, E\}$ and the fixed point A, not lying on ε. Show that: (1) The circles κ with diameter ΔE pass through two fixed points $\{B, B'\}$. (2) The circles λ pass also through two fixed points $\{A, A'\}$. (3) The lines $\{AA', BB'\}$ intersect at Z (See Figure 4.7-II).

4.2 Golden section and regular pentagon

> Such is always the pursuit of knowledge. The celestial fruits, the golden apples of the Hesperides, are ever guarded by a hundred headed dragon which never sleeps, so that it is an Herculean labor to pluck them.
>
> Henry David Thoreau, Writings v.9

Two famous numbers, which satisfy a quadratic equation are: $\sqrt{2}$, which satisfies the equation $x^2 - 2 = 0$, and the **golden section** or **golden ratio** $\phi = \frac{\sqrt{5}+1}{2}$, which satisfies equation $x^2 - x - 1 = 0$. Geometrically, the number ϕ is defined indirectly from the division of a line segment into two parts, such that:

The ratio of the whole to the larger one is equal to the ratio of the larger to the smaller one.

This is traditionally expressed using the terminology of the next construction.

Construction 4.1 (Golden section) *On a given line segment AB find a point Γ, which divides it into mean and extreme ratio, in other words, such that*

$$\frac{|AB|}{|A\Gamma|} = \frac{|A\Gamma|}{|B\Gamma|} \iff |A\Gamma|^2 = |AB||\Gamma B|.$$

Construction: Construct the right triangle ABE with $|BE| = \frac{\delta}{2}$, where $\delta = |AB|$ (See Figure 4.8). On the hypotenuse EA consider the point Δ: $|E\Delta| = |EB|$. The circle κ with center A and radius $A\Delta$ intersects AB at the wanted point Γ. This results by computing the power of A relative to the circle $\kappa(E, |E\Delta|)$.

$$|AB|^2 = |A\Delta||AZ| \iff \delta^2 = x(x+\delta),$$

where $x = |A\Delta| = |A\Gamma|$. We therefore have the equation with respect to x:

$$x^2 + x\delta - \delta^2 = 0,$$

which has one non acceptable negative solution and one positive:

4.2. GOLDEN SECTION AND REGULAR PENTAGON

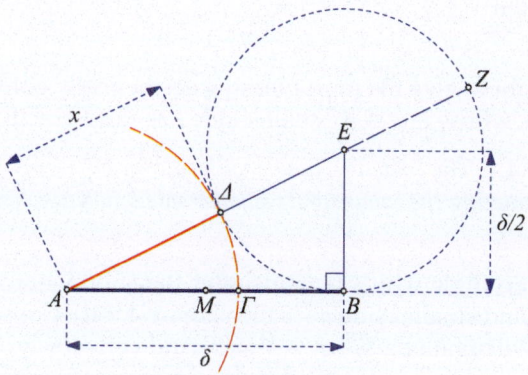

Fig. 4.8: Golden section

$$x = \frac{\sqrt{5}-1}{2}\delta \sim (0.61803398874989484820...)\cdot \delta,$$

for which we observe that it satisfies the requested condition

$$\frac{\delta}{x} = \frac{x}{\delta - x} \quad \Leftrightarrow \quad \frac{|AB|}{|A\Gamma|} = \frac{|A\Gamma|}{|B\Gamma|}.$$

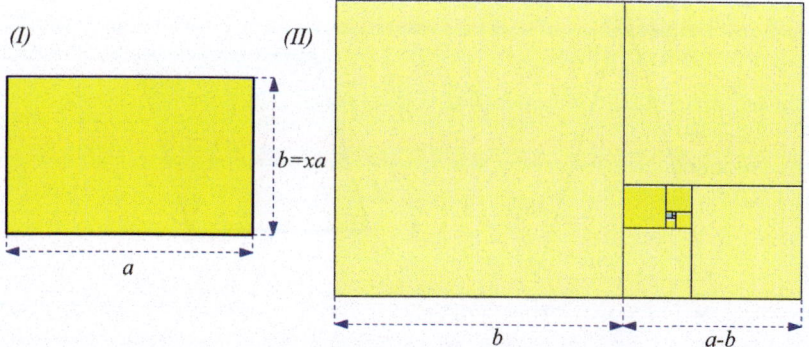

Fig. 4.9: "Golden" rectangle   Square + similar to original

Remark 4.1. The positive number $x = \frac{\sqrt{5}-1}{2}$, which is the root of the equation

$$x^2 + x - 1 = 0 \quad \Leftrightarrow \quad \frac{1}{x} = 1 + x$$

leads to the so called **golden ratio** or **golden section** traditionally denoted by the Greek letter : $\phi = 1 + x = \frac{1+\sqrt{5}}{2}$ and coinciding with the positive root

of the equation
$$\phi^2 - \phi - 1 = 0.$$
This is thought to give the most pleasing aesthetically **golden rectangle** (See Figure 4.9-I), whose larger to smaller side has ratio exactly ϕ.

Proposition 4.1. *The golden rectangle with side ratio $\frac{b}{a} = x$ is characterized by the fact that subtracting the square of its smaller side b, what remains is similar to the original.*

Proof. The rectangle which remains after the subtraction of the square has great side b and smaller side $a - b$ (See Figure 4.9-II). Consequently the ratio of its sides will again be $\frac{a-b}{b} = \frac{a-xa}{xa} = x$. And conversely, if what remains is similar to the original, then the following holds $\frac{a-b}{b} = \frac{b}{a} \Leftrightarrow b^2 + ab - a^2 = 0$, which is the equation of the golden section and gives $b = xa$.

Figure 4.9-II shows the repeated division of the golden rectangle into a square and a smaller golden rectangle. What results is a sequence of rectangles, all of which contain in their interior a unique point O. A property of this point is discussed in Exercise 2.48.

Another usage of the golden ratio is in the construction of the isosceles triangle with apical angle of 36 degrees, which, in turn, leads to the construction of the regular pentagon.

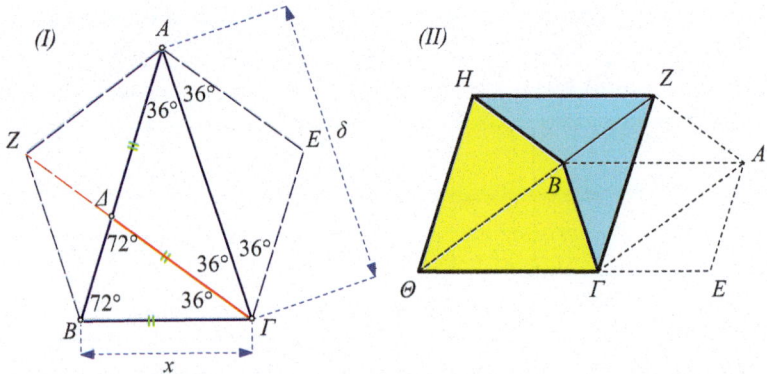

Fig. 4.10: Isosceles with apical angle 36 degrees and Penrose tiles

Construction 4.2 *Construct an isosceles triangle with apical angle of 36 degrees.*

Construction: From the fact that the other two angles sum up to $180° - 36° = 144°$, follows that the two angles of the base of the isosceles will be 72 degrees, which is twice 36 (Figure 4.10-I). Consequently the bisector $\Gamma \Delta$ will define also an isosceles $\Gamma \Delta B$ similar to $AB\Gamma$. Also it will be $|\Delta A| = |\Delta \Gamma| = |B\Gamma|$. From the similar triangles therefore we conclude, setting $|A\Gamma| = \delta$ and $|B\Gamma| = x$:

4.2. GOLDEN SECTION AND REGULAR PENTAGON

$$\frac{|B\Gamma|}{|A\Gamma|} = \frac{|B\Delta|}{|B\Gamma|} \Leftrightarrow \frac{x}{\delta} = \frac{\delta-x}{x} \Leftrightarrow x^2 + \delta x - \delta^2 = 0,$$

which means that the point Δ divides AB into mean and extreme ratio (Construction 4.1). Therefore, given the length $\delta = |AB|$ the length of $|B\Gamma|$ is found from the preceding construction and the isosceles is constructed as a triangle whose side lengths are known.

Exercise 4.8. Let $AB\Gamma$ be an isosceles with apical angle $\widehat{A}$ of measure 36 degrees. Show that, drawing from its vertices lines with 36 degree inclination towards its legs (See Figure 4.10-I), forms a regular pentagon.

Remark 4.2. Using the triangles $AB\Gamma$ and $B\Gamma Z$ of figure 4.10-I, we construct also the two quadrilateral *tiles of Penrose* (See Figure 4.10-II), which are used in the example of II-§ 2.10. One of them ($B\Gamma\Theta H$) results by attaching two triangles congruent to $AB\Gamma$ and the second ($B\Gamma ZH$) by attaching two triangles congruent to $B\Gamma Z$.

Exercise 4.9. Construct a regular pentagon and a regular decagon with side length α.

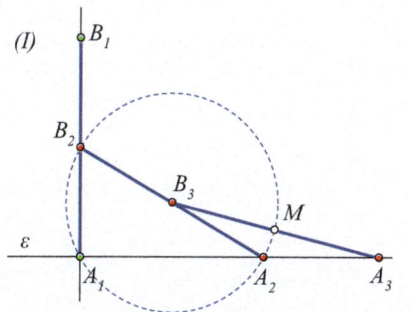
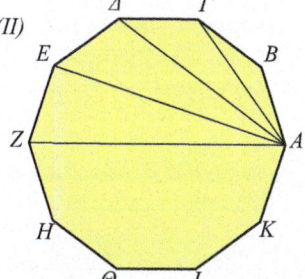

Fig. 4.11: Golden section construction   Golden section in decagon

Exercise 4.10. Starting with a line segment A_1B_1 orthogonal to line ε, construct two equal to it A_2B_2, A_3B_3, each of which rests on the middle of the other (See Figure 4.11-I). Show that the points A_1, A_2, A_3 define the golden ratio $\frac{|A_1A_2|}{|A_2A_3|} = \frac{1+\sqrt{5}}{2}$ ([104]).

Exercise 4.11. Show that the diagonals of the regular decagon (See Figure 4.11-II) are:

$$\frac{|AB|}{|AZ|} = \frac{x}{2}, \quad \frac{|A\Gamma|}{|AE|} = x, \quad \frac{|A\Delta|}{|AZ|} = \frac{1+x}{2} \quad \text{with} \quad x = \frac{\sqrt{5}-1}{2}.$$

Remark 4.3. Experimentally, the regular pentagon is constructed by making a simple knot with a tape, pressing and squeezing it, so as to become flat by folding without the tape becoming torn, as in figure 4.12-I.

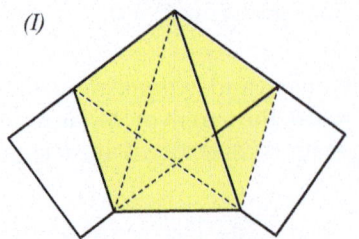
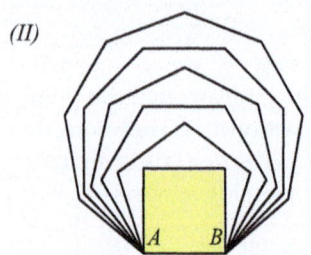

Fig. 4.12: Tape knot   Regular polygons

Exercise 4.12. On the line segment AB and towards the same side we construct regular polygons p_n with sides $n > 3$ (See Figure 4.12-II). Show that for $n > m$, p_n contains p_m in its interior.

Exercise 4.13. Show that the trigonometric numbers of the angles $18°$, $36°$, $54°$, $72°$ are given by the expressions of the following table.

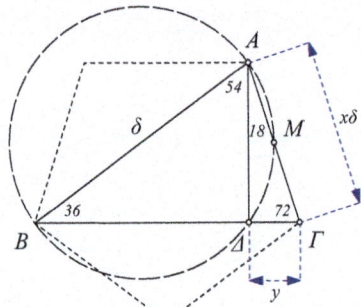

angle	sin		cos
$18°$	$\frac{1}{4}(\sqrt{5}-1)$	$= \frac{\phi-1}{2}$	$= \sin(72°)$
$36°$	$\frac{1}{4}\sqrt{10-2\sqrt{5}}$	$= \frac{\sqrt{3-\phi}}{2}$	$= \sin(54°)$
$54°$	$\frac{1}{4}(1+\sqrt{5})$	$= \frac{\phi}{2}$	$= \sin(36°)$
$72°$	$\frac{1}{4}\sqrt{10+2\sqrt{5}}$	$= \frac{\sqrt{2+\phi}}{2}$	$= \sin(18°)$

Fig. 4.13: Trigonometric numbers of angles connected to the golden section

Hint: Use the isosceles triangle with angles $36°$, $72°$, $72°$ (See Figure 4.13), for which holds $y = \frac{x^2 \delta}{2}$, $|A\Delta| = \frac{x\delta\sqrt{10+2\sqrt{5}}}{4}$, $x = \frac{\sqrt{5}-1}{2} = \phi - 1$.

Exercise 4.14. The points $\{Z, \Delta\}$ are projections of the endpoints of the chord $E\Gamma$ of the circle κ onto its tangent κ at its point A and B is the projection of A onto $E\Gamma$ (See Figure 4.14-I). Show that $a = |AB|$ is the mean proportional of $\{x = |EZ|, y = |\Gamma\Delta|\}$, i.e. it holds $a^2 = x \cdot y$.

Hint: The right triangles $\{OZE, OAB, O\Gamma\Delta\}$ are similar.

Exercise 4.15. The isosceles trapezium $AB\Gamma\Delta$ is circumscribed to a circle (See Figure 4.14-II). Show that the segments $\{A\Delta, \Delta E, \Delta Z\}$ represent correspondingly the arithmetic, geometric and harmonic mean of the parallel sides $\{AB, \Gamma\Delta\}$.

4.3. RADICAL AXIS, RADICAL CENTER

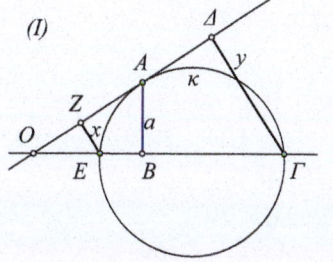
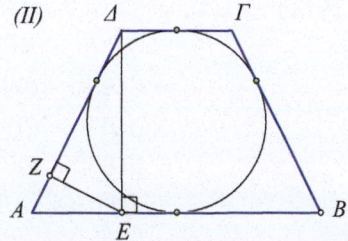

Fig. 4.14: $a^2 = x \cdot y$ Arithm.($A\Delta$), geom.(ΔE), harmon.(ΔZ) mean

4.3 Radical axis, radical center

> It is well known that the most radical revolutionary will become a conservative on the day after the revolution.
>
> H. Arendt, New Yorker, 12 Sept. 1970

The **radical axis** of two non concentric circles κ and λ is defined as the geometric locus of points X which have equal powers relative to the circles κ and λ. The existence of this line ε follows from next proposition.

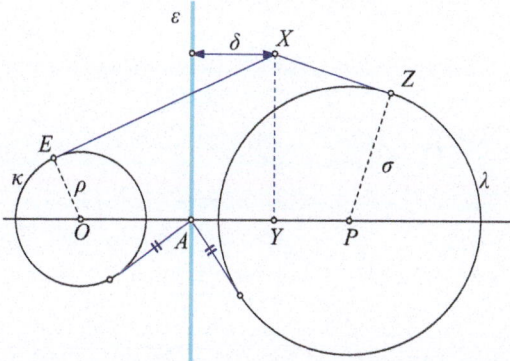

Fig. 4.15: Difference of powers relative to two circles

Proposition 4.2. *The difference of powers* $|p_\kappa(X) - p_\lambda(X)|$ *of a point X relative to two non-concentric circles* $\kappa(O, \rho)$ *and* $\lambda(P, \sigma)$ *is given by the formula*

$$|p_\kappa(X) - p_\lambda(X)| = 2|OP| \cdot \delta,$$

where δ is the distance of X from the radical axis.

Proof. Indeed, using the expression of the power trhough the radii (theorem 4.2), we have (See Figure 4.15)

$$\begin{aligned}
p_\kappa(X) - p_\lambda(X) &= (|XO|^2 - \rho^2) - (|XP|^2 - \sigma^2) \\
&= (|XO|^2 - |XP|^2) - (\rho^2 - \sigma^2) \\
&= (|OY|^2 - |PY|^2) - (\rho^2 - \sigma^2) \\
&= (|OY|^2 - |PY|^2) - ((|OA|^2 - p_\kappa(A)) - (|PA|^2 - p_\lambda(A))) \\
&= (|OY|^2 - |PY|^2) - (|OA|^2 - |PA|^2) \\
&= |OP|(|OY| - |PY|) - |OP|(|OA| - |PA|) \\
&= |OP|(|OY| - |OA| + |PA| - |PY|) = \pm 2|OP| \cdot \delta.
\end{aligned}$$

Here Y is the projection of X onto the center-line and A is the intersection of the center-line with the radical axis, for which $p_\kappa(A) = p_\lambda(A)$. The result follows by taking absolute values. The proof, apparently valid for non-intersecting circles is valid also for intersecting or tangent circles.

Proposition 4.3. *The radical axis of two circles is a line orthogonal to the center-line of the circles.*

Proof. This follows directly from proposition 4.2 in which the hypothesis implies $\delta = 0$. This shows that the geometric locus of points having equal power w.r.t. the two circles is the line ε orthogonal to the center-line at A (See Figure 4.15).

Figures 4.16 and 4.17 show the radical axis of intersecting, non-intersecting and tangent circles.

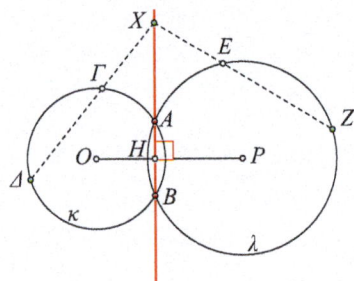

Fig. 4.16: Radical axis of intersecting circles

Corollary 4.3. *The power $p_\kappa(X)$ of the point X of circle $\lambda(P, \sigma)$ relative to circle $\kappa(O, \rho)$ is equal to $p_\kappa(X) = \pm 2|OP|\delta$, where δ the distance of X from the radical axis of the two circles.*

Exercise 4.16. The radical axis of two circles κ and λ which intersect at two different points A and B is the line AB.

Exercise 4.17. Show that if the two circles are tangent, then the radical axis coincides with their common tangent at their point of contact.

4.3. RADICAL AXIS, RADICAL CENTER

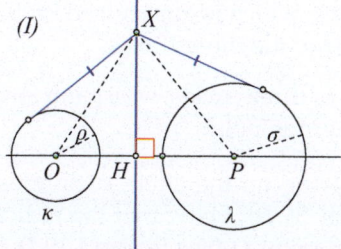
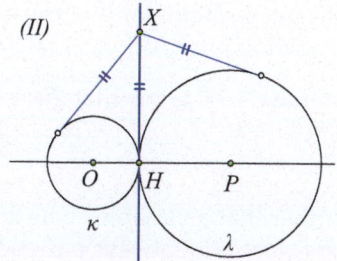

Fig. 4.17: Radical axis of non intersecting,   ... and tangent circles

Exercise 4.18. Given are two circles κ and λ. A third circle μ intersects κ at two points A and B and λ at two points Γ and Δ. Show that the lines AB and $\Gamma\Delta$ intersect on the radical axis of κ and λ (See Figure 4.18-I).

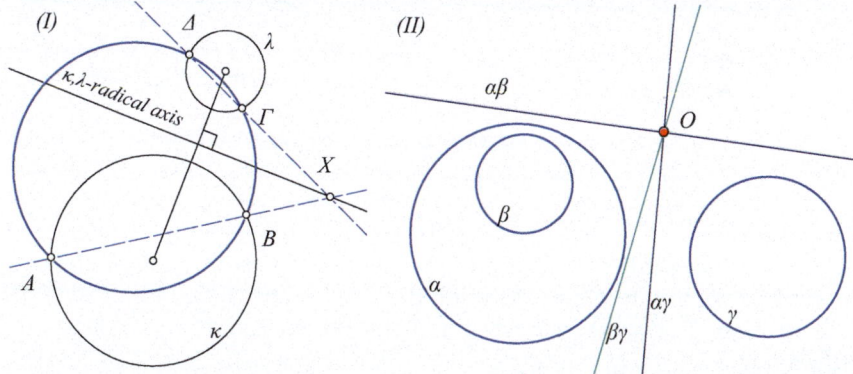

Fig. 4.18: Radical axis construction   Radical center of three circles

Hint: The intersection point X of AB and $\Delta\Gamma$ has power relative to κ and λ respectively $|XA||XB|$, $|X\Gamma||X\Delta|$. These products are, however, equal (Proposition 4.3).

Theorem 4.5. *The radical axes of three circles, whose centers are not collinear, pass through a common point.*

Proof. Let O be the intersection point of the radical axes $\alpha\beta$ and $\beta\gamma$ of the pairs of circles (α, β) and (β, γ) respectively (See Figure 4.18-II). The point O has the same power relative to α and β as well as also relative to β and γ, therefore also the same power relative to α and γ. Consequently it is also contained in the radical axis $\alpha\gamma$ of the pair of circles (α, γ). Therefore the radical axis $\alpha\gamma$ of this pair also passes through O. Note that the requirement of non collinearity of the three centers ensures the existence of O. If the three centers were collinear, then their radical axes would be pairwise parallel, being orthogonal to their center-line.

The intersection point of the three radical axes, which is guaranteed by the preceding proposition, is called **radical center** of the three circles.

Exercise 4.19. Show that the common chords of three circles, which intersect pairwise at two points, pass all through a common point.

Exercise 4.20. Let $\{A,B\}$ be two points not lying on the circle κ. Suppose further that a variable circle λ passes through $\{A,B\}$. Show that the radical axis of κ and λ always passes through a fixed point Ω (See Figure 4.19-I).

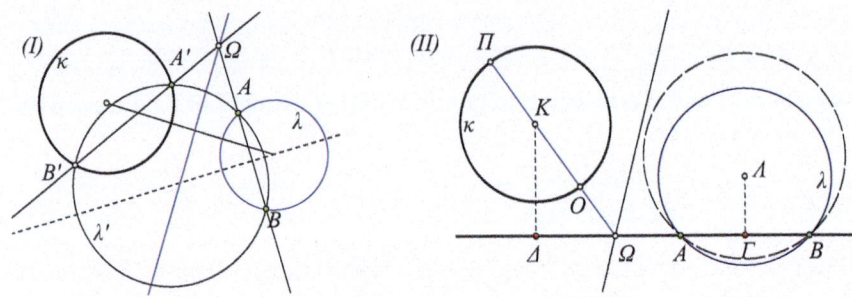

Fig. 4.19: Variable radical axis

Hint: Consider the circle λ' passing through from A, B and intersecting κ at points A', B'. According to Theorem 4.5 the radical axes of the three circles κ, λ and λ', taken by two, pass through the fixed point Ω (See Figure 4.19-II). The position of Ω is not changed if we consider the variable circle λ passing through A, B.

For an analytic calculation of the position of Ω, observe first that the projections Δ and Γ of the centers K, Λ of κ and of the variable circles λ are fixed points determined from the given data of the exercise (See Figure 4.19-II). Calculate next the powers $p_\kappa(\Omega)$, $p_\lambda(\Omega)$ of Ω relative to κ and λ respectively.

$$p_\kappa(\Omega) = |\Omega O||\Omega \Pi| = (|\Omega K| - |KO|)(|\Omega K| + |KO|) = |\Omega K|^2 - |KO|^2,$$

where O, Π the points of intersection of ΩK with κ. From Pythagoras' theorem applied to $K\Delta\Omega$ we have finally

$$p_\kappa(\Omega) = |\Omega K|^2 - |KO|^2 = |K\Delta|^2 + |\Delta\Omega|^2 - |KO|^2.$$

Also the power of Ω relative to λ is

$$p_\lambda(\Omega) = |\Omega A||\Omega B| = (|\Omega \Gamma| - |\Gamma B|)(|\Omega \Gamma| + |\Gamma B|) = |\Omega \Gamma|^2 - |\Gamma B|^2.$$

From the equality of powers follows that

4.3. RADICAL AXIS, RADICAL CENTER

$$|\Omega\Delta|^2 - |\Omega\Gamma|^2 = |KO|^2 - |K\Delta|^2 - |\Gamma B|^2.$$

The right side of the equality is fixed and independent of Ω, consequently the position of Ω will be defined uniquely from the given data. The position of Ω is found from the last equation. For this, define $x = |\Omega\Delta|$, $y = |\Omega\Gamma|$ and set $d = |KO|^2 - |K\Delta|^2 - |\Gamma B|^2$ and $e = |\Delta\Gamma|$. Then the equation becomes $x^2 - y^2 = d$, hence (if Ω is between points Δ and Γ) $x + y = e$. Solving we find $(x-y)(x+y) = d \Rightarrow x - y = \frac{d}{e}$ and consequently $x = \frac{1}{2}(e + \frac{d}{e})$, $y = \frac{1}{2}(e - \frac{d}{e})$ (see also Proposition 4.9).

Exercise 4.21. Investigate the different cases which result for the position of Ω between Γ and Δ in the preceding problem. Examine especially the case where Γ and Δ coincide. What happens then with the radical axes of the circles κ and λ?

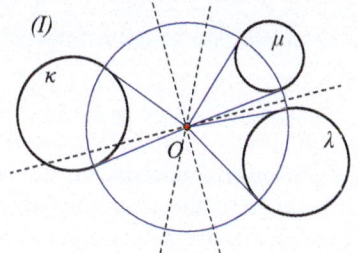

Fig. 4.20: Equal tangents   Tangent circle through two points

Exercise 4.22. Given three circles with non collinear centers, find a point from which equal tangents to them can be drawn (See Figure 4.20).

Hint: If there is a point O from which are drawn equal tangents to the three circles, then this point will also be in the three radical axes of the circles, taken by two. Consequently it will coincide with the radical center of the three circles. In order to draw from it tangents to all, the center will have to be in the exterior of all circles.

Exercise 4.23. Given is a circle κ and two points A, B in its exterior. Find a second circle λ passing through A, B and tangent to κ.

Hint: Suppose that the wanted circle λ has been constructed and is tangent to κ at point X (See Figure 4.20-II). Relying on Exercise 4.20, the common tangent to the two circles at X, which is also their radical axis (Exercise 4.17) will pass through a fixed point Ω of AB, constructible from the given data. Construct therefore point Ω and draw from it the tangents ΩX and ΩY towards κ. There are two solutions of the problem λ, λ', which correspond to circles passing through A, B and respectively also through X or Y (Proposition 2.1).

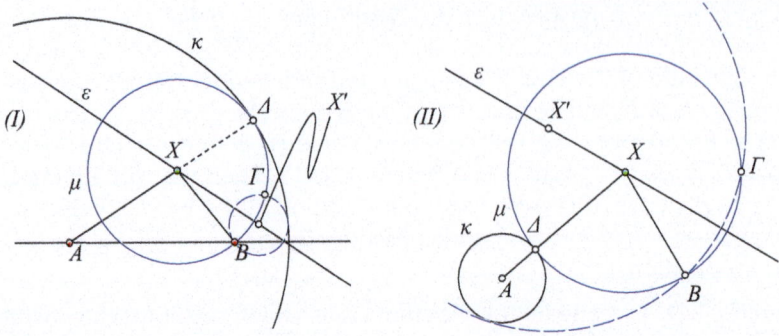

Fig. 4.21: Ellipse-line intersection   Hyperbola-line intersection

Exercise 4.24. Consider two points A, B, a line ε, and a constant λ. Find a point X of the line ε such that $|XA| + |XB| = \lambda$. Similarly find a point X on line ε such that $|XA| - |XB| = \lambda$.

Hint: Suppose that the requested point X has been found. Extend AX to Δ so that $|A\Delta| = \lambda$ (See Figure 4.21-I). The circle $\kappa(A, \lambda)$ is known and X is also the center of circle μ, which is tangent to κ, passes through B and its symmetric Γ relative to line ε. The problem therefore reduces to the construction of such a circle μ (Exercise 4.23). The construction of X with $|XA| - |XB| = \lambda$ is similar (See Figure 4.21-II).

Remark 4.4. The last exercise shows that the intersection points of a line and an ellipse/hyperbola, of which are given the focal points and the constant λ, are constructible using ruler and compass (Theorem II-6.7).

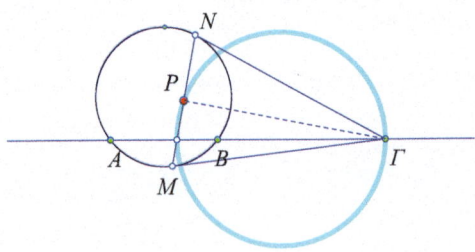

Fig. 4.22: dd

Exercise 4.25. Given are three collinear points $\{A, B, \Gamma\}$ point Γ lying outside the segment AB. We consider all circles through the points $\{A, B\}$ and draw the tangents $\{\Gamma M, \Gamma N\}$ from Γ. To determine the geometric locus of the middle P of the variable chord MN of contacts.

4.4 Apollonian circles

> The poets did well to conjoin music and medicine in Apollo, because the office of medicine is but to tune this curious harp of man's body and to reduce it to harmony.
>
> F. Bacon, Advancement of Learning, 2nd Book p. 51

The *medial line* of a segment AB is the geometric locus of the points X, for which the ratio of distances satisfies the equation $\frac{|XA|}{|XB|} = 1$. The *Apollonian circles* answer to the more general problem of the geometric locus of points, for which the equation is $\frac{|XA|}{|XB|} = \kappa \neq 1$. Next theorem shows, that for every such κ we have a circle.

Theorem 4.6. *Given a line segment AB and a positive number $\kappa \neq 1$, the geometric locus of points X, for which the ratio of distances $\frac{|XA|}{|XB|} = \kappa$, is a circle centered on line AB (See Figure 4.23).*

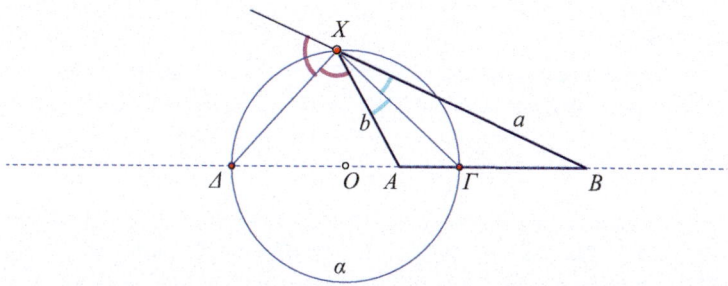

Fig. 4.23: An Apollonian circle of the segment AB

Proof. We show first that, if a point X has the aforementioned property relative to the line segment AB, then it lies on a specific circle. Indeed, let us consider the triangle AXB and its two bisectors at A, the internal $A\Gamma$ and the external $A\Delta$. The last one exists, in other words $X\Delta$ is not parallel to AB, because we supposed that $\kappa = \frac{|XA|}{|XB|} \neq 1$. As we proved in Theorem 3.3 and Exercise 3.12 respectively, the points $\{\Gamma, \Delta\}$ are harmonic conjugate to $\{A,B\}$ (§ 1.17) and divide AB into ratio equal to the ratio of XA and XB, in other words κ. Consequently, the position of these points is determined completely from A, B and κ (equations (1)-(3) of section 1.17). However the angle formed by $X\Gamma$ and $X\Delta$ is right, as half of a flat one. Therefore X sees the line segment $\Gamma\Delta$ under a right angle and consequently (Corollary 2.7) lies on the circle with diameter $\Gamma\Delta$.

As with every geometric locus, also here we must show the converse. In other words, that every point of the circle with diameter $\Gamma\Delta$ is a point of the locus. We suppose then that the point X lies on the circle with diameter $\Gamma\Delta$, therefore angle $\widehat{\Gamma X \Delta}$ is right. We draw line XA and we find on line AB point B' such that the angle $\widehat{\Gamma X B'}$ is equal to $\widehat{AX\Gamma}$. $X\Gamma$ is therefore, by construction, the bisector of the angle $\widehat{AXB}$. Therefore points $\{\Gamma, \Delta\}$ are harmonic conjugate relative to $\{A, B'\}$ and consequently (Exercise 1.77) points $\{A, B'\}$ are harmonic conjugate relative to $\{\Gamma, \Delta\}$. The same happens however also with points $\{A, B\}$, therefore points B and B' coincide and because of the bisector (Theorem 3.3) $\frac{|XA|}{|XB|} = \frac{|\Gamma A|}{|\Gamma B|} = \kappa$.

The circles which are defined for the line segment AB for different ratios $\kappa > 0$, through the preceding proposition, are called **Apollonian circles** of the segment AB (Apollonius 262-190 B.C.).

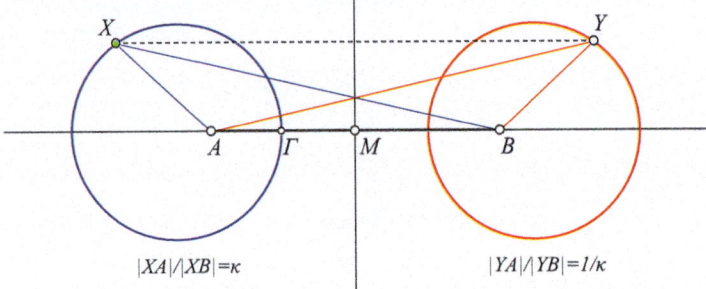

Fig. 4.24: Symmetry of the system of Apollonian circles

Proposition 4.4. *The distance $|A\Gamma|$, of the intersection Γ with AB, of the Apollonian circle of AB, relative to $\kappa \neq 1$, and the diameter d_κ of this circle are correspondingly*

$$|A\Gamma| = |AB|\frac{\kappa}{1+\kappa}, \qquad d_\kappa = |AB|\frac{2\kappa}{|1-\kappa^2|}.$$

The Apollonian circles of AB relative to the ratios κ and $\frac{1}{\kappa}$ have equal radii and lie symmetrically relative to the medial line of AB (See Figure 4.24).

Proof. The formula for the diameter is this of Corollary 1.43. Substituting the expression for the diameter for κ with $\frac{1}{\kappa}$ we see that its value doesn't change, hence the equality of the radii. The fact that the circles corresponding to ratios κ and $\frac{1}{\kappa}$ lie symmetrically relative to the medial line, results from the fact that for every point X which satisfies $\frac{|XA|}{|XB|} = \kappa$ its symmetric Y relative to the the medial line of AB will satisfy $\frac{|YA|}{|YB|} = \frac{1}{\kappa}$.

Proposition 4.5. *Consider the line segment AB and circle $\alpha(O, \rho)$ with diameter $\Gamma\Delta$ on the line AB (See Figure 4.25-I). Next propositions are equivalent:*

1. *The circle α is an Apollonian circle of the line segment AB.*

4.4. APOLLONIAN CIRCLES

2. For an arbitrary point X of the circle α, $X\Gamma$ and $X\Delta$ are bisectors of the angle $\widehat{AXB}$.
3. Triangles OXA and OBX are similar.
4. It holds $\rho^2 = |OA||OB|$ and OX is tangent to the circumcircle (ABX) at X.
5. (A,B) and (Γ,Δ) are pairs of harmonic conjugate points.

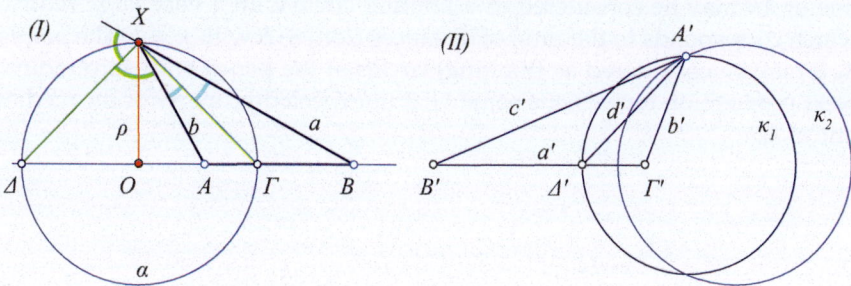

Fig. 4.25: Similar triangles: OXA and OXB Triangle construction

Proof. $(1) \Leftrightarrow (2)$: Was proved in Theorem 4.6.

$(2) \Leftrightarrow (3)$: If $X\Gamma$, $X\Delta$ are bisectors of angle $\widehat{AXB}$, then

$$\widehat{OXA} = \widehat{OX\Gamma} - \widehat{AX\Gamma}, \quad \widehat{OBX} = \widehat{O\Gamma X} - \widehat{\Gamma XB}.$$

However in the two differences the minuends are equal and the subtrahends are also equal, therefore the differences are equal. The two triangles, therefore, have $\widehat{AOX}$ common and angles $\widehat{OXA}$ and $\widehat{OBX}$ equal, therefore they are similar. The argument here is reversible and shows that (3) implies (2).

$(3) \Leftrightarrow (4)$: From the similarity of the triangles follows

$$\frac{|OX|}{|OB|} = \frac{|OA|}{|OX|} \Rightarrow \rho^2 = |OX|^2 = |OA||OB|.$$

The argument here is reversible and shows that (4) implies (3).

$(4) \Leftrightarrow (5)$: Follows directly from Proposition 1.15.

Exercise 4.26. Construct a triangle $AB\Gamma$ from its sides $\{b = |A\Gamma|, c = |AB|\}$ and its bisector $d = |A\Delta|$.

Hint: Consider an arbitrary segment $B'\Gamma'$ of length a' (See Figure 4.25-II). Construct its Apollonian circle κ_1, for the ratio $\lambda = b/c$, which intersects $B'\Gamma'$ at its inner point Δ'. Construct then the Apollonian circle κ_2 of the segment $\Delta'\Gamma'$ for the ratio $\mu = d/b$. If A' is an intersection point of the circles $\{\kappa_1, \kappa_2\}$, then the triangle $A'B'\Gamma'$ is similar to the requested one, with a known similarity ratio (alternatively use exercise 3.129).

Remark 4.5. Figure 4.26 shows the Apollonian circles of a line segment *AB* for different ratios κ. These circles constitute a **pencil** (§ 4.5) of non intersecting circles, which is divided by the medial line of *AB* into two subsystems of circles. The first system consists of the circles for $\kappa < 1$ and has point *A* in the interior of all its circles. The second subsystem is characterized by the circles for $\kappa > 1$ and has point *B* in the interior of all its circles. The medial line of *AB* may be considered as a limiting circle with a very large radius, which corresponds to the ratio of distances from *A*, *B* with $\kappa = 1$. The points *A*, *B* may be considered as (limiting) circles of the pencil with zero radius. Next proposition reveals one somewhat unexpected property of the medial line of *AB*.

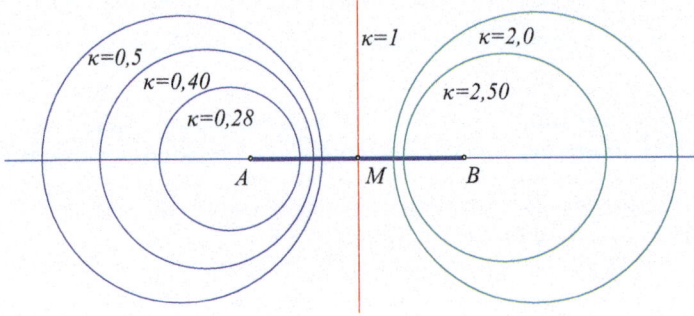

Fig. 4.26: The pencil of Apollonian circles of *AB*

Proposition 4.6. *For every point P of the medial line of the segment AB, the tangents PΣ from P to the Apollonian circles of AB, have the same length $t_p = |P\Sigma|$, which is equal to the distance $|PA| = |PB|$ (See Figure 4.27-I).*

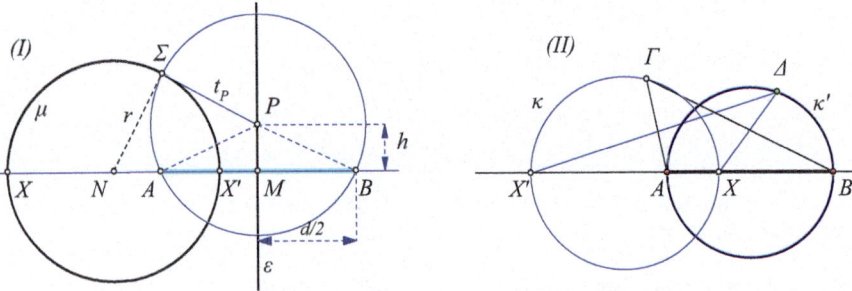

Fig. 4.27: Property of the medial line   Pairing of Apollonian circles

Proof. We must show that for the random point *P* of the medial line and for one Apollonian circle of *AB* for random ratio κ, the length of the tangent

4.4. APOLLONIAN CIRCLES

is independent of κ. Let us suppose that $\kappa < 1$ (the case $\kappa > 1$ is proved similarly). If X, X' are the intersection points of the Apollonian circle with AB and N is its center, then from Pythagoras' theorem for triangles $NP\Sigma$ and NPM, we have

$$t_P^2 = |PN|^2 - r^2 = |MN|^2 + h^2 - r^2,$$

where r the radius of the circle and $h = |MP|$ the distance of P from the middle M of AB. From Corollary 1.43 we have $r = \frac{d\kappa}{1-\kappa^2}$. We have $|MN| = |MA| + |AN| = \frac{d}{2} + d\frac{\kappa^2}{1-\kappa^2}$, according to Exercise 1.78. Setting $d = |AB|$ and doing the calculations, we find that $|MN|^2 - r^2 = \frac{d^2}{4} \Rightarrow |t_P|^2 = \frac{d^2}{4} + h^2 = |PA|^2$.

Remark 4.6. The proposition shows that all the Apollonian circles of the line segment AB have, pairwise, the same radical axis, which coincides with the medial line of AB.

Exercise 4.27. Show that if X, X' are diametrically opposite points on the line AB of the Apollonian circle κ of the points for which $\frac{|\Gamma A|}{|\Gamma B|} = k$, then also the circle κ' with diameter AB is an Apollonian circle relative to the line segment XX' and of ratio $\frac{|\Delta X|}{|\Delta X'|} = \frac{1-k}{1+k}$ (See Figure 4.27-II).

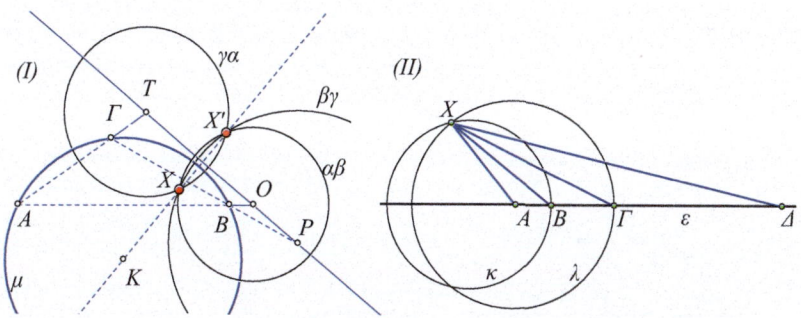

Fig. 4.28: Point X: $\frac{|XA|}{|XB|} = \kappa$, $\frac{|XB|}{|X\Gamma|} = \lambda$ $\widehat{AXB} = \widehat{BX\Gamma} = \widehat{\Gamma X\Delta}$

Exercise 4.28. Given the triangle $AB\Gamma$, to determine a point X, which has ratio of distances $\frac{|XA|}{|XB|} = \kappa$ and $\frac{|XB|}{|X\Gamma|} = \lambda$. Show that for given κ, λ, there exist in general two points X, X', whose line XX' passes through the center K of the circumcircle $\mu(K, r_\mu)$ of the triangle $AB\Gamma$.

Hint: Let X be one of the intersection points of the two Apollonian circles $\alpha\beta$, $\beta\gamma$, whose points satisfy respectively $\frac{|XA|}{|XB|} = \kappa$ and $\frac{|XB|}{|X\Gamma|} = \lambda$ (See Figure 4.28-I). Then it will also hold $\frac{|XA|}{|X\Gamma|} = \kappa \cdot \lambda$ and the corresponding Apollonian circle for $A\Gamma$ relative to the ratio $\kappa\lambda$ will also pass through points X, X'. From proposition 4.5-(4) follows that the power of the center O of $\alpha\beta$ relative to the circle μ is equal to the square $r_{\alpha\beta}^2$. From this also follows that the power of the center K of μ relative to the circle $\alpha\beta$ is equal to r_μ^2. The same property is proved similarly also for the circles $\beta\gamma$ and $\gamma\alpha$ and this has as a consequence K to be contained in the common radical axis of $\alpha\beta$, $\beta\gamma$, $\gamma\alpha$. From the fact that the power of K relative to $\alpha\beta$ is equal to r_μ^2, also follows that points X, X' satisfy $|KX| \cdot |KX'| = r_\mu^2$. This shows that the two points X, X' are coincident only in the case when the circles $\alpha\beta$, $\beta\gamma$, $\gamma\alpha$ are tangent at a point X contained in the circumcircle μ of $AB\Gamma$ and then the common tangent to these circles coincides with XK (also see exercise 4.40).

Exercise 4.29. Given are three successive line segments AB, $B\Gamma$, $\Gamma\Delta$ on line ε. To find a point X, from which these segments are seen under equal angles.

Hint: If there is such a point X, considering the first two segments, point X must be on the Apollonian circle κ of segment $A\Gamma$, relative to ratio $\tau = \frac{|AB|}{|B\Gamma|}$ (See Figure 4.28-II). Similarly, considering the segment $B\Delta$, point X must be contained in the Apollonian circle λ relative to the ratio $\sigma = \frac{|B\Gamma|}{|\Gamma\Delta|}$. It follows that X must be coincident with the intersection point of these two circles. The problem may not be solvable and this will happen exactly when the two circles κ and λ do not intersect. The investigation of the problem and the expression of the condition which ensures the existence of a solution is interesting.

Proposition 4.7. *Let H be a point lying outside the line η and $\kappa > 0$ a constant. Let also A be a variable point of the α. There exist at most two positions of A, for which $|AH| = \kappa|AX|$, where X is the projection of A onto η (See Figure 4.29-I).*

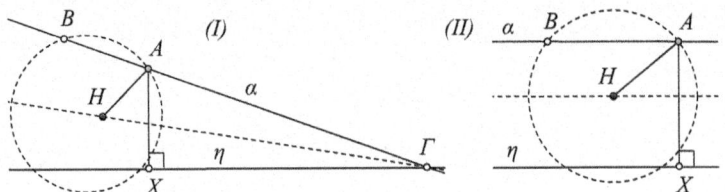

Fig. 4.29: Intersections of a line and a conic

Proof. Suppose that line α intersects line η at point Γ. Then, for every point A on α the ratio $\lambda = \frac{|AX|}{|A\Gamma|}$ will be fixed. Consequently, for the requested positions of A, the ratio $\mu = \frac{|AH|}{|A\Gamma|} = \frac{|AH|}{|AX|} \cdot \frac{|AX|}{|A\Gamma|} = \kappa\lambda$ will be known. It follows that the requested points A coincide with the intersection points of line α and the Apollonian circle of segment $H\Gamma$ relative to ratio μ. In the case where α is parallel to η, then $|AX|$ is a known fixed length and consequently length $|AH| = \kappa|AX|$ is also known (See Figure 4.29-II). In this case then, the requested points are again intersections of a circle and a line.

Remark 4.7. Last proposition gives the geometric proof of the fact that a line (α) intersects a conic in at most two points (Theorem 6.7) ([101, p.42]).

4.5 Circle pencils

> Philosophy is written in that great book which ever is before our eyes – I mean the universe – but we cannot understand it if we do not first learn the language and grasp the symbols in which it is written. The book is written in mathematical language, and the symbols are triangles, circles and other geometrical figures, without whose help it is impossible to comprehend a single word of it; without which one wanders in vain through a dark labyrinth.
>
> *Galileo Galilei*, The Assayer

Circle pencils are the analogue of line pencils for circles. As we did there (§ 3.8), distinguishing between two families of pencils of lines, we do here too and define as **circle pencil** one of the following three families of circles:

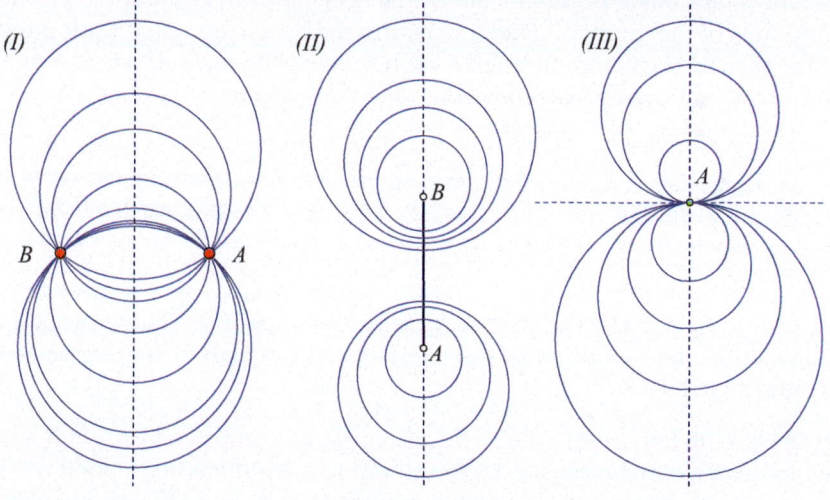

Fig. 4.30:   Intersecting   non-intersecting   tangent pencil

1. The family of all circles which pass through two fixed points A and B (See Figure 4.30-I),
2. The entire family of all Apollonian circles of a line segment AB (See Figure 4.30-II),
3. The family of all circles pairwise tangent at a fixed point A (See Figure 4.30-III).

Next theorem expresses the common characteristic of these three kinds of circle pencils, which we respectively call **intersecting pencil** or *elliptic pencil*, **non intersecting pencil** or *hyperbolic pencil* and **tangent pencil** or *parabolic pencil*. Collectively, all three kinds are often called **coaxal systems of circles** ([106, p. 106]).

Theorem 4.7. *For every circle pencil there is a line ε, such that any pair of different circles of the pencil has its radical axis identical with ε. For intersecting pencils this is the line through their common points. For non intersecting pencils ε coincides with the medial line of AB, and for tangent pencils ε coincides with the common tangent to the circles.*

Proof. For pencils of intersecting circles the conclusion follows from Exercise 4.16, for tangent pencils from Exercise 4.17 and for pencils of non intersecting circles from Proposition 4.6.

The common radical axis of all pairs of circles of a pencil is called the **radical axis** of the pencil. The line containing the centers of the pencil circles is called **center-line** of the pencil. This line is orthogonal to the radical axis of the pencil.

For intersecting pencils, the common points A, B of all circles of the pencil, are called **basic points** of the pencil. For non intersecting pencils, the endpoints of the interval AB which defines the pencil are called **limit points** of the pencil. The common point A of the circles in a tangent pencil may be considered simultaneously as a basic and a limit point.

Corollary 4.4. *The limit points A, B of a non intersecting pencil are inverse relative to every circle κ of the pencil. They are also harmonic conjugate relative to the diametrically opposite points $\{X,Y\}$ of κ, which are defined from its intersection with the center-line of the pencil.*

Proof. The term *inverse*, will be defined below (§ 4.8). Here it means, that for every circle $\kappa(O,r)$ of the pencil, it is $|OA| \cdot |OB| = r^2$. These properties follow from the definition of the circles as Apollonian of the segment AB (Proposition 4.5).

Exercise 4.30. Let κ be a circle of a non intersecting pencil $\mathscr{D}$ with limit points $\{A,B\}$, containing A in its interior. If Γ is an intersection point of this circle with the line ε, which is orthogonal to AB at A, show that its tangent at Γ passes through B.

Proposition 4.8. *Every pair consisting of a circle $\mu(N,\rho)$ and a line ε, not passing through the center N of the circle, defines exactly one pencil of circles which contains the circle μ and has corresponding radical axis the line ε (See Figure 4.31-I).*

Proof. If the line ε and the circle μ intersect at two points A and B, then obviously the set of circles which pass through points A and B is the requested pencil. If the line ε and the circle μ are tangent at A, then the pencil consists of all the circles which are tangent to ε at A. If the circle and the line do not intersect, then we can easily construct a segment AB whose medial line coincides with ε and whose μ is an Apollonian circle relative to an appropriate ratio of lengths κ.

Indeed, we draw initially the orthogonal to ε line NM from the center of the given circle. Then, from an arbitrary point P of ε we draw a tangent

4.5. CIRCLE PENCILS

$P\Sigma$ to the circle μ. With center P and radius $P\Sigma$ we draw a circle which intersects the line NM at points A and B. The segment AB is the requested one (Proposition 4.6) and the ratio κ, with respect to which μ is an Apollonian circle of the segment AB, is given by the quotient $\kappa = \frac{|\Sigma A|}{|\Sigma B|}$.

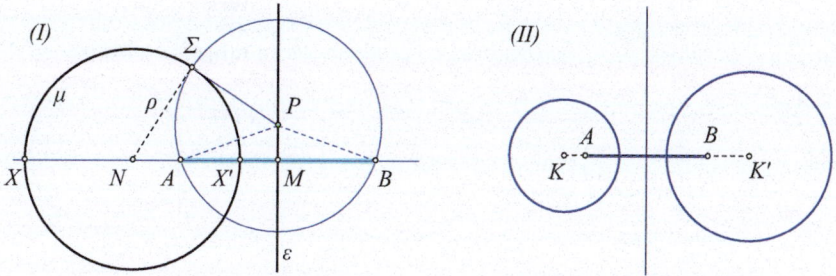

Fig. 4.31: Determination of AB Non intersecting pencil

Corollary 4.5. *Every pair of non concentric circles defines exactly one circle pencil which contains them as members.*

Proof. We consider one of the two circles and the radical axis of the two circles and we apply the preceding proposition. Be careful in the case of two non-intersecting circles. The line segment AB defined by the limit points $\{A, B\}$ of the pencil, does not coincide with the segment KK' of the centers of the circles (See Figure 4.31-II).

The pencil defined from two circles, as in the preceding corollary, we call **pencil generated** from the two circles. Usually, to a pencil we count as a member also its radical axis, considering it as a circle with infinite radius. In the case of non-intersecting and tangent pencils we also include as special members their limit points, considering them as circles of zero radius. The preceding corollary is extended also for pairs of circles, of which one or both coincide with points. These points are then the limit points of the pencil. Next corollaries are valid for this extended notion of pencil.

Corollary 4.6. *Every pair of two points or a point and circle/line defines a circle pencil which contains these given elements as members.*

Proof. If two points are given, then this defines one non intersecting pencil which has limit points the given points. If a circle $\kappa(O, \rho)$ and point A (different from the circle center) are given, then two things happen: (a) A belongs to the circle κ and this defines a tangent pencil. (b) A does not belong to the circle and the relation $\rho^2 = |OA||OB|$ defines on OA a second point B, such that the circle $\kappa(O, \rho)$ is an Apollonian one relative to segment AB (Proposition 4.5). Finally, if a line ε and a point A are given, there exist two alternatives: (i) A belongs to ε and this defines a tangent pencil with radical axis ε and limit point A. (ii) Point A does not belong to ε, and this defines the symmetric of B relative to ε and one non intersecting pencil with limit points A, B and radical axis the line ε.

Corollary 4.7. *Given a pencil of circles, for every point X of the plane there exists exactly one pencil-member κ which passes through it (See Figure 4.32).*

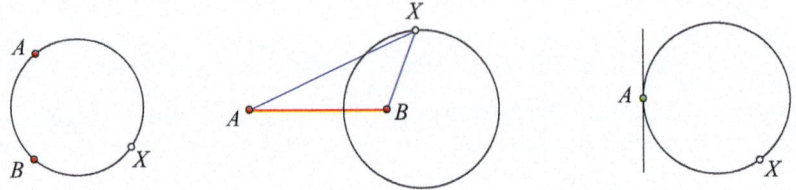

Fig. 4.32: The unique pencil circle through X

Proof. In the case of an intersecting pencil, circle κ is exactly the circle which passes through the two basic points of the pencil A and B and the point X. In the case of a non intersecting pencil, point X determines the ratio $\kappa = \frac{|XA|}{|XB|}$ and this defines uniquely the circle of the pencil. Finally, in the case of a tangent pencil, the circle is the one tangent to the radical axis at A and passing through point X.

Proposition 4.9. *Given a circle pencil and a circle κ, which does not belong to the pencil and does not have its center at the center-line of the pencil, the radical axes of the pairs of circles (κ, μ), where μ is a circle that belongs to the pencil, pass all through a fixed point Γ of the radical axis of the pencil (See Figure 4.33-I).*

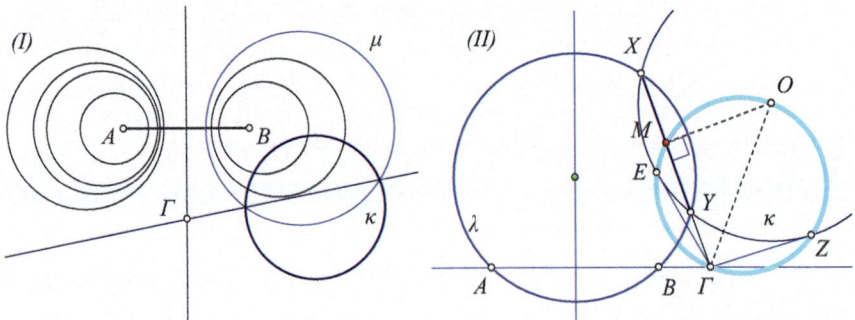

Fig. 4.33: Radical axes with pencil circles   Chord middles of pencil circles

Proof. Let Γ be the intersection point of the radical axis of the pencil and the radical axis of κ with a member μ of the pencil. Because every point of the radical axis of the pencil has the same power relative to all the members

4.5. CIRCLE PENCILS

of the pencil, if Γ has power δ relative to κ and μ, it will have the same power δ also relative to every other circle μ' of the pencil, therefore it will be contained also in the radical axis of κ and μ'. The requirement of κ to not have its center on the axis of the pencil ensures the existence of point Γ.

Remark 4.8. Last proposition, in the case of an intersecting pencil, gives a different aspect of Exercise 4.20 and generalizes its conclusion for all kinds of pencils.

Exercise 4.31. Given a circle pencil $\mathcal{D}$ and circle $\kappa(O, \rho)$, which does not belonging to the pencil and does not have its center on the central axis of the pencil, find the geometric locus of the middles M of the chords XY, which are excised by the circles λ of the pencil $\mathcal{D}$ from the circle κ.

Hint: All the lines XY pass through the fixed point Γ of the radical axis of the pencil (Proposition 4.9) (See Figure 4.33-II). Point M sees the fixed line segment $O\Gamma$ under a right angle. Therefore the locus is the arc $\widehat{EOZ}$ of the circle with diameter $O\Gamma$ which is contained in the circle κ.

Exercise 4.32. Show that the unique pencil, which has a circle of minimal diameter $\delta > 0$ is the intersecting pencil. How is this δ related to the basic points A, B of the pencil?

Exercise 4.33. Given a circle pencil $\mathcal{D}$ and a line ε, find the circles of the pencil tangent to ε (See Figure 4.34-I).

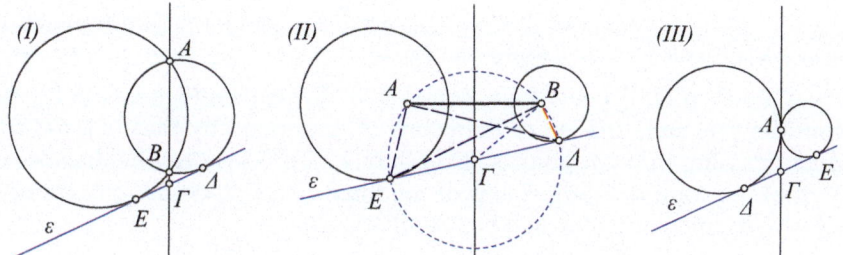

Fig. 4.34: Pencil members tangent to line

Hint: Let Γ be the intersection point of ε with the radical axis of the pencil. In the intersecting pencil, besides A, B, the position of the point of contact Δ of the requested circle can be calculated from $|\Gamma\Delta|^2 = |\Gamma A||\Gamma B|$. In the non intersecting pencil, the position of the point of contact is calculated directly from $|\Gamma\Delta| = |\Gamma A| = |\Gamma B|$ (See Figure 4.34-II) and in the tangent pencil respectively from $|\Gamma\Delta| = |\Gamma A|$ (See Figure 4.34-III). There are two solutions in general. Examine the special cases, in which ε is parallel to the radical axis of the pencil or when it passes through its basic points.

Exercise 4.34. Given a circle pencil $\mathcal{D}$ and a circle ω, which does not belong to the pencil, find the circles of the pencil which are tangent to ω.

Hint: If κ is the requested circle, the common tangent $E\Gamma$ of κ and of ω (See Figure 4.35-I) will pass through point Γ through which pass also all

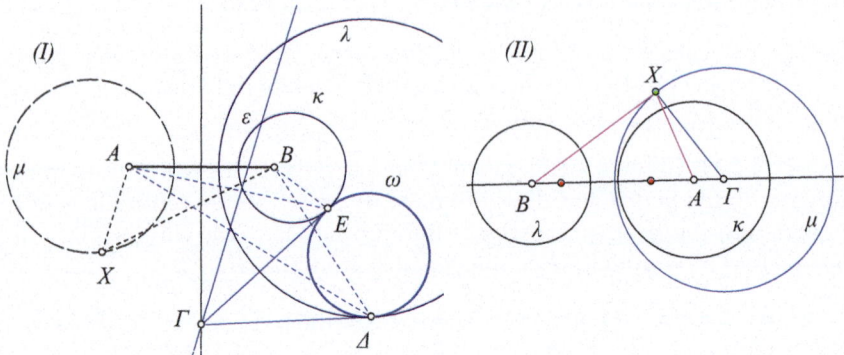

Fig. 4.35: Members tangent to ω Locus of points of fixed ratio of powers

the radical axes of the pairs of circles consisting of ω and a member μ of the pencil (Proposition 4.9). Construct then an arbitrary member μ of the pencil and find the intersection point Γ of the radical axis of $\{\omega,\mu\}$ with the radical axis of the pencil. From Γ then draw the tangents to ω and define the points of contact Δ and E. The requested circles are the circles of the pencil which pass through points Δ and E. Two solutions in general.

Theorem 4.8. *Given the circles $\{\kappa(A,\alpha),\lambda(B,\beta)\}$, the geometric locus of points X for which the ratio of powers relative to these two circles is constant is a circle μ, which belongs the the pencil defined by $\{\kappa,\lambda\}$.*

Proof. ([103, p.178]) Let X be a point of the locus (See Figure 4.35-II). We consider the circle $\mu(\Gamma,\gamma)$ of the pencil $\mathscr{D}$, generated by the circles κ and λ and passing through point X (Corollary 4.7). If δ denotes the distance of X from the common radical axis of the three circles, then the powers of X relative to circles κ and λ will be (Corollary 4.3)

$$p_\kappa(X) = \pm 2|A\Gamma|\delta, \qquad p_\lambda(X) = \pm 2|B\Gamma|\delta \quad \Rightarrow \quad \frac{p_\kappa(X)}{p_\lambda(X)} = \pm\frac{|A\Gamma|}{|B\Gamma|}.$$

This shows that point X is contained in μ, and simultaneously, that each point of μ is contained in the locus. Therefore the locus coincides with the circle μ.

Corollary 4.8. *The circle μ belongs to the pencil $\mathscr{D}$ generated by the circles $\{\kappa,\lambda\}$ if and only if, for every point X of μ the ratio of powers of X relative to the circles $\{\kappa,\lambda\}$ is constant.*

Corollary 4.9. *Given the circles $\kappa(A,\alpha)$ and $\lambda(B,\beta)$, the geometric locus of points X for which the ratio of the lengths of the tangents to the two circles κ and λ is fixed, is a circle μ which belongs to the pencil defined by κ and λ.*

4.5. CIRCLE PENCILS

The corollary, in the case of the non intersecting and tangent pencil (Figures 4.36-II,-III), has interesting special cases. In the case of non intersecting cir-

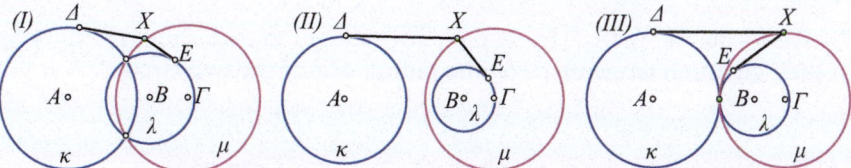

Fig. 4.36: Locus of points with fixed ratio of tangents

cles κ and λ, the corollary generalizes Theorem 4.6 of Apollonius, which can be considered as a special case in which the two circles κ and λ have zero radius and coincide with the limit points of the pencil. The cases in which only one of the two circles is coincident with one limit point are also interesting.

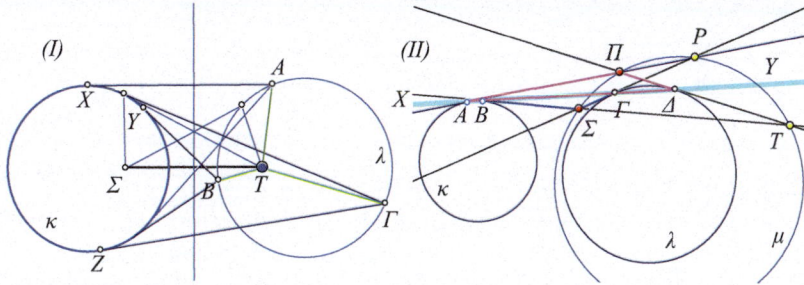

Fig. 4.37: $\frac{|AX|}{|AT|} = \frac{|BY|}{|BT|} = \frac{|\Gamma Z|}{|\Gamma T|}$ Line intersecting two circles

Corollary 4.10. *Given the circle $\kappa(O,\rho)$ and a point T, different from the center O of the circle, the geometric locus of the points A for which the ratio of the lengths $\frac{|AT|}{|AX|} = \sigma$ is constant, where AX is the tangent to κ from A, is a circle λ of the pencil generated by κ and the point T (See Figure 4.37-I).*

Corollary 4.11. *Given the circle $\kappa(O,\rho)$ and a point T different from O, three other points $\{A, B, \Gamma\}$ have the same ratio of lengths of tangents to their distance from T*

$$\frac{|AX|}{|AT|} = \frac{|BY|}{|BT|} = \frac{|\Gamma Z|}{|\Gamma T|}$$

if and only if, the circumscribed circle λ of triangle $AB\Gamma$ belongs to the pencil generated by the circle κ and the point T (See Figure 4.37-I).

Exercise 4.35. Line XY intersects the circles κ and λ respectively at points $\{A, B\}$ and $\{\Gamma, \Delta\}$. Show that the tangents to κ at $\{A, B\}$ and the tangents to λ at $\{\Gamma, \Delta\}$ intersect at four points $\{\Pi, P, \Sigma, T\}$ contained in a circle μ of the pencil $\mathscr{D}$ generated by κ and λ (See Figure 4.37-II).

Hint: ([27, p.205]) Triangles $(A\Pi\Delta, B\Sigma\Gamma)$ and $(TB\Delta, PA\Gamma)$ are pairs of similar triangles. Corollary 3.17 is applied to triangles $B\Sigma\Gamma$ and $TB\Delta$. The proof results by combining these with Theorem 4.8.

Exercise 4.36. Show that, for two circles external to each other the middles of their common tangents lie on the radical axis of the two circles.

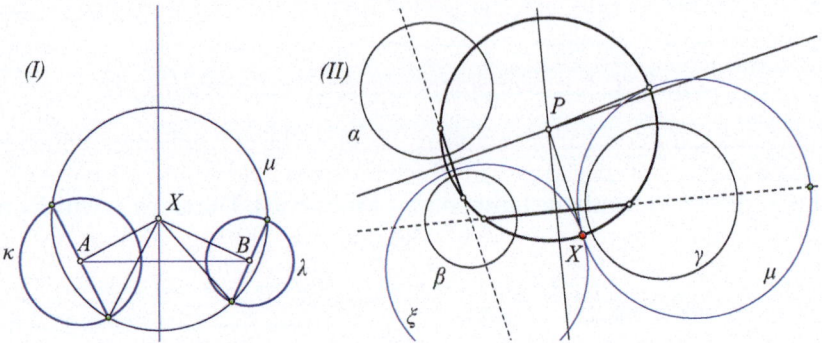

Fig. 4.38: Intersection at diameter   Contact of circles $\{\xi, \mu\}$

Exercise 4.37. The geometric locus of the centers X of the circles, which intersect two given circles $\{\kappa(A,\alpha), \lambda(B,\beta)\}$ along diameters of these circles, is a line ε perpendicular to the center-line AB (See Figure 4.38-I).

Hint: Calculate the expression $|XA|^2 - |XB|^2$ (Corollary 3.12).

Exercise 4.38. Given three circles $\{\kappa, \lambda, \mu\}$, to construct a circle ν intersecting each of them along a diameter of it.

Exercise 4.39. Given three circles $\{\alpha, \beta, \gamma\}$ external to each other, consider two tangent circles $\{\xi, \mu\}$ of the pencils generated by $\{\alpha, \beta\}$ and $\{\beta, \gamma\}$ respectively. Show that the points of contact X of $\{\xi, \mu\}$ are contained in the circle which passes through the limit points of the two pencils and has center the radical center P of $\{\alpha, \beta, \gamma\}$ (See Figure 4.38-II).

Exercise 4.40. Relate the results of the last exercise to those of exercise 4.28.

Exercise 4.41. Given are points A and B on the two sides of line ε and a point X moving on the line. Find the position of X for which the expression $|XA|^2 + |XB|^2$ is minimized.

Exercise 4.42. Show that if $\{A, B\}$ are positive numbers and have fixed product $x \cdot y = d^2$, then the sum $x + y$ becomes minimal, when $x = y$.

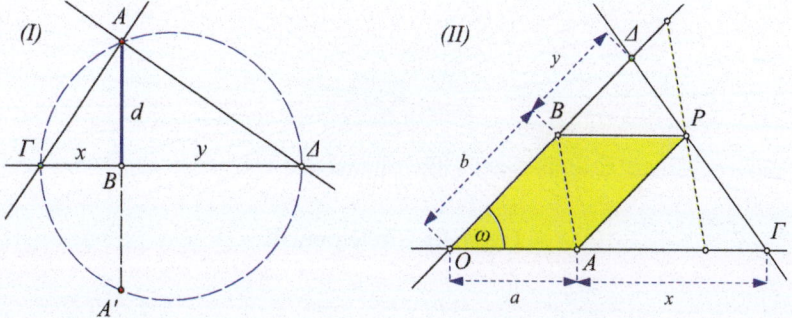

Fig. 4.39: Minimize $x+y$ under $xy = d^2$ Minimize area

Hint: Consider a segment AB with $|AB| = d$, the orthogonal to it at B line ε and points $\{\Gamma, \Delta\}$ on it and on the two sides of B such that $|B\Gamma| = x$, $|B\Delta| = y$ (See Figure 4.39-I). Then triangle $A\Gamma\Delta$ is right (Proposition 3.3). The circles with diameter $\Gamma\Delta$ are members of an intersecting circle pencil with basic points point A and its symmetric A' relative to ε. The numbers $x + y$ are lengths of diameters of such circles and the minimal of the pencil is the circle with diameter AA'.

Exercise 4.43. Show that if $\{a, b, x, y\}$ are positive, with fixed product $x \cdot y = d^2$, then $ax + by$ becomes minimal, when $ax = by$.

Hint: $x' = ax$, $y' = by$ also have fixed product $x' \cdot y' = abd^2$ and $x' + y' = ax + by$ becomes minimal when $x' = y'$ (Exercise 4.42).

Exercise 4.44. Given are a parallelogram $OAPB$ and a variable line passing through P and intersecting the extensions of OA, OB respectively at Γ, Δ (See Figure 4.39-II). Show that the area of the triangle $O\Gamma\Delta$ is minimized when $\Gamma\Delta$ is parallel to AB.

Hint: The area is minimized when the sum of the areas $\varepsilon(PA\Gamma) + \varepsilon(PB\Delta)$ is. The latter is $a\sin(\omega)y + b\sin(\omega)x$ and $xy = ab$ (see the remark 3.15). The conclusion follows applying Exercise 4.43.

Remark 4.9. Besides the pencils we saw, there exist also three other kinds, which we accept as non conventional circle pencils. In the first of these cases they are not even circles, but lines, and specifically all the lines which pass through a specific point A. We think of it as an intersecting pencil, which consists of very large circles which pass through the two basic points A and B. Point B however is far away at infinity. Figure 4.40 could be also a section of a genuine pencil of intersecting circles, whose second basic point is very far away, so that, the circles will be very large and indistinguishable from lines.

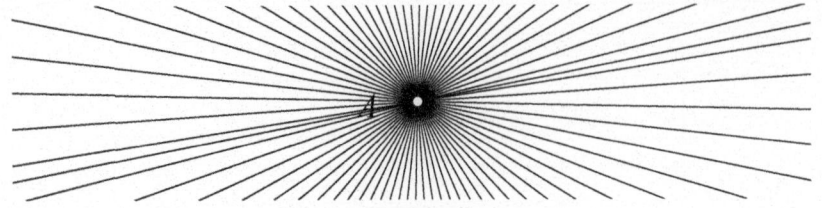

Fig. 4.40: Circle pencil!

The second kind of non conventional pencil consists of all the lines which are parallel to a given line ε. We think of it as a tangent circle pencil with a basic point A, which however is very far away at infinity.

The third non conventional circle pencil consists of all concentric circles with center a specific point A (See Figure 4.41). We think of it as a non in-

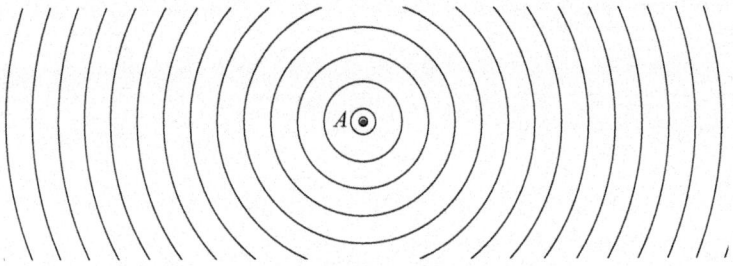

Fig. 4.41: Pencil of concentric circles

tersecting pencil with limit points A and B. Point B however is also very far away at infinity. The problem with these three pencils is that, contrary to the other cases, their radical axis is not defined. Their usefulness lies on the fact that from these we can produce all other pencils through a procedure, the *inversion*, which we'll examine below (§ 4.8).

Exercise 4.45. Show that if the points $\{A, B\}$ have the same power with respect to circles $\{\alpha, \beta, \dots\}$, then all these circles belong to the same pencil of circles.

Exercise 4.46. Point P lies on the radical axis of two circles $\{\alpha, \beta\}$ lying outside each other. How many circles there are, which pass through P and are tangent to $\{\alpha, \beta\}$? To find the geometric locus of the contact points of such a circle with one of the $\{\alpha, \beta\}$, when P moves on the radical axis.

Exercise 4.47. The circle $\kappa(E)$ intersects two circles $\{\alpha(Z), \beta(\Delta)\}$ lying outside each other along diameters (See Figure 4.42-I). Show that the intersection Γ of these diameters is contained in the radical axis $\alpha\beta$ of the circles

4.5. CIRCLE PENCILS

$\{\alpha, \beta\}$. Show also that the centers of these circles κ lie on a line δ which coincides with the symmetric of $\alpha\beta$ relative to the middle M of the segment $Z\Delta$.

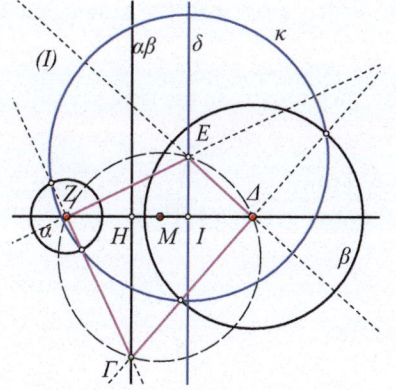
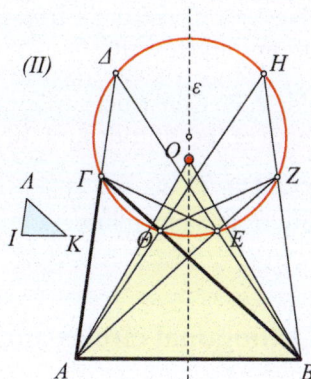

Fig. 4.42: Intersection along diameters   Non-intersecting pencil

Exercise 4.48. Given is a triangle $IK\Lambda$ and a line segment AB. With one side the segment AB, and in all possible ways we construct triangles similar to $IK\Lambda$ towards the same side of AB (See Figure 4.42-II). Show that the six vertices of the resulting triangles lie on a circle κ. Show also that all these circles κ, which result by varying the shape of the triangle $IK\Lambda$, are members of a non-intersecting pencil with limit points the vertex O of the equilateral with basis AB and its symmetric O' with respect to AB.

Exercise 4.49. Inversely to the preceding exercise, consider a circle κ of a non-intersecting pencil, which contains in its interior the limit point O of the pencil (See Figure 4.42-II). Consider also the equilateral triangle OAB with side AB on the radical axis of the pencil. Show that for every point Γ of the circle κ it is defined the second intersection point Θ of $B\Gamma$ and the second intersection point H of $A\Theta$ with κ and the triangles $\{AB\Gamma, AB\Theta, ABH\}$ are similar.

Exercise 4.50. Continuing the preceding exercise, for the circle κ of the non-intersecting pencil and the corresponding triangles $\{AB\Gamma, AB\Theta, ABH\}$, determine the position of Γ on κ, for which these triangles have an angle of the maximal/minimal possible measure.

Exercise 4.51. Construct a circle κ passing through two given points A and B and intersecting a given circle λ at two diametrically opposite points.

Exercise 4.52. Construct a triangle, for which are given the side $a = |B\Gamma|$, the altitude from the opposite side υ_A and the trace of the bisector (internal or external) from A.

Exercise 4.53. Construct a triangle, for which are given the side $a = |B\Gamma|$, the angle $\alpha = \widehat{BA\Gamma}$ of the opposite vertex and the trace of the bisector (internal or external) from A.

Exercise 4.54. Construct a triangle, for which are given the side a, the altitude v_A and the bisector δ_A.

Exercise 4.55. Construct a triangle, for which are given the side $a = |B\Gamma|$, and the traces of the altitude and the bisector on that side.

Exercise 4.56. Construct a circle κ passing through given points A and B and excising from a given line ε a line segment of given length λ.

4.6 Orthogonal circles and pencils

> Farming looks mighty easy when your plow is a pencil, and you're a thousand miles from the corn field.
>
> D. Eisenhower, Peoria, 1956

Orthogonal circles are formed by means of tangents to circles from a given point. The tangents $\{PA, PB\}$ to the circle $\kappa(O, \rho)$ from a point P are equal, therefore they define a circle $\lambda(P, \rho')$, which has these tangents as radii (See

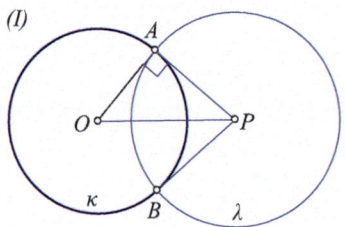
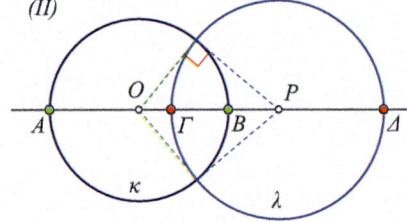

Fig. 4.43: Right circles   Harmonic quadruple $(AB;\Gamma\Delta) = -1$

Figure 4.43-I). At the intersection point A of the two circles the angle between their radii is a right one. This is a symmetric relation. The circle κ can be considered that it results the same way, from the tangents to λ from O. Two intersecting circles, whose radii at the intersection points are orthogonal are called **orhogonal circles**. By definition, therefore, this is equivalent to:

At each of their intersection points, the radius of one is tangent to the other.
From this characteristic property follow also the two next corollaries.

Corollary 4.12. *Two circles $\{\kappa(O,\rho), \lambda(P,\sigma)\}$ are orthogonal, if and only if*

4.6. ORTHOGONAL CIRCLES AND PENCILS

$$|OP|^2 = \rho^2 + \sigma^2.$$

Corollary 4.13. *Two circles $\{\kappa, \lambda\}$ are orthogonal, if and only if the diametrically opposite points $\{A, B\}$ of one and $\{\Gamma, \Delta\}$ of the other, on their center-line, form a harmonic quadruple (See Figure 4.43-II).*

Proof. It follows directly from the characteristic property of a harmonic quadruple of points (Proposition 1.15).

Exercise 4.57. Construct a circle λ, orthogonal to a given circle $\kappa(O, \rho)$ and having its center at a given point P external to κ.

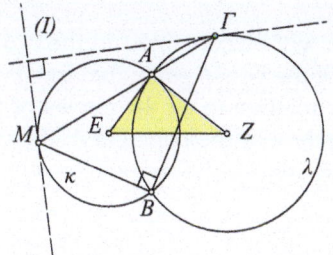 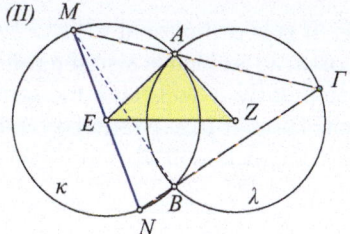

Fig. 4.44: Similar triangles AEZ, $BM\Gamma$ — Diametrically opposite points M, N

Exercise 4.58. Show that the circles $\{\kappa, \lambda\}$, intersecting at points $\{A, B\}$ are orthogonal, if and only if one of the following conditions holds:

1. Every line through point A defines points $\{M, \Gamma\}$ on circles $\{\kappa, \lambda\}$ such that the angle $\widehat{\Gamma BM}$ is a right one.
2. For every point Γ of λ the lines $\{\Gamma A, \Gamma B\}$ intersect again the circle κ at diametrically opposite points $\{M, N\}$.
3. The tangents to the circles at $\{M, \Gamma\}$ intersect orthogonally.

Hint: (1) The triangles EAZ and $BM\Gamma$ are similar even when the circles are not orthogonal (See Figure 4.44-I).
 (2) Draw BM and use (1) (See Figure 4.44-II).

Theorem 4.9. *The circles μ, which are orthogonal to both non concentric circles $\{\kappa, \lambda\}$, have their centers on the radical axis of $\{\kappa, \lambda\}$ (See Figure 4.45-I).*

Proof. If the circles $\mu(\Sigma, \rho)$ and $\kappa(O, \rho')$ are orthogonal, then their radii at one of their intersection points A will be orthogonal (See Figure 4.45-I), therefore line ΣA will be tangent to κ. The same will happen also with circles μ and λ. Consequently the tangents from point Σ towards the given circles will be equal and point Σ will lie on the radical axis of κ and λ.

Corollary 4.14. *A circle μ orthogonal to two other cirlces $\{\kappa, \lambda\}$ is simultaneously orthogonal also to every circle of the pencil generated by $\{\kappa, \lambda\}$.*

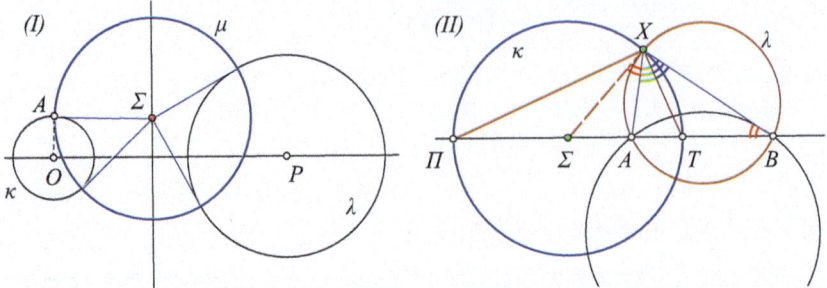

Fig. 4.45: Orthogonal to two circles Orthogonal to intersecting pencil

Proof. If μ is orthogonal to $\{\kappa, \lambda\}$ its center will be contained in the radical axis of $\{\kappa, \lambda\}$, which is also the radical axis of the pencil generated by $\{\kappa, \lambda\}$. Consequently μ will have the same power relative to all the circles of this pencil. Thus, if $\mu(\Sigma, \rho)$ intersects a third circle ν of the pencil at A, then the radius ΣA of μ will also be tangent to ν, therefore the two circles will be orthogonal.

Theorem 4.10. *The set of circles which are simultaneously orthogonal to all the circles of an intersecting, non-intersecting or tangent pencil $\mathscr{D}$, is, respectively, a non-intersecting, intersecting or tangent pencil $\mathscr{D}'$.*

Proof. In the case where the circle $\kappa(\Sigma, \rho)$ is orthogonal to the circles of an intersecting pencil with basic points A, B (See Figure 4.45-II), from an arbitrary point X of κ passes a circle λ of the pencil $\mathscr{D}$ and holds $|\Sigma X|^2 = |\Sigma A||\Sigma B|$. According to Proposition 4.5 this is equivalent to the fact that κ is Apollonian relative to AB. Consequently, all circles which are orthogonal to the circles of the intersecting at $\{A, B\}$ pencil $\mathscr{D}$ belong to the non intersecting pencil $\mathscr{D}'$ which is defined from the line segment AB. In the case where the cir-

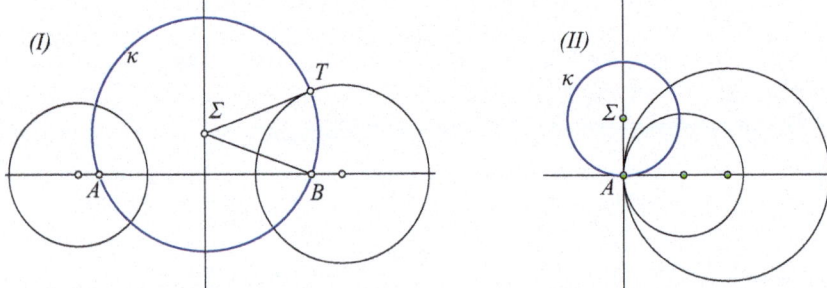

Fig. 4.46: Orthogonal to non intersecting Orthogonal to tangent pencil

cle $\kappa(\Sigma, \rho)$ is orthogonal to the circles of a non intersecting pencil which is defined from the line segment AB (See Figure 4.46-I), the tangent $|\Sigma T|$ to a member of the pencil is equal to $|\Sigma A| = |\Sigma B|$ (Proposition 4.6). Therefore the

4.6. ORTHOGONAL CIRCLES AND PENCILS

circle passes through points A, B and consequently is contained in the pencil with these as basic points.

In the case when circle $\kappa(\Sigma, \rho)$ is orthogonal to the circles of a tangent pencil with common point A, its radius ΣA will be tangent to all the circles and consequently the circle κ will be tangent at A to the central axis of the pencil (See Figure 4.46-II).

The pencil $\mathscr{D}'$, whose existence guarantees the preceding proposition, is called **orthogonal pencil** of $\mathscr{D}$.

Corollary 4.15. *Every circle pencil may be considered as the set of all circles which are orthogonal to two fixed circles α and β.*

Exercise 4.59. Construct a circle κ which is orthogonal to all the circles of a pencil $\mathscr{D}$.

Hint: Find a point P of the radical axis of $\mathscr{D}$ and draw a tangent PT to an arbitrary circle-member of $\mathscr{D}$. The circle $\kappa(P, |PT|)$ has the required property.

Exercise 4.60. Construct the limit points A and B of a non intersecting pencil $\mathscr{D}$, for which are given two circles α and β.

Hint: Construct the radical axis ε of α and β. From an arbitrary point P of ε (from where this is possible) draw a tangent PY to the circle α. The intersection points of the center-line of $\{\alpha, \beta\}$ along with the circle $\kappa(P, |PY|)$ are the wanted limit points of the pencil.

Exercise 4.61. Construct a circle κ orthogonal to two other non-concentric circles $\{\alpha, \beta\}$ and passing through a point X.

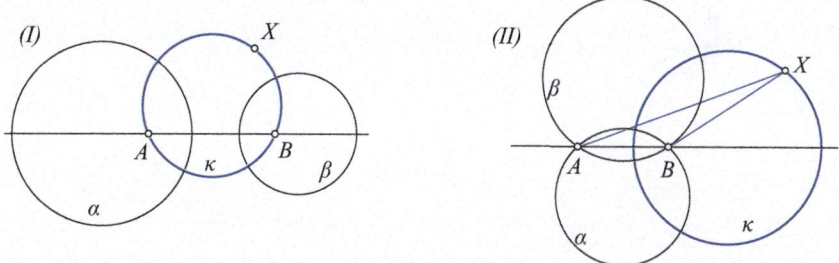

Fig. 4.47: κ orthogonal to $\{\alpha, \beta\}$ passing through X

Hint: If the given circles do not intersect (See Figure 4.47-I), determine first the limit points $\{A, B\}$ of the generated pencil (preceding exercise). The requested circle is the one passing through the three points $\{X, A, B\}$. If the given circles intersect at points $\{A, B\}$, then the requested one belongs to the non intersecting pencil with limit points $\{A, B\}$. Consequently it is the Apollonian circle of AB with ratio $\lambda = \frac{|XA|}{|XB|}$ (See Figure 4.47-II).

Exercise 4.62. Construct a circle κ orthogonal to a given circle α and passing through the given points {X,Y}.

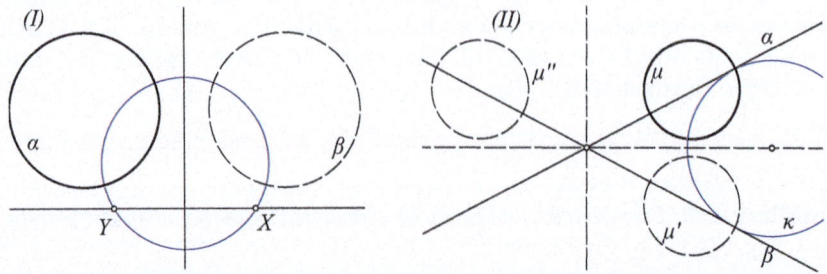

Fig. 4.48: Orthogonal α + through {X,Y}, Orthogonal μ + tangent α, β

Hint: Consider the symmetric circle β of the circle α relative to the medial line of XY (See Figure 4.48-I). If the requested circle is orthogonal to α, it will also be orthogonal to β. Therefore it coincides with the circle which is simultaneously orthogonal to α and β and passes through one of the two points, for example X (Exercise 4.61).

Exercise 4.63. Construct a circle κ orthogonal to a given circle μ and tangent to two lines α and β (See Figure 4.48-II).

Hint: κ will be orthogonal also to the symmetric μ' relative to the bisector of the angle of α, β (if they intersect). Therefore it will be a member of the pencil which is orthogonal to {μ,μ'} and simultaneously will be tangent to α (Exercise 4.33). What happens when α and β are parallel?

Exercise 4.64. Construct a circle λ tangent to a given line ε at a given point A of the line and simultaneously orthogonal to a given circle κ(O,ρ).

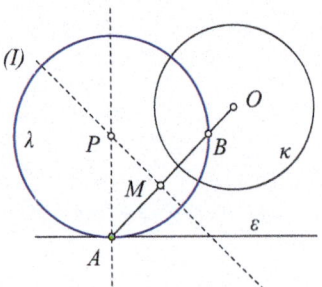 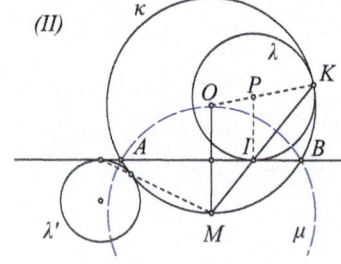

Fig. 4.49: λ orthogonal to κ   Collinear points {K,I,M}

Hint: Let λ be the requested circle and B be its other intersection point with OA. Because of the orthogonality of the two circles we have $|OA||OB| = \rho^2$. Therefore B may be constructed from the given data and the center P of λ is determined as an intersection point of the orthogonal to ε at A and the medial line of AB (See Figure 4.49-I).

4.6. ORTHOGONAL CIRCLES AND PENCILS

Exercise 4.65. Circle λ is tangent to circle κ internally and to the chord AB of κ. Show that the line IK, which joins the contact points, passes through the middle M of one of the arcs, which are defined by the chord. Show also, that the circle $\mu = M(|MA|)$ is orthogonal to λ. Show the analogous property for circle λ', which is tangent to the chord AB and to the circle κ externally.

Hint: The first part is Exercise 2.161 unchanged. For the second show that the triangles AMK, IBK and AMI are similar (See Figure 4.49-II).

Exercise 4.66. Circle λ is tangent to circle κ internally at point K and chord AB of circle κ at point I (See Figure 4.49-II). Show that the line KI is a bisector of the angle AKB.

Exercise 4.67. Show that, the geometric locus of the centers of circles λ, which pass through a fixed point B and intersect a fixed circle $\kappa(A,\rho)$ at diametrically opposite points relative to λ, is a circle.

Exercise 4.68. Chord $\Gamma\Delta$ is parallel to the diameter AB of the circle κ. Show that for every point X of the diameter AB holds $|X\Gamma|^2 + |X\Delta|^2 = |XA|^2 + |XB|^2$.

Exercise 4.69. Construct a circle κ simultaneously orthogonal to three given circles α, β and γ.

Hint: Exercise whose solution coincides with that of Exercise 4.22.

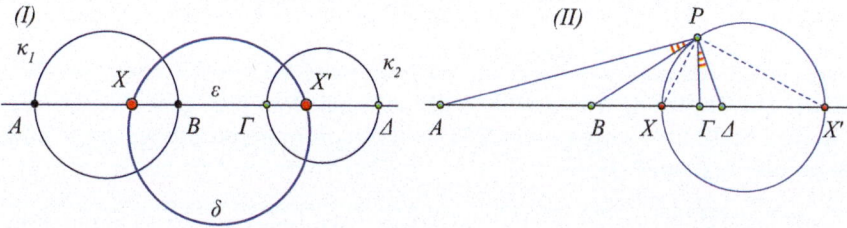

Fig. 4.50: Common harmonics of (A,B), (Γ,Δ) Locus of equal angles

Theorem 4.11. *Given four points $\{A, B, \Gamma, \Delta\}$ on line ε, there exist exactly two points X, X', which are simultaneously harmonic conjugate relative to the pairs (A,B) and (Γ, Δ), if and only if the intervals AB, $\Gamma\Delta$ have no common points or one contains the other in its interior (See Figure 4.50-I).*

Proof. According to Exercise 1.77, if such X, X' exist, then (A,B) and (Γ, Δ) will be pairs of harmonic conjugates relative to X, X'. Therefore the circles with diameter AB and $\Gamma\Delta$ respectively, will be Apollonian circles of the interval XX' (Proposition 4.5). Consequently these circles will belong to the non intersecting pencil $\mathcal{D}$ of the Apollonian circles of XX' and therefore the intervals AB, $\Gamma\Delta$ will either be external to each other or one of them will be contained in the other. Conversely, if this condition holds, then the two circles with diameter AB and $\Gamma\Delta$, respectively, will be non-intersecting and the pencil defined by them will be non intersecting. The limit points $\{X, X'\}$ of the pencil will be then the required by the theorem.

For intervals $\{AB, \Gamma\Delta\}$ of the line ε, which satisfy the requirement of the preceding theorem, we say that they do **not separate** each other. The points $\{X, X'\}$ of the theorem are called **common harmonics** of the pairs $\{(A, B), (\Gamma, \Delta)\}$. Under these terms, the theorem says that *the pairs (A, B) and (Γ, Δ) have common harmonics, exactly when the intervals $\{AB, \Gamma\Delta\}$ do not separate each other*. The proof of the theorem shows, that the limit points $\{X, X'\}$ of a non intersecting pencil are the common harmonics of pairs $\{(A, B), (\Gamma, \Delta)\}$ of diametrically opposite points of any two circles of the penicl. Figure 4.50-I shows the method of finding the common harmonics. We draw a circle δ intersecting orthogonally the circles κ_1, κ_2 with diameter, respectively, AB, $\Gamma\Delta$. The intersection points of δ with ε are the requested common harmonics of the pairs (A, B), (Γ, Δ). Next corollary formalizes these observations.

Corollary 4.16. *The pairs of collinear points $\{(A, B), (\Gamma, \Delta)\}$ have common harmonics $\{X, X'\}$, if and only if they are diametrically opposite points of two circles of a non-intersecting pencil. Points $\{X, X'\}$ coincide then with the limit points of the pencil.*

Exercise 4.70. For two mutually non-intersecting intervals $\{AB, \Gamma\Delta\}$ of a line ε, show that the geometric locus of the points from which these are seen under equal angles is the circle with diameter which has endpoints the common harmonics of these two intervals.

Hint: The angles $\widehat{AP\Delta}$ and $\widehat{BP\Gamma}$ will have common bisectors PX, PX' (See Figure 4.50-II), therefore points X, X' will be common harmonics of the pairs of points (A, B) and (Γ, Δ).

Exercise 4.71. Show that the radical axis of two circles does not change if the squares of their radii are modified by the same quantity.

Exercise 4.72. Show that the geometric locus of the points, whose difference of powers from two given circles is constant, is a line parallel to the radical axis of the two circles.

Exercise 4.73. Show that the line of the altitude AY of the triangle $AB\Gamma$ is the radical axis of the two circles with diameters the medians from the other vertices $\{B, \Gamma\}$.

Exercise 4.74. Show that the circles $\{\kappa, \lambda\}$ with diameters correspondingly the inner bisectors $\{B\Delta, \Gamma E\}$ of the triangle $AB\Gamma$, intersect the opposite sides $\{A\Gamma, AB\}$ at second points $\{\Delta', E'\}$ and the lines $\{B\Delta', \Gamma E'\}$ pass through the orthocenter of the triangle. Also the radical axis of $\{\kappa, \lambda\}$ passes through the orthocenter of the triangle (generalizes to exercise 4.116).

4.6. ORTHOGONAL CIRCLES AND PENCILS

Exercise 4.75. Show that the area of the triangle is $E = R\tau'$, where R the circumradius and τ' the half-perimeter of the orthic triangle.

Hint: Use exercise 3.123 and corollary 3.20.

Exercise 4.76. The circles $\{\kappa, \lambda\}$ are orthogonal and have respectively as chords the sides $\{AB, A\Gamma\}$ of the triangle $AB\Gamma$. Show that their second intersection point Δ is contained in a circle $\mu_A(O_A)$, which passes through $\{B, \Gamma\}$ and intersects sides $\{AB, A\Gamma\}$ for a second time at points $\{B', \Gamma'\}$. Show that the lines $\{B\Gamma', \Gamma B'\}$ are respectively orthogonal to $\{AB, A\Gamma\}$, hence $B'\Gamma'$ is a diameter of μ_A and AO_A is the symmedian of the triangle from A. Show that O_A and the analogous points $\{O_B, O_\Gamma\}$, relative to the other vertices, are the vertices of the tangential triangle of $AB\Gamma$.

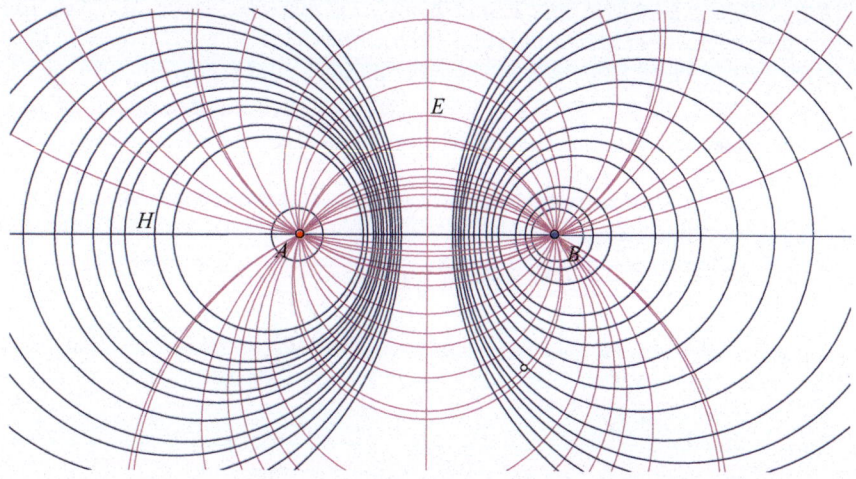

Fig. 4.51: Electromagnetic field

Remark 4.10. Circle pencils have applications in many fields. Figure 4.51 shows the *electromagnetic field*, which is created from two conductor wires $\{A, B\}$ of electric current. More precisely, the figure shows a plane section of the field, where the paper page coincides with the plane which is orthogonal to the wires. The field consists of two orthogonal circle pencils. The intersecting pencil E, with basic points $\{A, B\}$, represents the electric field. An electron, which is free to move under the influence of the electric field, will accept a tangential force relative to one circle of the pencil E. The orthogonal to E, non intersecting pencil H, represents the magnetic field. A magnetic compass, left under the influence of the field, will reorient and will take a tangential position relative to one circle of the pencil H.

4.7 Similarity centers of two circles

> He saw with his own eyes the moon was round,
> Was also certain that the earth was square,
> Because he had journey'd fifty miles and found
> No sign that it was circular any where;
>
> Lord Byron, Don Juan, canto V

Two non congruent and non concentric circles, besides the radical axis and the common tangents (when these exist), possess some additional interesting features, among which are the two *similarity centers*, guaranteed by the following theorem.

Theorem 4.12. *Given two non concentric and non congruent circles $\kappa(A, \alpha)$ and $\lambda(B, \beta)$, there exist two points Γ and Δ with the following property. Every line passing through one of these points and intersecting the two circles, defines on them points towards which their radii are pairwise parallel. And conversely, each pair of parallel radii AX and BY defines a line XY passing through one of the points $\{\Gamma, \Delta\}$.*

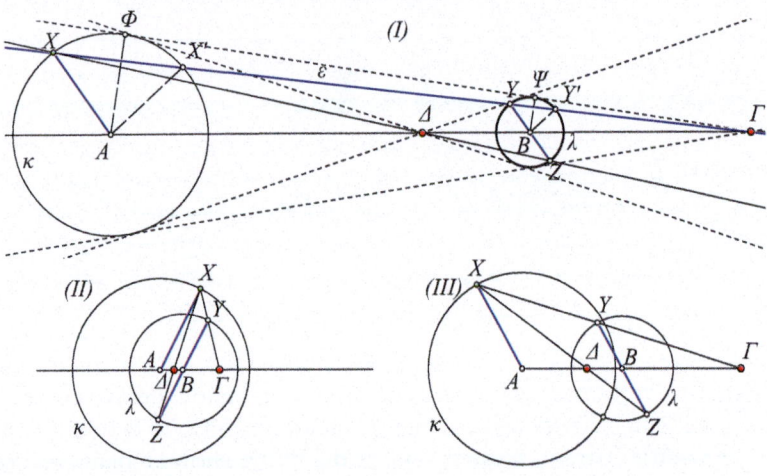

Fig. 4.52: Similarity centers (Γ, Δ) of two circles κ, λ

Proof. Consider the points Γ and Δ, external and internal respectively of the line segment AB, which divide it into segments of ratio equal to the ratio of the radii α/β (Figure 4.52 shows $\{\Gamma, \Delta\}$ for corresponding different relative positions of the circles):

$$\frac{|\Gamma A|}{|\Gamma B|} = \frac{|\Delta A|}{|\Delta B|} = \frac{\alpha}{\beta}.$$

4.7. SIMILARITY CENTERS OF TWO CIRCLES

Let also ε be an arbitrary line passing through Γ and intersecting the circle κ at X. From point B draw the parallel towards AX intersecting the line ε at the point Y. Triangles ΓAX and ΓBY are similar, because they have by construction their angles equal. Therefore

$$\frac{|AX|}{|BY|} = \frac{\alpha}{|BY|} = \frac{|\Gamma A|}{|\Gamma B|} = \frac{\alpha}{\beta} \Rightarrow |BY| = \beta,$$

which means that Y is a point of the circle. Similarly it is proved that line ΔX also intersects the circle λ at points, one of which defines a radius BZ parallel to AX. Conversely, if the radii AX and BY (resp. BZ) are parallel and equally (resp. inversely) oriented, then the triangles $XA\Gamma$ and $YB\Gamma$ (resp. $XA\Delta$ and $ZB\Delta$) are similar and consequently the points X, Y and Γ (resp. X, Z and Δ) are collinear.

Corollary 4.17. *The common tangent (if it exists) of two not congruent circles passes through one of the points $\{\Gamma, \Delta\}$ of the preceding theorem.*

Proof. The radii $A\Phi$ and $B\Psi$ to the contact points of a common tangent are parallel (See Figure 4.52-I), as orthogonals to the same line, therefore, according to the preceding theorem, line $\Phi\Psi$ will pass through one of the two points $\{\Gamma, \Delta\}$.

Points Γ and Δ, defined in the preceding theorem, are called **similarity centers** of the two circles κ and λ. They are harmonic conjugate (§ 1.17) relative to the centers A and B of the two circles. Point Γ (external to AB) is called **external center** of similarity and point Δ (internal to AB) is called **internal center** of similarity of the two circles (See Figure 4.52-I). Every line, which passes through one center of similarity and intersects each one of the two circles at two points, defines two pairs of points (X,Y), (X',Y') towards which the corresponding radii are parallel and are called **homologous points**, while the pairs of points (X,Y'), (X',Y) are called **antihomologous points** of the two circles ([102, p.19]).

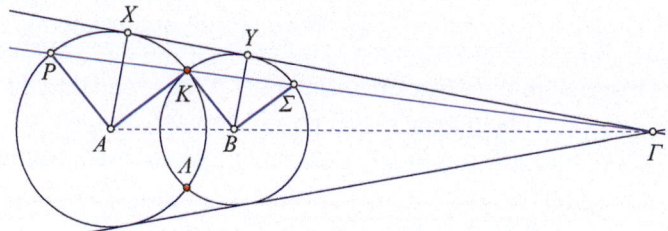

Fig. 4.53: Coincident antihomologous points

In the rest of this section we suppose, as in the theorem, that the circles $\kappa(A,\alpha)$ and $\lambda(B,\beta)$ are not concentric and not congruent. In the case where the line ε is a common tangent to the two circles, points X and X', as well as Y and Y' coincide and we have only one pair of points (X, Y) which are simultaneously homologous and antihomologous. Finally, in the case of the two circles which intersect at points K and Λ the lines ΓK and $\Gamma \Lambda$ intersect the two circles at three points, for example, the line ΓK at P, K and Σ and we have the pairs of homologous points (P, K), (K, Σ) and the pairs of antihomologous (K, K) and (P, Σ), in the first of which the two points coincide (See Figure 4.53).

The homologous points are characterized by the fact that their radii are parallel. Also a line which passes through two homologous points intersects the circles at two more points respectively, which are also homologous. Something similar also happens with antihomologous points, where, however, the intersecting lines are replaced by circles.

Exercise 4.77. Given two non congruent and non concentric circles $\{\kappa, \lambda\}$, show that if a line ε passes through one similarity center and intersects one of the circles, then necessarily intersects also the other circle.

Proposition 4.10. *For every pair of antihomologous points X, Y, of two circles $\kappa(A, \alpha)$ and $\lambda(B, \beta)$ and every circle μ passing through them, the second intersection points X', Y' of μ with κ and λ are also antihomologous points and the lines XY and $X'Y'$ pass through the same center of similarity.*

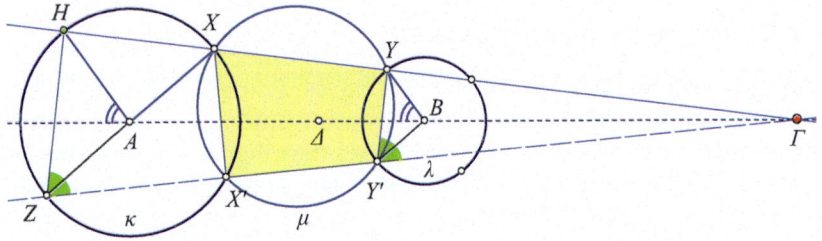

Fig. 4.54: Circle through antihomologous re-intersects in antihomologous

Proof. Let H and Z be the second intersection points respectively of XY and $X'Y'$ with circle κ (See Figure 4.54). Because the quadrilateral $XYY'X'$ is cyclic, its external angle at Y' is equal to the internal and opposite at X, which in turn, as external to the cyclic quadrilateral $X'XHZ$ is equal to the internal and opposite at Z. This implies that lines ZH and $Y'Y$ are parallel. Because by assumption lines AH and BY are parallel, it follows that the angles $\widehat{AHZ}$ and $\widehat{BYY'}$ are equal. Because of the isosceli triangles AHZ and BYY', it also follows that angles $\widehat{AZH}$ and $\widehat{BY'Y}$ are equal and they have corresponding sides parallel. Hence AZ is parallel to BY'. This means that points X' and Y' are antihomologous. The second claim follows from the fact that, the sides of the isosceli triangles AHZ and BYY' are parallel, which implies that the lines, which join their corresponding vertices, will pass through the same point (homothety center of two triangles).

4.7. SIMILARITY CENTERS OF TWO CIRCLES

Proposition 4.11. *From every pair of antihomologous points X, Y of two circles, κ(A, α) and λ(B, β), passes exactly one circle μ tangent at these points to κ and λ respectively (See Figure 4.55).*

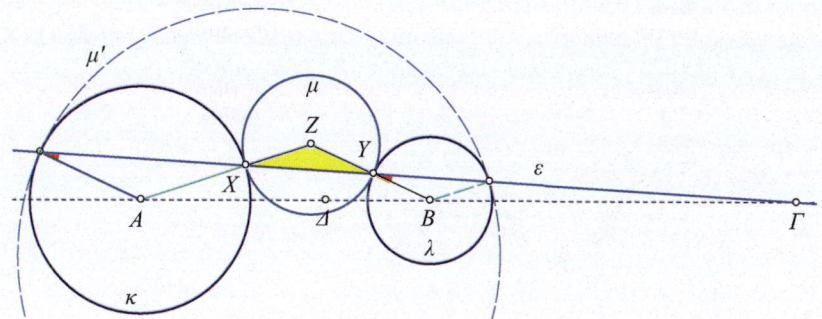

Fig. 4.55: Circle $\mu(\mu')$ tangent at two antihomologous points

Proof. Because of the hypothesis, the extensions of AX and BY will form a triangle ZXY with equal base angles, therefore an isosceles, which defines a circle μ tangent to κ, λ at X and Y respectively, as required. The uniqueness follows trivially.

Proposition 4.12. *Two points X, Y of two circles κ(A, α) and λ(B, β) are antihomologous, if and only if they are the contact points with a circle μ, which is tangent to both circles (See Figure 4.55).*

Proof. The first claim coincides with the preceding proposition. The converse coincides with Exercise 2.18.

Proposition 4.13. *Given the circles κ and λ, two different pairs (X,Y) and (X',Y') of antihomologous points, whose corresponding lines XY and X'Y' pass through the same center of similarity, are vertices of a cyclic quadrilateral (See Figure 4.56).*

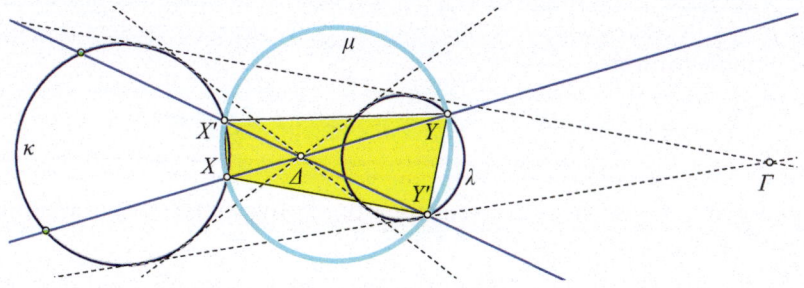

Fig. 4.56: Inscriptible quadrilateral of two pairs of antihomologous points

Proof. Consider the circle (XYX'). This circle, according to Proposition 4.10, since it passes through the antihomologous points X and Y, will intersect a second time the circles κ, λ at antihomologous points, one of which is X'. Therefore the other one will necessarily be Y'.

Proposition 4.14. *Given the circles $\kappa(A,\alpha)$ and $\lambda(B,\beta)$, for every line ε passing through a center of similarity Z and defining two antihomologous points $\{X,Y\}$, the product $|ZX||ZY|$ is constant and equal to $||ZA||ZB| - \alpha\beta|$.*

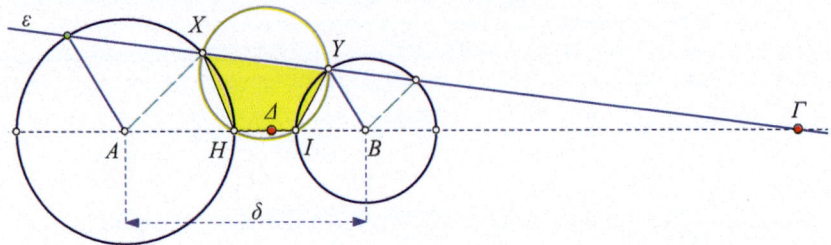

Fig. 4.57: The products $|\Gamma X||\Gamma Y|$

Proof. Let us see the case where Z coincides with the external center of similarity Γ (See Figure 4.57). The case of the internal center Δ is similar. Let H and I be two antihomologous points, which are also intersection points of κ and λ respectively with their center-line AB. According to Proposition 4.13, the quadrilateral $XYIH$ is inscriptible therefore

$$|\Gamma X||\Gamma Y| = |\Gamma H||\Gamma I| = (|\Gamma A| - \alpha)(|\Gamma B| + \beta) = |\Gamma A||\Gamma B| - \alpha\beta.$$

The last equality holds because, expanding the second to last product, we get $|\Gamma A|\beta - |\Gamma B|\alpha = 0$, since for the center of similarity holds the equivalent equality $\frac{|\Gamma A|}{|\Gamma B|} = \frac{\alpha}{\beta}$.

Exercise 4.78. Show that the products $|\Gamma X||\Gamma Y|$ and $|\Delta X||\Delta Y|$ of the preceding proposition (See Figure 4.57), are expressed as functions of the radii $\alpha > \beta$ of the circles κ, λ and the distance between their centers $\delta = |AB|$, through the formulas

$$|\Gamma X||\Gamma Y| = \delta^2 \frac{\alpha\beta}{(\alpha-\beta)^2} - \alpha\beta, \quad |\Delta X||\Delta Y| = \delta^2 \frac{\alpha\beta}{(\alpha+\beta)^2} - \alpha\beta.$$

Exercise 4.79. Using the notation of the preceding exercise, and assuming that $\alpha > \beta$, show that (See Figure 4.57):

$$|B\Gamma| = |AB|\frac{\beta}{\alpha-\beta}, \quad |B\Delta| = |AB|\frac{\beta}{\alpha+\beta}, \quad |\Gamma\Delta| = |AB|\frac{2\alpha\beta}{\alpha^2-\beta^2}.$$

4.7. SIMILARITY CENTERS OF TWO CIRCLES

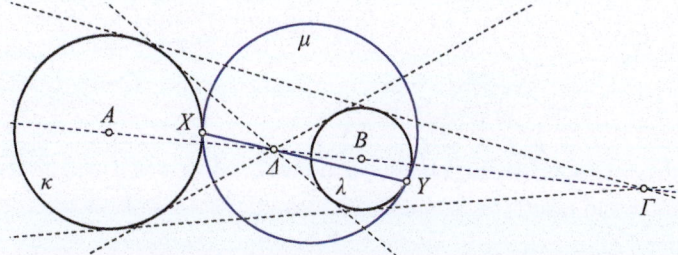

Fig. 4.58: $|\Delta X||\Delta Y|$ independent from tangent circle μ

Corollary 4.18. *For every circle μ, tangent to two others $\kappa(A, \alpha)$ and $\lambda(B, \beta)$, the chord XY of the contact points between the circles passes through one center of similarity Z of κ and λ. The kind of Z is determined from the way μ is tangent to κ and λ (internally-externally). The product $|ZX||ZY|$ is independent of μ and depends only on the size and the relative position of κ and λ (See Figure 4.58).*

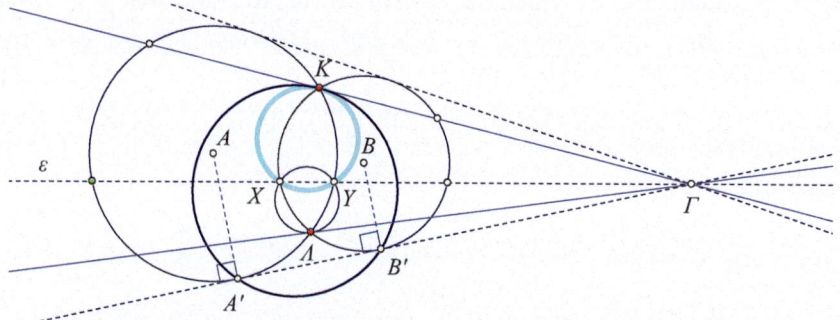

Fig. 4.59: Antihomologous points in intersecting circles

Corollary 4.19. *Let the circles $\kappa(A, \alpha)$ and $\lambda(B, \beta)$ intersect at points K and Λ and let X and Y be antihomologous points on line ε passing through their external center of similarity Γ. Then the circle (KXY) is tangent to ΓK and circle (ΛXY) is tangent to $\Gamma \Lambda$. In particular, the circle ($KA'B'$), where A', B' are the contact points of a common tangent through Γ, is tangent to line ΓK (See Figure 4.59).*

Proof. In the case of intersecting circles κ and λ, the fixed product $|\Gamma X||\Gamma Y|$ is equal to $|\Gamma K|^2$ and the conclusion follows from Proposition 4.4.

Exercise 4.80. Show that a line, which intersects two non concentric and unequal circles κ and λ, forms with the radii at the intersection points equal angles, if and only if it passes through one center of similarity of the two circles. How does this property change for congruent circles?

For two non-congruent circles $\kappa(A, \alpha)$ and $\lambda(B, \beta)$, we call **circle of similitude** the circle μ with diameter $\Gamma \Delta$, where Γ and Δ are the centers of similarity of the two circles (See Figure 4.60-I).

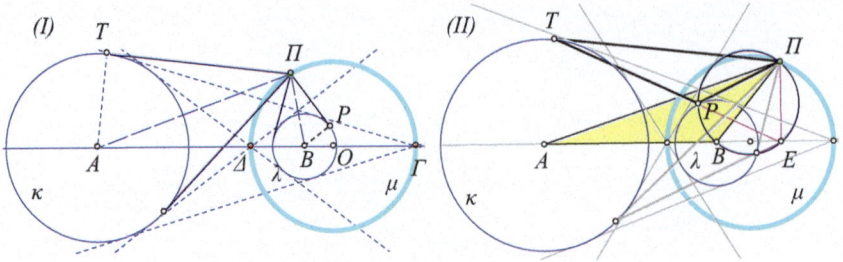

Fig. 4.60: Circle of similitude μ of $\kappa(A, \alpha)$ and $\lambda(B, \beta)$

Exercise 4.81. Show that for every point X of the circle of similitude μ the ratio of the powers relative to $\kappa(A, \alpha)$ and $\lambda(B, \beta)$ is equal to the ratio of the squares of their radii. Conclude that the circle of similitude μ of $\{\kappa, \lambda\}$ belongs to the pencil generated by the two circles.

Hint: The points Π, for which the ratio of powers to κ and λ is $\frac{p_\kappa(\Pi)}{p_\lambda(\Pi)} = \frac{\alpha^2}{\beta^2}$, are points of one circle of the pencil of κ and λ (Theorem 4.8). Points Γ and Δ also possess this property.

Exercise 4.82. Show that for two mutually external circles κ, λ, the geometric locus of the points Π which see them under equal angles is their circle of similitude μ.

Hint: The triangles ΠTA and ΠPB are similar, therefore $\frac{|\Pi A|}{|\Pi B|} = \frac{\alpha}{\beta}$ (See Figure 4.60).

Exercise 4.83. Given are two circles $\{\kappa(A), \lambda(B)\}$ lying external to each other, and a point Π outside both. Show that the following properties are equivalent (See Figure 4.60-II):

1. Two tangents $\{\Pi T, \Pi P\}$, to the circles, respectively $\{\kappa, \lambda\}$, separating the centers (i.e. one, at most of the centers, lying in $\widehat{T\Pi P}$), define a triangle $T\Pi P$ similar to $A\Pi B$.
2. The projection E of Π onto the line of centers AB is contained in the line TP.
3. The point Π is contained in the circle of similitude of the two circles $\{\kappa, \lambda\}$.

Hint: $(1 \Rightarrow 2)$ If the triangles $\{T\Pi P, A\Pi B\}$ are similar and $E = (AB, TP)$, then the quadrangle $PBE\Pi$ is cyclic and the angle $\widehat{BP\Pi}$ is a right one.

$(2 \Rightarrow 3)$ If the points $\{T, P, E\}$ are collinear, then the quadrangle $PBE\Pi$ is cyclic and the right angled triangles $\{\Pi TA, \Pi PB\}$ are similar.

$(3 \Rightarrow 1)$ If point Π is on the circle of similitude, then the right-angled triangles $\{\Pi TA, \Pi PB\}$ are similar. It follows that the triangles have the angles at Π equal and the adjacent sides, respectively, proportional.

4.7. SIMILARITY CENTERS OF TWO CIRCLES

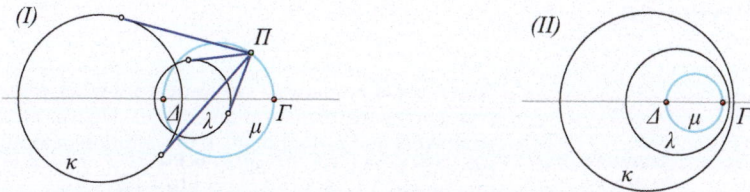

Fig. 4.61: Circle of similitude μ of two circles $\{\kappa, \lambda\}$

Exercise 4.84. How is modified the conclusion of the exercise 4.82 for intersecting circles (See Figure 4.61-I)? The same question in the case one of the circles is entirely inside the other (See Figure 4.61-II).

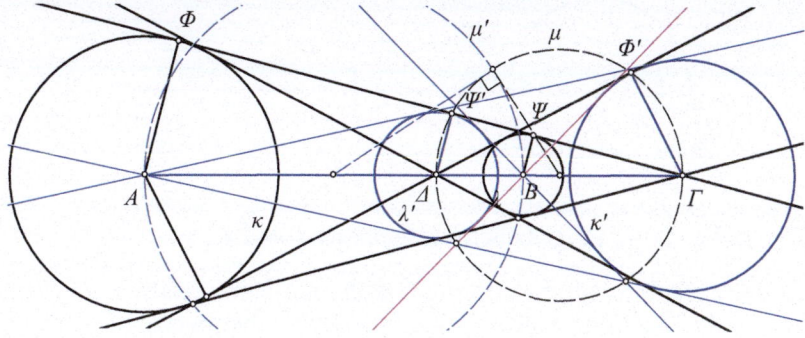

Fig. 4.62: Conjugate pairs of circles (κ, λ) and (κ', λ')

Remark 4.11. The pairs of non congruent and non concentric circles (κ, λ) appear ... in pairs. In the shape consisting of two circles κ, λ, mutually external (See Figure 4.62), this is seen by drawing the common tangents and defining the circles κ', λ' with radii, respectively, the perpendiculars $\Gamma\Phi'$, $\Delta\Psi'$ to the tangents from the centers of similarity Γ, Δ of κ, λ. We easily see that the centers of similarity of κ', λ' are the centers of circles κ, λ and the circle of similitude μ' of κ', λ' is orthogonal to the circle of similitude μ of κ, λ. The pairs (κ, λ) and (κ', λ'), then, determine each other the same way. In the cases where there are no common tangents of (κ, λ), we can again determine the circles (κ', λ') through their radii, which are calculated in the next exercise.

Exercise 4.85. Assuming that the circles κ, λ have respective radii $\alpha > \beta$ (See Figure 4.62), show that the circles κ', λ' have respective radii α', β' given by the formulae:

$$\alpha' = \frac{2\alpha\beta}{\alpha - \beta}, \qquad \beta' = \frac{2\alpha\beta}{\alpha + \beta}.$$

4.8 Inversion

> 'You must always invert', as Jacobi said when asked the secret of his mathematical discoveries. He was recalling what Abel and he had done. If the solution of a problem becomes hopelessly involved, try turning the problem backwards, put the quaesita for the data and vice versa.
>
> E.T. Bell, Men of Mathematics, v. II p. 355

Given a circle $\kappa(O,\rho)$, the **inversion** relative to κ, is a correspondence between points in the interior and exterior of the circle. To every point X different from the center O, we correspond a point Y on the half line OX, such that $|OX||OY| = \rho^2$ (See Figure 4.63). The definition establishes a kind of

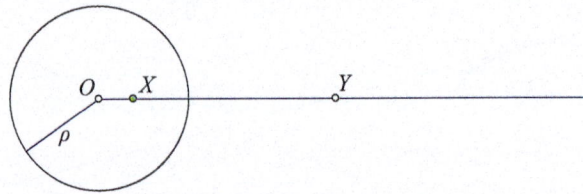

Fig. 4.63: Inverse points X and Y

symmetric relation, somewhat similar to the symmetry relative to a line (reflection). X is the inverse of Y, if and only if Y is the inverse of X. We then often say that such two points are **inverse points** relative to the circle κ. The points of the circle κ are characterized by being coincident to their inverse. We call these points the **fixed points** of the inversion. The circle κ is called **circle of inversion**, its center is called **center of inversion** and ρ^2 is called **power of inversion**.

Construction 4.3 *Given a circle $\kappa(O,\rho)$, construct the inverse Y of a given point X relative to κ.*

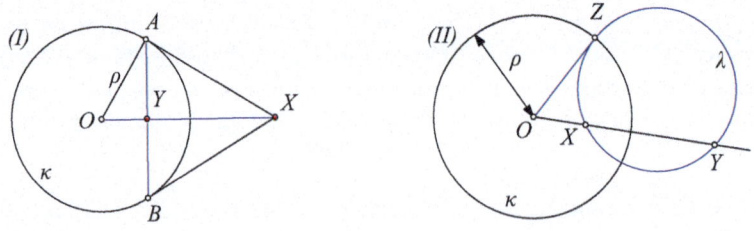

Fig. 4.64: Construction of inverses X, Y Orthogonal λ on κ

Construction: If the point X is external to the circle κ (See Figure 4.64-I), then the intersection Y of the chord AB of contacts of the tangents from X and the

4.8. INVERSION

line OX is the requested point, since from the right triangle OAX, we'll have (Proposition 3.3)
$$\rho^2 = |OX||OY|$$

The same relation shows that X is also the inverse of Y. Therefore, for internal points of the circle, we perform the reverse procedure: we raise the orthogonal AB on OY and draw the tangents of κ at its intersection points A, B with the circle. X is the intersection point of these tangents. The inverse of an X on κ is X itself.

Theorem 4.13. *Every circle λ, which passes through two inverse points $\{X,Y\}$ relative to the circle κ, is orthogonal to κ.*

Proof. Indeed, if $\{X,Y\}$ are inverse points relative to κ and Z is an intersection point of the circle λ which passes through X, Y (See Figure 4.64-II), then, according to the definition of inversion, we'll have $|OZ|^2 = \rho^2 = |OX||OY|$, which shows that OZ is tangent to the circle λ (Proposition 4.4).

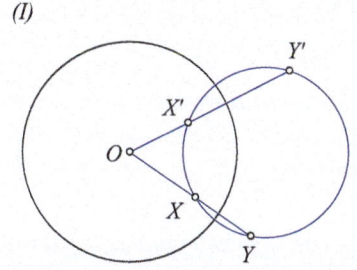
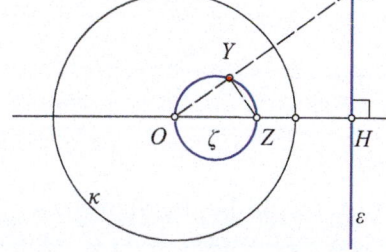

Fig. 4.65: Pairs of inverse points   Inverse of circle through O

Proposition 4.15. *Two different pairs of inverse points (X,Y) and (X',Y') relative to the circle $\kappa(O,\rho)$ define four concyclic points.*

Proof. Obviously, since the relation which connects them $\rho^2 = |OX||OY| = |OX'||OY'|$ (See Figure 4.65-I), means that $\{X,Y,X',Y'\}$ are concyclic (Proposition 4.3).

Proposition 4.16. *If the point X describes a line ε, then its inverse Y relative to circle $\kappa(O,\rho)$ describes a circle ζ passing through the center O of the circle κ. Conversely if the point Y describes a circle ζ passing through O, then its inverse X relative to κ, describes a line ε.*

Proof. Let H be the projection of O on line ε and Z the inverse of H relative to κ (See Figure 4.65-II). Pairs (X,Y) and (H,Z) are concyclic points (Proposition 4.15). Consequently, in the quadrilateral $ZHXY$ the angle at Y will be right, as supplementary to the angle at H, which is right. Consequently Y will be on the circle ζ with diameter OZ.

The second claim follows using the same argument. If Y lies on circle ζ with diameter OZ, consider the line ε, which is orthogonal to OZ at H, which is the inverse of Z relative to κ. For an arbitrary point Y of ζ, let X be the intersection of ε with OY. The quadrilateral $XHZY$ is by construction cyclic, because of the right angles at opposite vertices H and Y. Consequently, we'll have $|OY||OX| = |OZ||OH| = \rho^2$, therefore point X will be inverse of Y.

Proposition 4.17. *If the point X describes the circle λ not passing through the center of inversion O, then its inverse Y relative to circle $\kappa(O,\rho)$ describes also a circle μ not passing through O.*

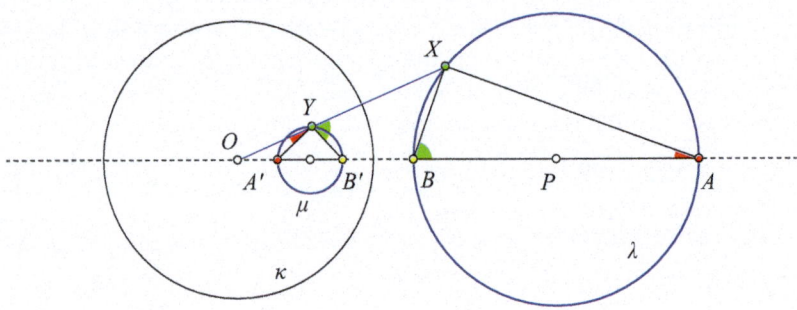

Fig. 4.66: Inverse of circle

Proof. Let AB be the diameter of circle λ on OP, where P is the center of λ (See Figure 4.66). If A', B' are respectively the inverses relative to κ of A and B, we show also that the inverse Y of the arbitrary point X of circle λ is contained in the circle μ with diameter $A'B'$. Indeed, the quadrilaterals $BB'YX$ and $AA'YX$ are cyclic (Proposition 4.15), therefore equal angles are formed:
$$\widehat{ABX} = \widehat{B'YX}, \quad \widehat{BAX} = \widehat{A'YO}.$$
This implies that the angle $\widehat{A'YB'}$ is right, therefore point Y lies on the circle with diameter $A'B'$.

For the conclusion of Proposition 4.16 we use often the formulation:

The inverse of a line relative to a circle is a circle through the center of inversion and the inverse of circle through the center of inversion is a line.

Similarly for the conclusion of Proposition 4.17 we use the formulation:

The inverse of circle not passing through the center of inversion is a circle.

Note the reciprocity of inverse circles. In figure 4.66 circle μ is the inverse of λ, but also circle λ is inverse of μ. Next proposition underlines the case in which circle λ and μ coincide, in other words the circle coincides with its inverse.

4.8. INVERSION

Corollary 4.20. *The inverses Y of points X of a circle λ are again points of λ, if and only if the circle λ is orthogonal to the circle of inversion $\kappa(O, \rho)$.*

Proof. If the same circle λ contains point X and its inverse Y, then it is orthogonal to the circle of inversion (Theorem 4.13). If again circle λ is orthogonal to the circle of inversion κ, then for every line ε through O which intersects λ at X and Y, we'll have the product $|OX||OY|$, equal to $|OZ|^2 = \rho^2$, where OZ is the tangent to λ from O.

For the conclusion of Corollary 4.20 we often use the formulation:

The circles which are orthogonal to the circle of inversion are precisely these, which remain invariant by the inversion.

Proposition 4.18. *Let Y be the inverse of the point X relative to the circle $\kappa(O, \rho)$, with X lying also on circle μ. If the circle or line λ is tangent to the circle μ at X, then the inverse λ' of λ and the inverse μ' of μ are tangent at Y.*

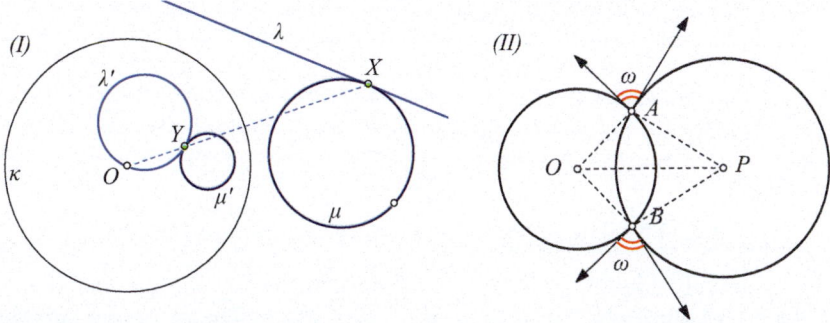

Fig. 4.67: Tangent inverted into tangent   Angle ω of two circles

Proof. λ and μ are tangent at X, precisely when X is their unique common point (See Figure 4.67-I). When this happens, then λ' and μ' will also have Y as a common point. If they had another common point Z, then the inverses of λ' and μ' which are λ and μ respectively would have also, besides X, the inverse Ω of Z in common, which contradicts the hypothesis. Therefore λ' and μ' have point Y as their unique common point, therefore they are tangent at this point.

We define the **angle between two circles** $\kappa(O)$, $\lambda(P)$, which intersect at two points A, B, to be the supplementary ω of the angle $\widehat{OAP} = \widehat{OBP}$, which is the angle of the tangents at their intersection point (See Figure 4.67-II). Next theorem shows that the inversion preserves these angles. This property is often mentioned with the phrase: *The inversion is a conformal transformation.*

Theorem 4.14. *Let Y be the inverse of point X relative to circle $\kappa(O,\rho)$. If circles λ and μ pass through X and form there the angle ω, then their inverses λ', μ' pass through point Y and form there the same angle ω (See Figure 4.68).*

Proof. Consider the circle μ'', which is tangent to μ at X and orthogonal to κ (Exercise 4.64). Consider also the circle λ'', which is tangent to λ at X and orthogonal to κ. Because the circles λ, μ are tangent to $\{\lambda'', \mu''\}$ respectively at X, the inverse circles $\{\lambda', \mu'\}$ will be tangent at Y of the inverses of λ'' and

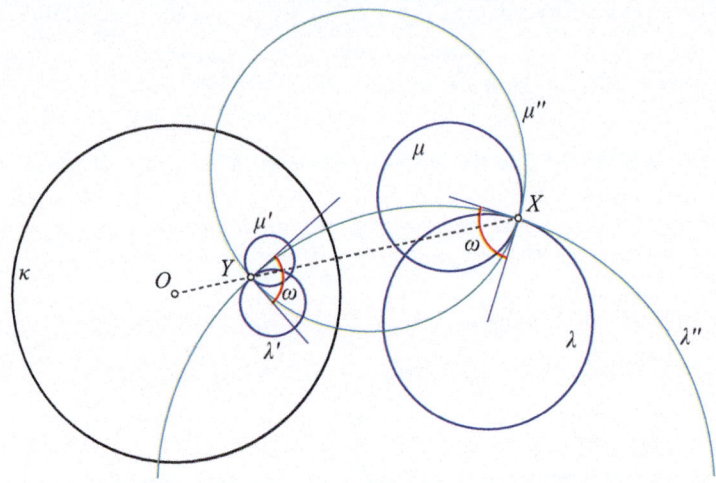

Fig. 4.68: The inversion is a conformal transformation

μ'' respectively. However, from the orthogonality to κ, the inverse of λ'' is λ'' itself and similarly the inverse of μ'' is μ'' itself. Therefore the tangents of λ' and μ' at Y will coincide respectively with the tangents of λ'' and μ'' at Y. Latter however, by symmetry, form at Y the same angle with the one formed by the tangents of λ'', μ'' at X, which by assumption make an angle of measure ω. Therefore the tangents of λ' and μ' at Y will form an angle of the same measure.

Theorem 4.15. *For a given pencil of circles $\mathcal{D}$ and a circle $\kappa(O,\rho)$, the inverses of circles λ of the pencil relative to κ form another pencil $\mathcal{D}'$.*

Proof. Suppose that the pencil $\mathcal{D}$ consists of all the circles which are orthogonal to two fixed circles μ and ν (Corollary 4.15). Let also μ' and ν' be the inverses of circles μ and ν relative to κ. By Theorem 4.14, every circle λ orthogonal to μ and ν will have an inverse λ' orthogonal to μ' and ν', therefore it will belong to the pencil $\mathcal{D}'$ of circles which are orthogonal to μ' and ν'. Similarly, every circle λ' orthogonal to μ' and ν' will have an inverse λ which is orthogonal to μ and ν, therefore it belongs to the pencil $\mathcal{D}$ and will have as inverse exactly circle λ'.

4.8. INVERSION 365

The pencil $\mathscr{D}'$ guaranteed by last theorem, is called **inverse pencil** of $\mathscr{D}$ relative to the circle κ (See Figure 4.69).

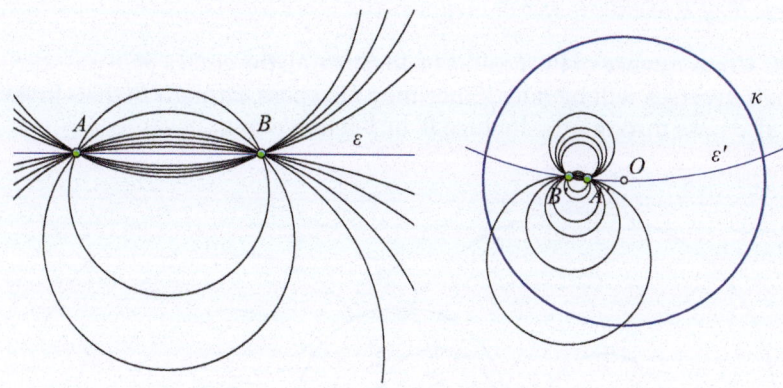

Fig. 4.69: Pencil $\mathscr{D}'$ of the inverses of circles of the pencil $\mathscr{D}$

In the preceding theorem, the inversion relative to κ, not only sends the pencil $\mathscr{D}$ onto a pencil $\mathscr{D}'$, but also preserves the quality of the pencil. In other words if the pencil $\mathscr{D}$ is intersecting, then pencil $\mathscr{D}'$ is intersecting too, if it is non intersecting then $\mathscr{D}'$ is also non intersecting etc. This however with a small concession. We must accept that, depending on the relative position of $\mathscr{D}$ and the circle of inversion κ, pencil $\mathscr{D}'$ may be non conventional (Remark 4.9). Next propositions show, for which positions of κ something like that may happen.

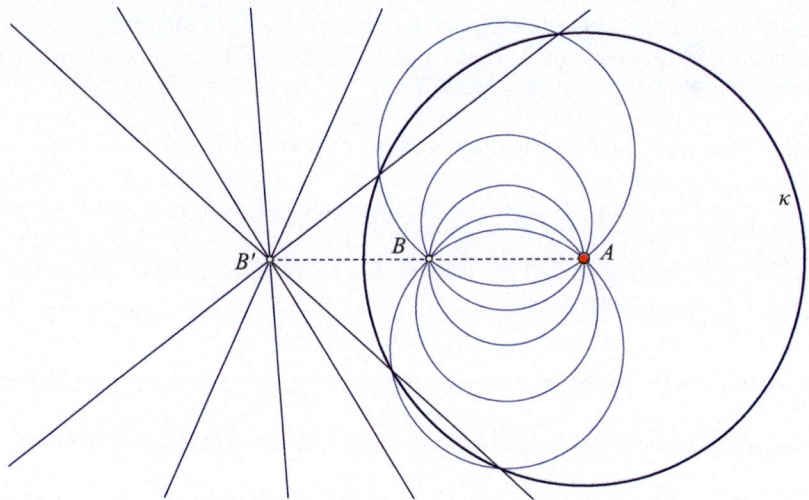

Fig. 4.70: Special inversion of intersecting pencil

Proposition 4.19. *Let $\mathscr{D}$ be an intersecting pencil and $\kappa(A, \rho)$ be a circle with center at one of the two basic points A and B of the pencil. Then the inverse pencil $\mathscr{D}'$ of $\mathscr{D}$ relative to κ is the non conventional pencil of the lines, which pass through the inverse B' of B relative to κ (See Figure 4.70).*

Proof. Consequence of Proposition 4.16. Since all the circles of $\mathscr{D}$ pass through A, their inverses will be lines. Since they also pass through B, their inverses will also pass through the inverse B' of B relative to κ.

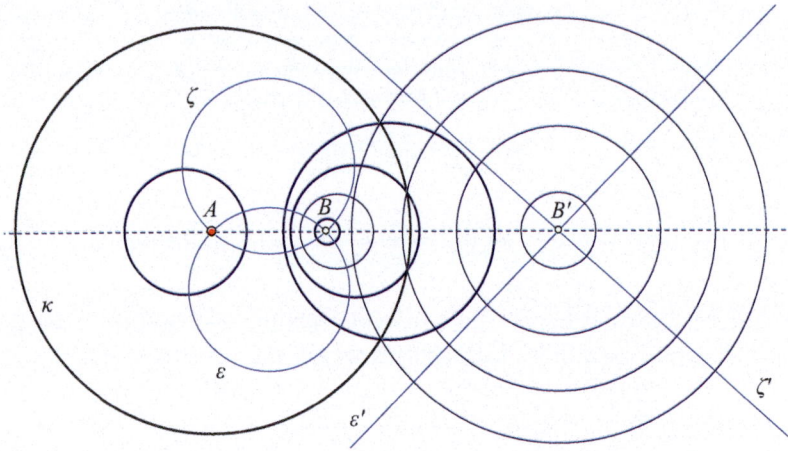

Fig. 4.71: Special inversion of non intersecting pencil

Proposition 4.20. *Let $\mathscr{D}$ be a non intersecting pencil and $\kappa(A, \rho)$ be a circle with center one of the limiting points A and B of the pencil. Then the inverse $\mathscr{D}'$ of $\mathscr{D}$, relative to κ, is the non conventional pencil of concentric circles with center the inverse B' of B relative to κ (See Figure 4.71).*

Proof. Consequence of Proposition 4.16, which we apply to two circles ε and ζ, passing through points A and B, which belong to the orthogonal pencil of $\mathscr{D}$. The inverses ε' and ζ' of these circles are lines through B'. Also because ε, ζ are orthogonal to every circle λ of $\mathscr{D}$, their inverses, which are lines ε' and ζ' will be orthogonal to the inverse λ' of λ (Theorem 4.14). Therefore λ' will be a circle with center B'.

Proposition 4.21. *Let $\mathscr{D}$ be a tangent pencil and $\kappa(A, \rho)$ be a circle with center the base point A of the pencil. Then the inverse $\mathscr{D}'$ of $\mathscr{D}$ relative to κ is the non conventional pencil of the lines, which are parallel to the radical axis ε of $\mathscr{D}$.*

Proof. Consequence of Proposition 4.16. Since all the circles of $\mathscr{D}$ pass through the point A, their inverses will be lines (See Figure 4.72-I). The inverse of the radical axis ε is itself. Every circle λ of the pencil has common with the

4.8. INVERSION

radical axis the unique point A coinciding with the center of the inversion. Consequently the inverse λ' of λ, which is a line, will have no common point with ε, therefore it will be parallel to it.

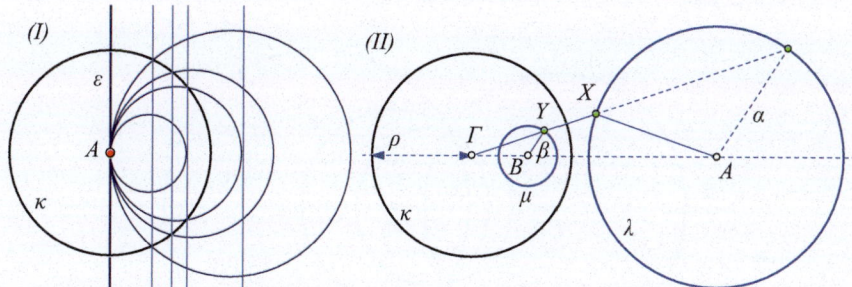

Fig. 4.72: Tangential pencil inverted   Inverting external circles

Theorem 4.16. *Given two non congruent circles $\lambda(A, \alpha)$ and $\mu(B, \beta)$, there exists an inversion relative to a circle κ, which interchanges the two given circles. In other words the inverse of λ relative to κ is μ and vice versa.*

Proof. Follows from Proposition 4.14. This proposition proves that, in the case where the circles λ and μ are mutually external or are externally tangent, the external similarity point Γ of the two circles has the desired property of the center of inversion (See Figure 4.72-II). Specifically, in this case the product of distances $|\Gamma X||\Gamma Y|$ of two antihomologous points from Γ is a fixed positive number equal to $|\Gamma A||\Gamma B| - \alpha\beta$. In this case, the desired circle has as center the external center of similarity Γ and radius

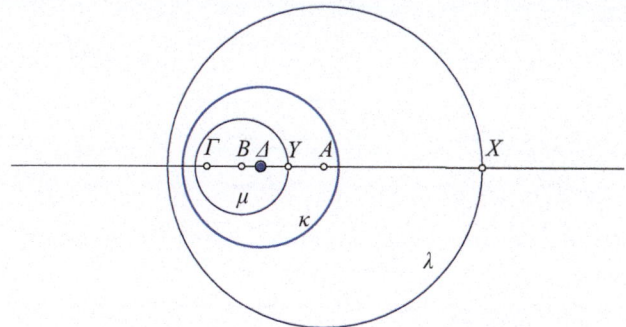

Fig. 4.73: Circle of inversion when λ contains μ

$\rho = \sqrt{|\Gamma A||\Gamma B| - \alpha\beta}$.

In the case where the circle $\mu(B, \beta)$ is internal to $\lambda(A, \alpha)$ or is internally tangent, the desired circle has as center the internal center of similarity Δ

and radius $\rho = \sqrt{\alpha\beta - |\Delta A||\Delta B|}$ (See Figure 4.73). Finally, in the case of two circles λ and μ intersecting at two points there exist two circles $\kappa(\Gamma, \rho)$ and $\kappa'(\Delta, \rho')$ with radii $\rho = \sqrt{|\Gamma A||\Gamma B| - \alpha\beta}$, $\rho' = \sqrt{\alpha\beta - |\Delta A||\Delta B|}$ and centers the centers of similarity Γ (external) and Δ (internal) of the circles λ and μ (See Figure 4.74).

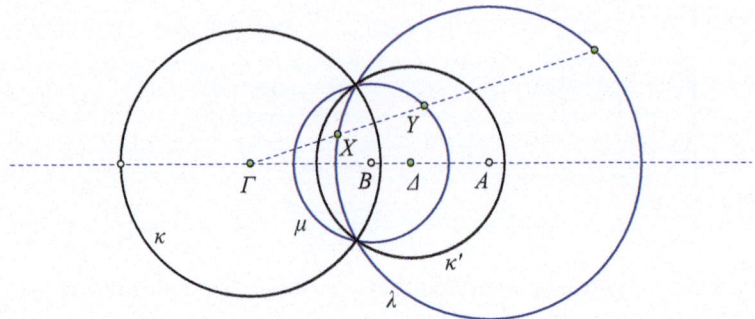

Fig. 4.74: Circles of inversion of two intersecting circles λ and μ

Exercise 4.86. Show that the circle of inversion of theorem 4.16, which interchanges the circles $\{\lambda, \mu\}$, belongs to the pencil of circles generated by $\{\lambda, \mu\}$.

Hint: Apply corollary 4.8, calculating the power of the center Γ of the inversion with respect to the two circles $\{\lambda, \mu\}$ for a secant line through Γ.

Exercise 4.87. Show that in the case of two intersecting circles λ and μ, the two circles of inversion which are ensured by Theorem 4.16 are orthogonal and pass through the intersection points of λ and μ.

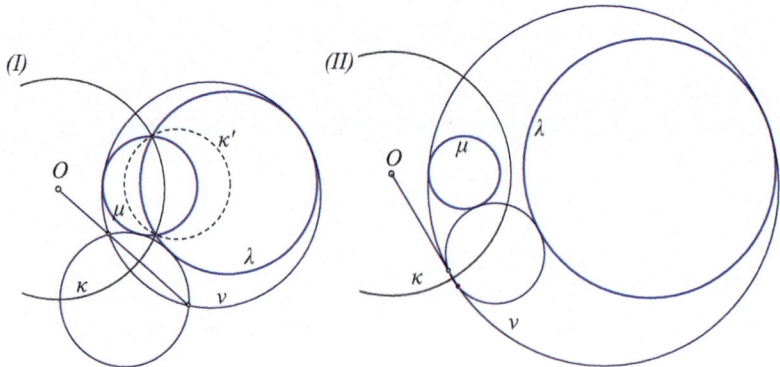

Fig. 4.75: Circle ν tangent to λ, μ is orthogonal to κ

4.8. INVERSION

Exercise 4.88. Let λ and μ be unequal circles and circle ν be tangent to them and containing either both in its exterior or both in its interior. Show that ν is orthogonal to one of the circles which interchanges circles λ, μ (there exist two, only in the case where λ, μ intersect at two points) (See Figure 4.75).

Hint: The center of circle κ, which defines the inversion interchanging the circles λ, μ, is one of the centers of similarity of λ, μ. Proposition 4.12 is applied.

Exercise 4.89. The circle $\kappa(A, a)$ does not pass through the center of the circle $\lambda(O, r)$. In the inversion relative to λ, the circle κ maps to $\kappa'(A', a')$. To find the center and the radius of κ' in dependence of $\{|OA|, a, r\}$.

Hint: The center A' of κ' is on OA at distance $|OA'| = (r^2|OA|)/(|OA|^2 - a^2)$. The radius of κ' is $a' = r^2 a/(|OA|^2 - a^2)$.

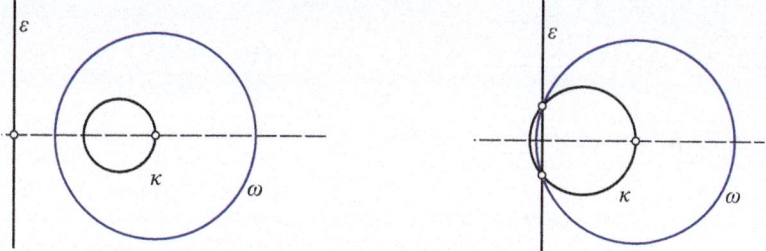

Fig. 4.76: Inversion interchanging circle and line

Exercise 4.90. Given a circle κ and a line ε, show that there exists a circle ω, defining an inversion which interchanges κ with ε (circle κ is the inverse of ε relative to ω and vice versa) (See Figure 4.76).

Exercise 4.91. Line ε does not pass through the center of the circle $\lambda(O, r)$. In the inversion relative to λ, the line ε maps to the circle $\kappa(A, a)$. To find the center and the radius of κ in dependence of r and the distance d of the point O from the line ε.

Hint: Let B be the projection of O on ε, so that $d = |OB|$. The center A of the circle κ is on the line OB at distance $|OA| = r^2/(2d)$. The quantity on the right is also the radius of κ.

Exercise 4.92. Given is a circle $\kappa(A, a)$ and a line ε not intersecting the circle κ. To investigate when there exists a circle $\lambda(O, r)$ defining an inversion, which maps κ and ε into two equal circles.

Exercise 4.93. Show that the length of the line segment $A'B'$, which has endpoints the inverses of the endpoints of the line segment AB, relative to the circle $\kappa(O, \rho)$ is equal to (See Figure 4.77-I)

$$|A'B'| = |AB| \cdot \frac{\rho^2}{|OA||OB|}.$$

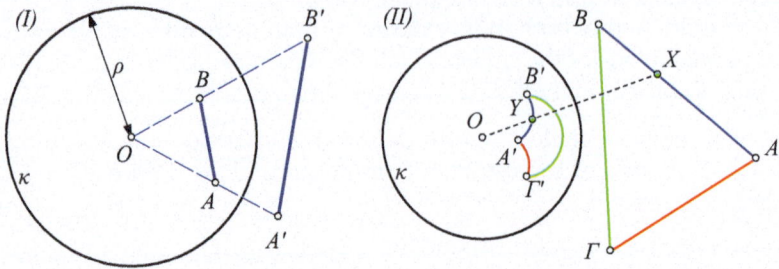

Fig. 4.77: Lengths of AB and $A'B'$ Inverse $A'B'\Gamma'$ of triangle $AB\Gamma$

Hint: The triangles OAB and $OB'A'$ are similar since

$$\rho^2 = |OA||OA'| = |OB||OB'| \Rightarrow \frac{|OA|}{|OB|} = \frac{|OB'|}{|OA'|}.$$

Consequently we have

$$\frac{|A'B'|}{|AB|} = \frac{|OB'|}{|OA|} = \frac{\frac{\rho^2}{|OB|}}{|OA|} = \frac{\rho^2}{|OA||OB|}.$$

Remark 4.12. On the occasion of the preceding exercise, let us note that it relates only the endpoints of the line segments $\{AB, A'B'\}$. The intermediate points X of AB are mapped under the inversion to the points Y of an arc of circle $\widehat{A'B'}$ (See Figure 4.77-II). When X traverses the perimeter of a triangle, its inverse Y will traverse a curvilinear triangle, whose sides are arcs of circles passing through O (Proposition 4.16).

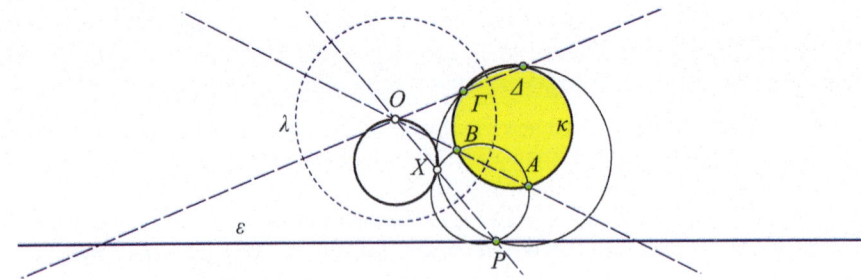

Fig. 4.78: Locus related to the radical center O of three circles

4.8. INVERSION

Exercise 4.94. Let $\{A, B, \Gamma, \Delta\}$ be four points on the circle κ and point P be a point of the line ε. Find the geometric locus of the second intersection points X of the circles (PAB) and $(P\Gamma\Delta)$ (See Figure 4.78).

Hint: The radical center O of the three circles κ, (PAB), $(P\Gamma\Delta)$ is a fixed point and center of a fixed circle λ orthogonal to all three circles. The locus is the circle, which is the inverse of line ε relative to λ.

Exercise 4.95. From a point X outside the circle κ we draw lines intersecting the circle at points $\{A, B\}$. To find the geometric locus of the middles of the chords AB.

Exercise 4.96. Given is a circle λ and a point B. Show that all the circles κ which pass through B and intersect λ under a fixed angle ω, are tangent to a fixed circle v (See Figure 4.79).

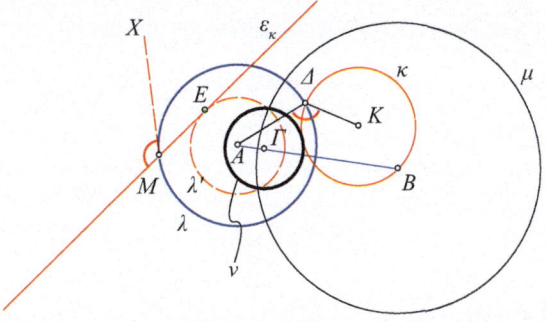

Fig. 4.79: Circles intersecting fixed circle under fixed angle

Hint: For a point B external to λ, consider the inversion relative to circle μ with center B, which is orthogonal to λ (See Figure 4.79). Under this inversion the circle λ maps to itself and the circle κ maps to a line ε_κ, which forms the same angle ω with λ. Such a line is tangent to a fixed circle λ', which is concentric with λ. It follows that the requested circles will be the inverses of the lines ε_κ, which are tangent to λ', therefore the circles κ will be tangent to the inverse v of λ' relative to μ.

Exercise 4.97. Construct a circle passing through two given points $\{\Gamma, \Delta\}$ and intersecting a given line ε at a given angle.

Hint: Triangle $\tau = OAB$ has known angles (See Figure 4.80-I). Therefore for arbitrary point X of the medial line η of $\Gamma\Delta$, a triangle AXY similar to τ can be constructed. Suppose Z is on $A\Delta$ and $|XZ| = |XY|$. From Δ draw the parallel to ZX intersecting η at O. Point O is the center of the requested circle.

Exercise 4.98. Given two circles κ and λ, construct a line ε intersecting them respectively under given angles α and β.

Hint: Line ε is a common tangent to two known circles κ' and λ' (See Figure 4.80-II).

Exercise 4.99. Given two circles κ and λ and a point O, construct a circle μ intersecting the circles respectively under given angles α and β and passing through point O.

Hint: Through an inversion relative to a circle with center O, reduce the problem to the construction of the preceding exercise.

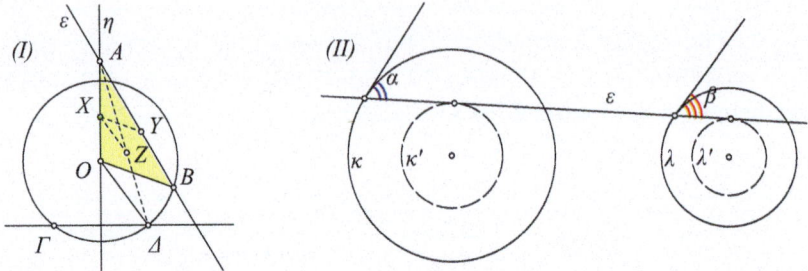

Fig. 4.80: Intersection at angle   Intersection of two circles by line

4.9 Polar and pole

> Like the meridians as they approach the poles, science, philosophy and religion are bound to converge as they draw nearer to the whole. I say "converge" advisedly, but without merging, and without ceasing, to the very end, to assail the real from different angles and on different planes.
>
> *J. Huxley, The Phenomenon of Man, p.30*

A circle $\kappa(O, r)$ defines a correspondence between points and lines on the plane, often called **polar reciprocity**: to every point $\Sigma \neq O$ corresponds the so called **polar line** $p(\Sigma)$ and to every line ε, not containing O, corresponds the so called **pole** $p(\varepsilon)$ of the line ε. A simple definition of this correspondence would be to consider the line $p(\Sigma)$ as the perpendicular to line $O\Sigma$ at the inverse Σ' of Σ relative to the circle κ (See Figure 4.81). Point Σ then is called the *pole* of the line $\varepsilon = p(\Sigma)$ relative to the circle. From the properties of inversion follows a reciprocity relation between *polar* and *pole*. Every line (not passing through the center O) is the polar of a suitable point Σ, which is constructed easily. And every point Σ (other than O) is the *pole* of a line which is also easily constructed. This relationship between lines and points

4.9. POLAR AND POLE

through a fixed circle κ, can be defined also in another way, described by the following theorem. The advantage of this method is that the corresponding definition can be transferred verbatim to the case in which the circle κ is replaced with a conic section (II-§ 6.1).

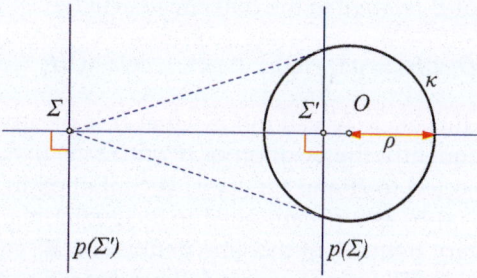

Fig. 4.81: Polar line $p(\Sigma)$ of Σ relative to circle κ

Theorem 4.17. *Given a circle $\kappa(O)$ and a point Σ, we consider all the lines through Σ intersecting the circle at points $\{A,B\}$. We define on every such line the harmonic conjugate T of Σ relative to $\{A,B\}$. The geometric locus of all points T is (part of) a line $p(\Sigma)$ orthogonal to $O\Sigma$ (See Figure 4.82).*

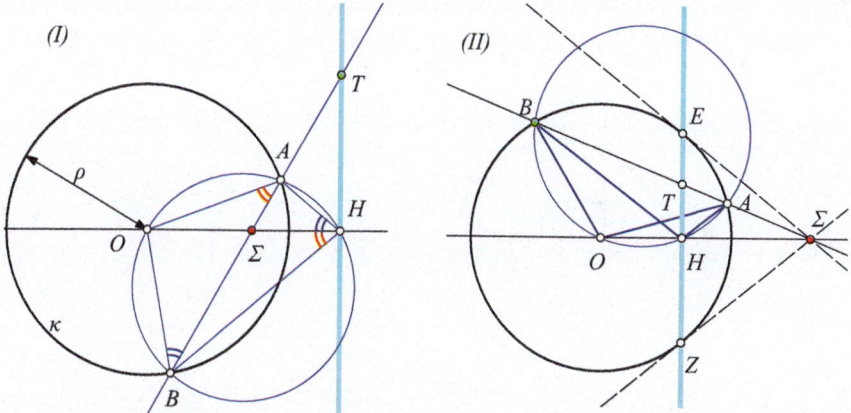

Fig. 4.82: Polar of Σ relative to κ Polar of external Σ

Proof. Consider the circle (OAB) passing through the center of κ and the intersection points $\{A,B\}$ of ΣT with κ (See Figure 4.82-I). Let H be the intersection of the circle with $O\Sigma$. The quadrilateral $BOAH$ is cyclic and the triangle AOB is isosceles, consequently

$$\widehat{OAB} = \widehat{OHB}, \ \widehat{OBA} = \widehat{OHA} \Rightarrow \widehat{OHB} = \widehat{OHA},$$

i.e. line ΣH is a bisector of the angle at H of triangle BHA. Because, by assumption, (A,B,T,Σ) is a harmonic quadruple, the point T will be the intersection point of AB with the external bisector of $\widehat{AHB}$, therefore TH will be orthogonal to OH. Because Σ lies on the radical axis of the two circles, which is AB, the power of Σ relative to the two circles will be

$$|\Sigma O||\Sigma H| = \rho^2 - |O\Sigma|^2 \quad \Leftrightarrow \quad |O\Sigma||OH| = \rho^2, \tag{4.1}$$

where ρ is the radius of circle κ. It follows that H is the inverse of Σ relative to circle κ. Consequently the position of H on $O\Sigma$ is fixed, all points T of the locus are projected to H and the locus is contained in the line which is orthogonal to $O\Sigma$ at H, as we mentioned in the beginning of this section.

Conversely, every point T of this line defines $\{A,B\}$ and (A,B,T,Σ) is a harmonic quadruple. The proof of this follows by reversing the preceding arguments. From the cyclic quadrilateral $AOBH$ follows again that OH is a bisector of the angle $\widehat{AHB}$ and consequently, because HT is orthogonal to OH, HT is an external bisector of $\widehat{AHB}$ and then (A,B,T,Σ) is a harmonic quadruple.

In the preceding proof we tacitly supposed that point Σ is in the interior of the circle κ and different from the circle's center (See Figure 4.82-I). With minor changes, the same proof works also for points Σ in the exterior of κ (See Figure 4.82-II).

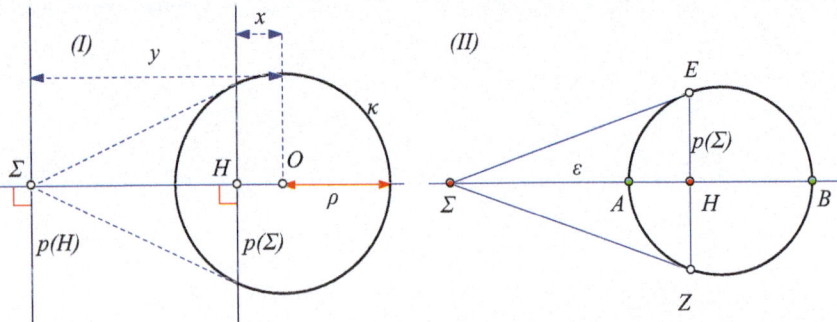

Fig. 4.83: $\Sigma \leftrightarrow H$ relation: $xy = \rho^2$ $\{\Sigma, H\}$ harmonic conjugate of $\{A, B\}$

Corollary 4.21. *The distances $|O\Sigma|$, $|OH|$ of the points Σ and H, at which the polar of Σ relative to the circle $\kappa(O,\rho)$ intersects line $O\Sigma$, satisfy the equation (See Figure 4.83-I)*

$$x \cdot y = \rho^2. \tag{4.2}$$

Corollary 4.22. *The polar line $p(\Sigma)$ of a point Σ relative to the circle $\kappa(O,\rho)$ coincides with the orthogonal to $O\Sigma$ at point H which is the inverse of Σ relative to the circle κ.*

4.9. POLAR AND POLE

Corollary 4.23. *For a circle κ and a point Σ in its exterior, the polar of Σ relative to the circle, coincides with the line which joins the contact points E, Z of the tangents to κ from Σ.*

Corollary 4.24. *The pairs of points $\{(\Sigma, H), (A, B)\}$ on a line ε are harmonic conjugate, if and only if the circle with diameter AB has one of them in its interior (H), the other (Σ) in the exterior of AB and the internal point coincides with the point of intersection of ε and the chord ZE of the contacts of the tangents from the external point (See Figure 4.83-II).*

From the last proposition (also see 4.2 follows also the extension of the notion of the polar for points Σ *on* the circle κ. When point Σ moves approaching externally the point Σ' of the circle, then the two tangents ΣE and ΣZ tend to coincide with the tangent to the circle at Σ'. We consider therefore that

At points Σ of the circle κ the polar $p(\Sigma)$, relative to κ, coincides with the tangent to κ at Σ.

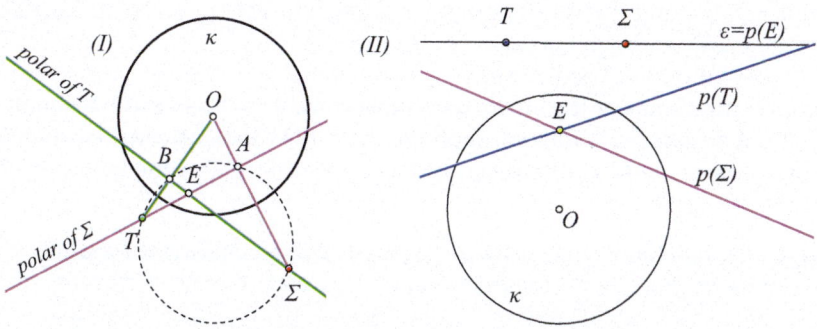

Fig. 4.84: $p(\Sigma) \ni T \Leftrightarrow p(T) \ni \Sigma$ The pole E of line TΣ

Corollary 4.25 (pole-polar reciprocity). *If the polar $p(\Sigma)$ of Σ relative to circle $\kappa(O, \rho)$ passes through T, then the polar $p(T)$ of T relative to κ passes through Σ.*

Proof. If the line ΣT intersects the circle, then this follows from the definition of the polar. If ΣT does not intersect the circle, one may think as follows. The polar ε of Σ relative to κ is the orthogonal to OΣ at the point A, which is the inverse of Σ relative to κ (Corollary 4.22). We draw then the perpendicular ΣB to TO (See Figure 4.84-I). This forms the quadrilateral ΣABT, which is cyclic, since points A and B see ΣT under a right angle. Then $|OB||OT| = |OA||O\Sigma| = \rho^2$, which means that B is also an inverse of T relative to κ, therefore line BΣ is the polar of T.

Corollary 4.26. *If the polars of the points Σ and T relative to circle κ intersect at E, then the line ΣT coincides with the polar $p(E)$ of E relative to κ.*

Proof. According to the preceding corollary, since the polar $p(\Sigma)$ passes through E, the polar $\varepsilon = p(E)$ of E will also pass through Σ (See Figure 4.84-II). Similarly, since the polar $p(T)$ of T passes through E, the polar ε of E will also pass through T.

Corollary 4.27. *Given a circle κ and a line ε, for every point Σ of ε the polar $p(\Sigma)$ passes through a fixed point E.*

Proof. According to the preceding corollary, for two points $\{\Sigma, T\}$ of ε, the corresponding polars $\{p(\Sigma), p(T)\}$ will intersect at a point E, such that ε coincides with the polar $p(E)$ of E (See Figure 4.84-II). Then (Corollary 4.25), also for any other point X of ε the polar $p(X)$ will pass through E.

Given the circle κ, the preceding corollary defines for every line ε, which does not pass through the center O of the circle, the point $E = p(\varepsilon)$, which we called *pole of the line ε* relative to the circle κ (See Figure 4.84-II). This correspondence between the pole E and the line ε is the inverse of the correspondence between the polar $\varepsilon = p(E)$ and the point E. This way an interesting relationship between points and lines is created, which is described arithmetically through equation (4.2).

This relation between pole-polar is extended also for lines which pass through the center of the circle, by corresponding to each line ε through O the point at infinity of the orthogonal ζ of ε at O. Also we consider that the polar of the center O coincides with the line at infinity (for the *line at infinity* see remark 5.19).

Exercise 4.100. Show that if the pole E, relative to circle κ, of the line ε is contained in line ζ, then the pole Z of line ζ, relative to κ, is also contained in ε.

Hint: The claim is equivalent to that of (Corollary 4.25).

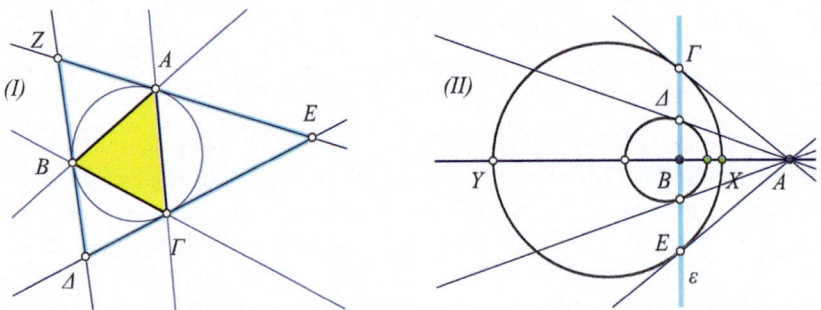

Fig. 4.85: Tangential triangle ΔEZ of $AB\Gamma$ \qquad Circles with same polar ε

Exercise 4.101. Show that the poles of the sides of triangle $AB\Gamma$ relative to its circumscribed circle κ are the vertices of the (tangential) triangle ΔEZ, whose sides are tangent to κ at the vertices of $AB\Gamma$ (See Figure 4.85-I).

4.9. POLAR AND POLE

Exercise 4.102. Given a line ε and a point A outside of it construct a circle κ such that the polar of A relative to κ will be ε. Show that all the circles which solve this problem build a non intersecting pencil (See Figure 4.85-II).

Hint: Let B be the projection of A onto ε. For every point X in the interior of AB and its harmonic conjugate Y relative to A, B, the circle κ with diameter XY has the required property. All these κ are Apollonian circles of the non intersecting pencil with limit points $\{A,B\}$

Exercise 4.103. Show that the poles, relative to circle κ, of the lines, which pass through a fixed point Σ, are contained in the polar $p(\Sigma)$ of Σ relative to circle κ.

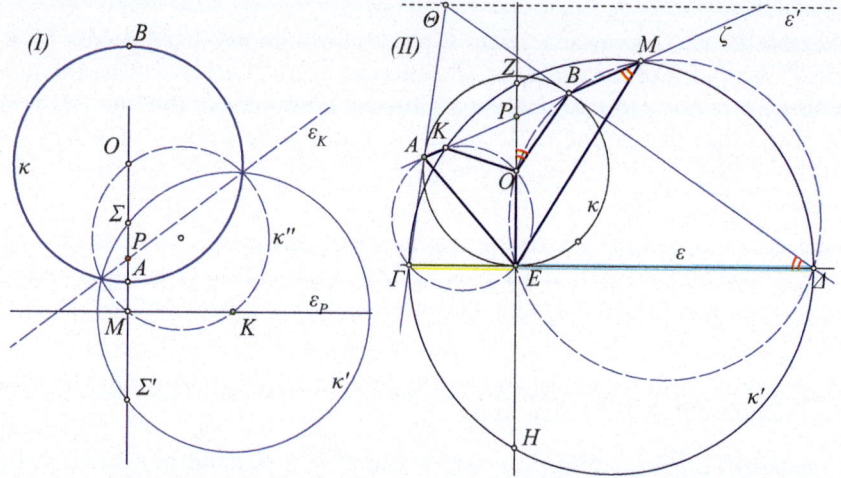

Fig. 4.86: Base points Σ, Σ' Line segments defined from tangents

Exercise 4.104. Let $\kappa(O,r)$ be a circle and $P \neq O$ a point in its interior whose polar is line ε_P. Let also $\kappa'(K,r')$ be a circle with center on ε_P, orthogonal to κ and Σ, Σ' be the points at which this circle intersects the line OP (See Figure 4.86-I). Show that $|O\Sigma| \cdot |O\Sigma'| = r^2$.

Exercise 4.105. Let $\kappa(O,r)$ be a circle and $P \neq O$ be a point in its interior. Let E be an endpoint of the diameter through P and ε be the tangent at E. Line ζ is rotated about P. Show that the tangents at A and B, where line ζ intersects the circle, excise from the line ε segments $E\Gamma$, $E\Delta$ having a fixed product of lengths. Show also that these tangents intersect at a point Θ of the polar ε' of P relative to circle κ (See Figure 4.86-II).

Hint: ([105, 16(Involution)]) Suppose that the circle κ' with diameter $\Gamma\Delta$ intersects OP at points Z, H. It suffices to prove that these points are fixed. This is equivalent to the fact, that $\sigma = |PZ||PH|$ is fixed. Comparing angles we see that the circles $(A\Gamma E)$, $(B\Delta E)$ intersect the line ζ at corresponding points K, M, which are also contained in the circle κ' and $\{\Delta M, AE\}$ are parallel and $MA = ME$. However $|PK||PA| = |PO||PE| = |PB||PM| = \sigma_1$ and $|PA||PB| = \sigma_2$ are constant and $\sigma = \sigma_1^2/\sigma_2$.

Exercise 4.106. From the point P of the circumcircle of the triangle $AB\Gamma$ we draw the lines $\{PB, P\Gamma\}$, which intersect correspondingly the sides $\{A\Gamma, AB\}$ at points $\{B', \Gamma'\}$. Show that the line $B'\Gamma'$ passes through the intersection point of the tangents to the circumcircle at $\{B, \Gamma\}$.

Exercise 4.107. Show that the polars of a similarity center of two non concentric circles, relative to these circles, are two parallel lines of whose the middle-parallel coincides with the radical axis of the two circles.

Exercise 4.108. The variable circle κ_t passes through two fixed points $\{A, B\}$ and Γ is a point of the segment AB different from $\{A, B\}$. Let P_t be the pole of line AB relative to κ and Δ_t be an intersection point of the line $P_t\Gamma$ with κ_t. Show that the geometric locus of Δ_t is the circle $\lambda(E)$, where E is the harmonic conjugate of Γ relative to $\{A, B\}$, and the λ is orthogonal to all the circles κ_t ([99, p.138]).

Exercise 4.109. Given three pairwise non concentric circles $\{\alpha, \beta, \gamma\}$, consider a variable point P and the three polars of it relative to these circles. Show that the locus of points P, for which the three polars are concurrent is the circle δ, which is orthogonal to the three given circles. For each point P on this locus, show that the concurrence point of the polars coincides with the diametral P' of P relative to δ.

Exercise 4.110. Show that for every polygon Π inscribed in a circle κ, the poles of the sides of Π are vertices of a polygon Π', which is circumscribed on κ and its sides are tangent to κ at the vertices of Π.

Exercise 4.111. Show that for every regular polygon Π inscribed in circle κ, the poles of the sides of Π are vertices of a regular polygon Π', which is circumscribed on κ and its sides are tangent to κ at the vertices of Π.

Exercise 4.112. Show that the polars relative to a circle κ of three collinear points are lines passing through the same point. Also the poles relative to circle κ of three lines, which pass through a common point, are three collinear points.

Exercise 4.113. Show that the polar lines $p(\Sigma)$ relative to circle κ, of the points Σ of a circle λ concentric to κ are tangents to the inverse λ' of λ relative to κ.

Exercise 4.114. Let κ be a circle and $\{A, B, \Gamma\}$ be three non collinear points lying on the exterior of κ. Construct a triangle circumscribed to κ, whose sides pass through the points $\{A, B, \Gamma\}$, respectively.

Hint: Consider the tangents of the points $\{A, B, \Gamma\}$ to κ. Taken by three, they form two triangles circumscribing κ and six triangle having κ as escribed circle.

4.10 Comments and exercises for the chapter

> Mathematics as an expression of the human mind reflects the active will, the contemplative reason, and the desire for aesthetic perfection. Its basic elements are logic and intuition, analysis and construction, generality and individuality.
>
> Richard Courant, *What is Mathematics*

Exercise 4.115. For a given circle $\kappa(O)$ and a point A outside of it, show that the centers of all circles $\lambda(P)$, which pass through A and are orthogonal to κ, lie on a line ε. Show also that the centers of all circles $\mu(\Sigma)$, which pass through A and intersect κ along a diameter of κ lie on a line ε', which is parallel to ε. Show finally that both lines $\{\varepsilon, \varepsilon'\}$ are parallel to the polar of A relative to κ.

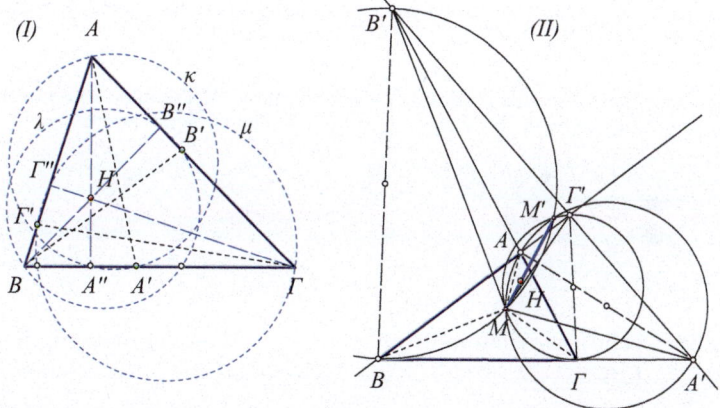

Fig. 4.87: Orthocenter as radical center   Pencil of 3 circles

Exercise 4.116. Show that the orthocenter H of triangle $AB\Gamma$ is the radical center of three circles which have diameters the altitudes $\{AA'', BB'', \Gamma\Gamma''\}$ of the triangle (See Figure 4.87-I). More general, show that for three points $\{A', B', \Gamma'\}$ on the sides $\{B\Gamma, \Gamma A, AB\}$ of the triangle, the radical center of the circles $\{\kappa, \lambda, \mu\}$ with diameters $\{AA', BB', \Gamma\Gamma'\}$ coincides with the orthocenter of the triangle.

Hint: For the first claim show that $|HA||HA''| = |HB||HB''| = |H\Gamma||H\Gamma''|$. See the related exercise 2.159. The second claim is a consequence of the first. The circle centered at the orthocenter H of the triangle and having radius ρ with $\rho^2 = |HA||HA''|$ is called **polar circle** of the triangle.

Exercise 4.117. Let M be a point not lying on a side-line of the triangle $AB\Gamma$. Consider also the segments $\{MA, MB, M\Gamma\}$ and the perpendicular lines to these at M, correspondingly, $\{MA', MB', M\Gamma'\}$, which intersect the side-lines $\{B\Gamma, \Gamma A, AB\}$ at points, correspondingly, $\{A', B', \Gamma'\}$. Show that the three circles with diameters $\{AA', BB', \Gamma\Gamma'\}$ are members of an intersecting pencil (See Figure 4.87-II).

Hint: Apply the preceding exercise for three points $\{A', B', \Gamma'\}$ on the corresponding side-lines. Conclude that the orthocenter of $AB\Gamma$ is the radical center of the three circles. After reading § 5.17 return to this figure and using exercise 5.133 show that $\{A', B', \Gamma'\}$ are collinear.

Exercise 4.118. Show that the inversion relative to a member-circle κ of the pencil $\mathscr{D}$, leaves this pencil invariant, i.e. it maps every circle κ' of $\mathscr{D}$ to a circle κ'' which also belongs to the same pencil. How does this inversion transforms the circles of the orthogonal pencil $\mathscr{D}'$ of $\mathscr{D}$?

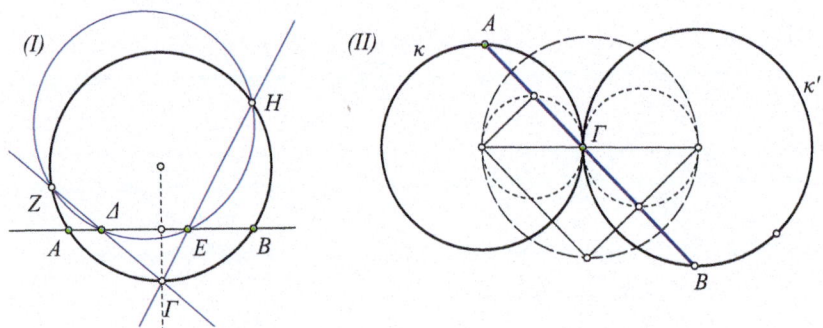

Fig. 4.88: Cyclic quadrilateral from Δ, E AB of definite length

Exercise 4.119. Two points Δ, E of a chord AB of a circle are joined with the point Γ of its arc $\widehat{AB}$. If the second intersection points of respectively $\{\Gamma\Delta, \Gamma E\}$ with the circle are $\{Z, H\}$, show that the quadrilateral ΔEHZ is cyclic, if and only if Γ is the middle of the arc $\widehat{AB}$ (See Figure 4.88-I).

Exercise 4.120. From the contact point Γ of two circles $\{\kappa, \kappa'\}$ draw a secant AB of given length λ (See Figure 4.88-II).

Exercise 4.121. Construct a right triangle with unknown side lengths $a > b > c$, but with given perimeter p, and such that $ac = b^2$ (See Figure 4.89-I).

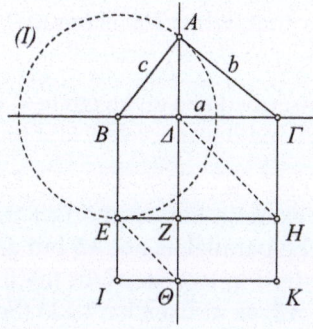 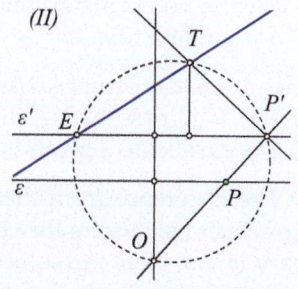

Fig. 4.89: Special rectangle   Locus for $|P'T| = |OP|$

Hint: Draw the altitude $A\Delta$ and extend it. form the rectangle $BEZ\Delta$ with $|BE| = c$ and show that the hypothesis $ac = b^2$ implies that this is a golden rectangle.

Exercise 4.122. From a point O, outside the strip of two parallels ε and ε', are drawn secants OPP' and orthogonal to it segments $|P'T| = |OP|$ always on the same side of OP. Find the geometric locus of T, as the secant rotates about O (See Figure 4.89-II).

Exercise 4.123. Let κ be a circle and $\{A, B\}$ be fixed points. Point M moves on κ and the lines MA, MB intersect the circle a second tima at the points

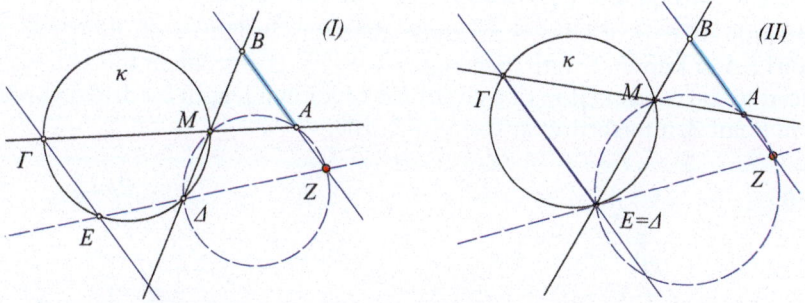

Fig. 4.90: Line passing through a fixed point

Γ and Δ. From Γ draw the parallel to the line AB which intersects again the circle at the point E. Show that the line ΔE passes through a fixed point Z of the line AB (See Figure 4.90-I).

Hint: The quadrilateral $AM\Delta Z$ is cyclic and the product $|BA||BZ|$ is equal to the power $p_\kappa(B)$ of B relative to κ, which is constant.

Exercise 4.124. Construct a triangle $M\Gamma\Delta$ inscribed in a given circle κ, whose sides $M\Gamma$, $M\Delta$ pass through fixed points A and B respectively and the third side $\Gamma\Delta$ is parallel to the line AB (See Figure 4.90-II).

Hint: Use the preceding exercise and show that, when ΓΔ is parallel to AB, then ZΔ is tangent to κ.

Exercise 4.125. Construct a triangle MΓΔ inscribed in a given circle κ, whose sides {MΓ, MΔ} pass through fixed points A and B respectively and the third side ΓΔ is parallel to a given line ε.

Hint: The difference from the preceding exercise lies on the fact that the line, to which we require the chord ΓΔ to be parallel, is not AB but another arbitrary line ε. Here the solution is organized in two steps. In the first we forget line ε and we construct point Z on AB, as in Exercise 4.123 (See Fig-

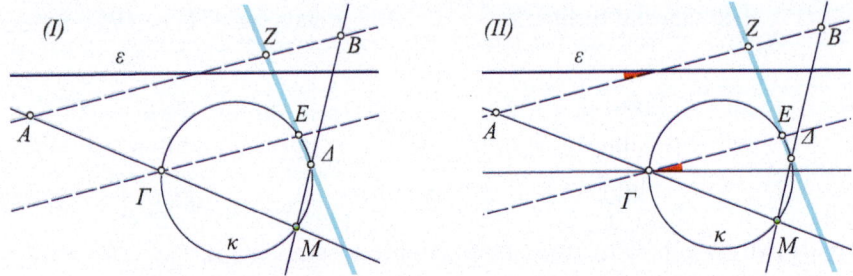

Fig. 4.91: Line passing through a fixed point II

ure 4.91-I). In the second step, we begin by assuming that the chord ΓΔ was constructed (parallel to ε). We join Δ with Z and we define the chord ΓE, which, according to the preceding exercise, will be parallel to AB. However the angle $\widehat{\Delta\Gamma E}$ is of known measure, therefore it defines on the circle κ a chord EΔ of known length (See Figure 4.91-II). The problem therefore is reduced to the construction of a chord EΔ of known length by drawing a line from point Z, intersecting κ (Exercise 3.201).

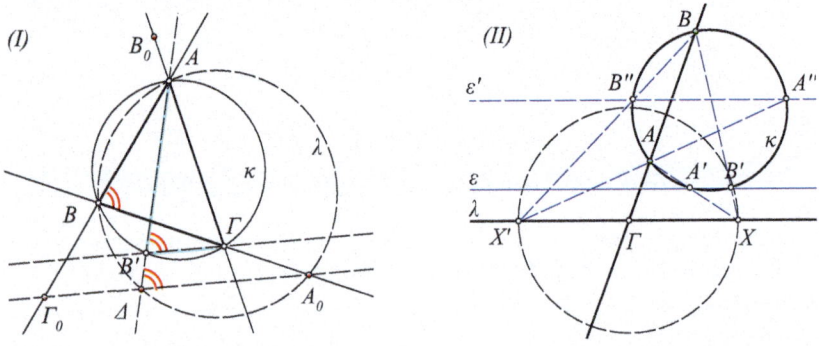

Fig. 4.92: Castillon's problem   Chords parallel to line λ

Exercise 4.126. Construct a triangle ABΓ inscribed in a given circle κ, whose sides {AB, BΓ, ΓA} pass through respective fixed points $\{\Gamma_0, A_0, B_0\}$.

4.10. COMMENTS AND EXERCISES FOR THE CHAPTER

Hint: ([9, p.136]) Suppose that $AB\Gamma$ has been constructed. Draw from Γ the parallel to $A_0\Gamma_0$, which intersects κ at B' and find the intersection Δ of $A_0\Gamma_0$ with AB' (See Figure 4.92-I). Point Δ is determined from the given data, because the quadrilateral $\Delta A_0 AB$ is cyclic in the circle λ and $p = |\Gamma_0\Delta||\Gamma_0 A_0|$ is the power of Γ_0 relative to κ. Then, the two sides $\{AB', A\Gamma\}$ of the triangle $AB'\Gamma$ pass through the two known points $\{\Delta, B_0\}$ and the third side $B'\Gamma$ is parallel to the given line $A_0\Gamma_0$ (Exercise 4.125).

The last problem is known as the **Castillon's problem**. A generalization and a more systematic treatment of this problem is discussed in § 5.21.

Exercise 4.127. Let κ be a circle non intersecting the line λ and A, B be two points of κ. Determine a point X on λ, such that the second intersection points A', B' of XA, XB with κ define a line $A'B'$ parallel to λ. With the same data, determine a point X on λ, such that the corresponding line $A'B'$ is orthogonal to λ.

Hint: Solution of the first part by the figure (See Figure 4.92-II). For the second part, suppose the position of X known and apply an inversion relative to the circle with center X and circle of inversion orthogonal to κ. Find the image of $A'B'$ relative to one such inversion.

Exercise 4.128. Show that in a line ε, the points $\{\Gamma, \Delta\}$ are harmonic conjugate of $\{A, B\}$, if and only if the points $\{\Gamma, \Delta\}$ are inverse points relative to the circle with diameter AB, and equivalently, when points $\{A, B\}$ are inverse relative to the circle with diameter $\Gamma\Delta$.

Exercise 4.129. Show that the polar of a point A of a circle κ, relative to the circle λ, which is orthogonal to κ, always passes through the diametrically opposite point B of A relative to κ. Conclude, that the polars of a point A relative to the circles λ of a pencil of circles $\mathscr{D}$ pass all through a fixed point.

Exercise 4.130. Given are three circles in general position. Show that the geometric locus of the points A, for which the polars relative to the three circles pass through a point B, is the simultaneously orthogonal (Exercise 4.69) to all three given circles, circle κ. Point B is the diametrically opposite of A relative to κ.

Exercise 4.131. Let $p = AB\Gamma\Delta$ be a parallelogram and $q = A'B'\Gamma'\Delta'$ be the parallelogram of its middles. Show that next propositions are equivalent:

1. An angle of p is equal to an angle of q.
2. Parallelograms p and q are similar.
3. A diagonal of p to one of its sides has ratio $\sqrt{2}$.

Hint: Through the figure (See Figure 4.93-I). Let M be the intersection point of the diagonals and suppose that the angles $\widehat{BA\Delta}$ and $\widehat{B'A'\Delta'}$ are equal. Then, the circle $(AM\Delta)$ is tangent to the line AB at A and the angle $\widehat{M\Delta A}$ is equal to $\widehat{MAB}$, which in turn is equal to the angle $\widehat{A'B'\Delta'}$. This shows the equivalence

1 ⇔ 2. The power of B relative to the circle is $|BA|^2 = 2|BM|^2$. This shows the direction 1 ⇒ 3. Reversing the argument, we have also the direction 3 ⇒ 1.

Exercise 4.132. Let $\{X, O, Y\}$ be fixed non collinear points and Z a point moving on line OY. Let also the circle $\kappa(K)$ be inscribed in the triangle OXZ. Show that the line AB of its contact points with $\{OZ, ZX\}$ passes through a fixed point Δ of the bisector of the angle $\widehat{XOY}$.

Hint: Consider the circle (ABZ) with diameter KZ and its intersections H, E with $A\Gamma$, $B\Gamma$ (See Figure 4.93-II). Show that: (i) HE is parallel to OX. (ii) ZE is the polar of Δ relative to κ. (iii) $X\Delta$ is the polar of E relative to κ.

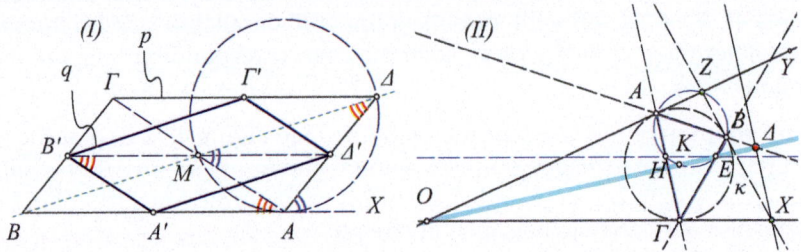

Fig. 4.93: Similar parallelograms p, q Passing through fixed point

Exercise 4.133. In an acute-angled triangle $AB\Gamma$ construct the circles κ_A, κ_B, κ_Γ, which are orthogonal respectively to the circles ν_A, ν_B, ν_Γ, which have respectively diameters the sides of the triangle $B\Gamma$, ΓA and AB (See Figure 4.94-I). Show that the circles κ_A, κ_B, κ_Γ have as radical axes the altitudes of

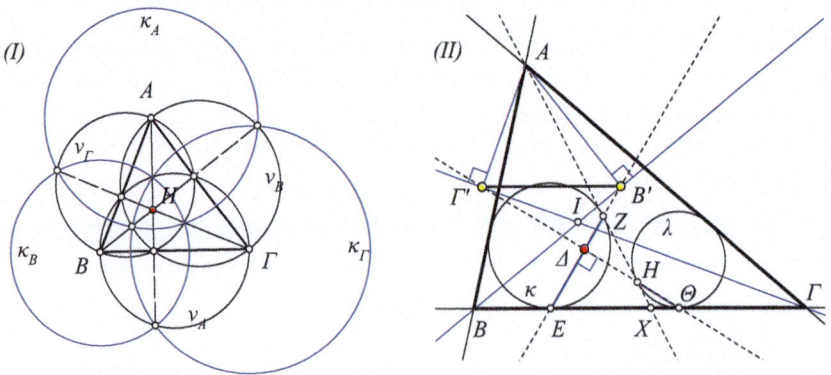

Fig. 4.94: Pencils from triangle altitudes   Chord intersection EZ, $H\Theta$

the triangle $AB\Gamma$. Show also that the circles ν_A, ν_B, ν_Γ have the same lines as radical axes. Finally show that the polar of the orthocenter H of the triangle $AB\Gamma$ relative to κ_A is line $B\Gamma$ and corresponding properties for κ_B and κ_Γ.

4.10. COMMENTS AND EXERCISES FOR THE CHAPTER

Exercise 4.134. Point X moves on the base $B\Gamma$ of triangle $AB\Gamma$. The circles κ, λ are the inscribed ones of triangles ABX, $AX\Gamma$, defined by AX. Show that the lines of the points of contact EZ, $H\Theta$ intersect orthogonally at a point Δ (See Figure 4.94-II). Also show that, for variable X on $B\Gamma$, the geometric locus of the corresponding Δ is a circle with diameter $B'\Gamma'$, where B', Γ' are the projections of A on the bisectors of $AB\Gamma$ from B and Γ.

Hint: Application of exercise 4.132.

Exercise 4.135. Find an inversion by which, three given circles α, β, γ are transformed into others whose centers are collinear.

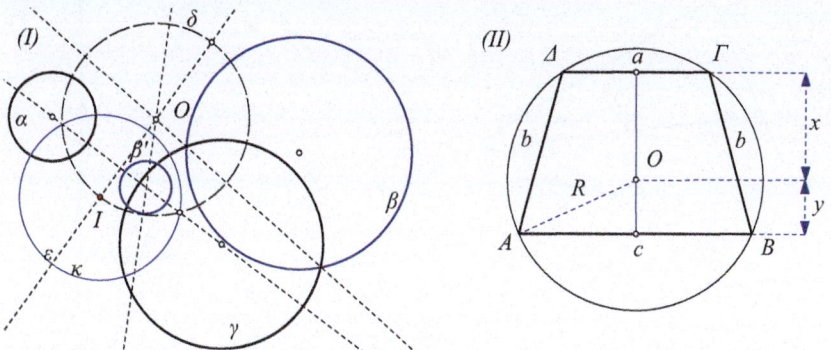

Fig. 4.95: $\{\alpha, \beta, \gamma\}$ with collinear centers   Isosceles trapezium

Hint: Let O be the radical center of α, β and γ and $\delta(O)$ be the circle simultaneously orthogonal to all three (See Figure 4.95-I). Consider the intersection point I of δ with the radical axis ε of α and γ. The inversion relative to circle $\kappa(I)$, which is orthogonal to α, therefore also to γ, inverts circle β into β', which has its center on the centerline of α, γ. This same inversion leaves α and γ invariant.

Exercise 4.136. Show that the circumscribed circle of an isosceles trapezium with side lengths a (small parallel), b, c (big parallel) has radius

$$R^2 = \frac{(ac+b^2)b^2}{4b^2 - (a-c)^2}.$$

Hint: Using the power of the middles of the parallel sides relative to the circumcircle show the formulas (See Figure 4.95-II): (i) $R^2 - x^2 - (a/2)^2 = 0$, (ii) $R^2 - y^2 - (c/2)^2 = 0$, as well as (iii) $(x+y)^2 - b^2 + ((a-c)/2)^2 = 0$. Solving the first two substitute x and y in the third.

Exercise 4.137. In a given triangle $AB\Gamma$ find points Δ, E on its sides, respectively AB, $A\Gamma$, such that the segments $|B\Delta| = |\Gamma E|$ and $\frac{|\Delta E|}{|B\Delta|} = \lambda$, where λ is a given constant (See Figure 4.96-I).

Hint: From the middle of the, supposedly constructed ΔE, draw the parallel and equal to ΔB, $E\Gamma$. The created isosceles triangle $ZH\Theta$, has the angle at Z equal to $\widehat{A}$, $BH\Gamma\Theta$ is a parallelogram, I is the middle of $H\Theta$ and ZI is parallel to the bisector of $\widehat{A}$. Therefore also the ratio $\frac{|I\Theta|}{|\Theta\Gamma|}$ will result from the given data and Θ will be an intersection of an Apollonian circle κ of the segment $I\Gamma$ and the line $I\Theta$, which also is determined from the given data (generalization of Exercise 3.191).

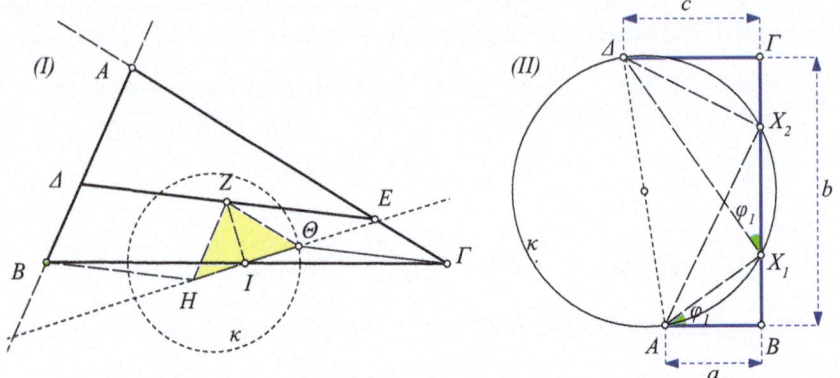

Fig. 4.96: Secant at given ratio   Graphical solution of quadratic

Exercise 4.138. Show that the next procedure produces the roots of the quadratic equation $ax^2 + bx + c = 0$ with $a > 0$, $bc \neq 0$ (See Figure 4.96-II).

1. Define segment $|AB| = a$.
2. Orthogonally to it at B, left if $b > 0$, right if $b < 0$, define $|B\Gamma| = b$.
3. Orthogonally to it at Γ, left if $c > 0$, right of $c < 0$, define $|\Gamma\Delta| = c$.
4. Define the points of intersection $\{X_1, X_2\}$ of the circle κ on diameter $A\Delta$ with line $B\Gamma$.
5. If these exist, then $x_1 = -\tan(\widehat{BAX_1})$, $x_2 = -\tan(\widehat{BAX_2})$ are the roots.

Hint: Put $\phi_1 = \widehat{BAX_1}$ and write $ax^2 + bx + c = (ax+b)x + c$. Then $(ax_1 + b) = -|AB|\tan(\phi_1) + |B\Gamma| = |X_1\Gamma|$ and $(ax_1+b)x_1 + c = -|X_1\Gamma|\tan(\phi_1) + |\Gamma\Delta| = 0$.

There are other simpler geometric methods to solve the quadratic equation, however the interest in this method is that it generalizes to equations of higher degree [107]. A simpler method is, assuming that $a \neq 0$, to write first $ax^2 + bx + c = 0$ equivalently as $x^2 + px + q = 0$, with $p = b/a$, $q = c/a$. Next we consider the fact that the the roots of the latter satisfy the equations

$$x_1 + x_2 = -p, \qquad x_1 \cdot x_2 = q, \qquad (4.3)$$

from which follow x_1, x_2, according to the next exercise.

4.10. COMMENTS AND EXERCISES FOR THE CHAPTER

Exercise 4.139. Show that the equations 4.3 are solved geometrically through one of the figures 4.97.

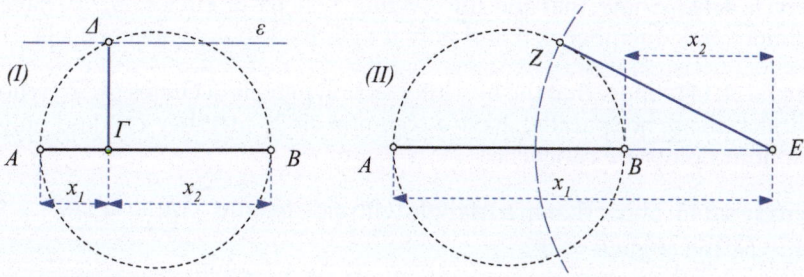

Fig. 4.97: $x_1 + x_2 = -p$, $x_1 \cdot x_2 = q > 0$ $x_1 + x_2 = -p$, $x_1 \cdot x_2 = q < 0$

Hint: Consider the circle with diameter the segment $|AB| = |p|$. In the case $q > 0$ draw a parallel ε at a distance $|A\Gamma| = \sqrt{q}$ (See Figure 4.97-I). If ε intersects the circle at Δ, then the projection Γ to AB defines the line segments $A\Gamma$, ΓB. If $p > 0$, then $x_1 = -|A\Gamma|$, $x_2 = -|\Gamma B|$. If $p < 0$, then $x_1 = |A\Gamma|$, $x_2 = |\Gamma B|$ are the wanted roots.

In the case $q < 0$ we draw the tangent EZ from point E of AB with $|EZ| = \sqrt{-q}$ (See Figure 4.97-II). If $p > 0$, then $x_1 = -|AE|$, $x_2 = |BE|$. If $p < 0$, then $x_1 = |AE|$, $x_2 = -|BE|$ are the wanted roots.

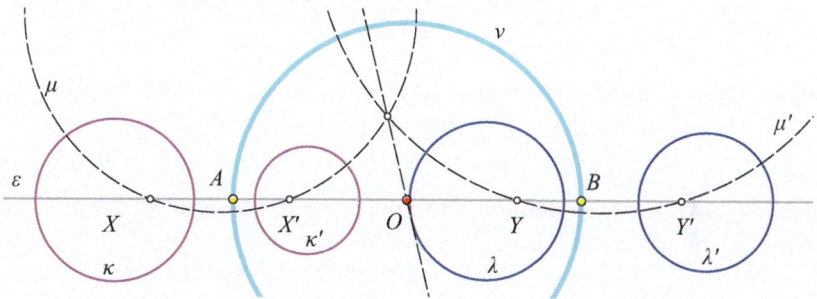

Fig. 4.98: Common circle member v of two pencils

Exercise 4.140. Let $\{\kappa, \kappa', \lambda, \lambda'\}$ be four pairwise non intersecting circles with centers on the line ε. Show that the pencil of circles $\mathscr{D}$ generated by the pair (κ, κ') and the pencil $\mathscr{D}'$ generated by the pair (λ, λ') have a common circle-member v, if and only if there exist the *common harmonics* $\{A, B\}$, of the pairs (X, X') and (Y, Y'), where (X, X') are the limit points of $\mathscr{D}$ and (Y, Y') are the limit points of $\mathscr{D}'$. Then, the circle v has diameter AB and from its center O pass the radical axes of all pairs (μ, μ') of circles orthogonal respectively to those of the pencils $\mathscr{D}$ and $\mathscr{D}'$ (See Figure 4.98).

Exercise 4.141. Examine when there exists a common circle member of an intersecting pencil $\mathscr{D}$ and a non intersecting one $\mathscr{D}'$.

Exercise 4.142. Is it possible that two different pencils of circles $\mathscr{D}$ and $\mathscr{D}'$ have more than one member-circles in common?

Exercise 4.143. Show that the intersecting pencils of circles $\mathscr{D}, \mathscr{D}'$ have a common circle-member κ, if and only if their basic points are concyclic.

Exercise 4.144. Show that the non intersecting pencils of circles $\mathscr{D}, \mathscr{D}'$, whose radical axes are not parallel, have a common circle-member κ, if and only if their limit points are concyclic.

Exercise 4.145. Show that in a triangle with side lengths $|AB| = 4k, |B\Gamma| = 5k$, $|\Gamma A| = 6k$ the angle $\widehat{B} = 2\widehat{\Gamma}$.

Exercise 4.146. To find a point O in the interior of the triangle $AB\Gamma$, such that the triangles OAB, $OB\Gamma$ and $O\Gamma A$ have the same area.

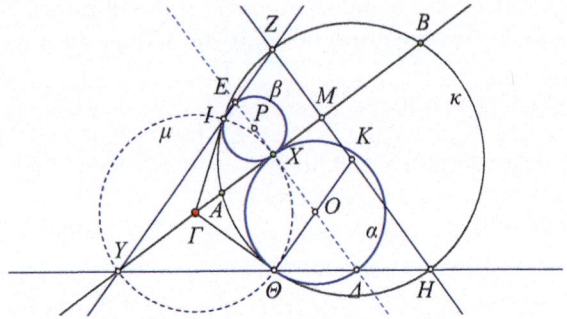

Fig. 4.99: Circles α, β tangent to κ and to the chord AB at X

Exercise 4.147. Let X be a point of the chord AB of the circle κ. Construct the circles $\{\alpha, \beta\}$, which are tangent to κ and to the chord at point X. Show that the ratio of the radii of $\{\alpha, \beta\}$ are independent of the position of X on AB (See Figure 4.99).

Hint: ([98, p.71]) The circles tangent to AB at X generate a circle pencil and the radical axis of κ with each member of the pencil passes through a fixed point Γ (Proposition 4.9). Drawing through it tangents to κ we find points $\{I, \Theta\}$ and the requested circles $\{\alpha, \beta\}$ (Exercise 4.34). The circle μ with center Γ, which contains points $\{I, \Theta\}$ is orthogonal to κ, the centerline ΔE of $\{\alpha, \beta\}$ is parallel to the diameter ZH of κ and lines $\{ZE, \Delta H\}$ pass through the symmetric Y of X relative to Γ. It follows that the ratio of the radii of α, β is equal to the ratio of $|MZ|, |MH|$.

Exercise 4.148. Construct a quadrilateral, for which are given the lengths of its sides and the length of the line segment which joins the middles of its diagonals (See Figure 4.100-I).

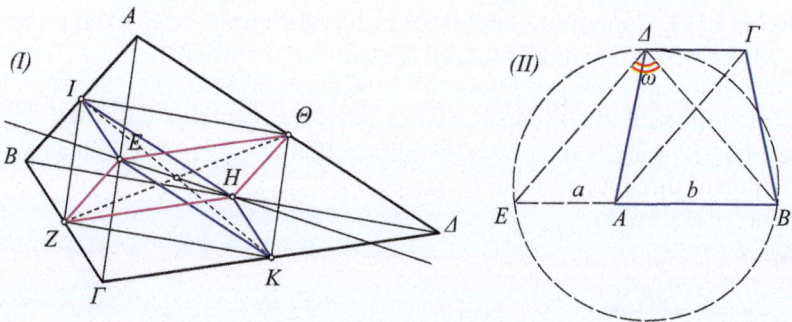

Fig. 4.100: Quadrilateral construction   Trapezium altitude

Exercise 4.149. Find the trapezium $AB\Gamma\Delta$ with maximal altitude, whose lengths of the bases are $a = |\Gamma\Delta|$, $b = |AB|$ and the angle between its diagonals is equal to ω (See Figure 4.100-II).

Exercise 4.150. From two given points $\{\Delta, E\}$ of a diameter of a circle to draw two equal chords, which intersect at a point A on the circle (See Figure 4.101-I).

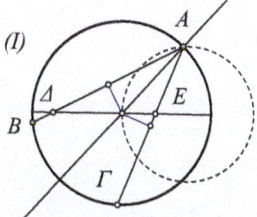 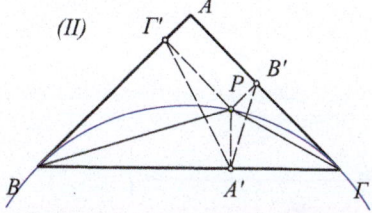

Fig. 4.101: Equal chords   Locus of points $|PA'|^2 = |PB'||P\Gamma'|$

Exercise 4.151. Find the geometric locus of the points P, for which the distance $|PA'|$ from the base $B\Gamma$ of an isosceles triangle $AB\Gamma$ is the mean proportional ($|PA'|^2 = |PB'||P\Gamma'|$) of its distances from the other sides (See Figure 4.101-II).

Hint: ([9, p.145]) Let P be a point of the locus. The triangles $PA'\Gamma'$ and $PB'A'$ are similar and, comparing angles, we see that $A'B\Gamma'$ and $B'\Gamma A'$ are also similar. Consequently the quadrilaterals $PA'B\Gamma'$ and $PB'\Gamma A'$ are similar. We conclude that $\widehat{BP\Gamma}$ is supplementary of $\widehat{AB\Gamma}$.

Exercise 4.152. Two billiard balls are placed at two arbitrary point $\{A, B\}$ of a diameter of a cyclic billiard. To which direction must move the ball at A, so that after one bounce against the wall it collides with the ball at B?

Exercise 4.153. Construct a circle κ, which is seen from three fixed points A, B, Γ under respective angles 2α, 2β, 2γ (See Figure 4.102-I).

Hint: Suppose that the circle $\kappa(O,r)$ has been constructed. Then $\frac{|OA|}{|OB|} = (r/\sin(\alpha))/(r/\sin(\beta))$ is known. It follows that O is constructible as an intersection of three Apollonian circles etc.

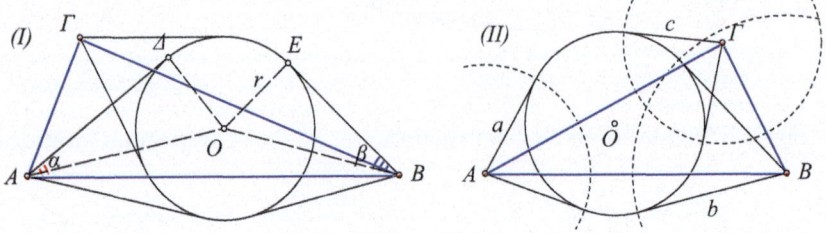

Fig. 4.102: Circle under 2α, 2β, 2γ Circle with tangents a, b, c

Exercise 4.154. Construct circle κ, towards which the tangents from three fixed points A, B, Γ have respective lengths a, b and c (See Figure 4.102-II).

Hint: The requested circle is centered at the radical center of three known circles and is orthogonal to them.

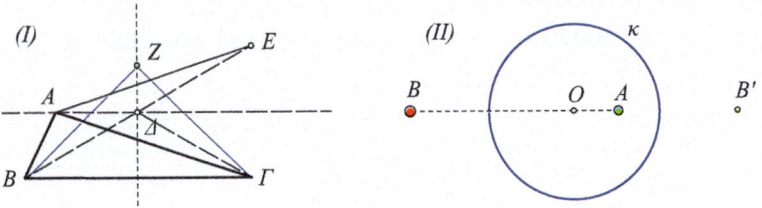

Fig. 4.103: Triangle of maximal area   Anti-inversion

Exercise 4.155. Show that from all triangles with fixed perimeter p, the equilateral has maximal area.

Hint: Suppose that the triangle $AB\Gamma$ has perimeter p and the maximal possible area. Then it must be isosceles with respect to the base $B\Gamma$ (See Figure 4.103-I). Otherwise, the isosceles $B\Delta\Gamma$ with the same base would have smaller perimeter. In fact, taking the symmetric E of Γ with respect to the parallel $A\Delta$ to the base $B\Gamma$, we would have $|AB| + |A\Gamma| = |AB| + |AE| > |B\Delta| + |\Delta\Gamma| = |BE|$, since the broken line has greater length than the line segment. Taking then Z on top of Δ, so that $BZ\Gamma$ has perimeter p, we would obtain a triangle with greater area, contradicting the hypothesis. The same argument shows that $AB\Gamma$ must be isosceles with respect to every one of its sides as base.

4.10. COMMENTS AND EXERCISES FOR THE CHAPTER

A slight modification of the concept inversion is the so called **anti-inversion**. It is defined, like the inversion, through a circle $\kappa(O,r)$ and corresponds to every point A of the plane, different from its center O, a point B on the same line OA, such that $|OA||OB| = r^2$ and such that OA, OB are opposite oriented (See Figure 4.103-II). It is easy to see that B is the symmetric, relative to O, of the inverse B' of A relative to the circle κ. The circle κ is called *circle of anti-inversion*, its center O *center of the anti-inversion* and its radius r *radius of the anti-inversion*. Next exercises show, that many properties of the *inversion* carry over to the *anti-inversion*. This gives a chance for a short review of the inversion's properties.

Exercise 4.156. Show that an anti-inversion maps the internal points of the circle to its external points and vice-versa. Also every point of the circle is mapped to its diametrically opposite. Conclude that the anti-inversion admits no fixed points.

Exercise 4.157. Show that for every pair of anti-inverse points $\{A,B\}$ relative to the circle κ and every circle λ passing through them, circles κ and λ intersect at diametral points of κ.

Exercise 4.158. Two different pairs of anti-inverse points (X,Y) and (X',Y') relative to the circle κ define four concyclic points.

Exercise 4.159. If the point X varies on a line ε, then its anti-inverse point Y relative to a circle $\kappa(O,r)$ varies on a circle ζ passing through the center O of κ. If the point Y varies on the circle ζ passing through point O, then its anti-inverse X relative to κ, varies on a line ε.

Exercise 4.160. If the point X varies on a circle λ not passing through O, then its anti-inverse Y relative to the circle $\kappa(O,r)$ varies on a circle μ.

Exercise 4.161. The anti-inverse points Y of points X of a circle λ are again points of λ, if and only if the circle λ intersects the anti-inversion circle κ at diametral points of κ.

Exercise 4.162. Let Y be the anti-inverse relative to the circle κ of the point X. If the circles λ and μ pass through X and intesect under the angle ω, then their anti-inverses λ', μ' pass through point Y and intersect under the same angle ω.

Exercise 4.163. For a given circle pencil $\mathscr{D}$ and a circle κ, the anti-inverses relative to κ, of the circles λ of the pencil, build another pencil of circles $\mathscr{D}'$.

Exercise 4.164. For a given intersecting pencil of circles $\mathscr{D}$ and a circle κ, show that the pencil $\mathscr{D}'$ of the anti-inverses relative to κ, of the circles λ of the pencil and the original pencil $\mathscr{D}$ have exactly one circle-member μ in common. Circle μ is characterized by the fact that it passes through the basic points of $\mathscr{D}$ as well as through the basic points of $\mathscr{D}'$.

Exercise 4.165. For a given non intersecting pencil of circles $\mathscr{D}$ and a circle κ, show that the pencil $\mathscr{D}'$ of the anti-inverses relative to κ, of the circles λ of the pencil and the original pencil $\mathscr{D}$ have exactly one circle-member μ in common. μ is characterized by the fact that it intersects κ at diametral points of κ.

Exercise 4.166. For a given tangent pencil of circles $\mathscr{D}$ and a circle κ, show that the pencil $\mathscr{D}'$ of the anti-inverses relative to κ, of the circles λ of the pencil and the original pencil $\mathscr{D}$ have exactly one circle-member μ common. What characterizes μ in this case?

Exercise 4.167. Given a circle κ and a line ε, show that there exists a circle ω, relative to which the anti-inversion interchanges κ with ε.

Exercise 4.168. Show that there is no anti-inversion interchanging two intersecting circles κ and λ.

Hint: If $\{\kappa, \lambda\}$ are tangent, then the anti-inversion should leave the point of contact fixed, which is impossible. If κ, λ intersect at two different points, then the anti-inversion should interchange them, therefore its center should be in the interior of the line segment defined by the intersection etc.

Exercise 4.169. Given two not congruent and not intersecting circles κ and λ, show that there exists exactly one circle ν, relative to which the anti-inversion interchanges κ and λ. The center of nu is the internal point of similarity Δ of κ, λ, the radius r of ν satisfies $r^2 = \alpha\beta \left(\frac{|AB|^2}{(\alpha+\beta)^2} - 1 \right)$ and the circle of inversion μ of the theorem 4.16, which also interchanges κ, λ, intersects the circle ν at two diametral points of ν.

Two circles like $\{\kappa, \lambda\}$, as well as the related to them circles $\{\mu, \nu\}$, defined in the preceding exercise, make an interesting system with many applications. Next exercise lists some basic properties of this system.

Exercise 4.170. Given two not congruent and not intersecting circles $\kappa(A, \alpha)$ and $\lambda(B, \beta)$ and the circles μ, ν (See Figure 4.104), defined in the preceding exercise, show that:

1. Two pairs of antihomologous points (X, X'), (Y, Y') whose lines XX', YY' pass through the exterior (resp. interior) center of similarity Γ (resp. Δ), are contained in a circle τ (resp. σ), which intersects orthogonally (resp. by diameter of ν) the circle μ (resp. ν).
2. Circle τ (resp. σ) is invariant with respect to the inversion (resp. anti-inversion) relative to μ (resp. relative to ν).
3. Conversely, every circle τ (resp. σ) which intersects circles κ, λ and is invariant with respect to the inversion relative to μ (resp. with respect to the anti-inversion relative to ν) intersects circles κ, λ by pairs of antihomologous points (X, X'), (Y, Y'), whose lines XX', YY' pass through point Γ (resp. Δ).

4. A circle tangent to κ, λ will be invariant with respect to the inversion relative to μ or it will be invariant with respect to the anti-inversion relative to ν. In the first case the line of the contact points with κ, λ will pass through Γ. In the second case this line will pass through point Δ.

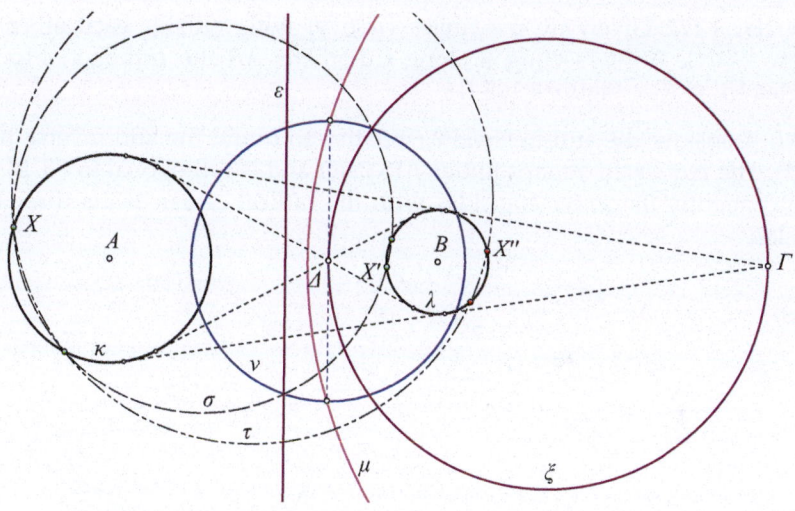

Fig. 4.104: The system of two non intersecting circles $\{\kappa, \lambda\}$

5. The inverse (resp. anti-inverse) of the circle of similitude ξ of κ, λ relative to μ (resp. ν), is the radical axis of κ, λ.

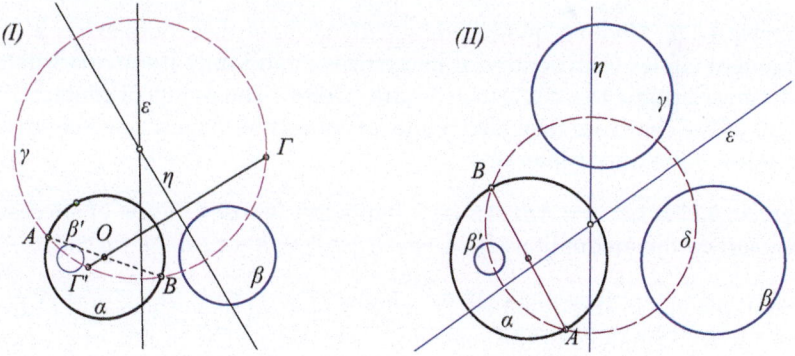

Fig. 4.105: Intersection along diameter   ... and orthogonal to β

Exercise 4.171. Let the circles $\{\alpha, \beta\}$ lie outside each other and the point Γ outside both. To construct a circle γ intersecting circle α along a diameter AB and orthogonal to β (See Figure 4.105-I).

Hint: Consider the anti-inversion f relative to α and the anti-inverse $\beta' = f(\beta)$, as well as the anti-inverse $\Gamma' = f(\Gamma)$ of the given point. The requested circle γ maps under f to itself, is orthogonal to $\{\beta, \beta'\}$ und passes through points $\{\Gamma, \Gamma'\}$. It follows that its center must be on the medial line η of $\Gamma\Gamma'$ and also on the radical axis ε of the two circles $\{\beta, \beta'\}$.

Exercise 4.172. Given are the circles $\{\alpha, \beta, \gamma\}$ lying outside each other. To draw a circle δ intersecting α along a diameter AB and orthogonal to the circles $\{\beta, \gamma\}$ (See Figure 4.105-II).

Hint: Consider the anti-inversion f relative to α and the anti-inverse $\beta' = f(\beta)$. The requested γ maps under f to itself and is orthogonal to $\{\beta, \beta', \gamma\}$. Concequently its center coincides with the radical center of the three last circles.

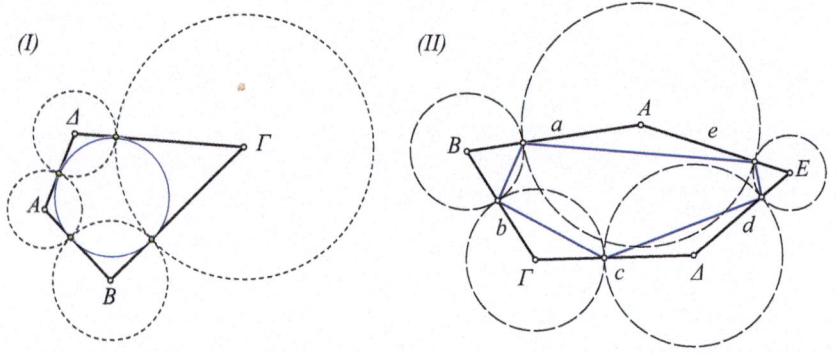

Fig. 4.106: Chain of tangent circles   Derived pentagon

Exercise 4.173. Given a quadrilateral $AB\Gamma\Delta$, show that there exists a chain of tangent circles with centers at its vertices, if and only if the quadrilateral is circumscriptible to a circle (See Figure 4.106-I). The points of contact of the circles of the chain are then also points of contact of the inscribed circle with the sides of the quadrilateral.

Exercise 4.174. Let $p = AB\Gamma\Delta E$ be a pentagon whose lengths of successive sides satisfy the inequalities $a - b + c - d + e > 0$, $b - c + d - e + a > 0$, $c - d + e - a + b > 0$, $d - e + a - b + c > 0$ and $e - a + b - c + d > 0$. Show that there is a unique chain of tangent circles with centers at its vertices and radii the corresponding halves of the preceding quantities. The points of contact of the circles of the chain define then a new pentagon, inscribed in p, which is called **derived** of p (See Figure 4.106-II).

Exercise 4.175. Let $AB\Gamma\Delta E$ be a pentagon and a variable point P on the medial line ε of its side EA (See Figure 4.107). Line PA intersects the medial line α of AB at H. Line HB intersects the medial line β of $B\Gamma$ at Θ. Line $\Theta\Gamma$ intersects the medial line γ of $\Gamma\Delta$ at I. Line HB intersects the medial line β of $B\Gamma$

4.10. COMMENTS AND EXERCISES FOR THE CHAPTER

at Θ. Line $\Theta\Gamma$ intersects the medial line γ of $\Gamma\Delta$ at I. Line $I\Delta$ intersects the medial line δ of ΔE at K. Line KE intersects PA at Z. Show that the geometric locus of Z is a circle passing through points A and E (angle $\widehat{AZE}$ is constant).

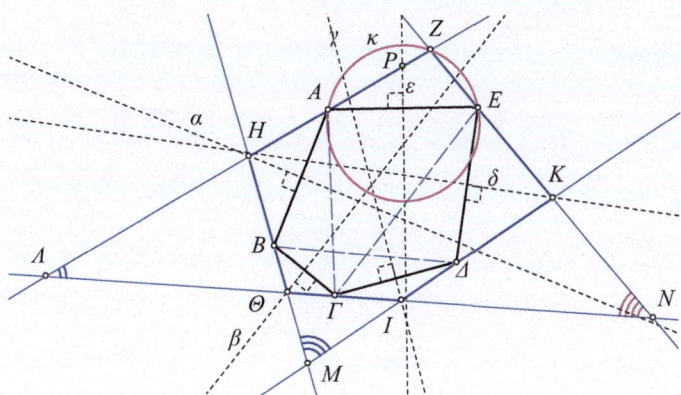

Fig. 4.107: Inverse construction of derived pentagon

Exercise 4.176. Using the notation of the preceding exercise, show that there exist exactly two positions of P on the medial line ε of side AE, for which Z coincides with P. These are the intersections of ε with κ. Conclude that for every pentagon $q = AB\Gamma\Delta E$ there are exactly two pentagons p, p', whose derived pentagon is q.

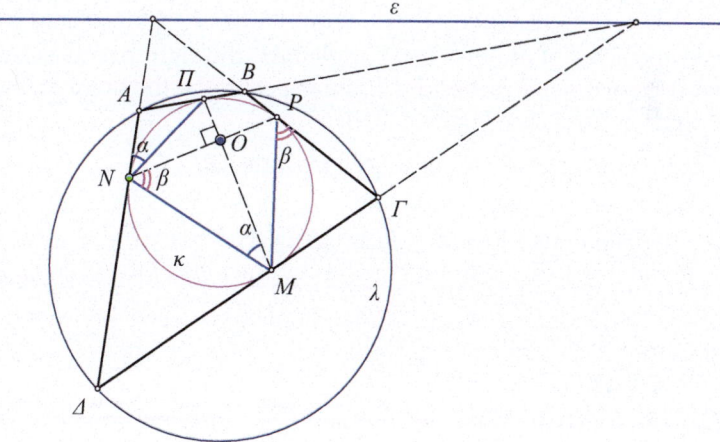

Fig. 4.108: Quadrilateral simultaneously inscribed and circumscribed

Exercise 4.177. Show that a quadrilateral simultaneously circumscribed in a circle κ and inscribed in a circle λ, has the lines NP and $M\Pi$, joining points of contact of opposite sides, orthogonal (See Figure 4.108). Conversely, if a right angle $\widehat{NOM}$ is rotated about the fixed point O and at the intersection points of its sides with the fixed circle κ we draw tangents, this forms a quadrilateral $AB\Gamma\Delta$, which is simultaneously circumscribed to κ and inscribed in another circle λ. Show also that circle λ is the same for every rotational position of the right angle and depends only on the position of O relative to κ (see also Exercise 3.35).

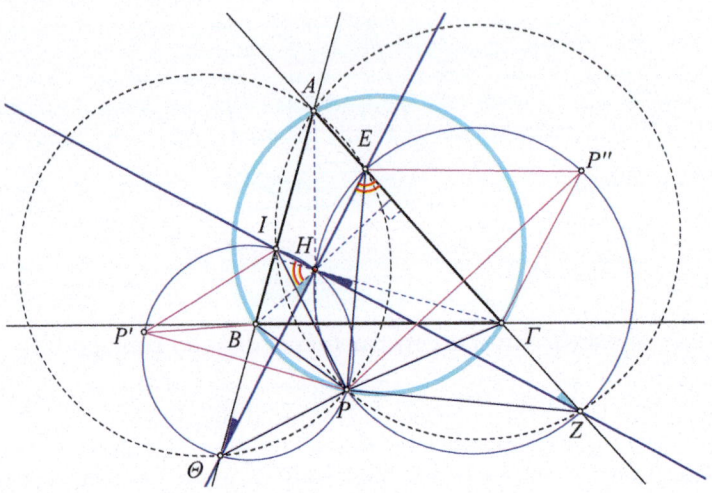

Fig. 4.109: Right angle rotating around H inside the angle $\widehat{BA\Gamma}$

Exercise 4.178. A right angle rotates about its fixed vertex H and its sides define on the sides of another fixed angle $\widehat{BA\Gamma}$ the right triangles $HI\Theta$ and HEZ, whose circumcircles, besides H, intersect also at the point P. Show the following properties (See Figure 4.109):

1. Point H is the orthocenter of a fixed triangle $AB\Gamma$.
2. The ratios $\frac{|BI|}{|B\Theta|}$ and $\frac{|\Gamma Z|}{|\Gamma E|}$ are equal.
3. The quadrilaterals $PEA\Theta$ and $PIAZ$ are cyclic.
4. The geometric locus of point P is the circumcircle of the triangle $AB\Gamma$.

Hint: For (1), draw from H the perpendiculars, respectively to AI, AE intersecting the other sides of the fixed angle, respectively at Γ and B. The triangle $AB\Gamma$ has H as its orthocenter.

For (2), observe that the angles $\widehat{I\Theta H}$ and $\widehat{\Gamma HZ}$ are equal having their sides, respectively orthogonal. Similarly angles $\widehat{BH\Theta}$ and $\widehat{HZ\Gamma}$ are equal and consequently the triangles $BH\Theta$ and ΓHZ are similar. Angles $\widehat{IHB}$ and $\widehat{HE\Gamma}$ also have their sides, respectively, orthogonal and are equal. It follows that the triangles BHI and ΓEH are similar. From the similar triangles follows the relation

$$\frac{|BI|}{|B\Theta|} = \frac{|BI|}{|BH|} \cdot \frac{|BH|}{|B\Theta|} = \frac{|H\Gamma|}{|\Gamma E|} \cdot \frac{|\Gamma Z|}{|\Gamma H|} = \frac{|\Gamma Z|}{|\Gamma E|}.$$

For (3), split the angle

$$\widehat{\Theta PE} = \widehat{\Theta PH} + \widehat{HPE} = \widehat{AIH} + \widehat{IZA} = 180° - \widehat{A}.$$

This shows that $PEA\Theta$ is inscriptible. Similar relations hold for $PIAZ$.

For (4), consider the symmetric points P', P'' of P, respectively, relative to AB, $A\Gamma$. From (3) follows that the isosceli triangles $PP'I$ and $PP''E$ have supplementary angles at their vertices $\widehat{P'IP}$ and $\widehat{PEP''}$. Combining with (2) we see that the isosceli triangles PBP' and $P\Gamma P''$ are similar hence the angles $\widehat{BP\Gamma}$ and $\widehat{P'PP''} = 180° - \widehat{A}$ are equal.

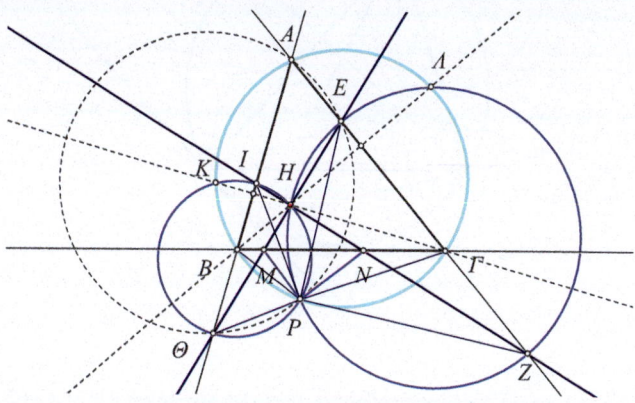

Fig. 4.110: Rotated right angle

Exercise 4.179. A right angle having its vertex at the orthocenter H of triangle $AB\Gamma$ rotates about H (See Figure 4.110). Its sides define on the sides of the triangle the right triangles $HI\Theta$, HMN and HEZ. Show that the circumcircles of these right triangles, besides H, pass also through a point P of the circumcircle of triangle $AB\Gamma$.

Hint: Suppose that angle $\widehat{A}$ is acute. The proof for obtuse or right angle is similar. The fact that the two circumcircles of triangles $IH\Theta$ and EHZ pass through point P of the circumcircle of $AB\Gamma$ follows from exercise 4.178. For the proof of the exercise, it suffices to show that angle $\widehat{MPN}$ is right. This follows from the fact that the quadrilaterals $PZ\Gamma N$ and $P\Theta BM$ are inscriptible. Indeed, assuming for the moment that this holds, we see that then $\widehat{NP\Gamma} = \widehat{NZ\Gamma}$ and $\widehat{MPB} = \widehat{M\Theta B}$. From this follows that $\widehat{BPM} + \widehat{NP\Gamma} = 90° - \widehat{A}$ and from the latter, because $\widehat{BP\Gamma} = 180° - \widehat{A}$ (Exercise 4.178), follows that $\widehat{MPN}$ is right. Now, the fact that $PZ\Gamma N$ is inscriptible follows from $\widehat{P\Gamma N} = \widehat{PAB} = \widehat{PE\Theta}$ (Exercise 4.178) and $\widehat{PE\Theta} = \widehat{PEH} = \widehat{PZN}$. Similarly follows that $PMB\Theta$ is also inscriptible.

Note that the last exercise shows that the three circumcircles $(HI\Theta)$, (HMN) and (HEZ) belong to an intersecting pencil whose one basic point is H and the other is P lying on the circumcircle of triangle $AB\Gamma$. This implies, that the centers of these circles are collinear. This conclusion is mentioned often as *theorem of Droz-Farny* (1856-1912) ([97], [96]).

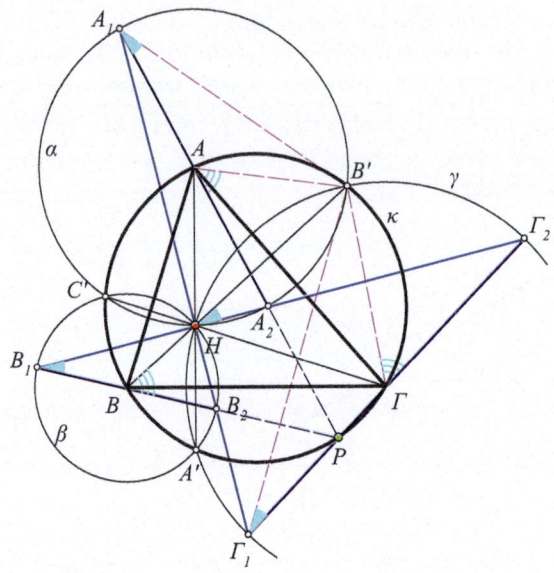

Fig. 4.111: Correspondence of right angle to diameters

Exercise 4.180. Consider the three circles $\{\alpha, \beta, \gamma\}$ with centers, respectively the vertices of the triangle $AB\Gamma$ and passing through the orthocenter H. Show that for every point P of the circumcircle, the diameters of these circles, which pass through point P, have their endpoints, by three, on two orthogonal lines which pass through H (See Figure 4.111).

Exercise 4.181. From a point A outside the circle κ draw a secant $AB\Gamma$ of it, such that the interior of the circle segment of $B\Gamma$ to its exterior segment AB are in a given ratio λ.

Exercise 4.182. Construct a circle κ passing through two given points and intersecting a given circle λ along a diameter of λ (or allong a diameter of κ if possible).

Exercise 4.183. Given are two circles $\{\alpha, \beta\}$ and a point Δ. Show that there are at most two lines through Δ intersecting the two circles at points at which the corresponding tangents are parallel. Determine the cases in which there are two, one or none such line.

4.10. COMMENTS AND EXERCISES FOR THE CHAPTER

Exercise 4.184. The circles $\{\beta,\gamma\}$ with centers, respectively, the vertices $\{B,\Gamma\}$ of triangle $AB\Gamma$, intersect at two points $\{\Delta,E\}$ of the altitude from A (See Figure 4.112) and intersect the other altitudes at points $\{I,K\}$ and $\{Z,\Theta\}$. Show that the points $\{K,I,Z,\Theta\}$ lie on a circle α, which has its center at A. Also show that the chords $\{KK',II',ZZ',\Theta\Theta'\}$, which are defined by the lines, re-

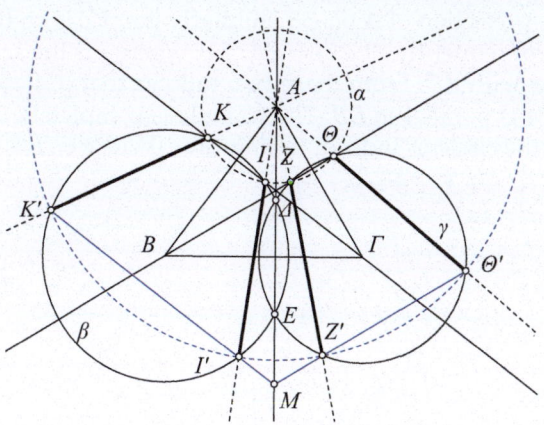

Fig. 4.112: Circles at vertices

spectively, $\{AK,AI,AZ,A\Theta\}$ on the two circles are equal. Finally show that the intersection point M of the lines $\{K'I',Z'\Theta'\}$ is on the line ΔE.

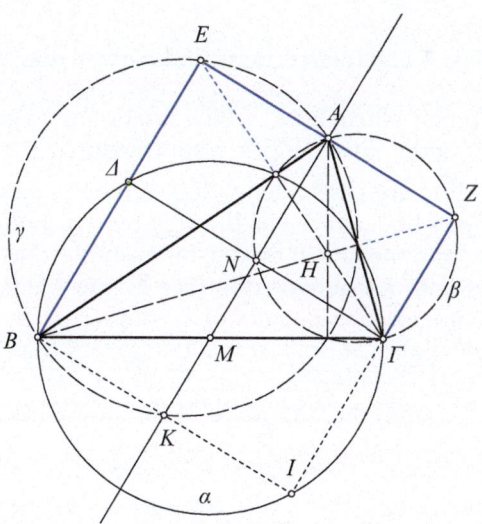

Fig. 4.113: Special triangles $AB\Gamma$

Exercise 4.185. Determine all the acute angled triangles $AB\Gamma$ with fixed base $B\Gamma$, for which the right angles $\widehat{AEB}$ and $\widehat{AZ\Gamma}$ with $\{E,Z\}$ on the extensions of the altitudes, respectively, $\{\Gamma H, BH\}$, have the points $\{E,A,Z\}$ collinear (See Figure 4.113).

Hint: If $\{\alpha, \beta, \gamma\}$ are respectively the circles with diameter the sides $B\Gamma, \Gamma A$, AB, then point E is on the radical axis of $\{\alpha, \beta\}$, therefore $EZ \cdot EA = E\Delta \cdot EB$, where Δ is the projection of A on EB. Similarly point Z is on the radical axis of $\{\alpha, \gamma\}$ and consequently $Z\Gamma \cdot ZI = ZA \cdot ZE$, where I is the second intersection point of α with $Z\Gamma$. From these two follows that $|EA| = |AZ|$ and that the median AM is parallel to EB. The triangle is completely determined by the point Δ of the circle α. Indeed, setting $\{d = |\Gamma\Delta|, g = |B\Delta|, x = |E\Delta|\}$, we find $EA \cdot EZ = \frac{d^2}{2} = E\Delta \cdot EB = x(x+g)$.

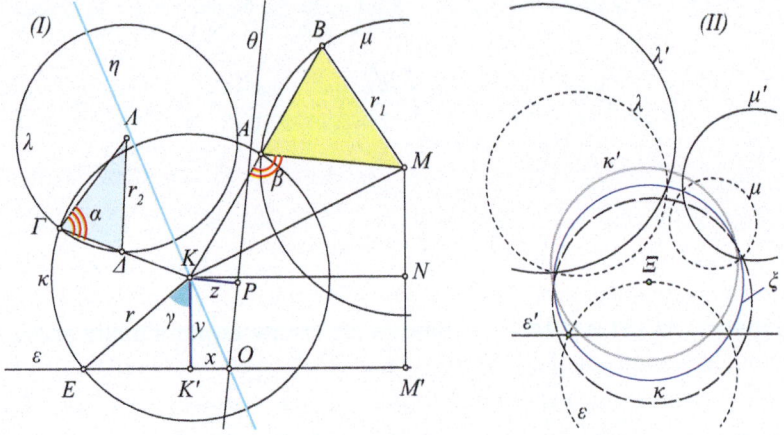

Fig. 4.114: Intersections under given angles

Exercise 4.186. To find all circles κ, which intersect two given circles λ, μ and a given line ε under given angles, respectively, $\{\alpha, \beta, \gamma\}$.

Hint: There are different cases, depending on the relative position of the given circles and the line, as well as the way by which the requested κ intersects the circles and the line. It can be, for example, that μ has the second intersection point with KA outside of κ (See Figure 4.114), while the corresponding second intersection point Δ of λ with $K\Gamma$ may be inside κ. Independently from the figure, it is a fact that the isosceli triangles $\{\Lambda\Gamma\Delta, ABM\}$ have sides determined by the given data. In the case of figure 4.114-I, where K is outside both circles, the difference of the powers of K relative to the circles $\{\lambda, \mu\}$ will be (Proposition 4.2)

$$||K\Gamma||K\Delta| - |KA||KB|| = |r(r-t_1) - r(r+t_2)| = r(t_1+t_2) = 2|KP||\Lambda M|,$$

where $\{t_1, t_2\}$ are the lengths of the bases of the isosceli triangles and $\{P, K'\}$ are, respectively, the projections of K on the radical axis θ of $\{\lambda, \mu\}$ and on

4.10. COMMENTS AND EXERCISES FOR THE CHAPTER

the line ε. Consequently, for $z = |KP|$, $y = |KK'|$, the ratio $\frac{z}{r} = \frac{z \cdot \cos(\gamma)}{y}$ will be fixed and consequently $z = k \cdot y$ and $x = |OK'| = n \cdot y$ for some constants, respectively, $\{k, n\}$, which are determined from the given data. The position of K is determined from the quadratic equation

$$|KM|^2 = r^2 + r_1^2 - 2rr_1 \cos(\beta) = (|OM'| + ny)^2 + (|MM'| - y)^2 \iff ay^2 + by + c = 0,$$

with coefficients $\{a, b, c\}$ which are determined from the given data.

Exercise 4.187. To find all circles κ, which intersect three circles $\{\lambda, \mu, \varepsilon\}$ under given angles, respectively, $\{\alpha, \beta, \gamma\}$.

Hint: Select a point Ξ on the circle ε and apply an inversion relative to the circle $\xi(\Xi)$ (the radius is immaterial). Through the inversion the circles $\{\lambda, \mu, \varepsilon\}$ map respectively to two circles $\{\lambda', \mu'\}$ and a line ε' (See Figure 4.114-II). Using exercise 4.186 construct a circle κ' which intersects $\{\lambda', \mu', \varepsilon'\}$ under the angles, respectively, $\{\alpha, \beta, \gamma\}$. The inverse κ of κ' is the required circle.

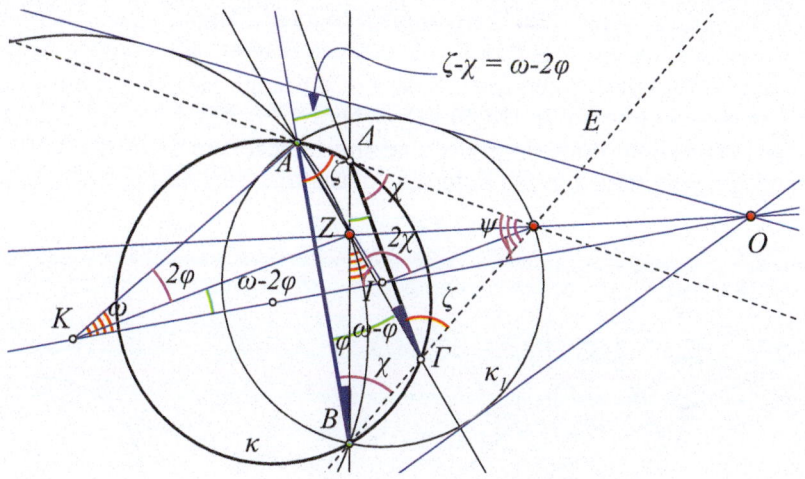

Fig. 4.115: Line through the similarity center

Exercise 4.188. On the circle κ the chord AB remains fixed and the chord $\Gamma \Delta$ varies without to change its length. Show that the intersection points of the lines $\{E = (A\Delta, B\Gamma), Z = (A\Gamma, B\Delta)\}$ move on two fixed circles respectively $\{\kappa_1, \kappa_2\}$ and the variable line EZ passes through a fixed point, coinciding with a similarity center O of these circles (See Figure 4.115).

Hint: The first claim is a consequence of the exercise 2.157. The second claim results by measuring the angles which are defined by various lines and by proving that the radii of the circles $\{IE, KZ\}$ are parallel.

Exercise 4.189. Let κ be a circle and the angle $\widehat{AB\Gamma}$ inscribed in it (See Figure 4.116). Show that there is a point O, such that the second intersection points $\{A', B', \Gamma'\}$ with the lines correspondingly $\{OA, OB, O\Gamma\}$, define an isosceles right triangle $A'B'\Gamma'$.

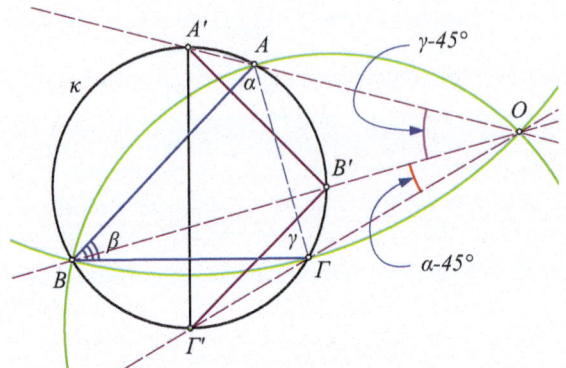

Fig. 4.116: From angle β to a right angle

Exercise 4.190. With centers the vertices of a triangle $AB\Gamma$ we draw circles with corresponding radii $\{ka, kb, kc\}$, where $\{a = |B\Gamma|, b = |\Gamma A|, c = |AB|\}$, where k a positive number. Show that the radical centers of these triples of circles are contained in the Euler line of the triangle (See Figure 4.117-I).

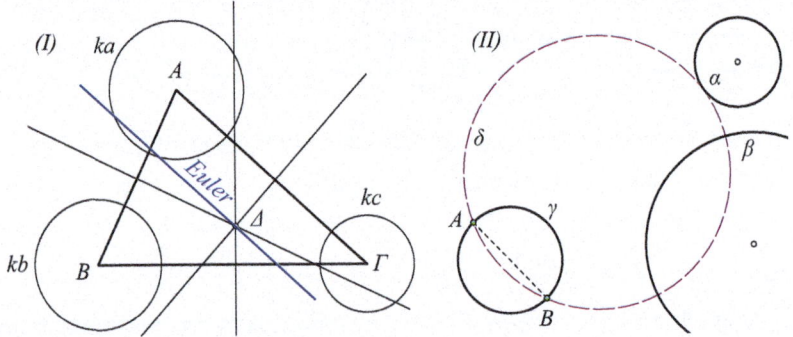

Fig. 4.117: Euler line characterization   Circle δ with 3 properties

Hint: Since the Euler line ε of the triangle passes through the centroid and the circumcenter, compute, for a point X of ε and its projection X' on a side e.g. $B\Gamma$, the ratio of distances $|X'B|/|X'\Gamma|$.

Exercise 4.191. Let $\{\alpha, \beta, \gamma\}$ be three circles lying outside each other. Construct a circle δ tangent to α, intersecting β orthogonally and intersecting γ along a diameter AB (See Figure 4.117-II).

4.10. COMMENTS AND EXERCISES FOR THE CHAPTER

Exercise 4.192. From a given point O to draw the tangents to the circle κ through three given points $\{A, B, \Gamma\}$ but whose center is outside the drawing sheet (See Figure 4.118).

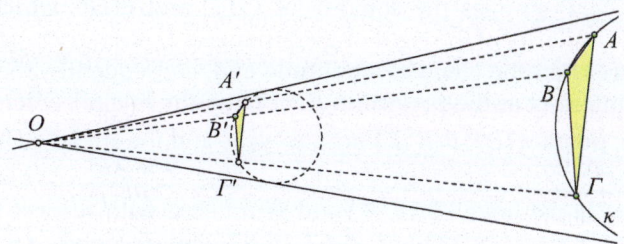

Fig. 4.118: Tangents from a point

Exercise 4.193. Let $\{\kappa(A, r_\kappa), \lambda(B, r_\lambda)\}$ be two circles lying outside each other and a point Π outside both. Show that the following properties are equivalent (See Figure 4.119):

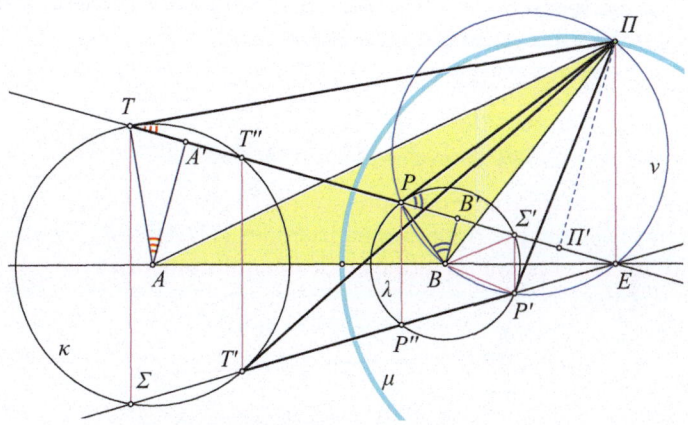

Fig. 4.119: Circle of similitude of two circles and equal triangles

1. The point Π is on the circle of similitude μ of $\{\kappa, \lambda\}$.
2. The tangents to the circles from Π, by two and such that they separate the centers (i.e. at most one center lies in the interior of the angle $\widehat{T\Pi P}$), define equal triangles $\{\Pi T P, \Pi T' P'\}$.
3. The base lines $\{TP, T'P'\}$ of these triangles are symmetric with respect to the line of centers.
4. The base lines $\{TP, T'P'\}$ cut from the circles equal chords.

Hint: (1 ⇒ 2) If the point Π lies on the circle of similitude, then these two triangles are similar to ΠAB (Exercise 4.83). The equality of the tangents $\{\Pi T, \Pi T'\}$, implies then the equality of the triangles.

(2 ⇒ 1,3) If the two triangles are equal, then the equality of the angles $\widehat{T\Pi P}$ and $\widehat{T'\Pi P'}$ implies the equality of $\widehat{T\Pi T'}$ and $\widehat{P\Pi P'}$, which, in turn, implies the similarity of the right-angled triangles $\{\Pi TA, \Pi PB\}$. This implies that point Π is on the circle of similitude and, consequently, the two lines $\{TP, T'P'\}$ intersect at the projection E, of Π on the line of centers (Exercise 4.83). Then, points $\{\Pi, P, B, P', E\}$ are concyclic and the angles $\widehat{P\Pi B} = \widehat{PEB} = \widehat{B\Pi P'} = \widehat{BEP'}$.

(3 ⇒ 1,4) If the lines $\{TP, T'P'\}$ are symmetric with respect to the line of centers and intersect at the point E of the line of centers AB, then they cut from the circles pairs of symmetric chords $(TT'', \Sigma T')$ and $(P\Sigma', P''P')$. In addition the quadrangle $PBP'E$ is inscriptible in a circle ν, because $\widehat{PBP'} = 2\widehat{PP''E} = 2\frac{180° - \widehat{PEP''}}{2} = 180° - \widehat{PEP''}$. It follows that Π is contained in this circle and that ΠE is orthogonal to AB. This implies that Π is on the circle of similitude of $\{\kappa, \lambda\}$ (Exercise 4.83) and consequently, by (1) $|PT| = |T'P'|$. Then, subtracting from equal segments $|T\Sigma'| = |\Sigma P'|$ the preceding equal segments, we find that $|P\Sigma'| = |\Sigma T'|$.

(4 ⇒ 1) If the segments $|TT''| = |P\Sigma'|$ (See Figure 4.119), then for the radii of the circles the following equation will be valid

$$r_\kappa \cdot \sin(\widehat{TAA'}) = r_\lambda \cdot \sin(\widehat{PBB'}) \Rightarrow \frac{r_\kappa}{r_\lambda} = \frac{\sin(\widehat{PBB'})}{\sin(\widehat{TAA'})} = \frac{|\Pi\Pi'|/|\Pi P|}{|\Pi\Pi'|/|\Pi T|} = \frac{|\Pi T|}{|\Pi P|},$$

where $\{A', B', \Pi'\}$ are the projections, respectively, of $\{A, B, \Pi\}$ on the line TP. This implies, that the right-angled triangles $\{\Pi TA, \Pi PB\}$ are similar and that the point Π is contained in the circle of similitude of $\{\kappa, \lambda\}$.

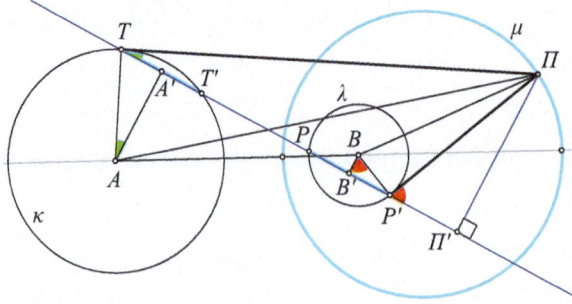

Fig. 4.120: Circle of similitude and equal chords

Exercise 4.194. Show that the last property of the preceding exercise is valid also in the case of two tangents from Π, which do not separate the centers (i.e. such that the centers of both circles lie in the interior of the angle formed

4.10. COMMENTS AND EXERCISES FOR THE CHAPTER

by the two tangents). In other words, given the circles $\{\kappa, \lambda\}$, each lying outside the other, and the tangents $\{\Pi T, \Pi P'\}$ from Π, which do not separate the centers, then Π lies on the circle of similitude of $\{\kappa, \lambda\}$, if and only if, the chords $\{TT', PP'\}$, cut by the line TP' of the contact points, are equal $|TT'| = |PP'|$ (See Figure 4.120).

Exercise 4.195. A right angle rotates about its vertex P, which is contained in the interior of a circle κ. Its sides intersect the circle respectively at A, Δ and B, Γ. From these points we draw parallels to the sides of the right angle, which form a rectangle. Show that

1. The diagonals of the rectangle intersect at the center O of the circle κ.
2. The geometric locus of the vertices of the rectangle is a circle κ' concentric to κ, of which calculate its radius as a function of the radius of κ and the distance $|PO|$.
3. Show that the diagonally-lying parallelograms, like, for example, these with diagonals $A\Gamma, B\Delta$ are similar.

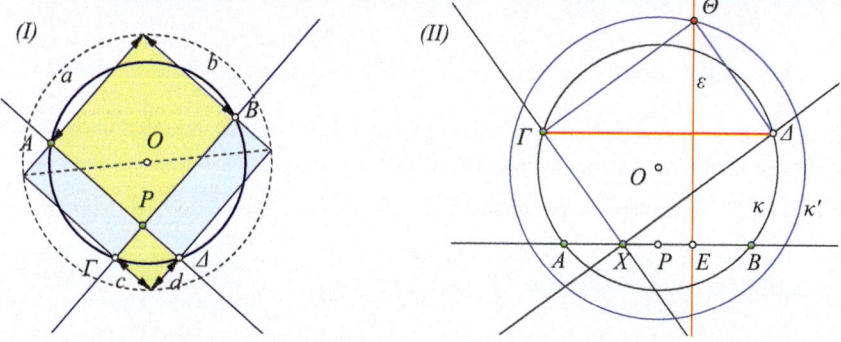

Fig. 4.121: Rotating right angle   Chord construction

Hint: Referring to figure 4.121-I, the square of the diagonal is equal to $(a+d)^2 + (b+c)^2$. Apply Exercise 3.36.

Exercise 4.196. Let X be a point on the chord AB of the circle κ. Construct a chord $\Gamma\Delta$ parallel to AB such that the angle $\Gamma X\Delta$ is right.

Hint: The requested chord defines a rectangle $X\Gamma\Theta\Delta$, whose vertex Θ belongs to the intersection of two geometric loci (See Figure 4.121-II): (i) of the circle κ', discussed in the preceding exercise and (ii) of the line ε which is the locus of vertices of parallelograms $\Gamma X \Delta \Theta$ for Γ moving on κ whith $\Gamma\Delta$ a chord of κ parallel to AB.

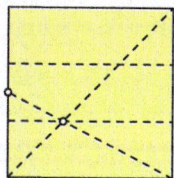

The segment $\Gamma\Delta$ could be determined also by computing its distance from AB. The preceding solution, though, gives a better insight into the geometric structure of the problem. Next exercise prepares a method of construction of the third root $\sqrt[3]{2}$ (Delian problem of doubling the cube) using a folding procedure of a square sheet of paper (the method of origami we encountered also in exercise 1.70). For this method of construction it is needed the preliminary division by appropriate folding of the sheet of paper in three equal strips suggested by the preceding figure on the right, the proof for this being supplied by exercise 2.86.

Exercise 4.197. Given are the square $AB\Gamma\Delta$, a variable point B' on the side $A\Delta$ and points $\{Z, I\}$ on side $B\Gamma$, which divide it in three equal parts. Show that there is a position of B', such that the symmetric Z' of Z relative to the medial line KM of BB' is contained in the parallel IH of AB (See Figure 4.122).

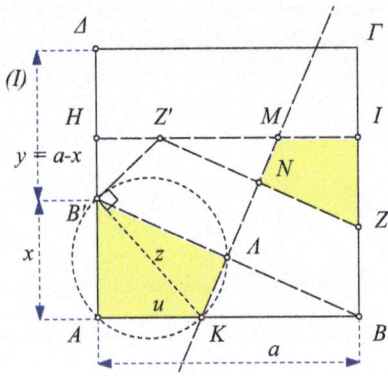
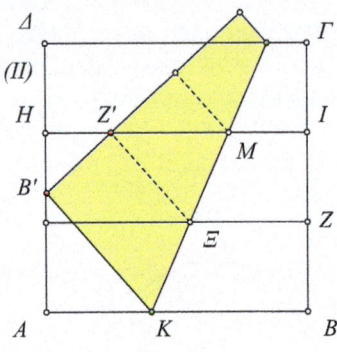

Fig. 4.122: Construction preparation   The $\sqrt[3]{2}$ with origami

Hint: With the notation of figure 4.122-I, where the position of B' is determined through the variable x, the following relations are valid.

$$w = |BB'| \Rightarrow w^2 = x^2 + a^2,$$
$$a(a-u) = w^2/2 \Rightarrow u = (a^2 - x^2)/(2a),$$
$$z = |B'K| = |KB| = a - u \Rightarrow z = (a^2 + x^2)/(2a),$$
$$|B'Z'| = |BZ| = a/3 \text{ and } B'AK \text{ similar to } Z'HB' \Rightarrow$$
$$|B'H|/|B'Z'| = u/z \Rightarrow |B'H| = (u/z)(a/3).$$

Assuming that $|AB'| + |B'H| = x + |B'H| = 2a/3$, and using the preceding relations, we find that x satisfies the equation

$$3x^3 - 3ax^2 + 3a^2x - a^3 = 0.$$

To complete the exercise we must find the real root of this equation which is $x_0 = \frac{a}{3}\left(\sqrt[3]{4} - \frac{2}{\sqrt[3]{4}} + 1\right)$, and see that this determines the position of B' with the requested properties. Next exercise shows a simpler method of determination of B'.

Exercise 4.198. With the notation and the value of x, determined in the preceding exercise, show that the ratio $\lambda = y/x = \sqrt[3]{2}$. Conclude that, by folding

4.10. COMMENTS AND EXERCISES FOR THE CHAPTER

the square sheet of paper, so that B falls onto $A\Delta$ and Z falls onto HI, the ratio $|AB'|/|B'\Delta| = \sqrt[3]{2}$.

Hint: The first claim results from the last equation. In fact, writing $\lambda = y/x = (a-x)/x$, from which follows $x = a/(1+\lambda)$ and replacing in that equation, we see that $2 - \lambda^3 = 0$.

The second claim results from the way we fold the paper (See Figure 4.122-II), the crease coincides with line KM and B passes exactly onto point B' of the preceding exercise.

Exercise 4.199. Consider an intersecting pencil of circles $\mathcal{D}$ at the points $\{O,A\}$ and two lines $\{\beta,\gamma\}$ through O. Show that for every circle κ of the pencil, its second intersection points $\{B,\Gamma\}$ with lines $\{\beta,\gamma\}$ define segments $\{x = |OB|, y = |O\Gamma|\}$, the lengths of whose satisfy a relation of the form $ax + by = c$, where $\{a,b,c\}$ are constants.

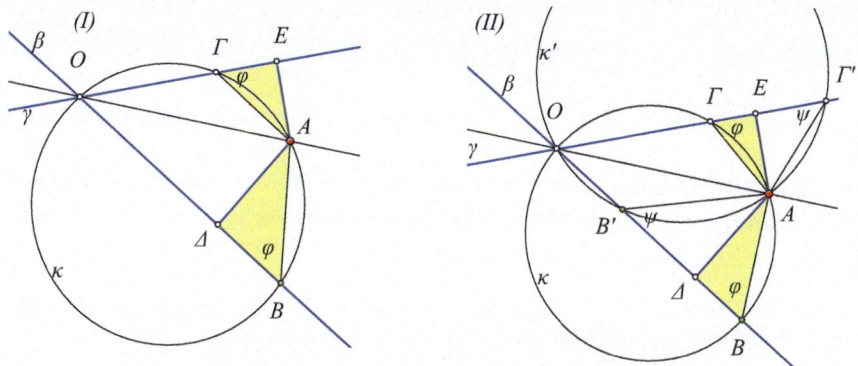

Fig. 4.123: Intersecting pencil of circles and lines

Hint: Draw the orthogonals $\{A\Delta, AE\}$ respectively to $\{\beta,\gamma\}$ (See Figure 4.123-I). The right triangles $AB\Delta$, $A\Gamma E$ are similar. Consequently,

$$\frac{|\Delta B|}{|A\Delta|} = \frac{|\Gamma E|}{|AE|} \Rightarrow \frac{x - |O\Delta|}{|A\Delta|} = \frac{|OE| - y}{|AE|} \Rightarrow ax + by = c,$$

where $\{a = 1/|A\Delta|, b = 1/|AE|, c = (|OE|/|AE|) + (|O\Delta|/|A\Delta|)\}$ does not depend on the special member-circle κ of the pencil $\mathcal{D}$.

Exercise 4.200. Show the inverse property of the one in the preceding exercise. That is, if on two lines $\{\beta,\gamma\}$ intersecting at the point O we take points respectively $\{B,\Gamma\}$, such that their distances $x = |OB|$, $y = |O\Gamma|$ satisfy a relation of the form $ax + by = c$, for some constants $\{a,b,c\}$, then the circles $\{\kappa = (OB\Gamma)\}$ are members of an intersecting pencil of circles and pass all through a second fixed point A.

Hint: Consider two circles $\{\kappa = (OB\Gamma), \kappa' = (OB'\Gamma')\}$, which intersect a second time at A (See Figure 4.123-II). The triangles $\{ABB', A\Gamma\Gamma'\}$ are similar. It follows that the ratios of their sides satisfy a relation

$$\frac{|BB'|}{|\Gamma\Gamma'|} = \frac{|OB'|-|OB|}{|O\Gamma'|-|O\Gamma|} = \frac{x'-x}{y'-y} = -\frac{b}{a},$$

which, because of the similarity, will be satisfied also from the corresponding altitudes $\frac{|A\Delta|}{|AE|} = \frac{|b|}{|a|}$. It follows (Proposition 3.7) that the position of the line OA is constant with respect to $\{\beta, \gamma\}$. It follows also that all the circles κ' pass through a common point A of the line OA and the circle κ.

Exercise 4.201. For a variable member-circle κ of the pencil $\mathscr{D}$ through the points $\{O, A\}$ of exercise 4.199, show that in the triangle $OB\Gamma$ of the figure 4.123-I, the second intersections of the altitudes from $\{B, \Gamma\}$ with κ move on two fixed lines $\{\beta', \gamma'\}$, which pass through A. Show also that the orthocenter of this triangle is moving on a line, which intersects the lines $\{\beta, \gamma\}$ at two points through which pass the preceding lines.

Hint: Examine the angles formed at A, and the orthocenter.

Exercise 4.202. Show that, for a given angle $\widehat{AB\Gamma}$ inscribed in the circle $\kappa(K)$, there is an inversion with respect to a circle $\nu(N)$ orthogonal to κ, which maps the three points $\{A, B\Gamma\}$ to three other $\{A', B', \Gamma'\}$, such that the angle $\widehat{A'B'\Gamma'}$ is a right one inscribed in the circle κ (Figure 4.124).

Hint: Consider the circle $\mu(M)$, which is orthogonal to κ and passes through points $\{A, \Gamma\}$. Any point N of this circle lying outside κ, defines a circle $\nu(N)$ orthogonal to κ, which has the aforementioned property.

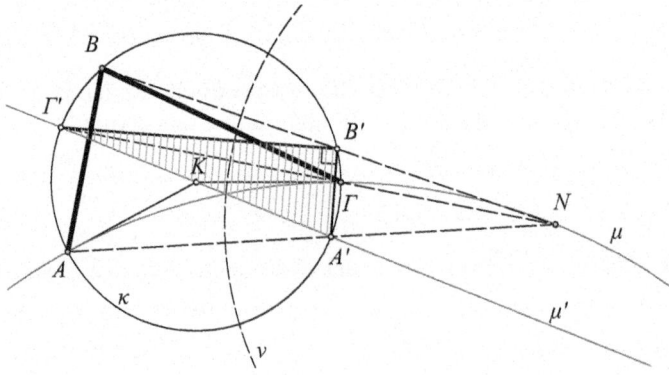

Fig. 4.124: Inverting any angle to a right one

Exercise 4.203. Show that, for a given quadrilateral $AB\Gamma\Delta$ inscribed in a circle $\kappa(K)$, there is an inversion with respect to a circle $\varepsilon(E)$, which maps the four points $\{A, B, \Gamma, \Delta\}$ to four other $\{A', B', \Gamma', \Delta'\}$, such that the $A'B'\Gamma'\Delta'$ is a rectangle inscribed in the circle κ (Figure 4.125).

4.10. COMMENTS AND EXERCISES FOR THE CHAPTER

Hint: Construct the circles $\{\mu(M), \nu(N)\}$, which are orthogonal to κ and contain respectively the pairs of points $\{(B,\Delta),(A,\Gamma)\}$. Show that these circles intersect at two points $\{E,Z\}$, one of which (E) is outside κ and defines a circle $\varepsilon(E)$ orthogonal to κ. Use the preceding exercise.

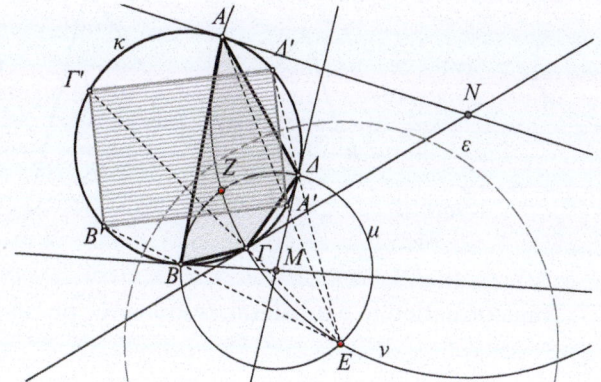

Fig. 4.125: Inverting any cyclic quadrilateral to a rectangle

Exercise 4.204. For a given triangle $AB\Gamma$, the circle κ_A is defined as locus of points $\left\{X : \frac{|XB|}{|X\Gamma|} = \frac{|AB|}{|A\Gamma|}\right\}$ and analogously are defined the other circles $\{\kappa_B, \kappa_\Gamma\}$, which are called **Apollonian circles** of the triangle $AB\Gamma$. Show that the three circles pass through two common points $\{I_1, I_2\}$ (See Figure 4.126).

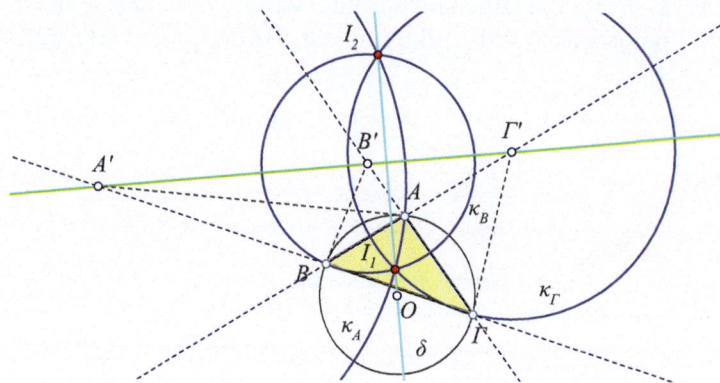

Fig. 4.126: Isodynamic points $\{I_1, I_2\}$ of the triangle

Hint: The circles are special Apollonian circles in the sense defined in section 4.4. They are orthogonal to the circumcircle δ of $AB\Gamma$ and $\{I_1, I_2\}$ which are called **Isodynamic points** of the triangle, are inverse relative to δ ([11, p.260]).

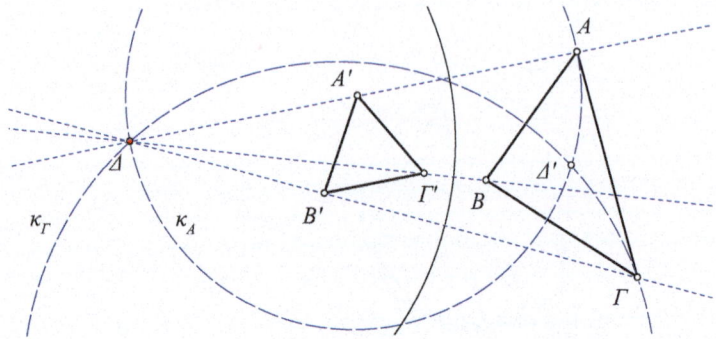

Fig. 4.127: Transformation to equilateral, isodynamic points $\{\Delta, \Delta'\}$

Exercise 4.205. For a given triangle $AB\Gamma$ to find an inversion mapping its vertices to three points which define an equilateral triangle (See Figure 4.127).

Hint: The center Δ of the inverting circle, coincides with an isodynamic point of the triangle. It is found using exercise 4.204. The radius can be arbitrary.

Exercise 4.206. Construct circles $\{\alpha, \beta, \gamma\}$ with diameters the medians AA', BB', $\Gamma\Gamma'$ of triangle $AB\Gamma$. Show that these circles intersect by pairs at points $\{A_1, A_2, B_1, B_2, \Gamma_1, \Gamma_2\}$ contained in the altitudes of the triangle, so that triangle $A_1A_2A_3$ is homothetic to $AB\Gamma$ and triangle $B_1B_2B_3$ is homothetic of the orthic of $AB\Gamma$.

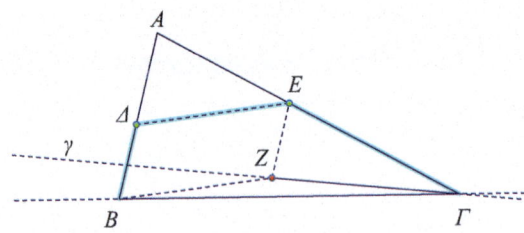

Fig. 4.128: Segments on the legs of given ratio

4.10. COMMENTS AND EXERCISES FOR THE CHAPTER

Exercise 4.207. On the sides $\{AB, A\Gamma\}$ of the triangle ABC find respectively points $\{\Delta, E\}$ such that $|B\Delta|/p = |\Delta E|/q = |E\Gamma|/r$, with $\{p, q, r\}$ given positive numbers (See Figure 4.128).

Hint: Draw ZE parallel and equal to $B\Delta$. Line $\gamma = \Gamma Z$ is determined from the give data. Also the ratio $|BZ|/|Z\Gamma|$ is determined from the data.

Next exercises discuss a generalization of the theorem of Thales for circles. Instead of a pencil of lines, we have here a pencil of circles $\{\kappa, \lambda, \mu, \ldots\}$ passing through two points $\{A, B\}$. On every line γ through A there are defined the second intersection points $\{\Gamma_\kappa, \Gamma_\lambda, \Gamma_\mu, \ldots\}$. The basic property is formulated in the next exercise (See Figure 4.129).

Exercise 4.208. With the preceding assumptions and notations, the ratios of the lengths $\{|\Gamma_\kappa \Gamma_\lambda|/|\Gamma_\lambda \Gamma_\mu|\}$ does not depend on the direction of the line γ passing through A. In other words, for another line δ through A will be defined corresponding points $\{\Delta_\kappa, \Delta_\lambda, \Delta_\mu, \ldots\}$ and the corresponding ratios will be equal

$$\frac{|\Gamma_\kappa \Gamma_\lambda|}{|\Gamma_\lambda \Gamma_\mu|} = \frac{|\Delta_\kappa \Delta_\lambda|}{|\Delta_\lambda \Delta_\mu|}.$$

Hint: Project the centers $\{I_\kappa, I_\lambda, I_\mu, \ldots\}$ of the circles to the corresponding points $\{P_\kappa, P_\lambda, P_\mu, \ldots\}$ of line δ and see that $|\Delta_\kappa \Delta_\lambda| = 2|P_\kappa P_\lambda| = 2|I_\kappa I_\lambda|\cos(\phi)$, where ϕ is the angle of the line of centers $\{I_\kappa\}$ with the line δ.

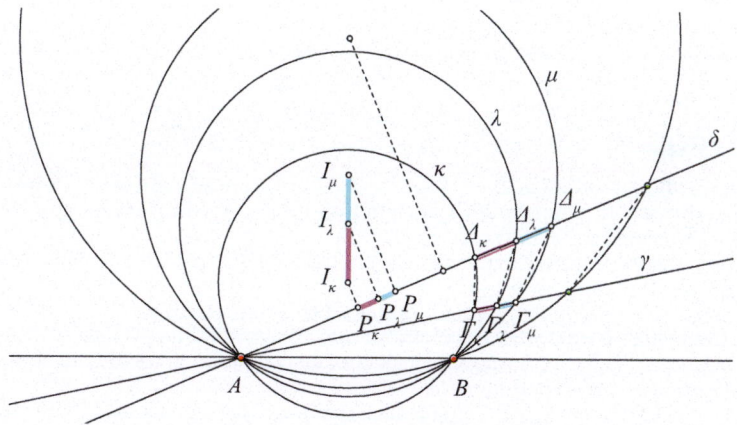

Fig. 4.129: Generalization of Thales for pencils of circles

Exercise 4.209. With the preceding assumptions and notations, show that, for constant lines $\{\gamma, \delta\}$ passing through A, the ratio $\frac{|\Delta_\kappa \Delta_\lambda|}{|\Gamma_\kappa \Gamma_\lambda|}$ does not depend on the special member-circles $\{\kappa, \lambda\}$ of the pencil and is the same for all pairs of circles passing through $\{A, B\}$ (See Figure 4.129).

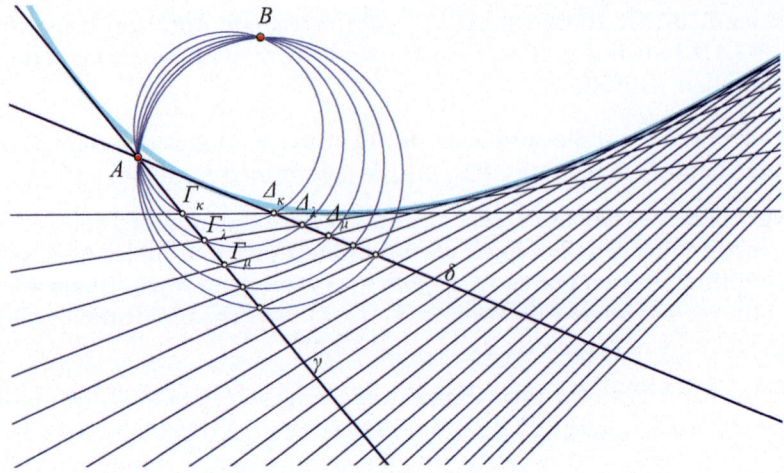

Fig. 4.130: Inverse of the generalization of Thales

Exercise 4.210. Show the inverse of the property of the exercise 4.208. If on two intersecting at A lines $\{\gamma, \delta\}$ the corresponding points $\{\Gamma_\kappa, \Gamma_\lambda, \Gamma_\mu\}$ and $\{\Delta_\kappa, \Delta_\lambda, \Delta_\mu\}$ satisfy the relation $\frac{|\Gamma_\kappa \Gamma_\lambda|}{|\Gamma_\lambda \Gamma_\mu|} = \frac{|\Delta_\kappa \Delta_\lambda|}{|\Delta_\lambda \Delta_\mu|}$, then the circles $\kappa = (A\Gamma_\kappa \Delta_\kappa)$, $\lambda = (A\Gamma_\lambda \Delta_\lambda), \mu = (A\Gamma_\mu \Delta_\mu)$ pass through a fixed point B.

In chapter 6 we'll see that the lines $\{\Gamma_\kappa \Delta_\kappa, \Gamma_\lambda \Delta_\lambda, \ldots\}$ are tangents to a parabola, which has its focus at B (See Figure 4.130).

References

96. T. Andreescu, C, Pohoata, 2012. Back to Eucldiean Geometry: Droz-Farny Demystified, *Mathematical Reflections*, 3:1-5
97. J. Ayme, 2004. A purely Synthetic Proof of the Droz-Farny Line Theorem, *Forum Geometricorum*, 4:219-224
98. T. Davis (2006) Geometry with Computers. Free internet edition
99. R. Deltheil, D. Caire (1989) Geometrie et complements. Editions Jaques Gabay, Paris
100. L. Gaultier, 1813. Sur le moyens generaux de construire graphiquement un cercle determine par trois conditions, *Journal de l'Ecole polytechnique*, 16:124-214
101. I. Ignazio, E. Suppa (2001) Il Problema Geometrico, dal compasso al cabri. interlinea editrice, Teramo
102. R. Johnson (1960) Advanced Euclidean Geometry. Dover Publications, New York
103. W. Mclelland (1891) A treatise on the Geometry of the Circle. Macmillan and Co, London
104. J. Niemeyer, 2011. A simple Construction of the Golden Section, *Forum Geometricorum*, 11:53
105. G. Papelier (1996) Exercices de Geometrie moderne. Editions Jacques Gabay, Paris
106. D. Pedoe (1990) A course of Geometry. Dover Publications, New York
107. M. Riaz, 1962. Geometric Solutions of Algebraic Equations, *American Mathematical Monthly*, 69:654-658

Chapter 5
From the classical theorems

5.1 Escribed circles and excenters

> We ... believe we are far more capable of reaching the
> center of things than of embracing their circumference.
>
> B. Pascal, Pensees, p. 61

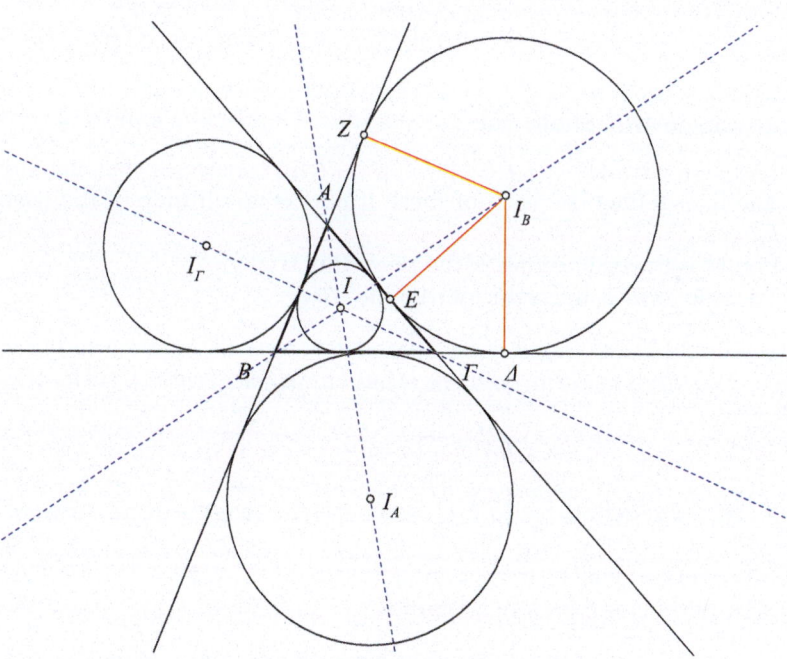

Fig. 5.1: The escribed (tritangent) circles of a triangle

Escribed or **tritangent** circles of the triangle $AB\Gamma$ are called the circles which are outside the triangle and are tangent to its sides. Their centers are called **excenters** of the triangle. The existence of these circles relies on next proposition, which is similar to the one for *inscribed* circles (Theorem 2.3).

Theorem 5.1. *The internal bisector of one triangle angle and the external bisectors of the other two angles pass through a common point (See Figure 5.1).*

Proof. We'll show that the internal bisector of angle $\widehat{B}$ and the external bisectors of $\widehat{A}$ and $\widehat{\Gamma}$ meet at a point I_B. Analogously are defined the points I_A and I_Γ. The proof is transferred almost verbatim from Theorem 2.3. Let I_B be the intersection point of two out of the three bisectors of the triangle and specifically of the external bisectors of the angles $\widehat{A}$ and $\widehat{\Gamma}$. We'll show that the internal bisector of angle $\widehat{B}$ passes also through I_B. Indeed, according to Corollary 1.35, the distances of I_B from the sides of angle $\widehat{A}$ are equal $|I_B Z| = |I_B E|$. Similarly, the distances of I_B from the sides of angle $\widehat{\Gamma}$ are equal $|I_B E| = |I_B \Delta|$. Consequently the three distances will all be equal $|I_B Z| = |I_B E| = |I_B \Delta|$, therefore the corresponding segments are radii of a circle κ with center I_B and radius $r_B = |I_B \Delta|$. The equality of the distances $|I_B Z| = |I_B \Delta|$ from the sides of angle $\widehat{B}$ shows that I_B is also on the bisector of angle $\widehat{B}$ (Corollary 1.35). Since the sides are orthogonal to these radii of κ (Corollary 2.10) latter is tangent to all three sides of the triangle.

Exercise 5.1. Let $\{I, I_A, I_B, I_\Gamma\}$ be respectively the incenter and the excenters of the triangle $AB\Gamma$. Show that:

1. Each of the triples (A, I, I_A), (B, I, I_B), (Γ, I, I_Γ) consists of collinear points.
2. The line defined by each of these triples is an altitude of the triangle $I_A I_B I_\Gamma$.
3. Point I is the orthocenter of the preceding triangle.
4. Triangle $AB\Gamma$ is the orthic of triangle $I_A I_B I_\Gamma$.

Proposition 5.1. *The length of the tangent $B\Delta$, from the vertex B to the corresponding escribed circle with center I_B, is equal to half the perimeter τ of the triangle $AB\Gamma$*

$$|B\Delta| = \frac{1}{2}(a+b+c) = \tau.$$

Proof. Here we denote by $\{a, b, c\}$, as usual, the lengths of the sides of the triangle. The proof follows directly from the equality of the tangents from B: $|BZ| = |B\Delta|$, as well as from A and Γ: $|AZ| = |AE|$, $|\Gamma E| = |\Gamma \Delta|$ (See Figure 5.1). The perimeter therefore is written

$$\begin{aligned} a+b+c &= (|B\Gamma|+|\Gamma E|) + (|AE|+|BA|) \\ &= (|B\Gamma|+|\Gamma \Delta|) + (|BA|+|AZ|) \\ &= 2(|B\Gamma|+|\Gamma \Delta|). \end{aligned}$$

5.1. ESCRIBED CIRCLES AND EXCENTERS

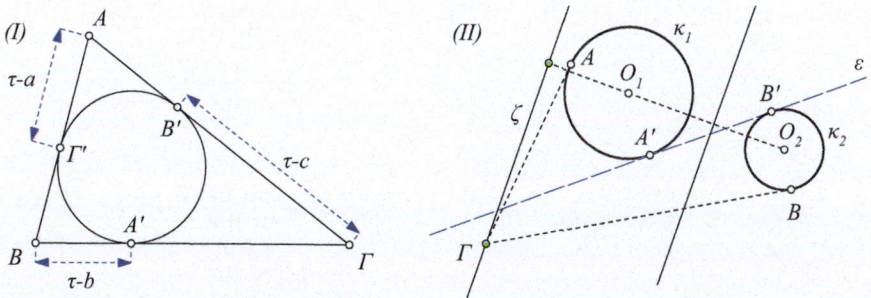

Fig. 5.2: Tangents from the vertices   Common tangent ε

Proposition 5.2. *The tangents AB', $A\Gamma'$ from the vertex A of triangle $AB\Gamma$ to its inscribed circle have length $|AB'| = |A\Gamma'| = \tau - a$, analogously $|B\Gamma'| = |BA'| = \tau - b$ and $|\Gamma A'| = |\Gamma B'| = \tau - c$, where $\tau = \frac{1}{2}(a+b+c)$ is the half-perimeter of the triangle.*

Proof. As the proof of the preceding proposition, so this one is also relying on the equality of the tangents from one point to a circle: $|AB'| = |A\Gamma'|$, $|B\Gamma'| = |BA'|$, $|\Gamma A'| = |\Gamma B'|$ (See Figure 5.2-I). It suffices therefore to write the perimeter as

$$a+b+c = 2\tau = 2(|AB'|+|B\Gamma'|+|\Gamma A'|) = 2(|AB'|+a),$$

from which the requested equality follows directly

Exercise 5.2. In the triangle $AB\Gamma$, with $|A\Gamma| \geq |AB|$, the circle κ is tangent to the sides $\{AB, A\Gamma\}$ and passes through the point A' of the base $B\Gamma$ (See Figure 5.2-I). Show that κ coincides with the inscribed circle of the triangle $AB\Gamma$, if and only if, it holds

$$|A'\Gamma| - |A'B| = |A\Gamma| - |AB|.$$

Then point A' coincides with the contact point of the circle with the base $B\Gamma$.

Hint: For $\{x = |A'B|, y = |A'\Gamma|\}$, the preceding relation is equivalent to $\{y - x = b - c, \; y + x = a\}$.

Exercise 5.3. Show that point Γ lies on the internal common tangent $\varepsilon = A'B'$ of the circles $\{\kappa_1(O_1), \kappa_2(O_2)\}$ which lie outside each other and are tangent to ε respectively at $\{A', B'\}$ and also lies on the exterior of segment $A'B'$, if and only if for the other tangents $\{\Gamma A, \Gamma B\}$ it is $|\Gamma A| - |\Gamma B| = |A'B'|$ (See Figure 5.2-II)

Hint: Obviously the relation holds if Γ is on ε. For the converse study the variation of Γ on a line ζ parallel to the radical axis of the two circles and show that the function $f(x) = ||\Gamma A| - |\Gamma B|| = k/(|\Gamma A| + |\Gamma B|)$ with k constant (Proposition 4.2) is decreasing with respect to the distance x of Γ from the line of centers.

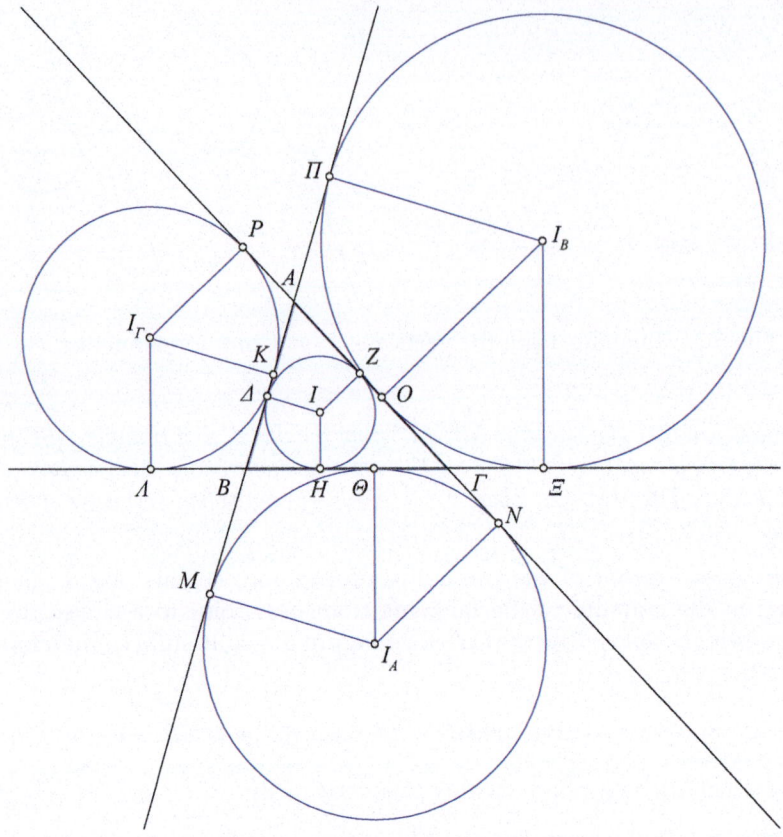

Fig. 5.3: Segments on the sides

Proposition 5.3. *Next table gives the centers of the inscribed and escribed circles of triangle $AB\Gamma$ as well as their respective projections on the sides AB, $B\Gamma$ and ΓA (See Figure 5.3).*

	AB	$B\Gamma$	ΓA
I	Δ	H	Z
I_A	M	Θ	N
I_B	Π	Ξ	O
I_Γ	K	Λ	P

The following relations are valid:

$$\tau - a = |A\Delta| = |AZ| = |\Gamma\Xi| = |\Gamma O| = |B\Lambda| = |BK|$$
$$\tau - b = |B\Delta| = |BH| = |\Gamma\Theta| = |\Gamma N| = |AK| = |AP|$$
$$\tau - c = |\Gamma Z| = |\Gamma H| = |A\Pi| = |AZ| = |BM| = |B\Theta|$$
$$|H\Theta| = |c - b|, \quad |OZ| = |a - c|, \quad |K\Delta| = |b - a|.$$

5.1. ESCRIBED CIRCLES AND EXCENTERS

Proof. In the preceding proposition we saw that $\tau - a = |A\Delta| = |AZ|$. For the other equalities on the same line write

$$|\Gamma\Xi| = |B\Xi| - |B\Gamma| = \tau - a.$$

Similarly follow also the equalities in the second and third line. Equality $|H\Theta| = |c - b|$ follows from the preceding ones

$$|H\Theta| = ||B\Gamma| - |BH| - |\Gamma\Theta|| = |a - (\tau - b) - (\tau - b)| = |b - c|.$$

Similarly follow also the two last equalities.

Exercise 5.4. Using Figure: 5.3 show that

1. $|\Lambda\Theta| = |KM| = b$, $|\Theta\Xi| = |ON| = c$, $|OP| = |\Pi K| = a$.
2. $|\Delta M| = |ZN| = a$, $|H\Xi| = |\Delta\Pi| = b$, $|ZP| = |H\Lambda| = c$.
3. $|PN| = a + c$, $|\Lambda\Xi| = b + c$, $|M\Pi| = a + b$.

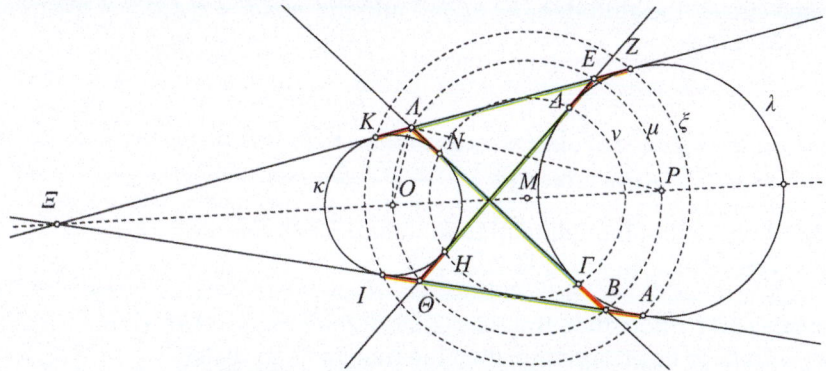

Fig. 5.4: Common tangents of two circles

Exercise 5.5. Let $\kappa(O)$ and $\lambda(P)$ be two non congruent and external to each other circles. Show the relations suggested by figure 5.4.

1. $|K\Lambda| = |\Lambda N| = |\Delta E| = |EZ|$ etc.
2. $|\Lambda E| = |N\Gamma|$ etc.
3. The circles $\nu = (\Gamma, \Delta, N, H)$, $\mu = (B, E, \Lambda, \Theta)$, $\xi = (A, Z, K, I)$ are concentric.
4. The circle μ passes through the centers O and P.

Exercise 5.6. Construct a triangle from its side $a = |B\Gamma|$ or its angle $\alpha = \widehat{BA\Gamma}$, its perimeter 2τ and its area $E = r \cdot \tau$, where r the inradius.

Exercise 5.7. Construct a triangle $AB\Gamma$ for which are given the radii of the inscribed r, escribed r_A and the difference of sides $|b - c|$.

Exercise 5.8. Construct a triangle $AB\Gamma$ for which are given the second intersection points of its bisectors with its circumcircle.

Exercise 5.9. Find a point M on the side AB of triangle $AB\Gamma$ (Figure 5.5-I), such that
$$|MA| + |A\Gamma| = |MB| + |B\Gamma|.$$

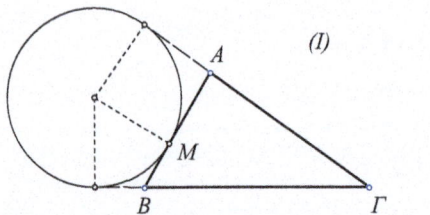
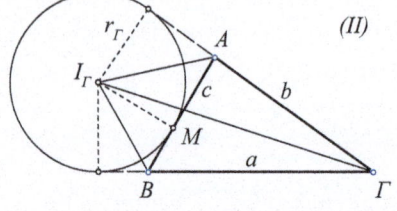

Fig. 5.5: M bisects the perimeter $\varepsilon(AB\Gamma) = \tfrac{1}{2}(a+b-c) \cdot r_\Gamma$

Exercise 5.10. Show that for the radii $\{r, r_A, r_B, r_\Gamma\}$ of the inscribed and escribed circles, the altitudes $\{h_A, h_B, h_C\}$ and the area $\varepsilon = \varepsilon(AB\Gamma)$ of the triangle $AB\Gamma$ holds
$$\varepsilon = r\tau = (\tau - a)r_A = (\tau - b)r_B = (\tau - c)r_\Gamma,$$
$$\frac{1}{r_A} + \frac{1}{r_B} + \frac{1}{r_\Gamma} = \frac{1}{r} = \frac{1}{h_A} + \frac{1}{h_B} + \frac{1}{h_C}.$$

Hint: For the first four equalities see figure 5.5-II. For the rest see that $\frac{1}{h_A} = \frac{a}{2\varepsilon}$, which implies that $\frac{1}{h_A} + \frac{1}{h_B} + \frac{1}{h_C} = \frac{\tau}{\varepsilon} = \frac{1}{r}$.

Exercise 5.11. Show that, if the incenter of the triangle $AB\Gamma$ coincides with its centroid or its orthocenter, then the triangle is equilateral.

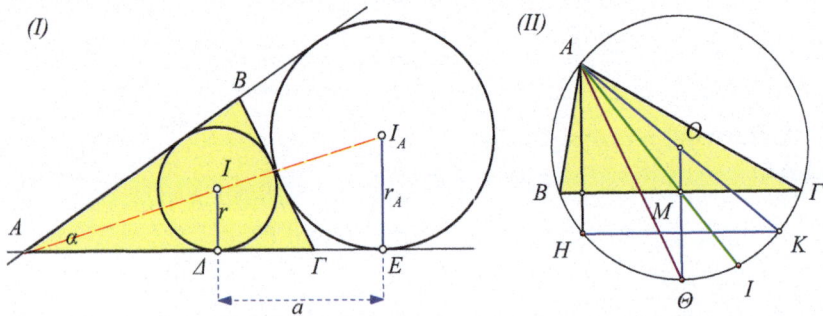

Fig. 5.6: $\{A, r(r_A), a\}$ triangle construction   Triangle from traces H, Θ, I

Exercise 5.12. Construct a triangle $AB\Gamma$ from its angle $\alpha = \widehat{BA\Gamma}$, its side $a = |B\Gamma|$ and the radius $r(r_A)$ of the inscribed (Escribed) circle.

Hint: Let Δ be the contact point of the inscribed circle $I(r)$ with side AB (See Figure 5.6-I). The right triangle $A\Delta I$ can be constructed, because we know the orthogonal $I\Delta$ of length r and the angle $\widehat{\Delta AI} = \frac{\alpha}{2}$. Extend $A\Delta$ by a until E. E is the contact point of AB with the escribed circle $I_A(r_A)$ (Exercise (2) 5.4). Therefore this too can be constructed by drawing the orthogonal to AB at E and finding the intersection point of I_A with AI. Draw next the common tangent $B\Gamma$ of the two circles (Construction 2.10). Triangle $AB\Gamma$ is the requested one.

Exercise 5.13. Construct a triangle $AB\Gamma$ for which are given the positions of the traces on its circumcircle of the altitude, bisector and median from A.

Hint: If $\{H, \Theta, I\}$ are the respective traces (See Figure 5.6-II), then these determine the circumcircle as well as the diametrically opposite K of A, on this circle. Then the right triangle AHK can be constructed, and from this the requested $AB\Gamma$.

5.2 Heron's formula

> One cannot escape the feeling that these mathematical formulas have an independent existence and an intelligence of their own, that they are wiser than we are, wiser even than their discoverers, that we get more out of them than was originally put into them.
>
> H. Hertz, in M.Kline's *Mathematics and the Search for Knowledge*, p. 144

The theorem (or **formula**) of Heron (approximately 10-75 A.D.) expresses the area of the triangle as a function of the lengths of its sides. With the help of one factorization exercise it can result as an application of the formula of Stewart (Theorem 3.27). In what follows I give this proof, as well as the more elegant one which is attributed to Heron himself ([14, p.160]).

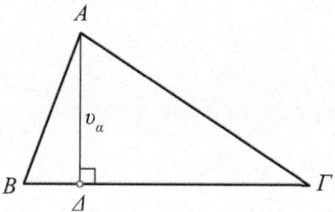

Fig. 5.7: $v_A^2 = \frac{1}{4a^2}(a+b+c)(b+c-a)(c+a-b)(a+b-c)$

Theorem 5.2. (*Heron's formula*) *Let* $\tau = \frac{1}{2}(a+b+c)$ *be the half perimeter of triangle* $AB\Gamma$. *Then the triangle's area E is given by the formula*

$$E = \sqrt{\tau(\tau-a)(\tau-b)(\tau-c)}.$$

Proof. We apply the formula for the altitude $v_A^2 = \frac{1}{4a^2}(a+b+c)(b+c-a)(c+a-b)(a+b-c)$ (Proposition 3.15), which holds even for obtuse triangles, provided we consider the altitude from its obtuse angle (assuming it to be at A), so that Δ falls inside $B\Gamma$ (See Figure 5.7). Then the factors which show up in the preceding proposition can be written

$$(a+b+c) = 2\tau$$
$$(b+c-a) = (b+c+a-2a) = 2(\tau-a)$$
$$(c+a-b) = (c+a+b-2b) = 2(\tau-b)$$
$$(a+b-c) = (a+b+c-2c) = 2(\tau-c)$$

Heron's formula then follows from the known formula for area $E = \frac{1}{2}v_A \cdot a$

The second proof of Heron's formula uses the calculation of the radii of the inscribed and escribed circles of the triangle (§ 5.1).

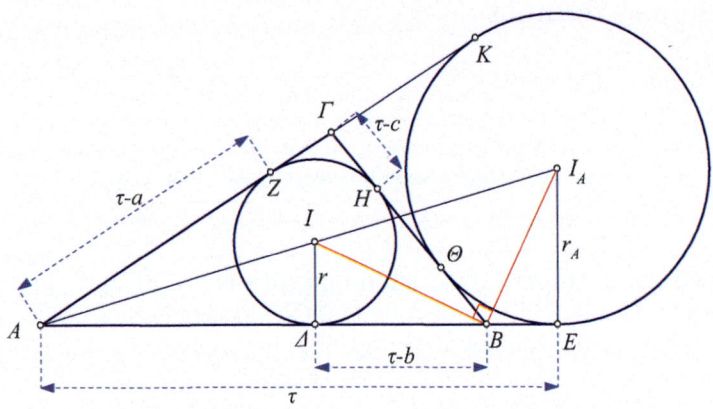

Fig. 5.8: r, r_A as functions of the sides

Proposition 5.4. *The radius r of the inscribed and r_A of the escribed circle of the triangle $AB\Gamma$ are given respectively by the formulas*

$$r^2 = \frac{(\tau-a)(\tau-b)(\tau-c)}{\tau} \quad r_A^2 = \frac{\tau(\tau-b)(\tau-c)}{\tau-a}.$$

Proof. We use the familiar figure 5.8 met in § 5.1. Two circles are shown, the inscribed $I(r)$ and the escribed $I_A(r_a)$. The proof uses the relations of the aforementioned section and the similarity between two pairs of triangles. The first pair of triangles is $(A\Delta I, AEI_A)$. The second pair is $(\Delta IB, EBI_A)$. Both pairs consist of right triangles and we have:

5.2. HERON'S FORMULA

$$\frac{|\Delta I|}{|EI_A|} = \frac{|A\Delta|}{|AE|} \Leftrightarrow \frac{r}{r_A} = \frac{\tau-a}{\tau},$$

$$\frac{|\Delta I|}{|\Delta B|} = \frac{|EB|}{|EI_A|} \Leftrightarrow \frac{r}{\tau-b} = \frac{\tau-c}{r_A}.$$

Solving the second relative to r_A and substituting into the first expression, we find the formula for r^2. Squaring the first formula and substituting with the found expression for r^2, we prove the second formula as well.

(I) 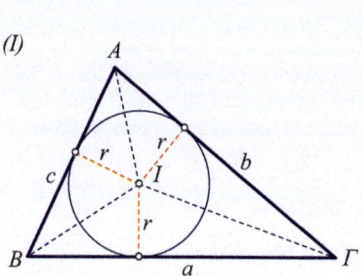 (II)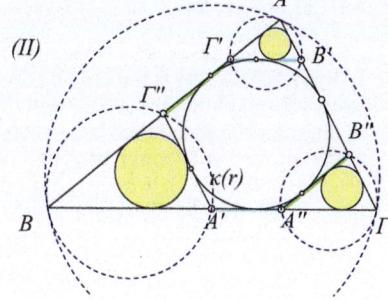

Fig. 5.9: Heron's proof   Symmetric inscribed hexagon

The second proof of Heron's formula follows by substituting in the formula for the area $\varepsilon(AB\Gamma) = r \cdot \tau$ (Corollary 3.4) the radius r through the formula of proposition 5.4 (See Figure 5.9-I).

Exercise 5.14. Let $\{B'\Gamma', \Gamma''A', A''B''\}$ be tangents of the inscribed circle $\kappa(r)$ of the triangle $AB\Gamma$, respectively parallel to the sides $\{B\Gamma, \Gamma A, AB\}$ (See Figure 5.9-II). Show that the hexagon $A'A''B''B'\Gamma'\Gamma''$ is symmetric. Show also that the sum of the inradii of the small circles $r_A + r_B + r_\Gamma = r$. Finally show that the circumcircles of the small triangles are tangent to the circumcircle of $AB\Gamma$.

Exercise 5.15. Let R be the radius of the circumcircle of the triangle with side lengths $\{a,b,c\}$. Prove the following formulas:

$$E = r \cdot \tau = r_A \cdot (\tau-a) = r_B \cdot (\tau-b) = r_\Gamma \cdot (\tau-c) \quad (1)$$

$$4R + r = r_A + r_B + r_\Gamma, \quad (2)$$

$$R^2 = \frac{a^2 b^2 c^2}{2(b^2c^2 + c^2a^2 + a^2b^2) - (a^4 + b^4 + c^4)}. \quad (3)$$

Hint: Use Heron's formula and Corollary 3.20 (see also exercise 5.159).

Exercise 5.16. Show that the area E of the triangle $AB\Gamma$ is expressed in terms of its altitudes h_A, h_B, h_Γ through formula:

$$\frac{1}{E^2} = \left(\frac{1}{h_A} + \frac{1}{h_B} + \frac{1}{h_\Gamma}\right) \cdot \left(-\frac{1}{h_A} + \frac{1}{h_B} + \frac{1}{h_\Gamma}\right) \cdot \left(\frac{1}{h_A} - \frac{1}{h_B} + \frac{1}{h_\Gamma}\right) \cdot$$
$$\left(\frac{1}{h_A} + \frac{1}{h_B} - \frac{1}{h_\Gamma}\right).$$

Remark 5.1. In the articles of Baker [110], [111] are contained 110 formulas for the area E of the triangle.

5.3 Euler's circle

> Euler (1707-1783) was not only by far the most productive mathematician in the history of mankind, but also one of the greatest scholars of all time. Cosmopolitan in the truest sense of the word - he lived during his first twenty years in Basel, was active altogether for more than thirty years in Petersburg and for a quarter of a century in Berlin - he attained, like only a few scholars, a degree of popularity and fame which may well be compared with that of Galilei, Newton, or Einstein.
>
> E. Fellmann, *Leonhard Euler, prologue xiv*

The Euler circle of the triangle, is the one which passes through the three middles of the sides. Its importance lies in the fact that it also passes through six noteworthy points of the triangle, that's why it is often called **nine point circle** of the triangle.

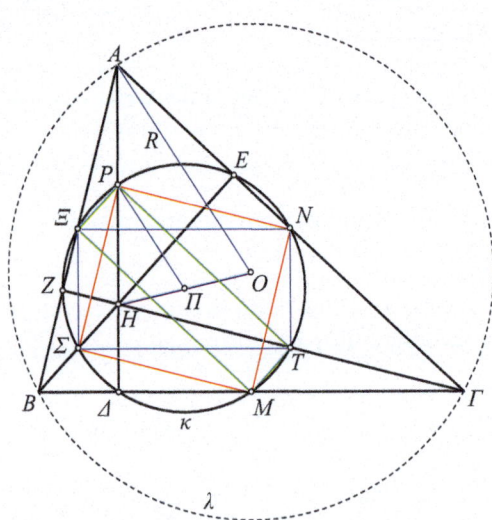

Fig. 5.10: Euler's circle κ of the triangle $AB\Gamma$

Theorem 5.3. *The circle κ through the middles M, N, Ξ of the sides of triangle $AB\Gamma$, has the following properties (See Figure 5.10):*

5.3. EULER'S CIRCLE

1. It passes also through the traces $\{\Delta, E, Z\}$ of the altitudes of the triangle.
2. It passes also through the middles $\{P, \Sigma, T\}$ of the line segments which join the vertices with the orthocenter H of the triangle.
3. Its center Π is the middle of the segment which joins the orthocenter H with the circumcenter O of the triangle.
4. Its radius $|\Pi P|$ is half that of the radius $R = |OA|$ of the circumscribed circle λ of the triangle.
5. Point H is a center of similarity of κ and the circumcircle of the triangle, with similarity ratio $1:2$.

Proof. The proof relies on the existence of three rectangles which have, by two, a common diagonal. The rectangles are $\Sigma T N \Xi$, $\Sigma M N P$ and $P \Xi M T$. First let us see that these rectangles exist. I show that $\Sigma T N \Xi$ is such a rectangle. The proof for rectangles $\Sigma M N P$ and $P \Xi M T$ is similar.

In $\Sigma T N \Xi$ then, $\Sigma \Xi$ joins the middles of sides of the triangle BHA. Therefore it is parallel and the half of HA. Similarly TN joins the middles of sides of triangle $AH\Gamma$. Therefore it is parallel and the half of HA. Consequently $\Sigma \Xi$ and TN, being parallel and equal, they define a parallelogram $\Sigma T N \Xi$. That this is actually a rectangle, follows from the fact that ΣT joins middles of sides of triangle $HB\Gamma$, therefore it is parallel and the half of $B\Gamma$. Since $A\Delta$ and $B\Gamma$ are mutually orthogonal, the same will happen also with their parallels $\Sigma \Xi$ and ΣT.

The three rectangles have by two a common diagonal, which is the diameter of their circumscribed circle. This implies that the three circumscribed circles of these rectangles coincide. This completes the proof of the first two claims of the proposition.

For the proof of the next two claims, it suffices to observe that in the triangle HOA the segment ΠP joins the middles of the sides of triangle HOA, therefore it is parallel and the half of OA. However OA is a radius of the circumscribed circle λ and ΠP is a radius of circle κ. The last follows from the fact $|AH| = 2|OM|$ (Proposition 3.11), therefore $PHMO$ is a parallelogram and its diagonals are bisected at Π. However PM, as seen previously, is a diameter of the circle κ. The last claim is a consequence of the two preceding ones.

Exercise 5.17. Show that the triangle $AB\Gamma$ coincides with the orthic triangle of triangle $I_A I_B I_\Gamma$, with vertices the excenters of $AB\Gamma$. Conclude that the circumcircle of the triangle $AB\Gamma$ coincides with the Euler circle of $I_A I_B I_\Gamma$ (See Figure 5.11).

Exercise 5.18. Construct a triangle $AB\Gamma$ for which are given the position of the vertex A, the position of the projection Δ of A on the opposite side $B\Gamma$ and the position of the center Π of its Euler circle.

Exercise 5.19. The orthocenter H of the triangle $AB\Gamma$ is projected on the bisectors of angle $\widehat{A}$, internal and external, at the points Σ and P. Show that the line ΣP passes through the middle of $B\Gamma$ and the center of its Euler circle.

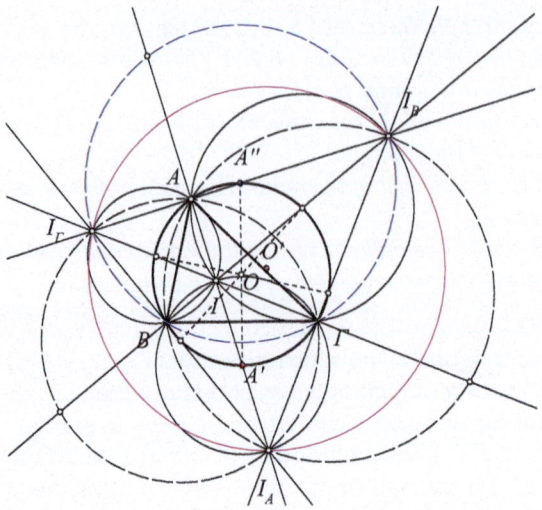

Fig. 5.11: Circumcircle of $AB\Gamma$ as Euler circle of $I_A I_B I_\Gamma$

Hint: The first part of the exercise is Exercise 3.212.

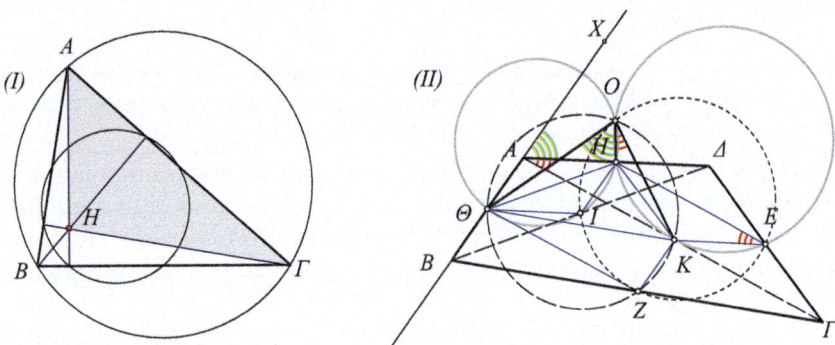

Fig. 5.12: Common Euler circle   Intersection of four Euler circles

Exercise 5.20. Show that for every triangle $AB\Gamma$ with orthocenter H, the Euler circles of the triangles $AB\Gamma$, ABH, $B\Gamma H$ and ΓAH coincide (See Figure 5.12-I). Conclude that these four triangles have circumscribed circles of equal radii.

Exercise 5.21. The diagonals of the quadrilateral $AB\Gamma\Delta$ define four triangles $AB\Gamma$, $B\Gamma\Delta$, $\Gamma\Delta A$, ΔAB. Show that the Euler circles of these triangles pass through a common point O (See Figure 5.12-II).

Hint: Consider the parallelogram $EZH\Theta$ of the middles of the sides of the quadrilateral and the middles $\{I, K\}$ of the diagonals. Let also O be the intersection of two of these Euler circles e.g. of $\{AB\Delta, A\Gamma\Delta\}$. Show that the other

5.3. EULER'S CIRCLE

Euler circles pass through the same point, proving f.e. that $\widehat{\Theta ZK}$ and $\widehat{\Theta OK}$ are supplementary angles.

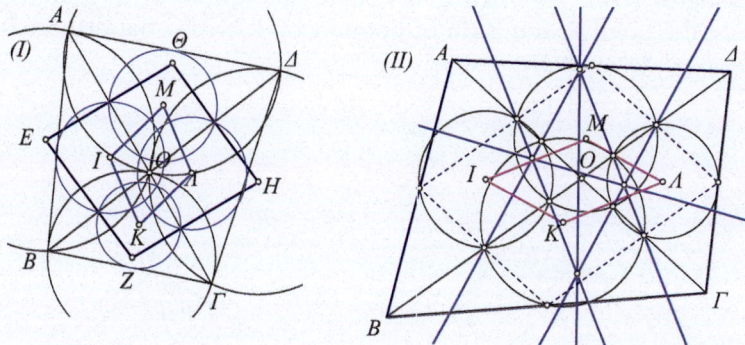

Fig. 5.13: Four circumscribed circles   Four Euler circles

Exercise 5.22. The intersection point O of the diagonals of the quadrilateral $AB\Gamma\Delta$ defines four triangles ABO, $B\Gamma O$, $\Gamma\Delta O$, ΔAO. Show that the circumcenters $\{E, Z, H, \Theta\}$ of these triangles are vertices of a parallelogram. Show the same also for the centers $\{I, K, \Lambda, M\}$ of the Euler circles of these triangles (See Figure 5.13-I).

Exercise 5.23. For the quadrilateral and the four triangles, defined in the preceding exercise, show that the radical axes of the Euler circles of these triangles define the sides and the diagonals of a parallelogram, which is similar to $IK\Lambda M$ (See Figure 5.13-II).

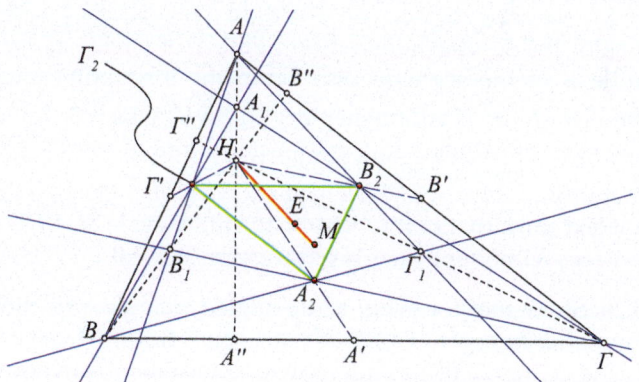

Fig. 5.14: Property of the center of the Euler circle

Exercise 5.24. For a triangle $AB\Gamma$ the middles of its sides are respectively $\{A', B', \Gamma'\}$, the traces of its altitudes are $\{A'', B'', \Gamma''\}$, the orthocenter, the center of mass and the center of its Euler circle are respectively $\{H, M, E\}$ (See

Figure 5.14). Also $\{A_1, B_1, \Gamma_1\}$ are respectively the middles of $\{HA, HB, H\Gamma\}$. Show that the triangle with vertices the centroids $\{A_2, B_2, \Gamma_2\}$ of the respective triangles $\{HB\Gamma, H\Gamma A, HAB\}$ is homothetic to $AB\Gamma$ with center of homothecy the point E and ratio of homothecy 1:3. Also the orthocenter of $A_2 B_2 \Gamma_2$ coincides with M.

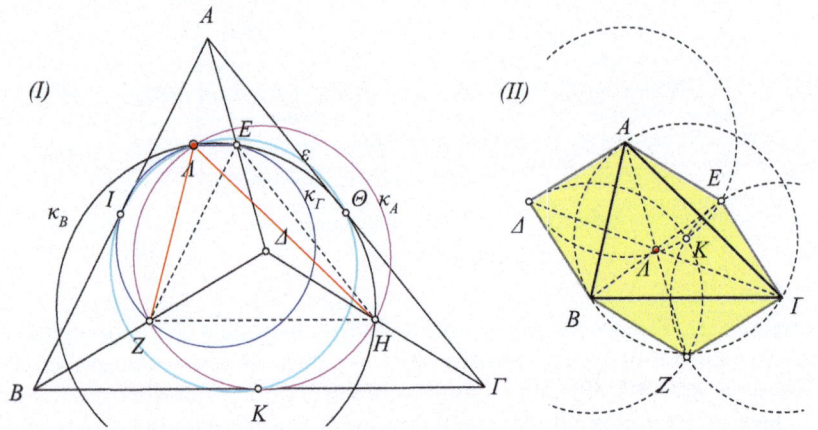

Fig. 5.15: Concurring Euler circles   Symmetric circumscribed hexagon

Exercise 5.25. Let Δ be a point not contained on the side-lines of triangle $AB\Gamma$. Show that the Euler circles $\{\kappa_A, \kappa_B, \kappa_\Gamma\}$ respectively of triangles $\{\Delta B\Gamma, \Delta \Gamma A, \Delta AB\}$ intersect at a point Λ contained in the Euler circle of the triangle $AB\Gamma$ (See Figure 5.15-I).

Hint: Consider the second intersection point Λ of circles κ_B, κ_Γ (the first is the middle E of the segment ΔA). From the inscriptible quadrilateral $\Lambda H \Theta E$, the angle $\widehat{E\Lambda H}$ is supplementary to $\widehat{E\Theta H}$, which is equal to $\widehat{E\Delta H}$, because $E\Theta H \Delta$ is a parallelogram. Similarly, $\widehat{Z\Lambda E}$ is equal to $\widehat{Z\Delta E}$ and $\widehat{Z\Lambda H}$ is supplementary to $\widehat{Z\Delta H} = \widehat{ZKH}$, which shows that circle κ_A will also pass through Λ. Next show that point Λ lies on the Euler circle of $AB\Gamma$ by proving that point Λ sees KI under an angle of measure $180° - \beta$.

Exercise 5.26. Show that for every acute-angled triangle $AB\Gamma$ there exists a convex symmetric hexagon $A\Delta BZ\Gamma E$ with equal sides, whose center coincides with the center Λ of its Euler circle. Show that, conversely in each convex symmetric hexagon with equal sides, the triangle which results by taking non successive vertices of it has as center of its Euler circle the center of symmetry of the hexagon (See Figure 5.15-II).

5.4 Feuerbach's Theorem

> Let alone the fact that the habit they will gain by first searching the easy and passing ever so slowly and step by step to other more difficult things will be of much more use to them than any of my other teachings would be.
>
> Descartes, Discourse on the Method, 6-th part

In 1822 Feuerbach (1800-1834), who was a high school teacher, published a small book, which, among other noteworthy theorems on the triangle ([102, p.190]), contained also the theorem we prove below. In this proof ([109, p.110], [13, p.117]), a second proof is given in Exercise 5.52) the key role is played by the circle κ with diameter $K\Lambda$, where $\{K, \Lambda\}$ are the projections on $B\Gamma$ of the incenter I and of the excenter Θ contained in the angle $\widehat{A}$ of triangle $AB\Gamma$. From proposition 5.3 we know that this circle has its center

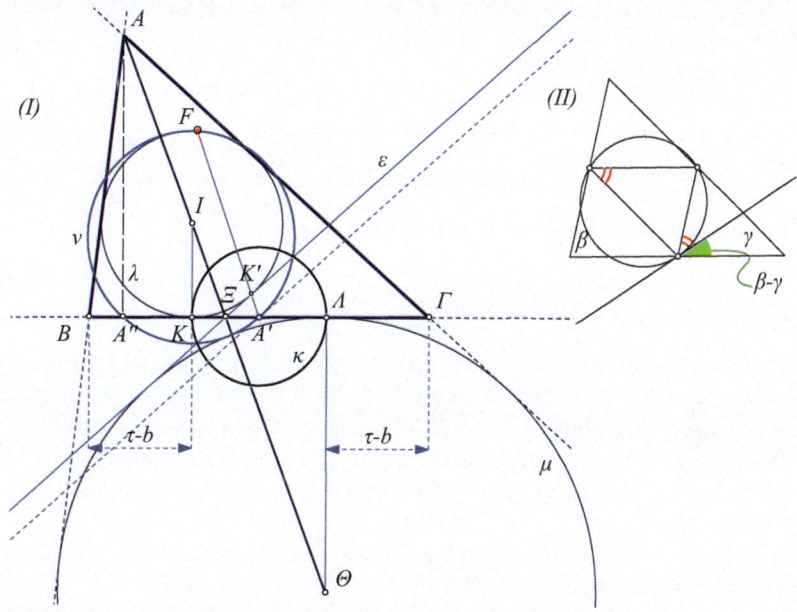

Fig. 5.16: The circle κ and its inversion

at the middle A' of $B\Gamma$ (See Figure 5.16). The proof of Feuerbach's theorem results from properties of the inversion relative to that circle. Next proposition formulates these properties, beginning with the acute triangle with $b > c$, and arguments which hold in all cases.

Proposition 5.5. *With the preceding definitions and notations, hold the properties (See Figure 5.16):*

1. *The circle κ is orthogonal to the inscribed circle λ as well as to the escribed μ.*

2. *The intersection point Ξ of the bisector AI with $B\Gamma$ is the inverse relative to κ of the trace A'' of the altitude.*
3. *The angles with $B\Gamma$, of the tangent ε to λ from Ξ and the tangent to the Euler circle at the middle A' of $B\Gamma$ are equal to $|\beta - \gamma|$.*
4. *The inversion relative to κ maps the Euler circle ν to line ε.*

Proof. (1) is obvious, since the radii of the circles at K and Λ, respectively, are orthogonal to the corresponding ones of κ (See Figure 5.16-I).

(2) follows from calculations we did in preceding sections. According to proposition 5.3, the radius of κ is equal to $r = |b-c|/2$. The length $|A'A''|$ is calculated by considering the power of B relative to the circle with diameter $A\Gamma$, and is found to be equal to $\frac{|b^2-c^2|}{2a}$. The length $|A'\Xi|$ is calculated through the ratio in which the bisector divides $B\Gamma$, and is found to be equal to $\frac{a|c-b|}{2(b+c)}$. The claim follows from the fact $r^2 = |A'\Xi||A'A''|$.

(3) follows, on one hand from the figure 5.16-II, which shows that the measure of the angle at A is equal to $|\beta - \gamma|$, and on the other from the fact, that the angle at Ξ will have measure $|180° - 2\widehat{B\Xi A}|$, which is also easily seen to be equal to $|\beta - \gamma|$.

(4) follows from the properties of inversion and the preceding claims. Indeed, since the Euler circle ν passes through the center of inversion and through A'', its image will be a line passing through Ξ and forming at Ξ an angle equal to that formed by the circle ν with $B\Gamma$, therefore coincident with ε. This conclusion follows from (3).

Theorem 5.4 (Feuerbach's theorem). *In every triangle the Euler circle is tangent to the inscribed, as well as, to its three escribed circles.*

Proof. With the notation of the preceding proposition, we consider the inversion relative to circle κ (See Figure 5.16-I). In it the inverse of the inscribed circle λ is itself (Corollary 4.20) and, according to the preceding proposition, the inverse of the Euler circle ν is the line ε, i.e. the second tangent to λ from Ξ. Since ε is tangent to λ, its inverse, which is the circle ν, will be also tangent to λ at a point F, which is the inverse of the contact point K' of ε with λ. Note that, because of the symmetry of λ relative to the bisector AI, point K' is the symmetric of K relative to AI.

The same argument is applied also to the escribed circle μ with center Θ. This circle, as well, is orthogonal to κ and has ε as a tangent. Therefore the inverse of this circle is itself and the inverse of the line ε, which is the Euler circle, will be tangent to it.

What we said proves then that the Euler circle is tangent simultaneously to the inscribed and escribed contained in the angle A. Similarly we prove that the Euler circle is tangent to the escribed circles contained in the other angles

The contact point F of the inscribed circle and the Euler circle is called **Feuerbach point** of the triangle.

Exercise 5.27. Examine the proof in the case where the triangle is obtuse at B. Finally, show that the point of contact of μ with ν is the inverse relative of κ of the symmetric Λ' of Λ relative to the bisector AI.

Exercise 5.28. Study the proof of the theorem of Feurbach in the case of the isosceles and equilateral triangle.

Exercise 5.29. Analogously to the circle $\kappa_A = \kappa$ of proposition 5.5 (See Figure 5.16-I), construct the circles κ_B, κ_Γ with centers the middles of sides, respectively, ΓA, AB and show that: (i) the incenter I is their radical center, (ii) the radical axis of κ_B, κ_Γ is the tangent to $\kappa = \kappa_A$ at the point of contact K of the inscribed with $B\Gamma$ and similar properties that hold for the radical axes of the other pairs of circles (κ_Γ, κ_A), (κ_A, κ_B), (iii) the radius of the larger of these circles is equal to the sum of the radii of the two other circles.

Exercise 5.30. Construct a triangle $AB\Gamma$ for which are given the position of the incenter (I), the position of the point of Feuerbach (F), as well as the position of the middle A' of the side $B\Gamma$.

Exercise 5.31. Show that the Euler circle coincides with the inscribed circle, if and only if the triangle is equilateral.

Exercise 5.32. Show, that in an isosceles triangle the line of centers IE of the inscribed circle $\kappa(I)$ and the Euler circle $\lambda(E)$ passes through the apex. Inversely, show that, if this line passes through a vertex A of the triangle $AB\Gamma$, then the triangle is either isosceles with apex at A, or it is non isosceles with angle $\widehat{A} = 60°$ and circumradius equal to $|AH|$, where H is the orthocenter of the triangle.

Exercise 5.33. Continuing the preceding exercise, show that all such non-isosceli triangles $AB\Gamma$, with collinear $\{A, I, E\}$, are created from an equilateral $AB'\Gamma$ by taking a point B on its basis AB'.

5.5 Euler's theorem

> Euler's formula or identity $e^{i\theta} = \cos(\theta) + i\sin(\theta)$, where $i = \sqrt{-1}$. The special case of $\theta = \pi$ gives $e^{i\pi} = -1$ or, as it is usually written, $e^{i\pi} + 1 = 0$, a compact expression that I think is of exquisite beauty.
>
> P. Nahin, Dr. Euler's Fabulous Formula, preface xxxii

Every triangle has a circumscribed circle λ and an inscribed κ. If we hide the triangle we see two circles (κ in the interior of λ) and the question rises, whether there are other triangles which have these two circles respectively as inscribed and circumscribed. More generally, for two circles $\{\kappa, \lambda\}$, the first of which is contained in the second, one may ask, whether there is a triangle circumscribed to the first and inscribed to the second.

Next theorem of Euler (one of many [150], [142]) leads to the expression of the power of the incenter *I* relative to the circumcircle of the triangle as a function of the radii of the circumcircle and the incircle. The formula gives also a quantitative criterion answering the preceding question of the existence of a triangle circumscribed resp. inscribed to two given circles.

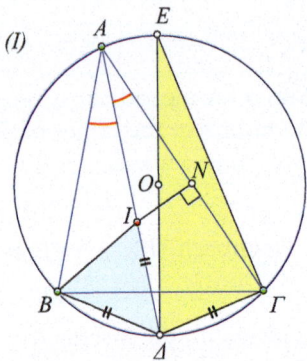
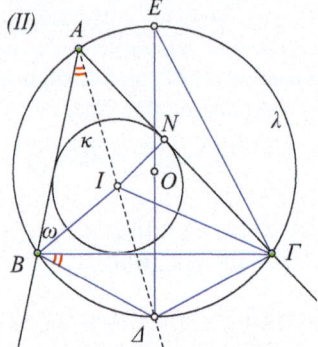

Fig. 5.17: Euler's theorem   Same incircle and circumcircle

Theorem 5.5 (Euler's theorem). *In every triangle the radius R of its circumcircle, the radius r of its incircle and the distance OI of the centers of these circles are related through the formula*

$$|OI|^2 = R(R-2r).$$

Proof. The formula reminds of the power $p(I) = |OI|^2 - R^2$ of the incenter *I* (center of the inscribed circle) relative to the circumcircle, especially if we write it as $|OI|^2 - R^2 = -2rR$. It suffices therefore to show that the power of the incenter *I* relative to the circumcircle is $-2rR$. The basic observations which lead to the proof are two.

First, that the extension of the bisector *AI* passes through the middle *Δ* of the arc $\widehat{B\Gamma}$ (Exercise 2.141) and the line segments *ΔB*, *IΔ*, *ΔΓ* are equal (See Figure 5.17-I). Segments *ΔB* and *ΔΓ* are of course equal, since *Δ* is the middle of the arc $\widehat{B\Gamma}$. Triangle *BΔI* is however isosceles, because its angle at *B* is the sum $\frac{\alpha+\beta}{2}$ and the same happens with its angle at *I*, as external of the triangle *BIA*. The power of *I* then is $|AI||I\Delta| = |AI||\Delta\Gamma|$.

The second observation is that the right triangles *AIN* and *EΔΓ* are similar. Here, *N* is the projection of *I* on *AΓ*, therefore its length is $|IN| = r$. Point *E* is the diametrically opposite of *Δ*. Obviously the triangles are similar because they have their acute angles at *A* and *E* equal. Then their sides are proportional:

5.5. EULER'S THEOREM

$$\frac{|AI|}{|IN|} = \frac{|E\Delta|}{|\Delta\Gamma|} \Rightarrow |AI||\Delta\Gamma| = |IN||E\Delta|,$$

which, with what we said, translates to the requested relation:

$$-p(I) = |AI||I\Delta| = |AI||\Delta\Gamma| = |IN||E\Delta| = r(2R).$$

Theorem 5.6. *If for two circles $\kappa(I,r)$ and $\lambda(O,R)$ holds the relation $|OI|^2 = R(R-2r)$, then for every point A of λ there exists a triangle $AB\Gamma$ inscribed in λ and cirumscribed to κ.*

Proof. To begin with, κ is inside circle λ. In fact, the relation $|OI|^2 = R(R-2r)$ implies $|OI| < R$ and I is inside λ. Also $(R-r)^2 = R^2 + r^2 - 2Rr > R^2 - 2Rr = |OI|^2$ implies $R > |OI| + r$ i.e. κ is located inside λ. Thus, we can draw from an arbitrary point A of λ the tangents to κ, which intersect again the circle λ at B and Γ (See Figure 5.17-II). It suffices to show that line $B\Gamma$ is tangent to κ. To see this, we draw the bisector of the angle $\widehat{BA\Gamma}$, which intersects the circle λ at the middle Δ of arc $\widehat{B\Gamma}$. Let E be the diametrically opposite of Δ and N be the projection of I on $A\Gamma$. The right triangles AIN and $EA\Gamma$ have their acute angles at A and E equal, therefore they are similar. Consequently

$$\frac{|AI|}{|IN|} = \frac{|E\Delta|}{|\Delta\Gamma|} \Rightarrow |AI||\Delta\Gamma| = |IN||E\Delta| = r(2R).$$

However, by hypothesis, the power of I relative to λ is also $-p(I) = |AI||I\Delta| = R^2 - |OI|^2 = 2Rr$. Therefore

$$|AI||I\Delta| = |AI||\Delta\Gamma| \Rightarrow |I\Delta| = |\Delta\Gamma|.$$

Thus, there are defined two isosceli triangles, $B\Delta I$ and $\Delta I\Gamma$. Let $\omega = \widehat{IBA}$. Taking into account that angles $\widehat{\Delta B\Gamma}$ and $\widehat{\Delta AB}$ are equal to $\alpha/2$, where $\hat{\alpha} = \widehat{BA\Gamma}$, we have

$$\widehat{BI\Delta} = \omega + \frac{\alpha}{2} = \widehat{IBA} = \widehat{IB\Gamma} + \frac{\alpha}{2} \Rightarrow \widehat{IB\Gamma} = \omega.$$

This means that BI is a bisector of the angle $\widehat{AB\Gamma}$.

Exercise 5.34. Show that the line $A\Delta$, passing through a vertex of triangle $AB\Gamma$ and intersecting the opposite side at Δ and the circumcircle at E is a bisector of the angle $\widehat{BA\Gamma}$, if and only if it holds $|E\Delta||EA| = |EB|^2 = |E\Gamma|^2$ (See Figure 5.18-I).

Hint: Point E is the middle of the arc $\widehat{B\Gamma}$ and the triangles ABE and $B\Delta E$ are similar.

Exercise 5.35. Show that for the isosceles triangle $AB\Gamma$ and the point Δ of its base $B\Gamma$ and with the notation $\{a = |AB|, d = |A\Delta|, x = |B\Delta|, y = |\Delta\Gamma|\}$, the relation $a^2 = d^2 + x \cdot y$ holds true.

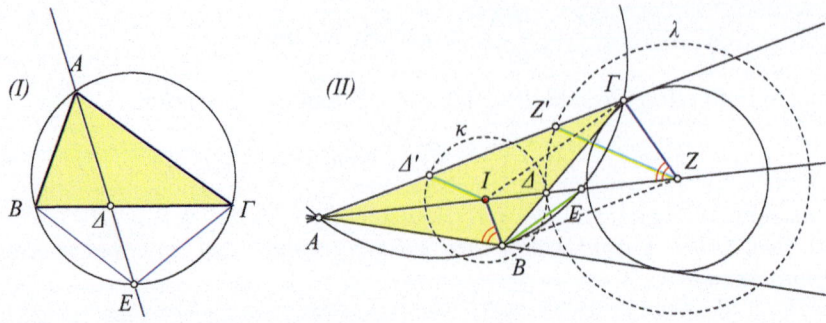

Fig. 5.18: Relation with bisector   Segments on the bisector

Hint: Almost identical to exercise 5.34 (See Figure 5.18-I).

Exercise 5.36. Construct a triangle from its elements $\alpha = |BA\Gamma|$, $a = |B\Gamma|$ and $\delta_A = |A\Delta|$, where $A\Delta$ is the bisector of angle $\widehat{A}$.

Hint: From the first two given elements the circumcircle of the requested triangle $AB\Gamma$ can be constructed and the position of B, Γ on it can be determined. Using figure 5.18-I, the conclusion of the preceding exercise, and setting $x = |EA|$, we have

$$x(x - \delta_A) = |BE|^2,$$

from which $|EA|$ is determined. With center E and radius $|EA|$ we draw a circle which intersects the previously constructed circumcircle at A.

Exercise 5.37. Let I be the incenter of triangle $AB\Gamma$ and Z be the excenter in the bisector from A (See Figure 5.18-II). Let also E be the intersection of AZ with the circumcircle and Δ be the intersection with $B\Gamma$. Show that the circles $\{\kappa, \lambda\}$ with centers, respectively, $\{I, Z\}$ and radii $\{|I\Delta|, |Z\Delta|\}$ define points $\{\Delta', Z'\}$ on $A\Gamma$, such that the triangles $\{ABE, A\Delta'I, A\Delta\Gamma, AZ'Z\}$ are similar. Also similar are the triangles $\{AIB, A\Gamma Z\}$, as well as the triangles $\{AI\Gamma, ABZ\}$ and the following relations hold:

1. $|AE| \cdot |I\Delta| = |EI| \cdot |AI|$.
2. $|AE| \cdot |A\Delta| = |AI| \cdot |AZ| = |AB| \cdot |A\Gamma|$.
3. $|IA| \cdot |Z\Delta| = |ZA| \cdot |I\Delta|$.

Exercise 5.38. In the triangle $AB\Gamma$, the middles of the sides are respectively A', B', Γ', the traces of the altitudes are A'', B'', Γ'', the orthocenter is H and A_1, B_1, Γ_1 are the middles of $HA, HB, H\Gamma$. Show that the points of intersection of the lines $\Delta = (A'\Gamma'', A_1B')$ and $E = (A_1\Gamma', A'B'')$ are contained in the parallel to $B\Gamma$ from A. Moreover line EA_1 is a bisector of the angle $\widehat{AEA'}$ and line $\Delta B'$ is a bisector of the angle $\widehat{A\Delta A'}$. Show that the circle with diameter AH is the inscribed circle of triangle $\Delta EA'$ (See Figure 5.19-I).

5.5. EULER'S THEOREM

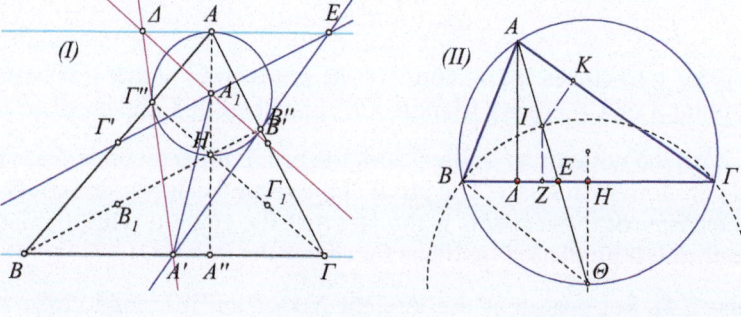

Fig. 5.19: $\{A, \Delta, E\}$ collinear   Traces $\{\Delta, Z, H\}$

Exercise 5.39. Let Δ, E, H be respectively the traces of the altitude, bisector and median on the side $B\Gamma$ of the triangle $AB\Gamma$ (See Figure 5.19-II). Let also Z be the projection of the incenter I on the side $B\Gamma$. Show that $|HZ|^2 = |H\Delta| \cdot |HE|$. Using $|ZH| = \frac{|b-c|}{2}$, construct the triangle, whose given are the lengths $\{v_A = |A\Delta|, \mu_A = |AH|, |b-c|\}$.

Hint: The relation follows from the equality of the first and last term of the equations

$$\frac{|\Delta Z|}{|ZH|} = \frac{|AI|}{|I\Theta|} = \frac{|AI|}{|B\Theta|} = \frac{|IK|}{|H\Theta|} = \frac{|IZ|}{|H\Theta|} = \frac{|ZE|}{|EH|} \Rightarrow |EH| \cdot |H\Delta - HZ| = |ZH| \cdot |ZE|.$$

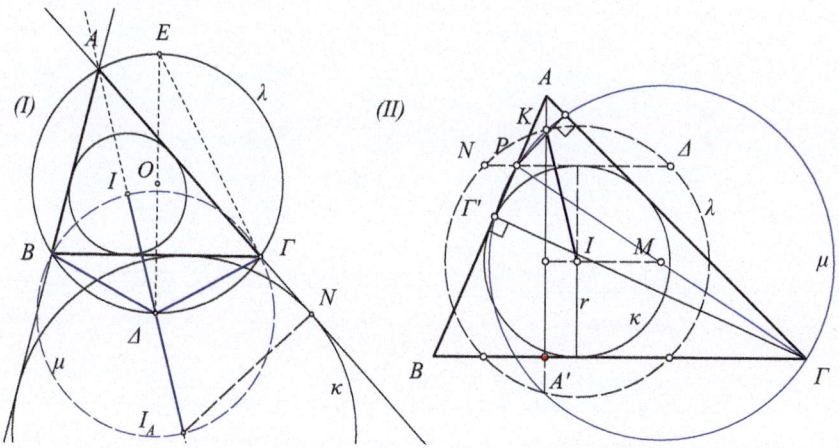

Fig. 5.20: Euler's theorem for escribed   Circle $\lambda(I, \sqrt{2} \cdot r)$

Exercise 5.40. Show that the Euler's theorem 5.5 as well as 5.6 holds analogously also for the escribed circle of triangle $AB\Gamma$. Specifically, if $I_A(r_A)$ is an escribed circle contained in the angle $\widehat{A}$ (See Figure 5.20-I), then

$$|OI_A|^2 = R(R+2r_A).$$

Also, if for two circles $\kappa(I,r)$ and $\lambda(O,R)$ holds $|OI|^2 = R(R+2r)$, then for every point A of λ there is a triangle $AB\Gamma$ inscribed in λ and escribed to κ.

Hint: With the notation of figure 5.20-I, the proof of the theorem 5.6 is transferred verbatim to this case. The proof of the second claim is similar. The key in the transfer of these proofs is the fact that the circle μ with diameter II_A passes through points B, Γ and has for center the point Δ.

Exercise 5.41. Suppose that the tangent $N\Delta$ of the inscribed circle $\kappa(I,r)$, which is parallel to the base $B\Gamma$ of the triangle $AB\Gamma$, intersects side AB at the point P. Show that the circle μ, with diameter $P\Gamma$, intersects the altitude AA' at a point K, whose distance from the incenter I is equal to $\sqrt{2}\cdot r$.

Hint: Show that A lies on the radical axis of circles $\lambda(I, \sqrt{2}\cdot r)$ and μ (See Figure 5.20-II), by calculating its two powers $\{\delta_1, \delta_2\}$ relative to the circles $\{\lambda, \mu\}$. If $\{\tau, r, \upsilon = |AA'|\}$ denote respectively the half perimeter, the radius of the inscribed circle and the altitude from A, these powers are (Proposition 5.4)

$$\delta_1 = (\tau-a)^2 - r^2 \;=\; \delta_2 = \frac{(\upsilon-2r)bc\cos(\alpha)}{\upsilon} \;=\; \frac{1}{2}(b^2+c^2-a^2)\frac{\tau-a}{\tau}.$$

Exercise 5.42. Using the notation of this section and figure 5.20-I, show the relations:

$$|\Gamma\Delta| = \frac{a}{2\cos\left(\frac{\alpha}{2}\right)} = 2R\sin\left(\frac{\alpha}{2}\right), \; |IB| = |II_A|\sin\left(\frac{\gamma}{2}\right) = 4R\sin\left(\frac{\alpha}{2}\right)\sin\left(\frac{\gamma}{2}\right),$$

$$r = 4R\sin\left(\frac{\alpha}{2}\right)\sin\left(\frac{\beta}{2}\right)\sin\left(\frac{\gamma}{2}\right),$$

$$\tau - a = |AI|\cos\left(\frac{\alpha}{2}\right) = 4R\sin\left(\frac{\beta}{2}\right)\sin\left(\frac{\gamma}{2}\right)\cos\left(\frac{\alpha}{2}\right),$$

$$\cos\left(\frac{\alpha}{2}\right) = \frac{\tau}{|AI_A|}, \; \widehat{A\Delta E} = \frac{|\beta-\gamma|}{2}, \; |A\Delta| = 2R\cos\left(\frac{|\beta-\gamma|}{2}\right),$$

$$|AE| = 2R\sin\left(\frac{|\beta-\gamma|}{2}\right),$$

$$\sin\left(\frac{\alpha}{2}\right) = \sqrt{\frac{(\tau-b)(\tau-c)}{bc}}, \; \cos\left(\frac{\alpha}{2}\right) = \sqrt{\frac{\tau(\tau-a)}{bc}},$$

$$\cot\left(\frac{\alpha}{2}\right) = \sqrt{\frac{\tau(\tau-a)}{(\tau-b)(\tau-c)}},$$

$$\tau^2 - r^2 - 4Rr = \frac{1}{2}(a^2+b^2+c^2), \; \tau^2 + r^2 + 4rR = ab+bc+ca.$$

5.5. EULER'S THEOREM

Hint: The relations in the first three lines result immediately from the figure and the rule of sine for triangles (§ 3.11). For the relations of the fourth line start from $\tan\left(\frac{\alpha}{2}\right) = \frac{r}{\tau - a}$ and the relations of proposition 5.4. The one before last relation follows from the left side, by expressing r^2 through proposition 5.4 and $4Rr$ through corollary 3.20 and doing some calculation. The last equation follows from the preceding one and the identity $2(ab + bc + ca) = (a + b + c)^2 - (a^2 + b^2 + c^2)$.

Exercise 5.43. Show that the lengths of the sides of the triangle $\{a, b, c\}$ satisfy the cubic equation

$$x^3 - 2\tau \cdot x^2 + (\tau^2 + r^2 + 4Rr)x - 4\tau Rr = 0. \tag{5.1}$$

Hint: Replace in $\sin^2\left(\frac{\alpha}{2}\right) + \cos^2\left(\frac{\alpha}{2}\right) = 1$ the corresponding expressions of the preceding exercise, which using corollary 3.20 are written as:

$$\sin^2\left(\frac{\alpha}{2}\right) = \frac{ar}{4R(\tau - a)}, \quad \cos^2\left(\frac{\alpha}{2}\right) = \frac{a(\tau - a)}{4Rr} \Rightarrow \frac{ar}{4R(\tau - a)} + \frac{a(\tau - a)}{4Rr} = 1.$$

This equation is equivalent to equation (5.1), for $x = a$ and will hold also for $\{x = b, x = c\}$, since the coefficients of equation (5.1) are independent of a.

Remark 5.2. $\{\tau, r, R\}$ are called **fundamental invariants** of the triangle ([108, p.110]). The converse problem, that of the existence of a triangle with given such three quantities, occupied Euler, in a slight variation ([88, p.7]) and led him to the above third degree equation. In order for its roots to be real, the known inequality ([117, p.71]) for the coefficients of the cubic equation must be satisfied. This is a condition, which in the present case reduces to

$$(\tau^2 + r^2)^2 + 4R[4rR(4R + 3r) - \tau^2(R + 5r) + 3r^3] \le 0.$$

Besides, in order for a triangle to exist with the given data, certain additional conditions must be satisfied, like for example the deduced from the Euler's theorem inequality, $R > 2r$ as well as the deduced from the preceding exercise inequality, $\tau > r$. If such a triangle exists, then the lengths of its sides are determined fully through the roots of the polynomial. However the construction of the triangle with these data, in general, is not possible using only a ruler and compass. The last inequality, considered with respect to τ^2 is quadratic and is satisfied when τ^2 is between the roots of the corresponding trinomial. This leads to the double inequality (of Blundon), [114]

$$2R(R + 5r) - r^2 - 2(R - 2r)\sqrt{R^2 - 2Rr} \le \tau^2$$
$$\le 2R(R + 5r) - r^2 + 2(R - 2r)\sqrt{R^2 - 2Rr}.$$

Exercise 5.44. To construct a right triangle with given length of hypotenuse and length of the bisector of one acute angle.

Exercise 5.45. Show that in every triangle the radius R of the circumcircle and the radius r of the incircle satisfy the inequality $R \geq 2r$ and the equality is valid precisely when the triangle is equilateral.

Exercise 5.46. Construct a triangle $AB\Gamma$ for which is given the position of the vertex A, of the incenter I and the position of the circumcenter O.

5.6 Tangent circles of Apollonius

> Meanwhile, there developed among the Greeks a quite special, one might say instinctive, sense of beauty which was peculiar to them alone of all the nations that have ever existed on earth; a sense that was fine and correct.
>
> A. Schopenhauer, *On Religion*

The problem of tangent circles of Apollonius consists in finding the circles which are simultaneously tangent to three given circles. Figure 5.21 shows the eight circles which solve the problem, in the case where the given circles κ, λ and μ are mutually external.

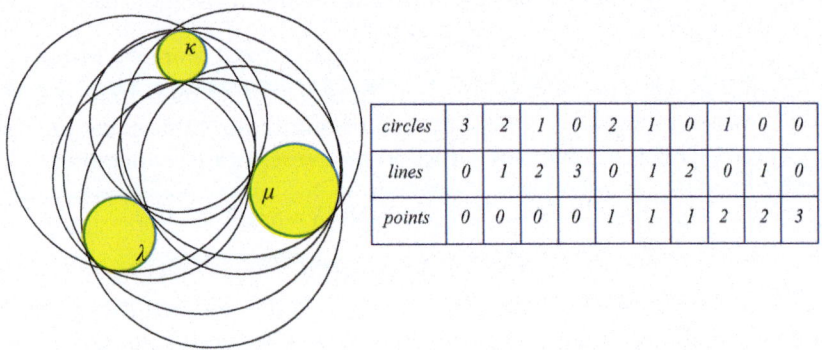

circles	3	2	1	0	2	1	0	1	0	0
lines	0	1	2	3	0	1	2	0	1	0
points	0	0	0	0	1	1	1	2	2	3

Fig. 5.21: Eight circles tangent to three given ones

Remark 5.3. Apollonius in his writing "contacts", which was lost, formulated more generally the problem of finding circles which are tangent to three *"things"*. The "things" may be circles, lines and points. The requested circles therefore may be tangent to given circles and/or lines and/or pass through given points. This results in 10 main categories for the problem where the "things" are expressed by the triples of numbers contained in the columns of the table on the right.

Some of these problems have been already discussed. The last for example, of the construction of a circle through three points (Theorem 2.1), the

5.6. TANGENT CIRCLES OF APOLLONIUS

fourth, of the construction of a circle tangent to three lines (which form a triangle, Theorem 2.3, Theorem 5.1), the eighth (Exercise 4.23) and the ninth (Exercise 4.1). In each category there are different cases and the individual problems have drawn the attention of many Mathematicians throughout the centuries. The first category shows the most special cases, depending on the position of the circles. This way, for example, for non intersecting circles we have figuratively the four possibilities of figure 5.22, from which the first two do not admit a solution. In this section we'll discuss the first and fifth of the ten categories, in the case where the "things" are mutually external.

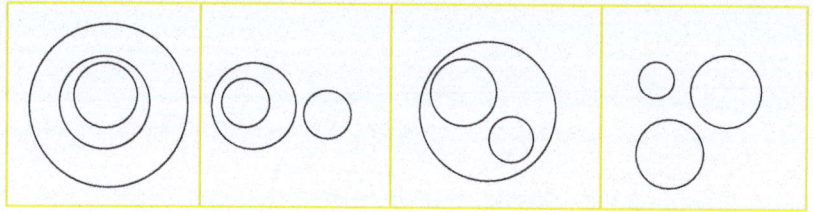

Fig. 5.22: Possible positions of three non intersecting circles

Construction 5.1 *Construct a circle μ simultaneously tangent to two mutually external given circles $\kappa(A, \alpha)$, $\lambda(B, \beta)$ and passing through a point Σ also external to the two circles.*

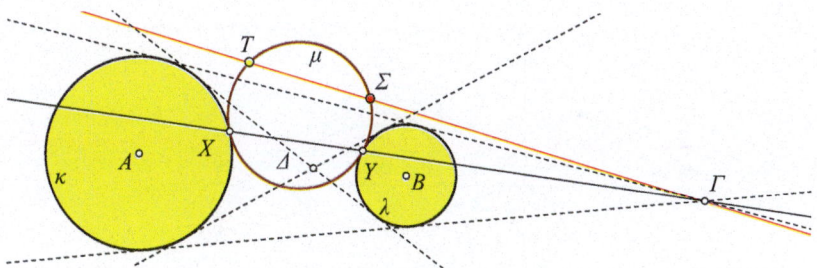

Fig. 5.23: μ tangent to κ, λ and passing through point Σ

Construction: Let us suppose that the requested circle μ is tangent externally to κ and λ respectively at points X and Y (Figure 5.23). According to Corollary 4.18, the line XY passes through a similarity center Γ of the two given circles and the product $\xi = |\Gamma X||\Gamma Y|$ will be fixed, independent of μ and determined from the given data. Consequently, if T is the second point of intersection of the line $\Gamma\Sigma$ with the circle μ, then also the product

$$|\Gamma\Sigma||\Gamma T| = |\Gamma X||\Gamma Y| = \xi,$$

and through it, the position of T, will be determined from the given data. Consequently the problem is reduced to the known problem of finding a circle μ, passing through two points $\{\Sigma, T\}$ and simultaneously tangent to one of the two circles, for example to $\kappa(A, \alpha)$. A problem which has in general two solutions and we already solved (Exercise 4.23).

In a similar way we get two solutions for a circle μ, which is tangent to one circle externally and the other internally. This time the line XY passes through the internal center of similarity Δ of circles κ and λ. Concluding therefore we get four circles μ which satisfy the requirements of the construction. Figure 5.24 shows the four circles-solutions of the problem. For

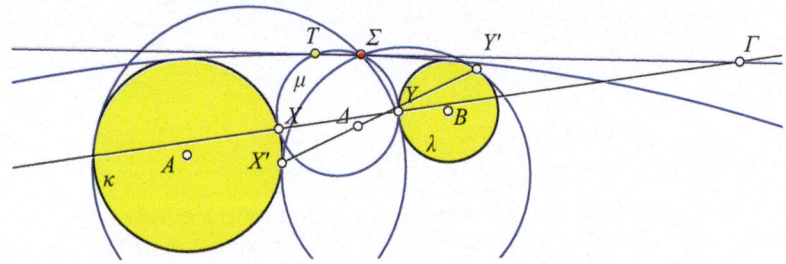

Fig. 5.24: Circles tangent to κ and λ and passing through Σ

two of them visible are also the chords of the contact points XY and $X'Y'$, which pass through the centers of similarity $\{\Gamma, \Delta\}$ of the given circles. Fig-

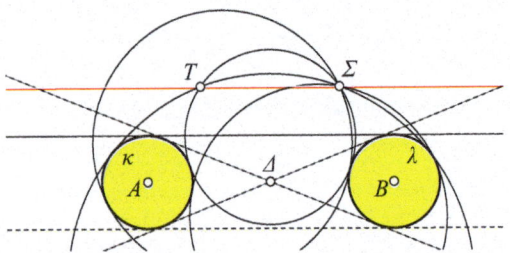

Fig. 5.25: Circles tangent to κ, λ and passing through Σ (II)

ure 5.25 shows the solutions of the problem in the case where the radii of κ and λ are equal. In this case the external center of similarity Γ lies at infinity and the second point T, which is the key to the solution, is the symmetric of the given Σ relative to the medial line of AB.

Construction 5.2 *Construct a circle ν simultaneously tangent to three given, mutually external, circles $\{\kappa(A, \alpha), \lambda(B, \beta), \mu(\Gamma, \gamma)\}$ (See Figure 5.26-I).*

Construction: We handle here the general case, in which the radii of the circles are pairwise different and satisfy $\alpha > \beta > \gamma$. We'll discuss the special cases

5.6. TANGENT CIRCLES OF APOLLONIUS

later. For the position of the requested circle v there are eight cases, which can be denoted with one symbol (***), consisting of three signs. In each case we put a + or a - depending on whether a solution-circle v has the given circle in its exterior or its interior. This way, (+++) means that v has all circles in its exterior, (---) that it has all circles in its interior, (-++) that it has κ in its interior and λ and μ in its exterior, as in figure 5.26-(I). In each of the eight cases we reduce the problem into that of the preceding construction using the following trick:

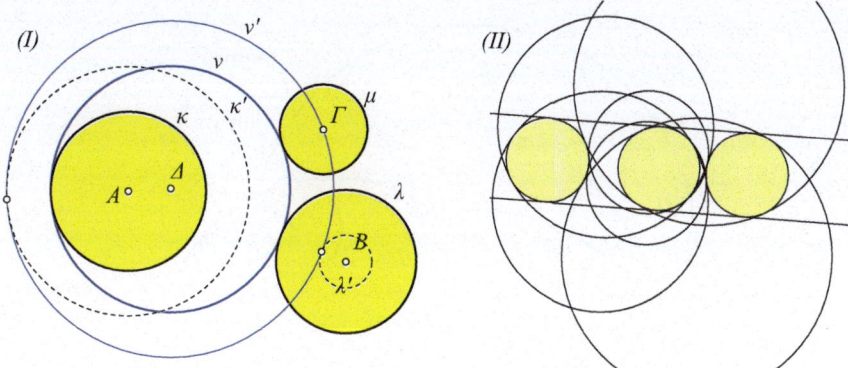

Fig. 5.26: Circle v with (-++) kind of contact   Tangent to three congruent

1. We replace the smaller circle (μ) with its center Γ.
2. We replace each of the two other circles κ, λ with a circle of the same center but radii respectively $\alpha \pm \gamma$ and $\beta \pm \gamma$.
3. This results in two new circles κ', λ' and instead of v we construct its concentric v', which passes through point Γ and is tangent to κ' and λ'.

The correct signs in (2) follow from this requirement. This way, in the problem of figure 5.26-I we are after the circle v' which passes through Γ and is tangent to $\kappa'(A, \alpha+\gamma)$ and $\lambda'(B, \beta-\gamma)$, in other words the circle which has them in its interior and exterior respectively. According to the preceding construction there exists exactly one such circle $v'(\Delta,\rho)$, constructible from the given data. The requested v will have the same center Δ with v' and radius $\rho - \alpha$. Solving therefore the problem for each one of the eight cases we also get the general solution to the problem.

In the cases where the two circles have the same radius, the solution is simplified a bit, since the initial problem reduces to the simpler constructions. Thus, for example in case $\alpha > \beta = \gamma$, the preceding process, in the case (-++), is reduced to the one of finding a circle v' which is tangent to $\kappa'(A, \alpha+\gamma)$ and passes through the two points B and Γ (Exercise 4.23).

Another example is the one where the three circles have the same radius and their centers are collinear (See Figure 5.26-II). Then the solutions include two lines symmetrically lying relative to their center-line. Also the rest of the solutions consists of three pairs of circles symmetrically lying relative to the center line of the three circles. The work therefore is reduced to one half.

Remark 5.4. Figure 5.27 shows the eight solutions for three mutually external circles with radii $\alpha > \beta > \gamma$. Noted on them is also their kind, as it is described in the preceding proof. I note that a detailed solution of the problem of Apollonius for all possible categories and all possible sub-cases is, if not difficult, rather painful (a list of all the cases is contained in [141, p.135], [116, p.97]).

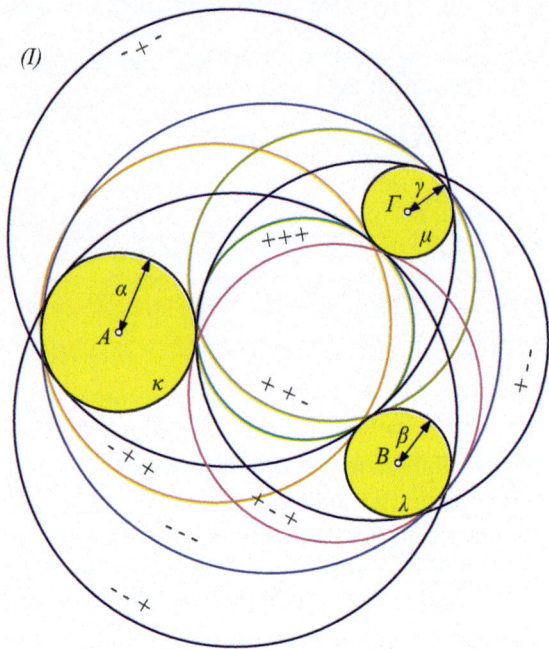

Fig. 5.27: The tangency kind

Remark 5.5. The solution in the last construction is due to Viete (1540-1603). There are several known solutions of the problem, geometric but also computational ([125, p.5], [131, p.15]). In certain special problems an inversion helps a lot. Through it the initial problem *A* is transformed into a simpler one *B* and after the solution of *B*, the solution to *A* is obtained by transforming back *B* to *A* through the same inversion.

We can see an example of such a transformation in the case of the two so called *circles of Soddy* of three externally pairwise tangent circles. This particular Apollonius problem has two solutions only: the internal (σ), and external (τ) *Soddy* circles of the three tangent circles $\{\kappa, \lambda, \mu\}$ (See Figure 5.28).

The construction problem of the Soddy circles σ and τ is reduced to a simpler one through the inversion relative to a suitable circle $\xi(\Xi, \rho)$ ([127,

5.6. TANGENT CIRCLES OF APOLLONIUS

p.191]). The center of this circle is taken to be one (of the three) contact points of the given circles, for example the contact point Ξ of κ and λ. The radius ρ of ξ is taken so, that this circle is orthogonal to the third circle (μ). The inverses relative to ξ, κ' and λ' of κ and λ are then two parallel lines tangent to μ (Proposition 4.16), while μ is the inverse of itself (Corollary 4.20). The inverses σ' and τ' of the requested circles are circles tangent to the two parallels and tangent also to μ, hence equal to μ and tangent to it. Thus $\{\sigma', \tau'\}$ are directly constructible and their inverses, relative to ξ, are the requested circles σ and τ.

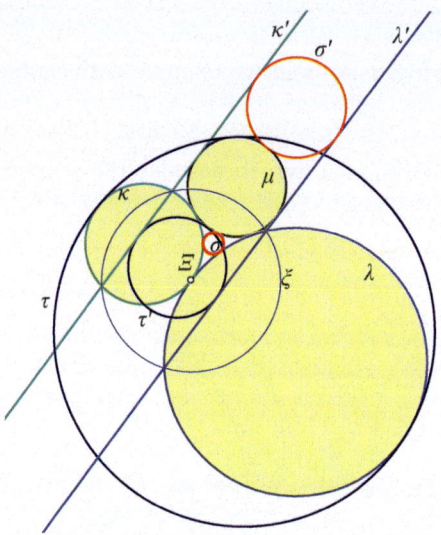

Fig. 5.28: Circles σ, τ of Soddy

Construction 5.3 *Construct a circle ν simultaneously tangent to three given circles κ, λ and μ, which pass through a common point A without being tangent there.*

Construction: Let Λ be another intersection point of the three circles, for example of κ and μ (See Figure 5.29). We consider the inversion relative to the circle $\xi(A, |A\Lambda|)$. The inverses of circles κ, λ and μ are three lines κ', λ' and μ' (Proposition 4.16), two of which pass through the point Λ. This forms triangle $K\Lambda M$ and the inverses $\{\nu'\}$ of the solutions to the initial problem will be circles tangent to the sides of the triangle (Proposition 4.18), therefore they will coincide with the four circles which are tangent to the sides of the triangle $K\Lambda M$ (inscribed and three escribed). Consequently the solutions to the original problem will be the inverses $\{\nu\}$ of $\{\nu'\}$ relative to the same inversion. There are therefore four solutions in this case.

Figure 5.30 shows in its left part the four circles $\{\nu'\}$ which are tangent to the sides of the triangle $K\Lambda M$. Its right part shows the inverses $\{\nu\}$ of these

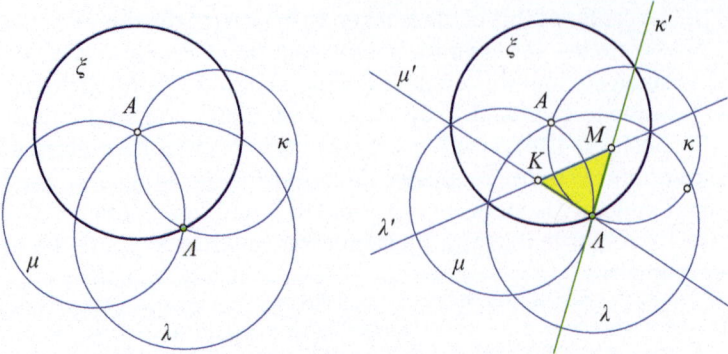

Fig. 5.29: Inversion relative to circle with center A

four circles (relative to circle ξ), which are also the solutions to the initial problem.

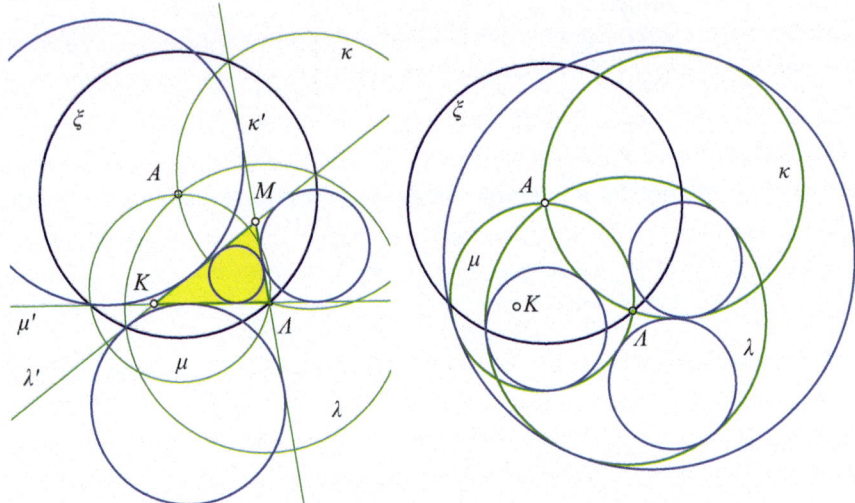

Fig. 5.30: The circles tangent to κ, λ and μ

Exercise 5.47. Show that in the case, where two circles κ and λ are tangent at A and a line ε passes through A without being tangent to the two circles, there exist exactly two circles simultaneously tangent to κ, λ and the line ε.

5.7 Theorems of Ptolemy and Brahmagupta

> A mind which has once imbibed a taste for scientific enquiry, and has learnt the habit of applying its principles readily to the cases which occur, has within itself an inexhaustible source of pure and exciting contemplations.
>
> J. F. W. Herschel, *Discourse on the Study of Natural Philosophy*

The theorems of Ptolemy (85-165) and of Brahmagupta (598-668) express simple characteristic properties of quadrilaterals inscriptible in a circle. All three theorems have the form of quantitative relations, which are useful in the investigation of more complex properties of quadrilaterals of this kind.

Theorem 5.7 (Ptolemy A). *For every convex inscribed in a circle quadrilateral $AB\Gamma\Delta$ the product of its diagonals is equal to the sum of the products of its opposite sides:*

$$|A\Gamma||B\Delta| = |AB||\Gamma\Delta| + |B\Gamma||\Delta A|.$$

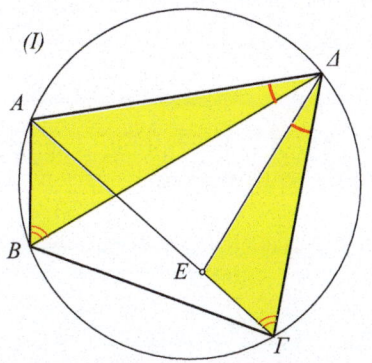
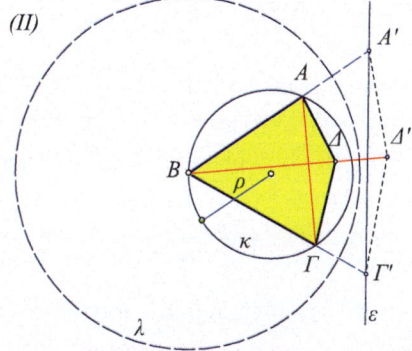

Fig. 5.31: Ptolemy's theorem   Ptolemy's inequality

Proof. (By Ptolemy [14, p.173]) Draw ΔE, so that the created angle is $\widehat{E\Delta\Gamma} = \widehat{A\Delta B}$ (See Figure 5.31-I). Then the triangles $\Delta E\Gamma$ and ΔAB are similar, having equal angles respectively at Δ and at B, Γ. From this similarity follows

$$\frac{|AB|}{|E\Gamma|} = \frac{|B\Delta|}{|\Delta\Gamma|} \Leftrightarrow |AB||\Delta\Gamma| = |B\Delta||E\Gamma|.$$

Also the triangles ΔAE and $\Delta B\Gamma$ are similar, having equal angles respectively at Δ and at A and B. From this similarity follows

$$\frac{|AE|}{|B\Gamma|} = \frac{|A\Delta|}{|B\Delta|} \Leftrightarrow |A\Delta||B\Gamma| = |B\Delta||AE|.$$

Adding the two equations by parts we have the requested

$$|AB||\Delta\Gamma|+|A\Delta||B\Gamma|=|B\Delta||E\Gamma|+|B\Delta||AE|=|B\Delta|(|E\Gamma|+|AE|)=|B\Delta||A\Gamma|,$$

Theorem 5.8. *For every convex quadrilateral $AB\Gamma\Delta$ the following inequality holds*

$$|AB||\Gamma\Delta|+|B\Gamma||\Delta A| \geq |A\Gamma||B\Delta|.$$

The equality holds, if and only if the quadrilateral is inscriptible in a circle.

Proof. The proof of this, more general than the preceding theorem, can be reduced to one application of inversion ([11, p.128], [106, p.111]). Using for center one vertex of the quadrilateral, B say, we draw a circle $\lambda(B,\rho)$, which contains the entire quadrilateral in its interior (See Figure 5.31-II). Next, we consider the circle κ, which passes through B and the two neighboring vertices A and Γ. We also consider the inverse of κ relative to λ, which is a line ε (Proposition 4.16). The inverses of A, Γ and Δ are the points A', Γ' and Δ'. From which, the two first are on the line ε, while the third Δ' is on the line ε, if and only if Δ lies on the circle κ, in other words when $AB\Gamma\Delta$ is inscriptible. Then, the triangle inequality holds

$$|A'\Delta'|+|\Delta'\Gamma'| \geq |A'\Gamma'|,$$

where the equality holds exactly in the case $AB\Gamma\Delta$ is inscriptible. The last inequality however, relying on (Exercise 4.93) is equivalent to

$$|A\Delta|\cdot\frac{\rho^2}{|BA||B\Delta|}+|\Delta\Gamma|\cdot\frac{\rho^2}{|B\Delta||B\Gamma|} \geq |A\Gamma|\cdot\frac{\rho^2}{|BA||B\Gamma|},$$

which is reducible to the inequality to show.

Theorem 5.9 (Ptolemy B). *For every convex, inscriptible quadrilateral $AB\Gamma\Delta$ the ratio of its diagonals is equal to*

$$\frac{|A\Gamma|}{|B\Delta|}=\frac{|AB||\Delta A|+|B\Gamma||\Gamma\Delta|}{|AB||B\Gamma|+|\Gamma\Delta||\Delta A|}.$$

Proof. The proof follows by applying the formula of Corollary 3.20 on the four triangles which are formed by the diagonals (See Figure 5.31-I).

$$\varepsilon(AB\Gamma)=\frac{|AB||B\Gamma||\Gamma A|}{4\rho}, \quad \varepsilon(A\Gamma\Delta)=\frac{|A\Gamma||\Gamma\Delta||\Delta A|}{4\rho},$$

$$\varepsilon(AB\Delta)=\frac{|AB||B\Delta||\Delta A|}{4\rho}, \quad \varepsilon(B\Gamma\Delta)=\frac{|B\Gamma||\Gamma\Delta||\Delta B|}{4\rho}.$$

We complete the proof, using these formulas in the obvious equality

$$\varepsilon(AB\Gamma)+\varepsilon(A\Gamma\Delta)=\varepsilon(AB\Delta)+\varepsilon(B\Gamma\Delta).$$

5.7. THEOREMS OF PTOLEMY AND BRAHMAGUPTA

and simplifying.

Remark 5.6. This proposition expresses a characteristic property of the inscribed in a circle convex quadrilateral as well. In fact, it can be proved, that conversely if the relation of the theorem holds, then the quadrilateral can be inscribed in a circle (see Exercise 5.56).

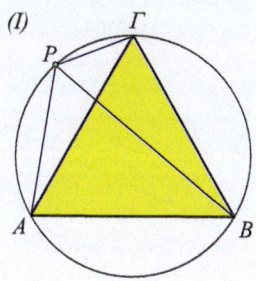
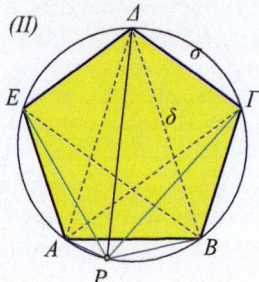

Fig. 5.32: Equilateral property   Regular pentagon property

Exercise 5.48. Show that for every point P of the circumcircle of the equilateral triangle $AB\Gamma$, the greatest of the line segments PA, PB and $P\Gamma$ is the sum of the two others.

Hint: Let PB be the greatest segment from the three and apply the theorem of Ptolemy on the quadrilateral $AB\Gamma P$ (See Figure 5.32-I): $|A\Gamma||PB| = |P\Gamma||AB| + |PA||B\Gamma|$, which, because of the equality of the sides, simplifies and gives the requested formula.

Exercise 5.49. Show that, for every point P of the arc AB of the circumcircle of the regular pentagon $AB\Gamma\Delta E$ holds

$$|PA| + |PB| + |P\Delta| = |P\Gamma| + |PE|.$$

Hint: Let σ and δ be the lengths of the sides and diagonals of the pentagon (See Figure 5.32-II). Apply the theorem of Ptolemy on the three quadrilaterals $APB\Delta$, $PE\Delta\Gamma$ and $AB\Gamma E$:

$$\sigma|P\Delta| = \delta|PA| + \delta|PB| \Rightarrow |PA| + |PB| = \frac{\sigma}{\delta}|P\Delta|,$$

$$\delta|P\Delta| = \sigma|PE| + \sigma|P\Gamma| \Rightarrow |PE| + |P\Gamma| = \frac{\delta}{\sigma}|P\Delta|,$$

$$\delta^2 = \sigma\delta + \sigma^2 \Rightarrow \frac{\delta}{\sigma} = 1 + \frac{\sigma}{\delta}.$$

The claim follows by combining the three last equations ([16, p.42]).

Exercise 5.50. Let $AB\Gamma\Delta$ be a quadrilateral inscribed in a circle κ. Show that, for an arbitrary point P of κ, the product of its distances from the two opposite sides of the quadrilateral is equal to that of the other opposite sides ($|PE||PZ| = |PH||P\Theta|$) (See Figure 5.33-I).

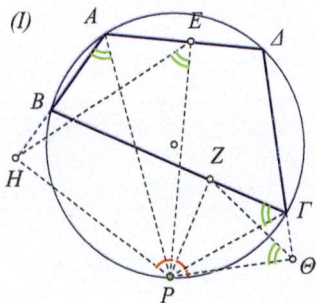
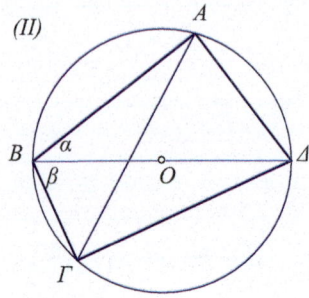

Fig. 5.33: Inscribed quadrilateral   Sine of sum of angles

Hint: Triangles PEH and $P\Theta Z$ are similar, because they have the angles $\widehat{EPH}$ and $\widehat{\Theta PZ}$ equal and angles $\widehat{PEH}$ and $\widehat{P\Theta Z}$ also equal. Consequently they have proportional sides: $\frac{|PE|}{|PH|} = \frac{|P\Theta|}{|PZ|}$.

Exercise 5.51. Show the sine formula for the sum of acute angles α and β

$$\sin(\alpha+\beta) = \sin(\alpha)\cos(\beta) + \sin(\beta)\cos(\alpha),$$

using Ptolemy's theorem.

Hint: Draw the angles α and β on either side of a diameter $B\Delta$ of the circle κ with arbitrary radius ρ (See Figure 5.33-II). Their second sides intersect the circle at points A and Γ, forming the inscribed convex quadrilateral $AB\Gamma\Delta$. From the sine formula we have

$$\frac{|A\Gamma|}{\sin(\alpha+\beta)} = |B\Delta| \Rightarrow |A\Gamma||B\Delta| = |B\Delta|^2 \sin(\alpha+\beta).$$

From the right triangles $AB\Delta$ and $B\Gamma\Delta$ we have

$$|B\Gamma||A\Delta| = (|B\Delta|\cos(\beta))(|B\Delta|\sin(\alpha)) = |B\Delta|^2 \sin(\alpha)\cos(\beta)$$
$$|AB||\Gamma\Delta| = (|B\Delta|\cos(\alpha))(|B\Delta|\sin(\beta)) = |B\Delta|^2 \cos(\alpha)\sin(\beta).$$

The claim follows from Ptolemy's theorem $|A\Gamma||B\Delta| = |AB||\Gamma\Delta| + |B\Gamma||A\Delta|$.

Proposition 5.6. *From the vertices of the triangle $AB\Gamma$ draw tangents $AX, BY, \Gamma Z$ to a circle κ. If one of the products $\{|AX||B\Gamma|, |BY||\Gamma A|, |\Gamma Z||AB|\}$ is equal to the sum of the other two, then the circle κ is tangent to the circumcircle λ of $AB\Gamma$.*

5.7. THEOREMS OF PTOLEMY AND BRAHMAGUPTA

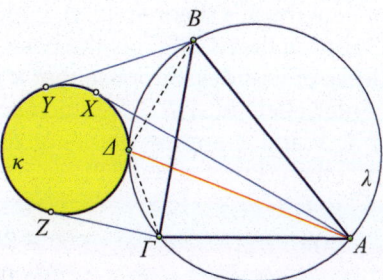

Fig. 5.34: Tangency criterion

Proof. ([27, p.206])(See Figure 5.34) Suppose that

$$|AX||B\Gamma| = |BY||\Gamma A| + |\Gamma Z||AB|.$$

Define on the arc $\widehat{B\Gamma}$ the point Δ, so that

$$\frac{|\Delta \Gamma|}{|\Delta B|} = \frac{|\Gamma Z|}{|BY|} \quad \Rightarrow \quad \frac{|BY|}{|B\Delta|} = \frac{|\Gamma Z|}{|\Gamma \Delta|} = \xi.$$

From the assumption and Ptolemy's theorem follows that

$$|AX||B\Gamma| = \xi|B\Delta||\Gamma A| + \xi|\Gamma \Delta||AB| = \xi(|B\Delta||\Gamma A| + |\Gamma \Delta||AB|)$$
$$= \xi \cdot |B\Gamma||A\Delta| \quad \Rightarrow \quad \frac{|AX|}{|A\Delta|} = \frac{|BY|}{|B\Delta|} = \frac{|\Gamma Z|}{|\Gamma \Delta|}.$$

The result follows from Corollary 4.11, according to which, when the last relations hold, then the circle λ belongs to the pencil of circles containing the circle κ and the limit point Δ. Because point Δ belongs to circle λ, the pencil is a tangential one and the two circles are tangent at Δ.

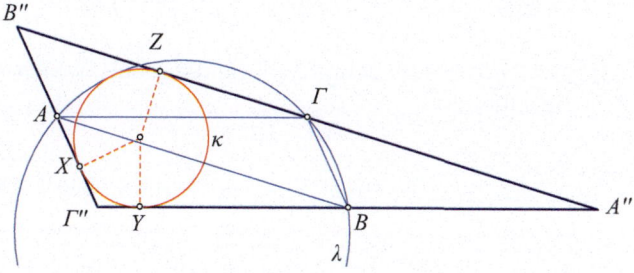

Fig. 5.35: Feuerbach through pencils and Ptolemy

Exercise 5.52 (Feuerbach). Using the preceding proposition, show that the Euler circle of a triangle is tangent to its inscribed, as well as to its escribed circles (See Figure 5.35).

Hint: This is a proof of Feuerbach's theorem (5.4), which reveals, I think, its structure a bit better. Suppose that $A''B''\Gamma''$ is the basic triangle and $AB\Gamma$ is the triangle of the middles of its sides. Suppose that κ is the inscribed circle of $A''B''\Gamma''$ and λ the Euler circle of $A''B''\Gamma''$, which coincides with the circumscribed of $AB\Gamma$. If X, Y and Z are the points of contact of the incircles with the sides of $A''B''\Gamma''$, then (Proposition 5.3)

$$|AX| = \frac{1}{2}|c-b|, \quad |BY| = \frac{1}{2}|a-c|, \quad |\Gamma Z| = \frac{1}{2}|b-a|,$$

where $\{a = |B''\Gamma''|, b = |\Gamma''A''|, c = |A''B''|\}$ are the lengths of the sides of $A''B''\Gamma''$. It follows

$$|AX||B\Gamma| = \frac{1}{4}|c-b|a, \quad |BY||\Gamma A| = \frac{1}{4}|a-c|b, \quad |\Gamma Z||AB| = \frac{1}{4}|b-a|c.$$

Taking into account the ordering of the lengths of the sides, so that we are rid of absolute values, for example, like in figure 5.35, in which $a < b < c$, we find that

$$|BY||\Gamma A| = |AX||B\Gamma| + |\Gamma Z||AB| \quad \Leftrightarrow \quad (c-a)b = (c-b)a + (b-a)c.$$

Applying Proposition 5.6 we see that the Euler circle λ of $A''B''\Gamma''$, is tangent to the inscribed κ of $A''B''\Gamma''$. The proof for the escribed circles of $A''B''\Gamma''$ is similar.

Theorem 5.10 (Brahmagupta). *Let $AB\Gamma\Delta$ be a cyclic quadrilateral, with lengths of successive sides $\{a = |AB|, b = |B\Gamma|, c = |\Gamma\Delta|, d = |\Delta A|\}$ and perimeter $2\tau = a+b+c+d$. Then its area is given by the formula*

$$\varepsilon(AB\Gamma\Delta) = \sqrt{(\tau-a)(\tau-b)(\tau-c)(\tau-d)}.$$

Proof. According to Exercise 3.61, the area of the quadrilateral is given by the formula

$$\varepsilon(AB\Gamma\Delta) = \frac{1}{4}\sqrt{4x^2y^2 - (a^2+c^2-b^2-d^2)^2},$$

where $x = |A\Gamma|$, $y = |B\Delta|$ are the lengths of the diagonals, which also satisfy (Theorem 5.7) the equation

$$xy = ac + bd.$$

The claim follows by substituting xy in the preceding to last equality and factoring.

Remark 5.7. Note that, by setting $a = 0$ in the preceding formula, we get Heron's formula for the area of triangle (Theorem 5.2). The converse also holds. The formula of Brahmagupta follows from that of Heron ([136, p.191]). The formula of Brahmagupta is also a special case of the next formula for the area of convex quadrilaterals, which was proved in 1842 by Bretschneider (1808-1878).

5.7. THEOREMS OF PTOLEMY AND BRAHMAGUPTA

Theorem 5.11 (Bretschneider). *The area of a convex quadrilateral $AB\Gamma\Delta$, with side lengths $\{a,b,c,d\}$, half perimeter τ and two opposite angles of measure α and γ, is equal to*

$$\varepsilon(AB\Gamma\Delta) = \sqrt{(\tau-a)(\tau-b)(\tau-c)(\tau-d) - abcd\cos^2\left(\frac{\alpha+\gamma}{2}\right)}.$$

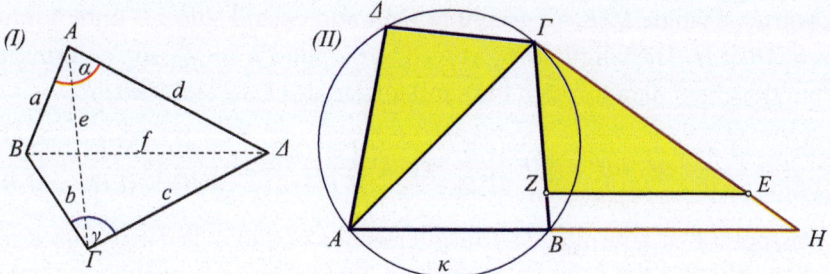

Fig. 5.36: Bretschneider   Construction of inscriptible quadrilateral

Proof. For the triangles $AB\Delta$ and $B\Gamma\Delta$, the formulas for the area and for the sines give (See Figure 5.36-I).

$$\varepsilon = \varepsilon(AB\Gamma\Delta) = \varepsilon(AB\Delta) + \varepsilon(B\Gamma\Delta) = \frac{1}{2}ad\sin(\alpha) + \frac{1}{2}bc\sin(\gamma) \Rightarrow$$

$$4\varepsilon^2 = (ad)^2\sin^2(\alpha) + (bc)^2\sin^2(\gamma) + 2abcd\sin(\alpha)\sin(\gamma)$$

$$|B\Delta|^2 = a^2 + d^2 - 2ad\cos(\alpha) = b^2 + c^2 - 2bc\cos(\gamma) \Rightarrow$$

$$\frac{(a^2+d^2-b^2-c^2)^2}{4} = (ad)^2\cos^2(\alpha) + (bc)^2\cos^2(\gamma) - 2abcd\cos(\alpha)\cos(\gamma).$$

Adding by parts the second and the fourth equality we have

$$4\varepsilon^2 + \frac{(a^2+d^2-b^2-c^2)^2}{4}$$
$$= (ad)^2 + (bc)^2 + 2abcd(\sin(\alpha)\sin(\gamma) - \cos(\alpha)\cos(\gamma))$$
$$= (ad)^2 + (bc)^2 - 2abcd\cos(\alpha+\gamma)$$
$$= (ad)^2 + (bc)^2 - 2abcd\left(2\cos^2\left(\frac{\alpha+\gamma}{2}\right) - 1\right)$$
$$= (ad+bc)^2 - 4abcd\cos^2\left(\frac{\alpha+\gamma}{2}\right).$$

The requested formula follows from the last equality and the easily proved identity

$$(ad+bc)^2 - \frac{(a^2+d^2-b^2-c^2)^2}{4} = 4(\tau-a)(\tau-b)(\tau-c)(\tau-d).$$

Exercise 5.53. Show the inverse of the theorem of Brahmagupta. That is, if the formula of the theorem 5.10 holds, then the quadrilateral is inscriptible in circle.

Exercise 5.54. Construct a quadrilateral inscriptible in a circle, for which are given the lengths of its four sides (See Figure 5.36-II).

Hint: Suppose that the requested quardilateral was constructed. On side $B\Gamma$ construct triangle ΓZE congruent to $\Gamma \Delta A$ and extend side ΓE until it intersects AB at H. The length $|BH|$, as well as the ratio $\lambda = \frac{|\Gamma A|}{|\Gamma H|}$, are determined from the given data through the similar triangles ΓZE and ΓBH:

$$\frac{|BH|}{|ZE|} = \frac{|\Gamma Z|}{|\Gamma B|} \Rightarrow |BH| = |ZE|\frac{|\Gamma Z|}{|\Gamma B|} = |A\Delta|\frac{|\Gamma A|}{|\Gamma B|}, \quad \lambda = \frac{|\Gamma A|}{|\Gamma H|} = \frac{|\Gamma E|}{|\Gamma H|} = \frac{|\Gamma Z|}{|\Gamma B|}.$$

Then, point Γ is determined as the intersection of the circle with center B and radius $|B\Gamma|$ and the Apollonian circle of the segment AH relative to ratio λ. From points A, B and Γ is determined the circumcircle κ of the quadrilateral and then the vertex Δ on it.

Exercise 5.55. Continuing the preceding exercise, show that, depending on the ordering of the side-lengths $\{a, b, c, d\}$ of the quadrilateral, there are three quadrilaterals with the given side-lengths. Show also that they have, in total, diagonals of only three different lengths.

Hint: Opposite to a there can be one of the sides $\{b, c, d\}$ and this determines the quadrilateral. For the lengths of the diagonals use the second theorem of Ptolemy ([11, 129]).

Exercise 5.56. Show that if the second theorem of Ptolemy (Theorem 5.9) is valid for the convex quadrilateral $AB\Gamma\Delta$, then this is inscriptible in a circle.

Hint: ([149]) Start with the relations (See Figure 5.36-I):

$$e^2 = c^2 + d^2 - 2dc\cos(\widehat{A}), \qquad e^2 = a^2 + b^2 - 2ab\cos(\widehat{B}),$$

multiply the first with ab and the second with cd and show, using a simple calculation that,

$$e^2 = \frac{(ac+bd)(ad+bc) - 2abcd(\cos(\widehat{B}) + \cos(\widehat{A}))}{ab+cd}.$$

Show analogously the equation

$$f^2 = \frac{(ab+cd)(ac+bd) - 2abcd(\cos(\widehat{\Gamma}) + \cos(\widehat{A}))}{ad+bc}.$$

Then, use these relations together with the hypothesis $\frac{e}{f} = \frac{ad+bc}{ab+cd}$, and show that

5.7. THEOREMS OF PTOLEMY AND BRAHMAGUPTA

$$\cos(\widehat{B}) + \cos(\widehat{\Delta}) = \cos(\widehat{\Gamma}) + \cos(\widehat{A}) = 0.$$

In the proof of these equations it is necessary to examine the signs of the two sums, which are opposite, because of the relations $\widehat{A} + \widehat{B} + \widehat{\Gamma} + \widehat{\Delta} = 360°$,

$$\cos(\widehat{B}) + \cos(\widehat{\Delta}) = 2\cos\left(\frac{\widehat{B}+\widehat{\Delta}}{2}\right)\cos\left(\frac{\widehat{B}-\widehat{\Delta}}{2}\right),$$

$$\cos(\widehat{A}) + \cos(\widehat{\Gamma}) = 2\cos\left(\frac{\widehat{A}+\widehat{\Gamma}}{2}\right)\cos\left(\frac{\widehat{A}-\widehat{\Gamma}}{2}\right)$$

$$= -2\cos\left(\frac{\widehat{B}+\widehat{\Delta}}{2}\right)\cos\left(\frac{\widehat{A}-\widehat{\Gamma}}{2}\right).$$

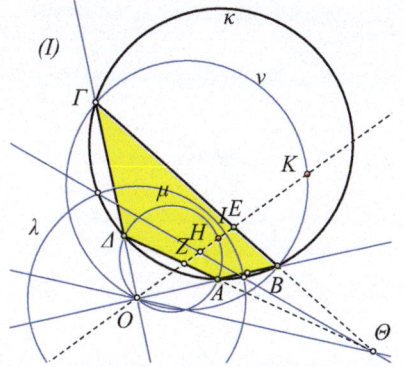
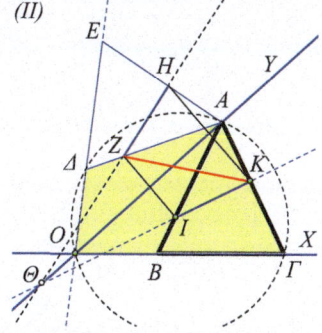

Fig. 5.37: Quadrilateral construction   Isosceles construction

Exercise 5.57. Inscribe in a given circle κ a quadrilateral $AB\Gamma\Delta$ for which the opposite sides AB and $\Gamma\Delta$ intersect orthogonally at a known point O and the side $B\Gamma$ passes through a known point E (See Figure 5.37-I).

Hint: Points A, B and Γ, Δ are pairs of inverse points relative to the circle λ with center O and orthogonal to κ. The inverse I of point E, relative to λ, is constructed and the circle with diameter ΔA (its center) is determined from the given data, as the intersection of the medial line of OI and the circle of Exercise 4.67.

Exercise 5.58. Given is an angle $\widehat{XOY}$ as well as two points I, K in its interior. Construct an isosceles triangle $AB\Gamma$, whose base $B\Gamma$ is contained in OX, the vertex A is contained in OY and the legs pass through points I and K.

Hint: Consider the symmetric $A\Delta E$ of $AB\Gamma$ relative to OY (See Figure 5.37-II). Show first that the quadrilateral $A\Delta O\Gamma$ is inscriptible. If Z is the symmetric of I relative to OY, then A sees ZK under a known angle ([157, I, p.44]).

5.8 Simson's and Steiner's lines

> A sentence should contain no unnecessary words, a paragraph no unnecessary sentences, for the same reason that a drawing should have no unnecessary lines and a machine no unnecessary parts.
>
> W. Strunk, *The elements of style, ch. 2*

Simson's lines (1687-1768) were discovered in reality by Wallace (1768-1843) ([138, p.61]). For this reason they are often mentioned as *Wallace-Simson lines*. These lines express a property of points of the triangle's circumcircle and their projections on the three sides of the triangle.

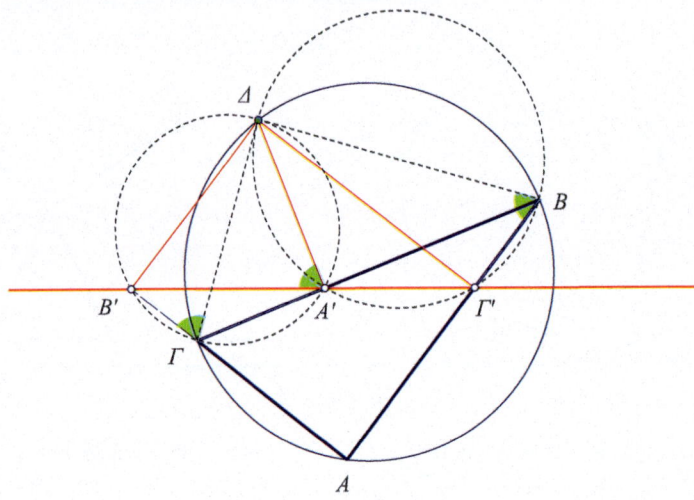

Fig. 5.38: Wallace-Simson line of $AB\Gamma$ relative to point Δ of the circumcircle

Theorem 5.12. *The projections $\{A', B', \Gamma'\}$ of a point Δ, of the triangle's $AB\Gamma$ circumcircle, on the side-lines of the triangle are collinear points (See Figure 5.38).*

Proof. The proof idea is to use three of the inscribed quadrangles, formed by Δ: $AB\Delta\Gamma$ and two from $\Delta A'\Gamma B'$, $\Delta B\Gamma'A'$, $\Delta B'A\Gamma'$, depending on the position of Δ in arcs $\widehat{B\Gamma}$, $\widehat{\Gamma A}$ and $\widehat{AB}$. In the case where Δ belongs to the arc $\widehat{B\Gamma}$ we use $\Delta A'\Gamma B'$, $\Delta B\Gamma'A'$. With their help we show that angles $\widehat{B'A'\Delta}$ and $\widehat{\Delta A'\Gamma'}$ are supplementary, therefore the three points A', B' and Γ' are collinear. $AB\Delta\Gamma$ is iscriptible by definition. The other two as well, because, on one hand $\Delta A'\Gamma B'$ has two opposite angles right (at A' and B'), on the other, $\Delta B\Gamma'A'$ has two successive vertices (A', Γ'), which see one of its sides (ΔB) under a right angle. From the inscriptible $\Delta B\Gamma'A'$ follows, that angle $\widehat{\Delta A'B'}$ is equal to the angle $\widehat{\Delta \Gamma B'}$. From the inscriptible $\Delta B\Gamma'A'$ follows, that angle $\widehat{\Delta A'\Gamma'}$ is supplementary to $\widehat{\Delta B\Gamma'}$. Finally, from the inscriptible $AB\Delta\Gamma$ follows, that $\widehat{\Delta B\Gamma'}$ is equal

5.8. SIMSON'S AND STEINER'S LINES

to $\widehat{\Delta \Gamma B'}$. The proof in the other cases, where Δ is in the arcs $\widehat{\Gamma A}$ and $\widehat{AB}$, is similar.

Exercise 5.59. Show that the converse of the theorem also holds: If the projections of the point Δ on the sides of a triangle are collinear points, then point Δ lies on the triangle's circumcircle.

Hint: In figure 5.38, if points A', B', Γ' are collinear, then $\widehat{\Gamma \Delta B} = \widehat{\Gamma \Delta A'} + \widehat{A' \Delta B} = \widehat{\Gamma B' A'} + \widehat{A' \Gamma' A} = 180° - \widehat{\Gamma' A B'}$. Consequently $AB\Delta\Gamma$ is an inscriptible quadrilateral.

Proposition 5.7. *Let point Δ be on the circumcircle of the triangle $AB\Gamma$ and E be its projection onto $B\Gamma$. Suppose also that Θ is the second intersection point of ΔE with the circumcircle. Then line $A\Theta$ is parallel to the Simson line of point Δ.*

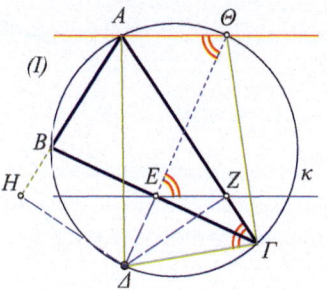
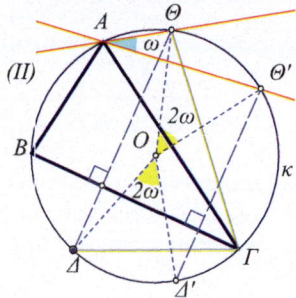

Fig. 5.39: Simson line direction   Angle of Simson lines

Proof. The quadrilateral $A\Delta\Gamma\Theta$ is inscriptible (See Figure 5.39-I), therefore $\widehat{A\Theta\Delta} = \widehat{A\Gamma\Delta}$. The quadrilateral $\Delta EZ\Gamma$, where Z is the projection of Δ onto $A\Gamma$, is also inscriptible, therefore $\widehat{Z\Gamma\Delta} = \widehat{\Theta EZ}$. Consequently, the lines $A\Theta$ and EZ form with ΘE internal and alternate angles equal.

Proposition 5.8. *The angle between two Simson lines, relative to points $\{\Delta, \Delta'\}$ of the circumcircle $\kappa(O)$ of the triangle $AB\Gamma$, is equal to half the central angle $\widehat{\Delta O \Delta'}$.*

Proof. According to the preceding proposition, the angle of the Simson lines is measured by the angle $\widehat{\Theta A \Theta'}$ formed by their parallels through A (See Figure 5.39-II). That however is half the central $\widehat{\Theta O \Theta'}$, which is equal to $\widehat{\Delta O \Delta'}$, because also lines $\Delta\Theta$, $\Delta'\Theta'$ are parallel and define congruent arcs on circle κ.

Exercise 5.60. Find a point Δ on the circumcircle of triangle $AB\Gamma$, for which the corresponding Simson line has a given direction.

Hint: Using Proposition 5.7 (See Figure 5.39-I), draw from A the parallel to the given direction, which intersects the circumcircle at a point Θ. From Θ draw an orthogonal to $B\Gamma$, which intersects the circumcircle at Δ. Δ is the requested point.

Theorem 5.13. *Let Δ be a point on the circumcircle of the triangle $AB\Gamma$ and A', B', Γ' be its projections on the sides. Extend until doubling $\Delta A'$, $\Delta B'$, $\Delta \Gamma'$ to $\Delta A''$, $\Delta B''$, $\Delta \Gamma''$. The points A'', B'', Δ'' are collinear and on a line, which passes through the orthocenter H of the triangle. The Simson line of Δ passes through the middle M of the segment ΔH (See Figure 5.40-I).*

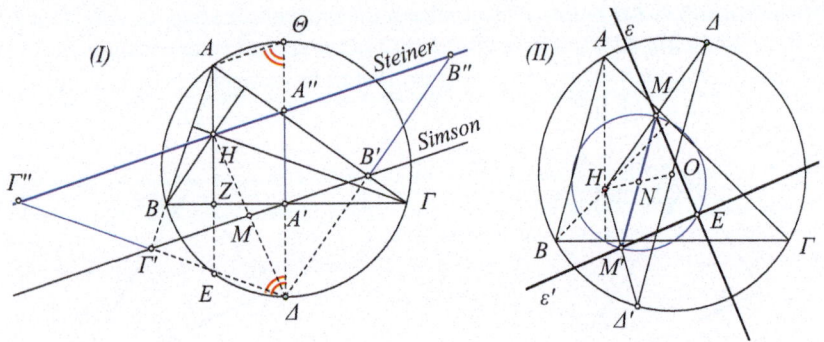

Fig. 5.40:   Steiner line of Δ Simson lines of diametrical $\{\Delta, \Delta'\}$

Proof. The fact, that the three points A'', B'', Γ'' are collinear and are contained in a parallel to the Simson line, follows from the corollary of Thales 3.15. To show that the line of these three points passes through H, we suppose that it intersects the altitude AH at another point H' and we show that point H' is coincident with H.

For this, we begin from the second intersection point Θ, of $\Delta A'$ with the circumcircle. According to Proposition 5.7, line $A\Theta$ is parallel to the Simson line $A'B'$. If E is the second intersection point of the altitude AZ with the circumcircle, because of the fact that AE and $\Theta \Delta$ are parallel, $AE\Delta\Theta$ is an isosceles trapezium. If H' is the intersection point of $A''B''$ with the altitude AH, then, because of the fact that ΘA, $A''B''$ are parallel, $A''H'E\Delta$ is also an isosceles trapezium and segment $B\Gamma$ will be the medial line of $\Delta A''$, therefore it will pass through the middle of EH'. However the symmetric of E relative to Z is the orthocenter H (Exercise 2.159), therefore $H' = H$. The fact, that the middle M of ΔH belongs to the Simson line, is a consequence of the fact that, according to the previously said, the Simson line contains all the middles N of the segments ΔX, where X is a point of the line $A''B''$.

The line defined by the preceding theorem is called **Steiner line** of the point Δ relative to the triangle $AB\Gamma$.

5.8. SIMSON'S AND STEINER'S LINES

Corollary 5.1. *The Simson ε and ε' of two diametrically opposite points Δ and Δ' of the circumcircle of triangle $AB\Gamma$ intersect orthogonally at point E of the Euler circle of the triangle (See Figure 5.40-II).*

Proof. That the lines ε, ε' are orthogonal follows from the fact that their angles are half that of the central $\widehat{\Delta O \Delta'} = 180°$ (Proposition 5.8). That their intersection point E is contained in the Euler circle, follows from the fact, that E sees a diameter MM' of the Euler circle under a right angle. Points M, M' are defined as middles of $H\Delta$, $H\Delta'$. Because points Δ, Δ' are diametrically opposite and point H is a center of similarity of the Euler circle and the circumcircle with similarity ratio $\frac{1}{2}$ (Theorem 5.3), it follows also, that M, M' will be diametrically opposite on the Euler circle. According to Theorem 5.13 points M, M' are points of ε, ε' respectively. Therefore point E sees the diameter MM' of the Euler circle under a right angle.

Exercise 5.61. Find the positions of the Simson line for special points of the circumcircle of triangle $AB\Gamma$, like: (i) the triangle's vertices, (ii) the intersections of the altitudes with the circumcircle, (iii) the intersections of the bisectors with the circumcircle, (iv) the diametrically opposite points of all these points.

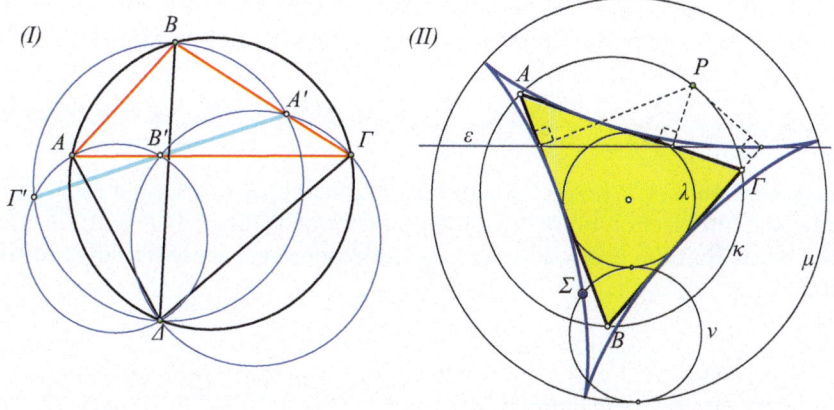

Fig. 5.41: Simson, other viewpoint   Deltoid as a line envelope

Exercise 5.62. Show that the circles with diameters three chords ΔA, ΔB, $\Delta \Gamma$ of the circle κ, intersect at three collinear points A', B', Γ' (See Figure 5.40-I).

Remark 5.8. It is noteworthy, that beginning from simple phenomena, we are led, relatively quickly, to more complex structures. One such case is presented by the **deltoid**, which is a curve intimately related to the Simson lines (See Figure 5.40-II). This curve is characterized by having for tangents all Simson lines $\{\varepsilon\}$ of a given triangle $AB\Gamma$ (the corresponding lines for all points of its circumscribed circle). It is, as we say, an *envelope* of all Simson lines of the triangle $AB\Gamma$. The triangle's sides and the Euler circle are also tangent to this curve ([70, p.226], [58, II, p.641]). The vertices of the deltoid are contained in a circle μ concentric with the Euler circle, having radius three times that of the latter. The same curve may also be produced from a fixed point Σ of a wheel (circle) ν, congruent to the Euler circle, which rolls without friction in the interior of μ.

Exercise 5.63. If x, y, z label the distances of point P from the sides of triangle $AB\Gamma$ with corresponding side lengths a, b, c and point P lies on the circumcircle of the triangle inside the arc which sees angle $\alpha = \widehat{BA\Gamma}$, then it holds $\frac{a}{x} = \frac{b}{y} + \frac{c}{z}$. Similar relations hold also for the arcs of the circumcircle which are opposite to angles β and γ of the triangle.

Hint: Combination of theorems of Ptolemy and Simson.

Exercise 5.64. Show that the Simson line of the second intersection point A' of the bisector $A\Delta$ of the triangle $AB\Gamma$ with the circumcircle of the triangle, is orthogonal to this bisector and passes through the middle A'' of the side $B\Gamma$. Show also that the Simson line of the second intersection point Δ of the parallel from A to $B\Gamma$, with the circumcircle, is parallel to the radius OA of the circumcircle.

Exercise 5.65. Show that the Simson line of the second intersection point A' of the symmedian $A\Delta$ of the triangle $AB\Gamma$, with the circumcircle of the triangle, is orthogonal to the median AM from A and, of course, passes through the projection A'' of A' on the side $B\Gamma$.

Exercise 5.66. Let P be a point of the circumcircle κ of triangle $AB\Gamma$. Consider the intersection point Δ of the diameter of κ through P with the side $B\Gamma$. Show that the projection E of Δ to the line PA is a point of the Simson line of P.

Exercise 5.67. Construct a triangle $AB\Gamma$ from the positions of the vertex A, the orthocenter H and the Simson line relative to the middle M of the of the arc of the circumcircle defined by $\widehat{A}$.

Exercise 5.68. The triangles $\{AB\Gamma, AB'\Gamma'\}$ have the same circumcircle κ and their sides $\{B\Gamma, B'\Gamma'\}$ are parallel. Show that their Simson lines relative to any point P of κ are parallel.

Exercise 5.69. The triangles $\{AB\Gamma, A'B\Gamma\}$ have the same circumcircle κ common the side $B\Gamma$. Show that their Simson lines relative to any point P of κ intersect on $B\Gamma$ and form a constant angle.

Exercise 5.70. Construct a triangle from a side and the two bisectors (internal, external) from some vertex.

5.9 Miquel point, pedal triangle

> We had two courses of three dishes each. In the first course there was a shoulder of mutton, cut into an equilateral triangle, a piece of beef into a rhomboid, and a pudding into a cycloid.
>
> J. Swift, Gulliver's Travels, part III, ch. 2

Miquel points and pedal triangles are related directly to the problem of inscribing a triangle inside another triangle, but also more generally to the inscription of a polygon into another polygon. In which way, for example,

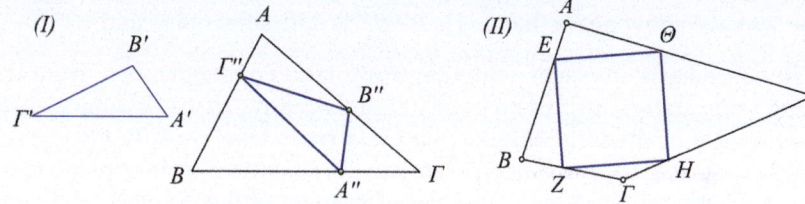

Fig. 5.42: Triangle inscribed   Square inscribed

can we inscribe a triangle $A''B''\Gamma''$, similar to $A'B'\Gamma'$, inside a given triangle $AB\Gamma$ (See Figure 5.42-I)? Or, in which way can we inscribe a square inside a given quadrilateral (See Figure 5.42-II)? How many ways are there to do something like that? Let us begin with the **Miquel point**, which is defined from three points on corresponding sides of a triangle, through the next theorem.

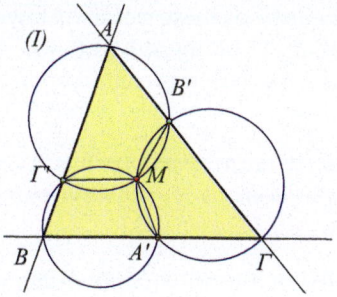
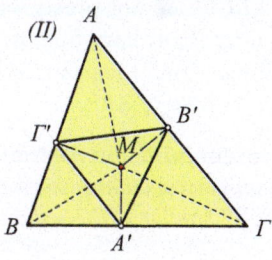

Fig. 5.43: Miquel point of $\{A', B', \Gamma'\}$ Pedal triangle $A'B'\Gamma'$ of M

Theorem 5.14. *For any three different points $\{\Gamma', A', B'\}$, lying respectively on the side-lines $\{AB, B\Gamma, \Gamma A\}$ of the triangle $AB\Gamma$, the three circles $(AB'\Gamma')$, $(B\Gamma'A')$, $(\Gamma A'B')$ intersect at a point M called Miquel point of the triple $\{A', B', \Gamma'\}$ relative to $AB\Gamma$ (See Figure 5.43-I).*

Proof. Let M be the other than Γ' point of intersection of the first two circles $(AB'\Gamma')$, $(B\Gamma'A')$. The created quadrilaterals: $AB'M\Gamma'$ and $BA'M\Gamma'$, are inscriptible and their angles at M are respectively $180° - \alpha$ and $180° - \beta$. Therefore angle $\widehat{A'MB'}$ will equal $360° - ((180° - \alpha) + (180° - \beta)) = \alpha + \beta$, in other words supplementary to γ. Therefore also the quadrilateral $\Gamma A'MB'$ will be inscriptible (Theorem 2.26), which means that point M will also be on the circle $(\Gamma A'B')$.

Corollary 5.2. *If A' is a point of the side $B\Gamma$ of the triangle $AB\Gamma$ and M is an arbitrary point, then the circles $(BA'M)$, $(\Gamma A'M)$ intersect again the sides AB, $A\Gamma$ respectively at Γ', B' and points A, B', M and Γ' are concyclic.*

Proof. The proof is similar to that of the theorem (See Figure 5.43-I). The same way we show again that angle $\widehat{B'M\Gamma'}$ is supplementary to α.

The corollary is, in essence, another formulation of the theorem. It shows, among other things, that there exist infinitely many triples of points A', B', Γ', which define the same Miquel point relative to triangle $AB\Gamma$. For a given point M, a special triple of such points are the *projections* of M onto the sides of $AB\Gamma$. These define the so called **pedal** triangle of the point M relative to $AB\Gamma$ (See Figure 5.43-II).

Remark 5.9. The pedal $A'B'\Gamma'$ of a point M is a genuine triangle for every point M of the plane of the triangle, with the exception of the case where M lies on the circumcircle of the triangle $AB\Gamma$. As we know from § 5.8, in this case, the three projections $\{A', B', \Gamma'\}$ of M on the sides of the triangle $AB\Gamma$ lie on a line, which is exactly the Simson line of M. The properties of the Miquel points, which we'll examine below, concern the cases where M **is not** on the circumcircle of $AB\Gamma$, consequently the pedal of M is a genuine triangle. We must, however, note that the definition of M from three points on the sides of $AB\Gamma$ does not presuppose the non-collinearity of these points. As we'll see later (Corollary 5.7), if the three points A', B', Γ' on the sides of $AB\Gamma$ are collinear, then the point M is defined and, as expected, it is contained in the circumcircle of $AB\Gamma$.

Theorem 5.15. *Given a triangle $AB\Gamma$ and a point M, every triple of points A'', B'', Γ'' on corresponding sides of $AB\Gamma$ whose Miquel point is M, defines a triangle $A''B''\Gamma''$ similar to the pedal triangle $A'B'\Gamma'$ of M (See Figure 5.44).*

Proof. The proof follows from the similarity of the three right triangles, which are formed from M and the points A', A'', B', B'' and Γ', Γ''. Let us see the similarity of $MA'A''$ and $MB'B''$. Quadrilateral $A'MB'\Gamma$ is inscriptible, therefore angle $\widehat{A'MB'}$ is supplementary to angle γ. Similarly angle $\widehat{A''MB''}$ is supplementary to γ, therefore the two angles are equal. It follows that the two right triangles $MA'A''$ and $MB'B''$ have their acute angles at M equal, therefore they are similar. The similarity of triangles $MA'A''$ and $M\Gamma'\Gamma''$ is proved analogously.

5.9. MIQUEL POINT, PEDAL TRIANGLE

The similarity of these right triangles implies the similarity of the triangles formed from their sides. Specifically, $(A'MB', A''MB'')$, $(B'M\Gamma', B''M\Gamma'')$ and $(\Gamma'MA', \Gamma''MA'')$ are pairs of similar triangles. In the first pair, for example, the angles at M are equal and the adjacent to M sides of the two triangles are proportional, because of the similarity of the corresponding right triangles $MA'A''$ and $MB'B''$. Consequently $A'MB'$ and $A''MB''$ are similar triangles. The similarity of the other two pairs is proved analogously.

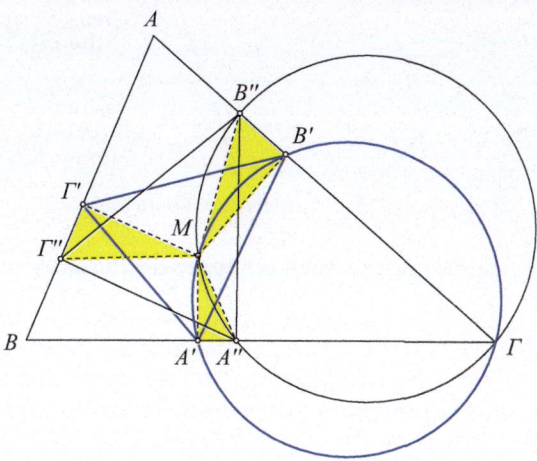

Fig. 5.44: Pedal $A'B'\Gamma'$ of M and $A''B''\Gamma''$ with Miquel point M

From the proved similarity of the pairs of triangles follows that the angles $\widehat{MA'B'}$ and $\widehat{MA''B''}$ are equal, as well as the fact that angles $\widehat{\Gamma'A'M}$ and $\widehat{\Gamma''A''M}$ are equal. Consequently also their sum $\alpha' = \widehat{\Gamma'A'M} + \widehat{MA'B'} = \widehat{\Gamma''A''M} + \widehat{MA''B''} = \alpha''$. The equality of the other two angles $\beta' = \beta''$ and $\gamma' = \gamma'$, is proved analogously.

Corollary 5.3. *If the points A'', B'', Γ'' on the sides of a triangle, define the Miquel point M, then the lines MA'', MB'', Γ'' form equal angles with the respective sides $B\Gamma$, ΓA, AB containing these points (See Figure 5.44).*

Proof. The proof is contained in that of the preceding theorem and follows from the similarity of the right triangles $MA'A''$, $MB'B''$, $M\Gamma'\Gamma''$.

Corollary 5.4. *From all the triangles $A''B''\Gamma''$ with the same Miquel point M, the pedal of M has the least perimeter and the least area.*

Proof. It follows from the fact that the ratio of similarity of $A''B''\Gamma''$ to $A'B'\Gamma'$ is equal to $\kappa = \frac{|MA''|}{|MA'|}$, which is the ratio of a hypotenuse to orthogonal in a right triangle (See Figure 5.44).

Corollary 5.5. *The angles under which the Miquel point M sees the sides of the triangle $A'B'\Gamma'$ are equal to the respective angles of $AB\Gamma$ or their supplementary.*

Proof. Through the figure. Because of the concyclicity of the points $AB'M\Gamma'$, if points A, M are on different sides of the line $B'\Gamma'$, then the angles $\widehat{\Gamma'AB'}$ and $\widehat{B'M\Gamma'}$ are supplementary (See Figure 5.45-I). If points A, M are on the same side of the line $B'\Gamma'$, then the angles are equal (See Figure 5.45-II).

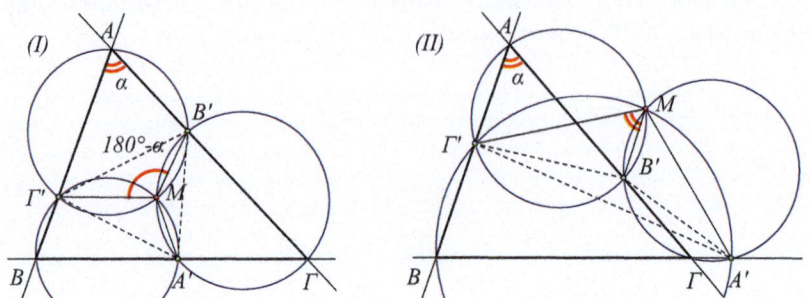

Fig. 5.45: Angles under which point M sees the sides of $A'B'\Gamma'$

Corollary 5.6. *The angles under which the Miquel point of three points A', B', Γ', on the respective sides of the triangle $AB\Gamma$, sees the sides of $AB\Gamma$ are expressed through the sums of corresponding angles of the triangles $AB\Gamma$ and $A'B'\Gamma'$.*

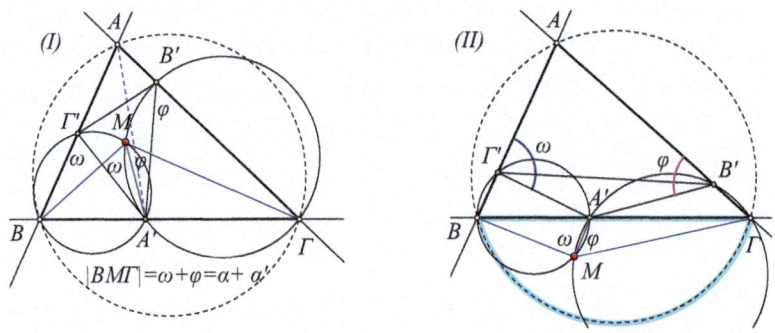

Fig. 5.46: Angles under which point M sees the sides of $AB\Gamma$

Proof. The formulation of the corollary needs a careful interpretation. As it is seen (and proved) with the help of figure 5.46, in all the cases except for the cyclic arc which is defined on the circumcircle by the angle $\widehat{BA\Gamma}$, the corresponding angle is $\widehat{BM\Gamma} = \alpha + \alpha'$ (See Figure 5.46-I). In the case where M is in the aforementioned cyclic arc the angle is $\widehat{BM\Gamma} = 360° - (\alpha + \alpha')$ (See Figure 5.46-II).

The last two corollaries have an important consequence which determines the way by which a triangle $A'B'\Gamma'$ with given angles can be inscribed by similarity (in other words, if not itself, one similar to it), in another triangle $AB\Gamma$, in such a way that A', B', Γ' will be respectively on the sides $B\Gamma$, ΓA and AB. From the corollaries it follows that the position of M relative

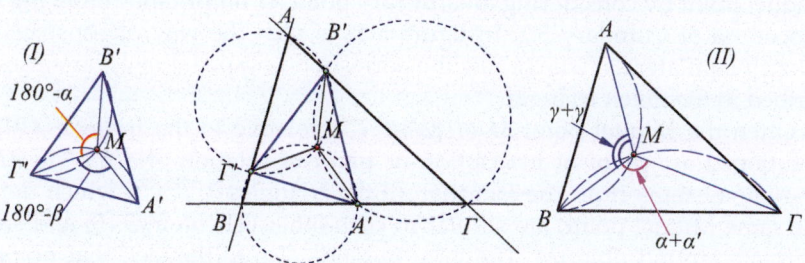

Fig. 5.47: Miquel point M relative to the inscribed/cicrumscribed triangle

to $A'B'\Gamma'$ is determined as the intersection of circular arcs whose points see the sides of $A'B'\Gamma'$ under supplementary angles to these of $AB\Gamma$ (or equal to these depending on the position of M) (See Figure 5.47-I). Correspondingly the position of M relative to $AB\Gamma$ is determined as the intersection of circular arcs whose points see the sides under equal angles with the sum (or the sum of the supplementary ones, depending on the position of M) of corresponding angles of the two triangles (See Figure 5.47-III). Thus, if we are given two triangles $AB\Gamma$ and $A'B'\Gamma'$ and we want to inscribe a similar $A''B''\Gamma''$ to $A'B'\Gamma'$ into $AB\Gamma$, we determine a point M, as in figure 5.47-II and we draw its pedal triangle $A''B''\Gamma''$ or any other triangle $A''B''\Gamma''$ which has M as its Miquel point (Theorem 5.15). In figure 5.48 we have an example application of this

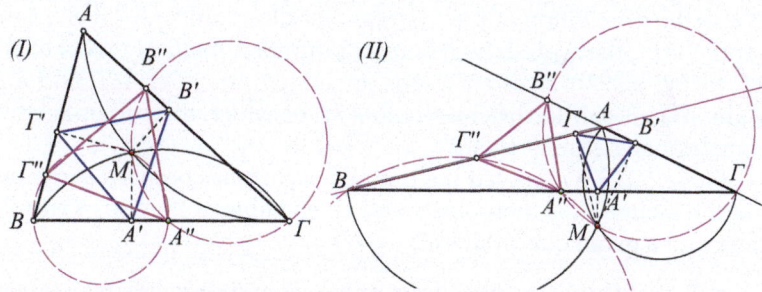

Fig. 5.48: Inscription of equilateral in triangle

method. The figure shows two cases of inscription of an equilateral $A'B'\Gamma'$, which is inscribed in an arbitrary triangle $AB\Gamma$. For this, using the preceding recipe, we construct a point M, the pedal $A'B'\Gamma'$ of which is an equilateral triangle. In figure 5.48-I, the point M is inside the triangle $AB\Gamma$ and we find it as the intersection of two arcs the points of which see respectively the sides $B\Gamma$, $A\Gamma$ under the angles $\alpha + 60°$ and $\beta + 60°$.

We follow the same recipe in the case of the obtuse triangle of figure 5.48-II. In this case the sum of the angles is $\alpha + \alpha' > 180°$ and point M is in the exterior of the triangle and sees the segment $B\Gamma$ under the angle $360° - (\alpha + \alpha')$. In this case point M is itself also outside the pedal triangle $A'B'\Gamma'$.

We can construct infinitely many other triangles $A''B''\Gamma''$ having the same Miquel point by considering an arbitrary point A'' on $B\Gamma$ and following the procedure of Corollary 5.2. In figure 5.48 is seen the way of construction of such equilaterals, which, like the corresponding pedal of M, are also inscribed in the given triangle.

Often the Miquel point M of $\{A', B', \Gamma'\}$ relative to the triangle $AB\Gamma$ is mentioned as **pivot** of inscription or **pivot** of rotation of $A'B'\Gamma'$ in $AB\Gamma$. The name stems from the fact that all the triangles $A''B''\Gamma''$, which define the same Miquel point, are similar to each other (Theorem 5.15) and seem to rotate around point M. The term reminds also to the fact, that point M is *invariant by similarity* or, as we use to say, *retains the same relative position* relative to the "rotated" triangle $A'B'\Gamma'$, as well as relative to the "enclosing" and immovable triangle $AB\Gamma$ (§3.10).

The procedure by which a polygon p varies, remaining similar to itself and, simultaneously, inscribed in another fixed polyon q, so that a point M remains relatively fixed with respect to both p and q, is called **pivoting** of the polygon p relative to q about point M. Thus, the various triangles $A'B'\Gamma'$ with the same Miquel point M, relative to triangle $AB\Gamma$, are snapshots of a specific pivoting of $A'B'\Gamma'$ about the point M, which has fixed relative position with respect to $A'B'\Gamma'$ but also relative to $AB\Gamma$.

In accordance with corollaries 5.5 and 5.6, this, relatively fixed position of M in the two triangles, is determined from the angles of the two triangles $AB\Gamma$, $A'B'\Gamma'$, as well as from the information, which vertex/angle of the "rotating" is moving on which side of the "immovable". Taking into account also the orientation of the triangles, it follows that, for given triangles $AB\Gamma$ and $A'B'\Gamma'$, there exist, in general, 12 different rotation pivots of $A'B'\Gamma'$ in $AB\Gamma$ ([102, p.297]). In other words, given two triangles $AB\Gamma$ and $A'B'\Gamma'$, there are, in general, 12 different points M, of which the pedals relative to $AB\Gamma$, are triangles similar to $A'B'\Gamma'$.

In exercise 5.173 it is solved the problem which has been set at the beginning of the section. It is there discussed an example of the use of *pivoting* of a square into a given quadrilateral.

Theorem 5.16 (Miquel point of four lines). *Four lines in general position (i.e. taken by three define a genuine triangle) form four triangles, whose circumscribed circles pass through a common point M.*

Proof. For the proof we apply the preceding theorem, which holds also for points A', B', Γ' contained in a line (collinear) (See Figure 5.49-I). For the proof then, we choose three out of the four given lines: e.g. the lines AB, $B\Gamma$, ΓA and think that the fourth intersects the sides (and/or their extensions) of the triangle $AB\Gamma$, formed by these lines, at A', B', Γ' respectively on $B\Gamma$, ΓA and AB. Applying Theorem 5.14, we conclude that the circles $(AB'\Gamma')$, $(B\Gamma'A')$ and $(\Gamma A'B')$ pass through a common point M. We choose now three

5.9. MIQUEL POINT, PEDAL TRIANGLE

other lines, e.g. $\Gamma A'$, $A'B'$ and $B'\Gamma$ and we think again, that the fourth line intersects the sides of the triangle $\Gamma A'B'$ at points B, Γ' and A. Again, from Theorem 5.14, we conclude that the circles (ΓAB), $(B'A\Gamma')$, $(A'\Gamma'B)$ pass through a common point M'. Because the two circles $(AB'\Gamma')$ and $(A'\Gamma'B)$ are contained in both triples of circles, points M and M' must coincide.

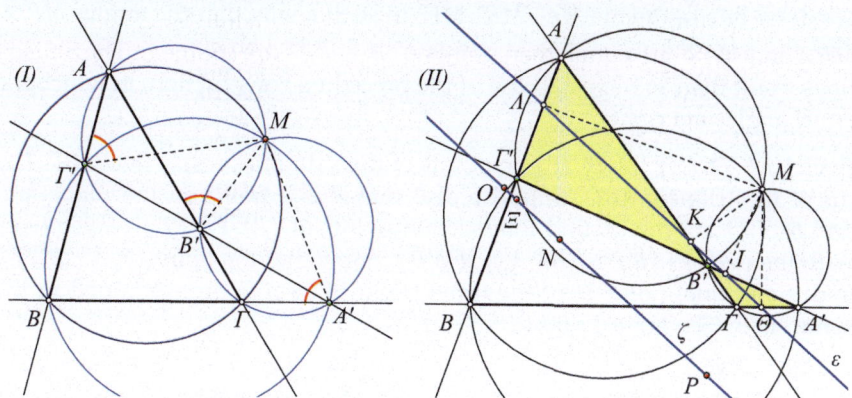

Fig. 5.49: Miquel point of 4 lines   Collinear orthocenters

The point M of the theorem 5.16 is called the **Miquel point** of the four lines.

Theorem 5.17. *The four triangles, formed by four lines in general position, have their orthocenters on a line (See Figure 5.49-II).*

Proof. Let M be the Miquel point of the four lines and $\{\Lambda, K, I, \Theta\}$ its projections on these lines. These points lie, by three, on the Simson line of a corresponding triangle with respect to M, hence these Simson lines coincide in one ε. By theorem 5.13, the orthocenters of the corresponding triangles are on a parallel ζ of ε at the double distance of that of ε from M.

Corollary 5.7. *The Miquel point of three collinear points A', B', Γ' on the sides of triangle, $B\Gamma$, ΓA, AB respectively, is contained in the circumcircle of the triangle $AB\Gamma$.*

Proof. Direct consequence of the preceding theorem, since the three of the circles of the theorem are those, through which point M is defined, and the fourth is the circumcircle of $AB\Gamma$.

Exercise 5.71. Complete the preceding figure (See Figure 5.49-I) by drawing the lines MA', MB', $M\Gamma'$ from the Miquel point of the four lines, and show the equality of the respective angles: $\widehat{M\Gamma'A} = \widehat{MB'A} = \widehat{MA'\Gamma}$.

Hint: The equality of the angles follows easily from the inscriptible quadrilaterals $MB'\Gamma'A$, $MB'\Gamma A'$. This property shows, that every line, which intersects the sides of a triangle at points A', B', Γ', defines through them a Miquel point contained in the circumcircle (Corollary 5.7) in such a way, that the aforementioned angles are equal.

Remark 5.10. A kind of converse to the preceding exercise is also valid. If from a point M of the circumcircle we draw line segments $M\Gamma'$ and MB', which form equal angles $\widehat{M\Gamma'A}$ and $\widehat{MB'A}$, then the line $\Gamma'B'$ will intersect the third side at point A', in such a way that angle $\widehat{MA'\Gamma}$ is also equal to the others (See Figure 5.49). This implies that if from point M of the circumcircle we draw line segments $MA', MB', M\Gamma'$ in such a way that the angles $\widehat{M\Gamma'A}$, $\widehat{MB'A}$ and $\widehat{MA'\Gamma}$ are equal, then points A', B' and Γ' are collinear. The Simson lines result then as a special case of the preceding construction, in which the equal angles are right.

Exercise 5.72. Let $\{A', B', \Gamma'\}$ be points of the side-lines respectively $B\Gamma, \Gamma A$, AB of the triangle $AB\Gamma$. Suppose also that P is a point on the plane and $\{A'', B'', \Gamma''\}$ are the second intersection points of $\{PA, PB, P\Gamma\}$ respectively with the circles $(AB'\Gamma'), (B\Gamma'A'), (\Gamma A'B')$. Show that the points P, A'', B'', Γ'' and the Miquel point M of the three triple $\{A', B', \Gamma'\}$ are contained in the same circle (See Figure 5.50-I).

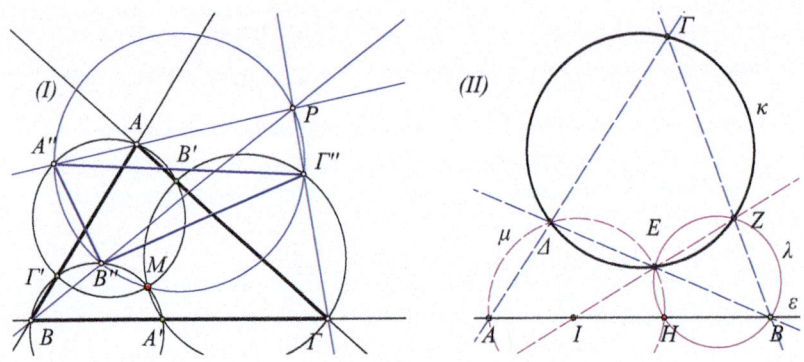

Fig. 5.50: Concyclic $\{A'', B'', \Gamma'', M, P\}$,   Variable Γ of κ+ points H, I fixed

Hint: Angle $\widehat{B''M\Gamma''}$ is equal to the sum of $\widehat{B''BA'}$ and $\widehat{A'\Gamma\Gamma''}$ and this sum is supplementary to angle $\widehat{BP\Gamma}$. It follows that quadrilateral $B''M\Gamma''P$ is inscriptible. With a similar analysis, follows also that $MB''A''P$ and $MA''P\Gamma''$ are inscriptible. For example, angle $\widehat{B'''MA''}$ is equal to the difference of angles $\widehat{B'MA''}$ and $\widehat{B'MB''}$, which are respectively equal to $\widehat{B'AP}$ and $\widehat{B''B\Gamma} + \widehat{B\Gamma B'}$, whose difference is $\widehat{BPA}$ etc.

Exercise 5.73. Let A, B be two points lying outside the circle κ. For every point Γ of the circle κ, the lines $A\Gamma, B\Gamma$ re-intersect the circle at Δ, Z and $B\Delta$ does so again at E. Show that the circle $\lambda = (BEZ)$ passes through a fixed point H of $\varepsilon = AB$ and the line EZ passes also through a fixed point I of ε.

Hint: According to theorem 5.15, applied to the triangle $AB\Gamma$, the circles $(\Gamma\Delta Z), (A\Delta H), (BZH)$ will pass through a common point (See Figure 5.50-II), which is E. The power of B relative to κ, which is a known constant

5.9. MIQUEL POINT, PEDAL TRIANGLE

depending from the given data, is calculated as

$$p_\kappa(B) = |BZ||B\Gamma| = |BE||B\Delta| = |BH||BA|,$$

from which it follows that $|BH|$ and, consequently, the position of B is fixed and independent of the position of Γ on κ. The constancy of I follows from that of H and Proposition 4.9 (for a different view see Exercise 5.134).

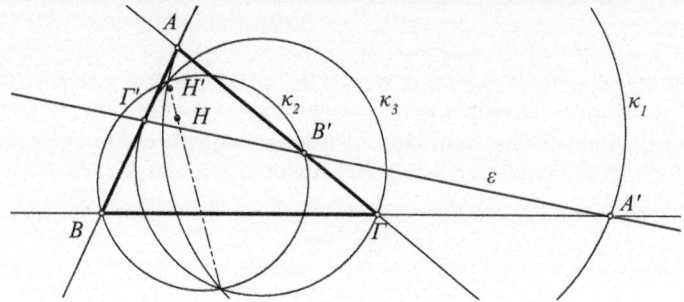

Fig. 5.51: A pencil of circles

Exercise 5.74. The collinear points $\{A', B', \Gamma'\}$ are respectively on the sides $\{B\Gamma, \Gamma A, AB\}$ of the triangle $AB\Gamma$. Show that the circles $\{\kappa_1, \kappa_2, \kappa_3\}$, with respective diameters $\{AA', BB', \Gamma\Gamma'\}$, belong to the same pencil of circles.

Hint: Apply the exercise 4.116 to the triangle $AB\Gamma$ and the secant line $A'B'$ and conclude that the orthocenter H of the triangle is a radical center of the three circles (See Figure 5.51). Then, apply again the exercise to the triangle $AB'\Gamma'$ and the secant line $B\Gamma$ and conclude that the orthocenter H' of the triangle $AB'\Gamma'$ is also a radical center of the three circles.

Exercise 5.75. The orthogonal to the base $B\Gamma$ of the triangle $AB\Gamma$ at its point A' intersects the sides $\{AB, A\Gamma\}$ correspondingly at points $\{\Gamma', B'\}$. Show that the circles with diameters $\{AA', BB', \Gamma\Gamma'\}$ define an intersecting pencil whose one base point is A' and whose radical axis passes through the orthocenter of $AB\Gamma$. Find the geometric locus of the other base point of the pencil as point A' moves on line $B\Gamma$.

Exercise 5.76. From the vertex A of the triangle $AB\Gamma$ we draw two lines lying symmetric relative to the bisector of $\widehat{A}$ and intersecting the side $B\Gamma$ at $\{\Delta, E\}$. Show that the projections of these points on the sides $\{AB, A\Gamma\}$ are four concyclic points. Show also that the product of the distance of $\{\Delta, E\}$ from AB equals the product of the distances of $\{\Delta, E\}$ from $A\Gamma$.

Exercise 5.77. Construct the pedals of the orthocenter, the centroid and the circumcenter of the triangle $AB\Gamma$ and determine other inscribed triangles $A'B'\Gamma'$ of $AB\Gamma$, which have these points as Miquel points.

5.10 Arbelos

> ... always make a definition or delineation of whatever presents itself to your mind, so that you can see distinctly what sort of thing it is when stripped down to its essence as a whole and in all its parts, and tell yourself its proper name, and the name of the elements from which it has been put together and into which it will be dissolved.
>
> Marcus Aurelius, Meditations, Book III, 11

The **arbelos** is a knife in a shape, which is defined by three semicircles with common endpoints. It was used by ancient shoe makers for the scratching and cutting of skin. The diameter $A\Gamma$ of the biggest circle is called **chord** of the arbelos. With the arbelos are connected the names of Archimedes,

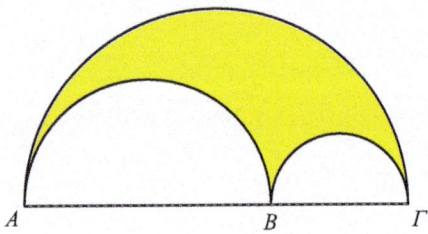

Fig. 5.52: Arbelos

Pappus and Steiner. Let us see first the theorem of Archimedes for the so-called **twin circles** ($\{\delta_1, \delta_2\}$ in figure 5.53).

Theorem 5.18. *The two circles $\{\delta_1, \delta_2\}$, which are tangent to the arcs of the arbelos and to the orthogonal to the chord at its intermediate point B, are equal.*

Proof. A proof can be done using simple calculations. Let us see, however, the subject with the help of inversion. The critical observation is, that the two inversions relative to circles $A(|AB|)$, $\Gamma(|\Gamma B|)$ map the great circle κ of the arbelos, with diameter $A\Gamma$, respectively to the lines κ_α and κ_γ, which are equidistant from the orthogonal ε to the chord at B (See Figure 5.53).

This follows by considering the circle λ, which is tangent to the chord at B and to κ at point E. This circle is orthogonal to the two circles, with respect to which we invert, consequently remains invariant by the respective inversions. Consequently, because it is tangent to κ, it will be also tangent to the inverses relative to these two inversions, which are the lines κ_α, κ_γ. Consequently these lines will be orthogonal to the chord at points A', Γ' equidistant from B. From this follows directly the theorem, because δ_1, being tangent to κ_1 and ε, is orthogonal to the circle of inversion $A(|AB|)$, which interchanges κ_1 with ε (Exercise 4.90), therefore is invariant by the inversion relative to it, and consequently, also tangent to the line κ_α, therefore tangent

5.10. ARBELOS

to the two parallels $κ_α$ and $ε$. Similarly, circle $δ_2$ will also be tangent to the parallels $κ_γ$ and $ε$. The congruence of the circles $δ_1$, $δ_2$ follows, then, from the equality of the distances of the parallels. From the proof follows also, that the radius of the twin circles is half that of the radius of $λ$.

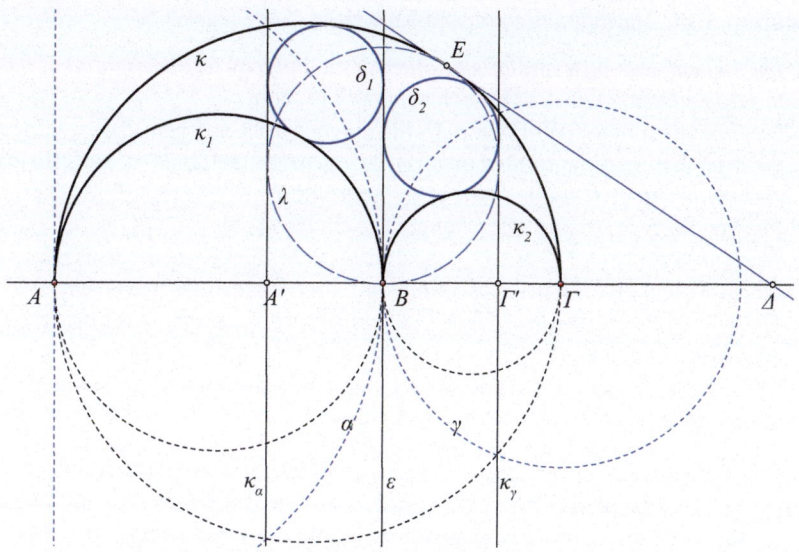

Fig. 5.53: The "twin" circles $δ_1$, $δ_2$ of Archimedes

There are infinitely many circles, intimately related to the arbelos and hidden in this figure ([156]), like the one of next theorem.

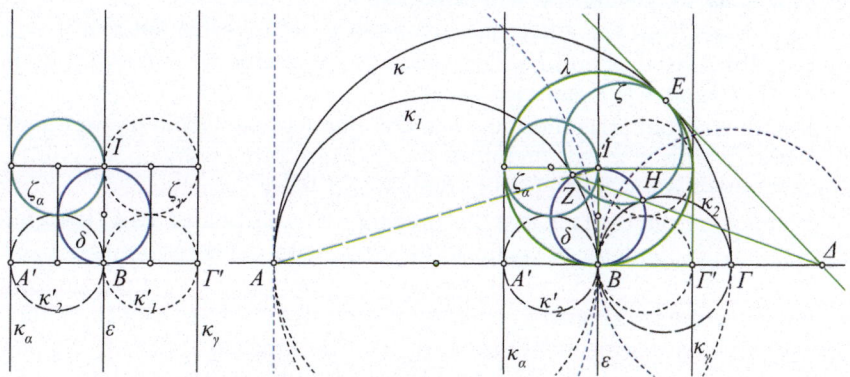

Fig. 5.54: Circle $δ$ congruent to the "twins" of Archimedes

The corresponding figure 5.54 is almost coincident with the preceding one. Besides the new circle $δ$, appears the circle $ζ$, which is tangent simultaneously to the three circles κ, κ_1, κ_2 of the arbelos and is called **inscribed circle of the arbelos**. Next theorem can also be proved by doing simple calculations. I prefer, however, to record the next proof, which reveals the structure of the subject and gives a chance to practice the inversion.

Theorem 5.19. *The following properties are valid (See Figure 5.54):*

1. *The radical axis of κ and of any other circle tangent to the chord at B passes through a fixed point Δ of the extension of the chord.*
2. *Let E be the contact point of κ with the tangent from Δ. Then $|B\Delta| = |E\Delta|$ and the inversion relative to the circle $\Delta(|\Delta E|)$, leaves the circle κ, but also every other circle tangent to the chord at B, invariant.*
3. *The preceding inversion interchanges the circles κ_1, κ_2.*
4. *The points of contact Z, H, of ζ respectively with κ_1, κ_2, are inverses relative to this inversion and the point of contact of ζ and κ coincides with point E.*
5. *The inversion relative to circle $A(|AB|)$ inverts the circle κ_2 into circle κ'_2, tangent to κ_α and ε.*
6. *The same inversion inverts the circle ζ into circle ζ_α, tangent to κ'_2 and tangent to κ_α and ε.*
7. *The circle $\delta = (BZH)$ is congruent to circles δ_1, δ_2.*

Proof. All the properties are, in essence, steps of the proof that the circle δ passes through the contact points of four congruent circles which have their centers at vertices of a square, as in figure 5.54, left, which is part of the wider figure on the right. Circle δ is characterized by the fact, that it is orthogonal to all four circles. Let us proceed, then, with the proof of the properties.

(1) Follows from Proposition 4.9, applied to the pencil $\mathscr{D}$ of the circles, which are tangent to the chord at B and the circle κ.

(2) follows from (1), since the tangents from Δ to the circles of the pencil $\mathscr{D}$, as well as, to the circle κ, will be equal.

(3) follows from (2), since this inversion must invert point A to Γ and leave point B fixed, therefore, the circle with diameter AB will invert to the one with diameter $B\Gamma$, and vice versa.

(4) follows from (3), since the circle ζ is characterized from its simultaneous contact with κ, κ_1, κ_2, therefore, by the preceding inversion, its point of contact with κ maps to itself and points Z and H are interchanged.

(5) follows from the fact, that this inversion inverts point Γ to A', therefore the circle with diameter $B\Gamma$ maps to the circle with diameter BA'.

(6) follows from the fact that the circle ζ is tangent to all three circles κ, κ_1, κ_2, therefore its inverse ζ_α will be tangent to the inverses of these three circles, which are respectively, the lines κ_α, ε and the circle κ'_2.

For (7) we first observe that the circle δ must be tangent to the chord at B, because, according to (4), it passes through the inverse points Z, H of the inversion relative to circle $\Delta(|\Delta E|)$. This implies, that δ is orthogonal to the circle of inversion, therefore is tangent to the chord at B. Circle δ, however, is also orthogonal to ζ and invariant relative to the inversion defined by

5.10. ARBELOS

the circle $A(|AB|)$. Therefore it will also be orthogonal to the inverse ζ_α of ζ relative to this inversion. Also, being orthogonal to κ_2, it will be as well orthogonal to κ'_γ and ζ_γ, which are defined similarly to circles κ'_α and ζ_α, respectively. In conclusion, then, circle δ will be orthogonal to the four congruent circles, as we analyzed in the beginning of the proof. Therefore it will have its center at the radical center of the four circles and will have radius equal to that of these four circles.

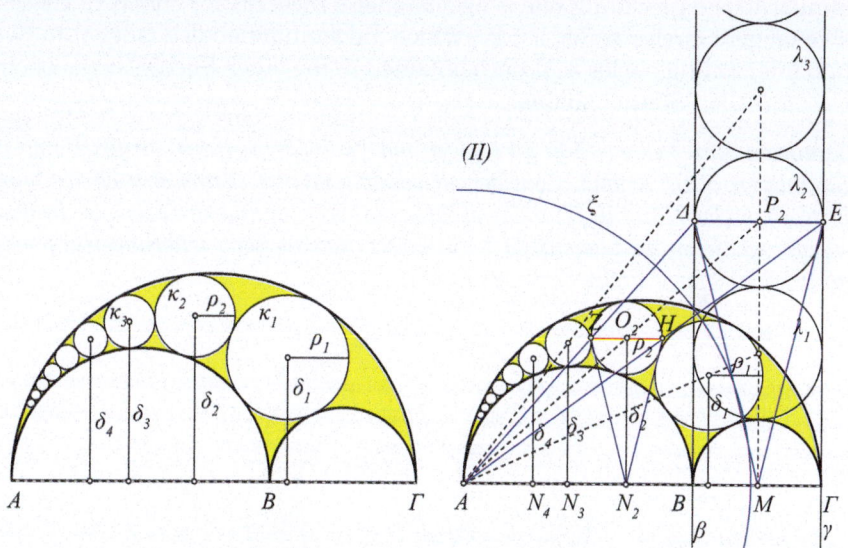

Fig. 5.55: Arbelos, theorem of Pappus ... and the proof

Next theorem is due to Pappus, while the proof, using inversion, is due to Steiner.

Theorem 5.20. *The ratios of distances δ_ν of the centers from the chord to the corresponding diameters $2\rho_\nu$ of the successive inscribed circles $\kappa_1, \kappa_2, \ldots,$ in an arbelos, form the progression of integers (See Figure 5.55-I)*

$$\frac{\delta_\nu}{2\rho_\nu} = \nu.$$

Proof. For the proof we consider the inversion relative to the circle ξ with center A (See Figure 5.55-II), which is orthogonal to the circle with diameter $B\Gamma$. This circle remains invariant by the inversion (Corollary 4.20), which interchanges points B and Γ. The two circles with diameters AB and $A\Gamma$, passing through the center of inversion (Proposition 4.16), invert respectively to lines β and γ. The successive tangent circles $\kappa_1, \kappa_2, \ldots$ invert to circles $\lambda_1, \lambda_2, \ldots$, which are pairwise congruent, successively tangent and also tangent to lines α and β (Proposition 4.18). For the circles λ_ν the to-be-proved relation holds obviously. It suffices therefore to show that the inversion preserves the ratio $\frac{\delta_\nu}{2\rho_\nu}$. However this follows directly from the fact that the center of inversion A is also a center of similarity of the corresponding circles κ_ν and λ_ν (Theorem 4.16). This implies that the diameters of κ_ν ($\nu = 2$ in figure 5.55) ZH, parallel to $A\Gamma$, correspond to diameters ΔE of λ_ν also parallel to $A\Gamma$ and the isosceli triangles ZHN_ν and ΔEM, where N_ν, M the projections of the centers of the circles to $A\Delta$, are similar.

Similar to the preceding one is also Steiner's theorem for chains of successive tangent circles κ_1, κ_2, ... , κ_ν, which are simultaneoulsy tangent to two non intersecting circles λ, μ one contained in the other. Such systems of circles are called **Steiner chains**.

Theorem 5.21. *Let $\nu \geq 3$ be an integer and $\{\lambda, \mu\}$ two circles, the first lying in the interior of the second. Consider all possible Steiner chains of ν circles lying between circles λ and μ. If for one of these chains the initial circle κ_1 and the final κ_ν intersect (resp. are tangent, do not intersect) then the same will happen also with all the chains for ν circles between λ and μ.*

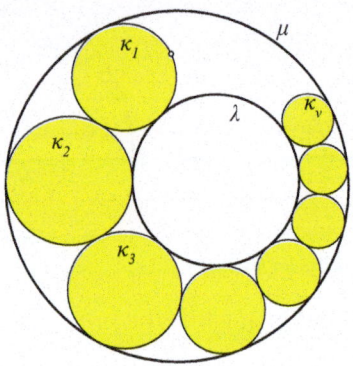

Fig. 5.56: Circle chains of Steiner

Proof. The preceding figure 5.56 displays a circle chain, whose initial (κ_1) and final circle (κ_ν) do not intersect. According to the preceding theorem the same will happen also with every other chain of ν circles between λ and μ. Note that one chain from ν circles is determined completely by the first circle κ_1, the number ν and the direction in which the circles are placed one after the other.

The proof is a simple application of the inversion. Indeed, the two containing the chain circles λ and μ define a non intersecting pencil, which has two limit points A and B. We consider therefore the inversion relative to a circle ξ of arbitrary radius, but with center coincident with one of these two points, A say. Under this inversion, the two circles λ and μ are inverted into two concentric, respectively λ' and μ' (Proposition 4.20). Consequently the chain of circles maps to another chain η_1, η_2, ..., η_ν contained between the

5.10. ARBELOS

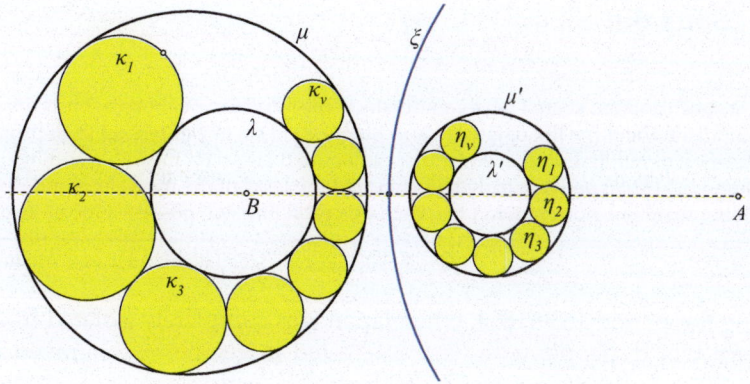

Fig. 5.57: Circle chains of Steiner, proof

concentric circles λ' and μ' (See Figure 5.57). For chains of concentric circles, however, the proposition is trivial, since the intersection/contact/non-contact of initial - final circle of the chain, because of the congruence of the circles of the chain, does not dependent on the position of the initial circle.

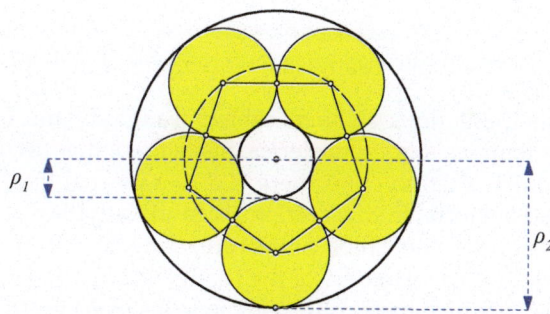

Fig. 5.58: Special Steiner chain

Exercise 5.78. Show that for two concentric circles with radii $\rho_1 < \rho_2$ there exists a Steiner chain with ν tangent circles, if and only if

$$\frac{\rho_2 - \rho_1}{\rho_2 + \rho_1} = \sin\left(\frac{180°}{\nu}\right).$$

Hint: Application of the formula of Exercise 3.58.

Exercise 5.79. Show that the inversion relative to one of two mutually orthogonal circles $\{\kappa(K), \mu(M)\}$ maps their radical axis α on the circle with diameter KM.

5.11 Sangaku

> At seventy-three years I partly understood the structure of animals, birds, insects and fishes, and the life of grasses and plants. And so, at eighty-six I shall progress further; at ninety I shall even further penetrate their secret meaning, and by one hundred I shall perhaps truly have reached the level of the marvelous and divine. When I am one hundred and ten, each dot, each line will possess a life of its own.
>
> Hokusai, afterword in "One Hundred Views of Mount Fuji"

During the period 1639-1854, when Japan was completely isolated from the western world, people who were interested in geometry, engraved prob-

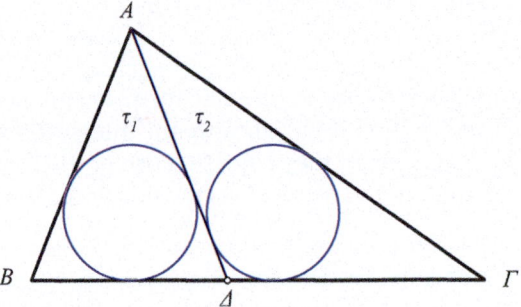

Fig. 5.59: Partition into triangles with congruent inscribed circles

lems or/and their solutions on small tablets (sangaku) and brought them, as offerings, to temples and monasteries. Since then, this circle of problems has attracted much attention and there is also a journal ([145]) devoted to Sangaku-related subjects. In this section we study a typical sangaku example from the temple in the Chiba prefecture, as well as its extensions. The problem asks to draw a secant $A\Delta$ of the triangle $AB\Gamma$, which divides it into two triangles τ_1, τ_2 which have congruent incircles (See Figure 5.59). We prepare the construction with the following proposition.

Theorem 5.22. *Let the secant line $A\Delta$ divide the triangle $\tau = AB\Gamma$, with incircle of radius r, into two triangles τ_1, τ_2 with incircles of radii, respectively r_1, r_2. Let also R, R_1, R_2 be respectively the radii of the escribed circles of the triangles τ, τ_1, τ_2, opposite vertex A (See Figure 5.60). Then it holds*

$$\frac{r}{R} = \frac{r_1}{R_1} \cdot \frac{r_2}{R_2}. \tag{5.2}$$

Proof. Let I, I_1, I_2 denote the incenters and Θ, Θ_1, Θ_2 respectively, the incenters of the triangles τ, τ_1, τ_2. We also define the angles $\omega = \widehat{IB\Gamma}$, $\phi = \widehat{I\Gamma B}$. Then $\omega = \beta/2$, $\phi = \gamma/2$ and the triangles $I_1B\Theta_1$ and $I_2\Gamma\Theta_2$ are similar right triangles. Latter follows immediately by considering their circumcircles, both

5.11. SANGAKU

passing through Δ. Then it holds

$$\frac{r_1}{R_1} = \frac{|I_1B|\sin(\omega)}{|\Theta_1B|\cos(\omega)}, \qquad \frac{r_2}{R_2} = \frac{|I_2\Gamma|\sin(\phi)}{|\Theta_2\Gamma|\cos(\phi)} \qquad \Rightarrow$$

$$\frac{r_1}{R_1} \cdot \frac{r_2}{R_2} = \frac{|I_1B|}{|\Theta_1B|} \cdot \frac{|I_2\Gamma|}{|\Theta_2\Gamma|} \tan(\omega)\tan(\phi) = \tan(\omega)\tan(\phi).$$

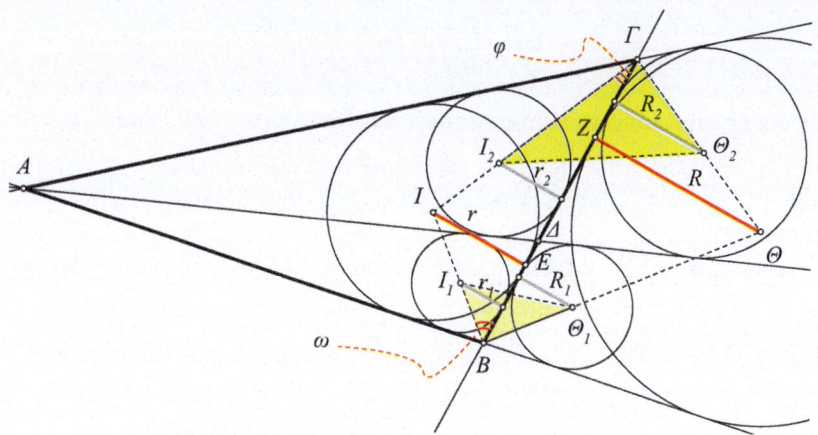

Fig. 5.60: Relation between inscribed/escribed circles

The last equality follows from the similarity of triangles $I_1B\Theta_1$ and $I_2\Gamma\Theta_2$. The last expression is equal to

$$\tan(\omega)\tan(\phi) = \frac{r}{|BE|} \cdot \frac{|Z\Gamma|}{R} = \frac{|Z\Gamma|}{|BE|} \cdot \frac{r}{R} = \frac{r}{R},$$

since $|BE| = |Z\Gamma|$ (Proposition 5.1, Proposition 5.2). This completes the proof.

Denoting by σ, σ_1, σ_2 the corresponding perimeters of the triangles τ, τ_1, τ_2 and with a, a_1, a_2 the lengths of their corresponding sides, opposite to vertex A ($a_1 + a_2 = a$), the above relation is equivalent to

$$\frac{\sigma - a}{\sigma} = \frac{\sigma_1 - a_1}{\sigma_1} \cdot \frac{\sigma_2 - a_2}{\sigma_2} \Leftrightarrow \left(1 - \frac{a}{\sigma}\right) = \left(1 - \frac{a_1}{\sigma_1}\right) \cdot \left(1 - \frac{a_1}{\sigma_2}\right). \quad (5.3)$$

Using the formulas for the area of the triangle $\varepsilon = \frac{1}{2}ha = \sigma r$, where h is the altitude from A, the last formula can be written

$$\left(1 - \frac{2r}{h}\right) = \left(1 - \frac{2r_1}{h}\right) \cdot \left(1 - \frac{2r_2}{h}\right). \quad (5.4)$$

Corollary 5.8. *If the secant line $A\Delta$ of the triangle $\tau = AB\Gamma$, of altitude h from A and inscribed of radius r, divides triangle τ into two others τ_1, τ_2 with congruent incircles of radius r', then*

$$r' = \frac{rh}{h+\sqrt{h^2-2rh}} = \frac{h-\sqrt{h^2-2rh}}{2}.$$

Proof. It follows from the formula 5.4, which for $r_1 = r_2 = r'$ becomes

$$\left(1-\frac{2r}{h}\right) = \left(1-\frac{2r'}{h}\right)^2.$$

The conclusion follows using a simple calculation

This formula leads directly to the solution of our initial sangaku problem, since, for given h, r, it can be solved for r' and determine the radius r' of the triangles τ_1, τ_2.

Exercise 5.80. Show that the length $d = |A\Delta|$ of the secant of the corollary 5.8, as well as the segments $a_1 = |B\Delta|$, $a_2 = |\Delta\Gamma|$ are given respectively by the formulas (σ is the half perimeter of $AB\Gamma$):

$$d^2 = \sigma(\sigma-a), \qquad a_1 = \frac{a(c+d)}{b+c+2d}, \qquad a_2 = \frac{a(b+d)}{b+c+2d}.$$

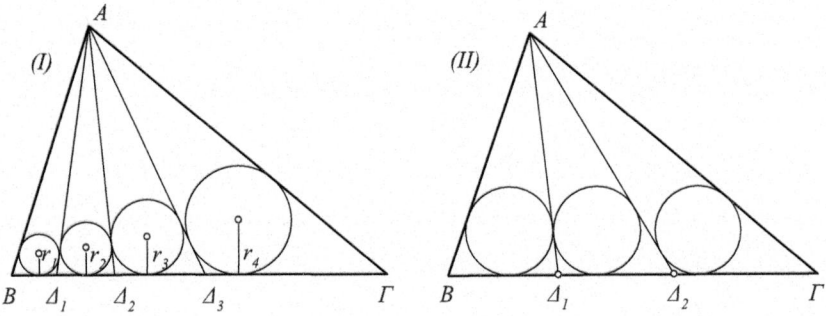

Fig. 5.61: Dividing into triangles with congruent inscribed circles

Let us note that formula (5.4), applied iteratively leads to the formula

$$\left(1-\frac{2r}{h}\right) = \left(1-\frac{2r_1}{h}\right)\cdot\left(1-\frac{2r_2}{h}\right)\cdots\left(1-\frac{2r_k}{h}\right), \qquad (5.5)$$

which connects the radii $\{r_1,\ldots,r_k\}$ of the k in number triangles, which are created by drawing secants $A\Delta_1, \ldots, A\Delta_{k_1}$ from vertex A (See Figure 5.61-I). Formula (5.5), putting $r_1 = \cdots = r_k = r'$, leads to the formula

$$\left(1 - \frac{2r}{h}\right) = \left(1 - \frac{2r'}{h}\right)^k, \tag{5.6}$$

which allows us, solving for r', to divide the triangle into a number k of triangles with congruent incircles (See Figure 5.61-II).

The same formula (5.6), also proves that if we divide the triangle into other triangles $\{\tau_1, \ldots, \tau_k\}$, with congruent inscribed circles, then considering together two successive τ_i, τ_{i+1} defines new triangles, we would say of order 2, which also have equal inscribed circles. Similarly triangles of order 3 also have congruent inscribed circles ([154, p.67]) (See Figure 5.62-I).

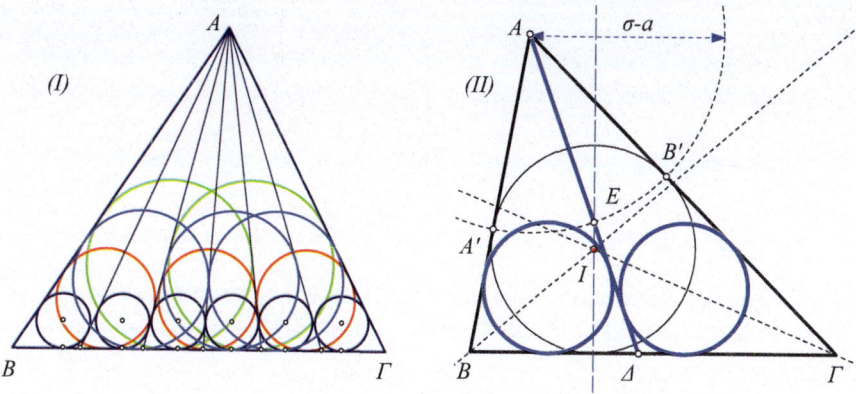

Fig. 5.62: Triangles of m order   Construction of secant $A\Delta$

Exercise 5.81. Show that the secant line $A\Delta$ of the triangle $AB\Gamma$, which defines two triangles with congruent inscribed circles is defined from the point of intersection E of the orthogonal from the incenter I with the circle $A(\sigma - a)$, where σ is the half perimeter and $a = |B\Gamma|$ (See Figure 5.62-II).

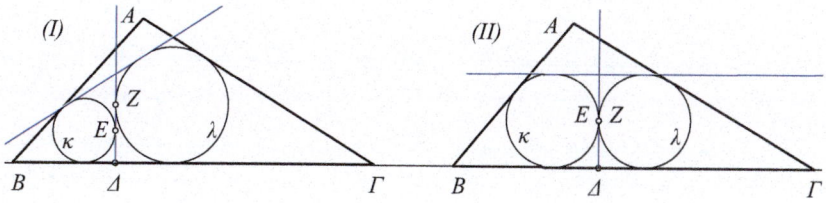

Fig. 5.63: Variation on a sangaku problem

Exercise 5.82. Consider a line orthogonal to the base $B\Gamma$ of the triangle $AB\Gamma$ at a point Δ and the two circles $\{\kappa, \lambda\}$, tangent to the base line and the adjacent to it sides. Locate the position of Δ, for which the two circles are equal (See Figure 5.63).

Let us close this section by presenting the figures for two more sangakus. The first (See Figure 5.64-I) concerns a relatively simple problem, while the second (See Figure 5.64-II) corresponds to a somewhat more difficult problem. In the first figure, triangle $AB\Gamma$ is right and contains three squares and three circles tangent to the squares and the sides of the triangle. We want the relation between the radii of the three circles.

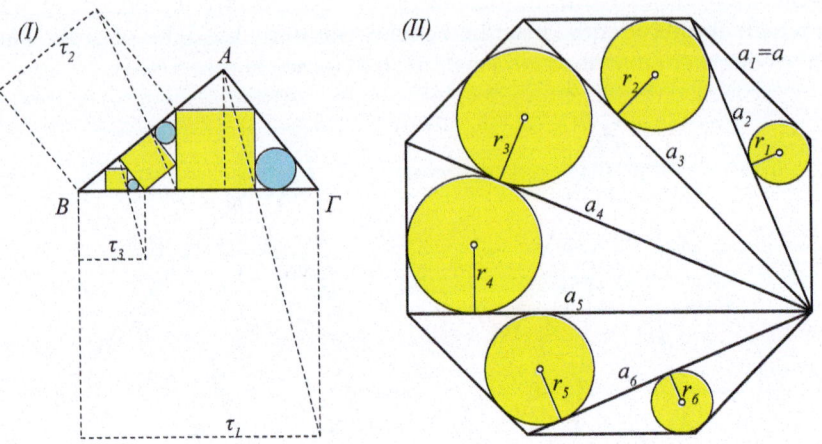

Fig. 5.64: Two typical sangaku problems

The dotted lines are mine and show the way to construct the figure. The answer is $r_2^2 = r_1 r_3$, where r_1, r_2, r_3 are the radii of the circles ordered by increasing magnitude. In the second figure we divide a regular ν-gon, of side-length a, into $\nu - 2$ triangles, through its diagonals $\{a_1, a_2, \dots\}$ from a vertex. We want the relations between the radii of the corresponding inscribed circles. It is proved ([130, p.267]) that the following relations hold:

$$a_{k+2} = e \cdot a_{k+1} - a_k, \qquad \text{where} \quad e = \frac{a_2}{a}, \tag{5.7}$$

$$r_{k+2} = e \cdot r_{k+1} - r_k - \frac{1}{2}t, \qquad \text{where} \quad t = (2-e)a\sqrt{\frac{2-e}{2+e}}. \tag{5.8}$$

Exercise 5.83. In the rectangle $ABCD$ with $|AB| > |AD|$ and center O, the incircle of the triangle ABC has inradius $r = \frac{3}{2}r'$, where r' the inradius of the incircle of the triangle OCD. Show that $|AD| = 3r$ ([146]).

5.12 Fermat's and Fagnano's theorems

> For the mind does not require filling like a bottle, but rather, like wood, it only requires kindling to create in it an impulse to think independently and an ardent desire for the truth.
>
> Plutarch, On listening to lectures, 48

The two theorems of Fermat and Fagnano concern problems of minimization connected with a triangle. The first searches for a point which minimizes a sum of distances and the second searches for a triangle of least perimeter inscribed in a given triangle. Next proposition is the key for the first theorem.

Proposition 5.9. *For every point Δ of the plane the sum of distances $\delta = |\Delta A| + |\Delta B| + |\Delta \Gamma|$ from the vertices of the triangle $AB\Gamma$, is greater than or equal to the length of the line segment AZ joining the vertex A with the vertex Z of the equilateral triangle $B\Gamma Z$ lying outside of $AB\Gamma$ (See Figure 5.65-I).*

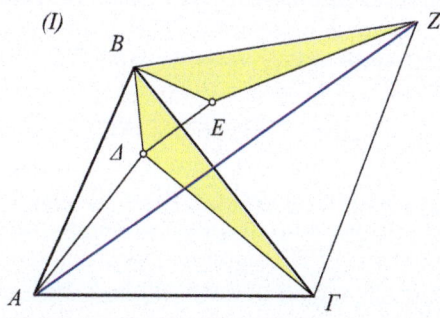
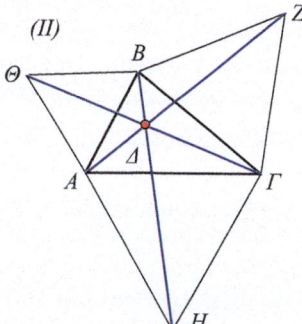

Fig. 5.65: Equilateral $B\Gamma Z$ on side $B\Gamma$ Point Δ of Fermat of $AB\Gamma$

Proof. We join the arbitrary point Δ with the vertices of the triangle and we construct two equilateral triangles $B\Gamma Z$ and $B\Delta E$ (See Figure 5.65-I). Because of the equilateral triangles, triangles $B\Delta\Gamma$ and BEZ are congruent (SAS-criterion). This way, the sum of distances $\delta = |\Delta A| + |\Delta B| + |\Delta \Gamma|$ transforms into the broken line of equal length $\delta = |A\Delta| + |\Delta E| + |EZ|$, which becomes minimal when points E and Δ are on AZ.

Theorem 5.23 (Fermat). *For every triangle $AB\Gamma$, with angles less than 120 degrees, the following properties are valid (See Figure 5.65-II):*

1. *The line segments AZ, BH and $\Gamma\Theta$, which join the vertices of the triangle with the vertices of the equilaterals which are constructed on the sides and outside it, are equal and pairwise form angles of 60 degrees.*
2. *Segments AZ, BH and $\Gamma\Theta$ pass through the same point Δ (**Fermat point** of the triangle).*

3. Δ *minimizes the sum δ of the distances of the point from the vertices of the triangle. This minimal value of δ is the common length of the segments AZ, BH and $\Gamma\Theta$.*

Proof. For (1): Define Δ to be the intersection point of AZ and BH. AZ and BH are equal because the triangles $A\Gamma Z$ and $B\Gamma H$ are congruent: $|B\Gamma| = |\Gamma Z|$, $|A\Gamma| = |H\Gamma|$ because of the equilaterals $B\Gamma Z$ and $A\Gamma H$. Also, $\widehat{B\Gamma H} = \widehat{A\Gamma Z}$, since each one is a sum of the angle at Γ and one of 60 degrees. Similarly we show that AZ and $\Gamma\Theta$ are equal, therefore all three segments are equal. Angle $\widehat{B\Delta Z}$ is 60 degrees. This follows from the fact that the quadrilateral $B\Delta\Gamma Z$ is inscriptible. Latter results from the congruence of triangles $B\Gamma H$ and $A\Gamma Z$, implying that angles $\widehat{\Delta B\Gamma}$ and $\widehat{\Delta Z\Gamma}$ are equal. Therefore B and Z see $\Delta\Gamma$ under equal angles and consequently $B\Delta\Gamma Z$ is inscriptible. $\widehat{B\Delta Z}$ then is equal to $\widehat{B\Gamma Z}$, which is 60 degrees. Because of the inscriptible $B\Delta\Gamma Z$, angle $\widehat{B\Delta\Gamma}$ will be 120 degrees, therefore $\widehat{\Gamma\Delta Z}$ will also be 60 degrees. Also, $\widehat{H\Delta\Gamma}$ as external and opposite to $\widehat{BZ\Gamma}$ will also be 60 degrees and $\widehat{A\Delta H}$, as vertical to $\widehat{B\Delta Z}$ will be itself also 60 degrees. It follows that $A\Delta B\Theta$ has the opposite angles at Θ and Δ supplementary, therefore is inscriptible. Then, however, $\widehat{\Theta\Delta B}$ will be equal to $\widehat{\Theta AB}$, therefore itself 60 degrees and finally $\widehat{\Theta\Delta A}$ will be also 60 degrees.

(2) is a consequence of the preceding equality of the angles. At Δ, which we defined as the intersection of BH and AZ, we have seen that all angles formed there, are 60 degrees, therefore points Θ, Δ and Γ will be collinear.

(3) is a consequence of the preceding proposition. If Δ exists, then it must lie on AZ, and similarly also on BH and also on $\Gamma\Theta$, therefore it will coincide with their intersection. The condition for the angles of the triangle, to be less

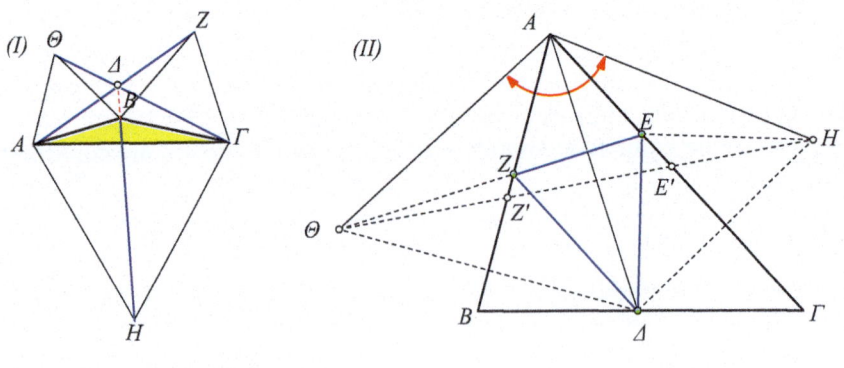

Fig. 5.66: $\widehat{AB\Gamma} > 120°$ \qquad Fagnano's theorem

than 120 degrees, guarantees that the line segments BH and AZ intersect.

5.12. FERMAT'S AND FAGNANO'S THEOREMS

Indeed, the existence of Δ is proved even without this restriction. We easily see however that in the case where the angle $AB\Gamma$ is greater than 120 degrees, point Δ is external to BH, while the minimum of δ occurs when Δ coincides with the vertex B (See Figure 5.66-I).

Theorem 5.24 (Fagnano 1715-1797). *The orthic triangle of an acute triangle $AB\Gamma$ has the least perimeter, among all triangles which are inscribed in $AB\Gamma$.*

Proof. Let us see first for a triangle ΔEZ, inscribed in $AB\Gamma$, how the perimeter varies, when we displace point Δ on $B\Gamma$ (See Figure 5.66-II). If Θ and H are respectively the reflected points of Δ relative to the sides AB and $A\Gamma$, then the perimeter of the triangle will be $\delta = |\Delta Z| + |ZE| + |E\Delta| = |\Theta Z| + |ZE| + |EH|$. The length of the broken line ΘZEH becomes minimal when points Z and E are on line ΘH. Therefore we can make the perimeter of ΔEZ smaller by replacing points Z, E with the intersection points Z' and E' of ΘH with AB and $A\Gamma$. Moreover the perimeter of $\Delta Z'E'$ will be equal to the length of ΘH.

We examine now for which position of Δ the length of ΘH becomes minimal. It suffices to note, that the isosceles $A\Theta H$ has its angle at the vertex always equal to 2α, where $\alpha = \widehat{BA\Gamma}$. Therefore the length of the base of $A\Theta H$ will become minimal when the length of the legs becomes minimal. This however happens when $A\Delta$ coincides with the altitude from A. Point Δ then, if there exists a triangle of least perimeter, must coincide with the trace of the altitude from A. A similar property however must be valid for every other vertex of ΔEZ, hence the truth of the theorem. The fact, that

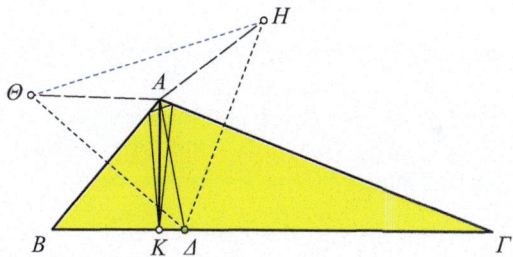

Fig. 5.67: Non existence of original triangle of least perimeter

the angle at A must be acute, is needed in order to guarantee that ΘH will intersect the sides AB and $A\Gamma$. In the case where the angle at A is obtuse (See Figure 5.67), we see easily that the triangle of least perimeter degenerates into taking twice the altitude AK from A (another proof: [148, p.19]).

Proposition 5.10 (Napoleon's theorem). *If on the sides of the triangle $AB\Gamma$ we construct equilateral triangles lying outside, then their centers are vertices of an equilateral triangle (See Figure 5.68-I).*

Proof. Triangle ΔEZ has for vertices the circumcenters of the equilateral triangles. The radius of the circumcircle of the equilateral is equal to $1/\sqrt{3}$ of its side. This implies that triangles AZE and ABB' are similar. Indeed

$$\frac{|AB|}{|AZ|} = \frac{|AB'|}{|AE|} = \sqrt{3}, \quad \widehat{ZAE} = \widehat{BAB'} = \widehat{BA\Gamma} + 60°,$$

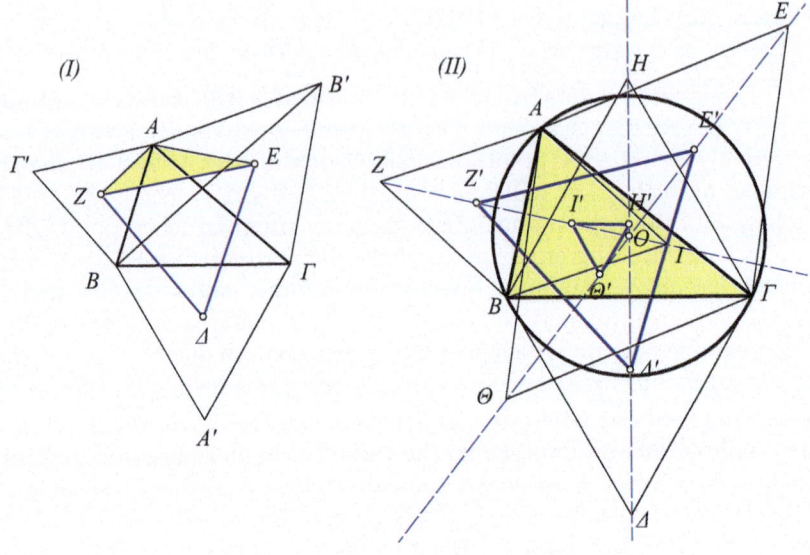

Fig. 5.68: Napoleon's theorem

which shows that the two triangles have two sides proportional and the contained in them angles equal. Similar relations will hold also for the other sides of triangle $E\Delta Z$. However, according to Theorem 5.23, $BB', \Gamma\Gamma', AA'$ are equal, therefore also $ZE, E\Delta, \Delta Z$, whose lengths have the same ratio $1/\sqrt{3}$ to the preceding, will be equal.

Exercise 5.84. Show that also in the case where the equilaterals are constructed on the sides of $AB\Gamma$ but on the same side as $AB\Gamma$, the triangle of their centers is again equilateral (See Figure 5.68-II).

Exercise 5.85. On the sides of the triangle $AB\Gamma$ we construct squares $B\Gamma E\Delta$, $A\Gamma ZH, AB I\Theta$ outside it with corresponding centers $\{M, K, \Lambda\}$ (See Figure 5.69-I). Show that:

1. The triangles $\{AB\Delta, IB\Gamma\}$ are equal and similar to ΛBM in ratio $\sqrt{2}$.
2. The lines $\{\Gamma I, \Delta A, BK, E\Theta\}$ pass through a common point N.
3. The lines $\{BK, \Lambda M\}$ are equal and orthogonal.
4. The lines $\{AM, BK, \Gamma\Lambda\}$ pass through a common point T.

The point T is called **Vecten point** of the triangle $AB\Gamma$. The triangle $K\Lambda M$ is called **Vecten triangle** of the triangle $AB\Gamma$. Notice that $\{K, \Lambda, M\}$ are apexes of similar isosceli constructed on the sides of $AB\Gamma$. It is more general true

5.12. FERMAT'S AND FAGNANO'S THEOREMS

that apexes of similar isosceli like $\{K, \Lambda, M\}$ define corresponding segments $\{BK, \Gamma\Lambda, AM\}$ passing through a common point T'. All these T', created from the various isosceli, are on the so called *rectangular hyperbola of Kiepert* (See Figure 5.69-II) of the triangle. This hyperbola passes through the vertices, as well as, through several other remarkable points of the triangle, like the orthocenter, the centroid and the Fermat point ([129]).

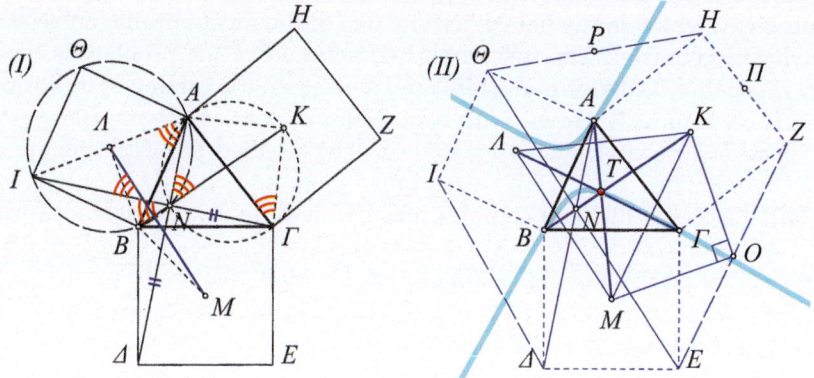

Fig. 5.69: Squares on the sides of the triangle

Exercise 5.86. Continuing the preceding exercise (See Figure 5.69-II), show that:

1. If O is the middle of EZ, then $\{OK, OM\}$ are equal and orthogonal.
2. The line $E\Theta$ passes through the intersection point N of $\{BK, A\Delta\}$.
3. The symmedian point of the triangle $AB\Gamma$ and of the triangle, which results by extending the sides of the squares $\{\Delta E, ZH, \Theta I\}$ coincide.

Exercise 5.87. In figure (See Figure 5.69-I) consider the intersection points $A' = (\Theta I, HZ)$, $B' = (\Delta E, I\Theta)$, $\Gamma' = (HZ, \Delta E)$. Show that lines $\{AA', BB', \Gamma\Gamma'\}$ are symmedians of the triangles $\{AB\Gamma, A'B'\Gamma'\}$. Construct the triangle $AB\Gamma$, given the triangle $A'B'\Gamma'$.

Exercise 5.88. On the sides of the triangle $AB\Gamma$ construct similar isosceli $\{A\Gamma'B, BA'\Gamma, \Gamma B'A\}$, all externally or all internally to $AB\Gamma$. Show that the lines $\{AA', BB', \Gamma\Gamma'\}$ pass through the same point.

Exercise 5.89. Consider the inscribed circle $\kappa(I, r)$ of the triangle $AB\Gamma$ and its contact points $\{A', B', C'\}$ with the sides $\{BC, CA, AB\}$. On the extensions of $\{IA', IB', IC'\}$ take equal segments $\{A'A'', B'B'', C'C''\}$. Show that the lines $\{AA'', BB'', CC''\}$ pass through the same point ([17, § 1242m]).

5.13 Morley's theorem

> The more ideas become mechanical and tool-oriented, the less one can see in them some thoughts with meaning in themselves.
>
> Max Horkheimer, *The eclipse of reason*

In 1899 Frank Morley (1860-1937), a prominent Mathematician, published a problem, which immediately became one of the most popular subjects of euclidean geometry. The first published solution of the problem is timed ten years after its exposition ([143]) and since then the catalog of its various proofs continuously grows, with contributions from Mathematicians of all calibers. The proofs which appeared can be classified into four categories, the trigonometric ones ([147]), the purely geometric ones ([133, p.103], [12, p.24]), the ones which use complex numbers and finally the algebraic proof [121] of Alain Connes (1947-), who was awarded the Fields metal (in 1982), which is the equivalent of the Nobel prize for Mathematics. In what follows I present a simple geometric proof, which relies on a trivial lemma for the bisectors of a triangle.

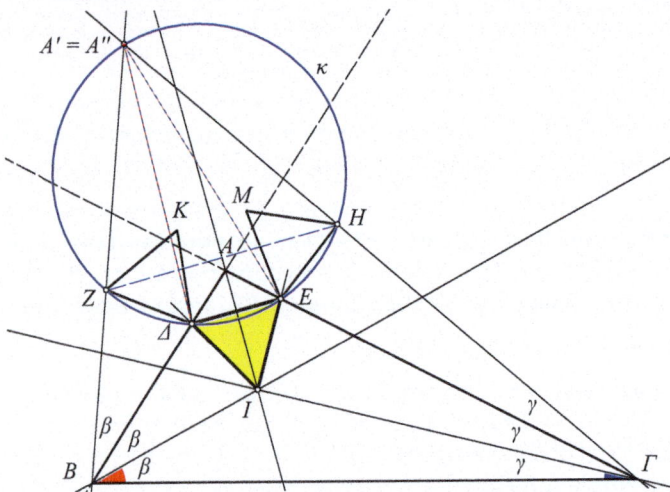

Fig. 5.70: Theorem of Morley

Lemma 5.1. *From the incenter I of the triangle ABΓ and on both sides of IA draw two lines with inclination 30° to IA, which intersect the other sides of the triangle at points Δ and E. The triangle IΔE is equilateral and ΔE is orthogonal to AI (See Figure 5.70).*

Proof. From the fact that I is equidistant from the sides, follows directly that IΔ and IE are equal, therefore IΔE is isosceles and IA is its bisector.

5.13. MORLEY'S THEOREM

Theorem 5.25 (Morley). *The successive trisectors of the angles of a triangle intersect at points which form an equilateral triangle.*

Proof. The trisectors are the lines dividing the triangle's angles into three equal parts. The proof relies on figure 5.70, in which starting from the triangle $A''B\Gamma$, with base angles 3β and 3γ, we define the triangle $AB\Gamma$ with base angles 2β and 2γ and reconstruct the triangle $A'B\Gamma = A''B\Gamma$ using the preceding lemma.

Indeed, using the results of the lemma, we reflect BI and $I\Delta E$ relative to line AB and we get line BA' and the equilateral ΔZK. A similar reflection relative to $A\Gamma$ gives $\Gamma A'$ and the equilateral EHM. From their definition we have

$$\widehat{\Gamma BA'} = 3\beta, \qquad \widehat{B\Gamma A'} = 3\gamma, \qquad \widehat{BA'\Gamma} = 180° - 3\beta - 3\gamma \qquad (*)$$

and the fact that A'' coincides with A'. On the other hand, the quadrilateral $Z\Delta EH$, because of its symmetry relative to AI, is an isosceles trapezium with three equal sides and its angle

$$\widehat{Z\Delta E} = 60° + 2(90° - \alpha/2) = 60° + 180° - \alpha = 60° + 2\beta + 2\gamma \Rightarrow$$
$$\widehat{ZHE} = 120° - 2\beta - 2\gamma.$$

Because of the arc congruence of the circumcircle κ of the trapezium, we conclude that

$$\widehat{ZH\Delta} = \frac{\widehat{ZHE}}{2} = 60° - \beta - \gamma.$$

Because of the arc congruence, the inscribed angles of κ, which view the arc $\widehat{Z\Delta EH}$, are the threefold of the angle $\widehat{ZH\Delta}$. Consequently, because of (*), the circumcircle of $Z\Delta EH$ passes through A'. This, implies again, because of the congruence of the arcs, that $A'\Delta$ and $A'E$ are trisectors of $\widehat{BA'\Gamma}$.

Exercise 5.90. On the triangle $AB\Gamma$ with respective trisectors AH, $A\Theta$, BI, BK, ΓM, $\Gamma \Lambda$ and Morley's triangle ΔEZ (See Figure 5.71), show that $\kappa_A = (A\Lambda ZEK)$, $\kappa_B = (BH\Delta ZM)$, $\kappa_\Gamma = (\Gamma IE\Delta\Theta)$ are three circles passing through the respective points in parentheses. The same happens also with circles $\lambda_A = (B\Gamma IM)$, $\lambda_B = (\Gamma A\Lambda\Theta)$, $\lambda_\Gamma = (ABHK)$. Also the lines $K\Lambda$, MH, ΘI are respectively parallel to the sides of ΔEZ and define an equilateral triangle $N\Xi O$, whose circumscribed circle is tangent to the circles λ_A, λ_B, λ_Γ.

Remark 5.11. Similar conclusions with those of the preceding theorem hold also for the intersection points of the trisectors of the external angles as well as intersections of internal with external trisectors (See Figure 5.72), for a total of five equilateral triangles ([133, p.103]).

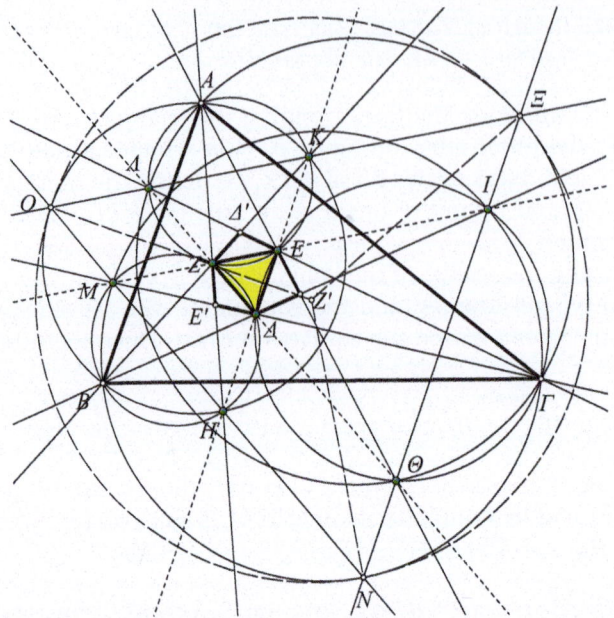

Fig. 5.71: Coincidences related to the theorem of Morley

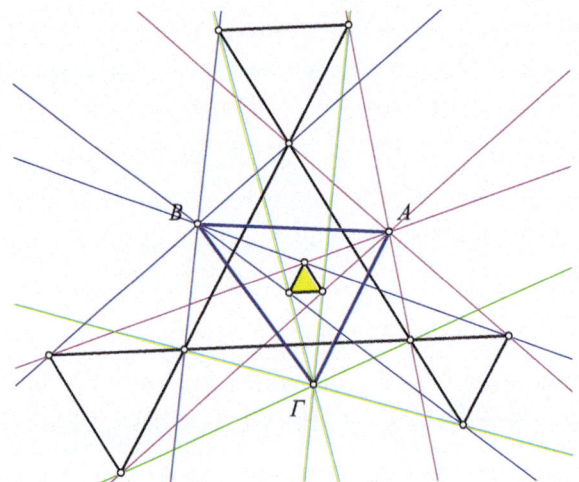

Fig. 5.72: Equilaterals from internal/external trisectors

Exercise 5.91. Show that the triangles $\{\Delta Z'E, E\Delta'Z, ZE'\Delta\}$ in figure 5.71, with bases the respective sides of the Morley triangle and lateral sides coinciding with trisectors of $AB\Gamma$, are isosceli (see figure 5.72 for other equilaterals formed by external trisectors of the angles of $AB\Gamma$).

5.14 Signed ratio and distance

> When we raise the question, " Supposing it possible to cease from heaping into this unconscionable flesh all these things from land and sea, what are we passing to do ? " it is because in our ignorance of noble things we are content with the life which our necessities impose.
>
> Plutarch, Dinner - party of the Seven Sages

The **signed ratio** is a slight modification of the simple (positive) ratio of two line segments of the same line, in which, besides the lengths of the segments, we also take into account their *orientation*.

$$\frac{XA}{XB} = \pm \frac{|XA|}{|XB|}$$

This coincides with the usual ratio of distances $\frac{|XA|}{|XB|}$, when XA and XB point to the same direction, while for opposite oriented XA and XB, it is equal to $-\frac{|XA|}{|XB|}$. By definition then, the signed and simple ratio are connected through the formula

$$\left|\frac{XA}{XB}\right| = \frac{|XA|}{|XB|}.$$

Next proposition shows the difference of the signed from the simple (positive) ratio.

Theorem 5.26. *Given two different points A and B of line ε, the position of a third point X on this line is completely determined through the signed ratio*

$$\lambda = \frac{XA}{XB}.$$

Proof. The theorem is a consequence of Theorem 1.16, according to which, given $\lambda > 0$, $\lambda \neq 1$, there exist exactly two points X, X' on the line AB with (common) ratio of distances $\frac{|XA|}{|XB|} = \lambda$, one of them internal to AB and the other external. Consequently, if we take into account the sign, the correspondence becomes unique. The internal points X of AB have negative signed ratio $\frac{XA}{XB}$, and the external points to AB have positive ratio.

Remark 5.12. Even though I define here the signed ratio, I don't use it except in the last sections of this chapter, which deal with subjects that are usually found outside the material of school-books.

Remark 5.13. Figure 5.73, shows how the signed ratio changes as a function of the position of X. When X is very far away from A and B, then XA and XB are almost equal and the ratio approaches the value 1, remaining however < 1 when X is far away on the left of A while it is > 1 for X far away on

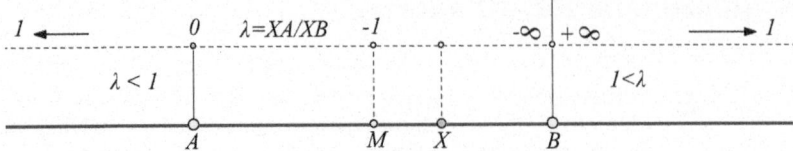

Fig. 5.73: Signed ratio

the right of B. As X approaches A from the left of AB, the ratio is positive and tends to 0. For $X = A$ the ratio becomes 0 and, continuing to the right, it becomes negative and tends to $-\infty$ as X approaches B. At the middle M of AB the ratio is -1. At B the value of the ratio is undefined, the same way it happens with fractions for which the denominator becomes zero. For X very close to B on the right of AB the signed ratio is a very big positive number and as X moves away from AB to the right it becomes less and less, remaining > 1, and tending to 1, as X moves further to infinity.

Corollary 5.9. *The points X and X' of the line ε are harmonic conjugate relative to A and B, if and only if the signed ratio satisfies*

$$\frac{XA}{XB} = -\frac{X'A}{X'B}.$$

The signed ratio is compatible with the **signed distance** of two points on a line $\varepsilon = AB$. For the definition of the signed distance we need the notion of **coordinate** on line ε. The latter is a simple mechanism, which assigns to every point on the line a positive or negative number. This way, if X has coordinate x and Y has coordinate y (we often write $X(x)$ for the point X and its coordinate x), then the signed distance is defined as the difference

$$XY = y - x.$$

It obviously holds

$$YX = x - y = -XY.$$

The definition of a system of coordinates on the line ε relies on the selection of two different fixed, but otherwise arbitrary points A and B of ε. We call A the **origin** of coordinates, and call **positive half line** the half line starting at A and containing point B. We call **negative half line** the opposite half line, starting at A and not containing point B. For every point X on ε we define then as coordinate of X the length of the segment $x = |AX|$, when point X is on the positive half line and the number $x = -|AX|$ when X is on the negative half line. This way, the coordinate of A is zero and the coordinate of B is the length $|AB|$.

Corollary 5.10. *For a system of coordinates, defined through the points A, B of the line ε, and three arbitrary points $X(x)$, $Y(y)$ and $Z(z)$ on ε we have*

5.14. SIGNED RATIO AND DISTANCE

$$x = AX, \quad XY = AY - AX = y - x, \quad XZ = XY + YZ.$$

Proposition 5.11. *The signed distance XY of two points does not depend on the special system of coordinates through which it is defined.*

Proof. Let us suppose that one system is defined from the two points A and B of ε and the other from A' and B'. Suppose that the same point X has relative to the first system coordinate x and relative to the second system x'. Suppose also α' the coordinate of A' relative to the first system (starting at A). Then, relying on the definitions and the corollary

$$x' = A'X = x - \alpha', \ y' = A'Y = y - \alpha' \Rightarrow XY = y' - x' = (y - \alpha') - (x - \alpha') = y - x.$$

Exercise 5.92. Show that for three arbitrary points A, B, Γ on line ε and their signed distances, holds
$$AB + B\Gamma + \Gamma A = 0.$$

With the help of a system of coordinates on the line ε, the signed ratio is written as a ratio of positive or negative numbers

$$\frac{XA}{XB} = \frac{a-x}{b-x},$$

where x, a, b are respectively the coordinates of points X, A and B. Figure 5.74 shows the graphical representation of the function $y = \frac{a-x}{b-x}$ and explains with an image the word-descriptions of the preceding remark 5.13.

The signed distance and the signed ratio simplify considerably the formulas of § 1.17, which uses exclusively positive ratios, and eliminates the case making and discrimination, made there between points in the interior and exterior of the line segment AB. The same happens with the formulas of Stewart (§ 3.12).

Next exercises begin with the substitution of formulas (1), (2) and (3) of § 1.17 through a unified formula and proceed gradually to the substitution of the rest of the formulas of that section. As a system of coordinates, we consider the one defined by the points A and B on the line ε. With $d = |AB|$ we denote the distance between A and B.

Exercise 5.93. Show that the signed ratio $t = \frac{XA}{XB}$ is given by the formula

$$t = \frac{x}{x-d} \quad \Leftrightarrow \quad x = \frac{t}{t-1}d, \qquad (d = |AB|)$$

where x is the coordinate of the point X of line ε.

Exercise 5.94. Show that the points $X(x)$ and $X'(x')$ are harmonic conjugate of A and B, if and only if

$$2xx' = d(x + x') \quad \Leftrightarrow \quad \frac{2}{d} = \frac{1}{x} + \frac{1}{x'}.$$

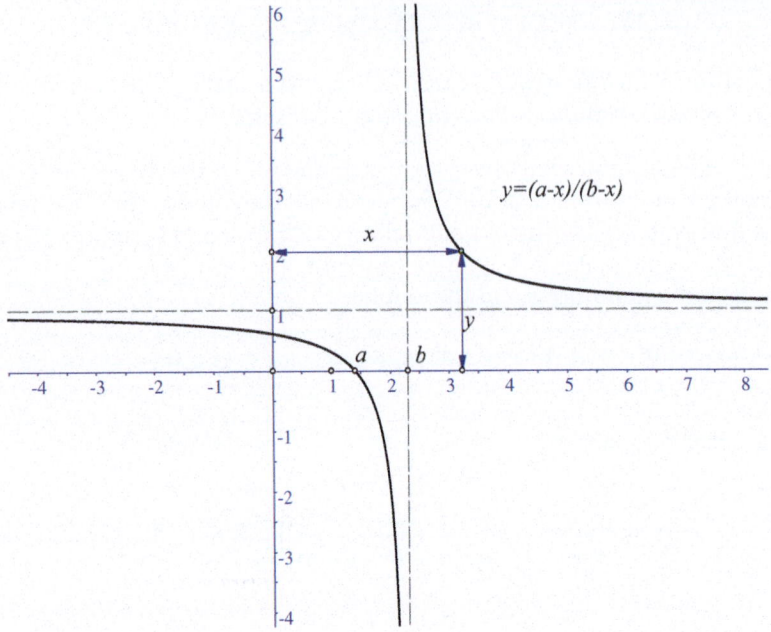

Fig. 5.74: The graphical representation of the ratio $y = \frac{XA}{XB} = \frac{a-x}{b-x}$

Exercise 5.95. Show that the coordinate v of the middle N of the line segment XX' which is defined from $X(x)$ and its harmonic conjugate $X'(x')$ relative to A, B, is equal to
$$v = \frac{t^2}{t^2 - 1}d,$$
where t is the signed ratio $t = \frac{XA}{XB}$.

Analogous are the simplifications and the unified form we achieve for the formula of Stewart. Next exercises give the corresponding unified formulas, when the ratios (positive), which are used in (§ 3.12), are replaced with the signed ones and the lengths through the corresponding signed lengths.

Exercise 5.96. Show that for a point Δ on the line $B\Gamma$ of the triangle $AB\Gamma$ with $d = |A\Delta|$ holds
$$\frac{c^2 - d^2 - \Delta B^2}{b^2 - d^2 - \Delta \Gamma^2} = \frac{\Delta B}{\Delta \Gamma}.$$

Exercise 5.97. Show that for a point Δ on the line $B\Gamma$ of the triangle $AB\Gamma$, with $d = |A\Delta|$ and $B\Delta = \kappa B\Gamma$ and $\Delta\Gamma = \lambda B\Gamma$ holds
$$d^2 = \kappa \cdot b^2 + \lambda \cdot c^2 - \kappa\lambda \cdot a^2.$$

5.14. SIGNED RATIO AND DISTANCE

Exercise 5.98. Show that for four real numbers $\alpha, \beta, \gamma, \delta$, the following identity holds

$$(\alpha-\delta)^2(\gamma-\beta) + (\beta-\delta)^2(\alpha-\gamma) + (\gamma-\delta)^2(\beta-\alpha) + (\beta-\alpha)(\gamma-\beta)(\alpha-\gamma) = 0.$$

```
    A(α)           Δ(δ)          B(β)      Γ(γ)
────●──────────────●─────────────●─────────●────────── ε
```

Fig. 5.75: Four points and their coordinates

Exercise 5.99. Show, that for four points A, B, Γ, Δ on line ε and their signed distances, the following identity holds (See Figure 5.75)

$$\Delta A^2 \cdot B\Gamma + \Delta B^2 \cdot \Gamma A + \Delta \Gamma^2 \cdot AB + AB \cdot B\Gamma \cdot \Gamma A = 0.$$

Hint: Consider any arbitrary system of cordinates on the line ε (it doesn't matter exactly which) and write the signed distances with the help of the coordinates of points $A(\alpha), B(\beta), \Gamma(\gamma)$ and $\Delta(\delta)$: $AB = (\beta - \alpha)$, $B\Gamma = (\gamma - \beta)$, ... etc. The proof follows from the identity of the preceding exercise.

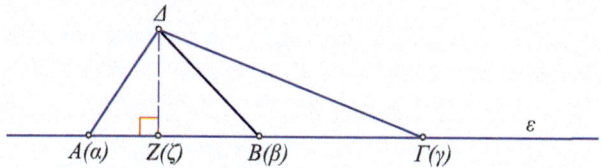

Fig. 5.76: Formula of Stewart

Exercise 5.100 (Stewart's formula). Show that the preceding equation holds even when points A, B, Γ are on the same line ε, yet point Δ is not contained in it (See Figure 5.76).

Hint: Project point Δ at point Z of ε and apply the Pythagorean theorem

$$\Delta A^2 \cdot B\Gamma + \Delta B^2 \cdot \Gamma A + \Delta \Gamma^2 \cdot AB + AB \cdot B\Gamma \cdot \Gamma A$$
$$= (\Delta Z^2 + ZA^2) \cdot B\Gamma + (\Delta Z^2 + ZB^2) \cdot \Gamma A + (\Delta Z^2 + Z\Gamma^2) \cdot AB + AB \cdot B\Gamma \cdot \Gamma A$$
$$= \left[ZA^2 \cdot B\Gamma + ZB^2 \cdot \Gamma A + Z\Gamma^2 \cdot AB + AB \cdot B\Gamma \cdot \Gamma A \right] + \left[\Delta Z^2 \cdot (B\Gamma + \Gamma A + AB) \right]$$
$$= 0 + 0 = 0.$$

The first bracket becomes zero because of Exercise 5.99 and the second because of Exercise 5.92.

Exercise 5.101. Show that, if the points A, B, Γ of the line ε have respective coordinates α, β and $\gamma = \frac{\lambda \cdot \alpha + \mu \cdot \beta}{\lambda + \mu}$, where λ, ν two numbers with $\lambda + \mu \neq 0$, then

$$\frac{\Gamma A}{\Gamma B} = -\frac{\mu}{\lambda}.$$

Fig. 5.77: Ratios of points on line

Exercise 5.102. Show that, if the points Γ, Δ are contained in the line segment AB and the ratios of the segments are $\frac{\Gamma A}{\Gamma B} = \kappa$, $\frac{\Delta A}{\Delta B} = \lambda$, then the following formulas hold (See Figure 5.77):

$$\Gamma A = \frac{\kappa}{1-\kappa}AB, \quad \Gamma B = \frac{1}{1-\kappa}AB, \quad \Gamma \Delta = AB\frac{\kappa-\lambda}{(\kappa-1)(\lambda-1)},$$

$$\frac{\Delta\Gamma}{\Delta B} = \frac{\kappa-\lambda}{\kappa-1}, \quad \frac{\Gamma A}{\Gamma \Delta} = \frac{\kappa(\lambda-1)}{\lambda-\kappa}.$$

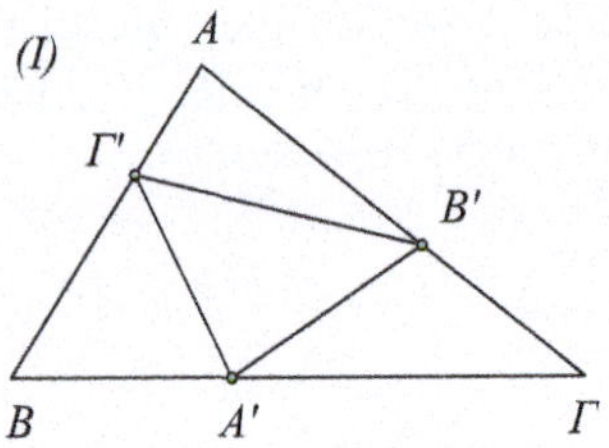

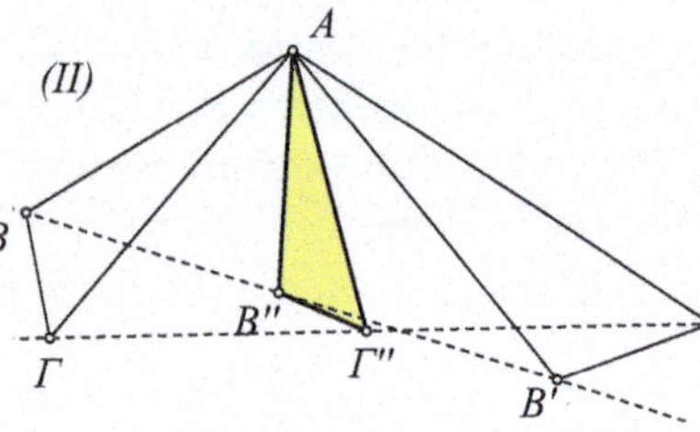

Fig. 5.78: Ratio of areas   Similar triangles produce similar

Exercise 5.103. Show that if the points A', B', Γ' are contained in sides $B\Gamma$, ΓA and AB of the triangle $AB\Gamma$ and divide them respectively into ratios

$$\frac{A'B}{A'\Gamma} = \kappa, \quad \frac{B'\Gamma}{B'A} = \lambda, \quad \frac{\Gamma'A}{\Gamma'B} = \mu,$$

then the ratio of areas of the triangles is

$$\frac{\varepsilon(A'B'\Gamma')}{\varepsilon(AB\Gamma)} = \frac{\kappa\cdot\lambda\cdot\mu-1}{(\kappa-1)(\lambda-1)(\mu-1)}.$$

Hint: According to Theorem 3.16, for the ratio of areas, $\frac{\varepsilon(AB'\Gamma')}{\varepsilon(AB\Gamma)} = \frac{A\Gamma'}{AB}\cdot\frac{AB'}{A\Gamma}$ (See Figure 5.78-I). From the preceding exercise the last ratio is calculated and is $\frac{\mu}{(\mu-1)(1-\lambda)}$. Similarly $\frac{\varepsilon(BA'\Gamma')}{\varepsilon(AB\Gamma)} = \frac{\kappa}{(\kappa-1)(1-\mu)}$ and $\frac{\varepsilon(\Gamma B'A')}{\varepsilon(AB\Gamma)} = \frac{\lambda}{(\lambda-1)(1-\kappa)}$. The formula follows by combining these ratios and the equality $\varepsilon(AB\Gamma) = \varepsilon(AB'\Gamma') + \varepsilon(B\Gamma'A') + \varepsilon(\Gamma A'B') + \varepsilon(A'B'\Gamma')$.

Proposition 5.12. *Given two similar and similarly oriented triangles $AB\Gamma$ and $AB'\Gamma'$ and on the lines BB', $\Gamma\Gamma'$ respective points B'', Γ'', such that $\frac{B''B}{B''B'} = \frac{\Gamma''\Gamma}{\Gamma''\Gamma'}$, triangle $AB''\Gamma''$ is similar to $AB\Gamma$, $A'B'\Gamma'$.*

5.14. SIGNED RATIO AND DISTANCE

Proof. The triangles ABB' and $A\Gamma\Gamma'$ are similar, because they have proportional sides and equal angles at A (See Figure 5.78-II). Also, by assumption, points B'', Γ'' divide the bases of these triangles into equal ratios, therefore the ratios will be $\frac{AB''}{A\Gamma''} = \frac{AB}{A\Gamma}$ and the angles $\widehat{BAB''}$ and $\widehat{\Gamma A\Gamma''}$ will be equal. It follows that triangles $AB\Gamma$ and $AB''\Gamma''$ have their angles at A equal and the corresponding adjacent sides proportional, therefore they are similar (see also Theorem 2.28).

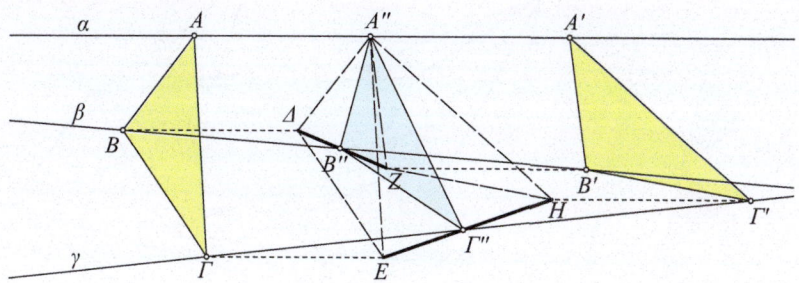

Fig. 5.79: Similar triangles with respective collinear vertices

Proposition 5.13. *Given two similar and similarly oriented triangles $AB\Gamma$ and $A'B'\Gamma'$ and on the lines $\alpha = AA'$, $\beta = BB'$, $\gamma = \Gamma\Gamma'$ respectively points A'', B'', Γ'' such that $\frac{A''A}{A''A'} = \frac{B''B}{B''B'} = \frac{\Gamma''\Gamma}{\Gamma''\Gamma'}$, the triangle $A''B''\Gamma''$ is similar to $AB\Gamma$, $A'B'\Gamma'$.*

Proof. Displace in a parallel way the triangles $AB\Gamma$, $A'B'\Gamma'$, so that their vertices A, A' become coincident with A'' (See Figure 5.79). The preceding proposition is applied to the resulting configuration. Indeed, $AB\Gamma$ takes the position $A''\Delta E$ and $A'B'\Gamma'$ the position $A''ZH$. Also, we easily see that $\frac{\Gamma''E}{\Gamma''H} = \frac{\Gamma''\Gamma}{\Gamma''\Gamma'} = \frac{B''B}{B''B'} = \frac{B''\Delta}{B''Z}$. The preceding proposition is applied to the triangles $A''\Delta E$, $A''B''\Gamma''$, $A''ZH$ and proves the claim.

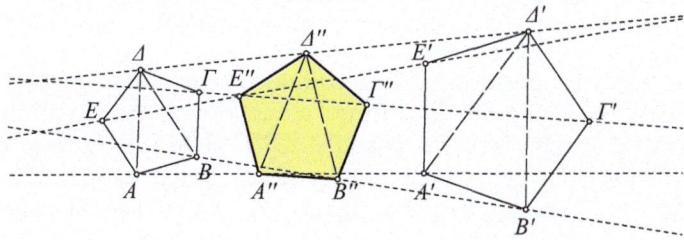

Fig. 5.80: Similar polygons with respective collinear vertices

Theorem 5.27. *Given two similar and similarly oriented polygons $AB\Gamma...$ and $A'B'\Gamma'...$ and on the lines $\alpha = AA'$, $\beta = BB'$, $\gamma = \Gamma\Gamma'$, ... respectively points A'', B'', Γ'', such that $\frac{A''A}{A''A'} = \frac{B''B}{B''B'} = \frac{\Gamma''\Gamma}{\Gamma''\Gamma'} = ...$, the polygon $A''B''\Gamma''...$ is similar to $AB\Gamma...$, $A'B'\Gamma'...$ (See Figure 5.80).*

Proof. Draw from respective vertices, like Δ, Δ', Δ'', the diagonals ΔA, ΔB, $\Delta \Gamma$, ..., $\Delta' A'$, $\Delta' B'$, $\Delta' \Gamma'$, ... etc. The polygons disassemble into similar triangles. Applying to them the preceding proposition, we prove that the triangles $\Delta'' A'' B''$, $\Delta'' B'' \Gamma''$, ... are similar to the corresponding triangles of the other two polygons, and from the similarity of all these triangles, we conclude the similarity of the polygons (Exercise 3.104).

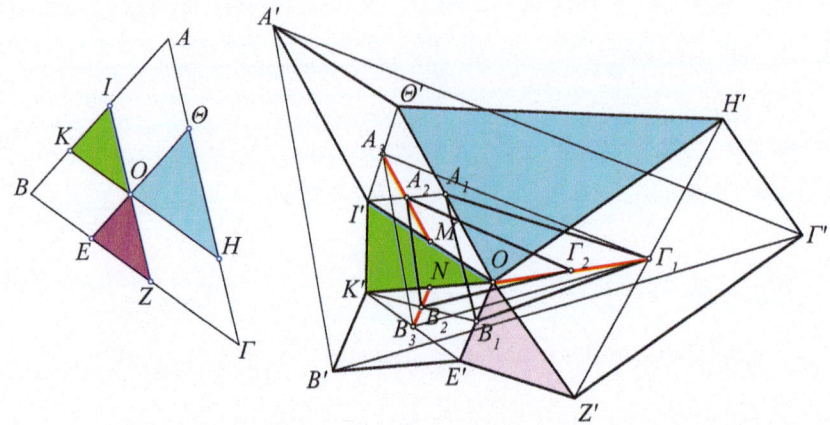

Fig. 5.81: Theorem of three bars

Theorem 5.28 (of three bars (Clifford 1845-1879, Cayley 1821-1895)). *From a point O, in the interior of a triangle $AB\Gamma$, we draw parallels to its sides and we define the similar to it triangles OEZ, $OH\Theta$, OIK. We suppose that the triangles remain fixed in magnitude, rotating about the fixed point O, while the rest of the segments $Z\Gamma$, ΓH, ΘA, AI, KB, BE remain fixed in length forming parallelograms. Then the vertices (different from O) of these parallelograms form a triangle $A'B'\Gamma'$ similar to $AB\Gamma$ (Figure 5.81 on the right magnified for better support of proof arguments).*

Proof. We apply Proposition 5.13 three times. The first time on the two similar triangles $OE'Z'$ and $O\Theta'H'$ (primed letters denote the variability of the positions of the triangles). The triangle $A_1 B_1 \Gamma_1$, which is formed from the middles respectively of $O\Theta'$, OE', $Z'H'$ will be similar to $AB\Gamma$. The second time on the two similar triangles $A_1 B_1 \Gamma_1$ and $I'K'O$. The triangle $A_2 B_2 \Gamma_2$, which is formed from the middles respectively $A_1 I'$, $B_1 K'$, $\Gamma_1 O$ is also similar to $AB\Gamma$. The third time we apply the proposition on the two similar triangles MNO and $A_2 B_2 \Gamma_2$, where M, N are respectively the middles of OI', OK'. Triangle $A_3 B_3 \Gamma_1$, is similar to $AB\Gamma$, as MA_3, NB_3, $O\Gamma_1$ are respectively double the segments MA_2, NB_2, $O\Gamma_2$. In total then $A_3 B_3 \Gamma_1$ is similar to $AB\Gamma$. However $A'B'\Gamma'$ is a homothety of $A_3 B_3 \Gamma_1$ with center of homothety the point O and ratio $\lambda = 2$ (alternatively [120], [137, p.129]).

5.15 Cross ratio, harmonic pencils

> As far as we can discern, the sole purpose of human existence is to kindle a light in the darkness of mere being.
>
> C. G. Jung, Errinerungen, Träume, Gedanken

A strange, on first sight, construction, which is proved to be rich in applications, is that of the cross ratio of four points on a line. For its definition we fix two of the four points and we consider the signed ratio (Theorem 5.26) of the other two relative to the first ([113]). We call **Cross ratio** of four points

```
———•————————•————————•————————————————•———
    A         Γ         B                Δ
```

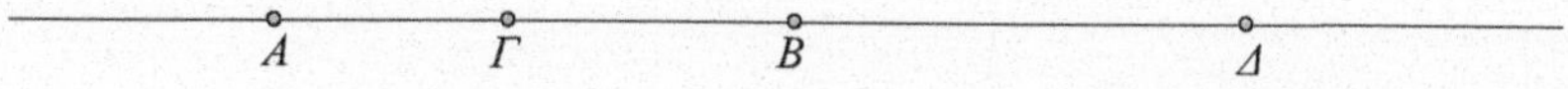

Fig. 5.82: Cross ratio $(AB;\Gamma\Delta) = \frac{\Gamma A}{\Gamma B} : \frac{\Delta A}{\Delta B}$

A, B, Γ and Δ on the same line ε, the quotient of the ratio $\frac{\Gamma A}{\Gamma B}$ to the ratio $\frac{\Delta A}{\Delta B}$. We write this as $\frac{\Gamma A}{\Gamma B} : \frac{\Delta A}{\Delta B}$ and we denote it by $(AB;\Gamma\Delta)$ (See Figure 5.82):

$$(AB;\Gamma\Delta) = \frac{\Gamma A}{\Gamma B} : \frac{\Delta A}{\Delta B} = \frac{\left(\frac{\Gamma A}{\Gamma B}\right)}{\left(\frac{\Delta A}{\Delta B}\right)}.$$

A special, but important case of cross ratio is that, which results from two points (A, B) and two others (Γ, Δ), which are harmonic conjugate to the first. This, by definition means, that the signed ratios are the same, except for the sign,

$$\frac{\Gamma A}{\Gamma B} = \lambda, \quad \frac{\Delta A}{\Delta B} = -\lambda \quad \Rightarrow \quad (AB;\Gamma\Delta) = \frac{\Gamma A}{\Gamma B} : \frac{\Delta A}{\Delta B} = -1.$$

Exercise 5.104. Show that given three different points $\{A, B, \Gamma\}$ on the line ε, the positin of a fourth point X on ε is completely determined through the cross ratio $k = (AB;\Gamma X)$.

Remark 5.14. The cross ratio reduces to the common signed ratio when point Δ tends to infinity. In fact, in this case $\frac{\Delta A}{\Delta B}$ tends to 1 and we admit, that in the limit, the cross ratio coincides with the simple signed ratio $\frac{\Gamma A}{\Gamma B}$. The importance of the cross ratio lies on the fact, that it is the unique arithmetic invariant of the "projective plane" (a kind of extension of the euclidean plane), underlying directly or indirectly, to all the arithmetic relations of the projective plane, in analogy with the "distance" of the euclidean plane, which underlies to all the arithmetic relations (formulas) of the euclidean plane.

Proposition 5.14. *Given four points A, B, Γ and Δ on a line ε and a point O outside ε, the cross ratio of the four points can be expressed, disregarding its sign,* through the angles of the lines which join the point O with the other four points (See Figure 5.83).

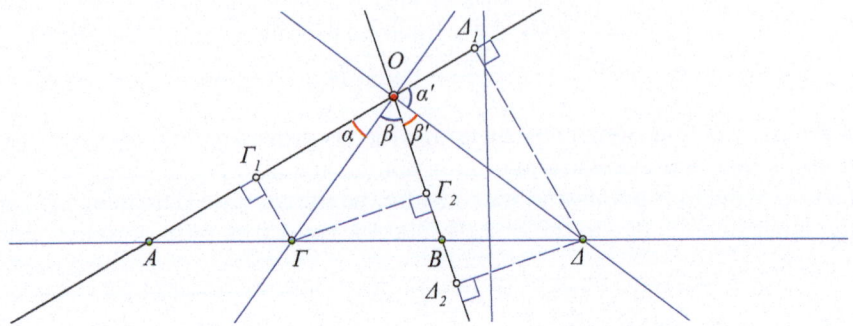

Fig. 5.83: Cross ratio $(AB;\Gamma\Delta) = \frac{\sin(\alpha)}{\sin(\alpha')} : \frac{\sin(\beta)}{\sin(\beta')}$.

Proof. Project the points Γ and Δ on the lines OA and OB at points Γ_1, Γ_2 and Δ_1, Δ_2 respectively. Because of the similar triangles $A\Gamma\Gamma_1$, $A\Delta\Delta_1$ and $B\Gamma\Gamma_2$, $B\Delta\Delta_2$, we have:

$$\frac{|\Gamma A|}{|\Gamma B|} : \frac{|\Delta A|}{|\Delta B|} = \frac{|\Gamma A|}{|\Delta A|} \cdot \frac{|\Delta B|}{|\Gamma B|} = \frac{|\Gamma\Gamma_1|}{|\Delta\Delta_1|} \cdot \frac{|\Delta\Delta_2|}{|\Gamma\Gamma_2|} = \frac{|O\Gamma|\sin(\alpha)}{|O\Delta|\sin(\alpha')} \cdot \frac{|O\Delta|\sin(\beta')}{|O\Gamma|\sin(\beta)},$$

which is simplified to

$$\frac{|\Gamma A|}{|\Gamma B|} : \frac{|\Delta A|}{|\Delta B|} = \frac{\sin(\alpha)}{\sin(\alpha')} : \frac{\sin(\beta)}{\sin(\beta')},$$

which proves the claim.

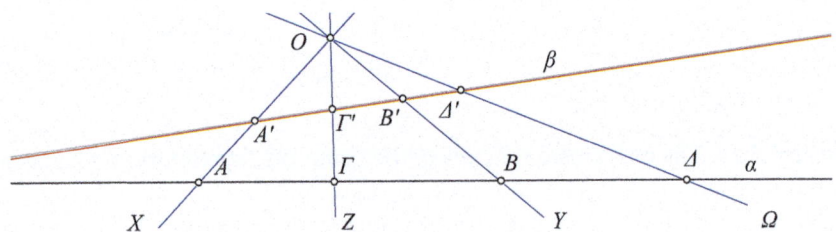

Fig. 5.84: Cross ratio of four lines passing through a point

Theorem 5.29. *Given four lines passing through the same point O and a fifth line ε, not passing through O, but intersecting the four lines, the cross ratio of the four intersection points on the line ε is the same for all such possible lines ε (See Figure 5.84).*

5.15. CROSS RATIO, HARMONIC PENCILS

Proof. Direct consequence of the preceding proposition, asserting that the cross ratio, except its sign, is expressed through the angles formed between the four lines, consequently being independent of the particular line ε, which intersects them. But even the sign of this cross ratio depends only on the position of the four lines and not on the special line which intersects them. In figure 5.84, for example, and for the two intersecting lines α, β the sign is negative. It would be positive, if point Γ was itself also, as is point Δ, outside of AB, in which case point Γ' would also be outside of $A'B'$ and the signs would again be coincident.

The preceding theorem allows us to define a number for each quadruple of lines passing through a common point: the cross ratio which the four lines define on any intersecting them line. According to the theorem, this cross ratio is independent of the particular intersecting line and, consequently, defines some characteristic of the four lines. We call this number **Cross ratio of four lines** which pass through the same point, or *cross ratio of the pencil of four lines*. A pencil consisting of four lines whose cross ratio equals -1 is called a **harmonic pencil** of four lines. Often a pencil of four lines OX, OY, OZ, $O\Omega$ is denoted by $O(X,Y,Z,\Omega)$.

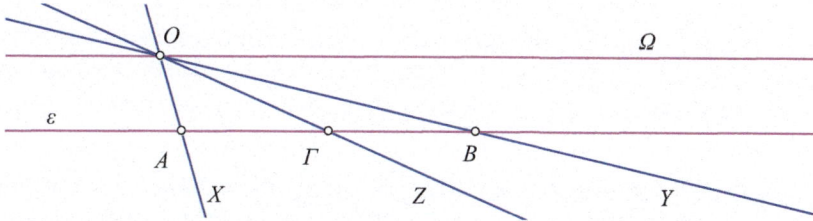

Fig. 5.85: Cross ratio of four lines on a parallel of $O\Omega$

Corollary 5.11. *The cross ratio of four lines OX, OY, OZ, $O\Omega$ is equal to the signed ratio $\frac{\Gamma A}{\Gamma B}$ of the three points A, B and Γ, which the three first lines excise on a line ε, which is parallel to the fourth line $O\Omega$ (See Figure 5.85).*

Proof. The proof follows from the remark made above, according to which, the cross ratio is reduced to the simple signed ratio $\frac{\Gamma A}{\Gamma B}$, in the case where the fourth point Δ tends to infinity.

Corollary 5.12. *A pencil of four lines $O(X,Y,Z,\Omega)$ is harmonic, if and only if a parallel ε to the fourth line $O\Omega$ intersects the three other lines respectively at three points A, B, Γ of which one (Γ) is the middle of the line segment (AB) which is defined by the other two (See Figure 5.86).*

Next proposition and the one after that examine the two main examples of harmonic pencils. We have met both examples in another form in Proposition 3.6.

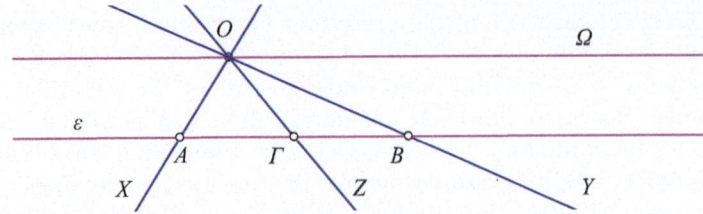

Fig. 5.86: Harmonic pencil $O(X,Y,Z,\Omega)$ and secant ε parallel to $O\Omega$

Proposition 5.15. *In every triangle OAB the two sides OA, OB and the bisectors $O\Gamma$, $O\Delta$ of the angle at O make a harmonic pencil (See Figure 5.87).*

Proof. As we saw in Theorem 3.3 and Exercise 3.12 for the triangle OAB with bisectors $O\Gamma$ and $O\Delta$, holds

$$\frac{|\Gamma A|}{|\Gamma B|} = \frac{|\Delta A|}{|\Delta B|} = \frac{|OA|}{|OB|} \quad \Rightarrow \quad \frac{\Gamma A}{\Gamma B} : \frac{\Delta A}{\Delta B} = -1.$$

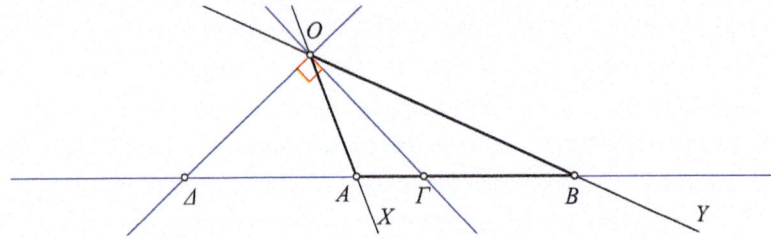

Fig. 5.87: Harmonic pencil of an angle and its bisectors

Exercise 5.105. Show that if in the harmonic pencil $O(X,Y,Z,\Omega)$ the two lines OX, OY are orthogonal, then these coincide with the bisectors of the angle $Z O \Omega$ (See Figure 5.88).

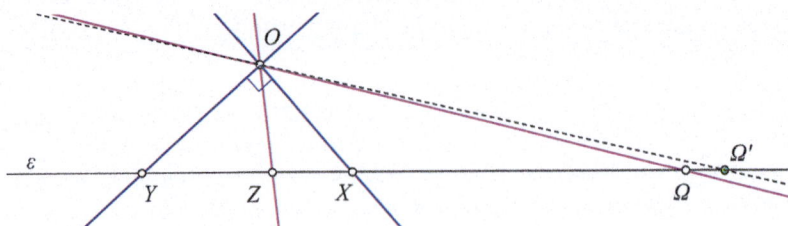

Fig. 5.88: Harmonic pencil with bisectors

Hint: Suppose that the intersection points of the pencil with a line ε satisfy $(XY;Z\Omega) = -1$. Consider the symmetric $O\Omega'$ of OZ relative to OX. Then the

5.15. CROSS RATIO, HARMONIC PENCILS

corresponding point Ω' on ε will satisfy also $(XY; Z\Omega') = -1$, therefore it will be coincident with point Ω.

Proposition 5.16. *In every triangle OEB the two sides OE, OB, the median OM and the parallel to the base EB from O make a harmonic pencil (See Figure 5.89).*

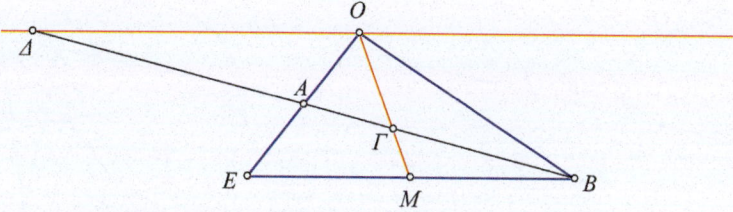

Fig. 5.89: Harmonic pencil of median and opposite side

Proof. If we draw the median OM and we extend the other median BA until it intersects the parallel to the base at Δ, this forms a quadruple of lines OA, OB, $O\Gamma$ and $O\Delta$ and holds

$$-\frac{\Gamma A}{\Gamma B} = \frac{\Delta A}{\Delta B} = \frac{1}{2},$$

from which follows that the pencil is harmonic (alternatively apply corollary 5.12).

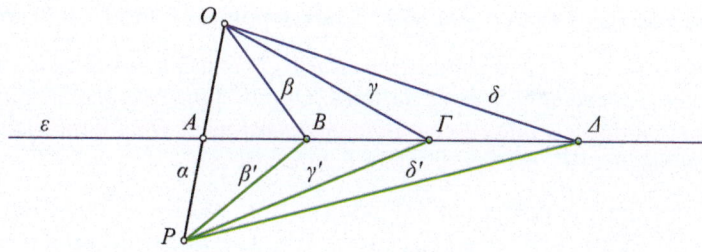

Fig. 5.90: Pencils with common line and the same cross ratio

Theorem 5.30. *Suppose that the two line pencils $O(\alpha, \beta, \gamma, \delta)$, $P(\alpha, \beta', \gamma', \delta')$ have the same cross ratio and common the line α, then the other corresponding lines intersect at three points $B = (\beta, \beta')$, $\Gamma = (\gamma, \gamma')$, $\Delta = (\delta, \delta')$ which are collinear (See Figure 5.90).*

Proof. Consider the line $\varepsilon = B\Gamma$ and the intersection point of A with α. On ε the two pencils will define the same cross ratio. If $\Delta' = (\varepsilon, \delta)$, $\Delta'' = (\varepsilon, \delta')$, then from the equality of the cross ratios $(AB; \Gamma\Delta') = (AB; \Gamma\Delta'')$, which the two pencils define on ε, follows that $\Delta' = \Delta''$, therefore also the third intersection point $\Delta' = \Delta'' = \Delta$ of δ and δ' will be on ε.

Theorem 5.31. *Suppose that on two intersecting lines at point O, are defined, respectively, points A, B, Γ and A', B', Γ', such that the cross ratios (OA; BΓ), (OA'; B'Γ') are equal. Then the lines AA', BB', ΓΓ' pass through a common point or are parallel (See Figure 5.91-I).*

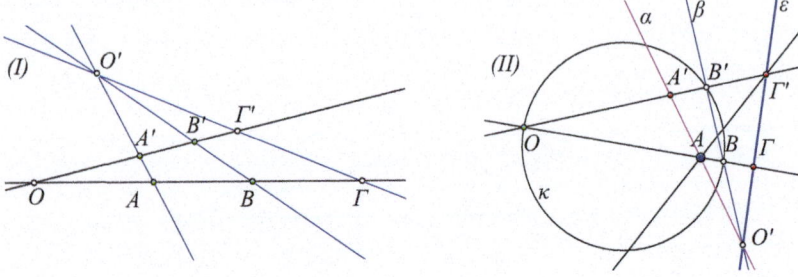

Fig. 5.91: Common cross ratio   Concurring lines

Proof. This theorem is a characteristic example of a "dual" of the preceding theorem. In "dual" theorems, in their formulation, and in the proof, we interchange the words "point" ↔ "line", as well as the words "intersection (of two lines)" ↔ "line (of two points)". This way the preceding proof is modified as follows.

Suppose that the lines are not parallel and that lines AA', BB' intersect at O' and Γ'' is the intersection of $O'\Gamma$ with $A'B'$. By hypothesis $(OA'; B'\Gamma') = (OA; B\Gamma)$. Because of the pencil $O'(O, A, B, \Gamma)$ and its intersection with the two lines (Theorem 5.29) will also hold $(OA'; B'\Gamma'') = (OA; B\Gamma)$. In total therefore it will hold also $(OA'; B'\Gamma') = (OA'; B'\Gamma'')$, hence points Γ' and Γ''' will coincide.

Corollary 5.13. *Let Γ' be a point of the polar ε of the point A relative to the circle κ. Then, for every point O of the circle, lines $\{OA, O\Gamma'\}$ intersect the circle at points $\{B, B'\}$, which define a line passing through a fixed point O' of ε coinciding with the pole of the line $O\Gamma'$ relative to κ (See Figure 5.91-II).*

Proof. Consider the pole O' of $A\Gamma'$ relative to κ and the intersection points $\{\Gamma, A'\}$ of the pairs of lines $\{(OA, \varepsilon), (O'A, O\Gamma')\}$ respectively. Line $\alpha = O'A$ is the polar of Γ' (Corollary 4.25), therefore the cross ratio will be $(OB'; A'\Gamma') = -1$. Similarly, because A is the pole of ε, the cross ratio will be $(OB; A\Gamma) = -1$. The conclusion follows by applying the preceding theorem.

Exercise 5.106. *Let Δ, E, Θ be three points on the base-line $B\Gamma$ of triangle $AB\Gamma$. From Θ we draw lines intersecting the sides $A\Gamma$, AB respectively at points X, Y. Show that the lines EX, ΔY intersect at a point Z contained in a fixed line, which passes through A (See Figure 5.92).*

Hint: The line pencils $A(B, \Gamma, A', \Theta)$ and $Z(\Delta, E, A', \Theta)$ define the same cross ratio $(YX; Z'\Theta)$ on line $B\Gamma$. Using a system of coordinates on the line $B\Gamma$ the equation $(B\Gamma; A'\Theta) = (\Delta E; A'\Theta)$ contains only one unknown, the coordinate x of A'. See also Exercise 5.183.

5.15. CROSS RATIO, HARMONIC PENCILS

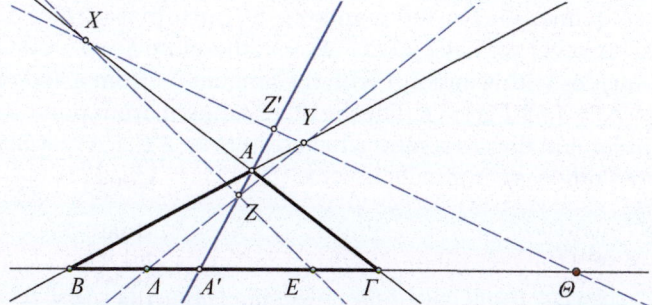

Fig. 5.92: Position determination from cross ratio

Theorem 5.32. *Given four points A, B, Γ, Δ on a circle κ, the pencil O(A,B,Γ,Δ) defined through a fifth point O of the circle, has a cross ratio independent of the position of O on the circle (See Figure 5.93-I).*

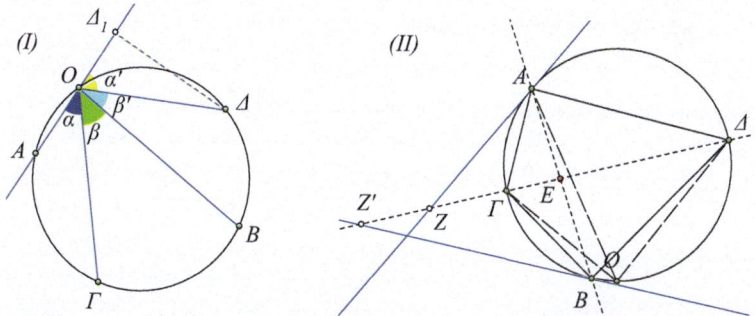

Fig. 5.93: Cross ratio of 4 circle points   cross ratio and tangent

Proof. Indeed, according to Proposition 5.14, the cross ratio of the pencil of the four lines *OA, OB, OΓ, OΔ* will be equal to

$$\frac{\sin(\alpha)}{\sin(\alpha')} : \frac{\sin(\beta)}{\sin(\beta')},$$

which, for fixed *A, B, Γ, Δ* on the circle, is independent of the position of *O* on the circle (Corollary 2.22).

Corollary 5.14. *Suppose that the cross ratio of four points of a circle is $(AB;\Gamma\Delta) = \kappa$. Suppose also that E is the point of intersection of the diagonals AB, ΓΔ and Z, Z' are the points of intersection of ΓΔ with the tangents at A and B respectively. Then $(ZE;\Gamma\Delta) = \kappa$ and $(Z'E;\Gamma\Delta) = \frac{1}{\kappa}$ (See Figure 5.93-II).*

Proof. Indeed, consider the point Z_1 for which holds $(Z_1E;\Gamma\Delta) = \kappa$ and suppose that AZ_1 intersects the circle at a second point *O*. Then the line pencil

$O(A,B,\Gamma,\Delta)$ defines on $\Gamma\Delta$ the points Z_1, E', such that $(Z_1 E'; \Gamma\Delta) = \kappa$. By hypothesis however we have $(Z_1 E; \Gamma\Delta) = \kappa$, therefore $E = E'$. Consequently point O coincides with A and $Z_1 A$ with the tangent ZA from A and we further have $(ZE; \Gamma\Delta) = (AB; \Gamma\Delta) = \kappa$. For the other tangent, from point B the argument is similar and the cross ratio which results is $(Z'E; \Gamma\Delta) = (BA; \Gamma\Delta)$. But the two cross ratios are mutually inverse $(BA; \Gamma\Delta) = \frac{\Gamma B}{\Gamma A} : \frac{\Delta B}{\Delta A} = (\frac{\Gamma A}{\Gamma B} : \frac{\Delta A}{\Delta B})^{-1} = (AB; \Gamma\Delta)^{-1}$,

Corollary 5.15. *Using the notation of the preceding corollary (See Figure 5.93-II), points Z, Z' coincide, if and only if the cross ratio is equal to $(AB; \Gamma\Delta) = -1$. For the corresponding quadrilateral then holds $|A\Gamma||B\Delta| = |B\Gamma||A\Delta|$. And conversely, if for the inscriptible quadrilateral the last relation holds, then $(AB; \Gamma\Delta) = -1$.*

Proof. The first claim is a direct consequence of the preceding corollary. For the second we compare the similar triangles $Z\Gamma A$ and $ZA\Delta$ (See Figure 5.93-II). We have $\frac{|A\Gamma|}{|A\Delta|} = \frac{|Z\Gamma|}{|ZA|}$ and $\frac{|B\Gamma|}{|B\Delta|} = \frac{|Z'\Gamma|}{|Z'B|}$ (*). If therefore points Z and Z' coincide, then we'll have equal tangents $|ZA| = |Z'B|$ and will hold $\frac{|A\Gamma|}{|A\Delta|} = \frac{|B\Gamma|}{|B\Delta|}$, which is the requested relation.

Conversely, if the claimed relation holds, then from (*) will follow $\frac{|Z\Gamma|}{|ZA|} = \frac{|Z'\Gamma|}{|Z'B|}$. It follows that for the squares of the tangents will hold

$$\frac{|ZA|^2}{|Z'B|^2} = \frac{|Z\Gamma|^2}{|Z'\Gamma|^2} \quad \Rightarrow \quad \frac{|Z\Gamma||Z\Delta|}{|Z'\Gamma||Z'\Delta|} = \frac{|Z\Gamma|^2}{|Z'\Gamma|^2} \quad \Rightarrow \quad \frac{|Z\Gamma|}{|Z'\Gamma|} = \frac{|Z\Delta|}{|Z'\Delta|}.$$

If points Z and Z' did not coincide, then, for the exterior points of the interval ZZ' and the two different points Γ and Δ, the preceding relation would mean that they have the same ratio relative to Z, Z', which is contradictory.

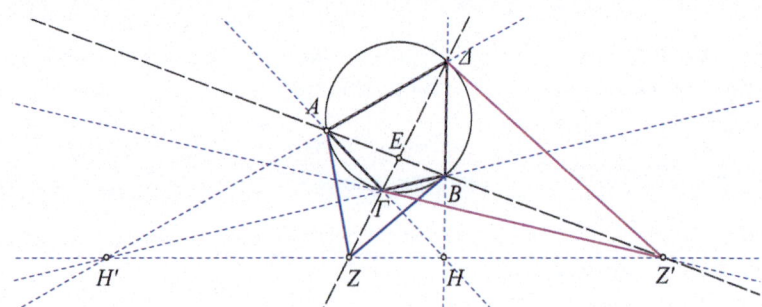

Fig. 5.94: Harmonic inscriptible quadrilateral

Inscriptible quadrilaterals which have the property of the preceding corollary are called **harmonic**. According to the corollary, these are characterized by the fact, that they have equal products of opposite sides, as well as the fact, that the tangents to their circumcircle at opposite vertices intersect on a diagonal (like HH' in the figure). This property gives also the general way of constructing a harmonic quadrilateral: From a point Z outside a circle κ we draw the tangents ZA, ZB (See Figure 5.94). Then, from an arbitrary point Γ of one of the arcs, defined by A, B, we draw a line $Z\Gamma$, which intersects again the circle at Δ. The quadrilateral $A\Gamma B\Delta$ is a harmonic one and every harmonic quadrilateral can be constructed in this way.

Remark 5.15. The intersection points H, H' of opposite sides, and the intersection points Z, Z' of the tangents at opposite vertices of an inscriptible quadrilateral (See Figure 5.94), are on a line, which coincides with the polar of the intersection point E of its diagonals (Exercise 5.128).

5.16 Theorems of Menelaus and Ceva

> Give me the best piano in Europe with an audience who understand nothing, desire to understand nothing and does not feel with me in what I play, and I would have no joy in it!
>
> W.A. Mozart, letter, May 1778

In this section we'll discuss the two main tools of detection of the coincidence: (i) of three points on a line, which is the theorem of Menelaus (70-140) and (ii) of three lines in a point (i.e. three lines passing through a point), which is the theorem of Ceva (1648-1734). In both cases the criteria rely on the *signed ratio* (signedratiosec).

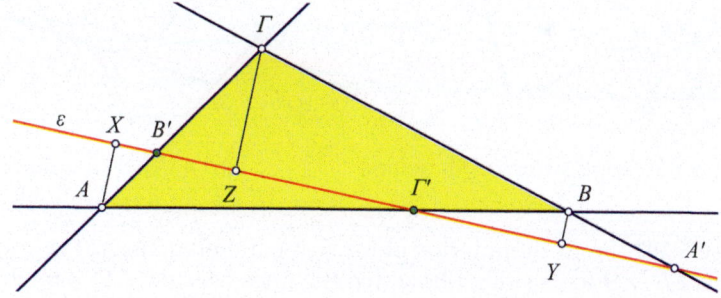

Fig. 5.95: Theorem of Menelaus

Theorem 5.33 (Menelaus). *A necessary and sufficent condition, so that the three points $\{A', B', \Gamma'\}$ on respective sides $\{B\Gamma, \Gamma A, AB\}$ of the triangle $AB\Gamma$, are on a line not containing some vertex of the triangle is the relation (See Figure 5.95)*

$$\frac{A'B}{A'\Gamma} \cdot \frac{B'\Gamma}{B'A} \cdot \frac{\Gamma'A}{\Gamma'B} = 1.$$

Proof. To begin with, observe that every line, which does not contain some triangle vertex, either intersects two sides of the triangle in their interior or none in its interior (Axiom 1.14). Therefore, from the three signed ratios which participate in the above relation, either two are negative or all are positive, hence the positive sign.

Now in the proof, we are careful to consider the right signs. Project the vertices A, B, Γ of the triangle onto the line ε, at points X, Y and Z respectively. This creates pairs of similar right triangles: $(A'BY, A'\Gamma Z)$, $(B'\Gamma Z, B'AX)$, $(\Gamma'AX, \Gamma'BY)$. The ratios of the sides of the similar triangles give:

$$\frac{A'B}{A'\Gamma} = \frac{|BY|}{|\Gamma Z|}, \quad \frac{B'\Gamma}{B'A} = -\frac{|\Gamma Z|}{|AX|}, \quad \frac{\Gamma'A}{\Gamma'B} = -\frac{|AX|}{|BY|}.$$

The claimed relation follows by multiplying the three preceding by parts and simplifying.

For the converse, let us suppose that the relation holds and that the line $A'B'$ intersects AB at Γ''. Then by the already proven part of the theorem will hold

$$\frac{A'B}{A'\Gamma} \cdot \frac{B'\Gamma}{B'A} \cdot \frac{\Gamma''A}{\Gamma''B} = 1, \quad \text{by hypothesis also} \quad \frac{A'B}{A'\Gamma} \cdot \frac{B'\Gamma}{B'A} \cdot \frac{\Gamma'A}{\Gamma'B} = 1.$$

From these two follows immediately $\frac{\Gamma''A}{\Gamma''B} = \frac{\Gamma'A}{\Gamma'B}$, which shows that Γ'' coincides with Γ' (Theorem 5.26).

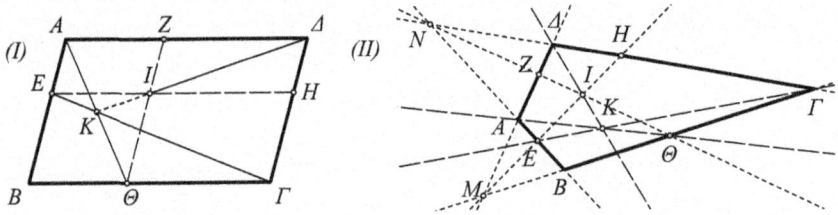

Fig. 5.96: Application of Menelaus ... and a generalization

Exercise 5.107. From the interior point I of parallelogram $AB\Gamma\Delta$ we draw parallels EH, $Z\Theta$ to its sides. Show that the lines $\{A\Theta, E\Gamma, I\Delta\}$ pass through a common point K (See Figure 5.96-I).

Hint: Suppose that K is the intersection point $K = (A\Theta, E\Gamma)$. It suffices to show that points $\{K, I, \Delta\}$ are collinear. This, by Menelaus, applied to the triangle $A\Theta Z$, is equivalent to showing $\frac{KA}{K\Theta} \cdot \frac{I\Theta}{IZ} \cdot \frac{\Delta Z}{\Delta A} = 1$. However, applying again the theorem of Menelaus, this time on $AB\Theta$ with the secant line $E\Gamma$, we have $\frac{KA}{K\Theta} \cdot \frac{EB}{EA} \cdot \frac{\Gamma\Theta}{\Gamma B} = 1$. Because the lines are parallel, it holds $\frac{I\Theta}{IZ} = \frac{EB}{EA}$ and $\frac{\Delta Z}{\Delta A} = \frac{\Gamma\Theta}{\Gamma B}$ (related exercise 3.111).

5.16. THEOREMS OF MENELAUS AND CEVA

Exercise 5.108. Generalize the preceding exercise as follows. Given the convex quadrilateral $AB\Gamma\Delta$ with opposite side intersections $M = (B\Gamma, A\Delta)$, and $N = (AB, \Gamma\Delta)$ draw two lines through these points intersecting the opposite sides respectively at the pairs of points $\{(E,H),(Z,\Theta)\}$ the lines themselves intersecting at $I = (EH, Z\Theta)$. Show that point $K = (A\Theta, E\Gamma)$ is on the line ΔI (See Figure 5.96-II).

Theorem 5.34 (Ceva). *A necessary and sufficent condition, so that the three points $\{A', B', \Gamma'\}$ on respective sides $\{B\Gamma, \Gamma A, AB\}$ of the triangle $AB\Gamma$, define three lines $\{AA', BB', \Gamma\Gamma'\}$ intersecting at the same point P not lying on the side-lines of $AB\Gamma$, is (See Figure 5.97)*

$$\frac{A'B}{A'\Gamma} \cdot \frac{B'\Gamma}{B'A} \cdot \frac{\Gamma'A}{\Gamma'B} = -1.$$

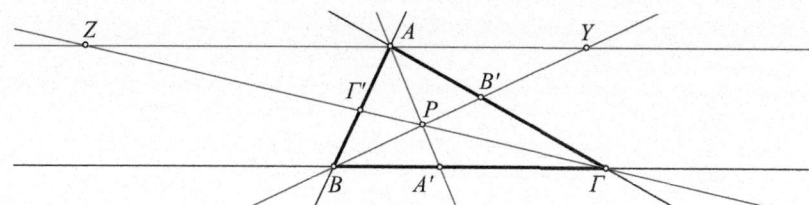

Fig. 5.97: Theorem of Ceva

Proof. Here again we first observe, that for every point P of the plane, not lying on the side-lines of the triangle, the intersections $\{A', B', \Gamma'\}$ of $\{AP, BP, \Gamma P\}$ with the respective opposite sides either all are contained in the interiors of the sides or exactly one of them is contained in the interior and all others are in the exterior, hence the sign -1.

Now to the proof, paying attention to the correct signs. Draw a parallel from one vertex, say from A to the base $B\Gamma$. This creates pairs of similar triangles: $(PB\Gamma, PYZ)$, $(B'B\Gamma, B'YA)$ and $(\Gamma'B\Gamma, \Gamma'AZ)$. From the side proportions of these similar triangles we have the equalities:

$$\frac{A'B}{A'\Gamma} = -\frac{|AY|}{|AZ|}, \quad \frac{B'\Gamma}{B'A} = -\frac{|B\Gamma|}{|AY|}, \quad \frac{\Gamma'A}{\Gamma'B} = -\frac{|AZ|}{|B\Gamma|}.$$

The claimed relation follows by multiplying these by parts and simplifying.

For the converse, suppose that the two lines AA', BB' intersect at point P and also suppose that Γ''' is the intersection point of ΓP with AB. Then, according to the proved part of the theorem, we'll have

$$\frac{A'B}{A'\Gamma} \cdot \frac{B'\Gamma}{B'A} \cdot \frac{\Gamma''A}{\Gamma''B} = -1, \quad \text{by assumption also} \quad \frac{A'B}{A'\Gamma} \cdot \frac{B'\Gamma}{B'A} \cdot \frac{\Gamma'A}{\Gamma'B} = -1.$$

From these two follows immediately $\frac{\Gamma''A}{\Gamma''B} = \frac{\Gamma'A}{\Gamma'B}$, which shows that Γ'' coincides with Γ' (Theorem 5.26).

Corollary 5.16. Let A', B' and Γ' be points on the sides of triangle ABC different from its vertices (See Figure 5.98). Then the relation of ratios of lengths:

$$\frac{|A'B|}{|A'\Gamma|} \cdot \frac{|B'\Gamma|}{|B'A|} \cdot \frac{|\Gamma'A|}{|\Gamma'B|} = 1$$

implies exactly one of the next two propositions:

1. points A', B' and Γ' are collinear,
2. the lines AA', BB', $\Gamma\Gamma'$ pass through a common point.
   Also, if Γ'' is the harmonic conjugate of Γ' relative to A, B, then,
3. when (1) occurs, the lines $\{AA', BB', \Gamma\Gamma''\}$, pass through a common point.
4. when (2) occurs, points $\{A', B', \Gamma'\}$, are collinear.

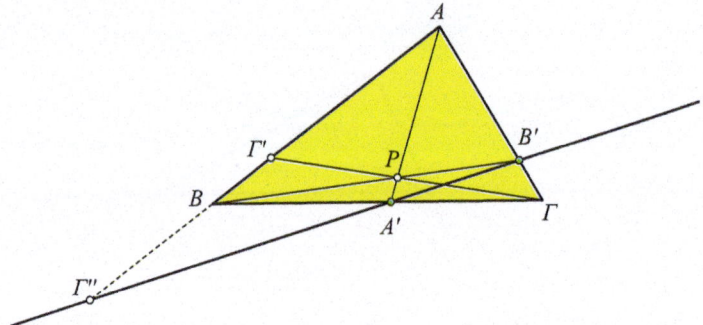

Fig. 5.98: Relation between theorems of Menelaus and Ceva

Proof. The proof follows directly from the preceding theorems. If the product of ratios of lengths is equal to 1, then the corresponding product of signed ratios will equal 1 or -1. In the first case we have (1). Because of the relation between harmonic conjugates Γ' and Γ'' relative to A, B

$$\frac{\Gamma'A}{\Gamma'B} = -\frac{\Gamma''A}{\Gamma''B},$$

(3) then will also hold. Similarly in the second case holds (2) and (4).

Note that similar properties hold also for the harmonic conjugates A'' of A' relative to B, Γ and B'' of B' relative to A and Γ.

Remark 5.16. The last corollary reveals that the two properties are intimately related. To see them in a unifying spirit and to include the symmetry implied in this relationship, we must, along with the three points $\{A', B', \Gamma'\}$ on the sides of triangle $AB\Gamma$, consider also their three harmonic conjugates $\{A'', B'', \Gamma''\}$ relative to the endpoints respectively on the sides $\{B\Gamma, \Gamma A, AB\}$. There results then the interesting figure 5.99, in which all the coincidences, besides those of $\{A'', B'', \Gamma''\}$ on a line, are consequences of the preceding propositions. The line which contains points $\{A'', B'', \Gamma''\}$ is called **trilinear polar** of P relative to triangle $AB\Gamma$ and point P is called **trilinear pole** of line $A''B''$ relative to the triangle $AB\Gamma$. These two notions occupy an important position in the so called *Geometry of the triangle* ([11, p.244], [40]). The fact, that points $\{A'', B'', \Gamma''\}$ are collinear, will be proved below (Corollary 5.18).

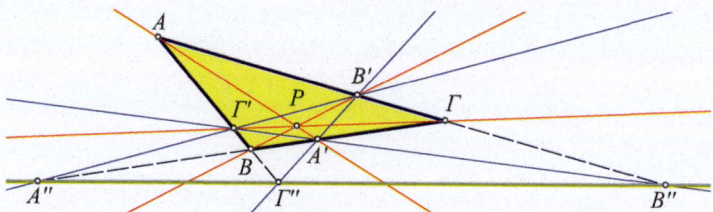

Fig. 5.99: Relation between theorems of Menelaus and Ceva II

Remark 5.17. Applying strictly our definitions, we can find the harmonic conjugate Γ'' relative to A, B for every point Γ' of the line AB, except for the three special points A, B and M, where M is the middle of AB. For the first two we accept that they are the only points which coincide with their conjugates. This is in agreement with the fact, that as X approaches point A or B from the exterior of AB, then the corresponding conjugate Y approaches the same point from the interior and vice-versa.

For the middle M, however, we are unable to make a correspondence with a conjugate point. We accept (define), therefore, an imaginary point on the line AB, which we name **point of the line at infinity** and we consider that this is, exactly, an *imaginary* point of the line, very far away at infinity, and consider this point as the harmonic conjugate of the middle M of AB. The point at infinity of the line AB fills the gap which is left by the conventional points of the line regarding the signed ratio. The ratio $\lambda = 1$ corresponds to no real point of the line AB. We consider therefore, that at the point at infinity corresponds the signed ratio $\lambda = 1$. Using these assumptions, we can also include in the theorem of Menelaus an old conclusion from the theorem of Thales. Drawing a parallel to the side $B\Gamma$ of the triangle $AB\Gamma$, ThalesFromMenelaos which intersects its sides AB, $A\Gamma$ respectively at

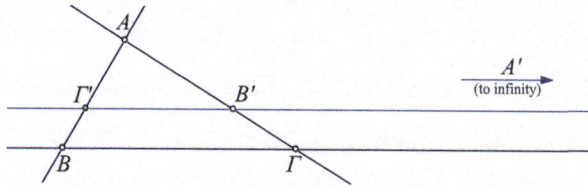

Fig. 5.100: Similarities relying on the theorem of Menelaus

points Γ' and B', we have according to Thales (See Figure 5.100):

$$\frac{\Gamma'A}{\Gamma'B} = \frac{B'A}{B'\Gamma} \Leftrightarrow \frac{\Gamma'A}{\Gamma'B} \cdot \frac{B'\Gamma}{B'A} = 1 \Leftrightarrow \frac{\Gamma'A}{\Gamma'B} \cdot \frac{B'\Gamma}{B'A} \cdot \frac{A'B}{A'\Gamma} = 1.$$

In the last equivalence we think exactly that $B'\Gamma'$ intersects $B\Gamma$ at its point at infinity A', for which (by assumption)

$$\frac{A'B}{A'\Gamma} = 1.$$

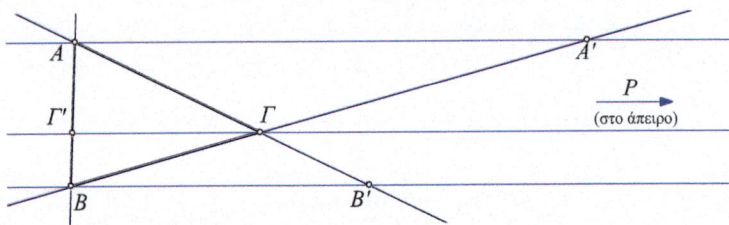

Fig. 5.101: Theorem of Ceva for parallels AA', BB', $\Gamma\Gamma'$

Remark 5.18. A similar to the preceding specialization holds also for the theorem of Ceva, in the case where P is a point at infinity, where the three lines AA', BB' and $\Gamma\Gamma'$ are three parallels, which we consider as concurring to this point (P at infinity)(See Figure 5.101). The theorem of Ceva follows, in this case, from that of Thales

$$\frac{A'B}{A'\Gamma} \cdot \frac{B'\Gamma}{B'A} \cdot \frac{\Gamma'A}{\Gamma'B} = -1.$$

Remark 5.19. This imaginary *point at infinity* of a line, whose existence we accept, unifies in some way, the relations that have to do with intersections of lines and removes the discrimination of lines betweem parallel and non-parallel. Two non-parallel lines intersect at their conventional point of intersection. Two parallel lines *intersect* at their point at infinity. We additionally consider that all these points at infinity form an *imaginary line*, which we call **line at infinity**. Every *conventional* line meets this imaginary line exactly at its point at infinity. With these assumptions, the theorems of Menelaus and Ceva hold for points $\{A', B', \Gamma'\}$ on the sides of the triangle, points, which may be conventional or even, some of them, may be points at infinity. Something similar holds, as we shall see, also for theorems in subsequent sections.

Exercise 5.109. Applying the theorem of Ceva, show that the medians of a triangle pass through the same point. Show in a similar way, that also the three internal bisectors pass through the same point. Finally, also that the three altitudes pass through a common point.

5.16. THEOREMS OF MENELAUS AND CEVA

Exercise 5.110. Let $\{\alpha(A, \rho_\alpha), \beta(B, \rho_\beta), \gamma(\Gamma, \rho_\Gamma)\}$ be three circles pairwise non-concentric and not congruent. Let also $\alpha\beta_1$, $\alpha\beta_2$ denote respectively the external and internal similarity centers of α and β. Then the following are triples of collinear points: $(\alpha\beta_1, \beta\gamma_1, \gamma\alpha_1)$, $(\alpha\beta_1, \beta\gamma_2, \gamma\alpha_2)$, $(\alpha\beta_2, \beta\gamma_1, \gamma\alpha_2)$, $(\alpha\beta_2, \beta\gamma_2, \gamma\alpha_1)$ (See Figure 5.102).

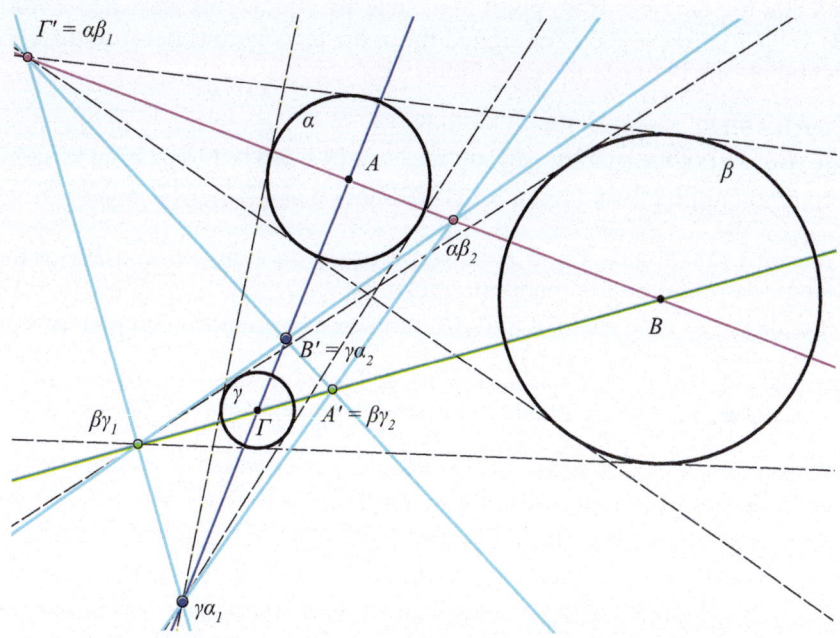

Fig. 5.102: Collinearities of similarity centers of three circles

Hint: We show that the points $\Gamma' = \alpha\beta_1$, $B' = \gamma\alpha_2$ and $A' = \beta\gamma_2$ are collinear, by applying Menelaus' theorem on the triangle $AB\Gamma$ of the centers of the circles:

$$\frac{\Gamma'A}{\Gamma'B} = \frac{\rho_\alpha}{\rho_\beta}, \quad \frac{A'B}{A'\Gamma} = -\frac{\rho_\beta}{\rho_\gamma}, \quad \frac{B'\Gamma}{B'A} = -\frac{\rho_\gamma}{\rho_\alpha} \Rightarrow \frac{\Gamma'A}{\Gamma'B} \cdot \frac{A'B}{A'\Gamma} \cdot \frac{B'\Gamma}{B'A} = 1.$$

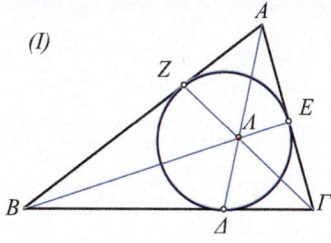

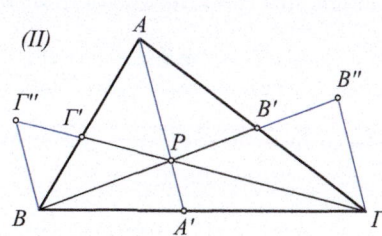

Fig. 5.103: Gergonne point   Ratios of segments

Exercise 5.111. (Point of Gergonne (1771-1859)) Show that the lines which join the vertices of a triangle with the opposite contact points of the incircle pass through a common point Λ (called *Gergonne point* of the triangle).

Hint: Apply the theorem of Ceva for the contact points $\{\Delta, E, Z\}$ of the incircle with the sides (See Figure 5.103-I). $\frac{AB}{A\Gamma} \frac{E\Gamma}{EA} \frac{Z A}{ZB} = -1$. This results from the fact that the tangents from point to a circle are equal. This way $|B\Delta| = |BZ|$, $|\Delta\Gamma| = |\Gamma E|$, $|AE| = |AZ|$. The signed ratios are all negative, because the contact points are between the endpoints of the sides.

Exercise 5.112. Suppose that the points $\{A', B', \Gamma'\}$ respectively on the sides $\{B\Gamma, \Gamma A, AB\}$ of the triangle $AB\Gamma$ define lines $\{AA', BB', \Gamma\Gamma'\}$ passing through a common point P (See Figure 5.103-II). Show that $\frac{|PA|}{|PA'|} = \frac{|\Gamma'A|}{|\Gamma'B|} + \frac{|B'A|}{|B'\Gamma|}$.

Exercise 5.113. Points Δ and E lie respectively on sides AB and $B\Gamma$ of the triangle $AB\Gamma$ and divide them into ratios $\frac{\Delta A}{\Delta B} = -2$, $\frac{EB}{E\Gamma} = -\frac{1}{2}$. Show that ΔE is parallel to $A\Gamma$ and that $\Gamma\Delta$ and AE intersect at a point Z, such that $\frac{AZ}{ZE} = 3$.

Exercise 5.114. Points Δ and E are respectively on the sides AB and $B\Gamma$ of the triangle $AB\Gamma$ and divide them into ratios $\frac{\Delta A}{\Delta B} = \lambda$, $\frac{EB}{E\Gamma} = \sigma$. Show that $\Gamma\Delta$ and AE intersect at point Z such that $\frac{ZA}{ZE} = \lambda \cdot (1-\sigma)$.

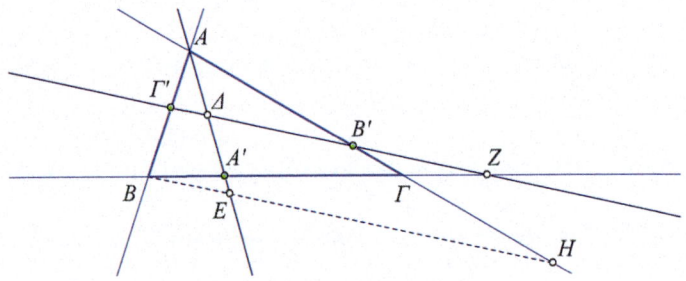

Fig. 5.104: Calculation of segment ratios

Exercise 5.115. Points $\{A', B', \Gamma'\}$ are respectively on sides $\{B\Gamma, \Gamma A, AB\}$ of the triangle $AB\Gamma$ and divide them respectively into ratios (See Figure 5.104)

$$\frac{A'B}{A'\Gamma} = \kappa, \quad \frac{B'\Gamma}{B'A} = \lambda, \quad \frac{\Gamma'A}{\Gamma'B} = \mu.$$

Line AA' intersects $B'\Gamma'$ at Δ. Show that Δ divides the respective segments into the following ratios:

$$\frac{\Delta\Gamma'}{\Delta B'} = \frac{\kappa\mu(1-\lambda)}{\mu - 1}, \quad \frac{\Delta A}{\Delta A'} = \frac{\mu(\kappa - 1)}{\kappa\lambda\mu - 1}.$$

5.16. THEOREMS OF MENELAUS AND CEVA

Hint: Draw the parallel from B to $B'\Gamma'$ which intersects $A\Gamma$ at H. Calculate first the ratio $\frac{A\Gamma}{AH}$ and next the ratio $\frac{A\Gamma'}{AB'} = \frac{EB}{EH}$, where E is the point at which AA' intersects BH.

$$\frac{B'A}{B'H} = \mu \Rightarrow B'H = \frac{1}{\mu}B'A = \frac{1}{\mu(1-\lambda)}\Gamma A \quad \text{(Exer. 5.102)}$$

$$HA = HB' + B'A = -\frac{1}{\mu(1-\lambda)}\Gamma A + \frac{1}{1-\lambda}\Gamma A = \frac{\mu - 1}{\mu(1-\lambda)}\Gamma A.$$

$$H\Gamma = HA + A\Gamma = \frac{\mu\lambda - 1}{\mu(1-\lambda)}\Gamma A \Rightarrow \frac{H\Gamma}{HA} = \frac{\mu\lambda - 1}{\mu - 1}.$$

The ratio $\frac{EB}{EH}$ is calculated from the theorem of Menelaus applied to the triangle $B\Gamma H$ with secant AA':

$$\frac{EB}{EH}\cdot\frac{AH}{A\Gamma}\cdot\frac{A'\Gamma}{A'B} = 1 \Rightarrow \frac{EB}{EH} = \frac{A\Gamma}{AH}\cdot\frac{A'B}{A'\Gamma} = \frac{\kappa\mu(1-\lambda)}{\mu - 1}.$$

From the theorem of Menelaus we also find the ratios

$$\frac{ZB}{Z\Gamma} = \frac{1}{\mu\lambda} \Rightarrow ZB = \frac{1}{\mu\lambda - 1}B\Gamma, \quad \frac{ZB}{ZA'} = \frac{\kappa - 1}{\kappa\lambda\mu - 1}.$$

Finally, one more application of the theorem of Menelaus on the triangle ABA' with secant $\Gamma'B'$ gives:

$$\frac{\Gamma'A}{\Gamma'B}\cdot\frac{ZB}{ZA'}\cdot\frac{\Delta A'}{\Delta A} = 1 \Rightarrow \frac{\Delta A}{\Delta A'} = \frac{\Gamma'A}{\Gamma'B}\cdot\frac{ZB}{ZA'} = \frac{\mu(\kappa - 1)}{\kappa\lambda\mu - 1}.$$

Exercise 5.116. On the sides AB and $A\Gamma$ of the triangle $AB\Gamma$ we consider respectively points Γ' and B', such that $|AB'| = |A\Gamma'|$. Show that the median AA' intersects the line segment $B'\Gamma'$ at Δ, in such a way as to have $\frac{|A\Gamma'|}{|AB'|} = \frac{|A\Gamma|}{|AB|}$.

Exercise 5.117. Show that for every point O, not lying on the side-lines of the triangle $AB\Gamma$, and the intersection points $\{A', B', \Gamma'\}$ respectively of $\{OA, OB, O\Gamma\}$ with $\{B\Gamma, \Gamma A, AB\}$, holds (compare with Exercise 3.84):

$$\frac{OA'}{AA'} + \frac{OB'}{BB'} + \frac{O\Gamma'}{\Gamma\Gamma'} = 1, \quad \frac{AO}{AA'} + \frac{BO}{BB'} + \frac{\Gamma O}{\Gamma\Gamma'} = 2.$$

Exercise 5.118. Given is a triangle $AB\Gamma$ and two points Δ, E. Point Z moves onto $B\Gamma$ and the lines $\Delta Z, EZ$ intersect $AB, A\Gamma$ respectively at points I and Θ. Show that the intersection point H of $\Delta\Theta, EI$ moves on a fixed line $B'\Gamma'$.

Hint: Apply twice Menelaus' theorem to $AB\Gamma$, for the secant lines $Z\Delta$ and $Z\Theta$ (See Figure 5.105). The following equalities result

$$\frac{IA}{IB} \cdot \frac{ZB}{Z\Gamma} \cdot \frac{K\Gamma}{KA} = 1, \quad \frac{ZB}{Z\Gamma} \cdot \frac{\Theta\Gamma}{\Theta A} \cdot \frac{\Lambda A}{\Lambda B} = 1 \quad \Rightarrow \quad \frac{IA}{IB} \cdot \frac{K\Gamma}{KA} = \frac{\Theta\Gamma}{\Theta A} \cdot \frac{\Lambda A}{\Lambda B}.$$

The last equality is written equivalently

$$(AB; I\Lambda) = \frac{IA}{IB} : \frac{\Lambda A}{\Lambda B} = \frac{KA}{K\Gamma} : \frac{\Theta A}{\Theta\Gamma} = (A\Gamma; K\Theta).$$

However, according to Theorem 5.29

$$(AB; I\Lambda) = (AB'; I'\Theta) \quad \text{and} \quad (A\Gamma; K\Theta) = (A\Gamma'; I\Theta') \quad \Rightarrow \quad (AB'; I'\Theta) = (A\Gamma'; I\Theta').$$

The last equality has as a consequence (Theorem 5.31) the concurrence of the lines $\{B'\Gamma', II', \Theta\Theta'\}$ at a point.

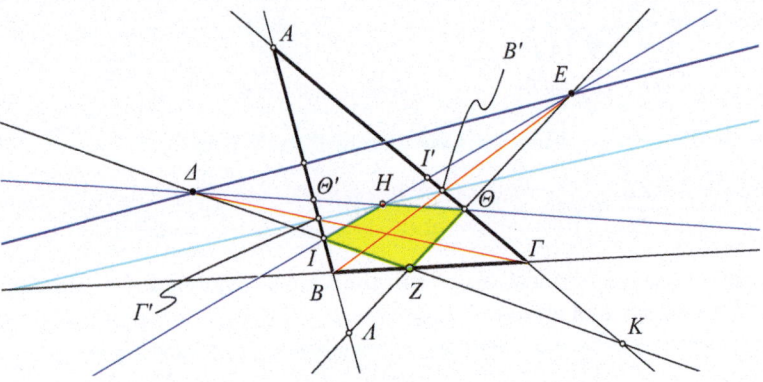

Fig. 5.105: Application of the cross ratio

Theorem 5.35. *Given is a triangle $AB\Gamma$ and a point P not lying on its side-lines. Suppose that the lines $PA, PB, P\Gamma$ intersect the opposite sides $B\Gamma, \Gamma A, AB$ at points A', B', Γ' (See Figure 5.106-I). Then, the symmetric points, respectively, A'', B'', Γ'' of A', B', Γ' relative to the middles of the sides, define three lines $AA'', BB'', \Gamma\Gamma''$, which also pass through a common point P'.*

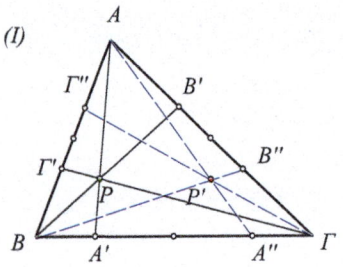
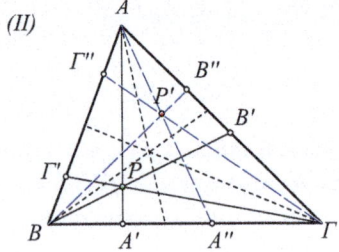

Fig. 5.106: Isotomic conjugacy   Isogonal conjugacy

5.16. THEOREMS OF MENELAUS AND CEVA

Proof. If the ratios $k_1 = \frac{A'B}{A'\Gamma}$, $k_2 = \frac{B'\Gamma}{B'A}$, $k_3 = \frac{\Gamma'A}{\Gamma'B}$ satisfy the equation of Ceva, then the ratios $\frac{1}{k_1} = \frac{A''B}{A''\Gamma}$, $\frac{1}{k_2} = \frac{B''\Gamma}{B''A}$, $\frac{1}{k_3} = \frac{\Gamma''A}{\Gamma''B}$ will also satisfy it.

Theorem 5.36. *Given is a triangle $AB\Gamma$ and a point P not lying on its side-lines. Suppose that the lines $PA, PB, P\Gamma$ intersect the opposite sides $B\Gamma, \Gamma A, AB$ at points A', B', Γ' (See Figure 5.106-II). Then, the symmetric lines $AA'', BB'', \Gamma\Gamma''$ of AA', $BB', \Gamma\Gamma'$, respectively, relative to the bisectors of angles $\{\widehat{A}, \widehat{B}, \widehat{\Gamma}\}$ pass also through a common point P'.*

Proof. According to exercise 3.79, if $\kappa = \frac{x}{y}$ is the ratio of distances of A' from the sides $AB, A\Gamma$, then the ratio of distances $\kappa' = \frac{x'}{y'}$ of A'' from the same sides will be $\kappa' = \frac{1}{\kappa}$. However, the ratios which participate in the equation of Ceva for the point P are expressed as functions of κ: $\mp\frac{A'B}{A'\Gamma} = \frac{x/\sin(\beta)}{y/\sin(\gamma)} = \kappa\frac{\sin(\gamma)}{\sin(\beta)}$, with analogous formulas for B' and Γ'. In these we get a negative sign when A' is inside the interval $B\Gamma$ and similarly the signs for B', Γ'. The equation of Ceva for P then takes the form

$$\kappa\frac{\sin(\gamma)}{\sin(\beta)} \cdot \lambda\frac{\sin(\alpha)}{\sin(\gamma)} \cdot \mu\frac{\sin(\beta)}{\sin(\alpha)} = 1 \quad \Leftrightarrow \quad \kappa \cdot \lambda \cdot \mu = 1.$$

Obviously then will also hold $\kappa' \cdot \lambda' \cdot \mu' = 1$, from which follows the equation of Ceva for point P'.

The points $\{P, P'\}$, defined by the last two theorems are called, respectively, **isotomic conjugates** and **isogonal conjugates** relative to the triangle $AB\Gamma$. The correspondence $P \mapsto P'$ is called, correspondingly, **isotomic conjugacy** and **isogonal conjugacy** with respect to the triangle $AB\Gamma$. The term *conjugacy* underlines the symmetry of the relation: *If P' is the isotomic/isogonal conjugate of P, then P is also the isotomic/isogonal conjugate of P'.*

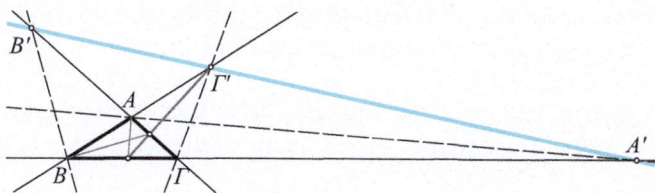

Fig. 5.107: $\{A', B', \Gamma'\}$: intersections of external bisectors with opposite sides

Exercise 5.119. Show that the external bisectors of the triangle $AB\Gamma$ intersect correspondingly the opposite sides of the triangle at three collinear points $\{A', B', \Gamma'\}$ (See Figure 5.107). Show the analogous property for two internal and one external bisector of the triangle.

Exercise 5.120. Consider three lines $\{\varepsilon_1, \varepsilon_2, \varepsilon_3\}$ passing through a point P, as well as three circles $\{\gamma_1, \gamma_2, \gamma_3\}$, tangent, each, to two of these lines (See Figure 5.108). Show that the other internal tangents $\{\varepsilon'_1, \varepsilon'_2, \varepsilon'_3\}$ of these circles pass also through a common point P'.

Hint: Consider the triangle $A'B'\Gamma'$ of the intersection points of the inner tangents ($A' = (\varepsilon_1, \varepsilon'_1)$, $B' = (\varepsilon_2, \varepsilon'_2)$, $\Gamma' = (\varepsilon_3, \varepsilon'_3)$). This triangle has for external bisectors the lines of centers of the circles. Hence the lines $\{\varepsilon'_1, \varepsilon_1\}$ are symmetric with respect to the bisector at A' of this triangle. The analogous result is valid for the pair of lines $\{\varepsilon'_2, \varepsilon_2\}$ and the pair $\{\varepsilon'_3, \varepsilon_3\}$. Hence, point P' is the isogonal conjugate of P with respect to the triangle $A'B'\Gamma'$.

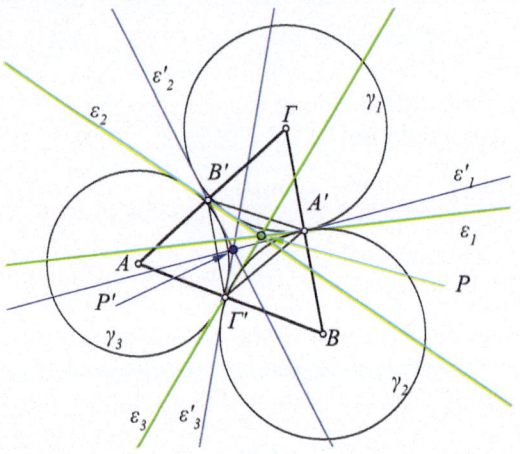

Fig. 5.108: A pair of isogonal points $\{P, P'\}$ w.r.t. $A'B'\Gamma'$

Exercise 5.121. Show that the centroid of the triangle $AB\Gamma$ is a fixed point of the isotomic conjugacy, i.e. it is identical with its isotomic conjugate. Show that the incenter of the triangle is a fixed point of the isogonal conjugacy, i.e. it is identical with its isogonal conjugate. Show finally that the symmedian point of the triangle is the isogonal conjugate of the centroid.

Exercise 5.122. Show that the isogonal conjugate of the orthocenter H of the triangle $AB\Gamma$ is the circumcenter of the triangle. Show also that the isotomic conjugate of the orthocenter is the symmedian point of the anticomplmentary triangle of $AB\Gamma$.

5.17 The complete quadrilateral

> Next above these [the straight line, the triangle, quadrilateral, and the pentagon] come the Nobility, of whom there are several degrees, beginning at six-sided Figures, or Hexagons, and from thence rising in the number of their sides till they receive the honorable title of Polygonal, or many-sided.
>
> E.A. Abbott, Flatland: A Romance of Many Dimensions

Complete quadrilateral is called the figure defined by four lines $\{\alpha, \beta, \gamma, \delta\}$ in general position. The lines are called **sides** of the complete quadrilateral. Each pair of sides defines a **vertex** of the complete quadrilateral ($A, B, \Gamma, \Delta, E, Z$ in figure 5.109). Two vertices not belonging to the same side are called

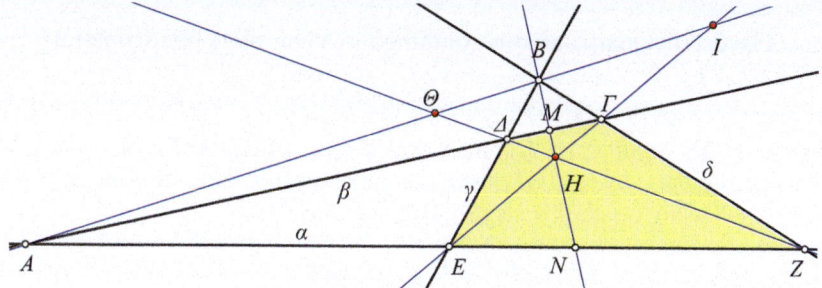

Fig. 5.109: Complete quadrilateral

opposite vertices of the complete quadrilateral. Each pair of opposite vertices defines a **diagonal** of the complete quadrilateral ($\{AB, \Delta Z, \Gamma E\}$ in the figure) ([12, p.231] , [128], [122]).

Remark 5.20. Notice that Exercise 5.110 shows that, the six centers of similarity of three circles, are vertices of a complete quadrilateral, whose diagonals coincide with the center-lines of the three circles.

Theorem 5.37. *For each vertex (B) of the complete quadrilateral, the two sides ($B\Gamma$, $B\Delta$), the diagonal (BA), which passes through it and the line (BH), which joins the vertex with the intersection of the two other diagonals make a harmonic pencil.*

Proof. For the vertex B the proof follows from Corollary 5.16, applied to the triangle BEZ (See Figure 5.109). According to this, point A is the harmonic conjugate of N relative to E, Z. Consequently (Theorem 5.29) the pencil of lines $B(A, N, E, Z)$ is harmonic. The proofs for the other vertices are similar.

Corollary 5.17. *All quadruples of collinear points in the figure 5.109 are harmonic.*

Proof. These quadruples are

$$(EZ; AN) = (\Delta\Gamma; AM) = (\Delta Z; \Theta H) = (E\Gamma; IH) = (MN; HB) = -1.$$

The conclusion follows from the preceding proposition and Theorem 5.29.

Exercise 5.123. Given is the complete quadrilateral $AB\Gamma\Delta EZ$. Find all the quadrilaterals of which the sides are contained in those of the complete quadrilateral (See Figure 5.110).

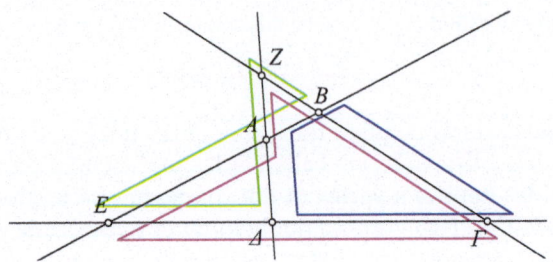

Fig. 5.110: Quadrilaterals contained in a complete quadrilateral

Theorem 5.38. *From vertex E of the complete quadrilateral $AB\Gamma\Delta\Theta E$ draw a line ε intersecting the sides ΘA, $\Theta \Gamma$, which do not contain point E, at points Z and H. Then the following equality holds (See Figure 5.111)*

$$\frac{ZA}{ZB} \cdot \frac{H\Gamma}{H\Delta} = \frac{EA}{E\Delta} \cdot \frac{E\Gamma}{EB}$$

and the products of ratios are independent of the position of line ε through E. Also holds

$$\frac{\Theta A}{\Theta B} \cdot \frac{\Theta \Gamma}{\Theta \Delta} = \frac{EA}{E\Delta} \cdot \frac{E\Gamma}{EB}.$$

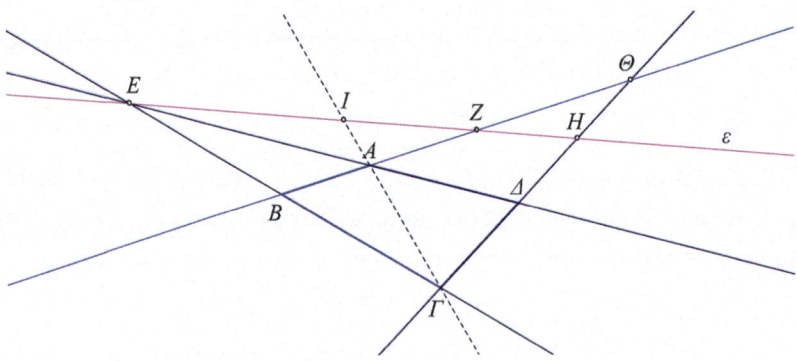

Fig. 5.111: Property of the complete quadrilateral

Proof. Draw the diagonal $A\Gamma$, which intersects line ε at the point I, and apply Menelaus' theorem on triangles $AB\Gamma$ and $A\Gamma\Delta$ for the secant line ε. This gives respectively

5.17. THE COMPLETE QUADRILATERAL

$$\frac{IA}{I\Gamma} \cdot \frac{E\Gamma}{EB} \cdot \frac{ZB}{ZA} = 1, \quad \frac{I\Gamma}{IA} \cdot \frac{EA}{E\Delta} \cdot \frac{H\Delta}{H\Gamma} = 1.$$

Multiplying the two relations, gives an equation equivalent to the first of the claimed. The second conclusion follows directly, since the right side of the first equation is independent of ε. The last conclusion follows when ε takes the position of $E\Theta$, in which case points Z and H coincide with Θ.

Exercise 5.124. Let $\{E, Z\}$ be opposite vertices of a complete quadrilateral and X be an arbitrary point, not lying on any of its sides. Let also N, Λ and K, M be the intersection points of EX, ZX respectively with the sides of the quadrilateral. Show that the lines MN and $K\Lambda$ intersect on the diagonal $B\Delta$.

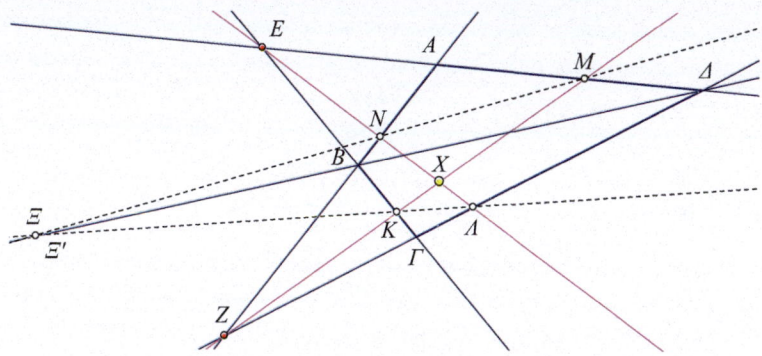

Fig. 5.112: Property of diagonals of the complete quadrilateral

Hint: (See Figure 5.112) Suppose that line MN intersects the diagonal at Ξ and line $K\Lambda$ at Ξ'. Apply Menelaus' theorem to triangles $AB\Delta$ and $\Delta\Gamma B$ with MN and $K\Lambda$ as intersecting lines:

$$\frac{\Xi B}{\Xi \Delta} \cdot \frac{M\Delta}{MA} \cdot \frac{NA}{NB} = 1 \Rightarrow \frac{\Xi B}{\Xi \Delta} = \frac{MA}{M\Delta} \cdot \frac{NB}{NA}, \quad \frac{\Xi'\Delta}{\Xi'B} \cdot \frac{KB}{K\Gamma} \cdot \frac{\Lambda\Gamma}{\Lambda\Delta} = 1 \Rightarrow \frac{\Xi'B}{\Xi'\Delta} = \frac{KB}{K\Gamma} \cdot \frac{\Lambda\Gamma}{\Lambda\Delta}.$$

According to Theorem 5.38 holds

$$\frac{EB}{E\Gamma} \cdot \frac{E\Delta}{EA} = \frac{ZB}{ZA} \cdot \frac{Z\Delta}{Z\Gamma} \Rightarrow \frac{E\Delta}{EA} \cdot \frac{ZA}{ZB} = \frac{Z\Delta}{Z\Gamma} \cdot \frac{E\Gamma}{EB}.$$

Multiplying the preceding by parts, we get respectively the expressions

$$\left(\frac{MA}{M\Delta} \cdot \frac{NB}{NA}\right)\left(\frac{E\Delta}{EA} \cdot \frac{ZA}{ZB}\right), \quad \left(\frac{KB}{K\Gamma} \cdot \frac{\Lambda\Gamma}{\Lambda\Delta}\right)\left(\frac{Z\Delta}{Z\Gamma} \cdot \frac{E\Gamma}{EB}\right),$$

which are respectively equal to the products of cross ratios

$$(A\Delta; ME)(AB; ZN), \quad (B\Gamma; KE)(\Delta\Gamma; Z\Lambda).$$

According to Theorem 5.29, the first factors in the two products are equal and the second factors are also equal. Therefore, the products are equal and, consequently, also $\frac{\Xi B}{\Xi \Delta}$, $\frac{\Xi' B}{\Xi' \Delta}$ will be equal, therefore Ξ and Ξ' will coincide (The exercise is a generalization of Exercise 3.111).

Exercise 5.125. Let $AB\Gamma\Delta$ be a convex quadrilateral with $\frac{EA}{E\Delta} \cdot \frac{E\Gamma}{EB} = \frac{\Theta A}{\Theta B} \cdot \frac{\Theta \Gamma}{\Theta \Delta} = \lambda^2$, where $E = (A\Delta, B\Gamma)$, $\Theta = (AB, \Gamma\Delta)$. Define respectively on its sides the points Z, H, K, I, such that $\frac{ZA}{Z\Delta} = \frac{H\Gamma}{H\Delta} = \frac{IA}{IB} = \frac{K\Gamma}{KB} = -\lambda$. Show that $ZHKI$ is a parallelogram, its sides are parallel to the diagonals of $AB\Gamma\Delta$ and its diagonals pass through points E, Θ respectively (See Figure 5.113).

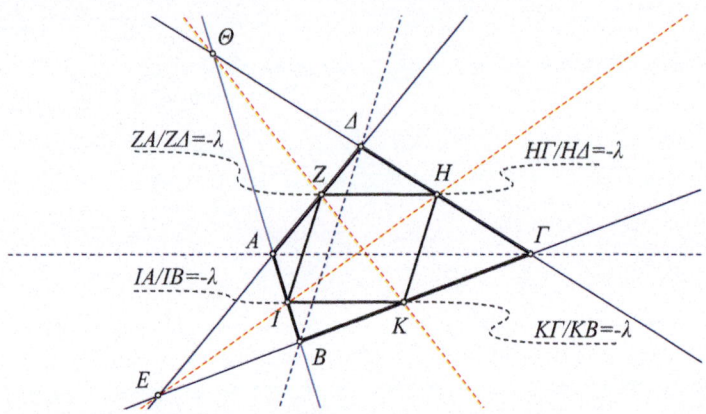

Fig. 5.113: Special parallelogram inscribed in a complete quadrilateral

Hint: The exercise relies on Theorem 5.38. From the equality of ratios, follows that ZH is parallel to the diagonal $A\Gamma$. Similarly also the other sides are parallel to the diagonals of $AB\Gamma\Delta$. The fact, that the diagonals of the parallelogram pass through points E, Θ, follows again from the aforementioned theorem. For example, suppose that ΘZ intersects $B\Gamma$ at K'. Then $\frac{ZA}{Z\Delta} \cdot \frac{K'\Gamma}{K'B} = \lambda^2$, which, because of $\frac{ZA}{Z\Delta} = -\lambda$, implies that $\frac{K'\Gamma}{K'B} = -\lambda$, therefore K' coincides with K.

Exercise 5.126. Continuing the preceding exercise, consider the harmonic conjugates $Z' = H(\Delta, \Gamma)$, $H' = Z(\Delta, A)$, $K' = I(B, A)$, $I' = K(B, \Gamma)$. Show that $Z'H'K'I'$ is a homothetic parallelogram of $ZHKI$ with homothety center at the middle M of $A\Gamma$ (See Figure 5.114).

Hint: The pencil $H'(\Delta, \Gamma, H, Z')$ is harmonic and line $A\Gamma$ is parallel to $H'Z'$. This shows, that line HH' passes through M. Similar properties are applied also for the other lines, which join the vertices ZZ', KK', II' and prove that these also pass through M.

5.17. THE COMPLETE QUADRILATERAL

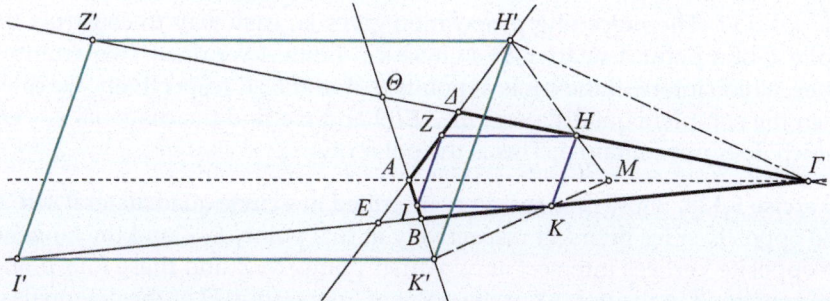

Fig. 5.114: Special parallelogram inscribed in a complete quadrilateral II

Proposition 5.17. *For every complete quadrilateral $B\Gamma\Delta EI\Theta$, whose four vertices are contained in a circle κ, the diagonal which doesn't pass through these four vertices is the polar of the intersection point of the two other diagonals (See Figure 5.115).*

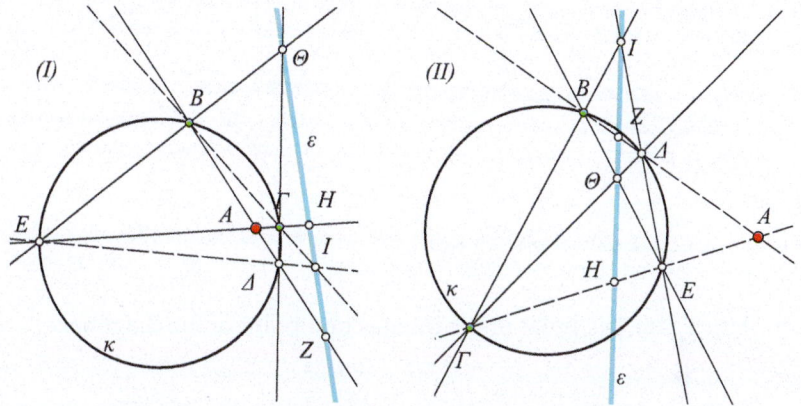

Fig. 5.115: Construction of the polar of a point

Proof. Suppose that the vertices B, Γ, Δ, E are contained in the circle κ. Then the diagonal $\varepsilon = \Theta I$ is the one which does not passes through the four vertices (Figure 5.115-I, the quadrilateral $B\Gamma\Delta E$ in Figure 5.115-II is non-convex). Suppose A is the intersection point of the other two diagonals ΓE and $B\Delta$. If Z, H are the intersection points of $B\Delta$ and ΓE with ε, then, according to Corollary 5.17, (B,Δ,A,Z) and (Γ,E,A,H) will be harmonic quadruples, therefore points H and Z will be points of the polar of A (Theorem 4.17).

Exercise 5.127. Show that in an inscriptible in a circle quadrilateral $AB\Gamma\Delta$ (See Figure 5.115-I), the bisectors of the angles $\{\widehat{E\Theta\Gamma}, \widehat{BIE}\}$, formed by the pairs of opposite sides intersect orthogonally.

Remark 5.21. The preceding proposition gives an easy way to construct the polar ε of a point A relative to a circle κ : From A we draw two arbitrary lines, which intersect the circle at points B, Δ and Γ, E respectively, we draw then the complete quadrilateral $B\Gamma\Delta EI\Theta$ and we find the diagonal $\varepsilon = \Theta I$, which does not contain A. This is the polar of A.

Exercise 5.128. Show that, for every inscribed in a circle quadrilateral $AB\Gamma\Delta$, the opposite sides intersect respectively at two points E, Z and the tangents to opposite vertices intersect at two other points N, Ξ and these four points are contained in the polar ε of the point of intersection Θ of the diagonals of the quadrilateral. Moreover points N, Ξ are harmonic conjugate relative to E, Z (See Figure 5.116-I).

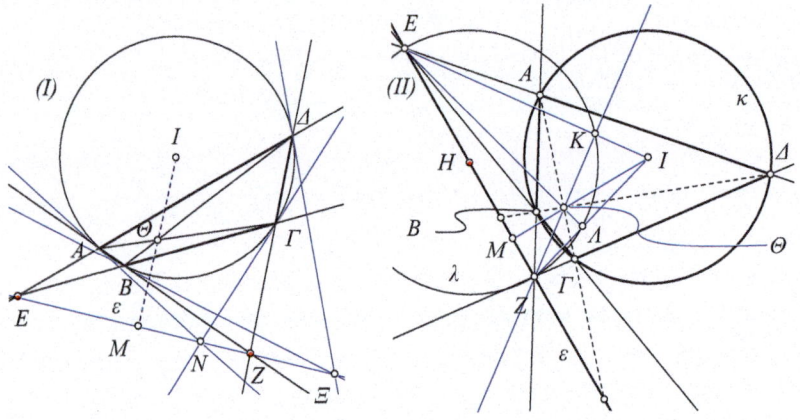

Fig. 5.116: The polar of the intersection point of the diagonals

Hint: According to the preceding proposition, the line $\varepsilon = EZ$ is the polar of the point Θ. Because the line $A\Gamma$ passes through Θ and is the polar of the point Ξ, point Ξ will also be contained in the polar ε of Θ. Similarly show that point N is contained in ε. For the last claim, find in the figure a complete quadrilateral whose ε is a diagonal and apply Corollary 5.17.

Theorem 5.39. *Let E, Z be the intersection points of the opposite sides of a convex quadrilateral $AB\Gamma\Delta$ inscribed in a circle κ. Then the circle λ, with diameter EZ, is orthogonal to the circumcircle κ of the quadrilateral (See Figure 5.116-II).*

Proof. Applying Proposition 5.17 to the quadrilateral $AB\Gamma\Delta$, we see that lines $\{Z\Theta, E\Theta, EZ\}$ are respectively the polars of points $\{E, Z, \Theta\}$. The line, which joins the center I with point E, is orthogonal to the polar of E, the triangle EZK is right and points E, K are inverse relative to κ (Corollary 4.22). According to Theorem 4.13 the circle λ, passing through the inverse points E, K, will be orthogonal to κ.

5.17. THE COMPLETE QUADRILATERAL

Exercise 5.129. In figure 5.116-II of the preceding theorem, show that each side of the triangle $EZ\Theta$ is the polar relative to κ of its opposite vertex. Show also that the circles with diameter the sides of this triangle are orthogonal to κ and the angle at Θ is obtuse.

Triangles like $EZ\Theta$ of the preceding exercise, for which there exists a circle, such that each of their sides is the polar relative to the opposite vertex, are called **autopolar** or **self-polar**.

Exercise 5.130. Show that, if the triangle $AB\Gamma$ is autopolar with respect to the circle κ, then it is obtuse, the center of κ coincides with the orthocenter of $AB\Gamma$, and κ coincides with the polar circle of the triangle (Exercise 4.116).

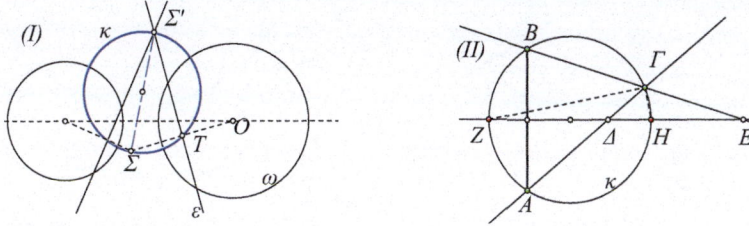

Fig. 5.117: Polars relative to pencil   Harmonic conjugates on diameter

Exercise 5.131. Show that the polar lines of a point Σ relative to the circles ω of one pencil of circles $\mathscr{D}$ pass through a point Σ'. Also show that Σ' is the diametrically opposite of Σ on the circle-member $\kappa(\Sigma)$, which passes through point Σ, of the circle pencil $\mathscr{D}'$, which is orthogonal to $\mathscr{D}$ (See Figure 5.117-I).

Exercise 5.132. Show that the lines, which join an arbitrary point of the circle with the endpoints of one of its chords intersect the diameter which is orthogonal to the chord at points which are harmonic conjugate relative to the endpoints of the diameter (See Figure 5.117-II).

Exercise 5.133. Show that the circles with diameters the diagonals $AE, BZ, \Gamma\Delta$ of a complete quadrilateral belong to a pencil of circles.

Hint: (See Figure 5.118-I) This follows immediately from exercise 5.74, applied to a triangle made by three, out of the four lines, and secant the fourth line. In figure 5.118-I these roles are played correspondingly by the triangle $AB\Gamma$ and the line ΔE.

The conclusion of the exercise is often referred to as the *theorem of Gauss-Bodenmiller* ([102, p.172]). Notice that this gives another proof of the fact, that the middles of the diagonals of a complete quadrilateral are on a line (the Newton line of the quadrilateral, see exercises 2.73 and 3.14).

Exercise 5.134. Given are two points A and B not lying on the circle κ and a chord ΔE of κ passing through B. If Z, H are the second intersection points of AE and AΔ respectively with κ, show that the chord ZH passes through a fixed point (See Figure 5.118-II).

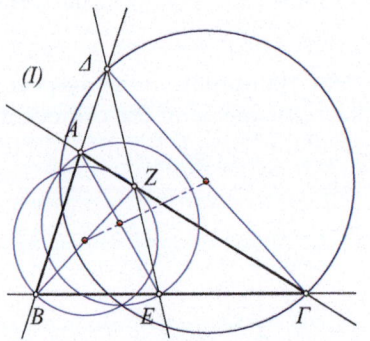
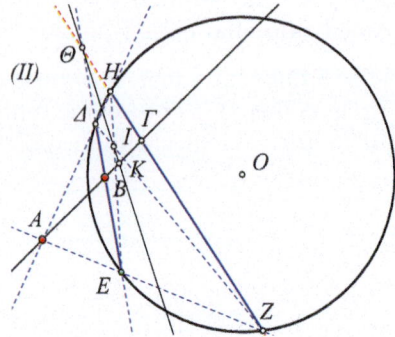

Fig. 5.118: Gauss - Bodenmiller   Inscribed with variable sides

Hint: (Alternatively see Exercise 5.73) The polar of A relative to κ is the diagonal ΘI of the complete quadrilateral ΔEZH (Proposition 5.17). If K and Γ are the intersection points of ΘI and ZH respectively with AB, then $(AK; B\Gamma) = -1$ (Corollary 5.17).

Exercise 5.135. Given is a circle κ(O), a point P lying outside it and a variable diameter AB of it. Let $\{A', B'\}$ be the second intersections with κ of the lines $\{PA, PB\}$. Show the following properties:

1. The circle (PAB) passes through a fixed point I and the line A'B' passes through a fixed point I'.
2. The circle (PA'B') passes through a fixed point K and is orthogonal to κ.
3. The pole of A'B' relative to κ is the center of κ'.
4. The segments $\{OI, OK\}$ are equal and it holds $(OI'; KP) = -1$.
5. The centroid Γ of the triangle PAB is a fixed point.
6. The symmedian point Δ of the triangle PAB is contained in a fixed circle.
7. The other intersection point P' of the circles $\{(PAB), (PA'B')\}$ is contained in the circle with diameter PO.

Exercise 5.136. In the triangle ABΓ the segments $\{AA', BB', \Gamma\Gamma'\}$ with points $\{A' \in B\Gamma, B' \in \Gamma A, \Gamma' \in AB\}$ intersect at the same point. Show that the three lines joining the middles of $\{AA', B\Gamma\}$, $\{BB', \Gamma A\}$ and $\{\Gamma\Gamma', AB\}$ intersect also at one point.

5.18 Desargues' theorem

> From all the works of men,
> most of all I love the used ones.
>
> *Bertold Brecht, From all the works*

The theorem of Desargues (1591-1661) represents the geometric foundation of *photography* and *perspectivity*, used by painters and designers in order to represent in paper objects of the space.

Two triangles are called **perspective relative to a point**, when we can label them $AB\Gamma$ and $A'B'\Gamma'$, in such a way, that lines $\{AA', BB', \Gamma\Gamma'\}$ pass through a common point P (See Figure 5.119). Points A, A' are then called **homologous**, and similarly points B, B' and Γ, Γ'. The point P is then called **perspectivity center** of the two triangles.

The two triangles are called **perspective relative to a line**, when we can label them $AB\Gamma$ and $A'B'\Gamma'$, in such a way, that the points of intersection of their sides $X = (AB, A'B')$, $Y = (B\Gamma, B'\Gamma')$ and $Z = (\Gamma A, \Gamma' A')$ are contained in the same line ε. Sides AB and $A'B'$ are then called **homologous**, and similarly sides $B\Gamma$, $B'\Gamma'$ and ΓA, $\Gamma' A'$. Line ε is called **perspectivity axis** of the two triangles.

Theorem 5.40 (Desargues). *Two triangles are perspective relative to a point, if and only if they are perspective relative to a line (See Figure 5.119).*

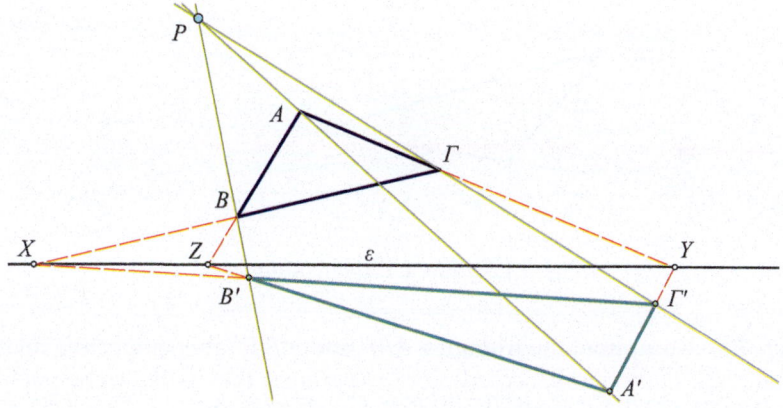

Fig. 5.119: Desargues' theorem

Proof. Suppose that the two triangles $AB\Gamma$ and $A'B'\Gamma'$ are perspective relative to a point P and apply three times Menelaus' theorem:

In triangle PAB with secant $ZB'A'$:
$$\frac{ZA}{ZB} \cdot \frac{A'P}{A'A} \cdot \frac{B'B}{B'P} = 1.$$

In triangle $PB\Gamma$ with secant $X\Gamma'B'$: $\quad \dfrac{XB}{X\Gamma} \cdot \dfrac{B'P}{B'B} \cdot \dfrac{\Gamma'\Gamma}{\Gamma'P} = 1.$

In triangle $P\Gamma A$ with secant $Y\Gamma'A'$: $\quad \dfrac{Y\Gamma}{YA} \cdot \dfrac{\Gamma'P}{\Gamma'\Gamma} \cdot \dfrac{A'A}{A'P} = 1.$

Multiplying by parts the three equalities and simplifying gives:

$$\dfrac{ZA}{ZB} \cdot \dfrac{XB}{X\Gamma} \cdot \dfrac{Y\Gamma}{YA} = 1.$$

According to Menelaus' theorem, this relation means that Z belongs to line XY.

Conversely, suppose that points X, Y and Z are collinear and apply the already proved part of the theorem on triangles ZBB' and $Y\Gamma'\Gamma$, which, by assumption now, are perspective relative to the point X. According to the proved part, the intersection points $A = (ZB, Y\Gamma)$, $A' = (ZB', Y\Gamma')$ and $P = (BB', \Gamma\Gamma')$ will be collinear. This is equivalent with the fact that the lines AA', BB' and $\Gamma\Gamma'$ pass through the same point, in other words, the triangles $AB\Gamma$ and $A'B'\Gamma'$ are perspective relative to a point ([152]).

Corollary 5.18 (Trilinear polar). *Suppose that point P does not belong to the side-lines of the triangle $AB\Gamma$. Consider also the intersection points of the pairs of lines $A' = (AP, B\Gamma)$, $B' = (BP, \Gamma A)$, $\Gamma' = (\Gamma P, AB)$ and their harmonic conjugates: $\{A'' = A'(B, \Gamma), B'' = B'(\Gamma, A), \Gamma'' = \Gamma'(A, B)\}$. Then points $\{A'', B'', \Gamma''\}$ are collinear (See Figure 5.120).*

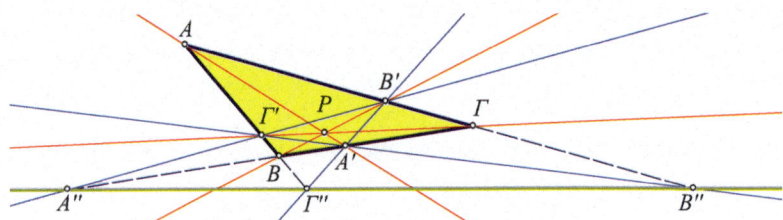

Fig. 5.120: The trilinear polar of P

Proof. By assumption, the triangles $AB\Gamma$ and $A'B'\Gamma'$ are perspective relative to the point P, therefore, according to Desargues, they will also be perspective relative to a line, in other words the pairs of respective sides $(AB, A'B')$, $(B\Gamma, B'\Gamma')$ and $(\Gamma A, \Gamma'A')$ will intersect on a line.

Remark 5.22. Desargues' theorem is a kind of generalization of Theorem 3.17 valid for homothetic triangles. If the sides of the triangles $AB\Gamma$ and $A'B'\Gamma'$ are respectively parallel, this means that the pairs of lines $(AB, A'B')$, $(B\Gamma, B'\Gamma')$ and $(\Gamma A, \Gamma'A')$ intersect at infinity (or, their points of intersection are contained in the *line at infinity*, see remark 5.19 in section § 5.16). As it was proved in the aforementioned proposition, then the lines AA', BB' and

5.18. DESARGUES' THEOREM

$\Gamma\Gamma'$ pass through a common point O (which we called homothety center of the two triangles) (See Figure 5.121). We can therefore consider that the perspectivity relative to a point is a generalization of the homothety (§ 3.9).

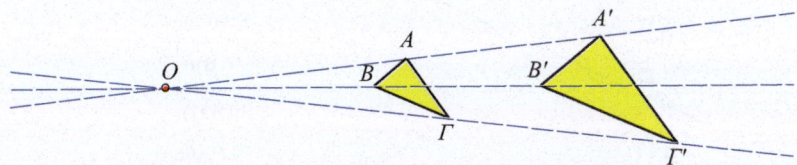

Fig. 5.121: Homothety and projectivity

Remark 5.23. A special application of Desargues' theorem is also shown in figure 5.122. In this, the two triangles $AB\Gamma$ and $A'B'\Gamma'$ have two respective sides $B\Gamma$ and $B'\Gamma'$ parallel to YZ, where $Y = (A\Gamma, A'\Gamma')$, $Z = (AB, A'B')$. The parallels are considered again as intersecting at a point at infinity X, through which passes YZ. According to Desargues then, the lines AA', BB' and $\Gamma\Gamma'$ will pass through a common point P.

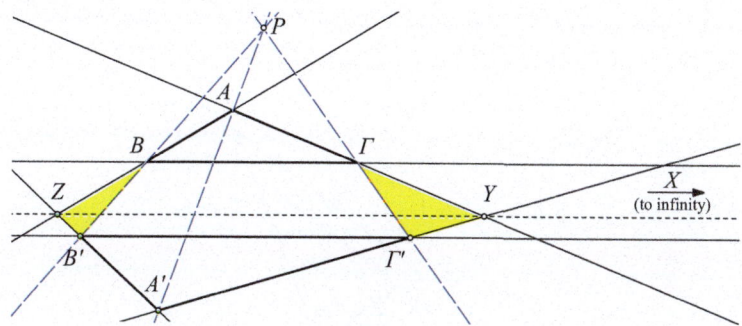

Fig. 5.122: Special case of application of Desargues' theorem

The same figure can also be interpreted in a different way, considering as main actors the triangles $BB'Z$ and $\Gamma\Gamma'Y$. In these triangles the lines $B\Gamma$, ZY and $B'\Gamma'$ are parallel and can be considered as passing through the same point X at infinity. According to Desargues, points $A = (BZ, \Gamma Y)$, $A' = (ZB', Y\Gamma')$ and $P = (BB', \Gamma\Gamma')$ are collinear. The conclusion in this case can be proved by applying the theorem of Thales.

Exercise 5.137. In a variable triangle $AB\Gamma$ the sides pass through three fixed points A'', B'', Γ'' which are collinear. Also its two vertices B and Γ move on fixed intersecting lines OB' and $O\Gamma'$ respectively. Show that its third vertex A moves on a line, which passes through the point O (See Figure 5.123).

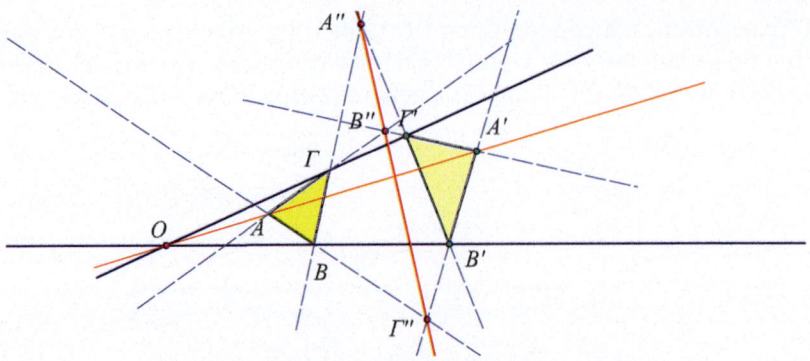

Fig. 5.123: Application of Desargues' theorem

Exercise 5.138. Let the triangles $\{AB\Gamma, A'B'\Gamma'\}$ have two pairs $(AB, A'B')$, $(A\Gamma, A'\Gamma')$ of sides parallel and the lines $\{AA', BB', \Gamma\Gamma'\}$ pass through a common point O. Then sides $\{B\Gamma, B'\Gamma'\}$ are also parallel.

Theorem 5.41. *Given two triangles $AB\Gamma$ and $A'B'\Gamma'$, the second of them, or a similar to it, can be placed in such a way, that the two triangles have common vertex A, the lines $\{BB', \Gamma\Gamma'\}$ are parallel to a line ζ and $Z = (B\Gamma, B'\Gamma')$ belongs to the orthogonal line ε to ζ at A (See Figure 5.124).*

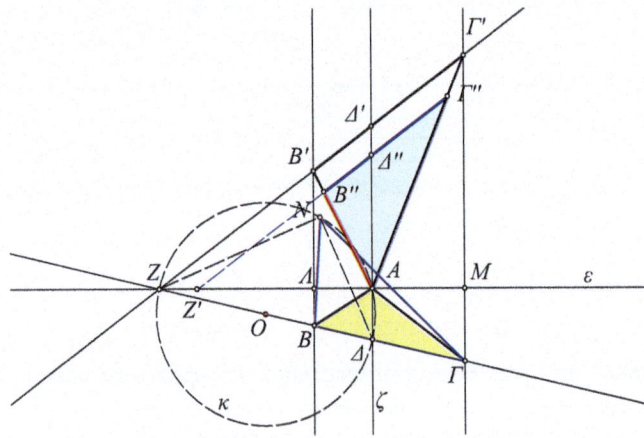

Fig. 5.124: Perspectivity relative to a point at infinity

Proof. Using the jargon of terminology, the property can be expressed briefly as follows:

Any two triangles, or their similars, can be placed in such a way that they are perspective from a point at infinity and the lines which join their homologous points are orthogonal to the axis of perspectivity.

5.18. DESARGUES' THEOREM

The lines which join homologous points are BB' and $\Gamma\Gamma'$. The axis of perspectivity is the line ε. For the proof let us use the method of analysis and synthesis ([70, p.305]). In the part of analysis, we then suppose that we have constructed such a perspectivity. Suppose also that $Z\Gamma$ is less than $Z\Gamma'$. Then, there exists a parallel $Z'\Gamma'''$ to $Z\Gamma'$ with length $|Z'\Gamma'''| = |Z\Gamma|$. Because of the parallels and the similarity, the triples $(ZB, B\Delta, \Delta\Gamma)$, $(ZB', B'\Delta', \Delta'\Gamma')$ and $(Z'B'', B''\Delta'', \Delta''\Gamma''')$ consist of proportional line segments and because, by construction of Z', holds $|Z'\Gamma'''| = |Z\Gamma|$, the first and third triples will consist of respectively equal segments. We place then triangle $Z'A\Gamma'''$ at a position $ZN\Gamma$, so that the base of $Z'\Gamma'''$ coincides with $Z\Gamma$. Then point Δ'' will coincide with Δ and B'' with B. Triangle $ZN\Delta$, being congruent to $Z'A\Delta''$ is right, as is also $ZA\Delta$, therefore points N and A will lie on the circle with diameter $Z\Delta$. Simultaneously, $NB\Gamma$ will be congruent to $B''A\Gamma$, which is similar to $B'A\Gamma'$.

Passing to synthesis, the construction of perspectivity, or equivalently of lines ε and ζ, is now easy and goes in reverse order to the preceding analysis, beginning from the circle κ with diameter $Z\Delta$. This circle is constructed by placing the similar $BN\Gamma$ of the given $A'B'\Gamma'$, in the way suggested by the figure, so that its base coincides with that of $AB\Gamma$. Circle κ is determined from its center O, which is the intersection of $B\Gamma$ with the medial line of AN, and its radius, which is OA. The requested lines are defined from the intersection points Z, Δ of κ with $B\Gamma$. They are $\varepsilon = AZ$ and $\zeta = A\Delta$. Triangle $ZN\Gamma$ is placed at position $Z'A\Gamma'''$ and the parallel $Z\Gamma'$ of $Z'\Gamma'''$ from Z defines, through its intersections with AB'', $A\Gamma'''$, respectively the points B', Γ', and with these, the triangle $AB'\Gamma'$. The verification, that all the preceding lead to two triangles which satisfy the theorem's requirements, follows from a simple application of the theorem of Thales.

An additional element, which is important in view of a future application of this theorem (Theorem 4.14), concerns the following ratio

$$\lambda = \frac{|A\Delta|}{|A\Delta'|} \quad \text{which, along with} \quad \frac{|A\Delta'|}{|A\Delta''|} = \frac{|AZ|}{|AZ'|} \Rightarrow |A\Delta'| = |A\Delta''|\frac{|AZ|}{|AZ'|}$$

$$= |N\Delta|\frac{|AZ|}{|NZ|} \quad \text{becomes} \quad \lambda = \frac{|A\Delta|}{|N\Delta|} \cdot \frac{|NZ|}{|AZ|} < 1.$$

The last inequality results from the Ptolemy's theorem (Theorem 5.7) for the quadrilateral $ANZ\Delta$.

5.19 Pappus' theorem

> All bodies, the firmament, the stars, the earth and its kingdoms, are not worth the least of minds. For mind knows all of these, and itself, and bodies know nothing.
>
> B. Pascal, Pensees

Pappus' theorem, like the theorem of Menelaus, is a tool in proving the collinearity of three points. In Menelaus' theorem the basic figure is a triangle and more generally, three lines in general position, each line carrying an additional point. The theorem decides when the triangle which is formed by these additional points is degenerate, that is when these three points are collinear. Pappus' theorem is of a different texture. The basic shape consists of two lines and three points on each. These, connected in a certain way, form a self-intersecting hexagon. The theorem warrants the collinearity of the points of intersection of the extensions of the three pairs of hexagon's opposite sides, that is, pairs of sides which are not successive.

Theorem 5.42 (Pappus). *Let points $\{A, \Gamma, E\}$ be on line α and points $\{B, \Delta, Z\}$ be on line β. Then the intersection points $X = (AB, \Delta E)$, $Y = (B\Gamma, EZ)$ and $\Theta = (\Gamma\Delta, ZA)$ are collinear (See Figure 5.125).*

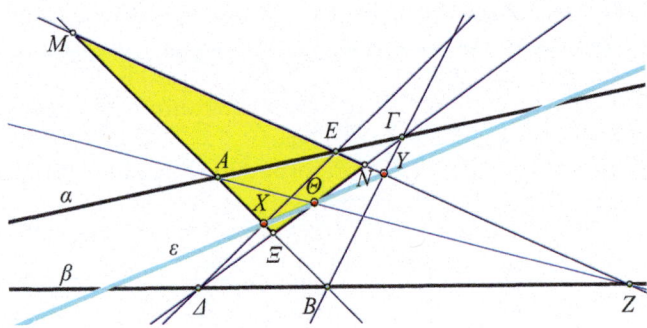

Fig. 5.125: Pappus' theorem

Proof. ([65, p.93]) In the triangle $MN\Xi$, where $\Xi = (AB, \Gamma\Delta)$, $M = (AB, EZ)$, $N = (EZ, \Gamma\Delta)$ we apply the theorem of Menelaus for the following 5 intersecting lines:

(1) intersecting line $E\Delta X$: $\quad \dfrac{EM}{EN} \cdot \dfrac{\Delta N}{\Delta \Xi} \cdot \dfrac{X\Xi}{XM} = 1,$

(2) intersecting line $Z\Theta A$: $\quad \dfrac{ZM}{ZN} \cdot \dfrac{\Theta N}{\Theta \Xi} \cdot \dfrac{A\Xi}{AM} = 1,$

5.19. PAPPUS' THEOREM

(3) intersecting line $Y\Gamma B$: $\quad \dfrac{YM}{YN} \cdot \dfrac{\Gamma N}{\Gamma \Xi} \cdot \dfrac{B\Xi}{BM} = 1,$

(4) intersecting line $E\Gamma A$: $\quad \dfrac{EM}{EN} \cdot \dfrac{\Gamma N}{\Gamma \Xi} \cdot \dfrac{A\Xi}{AM} = 1,$

(5) intersecting line $Z\Delta B$: $\quad \dfrac{ZM}{ZN} \cdot \dfrac{\Delta N}{\Delta \Xi} \cdot \dfrac{B\Xi}{BM} = 1.$

Multiplying by parts the three first and the two last equations we get

$$\frac{EM}{EN} \cdot \frac{\Delta N}{\Delta \Xi} \cdot \frac{X\Xi}{XM} \cdot \frac{ZM}{ZN} \cdot \frac{\Theta N}{\Theta \Xi} \cdot \frac{A\Xi}{AM} \cdot \frac{YM}{YN} \cdot \frac{\Gamma N}{\Gamma \Xi} \cdot \frac{B\Xi}{BM} = 1,$$

$$\frac{EM}{EN} \cdot \frac{\Gamma N}{\Gamma \Xi} \cdot \frac{A\Xi}{AM} \cdot \frac{ZM}{ZN} \cdot \frac{\Delta N}{\Delta \Xi} \cdot \frac{B\Xi}{BM} = 1.$$

Dividing by parts the first by the second, we get

$$\frac{X\Xi}{XM} \cdot \frac{\Theta N}{\Theta \Xi} \cdot \frac{YM}{YN} = 1,$$

which shows that the points X, Θ and Y are collinear ([140]).

Remark 5.24. The way we select the points, in order to define the lines and their intersections in Pappus' theorem, is the following:

1. We place $A, B, \Gamma,...$ alternatively on lines α and β and we form the "inscribed" in the system of two lines $\{\alpha, \beta\}$ hexagon $AB\Gamma\Delta EZ$.
2. Starting from any vertex, for example A, we take the next (B), we leave the next (Γ) and take the one after the next (Δ) and its next (E) and we form the intersection $(AB, \Delta E)$. This recipe determines the intersection of two *opposite* sides of the hexagon.
3. In the symbol $(AB, \Delta E)$ we substitute cyclically the letters $A \to B \to \Gamma \to \Delta \to E \to Z \to A$, and we determine the intersections of the other two pairs of *opposite* sides $(B\Gamma, EZ)$ and $(\Gamma\Delta, ZA)$.

Remark 5.25. Similar special cases, to those we met with the theorems of Menelaus, Ceva and Desargues, exist also for Pappus' theorem. These cases result, when some of the intersection points $\{X, Y, \Theta\}$, which appear in the theorem, go to infinity. One such case is seen in figure 5.126, in which the two lines AB and ΔE intersect at a point at infinity X, in other words they are parallel. Then the collinearity of X, $Y = (B\Gamma, EZ)$ and $\Theta = (\Gamma\Delta, AZ)$ means, that ΘY is also parallel to AB and ΔE.

Another special case is the one, where two of $\{X, Y, \Theta\}$ are points at infinity, for example, points X and Θ (See Figure 5.127). Then, from the collinearity of $\{X, \Theta, Y\}$ follows that the third point Y will also be contained in the line of X, Θ, which is the *line at infinity*. In other words, lines EZ and $B\Gamma$ will also be parallel.

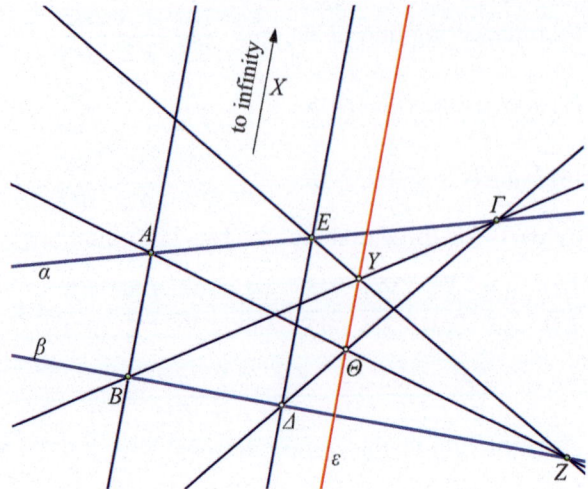

Fig. 5.126: Special case of application of of Pappus' theorem

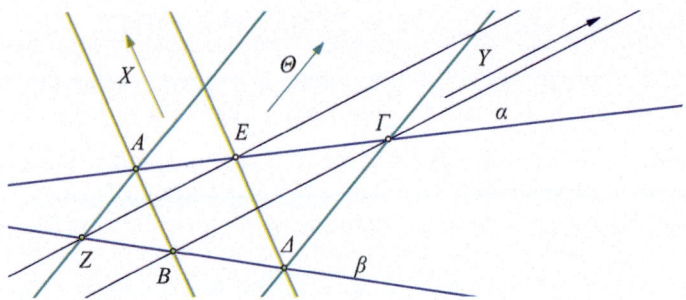

Fig. 5.127: Three intersection points at infinity

Exercise 5.139. Given three points on line α and three other points on line β, show that there are six different possibilities to build hexagons with the preceding recipe, and each such hexagon defines, through Pappus' theorem, a respective line.

Hint: One vertex on line α, for example (1) (See Figure 5.128) is connected with two others on line β. We therefore have three possibilities to choose the two others. After the selection of these three points, we have two possibilities to choose the fourth vertex on line α and this choice determines completely the polygon.

Exercise 5.140. Construct the line of Pappus for each hexagon of the figure 5.128.

Remark 5.26. The preceding and this section are a small introduction to the so called *Projective Geometry* ([126]), in which one examines coincidence prop-

5.19. PAPPUS' THEOREM

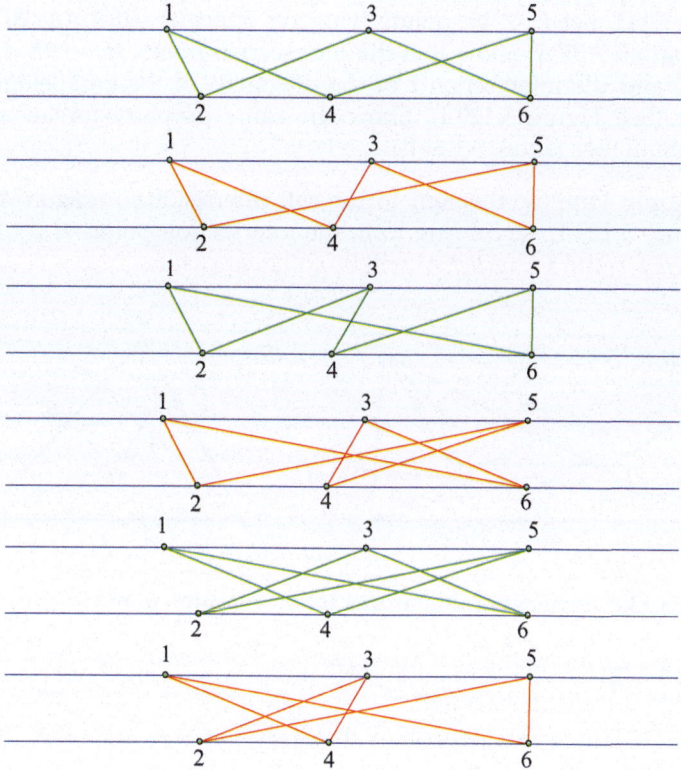

Fig. 5.128: The six possibilities for Pappus hexagons

erties of points in lines and lines on points (that is, passing through points). The theorems of Menelaus, Pappus, Ceva and Desargues are of fundamental importance in this Geometry, but they also have useful applications in the realm of Euclidean Geometry.

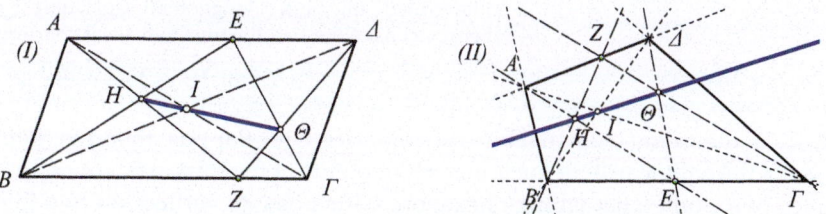

Fig. 5.129: Application of Pappus' theorem

Exercise 5.141. Let E, Z be points lying on opposite sides $A\Delta$, $B\Gamma$ of the parallelogram $AB\Gamma\Delta$. Show that the intersection points $H = (BE, AZ)$, $\Theta = (\Gamma E, \Delta Z)$ and the intersection I of the diagonals of the parallelogram are collinear (See Figure 5.129-I). Show the same property for an arbitrary quadrilateral (See Figure 5.129-II).

Hint: Apply Pappus' theorem to the self intersecting hexagon $AZ\Delta BE\Gamma$ (See Figure 5.129-I). According to it, the intersection points $H = (AZ, BE)$, $\Theta = (Z\Delta, E\Gamma)$ and $I = (\Delta B, \Gamma A)$ will be collinear points.

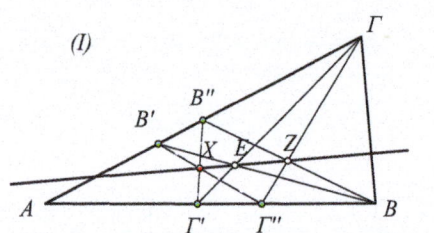
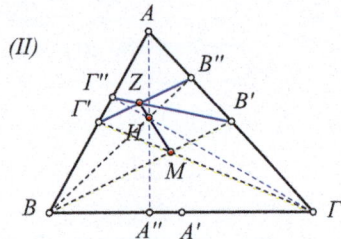

Fig. 5.130: Application of Pappus II Points on the Euler line

Exercise 5.142. Given is a triangle $AB\Gamma$ and two points E, Z not contained in its side-lines. Consider the intersection points $B' = (BE, A\Gamma)$, $B'' = (BZ, A\Gamma)$, $\Gamma' = (\Gamma E, AB)$, $\Gamma'' = (\Gamma Z, AB)$. Show that the point $X = (B'\Gamma'', \Gamma'B'')$ is on the line EZ (See Figure 5.130-I).

Exercise 5.143. Let $\{A', B', \Gamma'\}$ be the middles of the sides of the triangle $AB\Gamma$. Let also $\{A'', B'', \Gamma''\}$ be the traces of its altitudes and $\{H, M\}$ be its orthocenter and centroid. Show that the lines $B'\Gamma''$, $\Gamma'B''$ intersect at point Z on the Euler line HM of the triangle (See Figure 5.130-II).

5.20 Pascal's and Brianchon's theorems

> I have made this (letter) longer than usual, only because I have not had the time to make it shorter.
>
> B. Pascal, *Lettres Provinciales no. 16*

Pascal's theorem (1623-1662) has a similarity with Pappus' theorem. Latter asserts the collinearity of the intersection points of the three pairs of opposite sides of a, somewhat strange hexagon, which has its vertices on two lines. Pascal's theorem claims the collinearity of the three pairs of opposite sides of a hexagon inscribed in a circle. Both of these theorems are special cases of a corresponding theorem for conic sections (II-§ 6.1). Brianchon's theorem (1783-1864) can be proved as an application of Pascal's theorem by the, so

5.20. PASCAL'S AND BRIANCHON'S THEOREMS

called, *duality principle*, which we'll not analyze in this lesson, but give instead an independent proof (see however Exercise 5.146). According to this principle, every theorem which talks about a coincidence of points on a line corresponds to a theorem about lines which pass through a point and vice versa.

Theorem 5.43 (Pascal's Mystic hexagram). *The three intersection points $I = (AB, \Delta E)$, $K = (B\Gamma, EZ)$, $\Lambda = (\Gamma\Delta, ZA)$ of the pairs of opposite sides of an inscribed in a circle hexagon $AB\Gamma\Delta EZ$, are collinear (See Figure 5.131).*

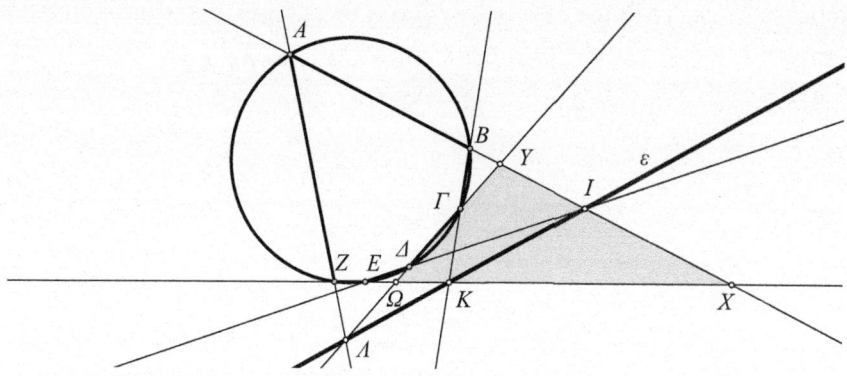

Fig. 5.131: Pascal's theorem

Proof. We apply three times the theorem of Menelaus on triangle $XY\Omega$, which is formed from the non consecutive sides of the hexagon $(AB, \Gamma\Delta, EZ)$.

(1) With the intersecting line $B\Gamma$: $\dfrac{BY}{BX} \cdot \dfrac{\Gamma\Omega}{\Gamma Y} \cdot \dfrac{KX}{K\Omega} = 1,$

(2) With the intersecting line ΔE : $\dfrac{IY}{IX} \cdot \dfrac{\Delta\Omega}{\Delta Y} \cdot \dfrac{EX}{E\Omega} = 1,$

(3) With the intersecting line AZ : $\dfrac{AY}{AX} \cdot \dfrac{ZX}{Z\Omega} \cdot \dfrac{\Lambda\Omega}{\Lambda Y} = 1.$

Multiplying the three equations by parts and noticing that

$$\frac{BY}{\Gamma Y} \cdot \frac{AY}{\Delta Y} = \frac{\Gamma\Omega}{E\Omega} \cdot \frac{\Delta\Omega}{Z\Omega} = \frac{EX}{BX} \cdot \frac{ZX}{AX} = 1,$$

because the products in the numerator and denominator express the power of the same point relative to the circle, we conclude by simplifying,

$$\frac{KX}{K\Omega} \cdot \frac{IY}{IX} \cdot \frac{\Lambda\Omega}{\Lambda Y} = 1,$$

which means that points $\{I, K, \Lambda\}$ are collinear.

Remark 5.27. The similarity between the theorems of Pappus and Pascal is reflected in their formulation, in the way we construct the points of intersection, but also in the proof as well. Indeed, the two theorems are special cases of a more general theorem, which is valid for hexagons inscribed in a conic section. This general theorem gives a necessary and sufficient criterion for the inscription of a hexagon in such a curve. The line, which contains the intersection points $\{I, K, \Lambda\}$ of the sides is called **Pascal line** of the inscribed hexagon.

Remark 5.28. Pascal's theorem leads to a property for pentagons, as it continues to hold also in the case where one of the points of the hexagon tends

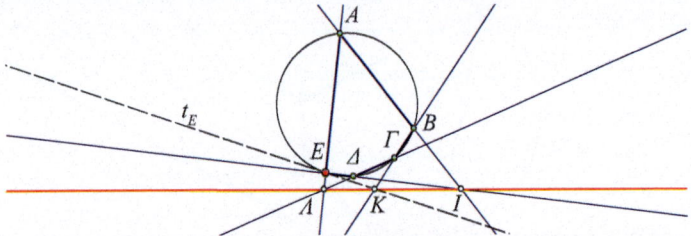

Fig. 5.132: Pascal's theorem for inscriptible pentagons

to one of its neighbors, for example as Z tends to coincide with E. Then the side EZ tends to coincide with the tangent to the circle at E and the theorem guarantees that the intersection points $I = (AB, E\Delta)$, $\Lambda = (AE, \Delta\Gamma)$ and $K = (t_E, B\Gamma)$, where t_E is the tangent to the circumscribed circle at E, are contained again in a line (See Figure 5.132).

The same process can be repeated with the pentagon and leads to a property for inscriptible quadrilaterals. For example, when Δ tends to coincide with Γ, then line $\Delta\Gamma$ tends to the tangent at Γ and Pascal's theorem guarantees that the intersection points $I = (AB, \Gamma E)$, $\Lambda = (AE, t_\Gamma)$, $K = (B\Gamma, t_E)$ are

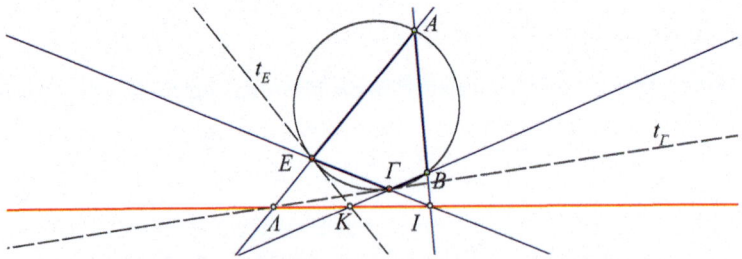

Fig. 5.133: Pascal's theorem for inscribed quadrilaterals

contained in a line (See Figure 5.133).

5.20. PASCAL'S AND BRIANCHON'S THEOREMS

Finally, applying the same process to inscriptible quadrilaterals we get a property for triangles. When B tends to coincide with A, then AB tends to the tangent at A and the theorem guarantees the known property of triangles, according to which (See Figure 5.134):

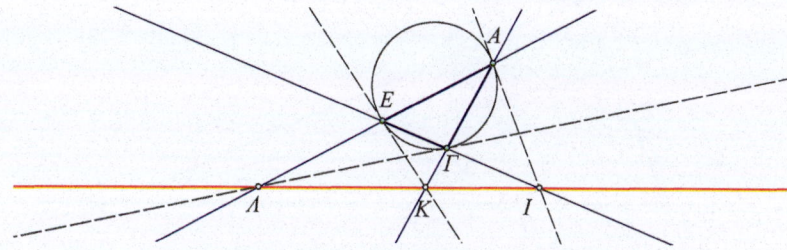

Fig. 5.134: Pascal's theorem for triangles

The tangents to the circumcircle at the vertices of the triangle intersect the opposite sides at collinear points.

Exercise 5.144. Show that the tangents of the circumcircle at the vertices of a triangle intersect the opposite sides at collinear points, using exercise 5.111 and Desargues' theorem.

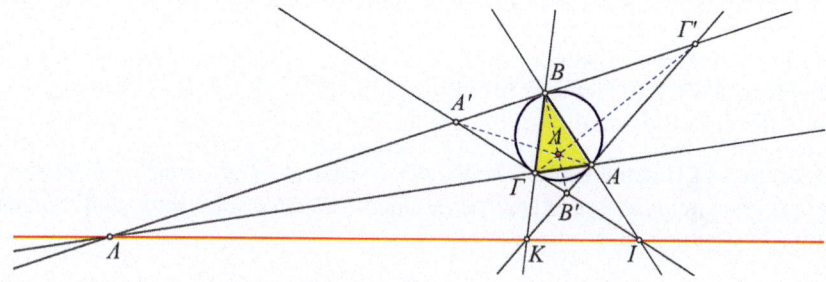

Fig. 5.135: The Lemoine line of a triangle

Hint: Drawing the tangents at the vertices of triangle $AB\Gamma$ creates the so called **tangential triangle** $A'B'\Gamma'$ of $AB\Gamma$ (See Figure 5.135), which has the circumcircle of $AB\Gamma$ as its inscribed circle. According to Exercise 5.111 the lines AA', BB' and $\Gamma\Gamma'$ pass through a common point Λ. This means that the triangles $AB\Gamma$ and $A'B'\Gamma'$ are perspective relative to point Λ, therefore, according to Desargues, they will also be perspective relative to a line, i.e. the intersection points of the lines $I = (AB, A'B')$, $K = (B\Gamma, B'\Gamma')$ and $\Lambda = (\Gamma A, \Gamma'A')$ will be collinear.

The line, which is defined in the preceding exercise and contains points $\{I, K, \Lambda\}$, is called **Lemoine line** (1840-1912) of triangle $AB\Gamma$ and plays a prominent role in the **Geometry of the Triangle** ([11, p.252], [40]).

Remark 5.29. The proof of Pascal's theorem we saw, makes use of the theorem of Menelaus and seems to be more complicated than the next, which uses only properties of inscriptible quadrilaterals (See Figure 5.136).

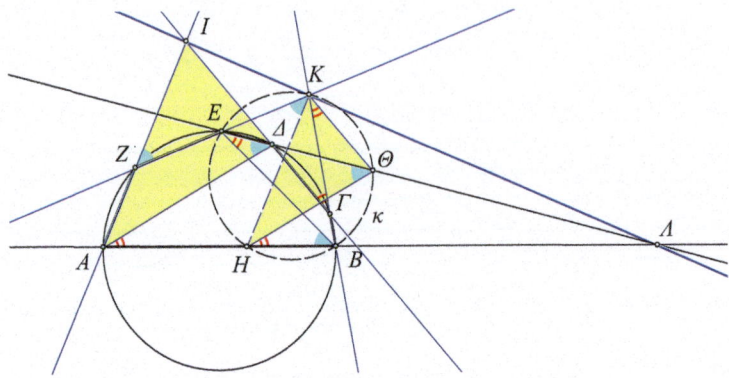

Fig. 5.136: Pascal's theorem, elementary proof

Let the hexagon $AB\Gamma\Delta EZ$ be inscriptible in a circle and $I = (AZ, \Gamma\Delta)$, $K = (B\Gamma, ZE)$ be the intersections of two pairs of opposite sides (See Figure 5.136). We show that the intersection point of the third pair of opposite sides $\Lambda = (AB, \Delta E)$ is contained in line IK. To this end, we consider the circle $\kappa = (EKB)$, which intersects a second time AB at H and $E\Delta$ at Θ. We show that the created triangles, $A\Delta I$ and $KH\Theta$, have their sides parallel, therefore the lines which join corresponding vertices pass through a common point Λ (Theorem 3.17). The fact that the sides of the triangles are parallel is proved easily through the equality of the angles shown in the figure 5.136 ([158]).

Theorem 5.44 (Brianchon). *The diagonals $\{A\Delta, BE, \Gamma Z\}$, which join opposite vertices of a circumscribed in a circle hexagon $AB\Gamma\Delta EZ$, pass through a common point (See Figure 5.137).*

Proof. Extend the sides of the hexagon and take equal segments from the contact points with the circle
$$|A_1 A_2| = |B_1 B_2| = |\Gamma_1 \Gamma_2| = |\Delta_1 \Delta_2| = |E_1 E_2| = |Z_1 Z_2| = \lambda.$$

Three other circles are defined: (κ) tangent to AB, ΔE at A_2, Δ_2, (μ) tangent to $\Gamma\Delta$, AZ at Γ_2, Z_2, (ν) tangent to EZ, ΓB at E_2, B_2 (See Figure 5.136). Because of the equality of the tangents from B, $|BA_1| = |BB_1|$, follows that $|BA_2| = \lambda - |BA_1| = \lambda - |BB_1| = |BB_2|$ and consequently point B lies on the radical axis of the circles κ and ν. Similarly, because of the equality of tangents from E: $|EE_1| = |E\Delta_1|$, follows that $|EE_2| = |E\Delta_2|$, therefore point E lies on the radical axis of κ and ν. This implies that line BE coincides with the radical axis of κ and ν. Similarly we show that line $A\Delta$ coincides with the radical axis of κ and μ and that ΓZ coincides with the radical axis of μ and μ. The theorem

5.20. PASCAL'S AND BRIANCHON'S THEOREMS

follows then from the fact that the radical axes of three circles pass through a common point O (Theorem 4.5).

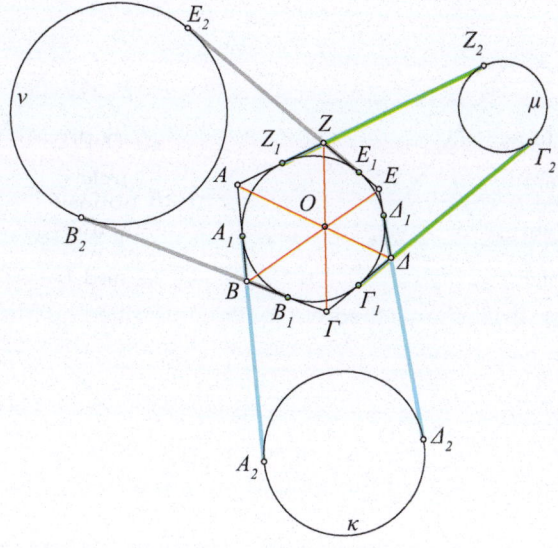

Fig. 5.137: Theorem of Brianchon

The intersection point of the diagonals O of the circumscribed hexagon is called **Brianchon point** of the circumscribed hexagon.

Remark 5.30. As in Pascal's theorem, so with that of Brianchon, in the case where one vertex of the hexagon tends to coincide with one of its neighbors, leads to a theorem for pentagons (See Figure 5.138):

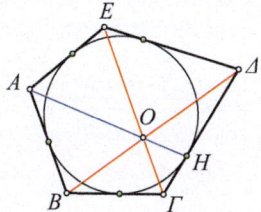

Fig. 5.138: Brianchon's theorem for circumscribed pentagons

Two diagonals $B\Delta$, ΓE of a pentagon circumscribed in circle and the line AH, which joins the fifth vertex with its opposite contact point, pass through a common point O.

In the preceding figure for the pentagon, letting E to coincide with A, transforms the pentagon into a quadrilateral (See Figure 5.139-I) and the theorem guarantees that:

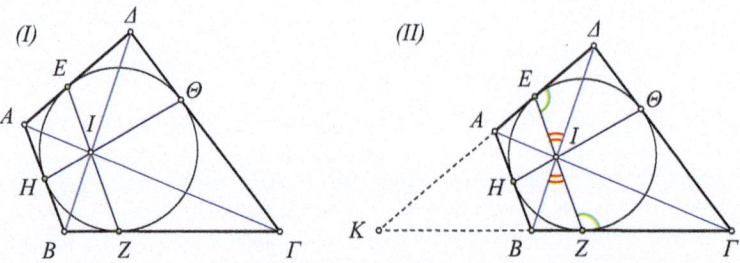

Fig. 5.139: Brianchon's theorem for circumscribed quadrilaterals

The diagonals of a quadrilateral circumscribed in circle and the lines which join points of contact of opposite lying vertices, pass through the same point.

Finally, letting in the preceding quadrilateral one vertex to coincide with one of its neighbors, we get the conclusion of Exercise 5.111, according to which:

The lines which join the vertices of a triangle with the opposite contact points of the inscribed circle pass through a common point (Gergonne point of the triangle).

Exercise 5.145. Show that in every circumscribed in a circle quadrilateral $AB\Gamma\Delta$, the diagonals and the lines which join the contact points of opposite sides pass through a common point (without using the theorem of Brianchon) (See Figure 5.139-II).

Hint: Suppose that line EZ, which joins the opposite contact points, intersects the diagonal $B\Delta$ at I (See Figure 5.139-II). Extend $A\Delta$ and $B\Gamma$ until they intersect at K. From the equality of tangents and the isosceles EZK, it follows that angles $\widehat{\Delta EI}$ and $\widehat{\Gamma ZI}$ are equal. Consequently triangles ΔEI and IBZ have their angles at I equal and those at E and Z supplementary. It follows (Corollary 3.17) that

$$\frac{|I\Delta|}{|IB|} = \frac{|E\Delta|}{|ZB|}.$$

Similarly, extending lines AB and $\Delta\Gamma$ and considering the intersection I' of $H\Theta$ with $B\Delta$, we find that

$$\frac{|I'\Delta|}{|I'B|} = \frac{|\Delta\Theta|}{|BH|}.$$

However, from the equality of the tangents $|BH| = |BZ|$, $|\Delta E| = |\Delta\Theta|$, follows that the ratios on the right side of the two preceding equalities are equal and consequently

$$\frac{|I\Delta|}{|IB|} = \frac{|I'\Delta|}{|I'B|}.$$

5.20. PASCAL'S AND BRIANCHON'S THEOREMS

From this follows that points I and I' coincide with the same point on ΔB. This shows that $B\Delta$ passes through the intersection point of EZ and $H\Theta$. Similarly we show, that the diagonal $A\Gamma$ passes also through the intersection point of EZ and $H\Theta$.

Remark 5.31. The theorems of Pascal and Brianchon are valid more generally ([153, p.111]) for the so called **conic sections**, which are precisely the intersection lines of a *circular cone* (II-§ 4.9) and a plane (II-§ 6.1). The circles are a special case of such conic sections. A fundamental property of the conic sections is that five points in general position always define exactly one conic section, which passes through them. Pascal's theorem, in its general form for conic sections, expresses a necessary *and sufficient condition* for *six* points to be contained in a conic section ([119, p.18]). This way, the converse of Theorem 5.43 guarantees that, if the opposite sides of a hexagon intersect at three collinear points, then the hexagon can be inscribed in a conic section, which is not, in general, a circle. From this viewpoint, Pascal's theorem is a tool similar to the theorem of Menelaus (Theorem 5.33), which gives a criterion for the collinearity of three points. The hexagon, consequently, is as "mystical" as the triangle.

A similar property holds also for the theorem of Brianchon, which, in the realm of *Projective Geometry* ([153]), is proved equivalent (*dual of*) with Pascal's theorem ([119, p.26]). The corresponding equivalence in the case of the circle is examined in exercise 5.146.

More generally it is proved that five lines in general position define a unique conic section tangent to them. The theorem of Brianchon gives a necessary and sufficient criterion for the existence of a conic section tangent to *six* given lines. This way, the converse of Theorem 5.44 guarantees that, if the lines which join opposite vertices of a hexagon pass all through a common point, then the sides of the hexagon are tangent to a conic section, which in general, is not a circle.

Remark 5.32. As is also apparent from the preceding remark, Pascal's theorem has to do mainly with the property of the six vertices of the hexagon to be on the same circle (conic). The hexagon itself is something of secondary importance and only auxiliary relative to this property. Indeed, six points on the circle can be connected and define a hexagon (convex or non-convex or self-intersecting) in 60 different ways ([151, p.379]). For each such hexagon results a corresponding Pascal line. Figure 5.140 gives an idea of the image of the six concyclic points and of the 60 corresponding (different in general) Pascal lines from the 60 corresponding inscribed polygons.

Exercise 5.146. Show that the poles of the sides of hexagon $AB\Gamma\Delta EZ$ inscribed in a circle κ are vertices of the hexagon $A'B'\Gamma'\Delta'E'Z'$ circumscribed in κ, whose contact points of its sides with the circle coincide with the vertices of $AB\Gamma\Delta EZ$ (See Figure 5.141). Also show that the Pascal line of

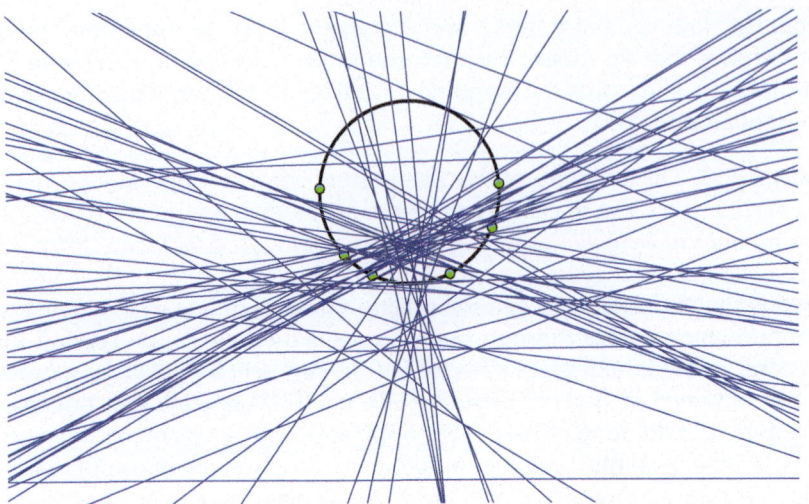

Fig. 5.140: Six concyclic points and the corresponding 60 Pascal lines

$AB\Gamma\Delta EZ$, which contains the intersection points X, Y, Z of the opposite sides of $AB\Gamma\Delta EZ$, coincides with the polar of the Brianchon point O of $A'B'\Gamma'\Delta'E'Z'$, which is the intersection point of the diagonals of this hexagon.

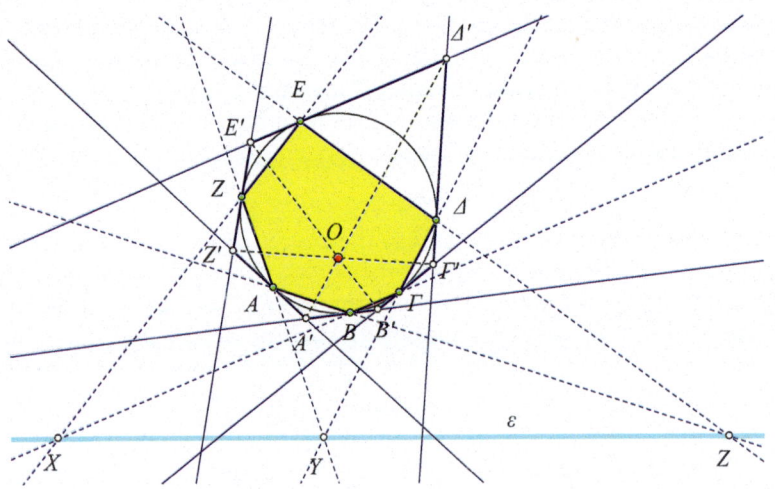

Fig. 5.141: Duality of theorems of Pascal and Brianchon

Hint: The first part, for the definition of $A'B'\Gamma'\Delta'E'Z'$ from $AB\Gamma\Delta EZ$, is Exercise 4.110. The key for the rest is the fact that each diagonal, like $A'\Delta'$ for example, is the polar of one of the three points (of Z) on the Pascal line. Indeed, point Δ' is the pole of $E\Delta$ and point A' is the pole of AB. Since then $A'\Delta'$ contains the poles of AB and ΔE, these lines themselves will also contain the pole of $A'\Delta'$, in other words their intersection Z will be exactly the pole of $A'\Delta'$. Similarly is proved, that points Y and X are respectively the poles of $B'E'$ and $\Gamma'Z'$. Because then line ε contains each one of these poles, the corresponding lines $A'\Delta'$, $B'E'$, $\Gamma'Z'$ will contain the pole O of ε, therefore they will intersect at this point.

Exercise 5.147. Give a proof for the theorem of Brianchon by reducing it to Pascal's theorem (See Figure 5.141).

Hint: If the given hexagon $A'B'\Gamma'\Delta'E'Z'$ is circumscribed to the circle κ, then its contact points with κ, taken with a suitable order, define an inscribed hexagon $AB\Gamma\Delta EZ$. Apply the preceding exercise.

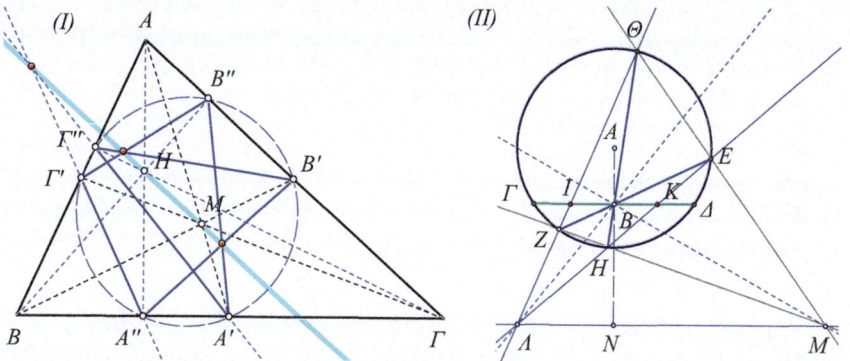

Fig. 5.142: Euler line as a Pascal line   Butterfly theorem

Exercise 5.148. Let $\{A', B', \Gamma'\}$ be the middles of sides of the triangle $AB\Gamma$. Let also $\{A'', B'', \Gamma''\}$ be the traces of the altitudes. Show that the Euler line coincides with the Pascal line of the hexagon $A'\Gamma''B'A''\Gamma'B''$, which is inscribed in the Euler circle of the triangle $AB\Gamma$ (See Figure 5.142-I).

Hint: Application of Exercise 5.143.

Exercise 5.149. (Butterfly theorem) The two chords ZE and $H\Theta$ of the circle κ pass through the middle B of the third chord $\Gamma\Delta$. Show that the lines $Z\Theta$ and EH intersect the chord $\Gamma\Delta$ respectively at points I and K lying symmetrically relative to B (See Figure 5.142-II).

Hint: The polar of the intersection point B of the diagonals of the quadrilateral $ZHE\Theta$ is the line ΛM parallel to $\Gamma\Delta$ (Proposition 5.17). Also the line pencil $\Lambda(ZHBM)$ is harmonic (Theorem 5.17), therefore it will excise a harmonic quadruple on every other line intersecting these four. Especially on $\Gamma\Delta$, which is parallel to ΛM, the three other lines of the pencil will excise equal segments (Corollary 5.12) (alternatively [13, p.45], [112]).

5.21 Castillon's problem, homographic relations

> Scientific discovery and scientific knowledge have been achieved only by those who have gone in pursuit of it without any practical purpose whatsoever in view.
>
> Max Planck, *Where is Science going?*

Castillon's problem (1704-1791) asks for the construction of a polygon with v sides, inscribed in a given circle (more generally in a conic section), so that its sides pass through v given fixed points. Initially the problem was formulated for triangles ([9, p.137], [70, p.144]) and various geometric solutions (like that of exercise 4.126) were given by known Mathematicians. Here, on the occasion of this problem, we'll examine the notion of **homographic relation**, which is at the root of of a systematic investigation and solution of the problem and which, with minor modifications, can be extended to the case of conic sections ([43, II, p.181]). A good starting point is to write the cross ratio of four points of a line relative to a coordinate system for that line (§ 5.14).

Proposition 5.18. *Let $\{A(a), B(b), X(x), Y(y)\}$ be collinear points on a line ε, where in parentheses are the corresponding coordinates relative to some coordinate system of ε. Let also $k = (AB;XY)$ be the corresponding cross ratio. The the following relation between $\{x,y\}$ is valid.*

$$y = \frac{mx+n}{px+q}, \quad \text{where} \quad m = ak-b, \quad n = ab(1-k), \quad p = k-1, \quad q = a-bk.$$

Proof. Using the coordinates, the cross ratio is written

$$k = \frac{XA}{XB} : \frac{YA}{YB} = \frac{a-x}{b-x} : \frac{a-y}{b-y}.$$

The relation follows by performing calculations and solving for y.

A relation between variables $\{x,y\}$, of the form $y = \frac{mx+n}{px+q} = f(x)$, with the restriction $mq - np \neq 0$, is called **homographic relation** ([118, p.100]). The preceding proposition shows, that such a relation holds for the coordinates $\{x,y\}$ of all the pairs of points $(X(x), Y(y))$ of the line ε, whose cross ratio, relative to two fixed points (A,B), is fixed. In this case the restriction $mq - np = k(a-b)^2 \neq 0$ is satisfied, when the points A, B are different and, of course, when $k \neq 0$. The homographic relation has a kind of symmetry relative to $\{x,y\}$, since x can be also expressed through y by a similar relation:

$$x = \frac{qy-n}{-py+m} = g(x).$$

5.21. CASTILLON'S PROBLEM, HOMOGRAPHIC RELATIONS

The definition of the homographic relation is extended from the coordinates to the points themselves. This way, we say that the variable *points* $\{X,Y\}$ of a line ε satisfy a homographic relation or are **homographically related**, when their corresponding coordinates $\{x,y\}$, relative to some system of coordinates, satisfy a homographic relation $y = f(x)$. We easily see that, if the coordinates $\{x,y\}$ of $\{X,Y\}$, relative to a specific system of coordinates, satisfy a homographic relation, then also the coordinates $\{x' = x - a, y' = y - a\}$, relative to any other system of coordinates, will also satisfy a similar homographic relation (see next exercise). This way, while the extension of the homographic relation from the coordinates to the points of a line, includes the use of a specific system of coordinates, in the end, the definition is independent of that special system.

Exercise 5.150. Show that if the points $\{X(x), Y(y)\}$ of a line ε are related homographically and $\{x' = x - a\}$ is a second system of coordinates, then the coordinates $\{x', y'\}$ of $\{X, Y\}$, relative to a second system of coordinates, will satisfy also a homographic relation.

Hint: The fact, that $\{X(x), Y(y)\}$ are related homographically, means that their coordinates will satisfy a relation of the form $y = \frac{mx+n}{px+q}$ with $mq - np \neq 0$. Another system of coordinates $\{x'\}$, is connected with $\{x\}$ through the equation $x' = x - a$, where a is a constant. Then $x = x' + a$ and by replacing in the formula we find that the corresponding $y' = y - a$ will satisfy $y' + a = \frac{m(x'+a)+n}{p(x'+a)+q}$, which is again of the same form:

$$y' = \frac{m'x' + n'}{p'x' + q'},$$

where $m' = m - pa$, $n' = ma + n - a(pa+q)$, $p' = p$, $q' = pa + q$, with $m'q' - n'p' = mq - np$.

Homographic relations lie at the root of many important theorems, like these of Pappus, Desargues and Pascal. In our presentation, however, we chose a more elementary way, which avoids their analytical usage. In what follows we examine some auxiliary propositions, which gradually lead to the solution of of Castillon's problem.

Exercise 5.151. Show that a homographic relation between $\{x, y\}$ results also in the cases, in which the variables are connected with a relation of the form:

1. $y = \frac{mx+n}{px+q}$ with $mq - np \neq 0$ (definition).
2. $axy + bx + cy + d = 0$ with $ad - bc \neq 0$.
3. $xy + ax + by + c = 0$ with $c - ab \neq 0$.
4. $(x-a) \cdot (y-b) = c$ with $c \neq 0$.

Theorem 5.45. *For points $\{X, Y, Z, \ldots\}$ of the line ε the following properties hold:*

1. If $\{X,Y\}$ are related homographically and $\{Y,Z\}$ are also related homographically, then $\{X,Z\}$ are related homographically too.
2. More generally, if the points $\{X_1,...,X_n\}$, by pairs: (X_1,X_2), (X_2,X_3), ... , and (X_{v-1},X_v), are related homographically, then the first and the last, (X_1,X_v), are related homographically too.

Proof. Let us denote, as usual, with corresponding small letters the coordinates of the points relative to a system of coordinates for the line ε. The fact that points $\{X,Y\}$ are related homographically means that the corresponding coordinates satisfy a relation $y = \frac{mx+n}{px+q}$, and similarly for the corresponding coordinates for $\{Y,Z\}$ will satisfy a relation of the form $z = \frac{m'y+n'}{p'y+q'}$. Substituting the second into the first one we find that

$$z = \frac{m'y+n'}{p'y+q'} = \frac{m'\frac{mx+n}{px+q}+n'}{p'\frac{mx+n}{px+q}+q'} = \frac{m''x+n''}{p''x+q''}, \text{ where}$$

$$m'' = mm'+pn', \quad n'' = nm'+qn', \quad p'' = mp'+pq', \quad q'' = np'+q'q,$$
$$m''q'' - n''p'' = (mq-np)(m'q'-n'p') \neq 0.$$

This shows (1). Applying (1) to $\{X_1,X_2\}$ and $\{X_2,X_3\}$, we see that $\{X_1,X_3\}$ are related homographically. Applying next (1) to $\{X_1,X_3\}$ and $\{X_3,X_4\}$, we see that $\{X_1,X_4\}$ are also related homographically. Continuing this way (inductively), we prove that $\{X_1,X_v\}$ are also related homographically.

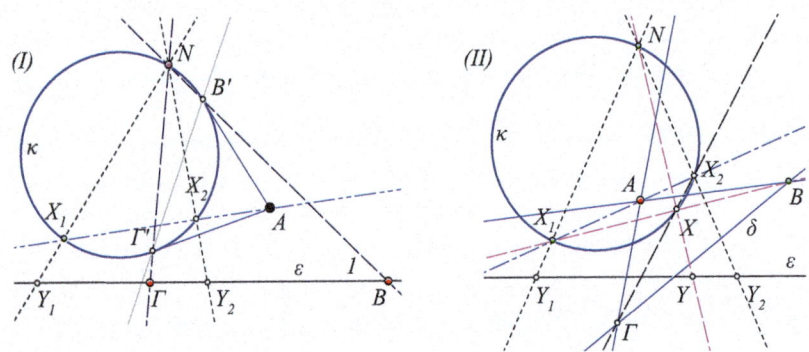

Fig. 5.143: Homographic relation through a circle κ and a point A

Theorem 5.46. *Let $\{\kappa,\varepsilon,N,A\}$ be respectively, a circle, a line, a point on κ and a point not contained in κ and ε. For every point X_1 of the circle κ we consider the other than X_1 intersection point X_2 with line AX_1. Then the intersection points $\{Y_1,Y_2\}$ of the lines $\{NX_1,NX_2\}$ with ε are related homographically.*

5.21. CASTILLON'S PROBLEM, HOMOGRAPHIC RELATIONS

Proof. In the case A is external to the circle (See Figure 5.143-I), we consider the points of contact $\{B', \Gamma'\}$ of the tangents from A and the corresponding intersection points $\{B, \Gamma\}$ of the lines $\{NB', N\Gamma'\}$ with ε. We observe that the cross ratio of the line pencil $N(B', \Gamma', X_1, X_2)$ is independent of the position of N on the circle (Theorem 5.32). When N coincides with B', then NB' coincides with the tangent at B' and $N\Gamma'$ with the polar of A. By the definition of the polar, the pencil $N(B', \Gamma', X_1, X_2)$ will be harmonic and consequently points $\{B, \Gamma, Y_1, Y_2\}$ make a harmonic quadruple. The conclusion, therefore, in this case results from proposition 5.18 for the value of the cross ratio $k = -1$.

In the case where A is internal to the circle (See Figure 5.143-II), we consider an arbitrary point B of the polar δ of A, relative to κ, as well as the pole Γ of AB, which is a point of δ. According to corollary 5.13, if X is the intersection point of $X_1 B$ with the circle, then $X X_2$ will pass through Γ. Suppose that Y is the intersection point of NX with ε. Point B is external to the circle and, by the preceding part of the proof, points $\{Y_1, Y\}$ are related homographically. According to theorem 5.45 points $\{Y_1, Y_2\}$ will be related homographically too.

Remark 5.33. The correspondence $X_1 \mapsto X_2$ defined on the circle κ through the point A, as it is described in the preceding theorem, coincides with the inversion relative to a circle λ, which has its center at point A and is orthogonal to κ. Using this view of the correspondence, one can give a solution to the general problem of Castillon, for polygons inscribed in circles (but not for polygons inscribed in conics) relying on the properties of inversion ([54, p.38]).

Remark 5.34. In the preceding theorem, it is worth noticing the characteristic difference of the two cases of this interesting homographic relation. When point A is external to the circle κ (See Figure 5.143-I), then, during the rotation of the secant line about A and the correspondence $X_1 \mapsto X_2$ points Γ' and B' are fixed points. Similarly points Γ and B are also fixed points of the correspondence $Y_1 \mapsto Y_2$. using the coordinates $\{x, y\}$ of $\{Y_1, Y_2\}$ relative to some system of coordinates of the line ε, this is equivalent to the fact that the homographic relation $y = f(x) = \frac{mx+n}{px+q}$ has two fixed points at the coordinates β of B and γ of Γ. In other words the numbers $\{\beta, \gamma\}$ and only these, satisfy the equation

$$\frac{mx+n}{px+q} = x \quad \Leftrightarrow \quad px^2 + (q-m)x - n = 0. \tag{1}$$

while for every other number x, we have $y = \frac{mx+n}{px+q} \neq x$. Then $\{\beta, \gamma\}$ are the roots of the quadratic polynomial of (1) and their existence means that $(q-m)^2 + 4pn \geq 0$.

It is easy to see that something like that does not happen in the case where A is internal to the circle. In this case we can ascertain that for every $Y_1(x)$ of ε and the corresponding $Y_2(y)$, we have $Y_2(y) \neq Y_1(x)$. In other words, in this case the homographic relation admits no fixed points or, equivalently, the homographic relation of their coordinates satisfies $y = \frac{mx+n}{px+q} \neq x$ for every value of the variable x. This again is equivalent to the fact that the quadratic equation of (1) does not have real roots and its discriminant satisfies $(q - m)^2 + 4pn < 0$.

Theorem 5.47. *Let $\{\kappa, \varepsilon, N, A_1, A_2, \ldots, A_\nu\}$ be respectively, a circle, a line, a point on κ and points not belonging to κ or ε. For every point X_1 of κ we consider the chain of points $\{X_1, X_2, \ldots, X_{\nu+1}\}$, where point X_{i+1} is equal to the other than X_i intersection of κ with the line $A_i X_i$, for $i = 1, \ldots, \nu$. Then the lines $\{NX_1, NX_{\nu+1}\}$ define, through their intersection with ε, points $\{Y_1, Y_{\nu+1}\}$ which are related homographically.*

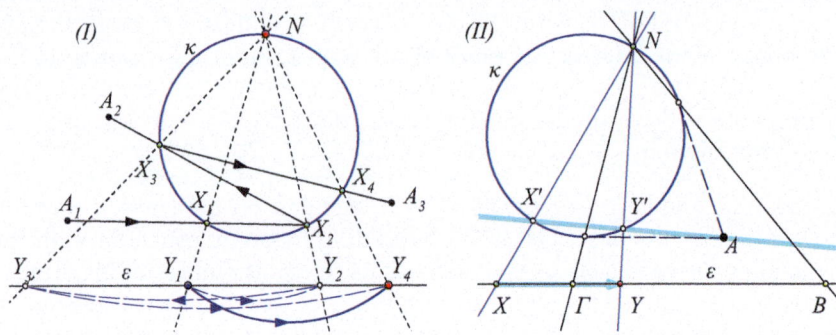

Fig. 5.144: $\{Y_1, Y_4\}$ in homogr. relation   Collinear points $\{X', Y', A\}$

Proof. Let $\{Y_1, \ldots, Y_{\nu+1}\}$ be the intersection points of the lines $\{NX_1, \ldots, NX_{\nu+1}\}$ with ε. Figure 5.144-I shows the process described by the theorem for $\nu = 3$. By the theorem 5.46, points (Y_1, Y_2) are related homographically. The same holds for the pairs of points (Y_2, Y_3), ..., $(Y_\nu, Y_{\nu+1})$. The conclusion follows by applying theorem 5.45.

Exercise 5.152. (Partial converse of theorem 5.46) For every point X of a fixed line ε we consider the harmonic conjugate Y relative to two fixed points $\{B, \Gamma\}$ of the line. Show that for an arbitrary circle κ and one of its points N, the second intersection points $\{X', Y'\}$ with the lines $\{NX, NY\}$ define a line $X'Y'$ passing through a fixed point A (See Figure 5.144-II).

Construction 5.4 (Castillon's problem) *In a given circle κ to inscribe a polygon with ν sides, so that each of these sides passes respectively through one of the given points $\{A_1, \ldots, A_\nu\}$.*

Construction: Let us see the construction in the case of quadrilaterals ($\nu = 4$), using a method which can be generalized directly for every value of ν. We consider a line ε and an arbitrary point N on κ (See Figure 5.145-I). Then, beginning from the arbitrary point X_1 on the circle κ, we define its second intersection point X_2 with $A_1 X_1$, next the second intersection point X_3 of $A_2 X_2$

with κ etc., creating the chain of points $\{X_1,...,X_5\}$ on the circle and the corresponding chain $\{Y_1,...,Y_5\}$ on line ε. Next, we apply the theorem 5.47, according to which, the points $\{Y_1(x), Y_5(y)\}$ are related homographically. This means that the coordinates of points $\{Y_1, Y_5\}$ are connected by a relation of the form $y = \frac{mx+n}{px+q}$, where the coefficients $\{m,n,p,q\}$ are determined from the given data, i.e. the circle κ, points $\{A_1, A_2, A_3, A_4\}$, point N, line ε and the selected system of coordinates of ε.

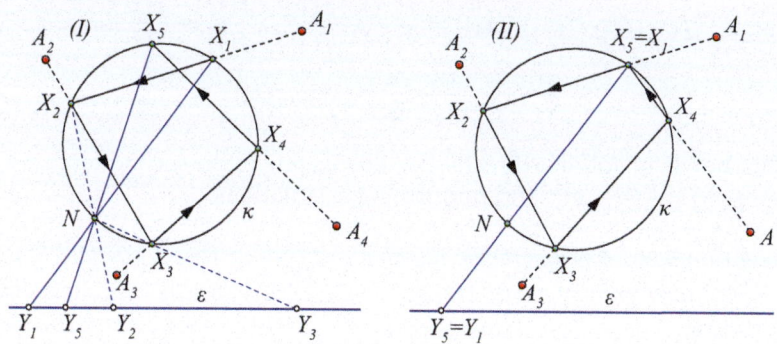

Fig. 5.145: Castillon's problem through homographically related $\{Y_1, Y_5\}$

The solution of the problem then corresponds to some particular positions of X_1 on the circle, for which the corresponding X_5 coincides with X_1. This again happens when on line ε, the corresponding Y_5 coincides with Y_1 (See Figure 5.145-II), or equivalently, when $y = \frac{mx+n}{px+q}$ satisfies $y = x$. In other words, x is the solution of the equation

$$x = \frac{mx+n}{px+q} \quad \Leftrightarrow \quad px^2 + (q-m)x - n = 0 \qquad (2)$$

It suffices then to construct $\{m,n,p,q\}$, from the given data of the problem and to find the roots of the quadratic polynomial of (2). The problem, depending on the sign of the discriminant of the polynomial, has two, one or no solutions.

Remark 5.35. The preceding proof leaves a feeling of incompleteness. How is $y = \frac{mx+n}{px+q} = f(x)$ determined from the given data of the problem? The question appears difficult, on first consideration. It can be dealt with, however, using a simple and unified way for any ν, if we exploit the basic properties of this function. Next propositions formulate these properties, which lead to the necessary complement of the preceding proof, that is the construction of the function, the determination of its fixed points and through them the determination of the position of X_1 on κ, equivalently of Y_1 on ε, which gives the solution to Castillon's problem.

Proposition 5.19. *A homographic relation,* $y = \frac{mx+n}{px+q} = f(x)$, *preserves the cross ratio of four numbers:*

$$(x_1 x_2; x_3 x_4) = \frac{x_1 - x_3}{x_2 - x_3} : \frac{x_1 - x_4}{x_2 - x_4}.$$

Proof. It suffices to show that for any quadruple of numbers $\{x_1, x_2, x_3, x_4\}$ holds

$$(f(x_1) f(x_2); f(x_3) f(x_4)) = \frac{f(x_1) - f(x_3)}{f(x_2) - f(x_3)} : \frac{f(x_1) - f(x_4)}{f(x_2) - f(x_4)} =$$

$$\frac{x_1 - x_3}{x_2 - x_3} : \frac{x_1 - x_4}{x_2 - x_4} = (x_1 x_2; x_3 x_4).$$

This, however, follows directly, by substituting in the second term $f(x_1) = \frac{mx_1+n}{px_1+q}$, $f(x_2) = \frac{mx_2+n}{px_2+q}$, etc. and performing calculations.

Proposition 5.20. *A homographic relation* $y = \frac{mx+n}{px+q} = f(x)$ *is completely determined from three pairs of points* $\{(x_1, y_1), (x_2, y_2), (x_3, y_3)\}$ *which are homographically related.*

Proof. Indeed by assumption, will hold $\{y_1 = f(x_1), y_2 = f(x_2), y_3 = f(x_3)\}$. Consequently, for arbitrary x and the corresponding $y = f(x)$, from the preservation of the cross ratio (Proposition 5.19), we'll have

$$\frac{y_1 - y_3}{y_2 - y_3} : \frac{y_1 - y}{y_2 - y} = \frac{x_1 - x_3}{x_2 - x_3} : \frac{x_1 - x}{x_2 - x}.$$

Performing calculations, we see that this relation can be written in the form

$$y = \frac{mx+n}{px+q} \quad \text{with} \quad m = ky_2 - y_1, \quad n = x_2 y_1 - kx_1 y_2, \quad p = k-1, \quad q = x_2 - kx_1$$

$$\text{where} \quad k = \frac{y_1 - y_3}{y_2 - y_3} : \frac{x_1 - x_3}{x_2 - x_3}.$$

Remark 5.36. In a continuation of the preceding remark, we can now use the last proposition to determine the function $y = f(x) = \frac{mx+n}{px+q}$, which gives the homographic relation of $\{Y_1(x), Y_5(y)\}$ in the proof of construction 5.4. For this, it suffices to consider three arbitrary points $\{X_1, X_1', X_1''\}$ on the circle κ and to determine the corresponding $\{Y_1, Y_1', Y_1''\}$ and their coordinates $\{x_1, x_2, x_3\}$ on ε. Subsequently, through the corresponding chains of points to determine the corresponding $\{Y_5, Y_5', Y_5''\}$ and their coordinates $\{y_1, y_2, y_3\}$. Applying the proposition 5.20 we determine $f(x)$, solving for y the equation

$$\frac{y_1 - y_3}{y_2 - y_3} : \frac{y_1 - y}{y_2 - y} = \frac{x_1 - x_3}{x_2 - x_3} : \frac{x_1 - x}{x_2 - x}.$$

5.21. CASTILLON'S PROBLEM, HOMOGRAPHIC RELATIONS

This process may be somewhat lengthy, especially for large v, but otherwise presents no other difficulty.

Exercise 5.153. Let $\{X(x), Y(y)\}$ be two variable points on the line ε. Show that if there exist two fixed points $\{A(a), B(b)\}$ of ε and a constant c, such that the signed distances satisfy the relation

$$AX \cdot BY = c, \tag{3}$$

then $\{x, y\}$ are related homographically. Under certain assumptions the converse holds as well, that is with the assumption that $\{x, y\}$ satisfy a homographic relation, then there exist points $\{A(a), B(b)\}$, which satisfy (3).

Hint: For the converse, suppose that the homographic relation has the form $xy + ax + by + c = 0$ and see that for $u = -b$, $v = -a$, $d = ab - c$, it takes the form $(x-u)(y-v) = d$.

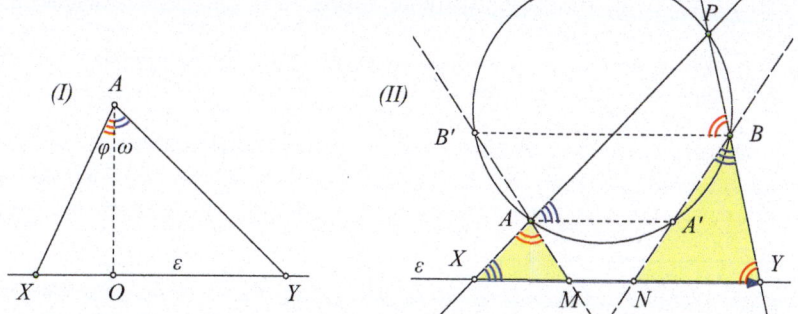

Fig. 5.146: Homographic rel. through angle ... and through circle

Exercise 5.154. The angle $\widehat{XAY}$ of fixed measure and with fixed vertex at point A rotates about A and defines the points of intersection $\{X(x), Y(y)\}$ of its sides with the fixed line ε (See Figure 5.146-I). Show that $\{X, Y\}$ are related homographically.

Hint: Consider a system of coordinates with origin at the projection O of A on ε. If the angle is not a right one, apply formula $\tan(\phi + \omega) = \frac{\tan(\phi) + \tan(\omega)}{1 - \tan(\phi)\tan(\omega)}$ and see that $\frac{(y-x)d}{d^2 + xy} = k$, where $d = |AO|$ and $k = \tan(\phi + \omega)$, where $\phi + \omega = \chi$ is the fixed measure of the rotating angle. If the rotating angle is right, then $x \cdot y = -d^2$ and the homographic relation is an **involution** (i.e. homography f satisfying $f^{-1} = f$).

Exercise 5.155. Point P is moving on a fixed circle and the lines $\{PA, PB\}$, which pass through two fixed points $\{A, B\}$ of the circle, intersect line ε at points $\{X(x), Y(y)\}$ (See Figure 5.146-II). Show that $\{X, Y\}$ are related homographically.

Hint: ([105, VII, p.24]) Consider the points $\{A', B'\}$, at which the parallels of ε from $\{A, B\}$ intersect the circle a second time. Define the points $\{M, N\}$, at which $\{AB', A'B\}$ intersect ε and show that the product of the signed distances $MX \cdot NY = MA \cdot NB$ is fixed. Finally apply exercise 5.153.

The homographic relation, the way we defined it previously, is a relation of two variable points $\{X, Y\}$ *of the same line* ε. The notion, however, can be naturally extended to points of two different lines $\{\varepsilon, \varepsilon'\}$. For this, it suffices to choose a system of coordinates on each one of the lines and say that the corresponding points $\{X(x), Y(y)\}$ of the lines $\{\varepsilon, \varepsilon'\}$ are related homographically, when, for their coordinates, holds a homographic relation $y = \frac{mx+n}{px+q}$ with $mq - pn \neq 0$. We easily see that the general properties of the homographic relation are carried over in this case too. For example, the definition is independent of the special systems we use (in this case exercise 5.150 still holds), the relation is symmetric relative to $\{X, Y\}$ and the analogue of theorem 5.45 holds for points $\{X_1, ..., X_v\}$ of the lines $\{\varepsilon_1, ..., \varepsilon_v\}$. Next exercises give examples of such homographic relations.

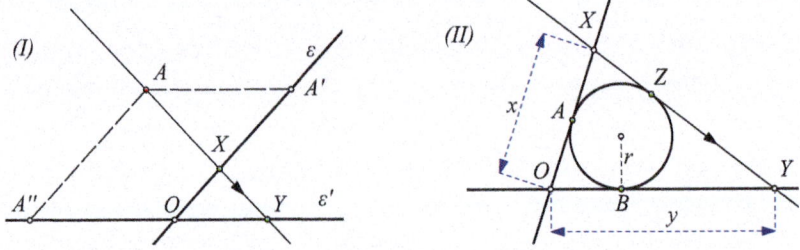

Fig. 5.147: Homographic relations between points of different lines

Exercise 5.156. Let $\{\varepsilon, \varepsilon', A\}$ be respectively two lines and a point not lying on them. For every point X of ε we define the point of intersection Y of AX with ε'. Show that $\{X, Y\}$ are related homographically.

Hint: Project point A parallel to the lines, to points $\{A', A''\}$ (See Figure 5.147-I). From the created similar triangles we have $\frac{OX}{XA'} = \frac{OY}{AA'} = \frac{OY}{A''O}$. This relation, using coordinates $\{x, y\}$ with origin at the intersection point O of the lines, translates to the homographic relation

$$\frac{x}{a'-x} = \frac{y}{-a''} \Leftrightarrow y = \frac{a''x}{x-a'}.$$

Exercise 5.157. Consider the tangents to a circle at its points $\{A, B\}$, intersecting at O. Point Z moves on the circle and the tangent to it intersects $\{OA, OB\}$ at points $\{X, Y\}$. Show that $\{X, Y\}$ are related homographically (See Figure 5.147-II).

5.22 MALFATTI'S PROBLEM

Hint: Use (proposition 5.4), according to which the radius r of the circle can be expressed through:

$$r^2 = \frac{|OA||AX||BY|}{|OA|+|AX|+|BY|} = \frac{a(x-a)(y-a)}{a+(x-a)+(y-a)} \Leftrightarrow y = (a^2+r^2)\frac{x-a}{ax-(a^2+r^2)},$$

where $a = |OA|$ and the systems of coordinates on $\{OA, OB\}$ have their origin at O.

5.22 Malfatti's problem

> A diagram is necessary to understand this obscure theorem; and when it is understood, the student says, - Of what service can it be to me? what does it matter? - He is disgusted with a science, of which he does not soon enough perceive the utility.
>
> *Voltaire, Dictionary, Geometry*

Malfatti's problem (1731-1807) asks for the construction of three circles $\{\kappa_1, \kappa_2, \kappa_3\}$ in the interior of a triangle. From the circles it is required that they are pairwise tangent and, each, be also tangent to two sides of the triangle (See Figure 5.148). There are various methods to solve this problem

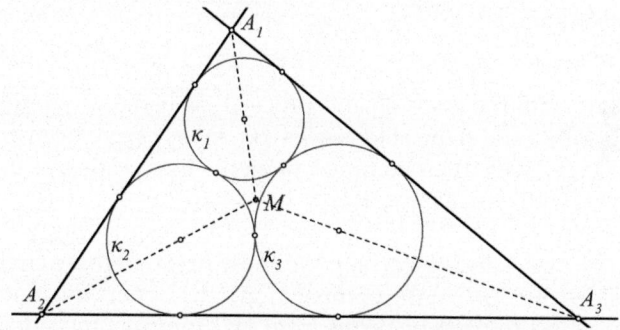

Fig. 5.148: Malfatti's problem

([70, p.147], [123, 17], [65]). Here we follow one, whose description comes from *Steiner*, whereas the complete proof is the work of Hart [134].

Construction 5.5 *Denoting the triangle by $A_1A_2A_3$ and its sides by $\{a_1, a_2, a_3\}$, instead of the usual $AB\Gamma$ etc., the construction of the solution can be achieved by the following procedure.*

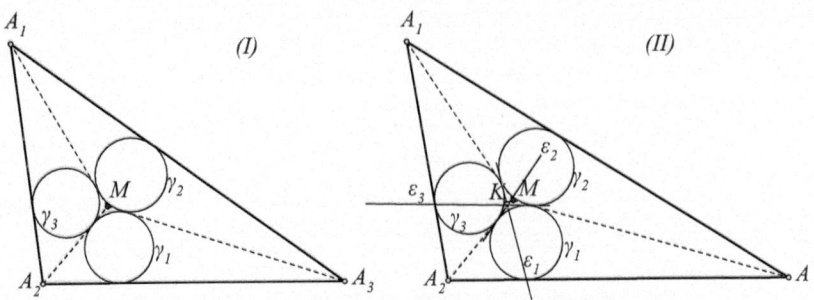

Fig. 5.149: Malfatti's problem, incenter M and point K

1. Consider the incenter M of the triangle and the inscribed circles $\{\gamma_1, \gamma_2, \gamma_3\}$ of the three triangles, respectively, $\{MA_2A_3, MA_3A_1, MA_1A_2\}$ (See Figure 5.149-I).
2. Line MA_1 is the common tangent to $\{\gamma_2, \gamma_3\}$, hence its symmetric ε_1, with respect to the line of centers of these circles, is also tangent to these circles. Analogously are defined the lines ε_2, common tangent to $\{\gamma_3, \gamma_1\}$ and ε_3, common tangent to circles $\{\gamma_1, \gamma_2\}$.
3. The three tangents $\{\varepsilon_1, \varepsilon_2, \varepsilon_3\}$ pass through a point K (See Figure 5.149-II).
4. The three tangents $\{\varepsilon_1 = \Delta_1 E_1, \varepsilon_2 = \Delta_2 E_2, \varepsilon_3 = \Delta_3 E_3\}$ pass also through the respective contact points $\{\Delta_1, \Delta_2, \Delta_3\}$ of the circles $\{\gamma_1, \gamma_2, \gamma_3\}$ with the respective sides of the triangle (Figure 5.150).
5. The quadrangles $\{A_1\Delta_2 K\Delta_3, A_2\Delta_3 K\Delta_1, A_3\Delta_1 K\Delta_2\}$ are circumscriptible and their corresponding inscribed circles $\{\kappa_1, \kappa_2, \kappa_3\}$ represent the solution to the problem.

Proof. We start with the analysis and proceed, as usual, to synthesis. Thus, let as suppose that the requested circles $\{\kappa_1, \kappa_2, \kappa_3\}$ have been constructed and $\{P_1 K, P_2 K, P_3 K\}$ are their common tangents intersecting at their radical center K, so that $|KP_1| = |KP_2| = |KP_3|$.

We define the circles $\{\gamma_1, \gamma_2, \gamma_3\}$, inscribed correspondingly to triangles $\{a_1, \varepsilon_2, \varepsilon_3\}$, $\{a_2, \varepsilon_3, \varepsilon_1\}$ and $\{a_3, \varepsilon_1, \varepsilon_2\}$. We show first that these circles are tangent to the sides of $\triangle AB\Gamma$ at the points $\{\Delta_1, \Delta_2, \Delta_3\}$, where the inner common tangents $\{\varepsilon_1, \varepsilon_2, \varepsilon_3\}$ to pairs of circles from $\{\kappa_i\}$ intersect the sides of $\triangle AB\Gamma$. Indeed, it holds

$$|E_3\Delta_1| - |E_2\Delta_1| = |E_3 B_1| - |E_3\Gamma_1| = |E_3 P_3| - |E_2 P_2| = |E_3 K| - |E_2 K|,$$

which means that Δ_1 is a contact point of the circle γ_1 with the side a_1 (Exercise 5.2). Analogously follows the corresponding property for Δ_2 and Δ_3.

Next, we consider the second inner tangents $\{\eta_1, \eta_2, \eta_3\}$ of the three circles $\{\gamma_1, \gamma_2, \gamma_3\}$. These are three lines, which, in view of exercise 5.120, pass through a point M. We prove that the three lines coincide with the bisectors

5.22. MALFATTI'S PROBLEM

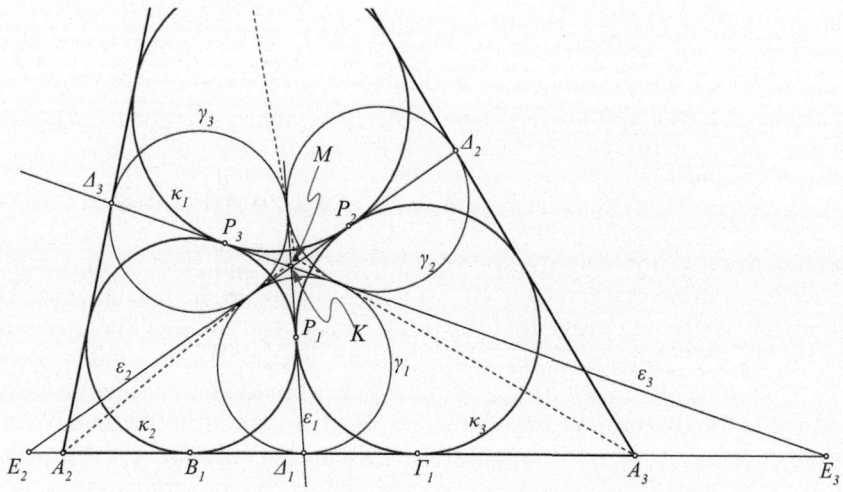

Fig. 5.150: Malfatti's problem, the other tangents $\{\varepsilon_1, \varepsilon_2, \varepsilon_3\}$

of the triangle $A_1A_2A_3$ and, consequently, point M coincides with the incenter of this triangle.

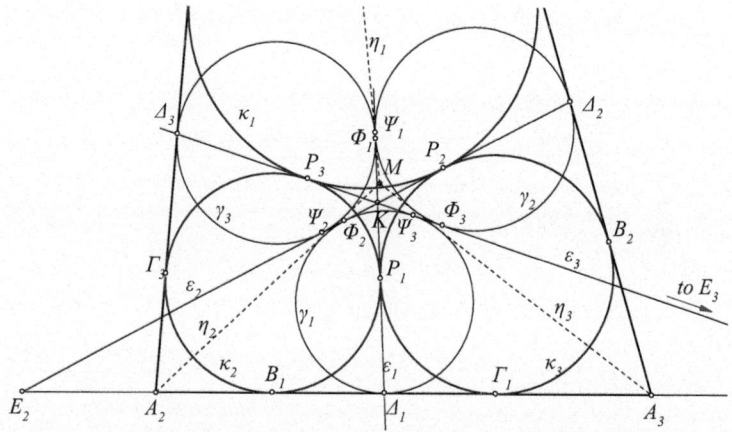

Fig. 5.151: Equal segments $|\Delta_1\Gamma_1| = |\Phi_2 P_2|$ ... on common tangents to circles

To show this, we consider the contact points (See Figure 5.151):
- $\{\Phi_1, \Psi_1\}$ of line ε_1 with circles, correspondinly, $\{\gamma_3, \gamma_2\}$.
- $\{\Phi_2, \Psi_2\}$ of line ε_2 with circles, correspondinly, $\{\gamma_1, \gamma_3\}$.
- $\{\Phi_3, \Psi_3\}$ of line ε_3 with circles, correspondinly, $\{\gamma_2, \gamma_1\}$.

We show first that the lines $\{\eta_i\}$ pass correspondingly through points A_i. To see this, say for point A_2 and line η_2, it suffices to show (Exercise 5.3) that

$$|\Phi_2\Psi_2| = |A_2\Delta_3| - |A_2\Delta_1|.$$

This, because, by the symmetry with respect to the line of centers of the circles $\{\gamma_1, \gamma_3\}$, the distance $|\Phi_2\Psi_2|$ equals to the distance of the contact points of these circles with η_2. But this equality follows from the relations:

$$|A_2\Delta_3| - |A_2\Delta_1| = |\Gamma_3\Delta_3| - |B_1\Delta_1| = |P_1\Phi_1| - |P_3\Psi_3| = |\Phi_2\Psi_2|.$$

Analogously is proved that η_1 passes through A_1 and η_3 passes through A_3.

Next we show that the lines $\{\eta_i\}$ are the bisectors of the triangle $A_1A_2A_3$. Thus, for A_2 say, we prove that it is contained in the circle of similitude of the circles $\{\gamma_1, \gamma_3\}$, hence it sees these two circles under equal angles (Exercise 4.82). In view of exercise 4.194, point A_2 lies on the circle of similitude of $\{\gamma_1, \gamma_3\}$, if the line joining the contact points $\Delta_3\Delta_1$ cuts from the two circles equal chords. Last condition is equivalent with the equality of the powers $p(\Delta_3, \gamma_1) = p(\Delta_1, \gamma_3)$. This, in turn, is equivalent with the equality of the tangents $|\Delta_3\Psi_3| = |\Delta_1\Phi_1|$ to the circles, which results from the equalities:

$$|\Delta_1\Phi_1| = |\Delta_1P_1| + |P_1\Phi_1| = |\Delta_1B_1| + |\Delta_3\Gamma_3| = |\Delta_3P_3| + |P_3\Psi_3| = |\Delta_3\Psi_3|.$$

Thus, line η_2 coincides with the bisector of the triangle at A_2. Analogous arguments identify the other bisectors with the lines η_1 and η_3.

The preceding arguments show that, if there is a solution to Malfatti's problem, then this can be delivered by the preceding procedure. It remains, though, the question of existence of a solution, which Hart addresses with a continuity argument, as follows: We consider a small circle γ_1, which is

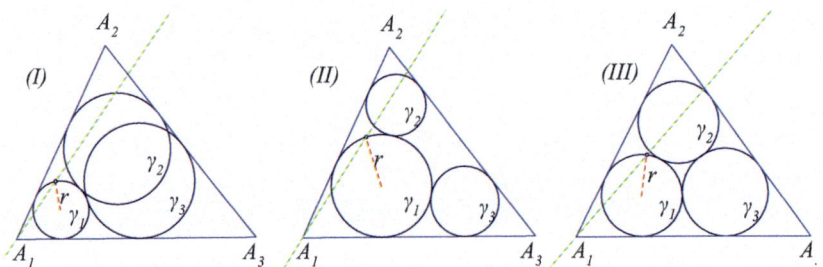

Fig. 5.152: Existence of solution to Malfatti's problem

tangent to the sides $\{A_1A_2, A_1A_3\}$ of the triangle and two additional circles $\{\gamma_2, \gamma_3\}$, which are tangent to γ_1 and, each, tangent to two sides of the triangle. These three circles *depend continuously* on the radius r of the circle γ_1. For small values of r the two circles $\{\gamma_2, \gamma_3\}$ will intersect (See Figure 5.152-I). For greater values of r, these circles will be non-intersecting (See Figure 5.152-

5.23. CALABI'S TRIANGLE

II). Hence, by continuity, there will be an intermediate value r_0, for which the two circles $\{\gamma_2, \gamma_3\}$ will be tangent (See Figure 5.152-III). Since, during this change of the radius r, the circles remain all the time tangent to γ_1, we obtain, this way, a solution to Malfatti's problem.

Remark 5.37. We should notice that the initial problem, formulated by Malfatti, was of a practical nature. It asked to cut three cylindrical columns of the maximal possible volume from a triangular prism of marmor. Equivalently, to place inside a given triangle three non overlapping discs with the maximal possible total area. Initially the problem was considered equivalent with the one we solved above. It was proved, though, relatively recently, that the above solution is never identical with that of the original problem ([144, p.145]). In figure 5.153 we see the difference of the solutions to the two problems in a pair of equal triangles. In the upper triangle is seen the solution of the initial problem, formulated by Malfatti, as well as the total area of the three solution circles. In the lower triangle is seen the solution of the problem discussed above and the total area of the corresponding three solution circles.

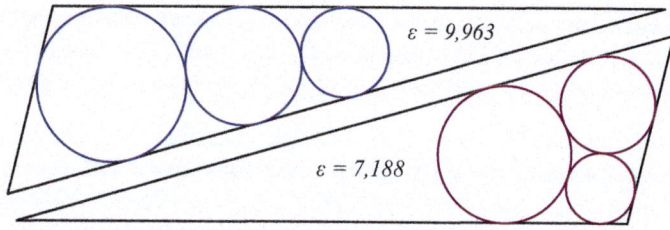

Fig. 5.153: Malfatti's initial problem

5.23 Calabi's triangle

> We must be the change we wish to see in the world.
>
> M. Gandhi, L.A. Times, 6/30/1989

This section, one would say, exaggerates the meaning of a detail, since it restricts to the study of a particular isosceles triangle. The subject though is appropriate for a review of some relations and ideas developed so far. At the same time it shows that a problem, that has been already solved, can be looked upon from another viewpoint and lead to new ideas and generalizations. The isosceles triangle we are talking about is related to exercise

3.106 and is called **Calabi's triangle** (1923-). As will be seen below, this triangle is the unique up to similarity, different from the equilateral, accepting in its interior three squares of maximal area. In all other cases of triangles we can place, in general, one only such maximal square, two if the triangle is isosceles and three if it is equilateral.

Our first proposition clarifies, how such a square of maximal area is placed inside a given triangle.

Theorem 5.48. *If a square of the maximal possible area is inscribed in a given triangle, then one of its sides is part of a side of the triangle.*

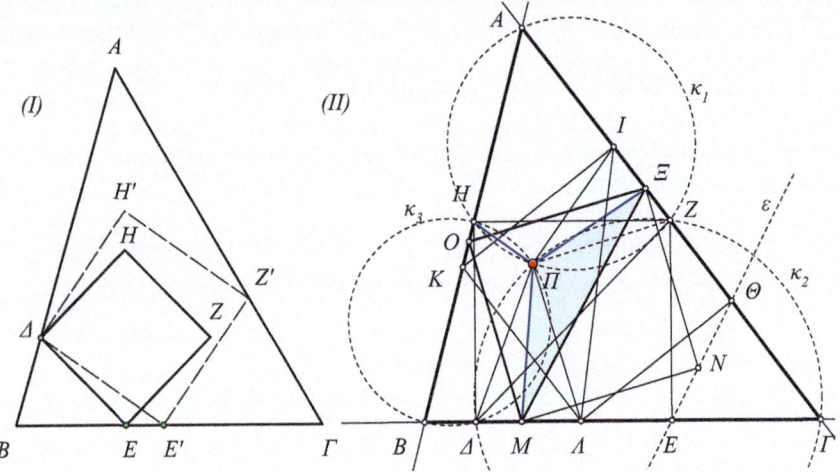

Fig. 5.154: Squares inscribed in a triangle

Proof. To begin with, we see easily, that if a square inscribed inside a triangle has at most two of its vertices on the sides of the triangle, then we can increase slightly its dimensions and obtain a bigger one still inscribed in the square (See Figure 5.154-I).

Thus, let us suppose that the inscribed square $MN\Xi O$ has at least three of its vertices on respective sides of the triangle (See Figure 5.154-II). Any such square defines through its diagonal $M\Xi$ an isosceles right triangle ΞOM also inscribed in the triangle. The triangle ΞOM defines in turn its pivot Π, so that every other inscribed in $AB\Gamma$ isosceles right triangle with corresponding vertices on the same sides which contain the vertices of ΞOM, has point Π at the same relative position with the corresponding in ΞOM. Thus, in the figure 5.154-II, which contains the inscribed similar triangles $\{\Xi OM, IK\Lambda, ZH\Delta\}$, visible are also the corresponding similar triangles with a vertex at Π, which are $\{\Xi\Pi M, I\Pi\Lambda, Z\Pi\Delta\}$. Obviously the greater square will correspond to a greater triangle $\Xi\Pi M$. Since all these triangles are similar, the greater of them is the one with the greater side ΠM. But the sides

5.23. CALABI'S TRIANGLE

ΠM, corresponding to squares totally contained inside the triangle $AB\Gamma$, are contained in the angle $\widehat{\Delta \Pi \Lambda}$ and the maximal one is obtained when ΠM coincides with one of the sides of this angle, in the case $AB\Gamma$ has acute angle $\widehat{B}$. But then, the corresponding square $MN\Xi O$ obtains the position of $ZH\Delta E$ or $IK\Lambda\Theta$, which are squares with one of their sides being part of a side of the triangle $AB\Gamma$.

In the case the angle $\widehat{B}$ is obtuse the construction of the exercise 3.106 does not deliver a square as we wish it, totally contained in the triangle (See Figure 5.155-I). The proof in this case proceeds along the same line with respect to the angle $\widehat{\Gamma}$, which then is acute (See Figure 5.155-II). In this case the maximal square is determined through the greater of the segments $\{\Pi\Theta, \Pi Z\}$, which corresponds to the square $IK\Lambda\Theta$ or the square ΔEZH, which has its vertex Δ coinciding with the vertex B of the triangle. This is also the case with Calabi's triangle, as will be seen below

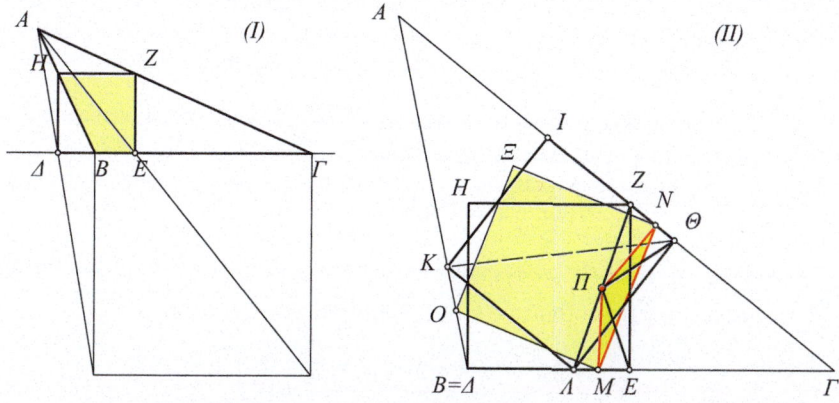

Fig. 5.155: The case of an obtuse angled triangle

Theorem 5.49. *There is exactly one, up to similarity, isosceles and non-equilateral triangle into which, a square of maximal area can be placed in three different ways. In this isosceles the ratio $t = x/y$ of its base $x = |B\Gamma|$ to the leg $y = |AB|$, satisfies the equation* $2t^3 - 2t^2 - 3t + 2 = 0$.

Proof. In exercise 3.106 it is constructed the square ΔEHI (See Figure 5.156-(1)). For its side $z = |\Delta E|$ a simple calculation gives

$$\frac{x}{z} = \frac{h+x}{h} = 1 + \frac{x}{h} = 1 + \frac{2x}{\sqrt{4y^2 - x^2}},$$

where h is the altitude of the triangle from A. If the squares $\{\Delta EHI, ZK\Lambda\Lambda\}$ are equal, then the triangle $A\Gamma\Delta$ is isosceles and it follows that $x - y = (x -$

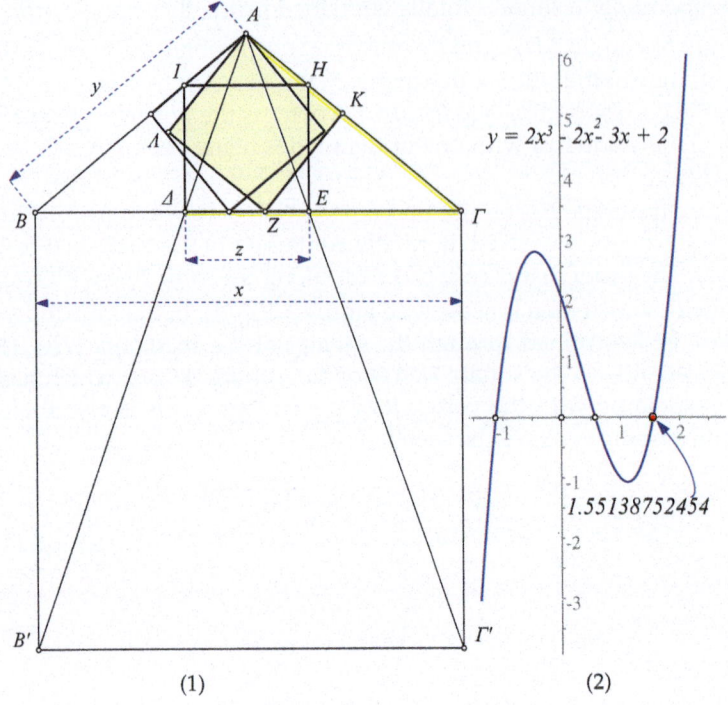

Fig. 5.156: Calabi's triangle

$z)/2$. Eliminating z from the two sides, we obtain the relation

$$\frac{4(2y^3 - 3xy^2 - 2x^2y + 2x^3)}{(2y-x)^2(2y+x)} = 0,$$

which implies the equation of third degree. Calabi's triangle (See Figure 5.156-(1)) results from the greater root of the third degree equation, which is approximately $t = 1,551388752454...$ (See Figure 5.156-(2)) and leads to an isosceles with angles approximately of 39:07:55 and 101:44:10 degrees. This triangle is not constructible by ruler and compass alone.

Exercise 5.158. Using figure 5.156-(1), show that triangles $AB'\Gamma'$ and $A\Gamma\Delta$ are similar and point Z, which determines the square $AZK\Lambda$, is the harmonic conjugate of Γ with respect to $\{\Delta, E\}$.

5.24 Comments and exercises for the chapter

> Would you have a man reason well, you must use him to it betimes; exercise his mind observing the connection between ideas, and following them in train. Nothing does this better than mathematics, which therefore, I think should be taught to all who have the time and opportunity, not so much to make them mathematicians, as to make them reasonable creatures;
>
> J. Locke, *Of the Conduct of the Understanding VI*

Exercise 5.159. Consider the circumcircle of the triangle $AB\Gamma$ and the sagittas $|P\Sigma|$ of the arc $\widehat{B\Sigma\Gamma}$ and $|P\Sigma'|$ of the arc $\widehat{B\Sigma'\Gamma}$ (See Figure 5.157). Show that

$$|P\Sigma| = \frac{1}{2}(r_A - r) \quad \text{and} \quad |P\Sigma'| = \frac{1}{2}(r_B + r_\Gamma).$$

Use these relations to give an alternative proof of exercise 5.15-(2):

$$4R + r = r_A + r_B + r_\Gamma,$$

where $\{R, r, r_A, r_B, r_\Gamma\}$ are respectively the circumradius the inradius and the exradii of the triangle (see also the related formulas in exercise 5.42).

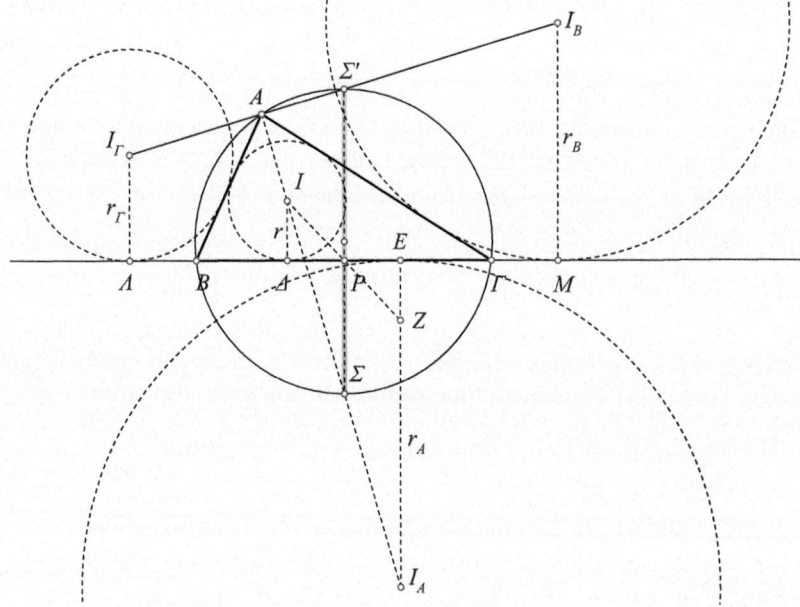

Fig. 5.157: Sagittas $|P\Sigma|$ of the arc $\widehat{B\Sigma\Gamma}$ and $|P\Sigma'|$ of the arc $\widehat{B\Sigma'\Gamma}$

Exercise 5.160. Construct a triangle $AB\Gamma$, whose given are the radii $\{R, r, r_A\}$.

Exercise 5.161. Show that the area of the triangle $AB\Gamma$ is $E = R\tau'$, where $\{R, \tau'\}$ are the circumradius and the semi-perimeter of the orthic triangle.

Hint: Use corollary 3.20 and exercise 3.123 to express the semi-perimeter of the orthic.

Exercise 5.162. Given two intersecting lines and a point P not lying on them, to draw a third line through P so that the triangle formed by the three lines has a given perimeter.

Exercise 5.163. Point E is the intersection of the hypotenuses of the two right triangles $\{AB\Gamma, B\Gamma\Delta\}$ and Z its projection on AB (See Figure 5.158). Show that $z = |EZ|$ does not depend on $|AB|$ but only on $\{x = |AB|, y = |\Gamma\Delta|\}$.

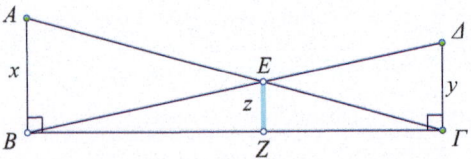

Fig. 5.158: z independent of $|AB|$

Exercise 5.164. Construct the bisector of two lines whose intersection point is outside the drawing sheet.

Exercise 5.165. Show that the similarity centers of the circumcircle and the Euler circle of the triangle $AB\Gamma$ are the orthocenter H and the centroid G of the triangle.

Exercise 5.166. Construct a triangle $AB\Gamma$ for which are given the position of the vertex A, the position of its orthocenter H and the position of the center Π of its Euler circle.

Exercise 5.167. Construct a triangle $AB\Gamma$ for which are given the lengths $a = |B\Gamma|$, $b + c = |A\Gamma| + |AB|$ and the radius r of the inscribed circle.

Hint: In this as well as in the next exercise use Proposition 5.3.

Exercise 5.168. Construct a triangle $AB\Gamma$ for which are given the lengths $a = |B\Gamma|$, $b - c = |A\Gamma| - |AB|$ and the radius r of the inscribed circle.

Exercise 5.169. If $\{AA', BB', \Gamma\Gamma', H, O, R\}$ are the altitudes, the orthocenter, the circumcenter and the circumradius of the triangle $AB\Gamma$, show that it holds (See Figure 5.159-I):

1. $|AH||HA'| = |BH||HB'| = |\Gamma H||H\Gamma'| = \frac{1}{2}(R^2 - |OH|^2)$.

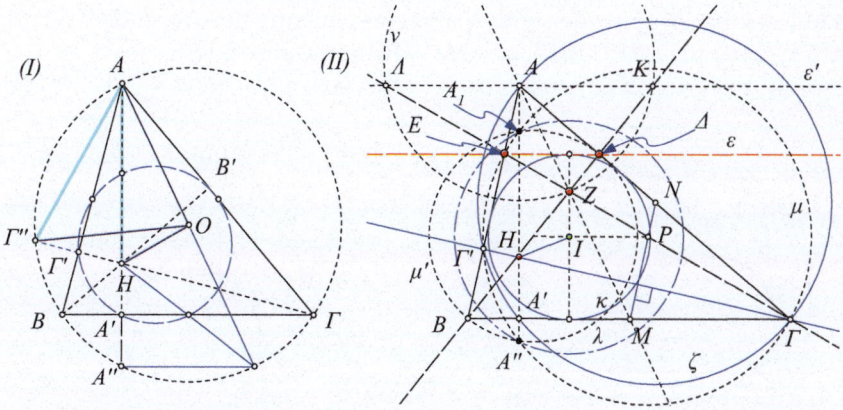

Fig. 5.159: Sides-circumcircle relations   Orthocenter-incenter relation

2. $\widehat{HAO} = |\widehat{B} - \widehat{\Gamma}|$.
3. $|AH| = 2R\cos(\widehat{A})$.
4. $|OH|^2 = 9R^2 - (a^2 + b^2 + c^2)$.

Hint: For (2) and (3) see figure 5.159-I. For (4) use the preceding and the rule of cosine, from which follows $a^2 + b^2 + c^2 = 2(\cos(\widehat{A}) + \cos(\widehat{B}) + \cos(\widehat{C}))$.

Exercise 5.170. ([139, p.124]) Let $\{AA', AM, H\}$ be, respectively, the altitude, the median and the orthocenter of the triangle $AB\Gamma$ (See Figure 5.159-II). Let also the parallels to $B\Gamma$ lines $\{\varepsilon = E\Delta, \varepsilon' = \Lambda K\}$ be respectively, the tangent to the incircle $\kappa(I, r)$ and the one passing through A. The points $\{E, \Delta\}$ are the intersections of ε with the sides $\{AB, A\Gamma\}$ and $\{\Lambda, K\}$ are the intersections of ε' with $\{E\Gamma, \Delta B\}$. Show the following properties:

1. Point A is the middle of ΛK.
2. $(B\Delta; ZK) = (\Gamma E; Z\Lambda) = -1$ and the triangle $Z\Lambda K$ is autopolar relative to the circle κ.
3. The circles $\{\mu, \mu'\}$ with diameters, respectively, $\{\Gamma E, B\Delta\}$ and $\lambda(I, \sqrt{2} \cdot r)$ are orthogonal to the circumcircle ν of the triangle ΛKZ and belong to a pencil of circles with radical axis the line AA'.
4. $|HA| \cdot |HA'| = 2r^2 - |HI|^2$.
5. $|HI|^2 = 2r^2 + 4R^2 - \frac{1}{2}(a^2 + b^2 + c^2)$, where R is the circumradius.

Hint: (1) and (2) are direct consequences of the properties of the trapezium $B\Gamma\Delta E$. (3) follows from the exercise 5.41. For (4) observe that $|HI|^2 - 2r^2$ is the power of H relative to the circle $\lambda(I, \sqrt{2} \cdot r)$. According to proposition 4.9 the radical axis of the circle ζ with diameter $A\Gamma$ and any one of the circles of the pencil $\mathscr{D}$ of (3) will pass through a fixed point H' contained in the radical axis AA' of the pencil. However the radical axis of ζ and of μ is their common chord of the altitude $\Gamma\Gamma'$, which intersects AA' at the orthocenter H.

It follows that H' coincides with H and consequently the power of H relative to ζ is equal to $-|H\Gamma|\cdot|H\Gamma'| = -|HA|\cdot|HA'|$ (Exercise 5.169).

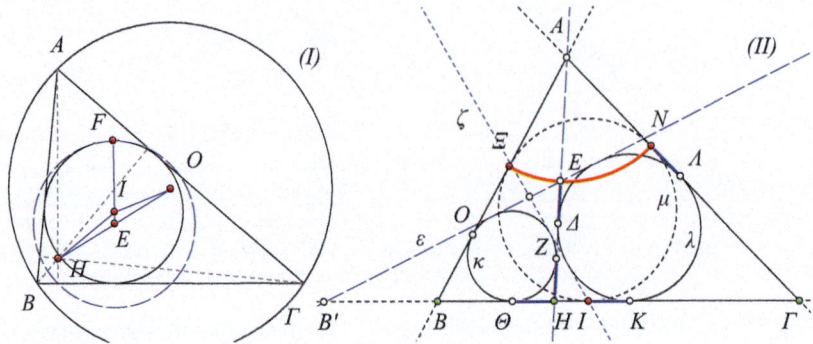

Fig. 5.160: Feuerbach again   Three inscribed circles

Exercise 5.171. Use Euler's theorem (§ 5.5), the preceding exercises, through which the lengths $|IH|$, $|OH|$ are determined, as well as the corollary for the length of the medians (3.24), to show, that $|IE| = \frac{R}{2} - r$, where $\{E, F, H, I, O\}$ are, respectively, the center of the Euler circle, the point of Feuerbach, the orthocenter, the incenter and the circumcenter (See Figure 5.160-I).

Exercise 5.172. Let H be a point of the base $B\Gamma$ of the triangle $AB\Gamma$. Let also κ and λ be the inscribed circles of the triangles ABH, $AH\Gamma$. Consider also the external tangent $\varepsilon \neq B\Gamma$ and the internal $\zeta \neq AH$. Show the following properties:

1. ζ passes always through the contact point I of the incircle μ of $AB\Gamma$.
2. $|AE| = |AN| = \tau - a$, where τ is half the perimeter of $AB\Gamma$ and $a = |B\Gamma|$.
3. $|N\Lambda| = |E\Delta| = |ZH| = |IK|$.
4. Similarly, $|O\Xi| = |EZ| = |\Theta I|$.
5. For variable H, the locus of the corresponding E is a circle with center at A.

Hint: Application of proposition 5.3, first on triangle EHB' to show $|E\Delta| = |HZ|$ (See Figure 5.160-II). Subsequently, application to $AH\Gamma$ and calculation of $|AE| = |A\Delta| - |E\Delta|$, leads to (2). (5) is a direct consequence. (2) implies also $|E\Delta| = |N\Lambda|$. Exercise 5.5 implies $|IK| = |HZ|$ and consequently $|IK| = |N\Lambda|$, therefore $|I\Gamma| = |N\Gamma|$ and I is also a contact point, as well as N.

Exercise 5.173. In a given quadrilateral $AB\Gamma\Delta$ inscribe a square.

Hint: Suppose that the requested square has been constructed and the point M found, about which the square is pivoting and its vertices X', Y' and Z'

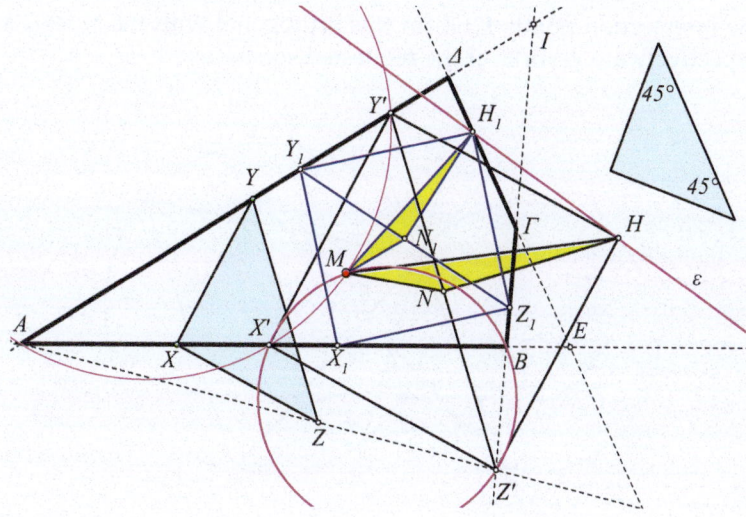

Fig. 5.161: Inscription of a square in a given quadrilateral using pivoting

move on the three sides respectively AB, $A\Delta$ and $B\Gamma$ of the given quadrilateral (Theorem 3.20) which form the triangle ABI (See Figure 5.161). Then the fourth vertex of the square will move on a fixed line ε, whose intersection with the fourth side $\Gamma\Delta$ of the given quadrilateral will determine one vertex of the square. It suffices then to find one such point M. The crucial observation is that M will also be the pivot for the inscription of the triangle $X'Y'Z'$, which is defined from one diagonal of the square, in triangle ABI. We determine, then, first a point M, as a pivot point of the right isosceles XYZ in triangle ABI.

The practical construction of such a pivot proceeds as follows. We place an arbitrary segment XY with endpoints on AB and $A\Delta$ and we construct the right isosceles XYZ having one side XY. Next we extend AZ until it intersects $B\Gamma$ at Z'. From Z' we draw parallels to XZ and ZY, which intersect respectively AB, $A\Delta$ at X', Y'. Triangle $X'Y'Z'$ is similar to XYZ and has its vertices on the sides of triangle ABI (or their extensions). We define next the Miquel point M of X', Y', Z' relative to ABI, which is also a pivot of $X'Y'Z'$ in ABI. Point M is determined as a second intersection point of the circles $(AX'Y')$ and $(BX'Z')$. Next we complete the isosceles triangle $X'Y'Z'$ to a square $X'Y'Z'H$ and we consider the pivoting of this square about M. During this pivoting, the square as much as the triangle MNH, where N is the middle of the diagonal $Y'Z'$, vary remaining similar to themselves and point H moves on a fixed line ε, which is determined from the given data. Let H_1 be the intersection point of $\Gamma\Delta$ with ε. We construct the triangle MN_1H_1, similar and similarly oriented to MNH, and we draw the orthogonal on N_1H_1 at

N_1. The intersection points Y_1, Z_1 of this orthogonal with the sides $A\Delta$ and $B\Gamma$ respectively are vertices of the requested square.

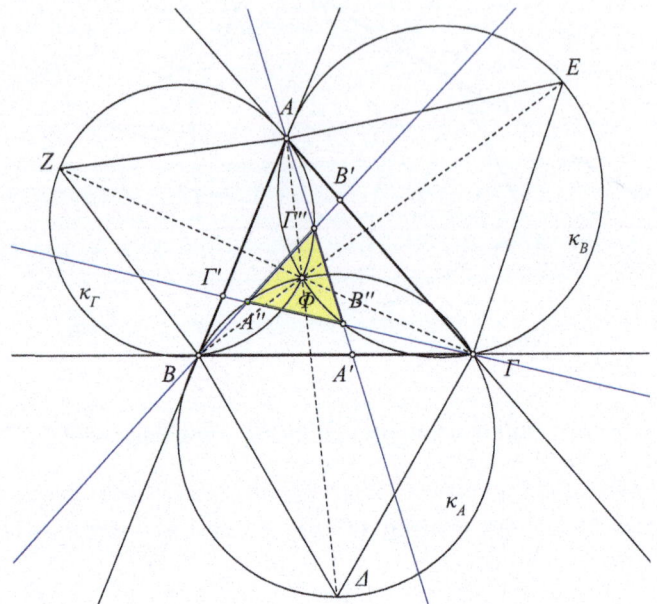

Fig. 5.162: Equilaterals $A''B''\Gamma''$ circumscribed to $AB\Gamma$

Exercise 5.174. Points $\{A', B', \Gamma'\}$ are respectively on the sides $\{B\Gamma, \Gamma A, AB\}$ of the triangle $AB\Gamma$ (See Figure 5.162). Show that the lines $\{AA', BB', \Gamma\Gamma'\}$ create an equilateral triangle $A''B''\Gamma''$, if and only if the points A'', B'', Γ'' are contained respectively in the circumcircles $\{\kappa_A, \kappa_B, \kappa_\Gamma\}$ of the equilaterals which are constructed on the sides of $AB\Gamma$.

Exercise 5.175. Three congruent circles with centers $\{A, B, \Gamma\}$ pass through point E. Show that E is the circumcenter of the triangle $AB\Gamma$, whose circumcircle is congruent to the three circles. Also show that the second intersection points of the three circles define a triangle $A'B'\Gamma'$ congruent to $AB\Gamma$, whose circumcenter is the orthocenter H of $AB\Gamma$. The two triangles have the same Euler circle and Euler line and are symmetric relative to the center of this circle (See Figure 5.163)

Exercise 5.176. Construct the triangle $AB\Gamma$, for which are given the intersection points $\{A', B', \Gamma'\}$ of its circumcircle with the inner bisectors (See Figure 5.164-I).

5.24. COMMENTS AND EXERCISES FOR THE CHAPTER 563

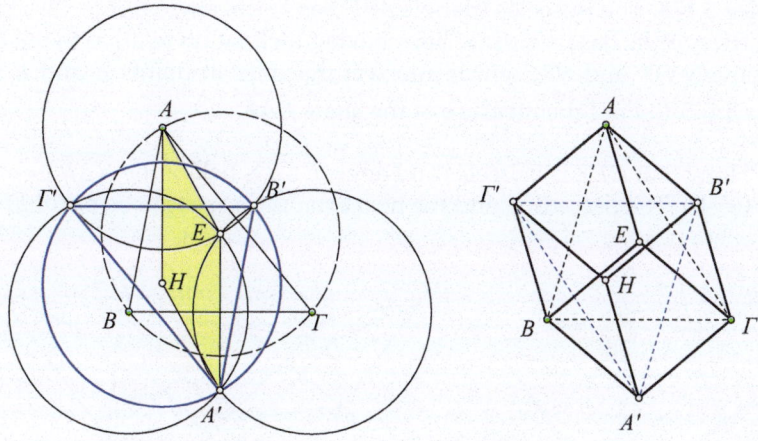

Fig. 5.163: Three intersecting congruent circles

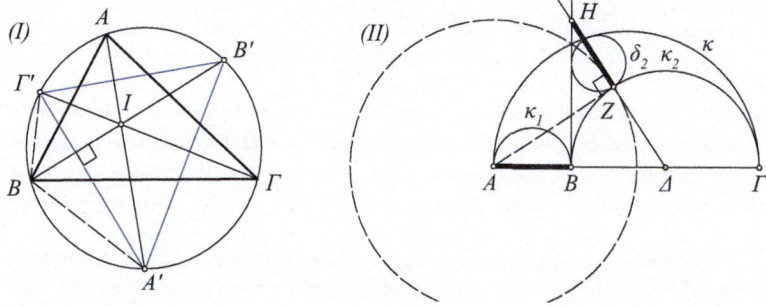

Fig. 5.164: Construction from A', B', Γ' Equal segments

Exercise 5.177. In figure 5.164-II show, that the inversion relative to circle $A(|AZ|)$, where AZ is the tangent to κ_2 from A, leaves the circles κ_2 and δ_2 invariant. Conclude that κ_2 and δ_2 are tangent at the point Z. Also conclude that $|AB| = |ZH|$ ([115]).

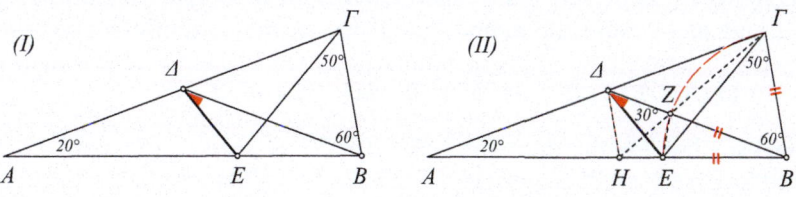

Fig. 5.165: A special triangle

Exercise 5.178. The isosceles triangle $AB\Gamma$ has apical angle $\widehat{BA\Gamma} = 20°$. From the vertices of its base we draw lines having inclination relative to the base respectively 50° and 60°, which intersect the sides at points Δ and E (See Figure 5.165-I). Find the measure of the angle $\widehat{E\Delta B}$.

Hint: Show that HEZ is isosceles and angle $\widehat{E\Delta B}$ is half of the $\widehat{H\Delta B}$ (See Figure 5.165-II). The preceding and the next exercise, appear to be peculiar rid-

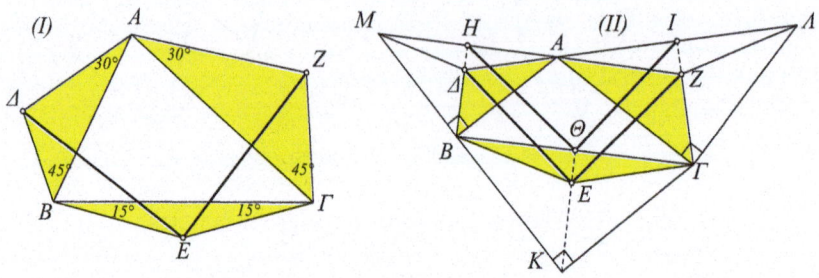

Fig. 5.166: A special triangle II

dles. Next exercise is difficult, if its connection with the natural environment where it belongs is not recognized ([132, p.6]). This difficulty is characteristic of many problems generally. A big part of the difficulty, if not the main, is to find the exact context to which belongs the specific problem.

Exercise 5.179. Given an arbitrary triangle $AB\Gamma$ we construct on its sides triangles with the specific angles of figure 5.166-I. Show that the line segments $\{E\Delta, EZ\}$ are equal and orthogonal at E.

Hint: On the sides of $AB\Gamma$ construct right isosceli triangles, as in the figure 5.166-II. Points $\{H, \Theta, I\}$ are the middles of the hypotenuse of these right triangles. To begin with, the triangle $H\Theta I$ is also a right isosceles. This follows from Theorem 5.27 applied to the two right isosceli triangles MBA and $A\Gamma\Lambda$. Respective vertices of these triangles are connected with the segments MA, $A\Lambda$, $B\Gamma$. Points H, I, Θ divide these segments, respectively, into the same ratio $v = \frac{1}{2}$. Therefore $H\Theta I$ is similar to MBA, $A\Gamma\Lambda$, in other words isosceles and right. We apply the same Theorem 5.27 now to the two right isosceli $H\Theta I$ and $BK\Gamma$. From the definitions of the points Δ, E and Z follows the equality of ratios $\frac{\Delta H}{\Delta B} = \frac{ZI}{Z\Gamma} = \frac{E\Theta}{EK}$. Consequently the triangle ΔEZ will also be similar to $H\Theta I$, $BK\Gamma$, in other words a right isosceles.

Exercise 5.180. Let the triangles $\{AB\Gamma, A'B'\Gamma'\}$ be similar and similarly oriented and ΔEZ another triangle. We construct similar to ΔEZ and similarly oriented triangles $A''AA'$, $B''BB'$ and $\Gamma''\Gamma\Gamma'$. Show that this defines a triangle $A''B''\Gamma''$ which is similar to $AB\Gamma$ and $A'B'\Gamma'$ (See Figure 5.167).

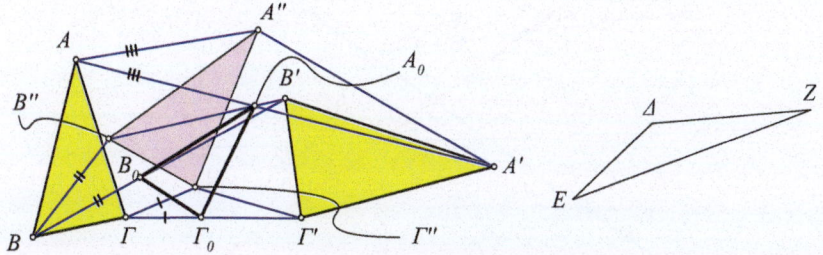

Fig. 5.167: Similar triangles

Hint: On the line AA' take the segment AA_0 equal to AA'' and analogously on the lines BB', $\Gamma\Gamma'$ the segments BB_0, $\Gamma\Gamma_0$ respectively equal to BB'', $\Gamma\Gamma''$. Because of the similarity of the triangles $A''AA'$, $B''BB'$, $\Gamma\Gamma''\Gamma'$, points A_0, B_0, Γ_0 divide AA', BB', $\Gamma\Gamma'$ respectively into the same ratio, therefore the triangle $A_0 B_0 \Gamma_0$ is similar to $AB\Gamma$, $A'B'\Gamma'$ (Proposition 5.13). Also, according to Exercise 3.179 triangles $A_0 B_0 \Gamma_0$ and $A''B''\Gamma''$ are similar.

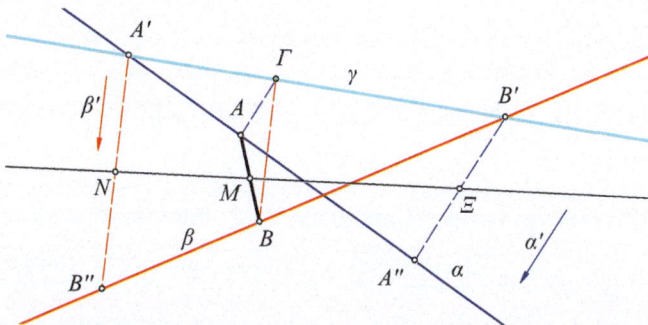

Fig. 5.168: Geometric locus of middle of moving segment

Exercise 5.181. Let $\{\alpha, \beta, \gamma\}$ be three lines in general position and α', β' be two directions. The points Γ of γ are projected parallel to α' to points A of α and parallel to β' to points B of β. Show that the middle M of the segment AB moves on a fixed line.

Hint: Let $A' = (\alpha, \gamma)$, $B' = (\beta, \gamma)$ and B'', A'' be the projections of these points respectively to β, α (See Figure 5.168). By Thales $\frac{AA'}{AA''} = \frac{\Gamma A'}{\Gamma B'} = \frac{BB''}{BB'}$. Apply the conclusion of Exercise 3.96.

Exercise 5.182. Let $\{\alpha, \beta, \gamma\}$ be three lines in general position and α', β' be two directions. Points Γ of γ are projected parallel to α' to points A of α and parallel to β' to points B of β. Drawing parallels to ΓA, ΓB we define the

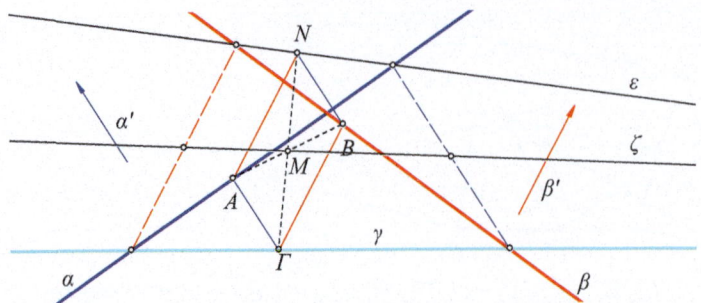

Fig. 5.169: Geometric locus of parallelogram vertex

parallelogram ΓABN. Show that its vertex N moves on a fixed line ε and the intersection point of its diagonals M moves on a fixed line ζ (See Figure 5.169).

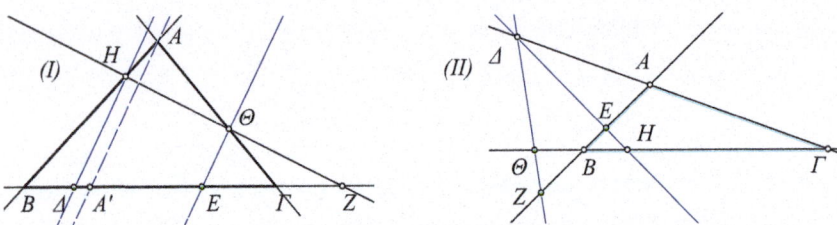

Fig. 5.170: Determination of parallels   Intersections on sides

Exercise 5.183. Let $\{\Delta, E, Z\}$ be three points on the base $B\Gamma$ of the triangle $AB\Gamma$. From point Z draw a line intersecting the sides $A\Gamma$, AB, respectively at points Θ, H, such that the lines ΔH, $E\Theta$ are parallel.

Hint: The cross ratios $(\Delta E; A'Z)$ and $(B\Gamma; A'Z)$ are equal (See Figure 5.170-I). The rest as in Exercise 5.106. The equality of the cross ratios leads to a quadratic equation in one unknown which apparently defines two solutions. There exists however only one and this because the ratios $x = \frac{HA}{HB}$ and $y = \frac{\Theta \Gamma}{\Theta A}$ are connected through the theorem of Menelaus:

$$x \cdot a \cdot y = 1, \qquad (1)$$

where $a = \frac{ZB}{Z\Gamma}$. Also because of the parallel lines holds

$$x = \frac{\Delta A'}{\Delta B}, \quad y = \frac{E\Gamma}{EA'}, \quad \Rightarrow \quad \Delta A' = x \cdot \Delta B, \quad EA' = \frac{E\Gamma}{y} \quad \Rightarrow \quad bx - \frac{c}{y} = d \qquad (2)$$

5.24. COMMENTS AND EXERCISES FOR THE CHAPTER

where $b = \Delta B$, $c = E\Gamma$, $d = \Delta E$ are numbers which result as signed lengths through a system of coordinates of the line $B\Gamma$. The solution to the system of equations (1),(2) is

$$x = \frac{d}{b-ac}, \quad y = \frac{b-ac}{ad}.$$

Exercise 5.184. Consider the points Θ, H on side $B\Gamma$ and E, Z on side AB of the triangle $AB\Gamma$. Show that the lines EH and ΘZ intersect on $A\Gamma$, if and only if the cross ratios satisfy $(B\Gamma; H\Theta) = (BA; ZE)$ (See Figure 5.170-II).

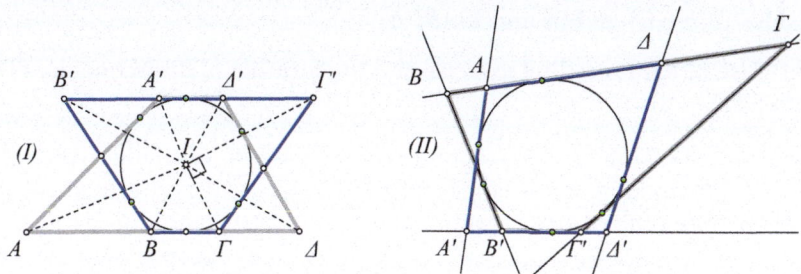

Fig. 5.171: Equal cross ratios $(AB; \Gamma\Delta) = (A'B'; \Gamma'\Delta')$

Exercise 5.185. The trapezia/quadrilaterals $\{A\Delta\Delta'A', B\Gamma\Gamma'B'\}$ are circumscriptible to the same circle with center I. In the case of trapezia (See Figures 5.171-I, II) show that the lines $\{IA, IB, I\Gamma, I\Delta\}$ are correspondingly orthogonal to $\{IA', IB', I\Gamma', I\Delta'\}$. In both cases show that the cross ratios are equal $(AB; \Gamma\Delta) = (A'B'; \Gamma'\Delta')$.

Hint: Apply exercise 5.157.

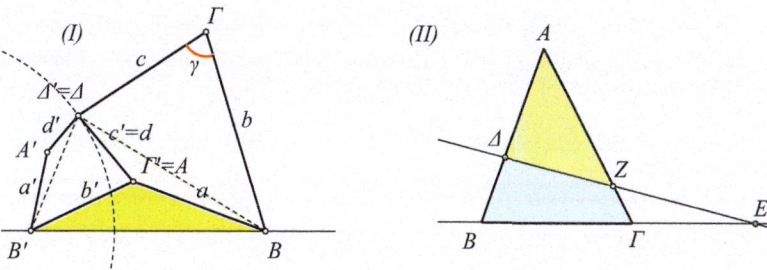

Fig. 5.172: Quadrilateral construction   Division of triangle

Exercise 5.186. Construct a convex quadrilateral $AB\Gamma\Delta$ from the lengths of its sides $\{a,b,c,d\}$ and the sum of the measures $\alpha + \gamma$ of two of its opposite angles.

Hint: Consider a quadrilateral $A'B'\Gamma'\Delta'$ similar to $AB\Gamma\Delta$ and place it on side $A\Delta$ so that it coincides with side $\Gamma'\Delta'$ (See Figure 5.172-I). The vertices A and Γ' coincide and at this point the angle of the triangle $AB'B$ is expressed through $\alpha + \gamma$. The similarity ratio is $\lambda = \frac{a'}{c} = \frac{d}{c}$. Triangle $\Gamma'B'B$ is constructible. Also $\frac{AB'}{AB} = \lambda$, therefore Δ is to be found on the Apollonian circle of ratio λ of segment $B'B$. Point Δ is also to be found on the circle $A(d)$.

Exercise 5.187. Construct a convex quadrilateral $AB\Gamma\Delta$ from the lengths of its sides $\{a,b,c,d\}$ and its area $\varepsilon(AB\Gamma\Delta)$.

Hint: Use of the preceding exercise and the formula of Bretschneider (Theorem 5.11).

Exercise 5.188. From the given point E of the base $B\Gamma$ of triangle $AB\Gamma$ draw a line $EZ\Delta$, dividing the triangle into a triangle $A\Delta Z$ and a quadrilateral $Z\Gamma B\Delta$, whose ratio of areas is a given number k (See Figure 5.172-II).

Hint: The product of $\{x = |AZ|, y = |A\Delta|\}$ results from the given data:

$$xy = \frac{\varepsilon(A\Delta Z)}{2\sin(\alpha)} = \frac{k_1 \varepsilon(AB\Gamma)}{2\sin(\alpha)}, \text{ where } k_1 = \frac{k}{1+k}.$$

In addition, the $\{x, y\}$ satisfy a known homographic relation (Exercise 5.156).

Exercise 5.189. The vertices of the polygon $A_1 A_2 ... A_v$ are projected onto the line ε at the points $B_1, B_2, ..., B_v$. Show that the signed lengths satisfy

$$B_1 B_2 + B_2 B_3 + ... + B_{v-1} B_v + B_v B_1 = 0.$$

Exercise 5.190. Let $AB\Gamma\Delta$ be a harmonic quadrilateral, $\{H, H'\}$ be the intersection points of opposite sides, $\{Z, Z'\}$ be the intersection points of the tangents of its circumcircle κ at its opposite vertices and E be the intersection point of its diagonals. Show that (See Figure 5.173):

1. The circles $\{\alpha = Z(|ZA|), \beta = Z'(|Z'\Gamma|)\}$ are mutually orthogonal and orthogonal to κ.
2. The radical axis of $\{\alpha, \beta\}$ passes through the center O of κ and point E.
3. The intersection points of $\{\alpha, \beta\}$ are harmonic conjugate relative to $\{O, E\}$.
4. The circles $\{\alpha', \beta'\}$ with respective diameters $\{ZZ', HH'\}$ belong to the pencil of $\{\alpha, \beta\}$.
5. The circles $\{\alpha', \beta'\}$ are orthogonal.

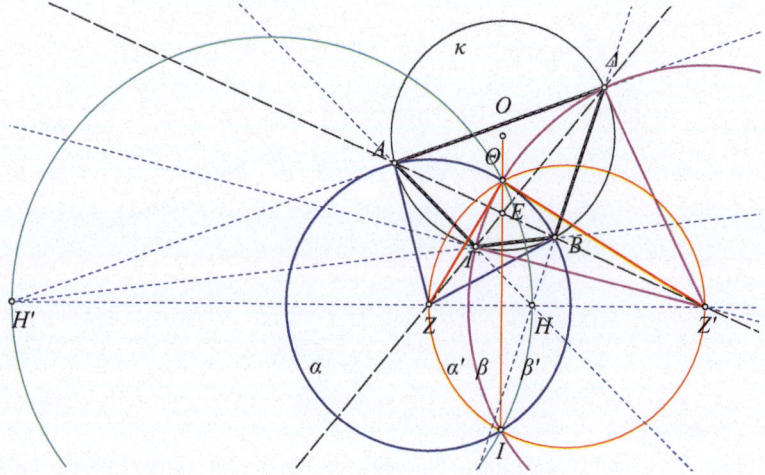

Fig. 5.173: Noteworthy circles associated with a harmonic quadrilateral

Hint: That circles $\{\alpha, \beta\}$ are orthogonal to κ, follows from the fact, that $\{Z\Alpha, Z'\Gamma\}$ are tangents to κ. From this follows also that O belongs to the radical axis of $\{\alpha, \beta\}$. Point E belongs to the radical axis of $\{\alpha, \beta\}$, because it possesses equal powers to them: $|E\Gamma||E\Delta| = |EA||EB|$. The fact that $\{\alpha, \beta\}$ are mutually orthogonal, results from the property of $\{A, B\}$, to be inverse relative to β, since $|Z'A||Z'B| = |Z'\Gamma|^2$. Therefore, every circle passing through $\{A, B\}$ is orthogonal to β (Theorem 4.13). The fact that $(ZZ'; HH') = -1$ implies, that the circles with diameters $\{ZZ', HH'\}$ respectively, are mutually orthogonal (Corollary 4.13). Because the circles $\{\alpha, \beta\}$ are orthogonal, the angle $\widehat{Z\Theta Z'}$ is right and the circle α', with diameter ZZ', passes through points $\{\Theta, I\}$, therefore belongs to the pencil of $\{\alpha, \beta\}$. Circle β' is also orthogonal to κ (Theorem 5.39), therefore itself also belongs to the pencil of $\{\alpha, \beta\}$ and passes through points $\{\Theta, I\}$. Finally, from the orthogonality of $\{\kappa, \beta\}$, follows that $\Gamma\Delta$ is the polar of O relative to β, therefore its intersection point E is the harmonic conjugate of O relative to $\{\Theta, I\}$.

Next exercise gives an application of **neusis**, which is defined as the placement of a line segment ΔE of given length a with its endpoints on two curves ε and ζ, and in such a way, that the line ΔE passes through a given point Γ. Figure 5.174 shows one such application of neusis, used for the construction of the cubic root of κ ([124, p.286]). The method of neusis was used by the ancient Greeks in some cases, when the construction of a line segment of given length was impossible using only ruler and compass ([135, c]).

Exercise 5.191. Construction of $\sqrt[3]{\kappa}$ using neusis. The side $A\Gamma$ of the isosceles triangle $AB\Gamma$, whose legs have the length 1 and the length of the base is $\kappa/4$, where $\kappa < 8$, is extended to its double up to H. This defines the line $\zeta = HB$

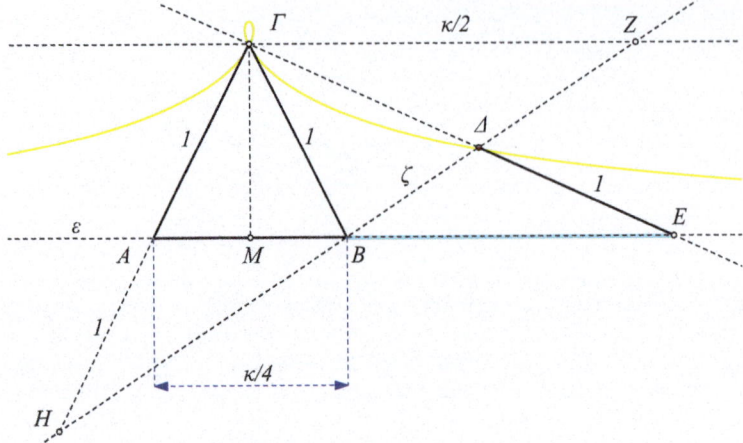

Fig. 5.174: Construction of $\sqrt[3]{\kappa}$ using neusis

and the segment ΔE, of length 1, is placed between the line of the base ε and ζ (neusis), in such a way, that $E\Delta$ passes through point Γ. Then BE has the length $\sqrt[3]{\kappa}$ (See Figure 5.174).

Hint: From the similar triangles $\Gamma\Delta Z$ and $E\Delta B$, follows $|\Gamma\Delta| = \kappa/(2|BE|)$. From the pythagorean theorem, follows

$$|\Gamma E|^2 = (1 + \frac{\kappa}{2|BE|})^2 = |\Gamma M|^2 + |ME|^2 = 1 - \left(\frac{\kappa}{8}\right)^2 + \left(\frac{\kappa}{8} + |BE|\right)^2 \Leftrightarrow$$
$$4x^4 + \kappa x^3 - 4\kappa x - \kappa^2 = (4x + k)(x^3 - k) = 0, \quad \text{where} \quad x = |BE|.$$

The only acceptable solution of the last equation is $x = \sqrt[3]{k}$. Why do we need the restriction $\kappa < 8$? How is the proof generalized for $\kappa > 8$?

Exercise 5.192. The sides of a convex quadrilateral $AB\Gamma\Delta$ are extended proportionaly to their lengths and to the same orientation, to create the quadrilateral $EZH\Theta$ with $\Delta E/\Delta A = AZ/AB = BH/B\Gamma = \Gamma\Theta/\Delta\Gamma = k$ (See Figure 5.175-I). Show that $EZH\Theta$ has the same center of mass(intersection point of the segments which join the middles of opposite sides) I as $AB\Gamma\Delta$ and calculate its area as a function of the area of $AB\Gamma\Delta$ and k.

Exercise 5.193. Let Θ be the point of the circumcircle α of triangle $AB\Gamma$, at which the external bisector AH intersects the circle. Let also β be the circle with center Θ, passing through $\{B, \Gamma\}$ and Δ be an arbitrary point on the bisector $A\Lambda$ (See Figure 5.175-II). Show that

1. The bisector $A\Delta$ is the polar relative to β of the point H of the line $B\Gamma$.
2. The line through $\{Z, E\}$, at which β intersects respectively ΔB and $\Delta \Gamma$ passes through H.

5.24. COMMENTS AND EXERCISES FOR THE CHAPTER

3. Points $\{\Delta, Z, A, \Gamma\}$ are concyclic. Similarly $\{\Delta, E, A, B\}$ are concyclic.
4. Points $\{E, Z, A, \Theta\}$ are concyclic.
5. The line through $\{K, I\}$ at which ΔB, $\Delta\Gamma$ respectively intersect α is parallel to ZE.

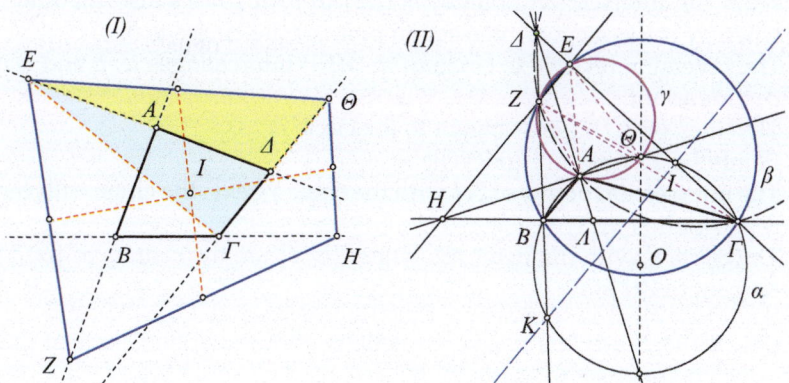

Fig. 5.175: Extending proportionally   Bisector as polar

Hint: (1) $\{H, \Lambda\}$ are harmonic conjugate of $\{B, \Gamma\}$ and $H\Theta$ is orthogonal to $A\Lambda$. (2) If H' is the intersection of HZ with $A\Lambda$, the harmonic conjugate E of Z relative to HH' belongs both to the circle β and to the line $\Delta\Gamma$. (3) The angle $\omega = \widehat{\Delta A \Gamma}$ is supplementary to $\widehat{BA\Gamma}/2$. We have $\widehat{\Delta Z\Gamma} = \widehat{\Delta Z E} + \widehat{EZ\Gamma} = \widehat{B\Theta E}/2 + \widehat{E\Theta \Gamma}/2 = \widehat{BA\Delta} = \omega$. (4) Because ΔA and HA are orthogonal and AZ, AE are harmonic conjugate of AH, $A\Delta$, it follows that $A\Delta$, AH are bisectors of $\widehat{ZAE}$. Then $\phi = \widehat{ZAE} = 2\widehat{ZA\Delta}$. And for $\widehat{ZA\Delta}$ relying on (3), $\widehat{Z\Gamma\Delta} = \widehat{Z\Theta E}/2$. (5) Follows directly that angles $\widehat{ZE\Gamma}$ and $\widehat{KI\Gamma}$ are equal.

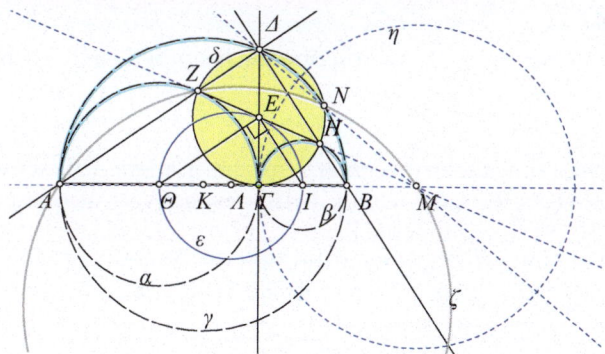

Fig. 5.176: The area of the arbelos

Exercise 5.194. Let $AB\Gamma$ be an arbelos with diameter AB. Besides the property of its area, which, according to Archimedes, is equal to the area of the circle δ with diameter $\Gamma\Delta$, where Δ is the intersection point of circle γ with the orthogonal at Γ (See Figure 5.176) show the relations:

1. The second intersection points Z, H of the circles α, β with δ define a line, which passes through the center E of δ and is tangent to α and β.
2. From the intersection point $M = (AB, ZH)$ AB passes also the line ΔN, where N is the second intersection point of the circles γ, δ.
3. The inversion relative to circle $\eta(M, |M\Gamma|)$ interchanges α, β and leaves δ, γ invariant.
4. The lines $\{\Theta Z, IH\}$ are tangent to the circle δ.
5. Points $\{A, Z, N, M\}$ are concyclic.
6. The intersection points of ZH with the circle γ lie on the circle with center Δ and radius $|\Delta\Gamma|$.
7. The circle ε, with diameter $I\Theta$, where $\{\Theta, I\}$ are the centers of $\{\alpha, \beta\}$, is tangent to ZH at E.

Exercise 5.195. Study the two constructions of the inscribed circle of the arbelos which are given in figure 5.177.

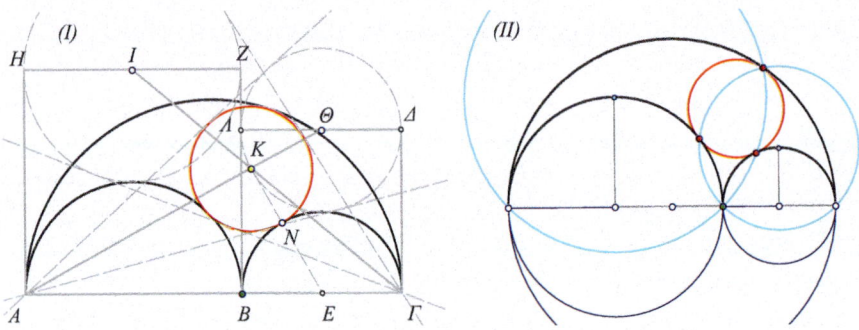

Fig. 5.177: Other ways of constructing the inscribed circle of the arbelos

Exercise 5.196. From the orthocenter H of the triangle $AB\Gamma$ we draw a line ε, which intersects the sides $\{B\Gamma, \Gamma A, AB\}$ respectively at points $\{A', B', \Gamma'\}$ (See Figure 5.178). Show that the lines $\{\varepsilon_1, \varepsilon_2, \varepsilon_3\}$, created by the reflection of ε on the sides of the triangle, pass through the point S_ε of the circumcircle of the triangle, whose corresponding Steiner line coincides with ε.

Hint: From the middle M of HS_ε draw the parallel to ε, which intersects the side AB at Γ''. Show that the triangle $\Gamma'\Gamma''S_\varepsilon$ is a right one etc.

Exercise 5.197. Construct a triangle $AB\Gamma$, for which are given the points of intersection $\{A', B', \Gamma'\}$ of its circumcircle with the external bisectors (See Figure 5.179-I)

5.24. COMMENTS AND EXERCISES FOR THE CHAPTER

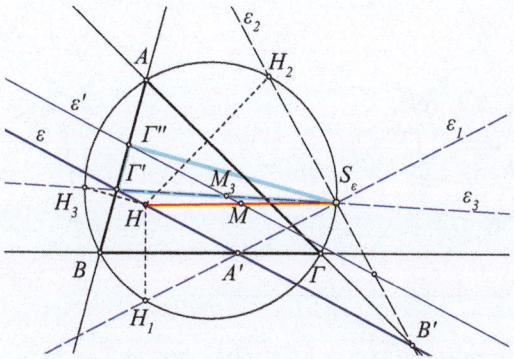

Fig. 5.178: Steiner's line, an inverse view

Exercise 5.198. Let $\{X, Y, Z\}$ be three points in general position lying outside the circle κ. Draw secants of κ: $\{XAB, Y\Gamma\Delta\}$, so that $A\Delta$, $B\Gamma$ intersect at Z (See Figure 5.179-II).

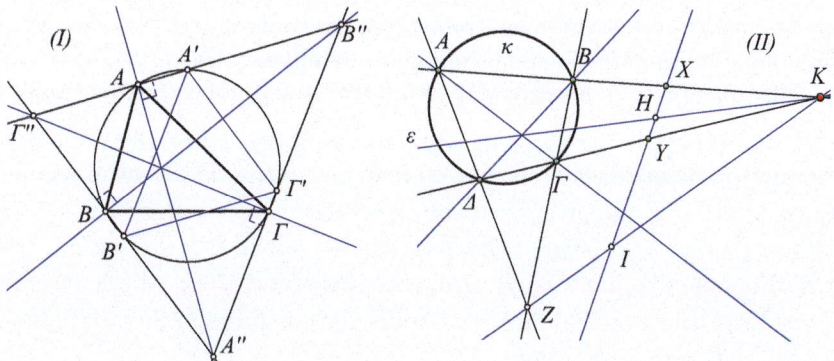

Fig. 5.179: Triangle construction   Construction of secants

Hint: Suppose the problem is solved and that lines XAB, $Y\Gamma\Delta$ intersect at point K. The polar ε of Z relative to κ intersects the line XY at point H and passes through K. The harmonic conjugate $I = H(X, Y)$ lies on line KZ. Point K is constructed as the intersection of ε and ZI.

Exercise 5.199. From a given point Δ, not contained in the side-lines of triangle's $AB\Gamma$, we draw lines which intersect the sides, making with them similarly oriented angles $\{\widehat{\Delta Z\Gamma}, \widehat{\Delta HA}, \widehat{\Delta\Theta B}\}$ equal to a fixed angle ω (See Figure 5.180-I). We consider subsequently the circle $\kappa = (ZH\Theta)$, which intersects for a second time the sides at points $\{I, K, \Lambda\}$. Through these points we

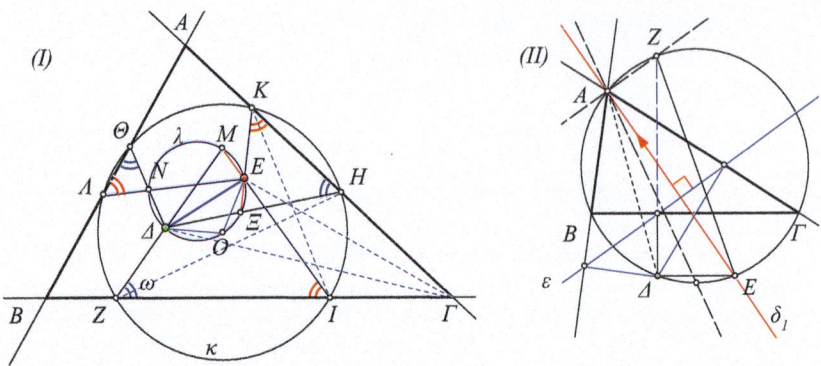

Fig. 5.180: Isogonal conjugates Δ, E Isogonal of point Δ

define lines which make angles $\{\widehat{EIZ}, \widehat{EK\Gamma}, \widehat{E\Lambda\Theta}\}$, opposite oriented and equal to the preceding. Show the following properties:

1. The lines $\{IE, KE, \Lambda E\}$ pass through a common point E.
2. The intersection points of the pairs of lines $\{M = (\Delta Z, IE), N = (\Delta\Theta, \Lambda E), \Xi = (\Delta H, KE)\}$ and Δ, E are contained in the same circle λ.
3. Circle λ passes through the center O of the circle κ.
4. Points Δ and E are isogonal conjugates relative to $AB\Gamma$.
5. The position of E depends only on that of Δ and not on the angle ω.

Hint: For (1) apply the theorem 5.14, observing that $\{BIE\Lambda, \Gamma KEI, A\Lambda EK\}$ are inscriptible quadrilaterals. For (2) note that $\{M, N, \Xi\}$ view the segment ΔE under the same angle ω' or its supplementary: $\widehat{MEN} = \widehat{MAN} = \widehat{B}$, $\widehat{MN\Xi} = \widehat{M\Delta\Xi} = \widehat{\Gamma}$ and $\widehat{NM\Xi} = 180° - (\widehat{B} + \widehat{\Gamma}) = \widehat{A}$.

For (3) note that the triangle IZM, is isosceles and MO is a bisector of the angle $\widehat{\Delta ME}$ hence medial line of ZI and will pass through the middle O of the arc $\widehat{\Delta OE}$ of the circle λ etc.

For (4) show that pairs like $(\Delta Z\Gamma, EK\Gamma)$ consist of similar triangles, therefore $\widehat{Z\Gamma\Delta} = \widehat{K\Gamma E}$.

(5) is a direct consequence of (4).

The isogonal Δ' of point Δ, besides the fact that it is defined only for points not contained in the line carriers of the sides of triangle $AB\Gamma$, can be at infinity. Next exercise shows exactly, which are the points with this property.

Exercise 5.200. Show that every point Δ on the circumcircle of the triangle $AB\Gamma$ has as isogonal conjugate the point at infinity, which corresponds to the direction which is orthogonal to the line of Δ (See Figure 5.180-II).

Hint: If Δ is a point of the circumcircle, then, according to theorem 5.36, the isogonal Δ' will coincide with the intersection point of lines δ_1, δ_2, δ_3 which are, respectively, symmetric of $\{\Delta A, \Delta B, \Delta \Gamma\}$ relative to the corresponding bisectors. One such, for example, is the symmetric $\delta_1 = AE$ of ΔA relative to the bisector of $\widehat{A}$. Because this bisector intersects arcs $\widehat{B\Delta\Gamma}$ and $\widehat{\Delta E}$ at their middle, ΔE is parallel to the base $B\Gamma$. Also, according to proposition 5.7, the orthogonal from Δ to $B\Gamma$ will intersect the circumcircle a second time at Z and AZ will be parallel to the Simson line ε of Δ. From the inscriptible quadrilateral ΔEZA follows that δ_1 is orthogonal to ε. The fact that δ_2 and δ_3 are also orthogonal to ε is proved similarly.

Exercise 5.201. Let P be a point not lying on the side-lines of the triangle $AB\Gamma$ with circumcircle κ. Let also $\{E, Z\}$ be the symmetrics of P respectively relative to its sides $\{B\Gamma, A\Gamma\}$. Define the circle $\lambda = (\Gamma EZ)$, which intersects κ a second time at K (See Figure 5.181-I). Show that:

1. If Θ is the symmetric of the orthocenter H relative to $B\Gamma$, then points $\{\Theta, E, K\}$ are collinear.
2. Line HP is the line of Steiner of K.

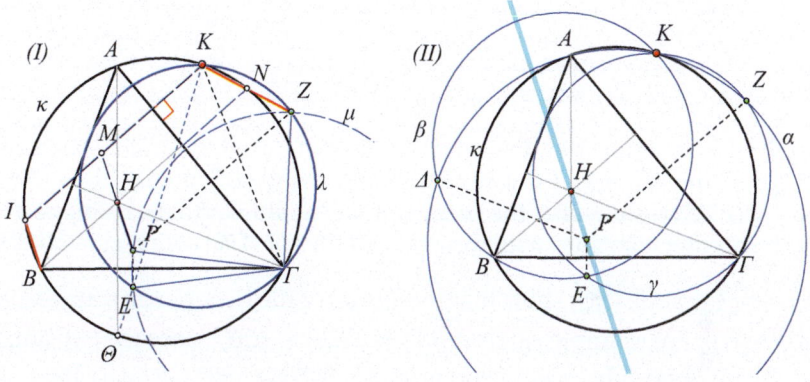

Fig. 5.181: Steiner line HP Concurrent circles

Hint: For (1) suppose that the line ΘE intersects circle κ at K' and λ at K''. Then we have $\widehat{\Theta A\Gamma} = \widehat{\Theta K'\Gamma}$. Also $\widehat{EK''\Gamma} = \widehat{EZ\Gamma}$. However, points $\{P, E, Z\}$ all lie on the same circle λ with center Γ. It follows that $\widehat{EZ\Gamma} = \widehat{PZ\Gamma} = \widehat{B\Gamma E} = \widehat{\Theta A\Gamma}$.

For (2) draw KI orthogonal to $A\Gamma$, which intersects PH at M. The symmetric N of H relative to $A\Gamma$ defines the line NZ, which, as in (1), is proved to pass through K. The trapezium $PZNH$ is isosceles and, consequently, $PZKM$ is also isosceles, as is also $BNKI$. The conclusion follows from the fact, that BI, PH are parallel, by applying proposition 5.7 and the fact that the Steiner line is parallel to the corresponding Simson line.

Exercise 5.202. Let P be a point not lying on the side-lines of the triangle $AB\Gamma$ with circumcircle κ. Let also $\{\Delta, E, Z\}$ be respectively the symmetrics of P relative to its sides $\{AB, B\Gamma, A\Gamma\}$. Show that the circles $(A\Delta Z), (BE\Delta)$, $(\Gamma ZE)\}$ and κ pass through a common point K (See Figure 5.181-II).

Hint: Application of exercise 5.201.

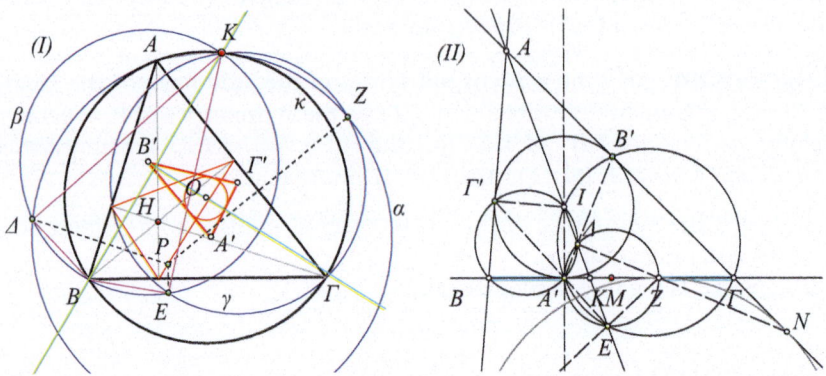

Fig. 5.182: $A'B'\Gamma'$ similar to orthic   Property of bisector AE

Exercise 5.203. Using the data of the preceding exercise and denoting by $\{A', B', \Gamma'\}$ the centers of the circles, respectively, $\{\alpha = (A\Delta Z), \beta = (BE\Delta)$ and $\gamma = (\Gamma Z E)\}$, show that the triangle $A'B'\Gamma'$ is similar to the orthic of $AB\Gamma$ and its incenter coincides with the circumcenter of $AB\Gamma$ (See Figure 5.182-I).

Hint: For the similarity, show that the angles of the two triangles are respectively equal. For example, angle $\widehat{A'B'\Gamma'}$ is equal to $\widehat{\Delta KE}$, since the two angles have orthogonal sides. On the other hand, $\widehat{\Delta KE} = 180° - \widehat{EB\Delta} = 180° - 2\widehat{B}$ and the last one is equal to one angle of the orthic. For the coincidence of centers notice that the bisector KB of $\widehat{\Delta KE}$ is orthogonal to the bisector of $\widehat{A'B'\Gamma'}$ and because KB is a chord of κ, this orthogonal will pass through the circumcenter O of $AB\Gamma$.

Exercise 5.204. If $\{A', B', \Gamma'\}$ are the points of contact of the incircle of triangle $AB\Gamma$ with its sides and the bisector AI intersects the circles $(BA'\Gamma')$, $(\Gamma B'A')$, respectively, at points $\{\Delta, E\}$, show that (See Figure 5.182-II):

1. (A', Δ, B') and (A', E, Γ') are triples of collinear points.
2. Circle $\kappa = (A'\Delta E)$ has its center M at the middle of $B\Gamma$.
3. The other intersection point Z of κ with $B\Gamma$ coincides with the contact point of $B\Gamma$ with the externally tangent and contained in angle $\widehat{A}$ circle of the triangle.

Hint: For (1) observe that $\widehat{\Delta A'\Gamma} = \widehat{B'A'\Gamma} = (\widehat{A}+\widehat{B})/2$.

For (2) observe first that $\widehat{A'\Delta Z}$ is right and subsequently that $|BA'| = |Z\Gamma|$, making use of the power of the intersection $K = (AI, B\Gamma)$ relative to the circles $(BA'\Gamma')$ and $(A'\Gamma B')$.

(3) is a consequence of proposition 5.3.

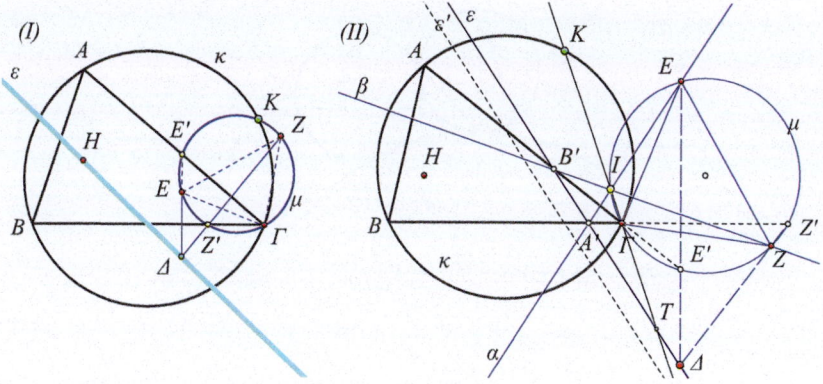

Fig. 5.183: Steiner line ε ε parallel to Simson $\varepsilon' = s(K)$

Exercise 5.205. Point Δ moves on the line ε passing through the orthocenter H of triangle $AB\Gamma$. Let $\{E,Z\}$ be the symmetrics of Δ respectively relative to the sides $\{B\Gamma, A\Gamma\}$ (See Figure 5.183-I). Show that:

1. Angle $\widehat{E\Gamma Z}$ is the double of $\widehat{\Gamma}$.
2. Circle $\mu = (E\Gamma Z)$ intersects $\{B\Gamma, A\Gamma\}$ at points $\{Z', E'\}$, lying respectively on lines $\{\Delta Z, \Delta E\}$.
3. Circle $\mu = (E\Gamma Z)$ passes through point K of the circumcircle of $AB\Gamma$, whose corresponding Steiner line $S(K)$ coincides with ε.

Exercise 5.206. Point Δ moves on line ε. Let $\{E,Z\}$ be the symmetrics of point Δ respectively relative to the sides $\{B\Gamma, A\Gamma\}$ of the triangle $AB\Gamma$ (See Figure 5.183-II). Show that:

1. Triangle $E\Gamma Z$ is isosceles and its vertices are contained in two fixed lines α and β.
2. Lines $B\Gamma$ and $A\Gamma$ bisect, respectively, the angles of the pairs of lines $\{(\alpha, \varepsilon), (\beta, \varepsilon)\}$.
3. The circle $\mu = (\Gamma EZ)$ passes through the intersection point I of lines α, β.
4. The second point of intersection K of the line ΓI with the circumcircle κ of triangle $AB\Gamma$ is the point at which the corresponding Simson line $\varepsilon' = s(K)$ is parallel to ε.

Hint: For (1) and (2), notice that E is contained in line α, which is the symmetric of ε relative to $B\Gamma$. Similarly β is the symmetric of ε relative to $A\Gamma$.

For (3), notice that $\widehat{EIZ} = \widehat{E\Gamma Z} = 2\hat{\Gamma}$.

For (4), notice that angle $\widehat{A\Gamma I}$ does not change when ε is relocated to a parallel to it. Consequently the position of K is the same with that which corresponds to the parallel of ε which passes through the orthocenter H. The conclusion follows from exercises 5.201 and 5.205.

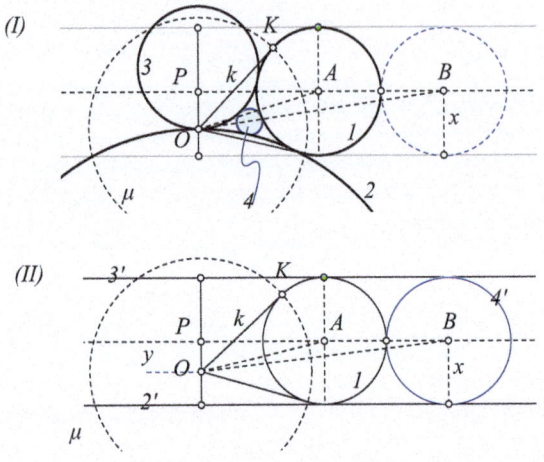

Fig. 5.184: Descartes formula for externally tangent circles

Exercise 5.207. Show that for the radii of four, pairwise tangent and external to each other circles, the following relation holds (Descartes 1596-1650)

$$2 \cdot \left(\frac{1}{r_1^2} + \frac{1}{r_2^2} + \frac{1}{r_3^2} + \frac{1}{r_4^2} \right) = \left(\frac{1}{r_1} + \frac{1}{r_2} + \frac{1}{r_3} + \frac{1}{r_4} \right)^2.$$

Hint: For the proof ([155], [125]) we use the inversion relative to circle $O(k)$ with center the common point of contact O of two out of the four tangent circles, which we denote by $\{1,2,3,4\}$ (See Figure 5.184-I). We select O to be the point of contact of the circles $\{2,3\}$ and for radius of the circle of inversion we take the length k of the tangent to circle 1 from O. We met this method in remark-2 of § 5.6, where we saw that, under the inversion, these circles map to two parallel lines $\{2',3'\}$ and two congruent tangent circles $\{1,4'\}$, which are tangent also to these two parallels (See Figure 5.184-II). We also use the formulas which express the radii of the inverses relative to the circle of inversion $O(k)$:

$$r' = \frac{rk^2}{d^2 - r^2}, \qquad r' = \frac{k^2}{2d}.$$

The first expresses the radius r' of the inverse of one circle of radius r, whose center is at distance d from O. The second expresses the radius r' of the circle

5.24. COMMENTS AND EXERCISES FOR THE CHAPTER

which represents the inverse of line ε at distance d from O. Both formulas are direct consequences of exercise 4.93. Relying on these formulas, we find for the radii of the four circles the expressions

$$r_1 = x, \quad r_2 = \frac{k^2}{2(x-y)}, \quad r_3 = \frac{k^2}{2(x+y)}, \quad r_4 = \frac{k^2 x}{y^2 + (z+2x)^2 - x^2},$$

where $k^2 = z^2 + y^2 - x^2$. Here $2x$ denotes the distance between the parallels, $y = |OP|$ and $z = |AP|$ (See Figure 5.184-II). Descartes formula results by substituting in it the above values of $\{r_1, r_2, r_3, r_4\}$ and performing calculations.

Exercise 5.208. Construct a circle δ, tangent to three others $\{\alpha, \beta, \gamma\}$, which are tangent to line ε and also tangent successively to each other (See Figure 5.185-I).

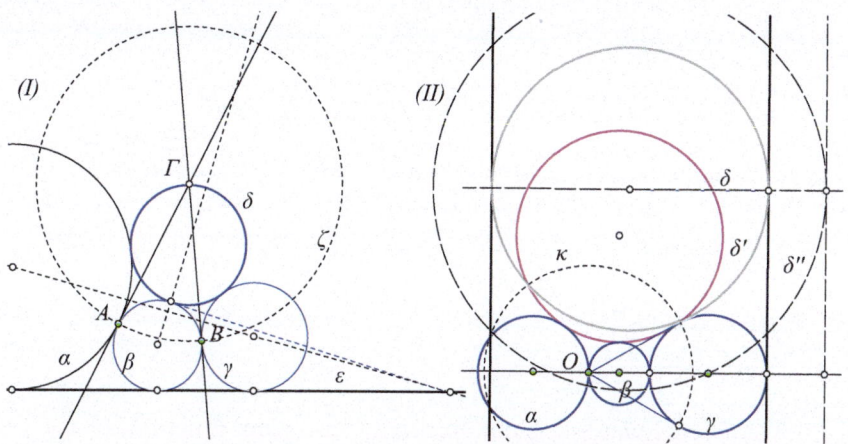

Fig. 5.185: Circle δ tangent to three others

Hint: The requested circle results from the inversion of the line ε relative to the circle ζ, which is orthogonal to the three given circles.

Exercise 5.209. Construct a circle δ, tangent to three others $\{\alpha, \beta, \gamma\}$, which are successively tangent to each other and have their centers on a line (See Figure 5.185-II).

Given the triangle $AB\Gamma$, next exercise defines a line, for every point Δ not contained in the sides of the triangle. This line, in the special case where point Δ is the orthocenter of the triangle, coinciides with the Euler line. (Proposition 3.11).

Exercise 5.210. Let $AB\Gamma$ be a triangle and Δ a point not lying on its side-lines. Consider the centroids $\{E, Z, H\}$ respectively of the triangles ΔAB, $\Delta B\Gamma$, $\Delta \Gamma A$. Suppose also that $\{\Theta, I\}$ are respectively the centroids of the triangles

$\{EZH, AB\Gamma\}$. Show that points $\{\Delta, \Theta, I\}$ are collinear and $|\Delta\Theta| = 2|\Theta I|$ (See Figure 5.186-I).

Exercise 5.211. Figure 5.186-II is that of the preceding exercise in the special case, in which point Δ coincides with the orthocenter of triangle $AB\Gamma$. Show that the three Euler lines of triangles ΔAB, $\Delta B\Gamma$, $\Delta\Gamma A$ pass through a common point K, which coincides with the center of the Euler circle of the triangle $AB\Gamma$. Also show that K is the center of the Euler circle of the triangle EZH. Show that the Euler line of the triangle $\Delta A\Gamma$ passes through the centroid H of $\Delta A\Gamma$, the vertex B, and the segment HB is divided by K into ratio 1:3. Show, finally, that point K is a center of antihomothety of triangles $AB\Gamma$ and ZHE with ratio 3:1 and K divides the segment $I\Theta$ into a ratio of 3:1.

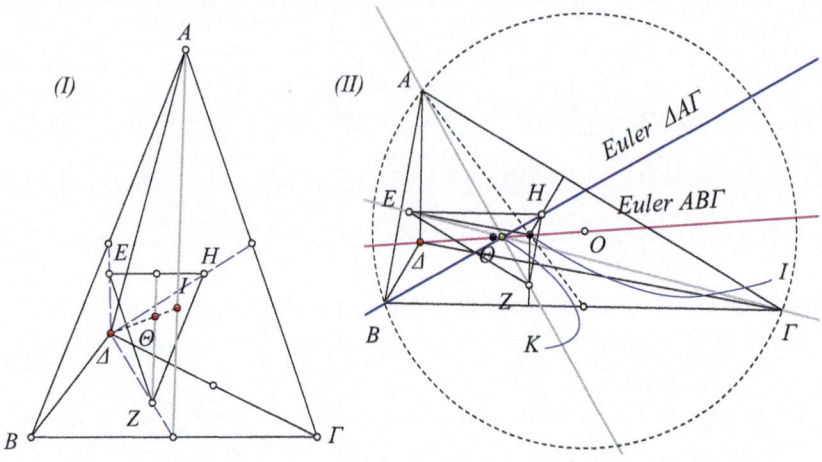

Fig. 5.186: Euler line generalization   Four Euler lines

Exercise 5.212. Let the lines $\{\alpha, \beta\}$ intersect at O, each with a coordinate system with origin at O. To every point $X(x)$ of α we correspond the point $Y(y)$ of β through the homographic relation $y = (ax+b)/(cx+d)$, where for the constants $ad - bc \neq 0$. Show that the lines $\{\varepsilon_x = XY\}$, defined by the corresponding points, pass all through a fixed point.

Hint: Consider the intersection point P of two lines $\{AA', BB'\}$ for arbitrary points $\{A, B\}$ of α and the corresponding homographic to them $\{A', B'\}$ on line β (See Figure 5.187-I). Consider also a third point Γ of α and its corresponding homographic Γ' of β. Let $k = (OAB\Gamma)$ the cross ratio of the four points. Since the homographic relation preserves the cross ratio (Proposition 5.19), it follows that the corresponding points on β will have the same cross ratio $k = (OA'B'\Gamma')$. From the theorem 5.31 follows that the three lines will pass through the same point. The conclusion results by holding $\{A, B\}$ fixed and varying the position of Γ.

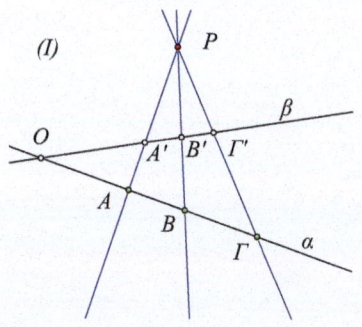
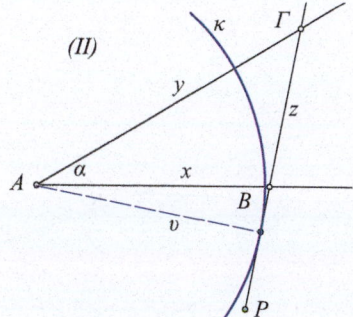

Fig. 5.187: Concurring lines   Triangle construction

Notice that the conclusion of the exercise is the reverse of the one of exercise 5.156.

Exercise 5.213. To construct a triangle $AB\Gamma$, whose sides $\{x = |AB|, y = |A\Gamma|\}$ satisfy a homographic relation $y = (ux+v)/(rx+s)$ and of whose is given the angle $\alpha = \widehat{BA\Gamma}$ and its altitude υ from A (See Figure 5.187-II).

Hint: On the sides $\{AB, A\Gamma\}$ of the angle α we define using coordinates the homographic relation $y = (ux+v)/(rx+s)$ and locate the fixed point P through which pass all the lines $\{B(x)\Gamma(y)\}$ (See Figure 5.187-II). With center at A and radius υ we draw a circle κ and draw the tangent to it from P.

Exercise 5.214. To construct a triangle $AB\Gamma$, whose sides $\{x = |AB|, y = |A\Gamma|\}$ satisfy the homographic relation $y = (ux+v)/(rx+s)$ and of whose is given the angle $\alpha = \widehat{BA\Gamma}$ and the internal, or external bisector, or the inradius.

Exercise 5.215. To construct a triangle $AB\Gamma$, whose sides $\{x = |AB|, y = |A\Gamma|\}$ satisfy the homographic relation $y = (ux+v)/(rx+s)$ and of which are given the angles.

Exercise 5.216. To construct a triangle $AB\Gamma$, whose sides $\{x = |AB|, y = |A\Gamma|\}$ satisfy the homographic relation of the form $ux+vy = w$ and of which is given also the angle $\alpha = \widehat{BA\Gamma}$ and the circumradius R or the median from A.

The homographic relation in the last exercise has not the general form of the preceding exercises. If the exercise is formulated using the general homographic relation $y = (ux+v)/(rx+s)$, then the corresponding problem is not solvable with ruler and compass only. For example, if the last relation is valid and given are also the angle $\omega = \widehat{BA\Gamma}$ and the circumradius R, then the side $z = |B\Gamma|$ is determined $z = 2R\sin(\omega)$ and the problem reduces to that of intersecting a known angle $\widehat{BA\Gamma}$ from P, by a secant $PB\Gamma$ through P, so that the segment $B\Gamma$ has a given length. Point P is determined from the angle ω and the homographic relation, as in the preceding exercises. In figure 5.188

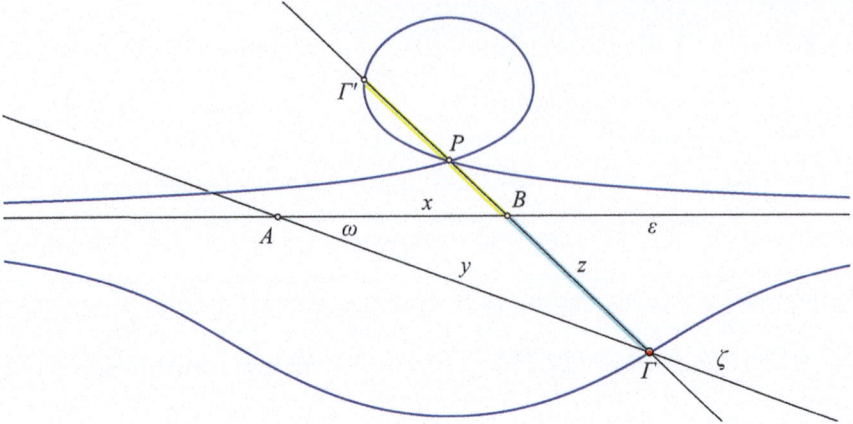

Fig. 5.188: Conchoid of Nicomedes

we see the geometric locus of points, which result from the fixed point P and the fixed line ε, taking on all lines through P segments $|B\Gamma| = |B\Gamma'| = z$. This is the known from ancient times **conchoid of Nicomedes** (about 200 b.C), which is intimately related to the problem of *neusis* (or verging) (Exercise 5.191) and was used to solve the problem of trisection of an angle and the doubling of the volume of the cube, equivalently the construction of the $\sqrt[3]{2}$. The requested Γ is precisely an intersection point of the other side ζ of the angle ω and the curve.

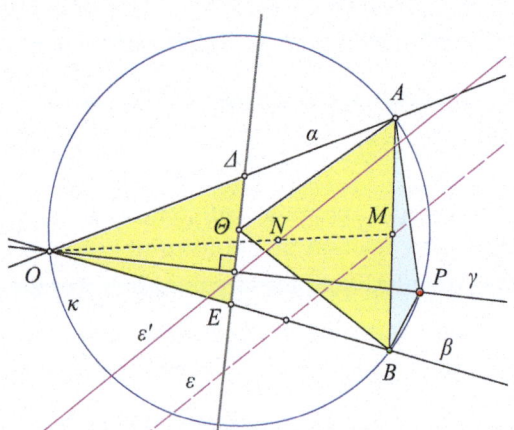

Fig. 5.189: Geometric locus of the middles

Exercise 5.217. Let $\{\alpha, \beta, \gamma\}$ be three lines through the point O and P a point of γ. We consider all the circles κ passing through $\{O, P\}$ and intersecting a

5.24. COMMENTS AND EXERCISES FOR THE CHAPTER

second time the lines $\{\alpha,\beta\}$ at corresponding points $\{A,B\}$. Show that the middle M of the segment AB is contained in a fixed line ε and the centroid N of the triangle OAB is contained in a line ε', which is parallel to ε (See Figure 5.189).

Hint: The triangle PAB has constant angles, its vertex P is fixed and points $\{A,B\}$ move on fixed lines. Apply theorem 2.28.

Exercise 5.218. In figure 5.189 of the preceding exercise consider the medial line of OP and its intersections $\{\Delta,E\}$ respectively with the lines $\{\alpha,\beta\}$ and the center Θ of the circle κ. Show that P is a Miquel point of $\{A,B,\Theta\}$ relative to the triangle $O\Delta E$ and a pivot of $AB\Theta$ relative to $O\Delta E$.

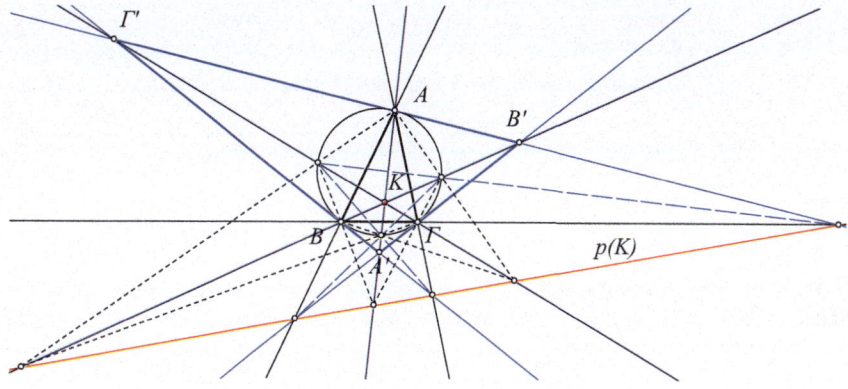

Fig. 5.190: Symmedian point K and the polar $p(K)$

Exercise 5.219. Let K be the symmedian point of the triangle $AB\Gamma$, whose tangential triangle is $A'B'\Gamma'$ (See Figure 5.190). Let also $p(K)$ be the polar of K with respect to the circumcircle of $AB\Gamma$, coinciding with the trilinear polar of K with respect to $AB\Gamma$ and with respect to $A'B'\Gamma'$. Study the various relations suggested by this figure. Find all harmonic quadruples of points contained in the lines of this figure.

Exercise 5.220. Construct a right triangle for which are given the length of the hypotenuse and the length of the bisector of one of its acute angles.

Exercise 5.221. Construct square $AB\Gamma\Delta$ for which is given the center O and two points Z, H on the sides, respectively, $\Gamma\Delta$ and ΔA, with $|ZO| \neq |HO|$.

Exercise 5.222. The vertices of polygon $A_1 A_2 ... A_\nu$ are projected on a line ε at points $B_1, B_2, ..., B_\nu$. Show that for the signed lengths holds

$$B_1 B_2 + B_2 B_3 + ... + B_{\nu-1} B_\nu + B_\nu B_1 = 0.$$

Exercise 5.223. Let P be a point on the circumcircle $\kappa(O,R)$ of the triangle $AB\Gamma$ and $\{A', B', \Gamma'\}$ be the projections respectively on lines $\{B\Gamma, \Gamma A, AB, s_P\}$, where s_P the Simson line of P. Prove the formulas ([28, p.14]):

1. $|PA||PA'| = |PB||PB'| = |P\Gamma||P\Gamma'| = 2R|PP'|$.
2. $|PA||PB||P\Gamma| = 4R^2|PP'|$.
3. $|PA'||PB'||P\Gamma'| = 2R|PP'|^2$.

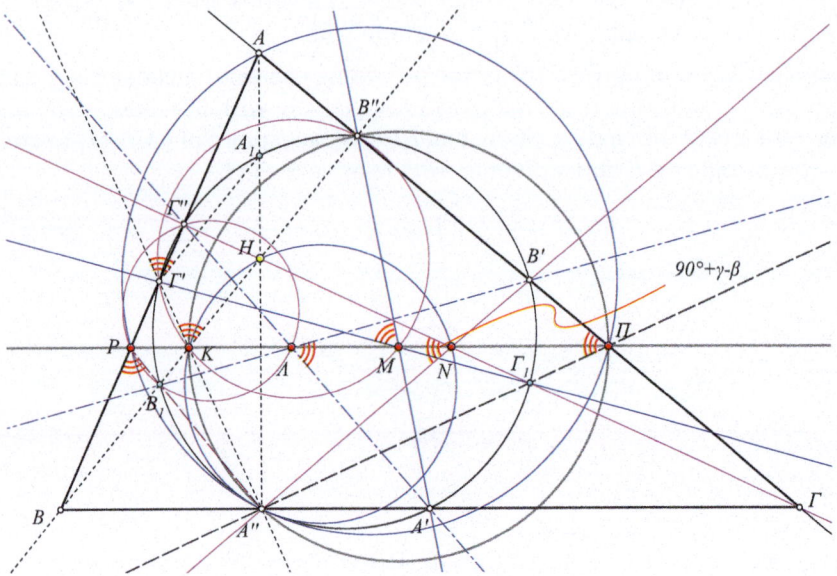

Fig. 5.191: Six collinear points

Exercise 5.224. Let A', B', Γ' be the middles of the sides and A'', B'', Γ'' be the traces of tje altitudes of the triangle $AB\Gamma$. Let also H be the orthocenter of $AB\Gamma$ and A_1, B_1, Γ_1 be the middles of HA, HB, $H\Gamma$. Show that next 6 points, which are defined as intersections of lines, are contained in a line parallel to $B\Gamma$: $P = (AB, A''B_1)$, $K = (BB'', A''\Gamma')$, $\Lambda = (A'\Gamma'', B'B_1)$, $N = (\Gamma\Gamma'', B'A'')$, $M = (A'B'', \Gamma'\Gamma_1)$ and $\Pi = (A''\Gamma_1, A\Gamma)$. Show also that point M is the center of the circle with diameter ΠP and that the quadrilaterals $A''KHN$, $A''P A\Pi$ are inscriptible and their circumcircles are tangent to the Euler circle at point A''.

Hint: Begin with the inscriptible quadrilaterals, for which the equality of the angles, suggested by the figure 5.191, is crucial.

Exercise 5.224 is a typical example of exercise, which results with the help of the computer. The specific one comes from a program, which starts from a given shape in which a few points have bin singled out, called *first generation* points (in this case they are 13: $\{A, B, \Gamma, A', B', \Gamma', A'', B'', \Gamma'', A_1, B_1, \Gamma_1, H\}$). The program calculates all the different lines defined by these points, the so called *first generation lines* (specifically 699). Subsequently, it determines points of *second generation*, which are the intersections of the first generation

5.24. COMMENTS AND EXERCISES FOR THE CHAPTER

lines (specifically 16470). Then, the program investigates how many lines contain more than two points. In the specific case there are 5286 such lines carrying 3, 4, 5, 6 and 7 points of second generation. Each such line defines an exercise, like the preceding one, where it is requested to *prove the collinearity* of certain points. In the specific case most of the 5286 lines carry 3 second generation points. The more collinears on a line we require, the less lines we find. This way, in the specific example, from the 5286 lines only 38 carry six points of a second generation (and consequently define 38 exercises similar to the preceding one), while there exist also 9, which carry 7 points each. In the specific example this is the maximum number of collinears. In other words, each of the 5286 lines contains no more that 7 second generation points.

Obviously the process could be continued ad infinitum, defining similarly *lines of second generation*, respectively *points of third generation* and so on and so forth, examining again the collinearities, finding the maximum number of collinear points in each generation etc. The actual program, for reasons of exhaustion of memory, stops at the collinearities of second generation points, but examines the corresponding process by substituting circles in the place of lines. These are defined from all possible triples of non-collinear points, the so called *first generation circles*. There the corresponding investigation concerns 4 or more concyclic points and in the specific example, among other things, determines the Euler circle, which contains 9 points. The following exercise resulted from a corresponding investigation for the same basic figure of 13 points of the preceding exercise. The exercise shows that the aforementioned process produces problems of non trivial geometric content. However, for reasons of consistency to historical evolution, I do not include in the book more problems produced this way. It is obvious that, by beginning with a small number of points from one shape, we can produce an infinity of problems. However, these problems do not seem to possess, generally speaking, an importance similar to that of the classical problems, which were developed in a long time evolution and are useful as tools for the solution of other problems.

Exercise 5.225. The middles of the sides of the triangle $AB\Gamma$ are respectively A', B', Γ' and the traces of its altitudes are A'', B'', Γ''. Let Z be the intersection point of the lines $B'\Gamma''$, $\Gamma'B''$, and K be the second intersection point of the circles $(B'B''Z)$, $(\Gamma'\Gamma''Z)$. Show that the lines $B'\Gamma'$, $B''\Gamma''$, EK, where E is the center of the Euler circle, pass through the same point Λ (See Figure 5.192).

Hint: Let $\Lambda = (\Gamma'B', \Gamma''B'')$ be the pole of the line AZ relative to the Euler circle of the triangle $AB\Gamma$ (Proposition 5.17). We easily see, that the triangles $A\Gamma'B''$, $AB'\Gamma''$ are similar isosceli. The intersection point $P = (B''\Gamma'', AZ)$ is the harmonic conjugate $P = \Lambda(\Gamma'', B'')$. Also the angles $\widehat{\Gamma''KP}$, $\widehat{PKB''}$ are equal, because of the inscriptible quadrilaterals $ZK\Gamma'\Gamma''$, $ZKB'B''$. It follows, that ΛK is the external bisector of triangle $KB''\Gamma''$, therefore it is orthogonal to AZ, which is the polar of Λ. The orthogonal from the pole Λ to the polar AZ

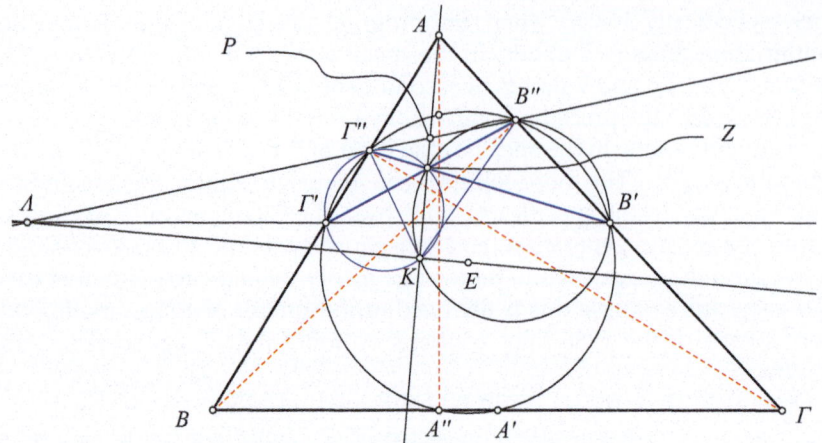

Fig. 5.192: Polar AZ of point Λ

passes through the center E. Note that line EZ is the Euler line of the triangle $AB\Gamma$ (Exercise 5.143).

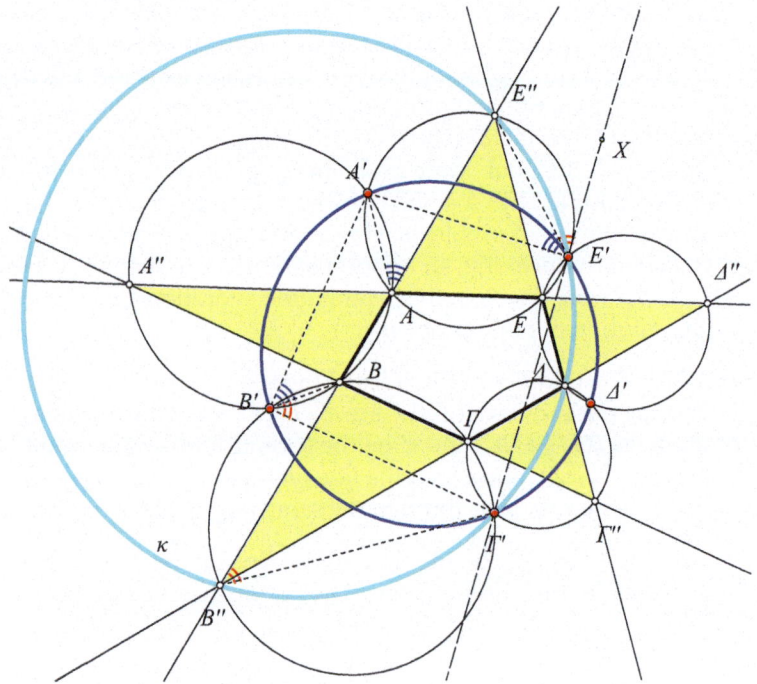

Fig. 5.193: Miquel's theorem for pentagons

Exercise 5.226 (Miquel's theorem for pentagons). We extend the sides of the convex pentagon $AB\Gamma\Delta E$, creating five triangles : $\{ABA'', B\Gamma B'', \Gamma\Delta\Gamma'', \Delta E\Delta'', EAE''\}$, the circumcircles of which intersect a second time at points $\{A', B', \Gamma', \Delta', E'\}$. Show that these points are concyclic (See Figure 5.193).

Hint: A solution results, by a, so called, *angle chasing*. With this we show that the resulting five quadrilaterals which result by omitting a vertex are inscriptible. For $A'B'\Gamma'E'$ say, resulting by omitting Δ', we show that the opposite angles $\{\widehat{B'}, \widehat{E'}\}$ are supplementary. We notice first, that pentagon $B''\Gamma'\Delta E'E''$ is inscriptible. In fact, consider the circumcirlce κ of the triangle $B''\Delta E''$ and its angles at Γ' :

$$\widehat{B''\Gamma'\Delta} = \widehat{B''\Gamma'\Gamma} + \widehat{\Gamma\Gamma'\Delta} = \widehat{\Gamma BA} + \widehat{\Gamma\Gamma''\Delta} = 180° - \widehat{AE''E}.$$

This shows that $B''\Gamma'\Delta E''$ is cyclic. Similarly we show that $B''\Delta E'E''$ is cyclic. Then, in $B'\Gamma'E'A'$ we notice that the opposite angles $\widehat{B'} = \widehat{A'B'B} + \widehat{BB'\Gamma'}$ and $\widehat{E'} = 180° - \widehat{A'E'E''} - \widehat{E''E'X}$, for which it is valid $\widehat{A'B'B} = \widehat{A'AE''} = \widehat{A'E'E''}$ and $\widehat{BB'\Gamma'} = \widehat{BB''\Gamma'} = \widehat{E''E'X}$.

References

108. T. Andreescu, D. Andrica (2006) Complex Numbers from A to ... Z. Birkhaeuser, Berlin
109. M. Audin (2002) Geometry. Springer, Heidelberg
110. M. Baker, 1885. A collection of formulae for the area of a plane triangle, *Annals of Mathematics*, 1:134-138
111. M. Baker, 1885. A collection of formulae for the area of a plane triangle, *Annals of Mathematics*, 2:11-18
112. L, Bankoff, 1988. The Metamorphosis of the Butterfly Problem, *Mathematics Magazine*, 60:195-210
113. L, Berman, G. Williams, B. Molnar, 2006. The Cross Ratio Is the Ratio of Cross Products!, *Mathematics Magazine*, 79:54-59
114. T. Birsan, 2015. Bounds for Elements of a Triangle Expressed by R,r and s, *Forum Geometricorum*, 15:99-103
115. H. Boas, 2006. Reflections on the Arbelos, *The American Mathematical Monthly*, 113:236-249
116. A. Bruen, J. Fisher, J. Wilker, 1983. Apollonius by Inversion, *Mathematics Magazine*, 56:97-103
117. W. Burnside (1886) Theory of equations. Longmans Green and Co., London
118. M. Chasles (1852) Traite de geometrie suprerieure.Bachelier, Paris
119. M. Chasles (1865) Traite de Sections Coniques. Gauthier-Villars, Paris
120. A. Cayley, 1876. On three-bar motion, *Proceedings of the London Mathematical Society*, 7:136-166
121. A. Connes, 1998. A new proof of Morley's theorem, *Les relations entre les mathematiques et la physique theorique, IHES*, 40:43-46
122. W. Clawson, 1919. The Complete Quadrilateral, *Annals of Mathematics*, 20:232-261
123. J. Coolidge (1980) A treatise on the circle and the sphere. Oxford University Press, Oxford

124. D. Cox (2012) Galois Theory, 2nd ed. Wiley., New York
125. H. Coxeter, 1968. The problem of Apollonius, *American Math. Monthly*, 75:5-15
126. H. Coxeter, 1949. Projective Geometry, *Mathematics Magazine*, 23:79-97
127. N. Dergiades, 2007. The Soddy Circles, *Forum Geometricorum*, 7:191-197
128. J. Ehrmann, 2004. Steiner's Theorems on the Complete Quadrilateral, *Forum Geometricorum*, 4:35-52
129. H. Eddy, R. Fritsch 1994. The conics of Ludwig Kiepert: A Comprehensive Lesson in the Geometry of the triangle, *Mathematics Magazine*, 67:188-205
130. H. Fukagawa (2008) Sacred Mathematics. Princeton University Press, Princeton
131. D. Gisch, J. Ribando, 2004. Apolloniu's Problem: A Study of Solutions and Their Connections, *American Journal of Undergraduate Research*, 3:15-26
132. S. Greitzer (1988) Arbelos, Special Geometry issue. Mathematical Association of America
133. L. Hahn (1994) Complex Numbers and Geometry. Mathematical Association of America
134. A. Hart, 1856. Geometrical investigation of Steiner's solution of Malfatti's problem, *Quarterly journal of pure and applied Mathematics*, 1:219-222
135. T. Heath (1897) The works of Archimedes. Cambridge University Press, Cambridge
136. A. Hess, 2012. A highway from Heron to Brahmagupta, *Forum Geometricorum*, 12:191-192
137. R. Honsberger (1997) In Polya's The Mathematical Association of America
138. R. Johnson, 1916. Relating to the Simson line or Wallace line, *American Math. Monthly*, 23:61-62
139. C. Lebosse, C. Hemery (1965) Geometrie. Fernand Nathan, Paris
140. E. Marchisotto, 2002. The Theorem of Pappus: A Bridge between Algebra and Geometry, *Amer. Math. Monthly*, 109:497-516
141. R. Muirhead, 1895. On the number and nature of the solutions of the Apollonian contact problem, *Proc. Edinburgh Math. Society*, 14:135-147
142. P. Nahin (2006) Dr. Euler's Fabulous Formula. Princeton University Press, Princeton
143. M. Naraniengar, 1909. Solution to Morley's problem, *Educational Times (New Series)*, 15:47
144. S. Ogivly (1969) Excursions in Geometry. Oxford University Press, Oxford
145. H. Okumura, 2023. A rare sangaku problem involving three congruent circles, *Sangaku Journal of Mathematics*, 7:1-8
146. H. Okumura, 2023. A configuration arising from Problem 2023-1-1, *Sangaku Journal of Mathematics*, 7:32-34
147. J. Peters, 1941. The theorem of Morley, *National Mathematics Magazine*, 16:119-126
148. H. Rademacher (1933) Von Zahlen und Figuren. Springer, Heidelberg
149. M. Rashid, A. Ajibade 2003. Two conditions for a quadrilateral to be cyclic, *International Journal of Mathematical Education in Science and Technology*, 34:739-799
150. D. Richeson (2008) Euler's gem. Princeton University Press, Princeton
151. G. Salmon (1917) A treatise on Conic Sections. Longmans, Green and Co., London
152. K. Tan, 1967. Different Proofs of Desargues' Theorem, *Mathematics Magazine*, 40:14-25
153. O. Veblen, J. Young (1910) Projective Geometry vol. I, II. Ginn and Company, New York
154. D. Wells (1991) Dictionary of Curious and Interesting Geometry. Penguin Books
155. J. Wilker, 1969. Four Proofs of a Generalization of the Descartes Circle Theorem, *American Mathematical Monthly*, 76:278-282
156. D. Woo, P. Yiu, 1999. Those Ubiquitous Archimedean Circles, *Mathematics Magazine*, 72:202-213
157. I. Yaglom (1962) Geometric Transformations I, II, III. Mathematical Association of America, New York
158. J. Yzeren, 1993. A Simple Proof of Pascal's Hexagon Theorem, *American Math. Monthly*, 100:930-931

Index

$O(\rho)$, 79, 85, 90
$\kappa(O,\rho)$, 79
$(AB\Gamma)$, 82

Absolute Geometry, 45, 48, 52
Acute, 17, 34
Acute triangle, 20, 157, 272, 275, 296, 384, 400, 427, 479
Additivity of areas, 182
Adjacent angles, 30, 53, 102, 127
Altitude, 22, 31, 32, 36, 38, 42, 72, 74, 75, 98, 102, 104, 118, 139, 147, 155, 157, 161, 163, 165, 166, 168, 172, 177, 188, 191, 202, 210, 238, 258, 268, 272, 273, 276, 281, 286, 291, 296–299, 309, 343, 350, 400, 408, 414, 418–421, 428, 474, 558, 559, 581, 584
Analysis, 95, 525, 550
Analysis, Synthesis, Discussion, 95
Angle, 12
Angle acute, 60
Angle central, 136, 140
Angle congruence, 23, 28, 37, 44
Angle exterior, 12
Angle inscribed, 139
Angle interior, 12
Angle measure, 13, 294
Angle non-convex, 12
Angle of circles, 344, 363, 371, 391
Angle sides, 12
Angle straight, 12
Angle sum, 14
Angle trisection, 100, 132
Angle vertex, 12
Angles adjacent, 14
Angles equal, 14
Angles vertical, 16

Anthyphairesis, 198
Anti-inversion, 391, 394
Anti-inversion interchanging, 392
Anticenter, 156
Anticomplementary, 116
Antihomologous points, 353, 357, 367, 392, 393
Antihomothetic, 241
Antihomothety, 580
Antiparallel, 277, 280, 283
Apex, 26, 37, 39, 42, 55, 72, 109, 118, 129, 176, 193, 211, 429, 481
Apollonian circle, 327–329, 333, 347, 386, 390, 450, 568
Apollonius, 327, 331, 339, 436, 439
Apothem, 87
Arbelos, 466, 571, 572
Arbelos-twins, 466
Arc, 135, 138
Arc complementary, 135
Arc of inscribed angle, 139
Archimedes, 146, 185, 466
Arcs congruent, 136
Area of parallelogram, 187, 188
Area of polygon, 181, 252
Area of quadrilateral, 220, 261
Area of rectangle, 185
Area of trapezium, 190
Area of triangle, 188, 259, 422
Area, axioms of, 182
Arithmetic mean, 209
ASA-criterion, 27, 28, 30–32, 55, 101, 125, 129, 130
Autopolar triangle, 519, 559
Axial symmetry, 62, 117
Axiom, 6, 9, 12, 20, 52, 185
Axiom of parallels, 52, 204, 218, 249

Axioms, 3, 4, 9, 20, 45, 48, 60, 136, 182, 249
Axis of symmetry, 62, 65, 66, 81, 91, 100, 121, 127

Bachmann, 4
Base, 26
Between, 3, 6, 7, 9
Bicentric, 159
Bicentric quadrilateral, 160, 219
Billiard, 211, 288
Birkhoff, 4
Bisector, 14, 22, 30, 47, 60, 71, 85, 94, 117, 129, 137, 144, 149, 155, 177, 192, 211, 268, 270, 272, 298, 344, 414, 428, 431
Bisector external, 570
Blundon, 435
Bodenmiller, 519
Bolyai, 48
Brahmagupta, 443, 448
Bretschneider, 448, 449, 568
Brianchon, 531, 534
Brianchon point, 535
Broken line, 41
Butterfly theorem, 539

Cairns, 4
Calabi, 554
Calabi's triangle, 554
Carpenter's square, 100
Carrier, 8
Castillon, 383, 540, 544
Cayley, 492
Center, 79, 121
Center line, 440
Center of mass = centroid, 114
Center of symmetry, 63
Center-line, 88, 286, 322, 334, 340, 345
Centers of similarity, 507
Centroid, 114, 117, 241, 285, 292
Ceva, 503, 506
Chord, 79, 135–137, 203, 466
Circle, 79
Circle construction, 437
Circle escribed, 414
Circle exterior, 79
Circle inscribed, 414, 415
Circle interior, 79
Circle of inversion, 360
Circle of similitude, 357, 393
Circle orthogonal, 347, 365
Circle pencil, 333, 345, 364, 380, 465
Circle squaring, 132
Circle tangent, 314
Circle tritangent, 414

Circumcenter, 82, 117, 431
Circumcircle, 82, 129, 154, 315
Circumradius, 129
Circumscircle, 121
Circumscribed, 129
Circumscribed circle = Circumcircle, 82
Circumscribed quadrilateral, 158
Circumscriptible, 158
Clifford, 492
Common harmonics, 350
Common tangents, 353, 417
Complementary, 18
Complementary triangle, 241
Complete quadrilateral, 513
Concentric, 88, 137
Conchoid of Nicomedes, 582
Concyclic, 313
Congruent circles, 80
Congruent polygons, 128
Congruent quadrilaterals, 105
Congruent shapes, 23
Congruent triangles, 29, 32, 73, 102, 154
Conic line intersection, 332
Conic sections, 537
Contact point, 84, 88
Convex, 105
Convex angle, 12
Coordinate, 486
Cosine rule, 258, 261
Cross ratio, 493
cross ratio of four lines, 495
Cyclic = Inscriptible, 152

Decatetragon, 132
Degrees, 13
Delian problem, 132
Deltoid, 455
Desargues, 521, 541
Descartes, 578
Descartes formula, 578
Diagonal, 101, 128, 129, 134, 182, 513
Diameter, 79, 83
Diametral, 146
Diametral = Diametrically opposite, 79
Diametrical, 343
Diametrically opposite, 79, 242, 260, 275, 430, 431
Direction, 7
Discussion, 96
Distance, 9
Distance from line, 43
Distance of parallels, 58, 188
Dodecagon, 132
Droz-Farny, 398

Index 591

Electromagnetic field, 351
Elliptic pencil = Pencil intersecting, 333
End, 7
Endpoint, 7
Envelope, 456
Equal segments, 10
Equality, 3, 23
Equation cubic, 435
Equilateral, 55, 72, 76, 93, 102, 117, 121, 130, 146, 147, 154, 176, 191, 195, 199, 295, 343, 390, 418, 429, 445, 461, 477, 479, 480, 482
Equilateral inscribed, 461
Escribed, 416, 433
Escribed circles, 419
Euler, 305, 422, 430
Euler circle, 422, 423, 426, 455, 539, 584, 586
Euler line, 242, 402, 530
Euler's theorem, 430
Excenter, 414
Exterior, 7
Exterior angle, 22
External angle, 33, 53
External bisector, 22, 192
External similarity center, 353

Fagnano, 477, 479
Fermat, 305, 477
Fermat numbers, 132
Feuerbach, 427, 428, 447
Feuerbach point, 428
Fixed point, 63
Fixed points, 62, 360
Flank, 194
Flat angle, 12
Full turn, 13
Fundamental invariants, 435

Gauss, 131, 519
Geometric locus, 30, 59, 83, 87, 120, 123, 138, 142, 161, 166, 172, 178, 201, 211, 232, 242, 243, 269, 274, 291, 292, 297–299, 315, 321, 322, 327, 337, 338, 340, 342, 349, 358, 371, 378, 381, 383, 385, 389, 395, 405, 451, 465, 560
Geometric mean, 208
Gergonne, 508, 536
Golden ratio section, 317
Golden rectangle, 318
Golden section, 130, 316

Half circumference, 136
Half line, 56, 84, 136
Half-line, 9

Half-perimeter, 19
Half-planes, 6
Harmonic conjugate, 68, 349, 373, 383, 518
Harmonic division = Harmonic quadruple, 67
Harmonic mean, 209, 285
Harmonic pencil, 495
Harmonic quadrilateral, 501
Harmonic quadruple, 67, 68, 231
Heron, 419, 421
Hexagon, 283, 528, 534
Hexagon regular, 130, 175, 195
Hilbert, 4
Homographic relation, 540, 541, 547, 581
Homologous, 241, 521
Homologous points, 353
Homothetic triangles, 240
Homothety center, 241, 249, 523
Homothety ratio, 241, 249
Hyperbolic Geometry, 45, 46, 48
Hyperbolic pencil = Pencil non-intersecting, 333
Hypotenuse, 26, 35, 97

Incenter, 86, 87, 414, 430, 431
Incidence angle, 45
Incircle, 86, 129
Incircle = Inscribed circle, 85
Inscribable = Inscriptible, 152
Inscribed, 129
Inscribed circle, 85, 158
Inscribed circle of the arbelos, 468
Inscribed quadrilateral, 152
Inscriptible, 152, 156, 160
Inscriptible quadrilateral, 260
Interior, 7
Internal similarity center, 353
Intervals separating each other, 350
Invariant by similarity, 462
Inverse pencil, 365
Inverse points, 360
Inverse points distance, 369
Inversion, 360, 383, 385, 408, 469, 470
Inversion center, 360
Inversion interchanging, 367
Isodynamic points, 410
Isogonal conjugacy, 511
Isogonal conjugate, 511, 512, 574
Isogonal conjugation, 574
Isometric shapes = congruent shapes, 23
Isometry, 23
Isosceles, 26, 29, 30, 35, 36, 42, 55, 60, 65, 73, 75, 80, 81, 87, 93, 98–100, 102, 104, 109, 114, 118, 145, 172, 191, 195, 205,

210, 238, 276, 318, 402, 429, 431, 451, 554
Isosceles trapezium, 155, 168, 192, 242, 243, 283, 297, 308, 320, 385, 454, 483
Isotomic conjugacy, 511
Isotomic conjugate, 511

Kiepert hyperbola, 481

Legendre, 48
Legs, 26
Lehmus, 104, 275
Lemoine, 279, 533
Lemoine circles, 280, 283
Lemoine line, 533
Length, 10
Length of broken line, 41
Line, 3, 6
Line at infinity, 376, 505, 506, 522, 527
Line-bisector, 27
Lobatsevsky, 48

Malfatti, 549
Maximal altitude, 389
Maximal angle, 235, 299, 343
Maximal area, 195, 246, 266, 296, 390, 553, 554
Maximal distance, 258
Maximal perimeter, 145, 147, 295
Maximal product, 245
Mean proportional, 208, 297
Measures, 182
Medial line, 27, 30, 80, 92, 137, 327, 330
Median, 22, 30, 42, 59, 114, 183, 189, 194, 268, 269, 309, 315, 559
Medians triangle, 115
Menelaus, 501, 532
Middle, 9, 92
Middle parallel, 111, 127
Middle-parallel, 59
Minimal angle, 343
Minimal area, 459
Minimal chord, 86
Minimal distance, 43, 44, 108, 109, 243, 258, 267
Minimal perimeter, 157, 191
Minimal sum, 286, 340, 478
Minute (of degree), 13
Miquel, 457, 458, 462, 562, 574, 587
Miquel pentagon theorem, 587
Miquel point, 457, 462, 463, 465, 583
Miquel point of four lines, 463
Morley, 482, 483
Morley's theorem, 482

Napoleon's theorem, 479
Negative half line, 486
Neusis, 132, 569, 582
Newton, 111
Newton line, 111, 193, 225, 270, 519
Nicomedes, 582
Nine point circle = Euler circle, 422
Non-convex angle, 12
Null angle, 13

Obtuse, 17, 34, 117
Obtuse triangle, 20
Opposite, 10, 19, 105, 513
Orientation, 24, 29
Oriented equally, 56
Oriented negatively, 24
Oriented opposite, 56
Oriented positively, 24
Origami, 63, 406
Origin, 486
Orthic, 157, 410, 576
Orthic triangle, 351, 414
Orthocenter, 22, 117, 144, 150, 162, 166, 175, 246, 291, 350, 379, 384, 396, 408, 418, 423, 426, 429, 432, 454, 463, 465, 481, 512, 519, 530, 558, 572, 583, 584
Orthocentric quadruple, 150
Orthodiagonal, 203, 395
Orthogonal, 17
Orthogonal circles, 344
Orthogonal pencil, 347, 366, 380
Orthogonal sides, 26, 34, 57, 196, 207, 247
Orthogonal to line, 43, 93

Pappus, 29, 175, 181, 199, 466, 469, 526, 541
Parabolic pencil = Pencil tangent, 333
Parallel, 8, 10
Parallel distance, 228
Parallel lines, 35, 45, 46, 52, 54, 56, 94
Parallelogram, 101, 102, 106, 110, 111, 118, 182, 187, 270, 383, 530
Pascal, 531, 541
Pascal line, 532
Pasch, 20
Pedal, 465
Pedal triangle, 458
Pencil center, 227
Pencil direction, 227
Pencil generated, 335
Pencil intersecting, 333, 377, 407
Pencil non-conventional, 341, 366
Pencil non-intersecting, 333
Pencil of circles, 330, 388
Pencil of lines, 227

Pencil orthogonal, 387
Pencil parallel, 227
Pencil tangent, 333
Pencil's basic points, 334
Pencil's limit points, 334, 388
Pencil's radical axis, 334
Penrose tiles, 319
Pentagon, 394, 532
Pentagon regular, 129, 319
Perimeter, 19, 97, 147, 171
Perpendicular = Orthogonal, 17
Perspective relative to line, 521
Perspective triangles, 521, 524
Perspectivity axis, 521
Perspectivity center, 521
Pick, 303
Pivot, 462, 583
Pivoting, 562
Plane, 3
Plane Geometry, 4
Playfair axiom, 54
Point, 3, 6
Point of line at infinity, 376, 527
Point of the line at infinity, 505, 506
Point symmetric, 64, 65, 119
Point symmetry, 63, 117
Polar, 372, 376, 378, 383, 518, 519
Polar circle, 380, 519
Pole, 376
Pole-polar reciprocity, 375
Polygon, 127
Polygon angles, 127
Polygon convex, 128
Polygon partition, 301
Polygon perimeter, 128
Polygon pivoting, 462
Polygon regular, 128, 129, 131, 135, 146, 147, 155, 167, 177, 291, 320
Polygon self-intersecting, 128
Polygon sides, 127
Polygons congruent, 249
Polygons homothetic, 249
Polygons similar, 247–249, 252
Positive half line, 486
Power, 311
Power of inversion, 360
Power ratio, 338
Projection, 43, 83
Projective geometry, 529
Ptolemy, 443
Pythagoras, 181, 196
Pythagorean theorem, 218

Quadrangle = Quadrilateral, 105

Quadratic equation, 387
Quadrilateral, 105, 109, 111, 112, 193, 219
Quadrilateral bicentric, 396
Quadrilateral complete, 513
Quadrilateral construction, 109, 127, 166, 173, 176, 258, 388, 450, 451, 568
Quadrilateral convex, 105, 153
Quadrilateral harmonic, 568, 569
Quadrilateral inscriptible, 444
Quardatic equation, 386

Radical axis, 321, 350, 471
Radical center, 324, 379, 385, 402, 535
Radius, 79
Ratio, 67
Rectangle, 118, 184, 408
Reduction to contradiction, 4, 8, 34, 48
Reflected relative to line, 62
Reflection angle, 45
Relative position, 250, 251
Rhombus, 123, 124, 126, 127
Right angle, 17, 83, 143
Right angle rotating, 138, 288, 396
Right triangle, 26, 30, 34, 53, 59, 72, 95, 122, 155, 166, 172, 188, 196, 199, 202, 205, 206, 215, 224, 253, 380
Rotation, 462

Saccheri, 46, 48
Sagitta, 87
Sangaku, 472, 476
SAS-criterion, 27, 30, 31, 37, 48, 59, 89, 97, 120
Scalene triangle, 20, 73
Secant, 8
Second (of degree), 14
Secondary elements, 22
Segment, 7
Self-intersecting, 105
Self-polar triangle, 519
Semi-perimeter, 19
Shape, 4
Side, 6
Side proportionality, 205
Sides, 101, 513
Signed distance, 486
Signed ratio, 272, 485
Similar polygons, 247, 248
Similar triangles, 205, 237, 458
Similarity center, 352, 378
Similarity centers, 353
Similarly invariant, 251
Simson, 452

Simson line, 452, 453, 455, 456, 458, 463, 574, 575, 583
Simson lines angle, 453
Sine rule, 258, 260
Soddy, 440
Space, 3
Space Geometry, 4
Spiral of Theodorus, 198
Square, 118, 120, 186, 194
Square inscription, 560
SSS-criterion, 20, 27, 28, 31
Start, 7
Start-point, 9
Steiner, 31, 104, 275, 466, 469
Steiner chains, 470
Steiner line, 454, 572, 575, 577
Stewart, 268, 419
Sum of squares, 270
Sum of triangle angles, 48
Supplementary, 17, 33, 54, 57
Sylvester, 170
Symmedian, 277, 280, 456, 512
Symmedian center, 279, 281
Symmetric relative to axis, 62
Symmetric relative to line, 62
Symmetry, 61, 108
Symmetry axis = Axis of symmetry, 62
Synthesis, 96

Tangent, 84, 95, 99, 312
Tangent circles, 88
Tangential triangle, 279, 351, 376, 533
Tangents common, 96
Tangram, 302
Taylor, 282
Taylor circle, 282
Thales, 181, 221
Thales theorem, 411
Theodorus of Kyrenia, 198
Theorem of Apollonius, 327
Theorem of Ceva, 503
Theorem of cosines, 261
Theorem of Desargues, 522
Theorem of Droz-Farny, 398
Theorem of Heron, 419
Theorem of Lhuilier, 525
Theorem of Menelaus, 501
Theorem of Pappus, 199, 526
Theorem of Pascal, 531
Theorem of Pick, 303
Theorem of Ptolemy, 443, 446, 525
Theorem of Pythagoras, 196, 224
Theorem of sines, 260
Theorem of Stewart, 268

Theorem of Thales, 221, 224
Theorem of three bars, 492
Tiles, 302
Transformation, 23, 66
Trapezium, 46, 123, 127, 137, 144, 163, 190, 192, 231, 285, 290, 291, 567
Trapezium isosceles, 123, 125, 138
Trapezoid = Trapezium, 123
Triangle, 19
Triangle angle sum, 53
Triangle circumscribed, 295
Triangle construction, 23, 92, 95, 98, 99, 118, 123, 147, 154, 155, 161, 165, 168, 172, 180, 194, 210, 212, 271, 274, 276, 286, 294, 295, 318, 343, 344, 378, 380, 381, 418, 419, 429, 432, 558, 581, 583
Triangle formulas, 284, 558
Triangle Geometry, 533
Triangle inequality, 40, 81, 90, 97, 109, 267
Triangle sides, 19
Triangle vertices, 19
Triangle's exterior, 20
Triangle's interior, 20
Triangles congruent, 20
Triangles equal = Triangles congruent, 20
Triangles isometric = Triangles congruent, 20
Triangles similar, 170, 289
Triantagon, 134
Trigonometric circle, 264
Trigonometric formulas, 263, 265, 320
Trigonometric functions, 213
Trigonometric functions: sin, cos, tan, cot, 213
Trilinear polar, 505, 522
Trilinear pole, 505
Trioker game, 76
Trisector, 482
Tritangent circle, 420
Tucker, 283
Tucker circles, 283

Undefined terms, 3

Van Aubel, 204
Vecten, 480
Vecten point, 480
Vecten triangle, 480
Vertex, 513
Vertical = Orthogonal, 17
Vertices, 101
Viete, 440

Wallace, 452

Wallace line = Simson line, 452

Zeno's paradox, 11

The manufacturer's authorised representative in the EU is Springer Nature Customer Service Centre GmbH, Europaplatz 3, 69115 Heidelberg, Germany. If you have any concerns regarding our products, please contact ProductSafety@springernature.com

Printed and bound by CPI Group (UK) Ltd, Croydon, CR0 4YY
26/03/2026
02078940-0005